High-Density Lipoproteins

Edited by
Christopher J. Fielding

1807–2007 Knowledge for Generations

Each generation has its unique needs and aspirations. When Charles Wiley first opened his small printing shop in lower Manhattan in 1807, it was a generation of boundless potential searching for an identity. And we were there, helping to define a new American literary tradition. Over half a century later, in the midst of the Second Industrial Revolution, it was a generation focused on building the future. Once again, we were there, supplying the critical scientific, technical, and engineering knowledge that helped frame the world. Throughout the 20th Century, and into the new millennium, nations began to reach out beyond their own borders and a new international community was born. Wiley was there, expanding its operations around the world to enable a global exchange of ideas, opinions, and know-how.

For 200 years, Wiley has been an integral part of each generation's journey, enabling the flow of information and understanding necessary to meet their needs and fulfill their aspirations. Today, bold new technologies are changing the way we live and learn. Wiley will be there, providing you the must-have knowledge you need to imagine new worlds, new possibilities, and new opportunities.

Generations come and go, but you can always count on Wiley to provide you the knowledge you need, when and where you need it!

William J. Pesce
President and Chief Executive Officer

Peter Booth Wiley
Chairman of the Board

High-Density Lipoproteins

From Basic Biology to Clinical Aspects

Edited by
Christopher J. Fielding

WILEY-VCH Verlag GmbH & Co. KGaA

The Editor

Prof. Dr. C. J. Fielding
University of California
Cardiovasc. Res. Inst.
4th and Parnassus
San Francisco, CA 94143-0130
USA

Cover
Parts of the cover picture are reprinted with kind permission from Robert O. Ryan (Chapter 1, Figure 3), Children's Hospital Oakland Research Institute, Oakland, CA, USA.:

Models of apoA-I alignment on discoidal nascent HDL. Proposed helical arrangement of apoA-I on the periphery of the HDL disc. ApoA-I has traditionally been presented as the "picket fence" (left) but recent evidence has eliminated this from consideration. Current data supports the "belt" conformation on HDL (second from left) and other models wherein apoA-I adopts and extended helical structure. Direct evaluation suggests that the "tandem hairpin" (second from rigth) and "z-scheme" (right) models are unlikely conformational states of apoA-I.

Library of Congress Card No.:
applied for

British Library Cataloguing-in-Publication Data
A catalogue record for this book is available from the British Library.

Bibliographic information published by the Deutsche Nationalbibliothek
Die Deutsche Nationalbibliothek lists this publication in the Deutsche Nationalbibliografie; detailed bibliographic data are available in the Internet at http://dnb.d-nb.de.

Typesetting Thomson Digital, Noida
Printing betz-druck GmbH, Darmstadt
Binding Litges & Dopf GmbH, Heppenheim
Cover Design Anne Christine Keßler, Karlsruhe
Wiley Bicentennial Logo Richard J. Pacifico

Printed in the Federal Republic of Germany
Printed on acid-free paper

ISBN: 978-3-527-31717-2

Contents

Preface XV
List of Contributors XVII

I **HDL Structure** 1

1 **Apolipoprotein A-I Structure** *3*
Michael N. Oda, Robert O. Ryan
1.1 Introduction and Perspective *3*
1.2 Lipid-Free ApoA-I *3*
1.3 The Role of Protein Structure in ApoA-I Function *4*
1.4 Full-Length ApoA-I X-Ray Structure *9*
1.5 Apolipoprotein A-I Structure – Lipid-Bound State *11*
1.6 Structural Studies of ApoA-I on Discoidal Nascent HDL *12*
1.7 Hinge Domain *14*
1.8 Computer Models *15*
1.9 Spherical HDL *16*
1.10 Future Directions *16*
1.11 Acknowledgments *16*

2 **Apolipoprotein A-II** *25*
Joan Carles Escolà-Gil, Jesús M. Martín-Campos, Josep Julve, Francisco Blanco-Vaca
2.1 Apolipoprotein A-II Structure and Plasma Level Regulation *25*
2.1.1 Structural Organization and Transcriptional Regulation of the Gene *25*
2.1.2 ApoA-II Mutations and Polymorphisms *27*
2.1.3 Protein Structure, Synthesis, Secretion, and Metabolism *28*
2.1.4 Other Determinants of ApoA-II Levels *33*
2.2 Pathophysiological Role of ApoA-II *33*
2.2.1 On HDL Metabolism *33*

High-Density Lipoproteins: From Basic Biology to Clinical Aspects. Edited by Christopher J. Fielding

ISBN: 978-3-527-31717-2

2.2.1.1 HDL Remodeling *33*
2.2.1.2 Reverse Cholesterol Transport Function *35*
2.2.1.3 Antioxidant and Anti-inflammatory Effects *36*
2.2.2 Role of ApoA-II in Triglyceride and Glucose Metabolism *37*
2.2.2.1 Triglyceride and Free Fatty Acid (FFA) Metabolism *37*
2.2.2.2 Glucose Homeostasis *38*
2.2.2.3 Role of ApoA-II in Dyslipidemia and Type 2 Diabetes Mellitus *38*
2.2.3 ApoA-II in Atherogenesis *41*
2.3 Acknowledgments *41*

3 Apolipoprotein E and Reverse Cholesterol Transport: The Role of Recycling in Plasma and Intracellular Lipid Trafficking *55*
Larry L. Swift, Alyssa H. Hasty, MacRae F. Linton, Sergio Fazio
3.1 Introduction *55*
3.2 Role of ApoE in Triglyceride-Rich Lipoprotein Metabolism *56*
3.2.1 ApoE and Lipoprotein Assembly and Secretion *56*
3.2.2 ApoE and Lipoprotein Processing and Catabolism *58*
3.2.3 ApoE and Lipoprotein Uptake *58*
3.3 Role of ApoE in HDL Metabolism *60*
3.3.1 HDL Formation and Maturation *60*
3.3.2 Cholesterol Efflux and RCT *62*
3.3.3 Roles of apoE in RCT and HDL Metabolism *63*
3.3.4 Critical Roles for ApoE and SR-BI in RCT *65*
3.4 ApoE and Recycling *66*
3.4.1 Evidence for apoE Recycling *66*
3.4.2 Regulation of ApoE recycling *70*
3.4.3 ApoE Recycling and Cholesterol Efflux *71*
3.4.4 Form of Recycled ApoE *72*
3.4.5 ApoE Recycling: Relevance to Atherosclerosis *73*
3.5 Summary *74*

4 ApoM – A Novel Apolipoprotein with Antiatherogenic Properties *89*
Ning Xu, Peter Nilsson-Ehle
4.1 Introduction *89*
4.2 Cloning and Characterization of Human ApoM *90*
4.3 Gene Location and Amino Acid Sequence of ApoM *91*
4.4 Protein Structure of ApoM *93*
4.5 Distribution of ApoM in Plasma Lipoproteins *94*
4.6 Tissue Distribution and Cellular Expression of ApoM *95*
4.7 Putative Functions of ApoM *97*
4.8 Regulation of ApoM *99*
4.9 Hormonal Regulation of ApoM *102*
4.10 Clinical Observations *104*
4.11 Conclusions and Perspectives *105*

II HDL Plasma Metabolic Factors 111

5 **ATP Binding Cassette Transporters** *113*
Laetitia Dufort, Giovanna Chimini
5.1 The Family of ABC Transporters – General Features *113*
5.2 ABC Transporters and Lipid Transport *118*
5.3 The ABCA Subfamily and the Handling of Excess Lipids *119*
5.3.1 ABCA1 *121*
5.3.1.1 ABCA1 Expression and its Control *121*
5.3.1.2 Structural Considerations *123*
5.3.1.3 ABCA1 and Cellular Effluxes of Phospholipid and Cholesterol – Facts and Speculation *124*
5.3.2 ABCA7 *126*
5.3.2.1 From the Gene to the Protein *126*
5.3.2.2 The ABCA7 Knockout Mouse – A Model with some Surprises *127*
5.4 The ABCG Subfamily and Sterol Trafficking *128*
5.4.1 ABCG1 *129*
5.4.1.1 Expression and its Regulation *129*
5.4.1.2 Physiopathology and Functional Predictions *130*
5.5 Acknowledgments *131*

6 **Reverse Cholesterol Transport – New Roles for Preβ_1-HDL and Lecithin:Cholesterol Acyltransferase** *143*
Christopher J. Fielding, Phoebe E. Fielding
6.1 Introduction *143*
6.2 Structure and Properties of Preβ_1-HDL *144*
6.3 Cell Membrane Origin of FC for RCT *146*
6.4 Role of ABCA1 in FC Transfer *147*
6.5 Mechanism of Lecithin:Cholesterol Acyltransferase (LCAT) *149*
6.6 Origins of FC and PL for LCAT Activity *150*
6.7 Further Metabolism of LCAT/Preβ_1-HDL Complexes *152*
6.8 ApoA-1 Recycling and the Production of Preβ_1-HDL *154*
6.9 Revised Model of RCT *155*

7 **HDL Remodeling by CETP and SR-BI** *163*
Christopher J. Harder, Ruth McPherson
7.1 Introduction *163*
7.2 Atheroprotective Function of HDL *163*
7.3 Cholesteryl Ester Transfer Protein (CETP) *164*
7.4 Selective Uptake *164*
7.5 SR-BI Function and Trafficking in Hepatocytes *165*
7.6 Cellular Mechanisms of SR-BI-Mediated Biliary Cholesterol Secretion *170*

7.7 Role of CETP in Direct Clearance of HDL-derived CE (Selective Uptake) *170*
7.8 Summary *173*

8 Human Plasma Phospholipid Transfer Protein (PLTP) – Structural and Functional Features *183*
Sarah Siggins, Kerry-Ann Rye, Vesa M. Olkkonen, Matti Jauhiainen, Christian Ehnholm
8.1 Introduction *183*
8.2 Lipid Transfer Proteins *184*
8.3 Phospholipid Transfer Protein (PLTP) *184*
8.3.1 Characteristics of PLTP *184*
8.3.2 Regulation of PLTP Gene Expression *186*
8.4 Functions of PLTP *187*
8.4.1 Phospholipid Transfer Activity *187*
8.4.2 HDL Remodeling *189*
8.4.3 Cellular Cholesterol Efflux *190*
8.4.4 PLTP in Human Plasma and Tissues *190*
8.4.5 Low-Activity (LA) and High-Activity (HA) PLTP *192*
8.5 PLTP and Atherosclerosis *193*
8.5.1 Insights from Mouse Models *193*

9 Use of Gene Transfer to Study HDL Remodeling *207*
Kazuhiro Oka, Lawrence Chan
9.1 Introduction *207*
9.2 Vectors for Gene Transfer *208*
9.2.1 Physical Delivery of Nonviral DNA *208*
9.2.2 Nonviral Vectors *209*
9.2.2.1 Phage Integrase *209*
9.2.2.2 Adeno-Associated Virus (AAV) Rep Protein-Mediated Integration *209*
9.2.3 Viral Vectors *210*
9.2.3.1 DNA Viral Vectors *210*
9.2.3.2 RNA Viral Vectors *213*
9.3 Roles of Lipases and HDL-Associated Enzymes in HDL Remodeling *214*
9.3.1 Endothelial Lipase (EL) *214*
9.3.2 Hepatic Lipase (HL) *216*
9.3.3 Lipoprotein Lipase (LPL) *217*
9.3.4 Structural and Functional Alterations of HDL by Enzymes *218*
9.3.5 Paraoxonases (PONs) *218*
9.3.6 Myeloperoxidase (MPO) *219*
9.4 Conclusion *220*
9.5 Perspective *220*
9.6 Acknowledgments *221*

III HDL Formation, Secretion, Removal 235

10 Regulation of Genes Involved in the Biogenesis and the Remodeling of HDL 237
Dimitris Kardassis, Costas Drosatos, and Vassilis I. Zannis
10.1 Introduction 237
10.2 The Coregulator Complexes 237
10.2.1 Nuclear Hormone Receptors 239
10.2.2 LXR, FXR, LRH-1, SHP 240
10.2.3 PPARα, -β, -γ 240
10.2.4 SF-1 240
10.2.5 HNF-4 240
10.2.6 ARP-1, EAR-3 241
10.2.7 SP1 Family of Proteins 241
10.2.8 SREBPs 241
10.3 Transcriptional Regulation of Genes Involved in HDL Biogenesis and Catabolism 242
10.3.1 Transcriptional Regulation of the Human ApoA-I Gene 242
10.3.1.1 *In vitro* Studies 242
10.3.1.2 *In vivo* Studies 244
10.3.1.3 Putative Mechanisms of ApoA-I Gene Transcription 245
10.3.1.4 *In vivo* Regulation of the ApoA-I Gene in Transgenic and Knockout Animal Models, as well as in Response to Pharmacological and Dietary Treatments 247
10.3.2 Transcriptional Regulation of the ABCA1 Gene 248
10.3.3 Transcriptional Regulation of the ABCG1 Gene 249
10.3.4 Transcriptional Regulation of the SR-BI Gene 251
10.3.5 Transcriptional Regulation of the CETP Gene 252
10.3.6 Transcriptional Regulation of the PLTP Gene 252
10.4 Conclusions 253
10.5 Acknowledgements 253

11 ApoA-I Functions and Synthesis of HDL: Insights from Mouse Models of Human HDL Metabolism 267
Vassilis I. Zannis, Eleni E. Zanni, Angeliki Papapanagiotou, Dimitris Kardassis, Christopher J. Fielding, Angeliki Chroni
11.1 Structures of ApoA-I and High-Density Lipoprotein (HDL) 267
11.2 ApoA-I Participates in the Biogenesis and Catabolism of HDL and Interacts with Several Other Proteins of the HDL Pathway 269
11.3 Functional Interactions between Lipid-Free ApoA-I and the ABCA1 Lipid Transporter Represent the First Step in the Biogenesis of HDL 270
11.3.1 *In vitro* Analysis of the Interactions between ApoA-I and ABCA1 that Lead to Cholesterol and Phospholipid Efflux through the Use of Targeted ApoA-I Mutations 270

11.3.2 Use of Adenovirus-Mediated Gene Transfer to Study how ApoA-I Mutations Affect the Biogenesis of HDL *274*
11.4 ApoA-I/ABCA1 Interactions: Intracellular Trafficking of the Complex *279*
11.5 Contribution of the HDL Pathway in the Reverse Transport of Cholesterol *279*
11.6 HDL Biogenesis can be Inhibited by Specific Mutations in ApoA-I that Inhibit Activation of LCAT *281*
11.7 Mutations in ApoA-I may Induce Hypertriglyceridemia and Other Dyslipidemias *283*
11.8 Remodeling of the HDL Particles after Interactions between HDL-Bound ApoA-I and SR-BI *284*
11.9 Remodeling of HDL Particles after Interactions with ABCG1 *287*
11.10 The Downstream Steps of the HDL Pathway *288*
11.11 Origins and Functions of HDL Subpopulations *288*
11.11.1 *De novo* Synthesis of Preβ HDL *288*
11.11.2 Generation of Preβ1 HDL from αHDL Particles *290*
11.12 Importance of HDL Subpopulations *291*
11.13 Role of ApoA-I and HDL in Atheroprotection *291*
11.14 Conclusions *292*
11.15 Acknowledgments *293*

12 Hepatic and Renal HDL Receptors *307*
Laurent O. Martinez, Bertrand Perret, Ronald Barbaras, François Tercé, Xavier Collet
12.1 Historical Background *307*
12.1.1 HDL Receptors: From Myth to Reality *307*
12.1.2 To be or not to be an HDL Receptor? *308*
12.2 SR-BI-Mediated Selective HDL Cholesterol Uptake *310*
12.2.1 The Selective HDL Cholesteryl Ester Uptake Pathway *310*
12.2.2 SR-BI as a Receptor for Selective Lipoprotein Cholesterol Uptake *311*
12.2.3 Role of Hepatic SR-BI in Murine Cholesterol Homeostasis and Atherogenesis *311*
12.2.4 Regulation of Hepatic SR-BI *312*
12.2.5 Mechanism of SR-BI-Mediated Selective Lipoprotein Cholesterol Uptake *314*
12.3 Ecto-F_1-ATPase: An Unexpected Role in HDL Endocytosis *316*
12.3.1 Two HDL Binding Sites Allocated to One HDL Endocytosis Pathway *316*
12.3.2 Ecto-F_1-ATPase as a High-Affinity ApoA-I Receptor *317*
12.3.3 The Ecto-F_1-ATPase/$P2Y_{13}$-Mediated HDL Endocytosis Pathway *319*
12.3.4 Extracellular ADP Levels as a Crucial Point of Regulation of HDL Endocytosis *319*
12.3.5 F_1-ATPase as a Moonlighting Protein Complex *320*

12.3.6 The Checkpoints for Regulation of the F_1-ATPase-Mediated HDL Endocytosis Pathway 322
12.4 Others Receptors involved in HDL Metabolism 323
12.4.1 Contribution of ABCA1 to Hepatic ApoA-I Secretion 323
12.4.2 The Role of Cubilin in Renal ApoA-I Catabolism 324
12.5 Relevance to Human Physiopathology 325

IV Biological Activities of HDL 339

13 HDL and Inflammation *341*
Mohamad Navab, Srinivasa T. Reddy, Brian J. Van Lenten, Georgette M. Buga, G. M. Anantharamaiah, Alan M. Fogelman
13.1 HDL as Part of the Innate Immune System *341*
13.2 The Role of Oxidized Phospholipids and HDL-Associated Enzymes *342*
13.3 A Connection between Reverse Cholesterol Transport and Inflammation *343*
13.4 The Inflammatory Properties of HDL in Humans *344*
13.5 Oxidative Stress and Inflammation *345*
13.6 Apolipoprotein Mimetic Peptides that Modulate HDL Function *346*
13.7 Summary *350*
13.8 Acknowledgments *350*

14 HDL in Health and Disease: Effects of Exercise and HIV Infection on HDL Metabolism *355*
Dmitri Sviridov, Michael I. Bukrinsky, Bronwyn A. Kingwell, Jennifer Hoy
14.1 Introduction *355*
14.2 HDL Metabolism *356*
14.2.1 HDL Formation *356*
14.2.2 HDL Maturation *356*
14.2.3 HDL Remodeling *357*
14.2.4 Efficiency of Reverse Cholesterol Transport *357*
14.3 Exercise and HDL Metabolism *360*
14.3.1 HDL Formation *360*
14.3.2 HDL Maturation *361*
14.3.3 HDL Remodeling *361*
14.3.4 Exercise and the Efficiency of Reverse Cholesterol Transport *362*
14.4 HIV Infection and HDL Metabolism *362*
14.4.1 HDL Formation *363*
14.4.2 HDL Maturation *363*
14.4.3 HDL Remodeling *364*
14.4.4 HIV Infection and the Efficiency of Reverse Cholesterol Transport *364*
14.5 Fixing HDL Metabolism *366*

15 **Endothelial Protection by High-Density Lipoproteins** *375*
Monica Gomaraschi, Laura Calabresi, Guido Franceschini
15.1 Endothelial Dysfunction and Cardiovascular Disease *375*
15.2 HDL and Regulation of Vascular Tone *377*
15.2.1 Nitric Oxide *377*
15.2.2 Prostacyclin *380*
15.3 HDL and Vascular Inflammation *381*
15.3.1 Cellular Adhesion Molecules *381*
15.3.2 Cytokines *383*
15.3.3 Platelet-Activating Factor *384*
15.4 HDL and Hemostasis *384*
15.4.1 Platelet Adhesion and Reactivity *384*
15.4.2 Coagulation *385*
15.4.3 Fibrinolysis *386*
15.5 HDL and Endothelial Monolayer Integrity *386*
15.5.1 Endothelial Cell Apoptosis *386*
15.5.2 Migration and Proliferation of Endothelial Cells *387*
15.6 Conclusions and Perspectives *388*

16 **The Effect of Nutrients on Apolipoprotein A-I Gene Expression** *399*
Arshag D. Mooradian
16.1 Introduction *399*
16.2 Effect of Caloric Intake *400*
16.3 Effects of Various Nutrients on HDLc and ApoA-I *400*
16.3.1 Effects of Lipids *400*
16.3.1.1 Effect of Fatty Acid Saturation *403*
16.3.1.2 Effects of Oxidized Fatty Acids *406*
16.3.1.3 Effect of *trans*-Fatty Acids *407*
16.3.1.4 Effects of Medium-Chain Fatty Acids *408*
16.3.1.5 Effects of ω-3-Fatty Acids *408*
16.3.1.6 Effect of Fat Metabolites: Ketones and Prostanoids *409*
16.4 Effects of Carbohydrates or their Metabolites *410*
16.4.1 Effect of Glucose *410*
16.4.2 Effects of Fructose *411*
16.4.3 Effect of Glucosamine *411*
16.5 Effects of Protein *412*
16.6 Effects of Micronutrients *412*
16.7 Effects of Alcohol *414*
16.8 Conclusions *414*

17 **Nutritional Factors and High-Density Lipoprotein Metabolism** *425*
Ernst J. Schaefer, Stefania Lamon-Fava, Bela F. Asztalos
17.1 Introduction *425*
17.2 High Density Lipoprotein Particles and their Metabolism *427*
17.3 Effects of Dietary Cholesterol *431*

17.4 Effects of Dietary Fatty Acids *431*
17.5 Effects of Exchange of Dietary Fat and Carbohydrate *432*
17.6 Effects of Diets Restricted in Cholesterol and Saturated Fat *433*
17.7 Effects of Alcohol Intake *434*
17.8 Effects of Weight Loss *434*
17.9 Effects of Exercise *434*
17.10 Conclusions *435*

V Clinical Aspects of HDL 443

18 HDL and Metabolic Syndrome *445*
Clara Cavelier, Arnold von Eckardstein
18.1 Introduction *445*
18.2 Altered Lipoprotein Metabolism in the Metabolic Syndrome *445*
18.3 Dysregulated Genes and Proteins of HDL Metabolism *447*
18.4 Biological Properties of Impaired HDL *452*
18.5 Treatment of Low HDL-C in Metabolic Syndrome *454*
18.6 Conclusions *456*

19 Genetics of High-Density Lipoproteins *465*
Jacques Genest, Zari Dastan, James C. Engert, Michel Marcil
19.1 Introduction *465*
19.2 Background *466*
19.2.1 Epidemiology of HDL and Risk of Coronary Artery Disease CAD *466*
19.2.2 HDL Metabolism *467*
19.3 Genetic Approaches *468*
19.3.1 Monogenic Disorders, Candidate Genes *468*
19.3.2 Genome-Wide Studies *476*
19.3.3 Segregation of HDL-C *478*
19.3.4 Heritability of HDL-C *479*
19.3.5 Genetic Association Studies *481*
19.4 Conclusions *482*
19.5 Future Considerations *482*

20 HDL and Atherosclerosis *491*
Philip J. Barter, Kerry-Anne Rye
20.1 Introduction *491*
20.2 HDL and Atherosclerosis: Cause and Effect or an Epiphenomenon? *491*
20.3 Potential Antiatherogenic Properties of HDL *492*
20.3.1 HDL and Cholesterol Efflux *492*
20.3.2 Antioxidant Properties of HDL *493*
20.3.3 Antiinflammatory Properties of HDL *494*
20.3.4 Antithrombotic Properties of HDL *495*

20.3.5 HDL and Endothelial Repair *495*
20.4 HDL Subpopulations and Atherosclerosis *496*
20.5 Evidence that Raising the Concentration of HDL Protects against Atherosclerosis *497*
20.5.1 Intervention Studies in Animals *497*
20.5.2 Intervention Studies in Humans *498*
20.6 Conclusions *499*

21 Therapeutic Targeting of High-Density Lipoproteins *507*
Daniel J. Rader
21.1 Introduction *507*
21.2 Current Clinical Approaches to Patients with Low HDL-C *507*
21.3 Approaches to Raising Levels of HDL-C and/or ApoA-I *508*
21.3.1 Increasing HDL/ApoA-I Production and Maturation *508*
21.3.2 Inhibiting ApoA-I Catabolism *509*
21.3.3 Inhibiting Cholesteryl Ester Transfer *510*
21.4 Approaches to Promoting Reverse Cholesterol Transport *511*
21.5 Full-Length ApoA-I and ApoA-I Mimetic Peptides *513*
21.6 Improving HDL Function *514*
21.7 Conclusions *514*

Index *523*

Preface

High-Density Lipoprotein – What it is and What it does

The level of cholesterol in serum high-density lipoprotein (HDL) is a parameter closely linked to the risk of human heart disease. While the "protective" effect of HDL is now widely accepted, the mechanism that supports this is still debated. The original proposal, and one supported by a wide range of evidence, is that HDL is a measure, albeit indirect, of 'reverse' cholesterol transport (RCT), which carries cholesterol from peripheral tissues back to the liver for catabolism. There have been significant recent advances in our understanding of the regulation and mechanism of this reaction sequence. Now, for the first time, serious efforts in drug design with the goal of raising circulating HDL levels are being made.

During RCT, successful formation of typical HDL particles – those α-migrating lipoproteins with densities spanning the 1.063–1.21 g mL^{-1} range – is a prerequisite for a second class of HDL functions now receiving intense scrutiny. Innate immunity, inflammation, and signaling are areas in which HDL plays a significant role. Mature HDL particles act as a scaffold on which proteins important in oxidativc protection are selectively adsorbed.

Convergence of this broad range of studies makes it an appropriate time to bring together chapters dealing with biochemical, physiological, and clinical studies of HDL.

Four chapters deal with the structures and properties of the HDL proteins apoA-I, -A-II, -E and -M. The first two provide important insights into the properties that support protein–lipid binding, ApoE has major significance in cholesterol homeostasis in disease and is the continuing subject of intensive research, and ApoM is a newly recognized apolipoprotein believed to play a key role in HDL recycling.

Five chapters deal with HDL biochemistry – the catalytic factors driving its formation and metabolism – and cover transporters, the enzyme lecithin:cholesterol acyltransferase, lipid transfer proteins, and receptors. These proteins regulate not only RCT but also the distribution of lipids, especially cholesterol, in HDL and other lipoproteins.

Three chapters deal with the secretion and removal of HDL. The cell biological pathways that determine circulating HDL levels are dealt with in detail. One of these chapters describes the use of animal models in HDL research.

High-Density Lipoproteins: From Basic Biology to Clinical Aspects. Edited by Christopher J. Fielding

ISBN: 978-3-527-31717-2

Four chapters describe the significance of HDL in different areas of normal metabolism – nutrition, exercise, and the endothelium.

The remaining five chapters review the rapid progress being made in the human pathology of HDL, and in the treatment of HDL diseases. One, on the abnormal HDL of inflammation, introduces the new field of apo AI-mimetic peptides, and their potential for atherosclerosis protection. The second deals with the new field of HDL and Metabolic Syndrome, the third reviews our current understanding of the human genetics of HDL, and the fourth surveys the structure and functions of HDL in atherosclerosis, the major human HDL-dependent disorder. The final chapter brings the reader up to date on the development of drugs that raise HDL, including both successes and failures, and the biological basis for these outcomes.

These chapters are the result of the thought and research of many investigators who have spent much of their professional lifetimes on the complex, but ultimately fascinating, subject of HDL. I would like to thank all my colleagues for their enthusiasm and dedication in making this volume possible, and hope that the book becomes a useful resource promoting further progress.

Christopher J. Fielding

List of Contributors

G. M. Anantharamaiah
University of Alabama at
Birmingham
Department of Medicine
Atherosclerosis Research Unit
1808 Seventh Avenue, South
Birmingham, AL 35294-0012
USA

Bela F. Asztalos
Tufts University School of Medicine
Lipid Metabolism Laboratory
711 Washington Street
Boston, MA 0211
USA

Ronald Barbaras
Paul Sabatier University
INSERM U563, IFR30
Lipoproteins and Lipid Mediators
department
BP 3028
31024-Toulouse cedex 3
France

Philip J. Barter
Heart Research Institute
114 Pyrmont Bridge Road
Camperdown
Sydney, NSW 2050
Australia

Georgette M. Buga
David Geffen School of Medicine at
UCLA
Division of Cardiology
1038 Le Conte Avenue
Room BH-307 CHS
Los Angeles, CA 90095-1679
USA

Michael I. Bukrinsky
The George Washington University
Department of Microbiology
Immunology and Tropical Medicine
2300 I St. NW, Ross Hall, Rm. 734
Washington, DC 20037
USA

Francisco Blanco-Vaca
Hospital de la Santa Creu i Sant Pau
Servei de Bioquímica
C/Antoni M. Claret 167
08025 Barcelona
Spain

Laura Calabresi
University of Milano
Faculty of Pharmacy
Department of Pharmacological
Sciences
Via Balzaretti 9
20133 Milano
Italy

Clara Cavelier
University Hospital of Zurich,
Institute of Clinical Chemistry and
University of Zurich
Center for Integrative Human
Physiology
Raemistrasse 100
8091 Zurich
Switzerland

Lawrence Chan
Baylor College of Medicine
Department of Medicine
Houston, TX 77030
USA

Giovanna Chimini
Centre d'Immunologie de Marseille
Luminy
INSERM, CNRS, Université de La
Méditerranée
Parc Scientifique de Luminy
Case 906
13288 Marseille Cedex 09
France

Angeliki Chroni
National Center for Scientific
Research "Demokritos"
Institute of Biology
Athens, 15310
Greece

Xavier Collet
Paul Sabatier University
INSERM U563, IFR30
Lipoproteins and Lipid Mediators
department
BP 3028
31024 Toulouse cedex 3
France

Zari Dastani
McGill University Health Center
Royal Victoria Hospital M4.72
687 Pine Avenue West
Montreal QC, H3A 1A1
Canada

Costas Drosatos
University of Crete Medical School
Heraklion
Crete, 71003
Greece

Laetitia Dufort
Centre d'Immunologie de Marseille
Luminy
INSERM, CNRS, Université de La
Méditerranée
Parc Scientifique de Luminy
Case 906
13288 Marseille Cedex 09
France

Christian Ehnholm
Heart Research Institute 135
Missenden Road Camperdown
Sydney, NSW 2050
Australia

James C. Engert
McGill University Health Center
Royal Victoria Hospital M4.72
687 Pine Avenue West
Montreal QC, H3A 1A1
Canada

Joan Carles Escolà-Gil
Hospital de la Santa Creu i Sant Pau
Servei de Bioquimica
C/Antoni M. Claret 167
08025 Barcelona
Spain

Sergio Fazio
Vanderbilt University School of Medicine
Departments of Medicine and Pathology
383 Preston Research Building
Nashville, TN 37232-6300
USA

Christopher J. Fielding
University of California
Cardiovascular Research Institute and
Departments of Physiology
4th and Parnassus
San Francisco, CA 94143-0130
USA

Phoebe E. Fielding
University of California
Cardiovascular Research Institute and
Departments of Medicine
4th and Parnassus
San Francisco, CA 94143-0130
USA

Alan M. Fogelman
University of California LA
David Geffer School of Medicine
Department of Medicine
Room 37–120 Center for Health Sciences
10833 Le Conte Avenue
Los Angeles, CA 90095-1736
USA

Guido Franceschini,
University of Milano
Faculty of Pharmacy
Department of Pharmacological Sciences
via Balzaretti 9
20133 Milano
Italy

Jacques Genest
McGill University Health Center
Royal Victoria Hospital M4.72
687 Pine Avenue West
Montreal QC, H3A 1A1
Canada

Monica Gomaraschi
University of Milano
Faculty of Pharmacy
Department of Pharmacological Sciences
via Balzaretti 9
20133 Milano
Italy

Christopher J. Harder
University of Ottawa
Heart Institute
40 Ruskin St – Lab H453
Ottawa, K1Y 4W7
Canada

Alyssa H. Hasty
Vanderbilt University
School of Medicine
Department of Molecular Physiology & Biophysics
813 Light Hall
Nashville, TN 37232-0615
USA

Jennifer Hoy
Monash University
Department of Medicine
Melbourne, Victoria 8008
Australia

Matti Jauhiainen
National Public Health Institute
Department of Molecular Medicine
Biomedicum
Haartmaninkatu 8
FI-00251 Helsinki
Finland

Josep Julve
Hospital de la Santa Creu i Sant Pau
Servei de Bioquímica
C/Antoni M. Claret 167
08025 Barcelona
Spain

Dimitris Kardassis
University of Crete Medical School
Heraklion Crete, 71003
Greece

Bronwyn A. Kingwell
Baker Heart Institute
P.O. Box 6492
St Kilda Rd Central
Melbourne, Victoria 8008
Australia

Stefania Lamon-Fava
Tufts University School of Medicine
Lipid Metabolism Lab.
711 Washington Street
Boston, MA 0211
USA

MacRae F. Linton
Vanderbilt University School of Medicine
Departments of Pharmacology and Medicine
383 Preston Research Building
Nashville, TN 37232-6300
USA

Michel Marcil
McGill University Health Center
Royal Victoria Hospital M4.72
687 Pine Avenue West
Montreal QC, H3A 1A1
Canada

Jesús M. Martín-Campos
Hospital de la Santa Creu i Sant Pau
Servei de Bioquimica
C/Antoni M. Claret 167
08025 Barcelona
Spain

Laurent O. Martinez
Paul Sabatier University
INSERM U563, IFR30
Lipoproteins and Lipid Mediators department
BP 3028
31024-Toulouse cedex 3
France

Ruth McPherson
University of Ottawa
Heart Institute
40 Ruskin St – Lab H453
Ottawa, K1Y 4W7
Canada

Arshag D. Mooradian
University of Florida
College of Medicine
Department of Medicine
653-1 West 8th Street
4th Floor - LRC
Jacksonville, FL 32209
USA

Arshag D. Mooradian
Saint Louis University School of Medicine
Department of Internal Medicine
1402 S. Grand Blvd.
St. Louis, MO 63104-1004
USA

Mohamad Navab
David Geffen School of Medicine at UCLA
Department of Medicine
Division of Cardiology
Room BH-307 CHS
10833 Le Conte Avenue
Los Angeles, Ca. 90095-1679
USA

Peter Nilsson-Ehle
Lund University, Medical faculty
Department of Laboratory Medicine
221 85 Lund
Sweden

Michael N. Oda
Children's Hospital Oakland Research Institute
Center for Prevention of Obesity, Diabetes and Cardiovascular Disease
5700 Martin Luther King Jr. Way
Oakland, CA 94609
USA

Kazuhiro Oka
Baylor College of Medicine
Department of Molecular and Cellular Biology Houston, TX 77030
USA

Vesa M. Olkkonen
Heart Research Institute
135 Missenden Road
Camperdown Sydney, NSW 2050
Australia

Angeliki Papapanagiotou
Boston University School of Medicine
Center for Advanced Biomedical Research
Whitaker Cardiovascular Institute
Molecular Genetics
Boston, MA 02118
USA

Bertrand Perret
Paul Sabatier University
INSERM U563, IFR30
Lipoproteins and Lipid Mediators
department BP 3028
31024-Toulouse cedex 3
France

Daniel J. Rader
University of Pennsylvania
School of Medicine, 654BRB II/III
421 Curie Boulevard
Philadelphia, PA 19104–6160
USA

Srinivasa T. Reddy
David Geffen School of Medicine at UCLA
Division of Cardiology
1038 Le Conte Avenue
Room BH-307 CHS
Los Angeles, CA 90095-1679
USA

Robert O. Ryan
Children's Hospital Oakland Research Institute
5700 Martin Luther King Jr. Way
Oakland, CA 94609
USA

Kerry-Ann Rye
Heart Research Institute
114 Pyrmont Bridge Road
Camperdown
Sydney, NSW 2050
Australia

Ernst J. Schaefer
Tufts University School of Medicine
Lipid Metabolism Lab.
711 Washington Street
Boston, MA 0211
USA

Sarah Siggins
Heart Research Institute
135 Missenden Road
Camperdown
Sydney, NSW 2050
Australia

Dmitri Sviridov
Baker Heart Institute
P.O. Box 6492
St Kilda Rd Central
Melbourne, Victoria 8008
Australia

Larry L. Swift
Vanderbilt University School of
Medicine
Department of Pathology
CC3327 Medical Center North
Nashville, TN 37232–2561
USA

François Tercé
Paul Sabatier University
INSERM U563, IFR30
Lipoproteins and Lipid Mediators
department
BP 3028
31024-Toulouse cedex 3
France

Brian J. Van Lenten
David Geffen School of
Medicine at UCLA
Division of Cardiology
1038 Le Conte Avenue
Room BH-307 CHS
Los Angeles, CA 90095-1679
USA

Arnold von Eckardstein
Universitätsspital Zürich
Institut für Klinische Chemie
Ramistrasse 100
8091, Zürich
Switzerland

Ning Xu
Lund University, Medical Faculty
Department of Laboratory Medicine
221 85 Lund
Sweden

Eleni E. Zanni
Boston University, School of Medicine
Center for Advanced Biomedical
Research
Whitaker Cardiovascular Institute
Molecular Genetics
Boston, MA 02118
USA

Vassilis I. Zannis
Boston University School of
Medicine
Center for Advanced Biomedical
Research
Whitaker Cardiovascular Institute
Molecular Genetics
Boston, MA 02118
USA

I
HDL STRUCTURE

High-Density Lipoproteins: From Basic Biology to Clinical Aspects. Edited by Christopher J. Fielding

ISBN: 978-3-527-31717-2

1
Apolipoprotein A-I Structure

Michael N. Oda, Robert O. Ryan

1.1
Introduction and Perspective

Human apolipoprotein A-I (apoA-I) is an abundant, multifunctional, exchangeable apolipoprotein whose plasma concentration is inversely correlated with the incidence of cardiovascular disease [1]. This effect is achieved through its ability to serve as the key physiological activator of the plasma enzyme lecithin:cholesterol acyltransferase (LCAT) and as an acceptor of cell membrane cholesterol in the context of the reverse cholesterol transport pathway [2,3]. An important aspect of apoA-I biology relates to its ability to exist in solution in alternate lipid-poor and lipid-associated states. Indeed, lipid-poor apoA-I has the capacity to promote efflux of cholesterol and phospholipid from cell membranes [4], forming pre-ß high density lipoprotein (HDL) particles that ultimately mature into spherical HDL. This process is of fundamental importance for maintenance of whole body cholesterol homeostasis and is accompanied by major structural alterations in apoA-I (see [5–9] for reviews). In an effort to integrate recent findings and to define better how the structure of apoA-I relates to its function, this chapter examines the molecular organization of lipid-free and lipid-associated apoA-I. As described, unique features of apoA-I primary sequence, secondary structure, and tertiary fold contribute to its versatility and conformational adaptability.

1.2
Lipid-Free ApoA-I

Human apoA-I is expressed in the liver and small intestine as a 267 amino acid preproprotein that is targeted to the endoplasmic reticulum for secretion. Following intracellular cleavage of an 18 amino acid signal peptide, proapoA-I is secreted into plasma, where an unknown protease removes a six amino acid propeptide. Mature apoA-I is a 243 amino acid nonglycosylated protein with a calculated molecular mass of 28 079. The protein lacks cysteine and isoleucine, yet contains 37 leucine residues,

High-Density Lipoproteins: From Basic Biology to Clinical Aspects. Edited by Christopher J. Fielding

ISBN: 978-3-527-31717-2

accounting for 15 % of its amino acid content. ApoA-I contains 46 negatively charged (glutamate plus aspartate) amino acids and 37 positively charged (arginine plus lysine) amino acids, giving rise to a theoretical isoelectric point of 5.27. A characteristic feature of the apoA-I amino acid sequence is the presence of 10 proline residues. The positioning of proline residues at regular intervals throughout the sequence, combined with the fact that the cyclic structure of proline imposes geometric constraints that influence secondary structure in proteins, has led to models of apoA-I organization [10]. With the exception of three proline residues that occur within the first seven amino acids of mature apoA-I, the remaining prolines, located at positions 66, 99, 121, 143, 165, 209, and 220, are spaced at regular intervals that have been predicted to demarcate individual amphipathic α helix segments. When considered in the light of spectroscopic studies and computer-based sequence analyses that indicate that the predominant secondary structure in apoA-I is α helix, it has been proposed that, in the absence of lipid, these segments may align with one another to form a globular bundle of α helices stabilized by hydrophobic helix–helix interactions in the bundle interior and endowed with water solubility through the presentation of polar and charged amino acid side chains to the exterior of the bundle [11,12]. Remarkably, as described below, these predictions proved to be an accurate description of the structural organization of lipid free apoA-I.

1.3 The Role of Protein Structure in ApoA-I Function

From a structural and functional perspective, apoA-I can be divided into three structure/function segments: the N-terminal region from residues 1–87, a central portion corresponding to residues 88–181, and a C-terminal segment encompassing residues 182–243. Interestingly, these regions also represent distinct structural aspects of both the lipid-free and the nascent HDL-associated forms of the molecule.

The extreme N-terminal 87 amino acids of apoA-I constitute a unique facet of apoA-I. These residues include the first 43 amino acids of the protein and two predicted helix segments (residues 46–65 and 66–87). Various predicted helix segments in apoA-I have been assigned to one of three classes of amphipathic α helix [13]: class A, with relatively high lipid affinity, class G, with a neutral affinity for lipid, or class Y, with a weak lipid affinity and a propensity for protein–protein interaction. Helical wheel projections (Fig. 1.1) illustrate the differences in charge distribution amongst the classes.

The first 43 amino acids of apoA-I are encoded by exon 3, whereas the remainder of the protein is encoded by exon 4 [14]. Exon 3 is the most highly conserved segment in apoA-I, suggesting that this region has a critical role distinct from the remainder of the protein. Furthermore, lipid-free apoA-I lacking residues 1–43 adopts a conformation resembling that of lipid-bound apoA-I [15,16]. This N-terminal deletion induces a significant decrease in apoA-I stability, as well as an increased number of solvent-exposed hydrophobic sites [17,18], from which it has been hypothesized that these residues serve to stabilize the compact lipid-free conformation by masking cryptic lipid-binding sites that are manifest only after initial lipid interaction [16]. Lipid affinity

Class G
(Residues 46 - 65)

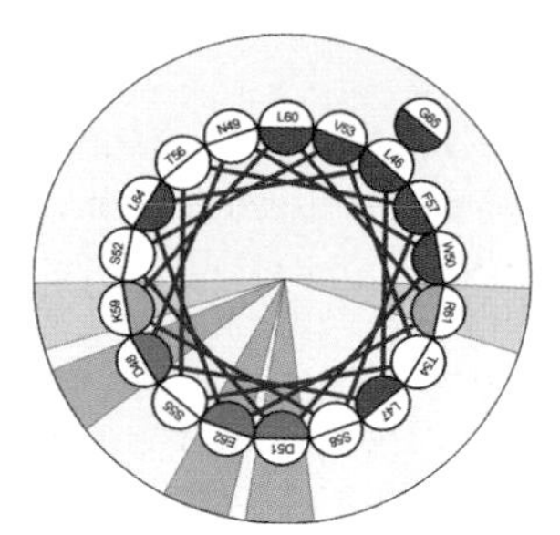

Class Y
(Residues 220- 243)

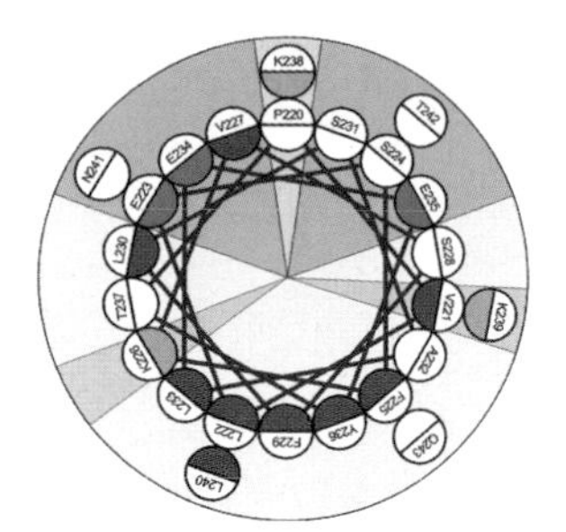

Class A
(Residues 143- 164)

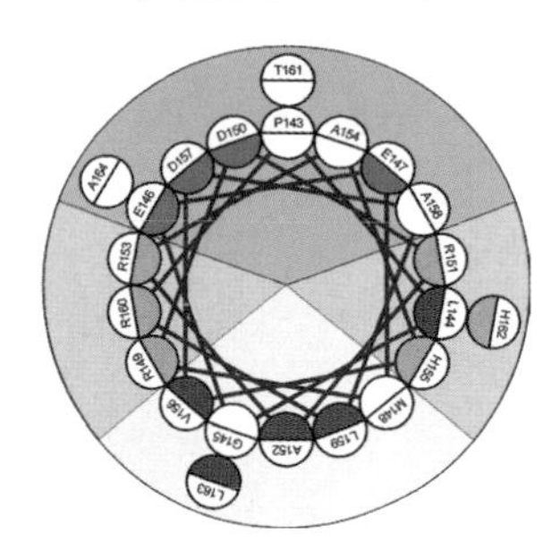

Fig. 1.1 Arrangement of charged residues found in apoA-I amphipathic α helices. Examples of class G (residues 46–65), class A (residues 143–164), and class Y (residues 220–243) helices from apoA-I. The distribution of hydrophobic (gray), negative (red), and positive (blue) residues are shown with the basis of the classification [10,13,106]. The helical classes share a common amphipathic character, but the distribution of charges imparts a differential protein lipid affinity, with class A helices possessing the greatest lipid affinity, class Y the greatest protein affinity, and class G intermediate affinity. Such regional distribution of affinity may play a significant role in imparting functional character to specific segments of apoA-I.

analysis of eight tandem 22 amino acid segments of apoA-I indicate that residues in the N and C termini (44–65 and 220–241) have the greatest lipid-binding affinity [19], suggesting that, if these residues are proximal, they may act in a cooperative fashion to draw apoA-I into contact with biological membranes or lipid complexes.

Further studies of (Δ1–43)apoA-I resulted in an X-ray crystal structure determination at 4 Å resolution [15]. This structure revealed a continuously curved, horseshoe-shaped series of 10 amphipathic α helices, with four truncated apoA-I molecules arranged as pairs of antiparallel dimers. The interaction of molecules A and B to form a stable dimer is due to electrostatic and hydrophobic contacts. The molecular organization of the A/B dimer produces a structure that presents two faces: one predominantly hydrophilic and the other strongly hydrophobic. Interactions between the hydrophobic surfaces of two dimers actually stabilize the dimer structure and give rise to the observed tetrameric organization. It is envisioned that apoA-I dimers, aligned as depicted in the X-ray structure, form an antiparallel "belt" around phospholipid bilayer discs or the circumference of spheroidal HDL (see below).

Interestingly, a majority of the naturally occurring mutations in apoA-I associated with autosomal dominant systemic amyloidosis are located in the first 87 residues, supporting the hypothesis that misfolding of the N-terminal third of the molecule leads to an accumulation of apoA-I in amyloid plaques [20]. These include the substitution mutants Gly26Arg [21], Thr50Arg [22], and Leu60Arg [23], together with the deletion mutant (Δ70–72)apoA-I [24], which are associated with increases in apoA-I ß strand content. Indeed, Andreola and colleagues [25] observed that misfolding of the N terminus to a predominantly ß strand conformer leads to amyloid fibril formation. *In vivo*, the predominant form of apoA-I present in amyloid plaques is a proteolytic fragment consisting of the first 80–93 residues [26], implicating the apoA-I N terminus as a potential contributing factor in systemic amyloidosis, as well as in the overall conformational stability of the molecule.

While the N-terminal residues play a significant role in maintaining the solubility of lipid-free apoA-I, Scott et al. [27] observed through deletion of the first 65 residues that this region is necessary for both *in vivo* and *in vitro* maturation of HDL particles. Collectively, these results suggest that the function of the apoA-I N terminus may be to establish an initial conformational state, which directs apoA-I down a structural transition pathway with minimal potential for conformational dead ends during lipid association and mature HDL formation.

Interestingly, while a majority of apoA-I on nascent HDL adopts α helical secondary structure (~80% by far-UV circular dichroism spectroscopy [28]) and is relatively immobile in relation to the lipid bilayer, the central residues (amino acids 88–183) represent a more conformationally dynamic region, which responds to changes in HDL lipid composition. Marcel and colleagues suggest that amino acids in the central region form a putative "hinge" domain [29,30]. These residues are attributed with apoA-I's ability to activate LCAT, suggesting that LCAT activation and regional conformational plasticity may be related. It is conceivable that cholesteryl ester dependent alterations in this region's conformation serve as a means by which apoA-I reports to LCAT the status of HDL's cholesterol cargo.

This hypothesis is supported by the observation that efficient LCAT activation is specific to HDL class. For instance, spherical (cholesteryl ester core containing) HDL has a LCAT affinity 36 times lower than that of discoidal HDL [28,31]. The differential affinity of LCAT for HDL subclasses may be due to differences in apoA-I conformation on discoidal HDL versus spherical HDL [32], particularly in residues 121–165 [33]. Increases in HDL particle phospholipid and cholesterol content each lead to specific conformational changes in the central region of apoA-I [29]. Curtiss and Banka [34] determined that these conformational changes are a result of tertiary and/or quaternary structural rearrangements and not the result of secondary structure content or rotation of α helices. These findings raise the possibility that the LCAT binding site on apoA-I is a structural element unique to the discoidal HDL complex formed by the tertiary or quaternary conformational adaptation of the "hinge" domain, a structural feature of nascent discoidal HDL.

A complementary set of LCAT recognition domains in apoA-I was identified from LCAT binding competition studies using monoclonal antibodies (Mabs) directed against specific segments of apoA-I. Banka and associates identified residues 95–121 [35] as part of the LCAT activation/binding site, whereas Uboldi et al. [36] point to residues 96–174. Interestingly, Uboldi and colleagues observed two Mabs, with specificity for noncontiguous segments of apoA-I (residues 124–128 and 144–148) [36]. Because of the effectiveness of one of these Mabs at disrupting LCAT activity, the authors proposed that the LCAT recognition site might be composed of nonsequential apoA-I segments that bind LCAT in a cooperative fashion. It is unclear whether this epitope is derived from an intermolecular or an intramolecular alignment of apoA-I segments on nascent HDL.

Engineered mutations have revealed that the amino acids within the central region are essential to LCAT activation. Deletion analyses of apoA-I have implicated residues 143–165 [37–42], together with the first 73 residues of apoA-I [27,43], as critical to LCAT activation. Point mutations corroborate deletion analyses implicating residues 144–164. Specifically, single amino acid substitutions at positions 149, 153, and 160 significantly disrupt apoA-I's ability to activate LCAT [18,44].

Similarly, naturally occurring single amino acid mutants – apoA-I_{FIN} (L159R) [45] and apoA-$I_{MARBURG}$/apoA-$I_{MUNSTER2}$ (ΔK107) [46] – are only 40 % as effective as wild-type apoA-I (apoA-I_{WT}) in activating LCAT. Interestingly, approximately 80 % of the identified apoA-I mutations associated with reduced HDL plasma levels are located in the region between residues 143–187 [47], supporting the hypothesis that the conformation of the central region of apoA-I is essential to the establishment of a stable cholesteryl ester core containing HDL particle.

The importance of apoA-I's C-terminal segment (residues 188–243) in apoA-I function was demonstrated by a dramatic decrease in lipid affinity and cellular association resulting from the deletion of a single amino acid: Glu235 [48,49]. Lipid affinity analysis of the eight tandem 22 amino acid segments of apoA-I indicated that residues in the C terminus (220–241) have the greatest lipid-binding activity [19], suggesting that this region may draw apoA-I into proximity to biological membranes or lipid complexes. Ji and Jonas [50] found that C-terminal truncation of apoA-I had a negative effect on the lipid-binding activity and self-association properties of the protein. In aqueous solution, (Δ193–243)apoA-I displayed an α helix content of 52 %, similar to that reported for full-length apoA-I. This concept is further supported by the observation that other C-terminal deletion mutants, including (Δ190–243)apoA-I [51], (Δ212–233)apoA-I, and (Δ213–243)apoA-I [39], are significantly impaired in lipid-binding ability. In contrast to the well known oligomerization of lipid-free apoA-I in solution [8,52], C-terminal truncated apoA-I exists predominantly as monomers and dimers. In terms of lipid interaction properties, compared to the full-length protein, (Δ193–243)apoA-I solubilized dimyristoylphosphatidylcholine liposomes slowly, yet retained the capacity to form reconstituted HDL complexes with palmitoyloleoylphosphatidylcholine or dipalmitoylphosphatidylcholine when prepared by the sodium cholate dialysis method [53]. The observation that (Δ193–243)apoA-I reconstituted HDL were comparable to full-length apoA-I HDL in terms of LCAT activation is consistent with studies by Sorci-Thomas et al. [40] and Frank et al. [54] defining the region between residues 143 and 164 as critical for LCAT activation (see above). These data suggest that C-terminal amino acids of apoA-I are crucial for self-association and initial lipid binding but are not specifically involved in LCAT activation. In keeping with this, Ren et al. [55] employed an iterative, site-directed mutagenesis strategy to engineer a monomeric apoA-I. These authors found that mutations in the C-terminal segment of apoA-I that replace six hydrophobic residues with either polar or smaller hydrophobic residues yielded a variant protein that was 90 % monomeric at 8 mg mL^{-1}. On the basis of the finding that "monomeric" apoA-I is indistinguishable from WT apoA-I in terms of secondary and tertiary structure and stability, this variant may be useful for high-resolution structural studies in solution.

Panagotopulos et al. [56], through a variety of recombinant methodologies (helix deletion, substitution, duplication), identified residues 220–243 as essential for interaction with ABCA1 transporter and efficient efflux of cholesterol from RAW mouse macrophages [56]. The sequence from residues 220–243 corresponds to a class Y amphipathic α helix, in which there is a lysine residue in the middle of the negative charge cluster of the polar face (see Figure 1.1). This charge distribution is important for proper association with lipid [19] and is a key element in ABCA1-mediated cholesterol efflux [56]. The positions and locations of lysine residues may also contribute to the lipid interaction and cholesterol efflux capacity of this region of apoA-I [57].

The differential recognition of apoA-I in various lipidation states by conformation-specific apoA-I antibodies [58–60] provides evidence that apoA-I undergoes a multi-step structural transition process during lipid association, explaining the variety of predicted conformational states. During apoA-I's transition from a compact to an extended structure [61] there is an increase in α helical content, from 40–50 % to 70–80 % [28], and a coincident decrease in random coil content. A majority of the secondary structure transition occurs at the C terminus and has been hypothesized to drive the overall conformational transition of apoA-I, analogously to that seen in viral fusion proteins such as hemagglutinin A2 [62]. For viral fusion proteins, this structural transformation is a key event during the fusion of viral envelope membranes with host cell membranes [63], as it serves to regulate the lipid fusing capability of these proteins tightly, to align functional domains properly, and to provide thermodynamic energy for the membrane fusion process [63]. Tanaka and colleagues [64] confirm that the random coil to α helix transition in apoA-I is an energy-yielding process.

Emerging from these studies is the concept that the N terminus of apoA-I functions to maintain the lipid-free conformation of the protein. In response to lipid surface availability, however, these stabilizing forces may be overcome and the protein may undergo a conformational change that results in substitution of helix–helix interactions for helix–lipid interactions. Spectroscopic studies examining this concept suggest that the N-terminal segment of apoA-I is stabilized by interactions with the C terminus of the protein [65,66]. Interestingly (considering the effect of the N-terminal truncation alone), a globular structure is retained when N- and C-terminal truncations [67] are combined, which is consistent with the concept that N-terminal–C-terminal interactions stabilize the lipid-free conformation of apoA-I. This has led to the concept of a lipid-sensitive molecular switch that functions as follows: in the absence of lipid, N- and C-terminal interactions serve to maintain the protein in a globular state. The extreme C terminus of the protein, serving as a lipid sensor, probably initiates contact with available lipid surface binding sites, altering its interaction with the N-terminal portion of the protein. This, in effect, triggers an extensive conformational change in which lipid-binding sites throughout the protein become exposed and available for interaction [62]. Combined, these findings lead to the conclusion that distinct structural elements in apoA-I function in a coordinated fashion to direct apoA-I from the lipid-free state to a biologically active lipid-bound state.

1.4
Full-Length ApoA-I X-Ray Structure

A long sought-after X-ray structure of lipid-free full-length apoA-I emerged in 2006 [68]. Consistently with early hydrodynamic studies, the molecule is asymmetric and consists of two antiparallel helical bundles, an N-terminal four-helix bundle, and a C-terminal two-helix bundle (Fig. 1.2). An interesting feature of this structure is the presence of helices longer than those anticipated from sequence analysis and the absence of proline punctuation for prolines 66, 121, and 165, which reside in the middles of helices. The helices in the bundle are connected primarily by Type IVb turns. The four-helix bundle is stabilized by shielding of >6500 Å^2 of accessible surface area from solvent. The hydrophobic core of the bundle is formed by side chains from each of the four helices in roughly equal proportions. In particular, the N-terminal domain (residues 1–43, Helix A) forms an integral part of the bundle, perhaps explaining the finding that deletion of this segment induces the formation of an extended "open" conformation [15]. Less clear from this structure, however, is the nature of potential N-terminal–C-terminal interactions predicted to function in maintenance of overall helix bundle stability. Overall, about 82% of the residues in this structure adopt α helix secondary structure. This value is higher than values

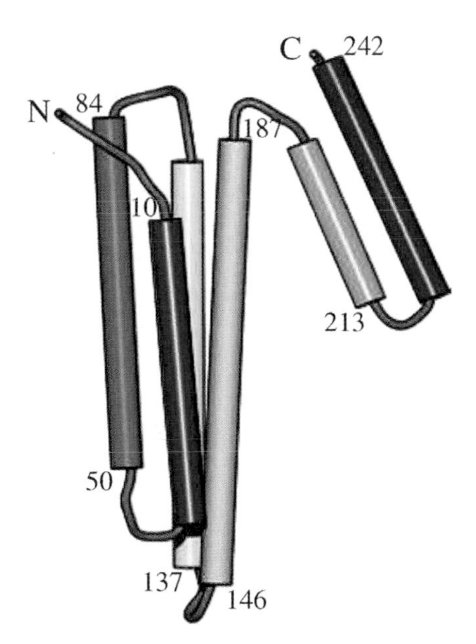

Fig. 1.2 Structural model of full-length lipid-free apoA-I. Helix segments are depicted as cylinders. Individual helices in the N-terminal four-helix bundle are colored blue (A), pink (B), yellow (C), and lavender (D), while the two C-terminal helices are colored cyan (E) and red (F). Numbers correspond to amino acid sequence. Adapted from Ajees et al. [68] with permission.

obtained by circular dichroism spectroscopy, but may be related to the high concentration of protein used in crystallization experiments. This discrepancy has prompted the suggestion that the reported crystal structure may actually correspond to one conformational state among a continuum that apoA-I adopts in solution as a metastable lipid surface seeking protein. Indeed, conformational flexibility is considered a hallmark feature of apoA-I and is probably related to the high lipid-binding activity observed for the protein and its general susceptibility to temperature or guanidine·HCl-induced unfolding [69]. Evidence that the crystal structure represents an accurate depiction of apoA-I solution structure was obtained by Silva et al. [70], who employed a combined mass spectrometry/molecular modeling approach to derive a model structure for apoA-I. This independent analysis yielded a three-dimensional model consisting of an elongated α helix bundle in which the predicted helix boundaries are in general agreement with those obtained by X-ray crystallography. The topology of the structural model also predicts that the N and C termini of the molecule should reside in close proximity to one another. The overall similarity of this model to that obtained by X-ray crystallography provides validation of this approach for structural analysis of members of the apolipoprotein family.

ApoA-I has long been considered to adopt a structure corresponding to a "molten globule" in solution, with the assumption of a metastable conformation that is poised to adapt to lipid surface availability. This feature of apoA-I probably contributed to the difficulty in crystallization of the protein. Thus, an important question that arises pertains to whether the X-ray structure corresponds to the true solution conformation of apoA-I. While it is likely that multiple conformations exist, it is reasonable to consider that the present structure represents one of the conformations that occur naturally in solution and exists in equilibrium with other, perhaps more metastable, conformations. In this manner, it may be considered that conventionally purified apoA-I, which normally involves denaturation and refolding steps, exists as a continuum of conformational states in equilibrium with one another. Regardless of the extent to which this structure accurately depicts the natural state of lipid-free apoA-I, it provides a wealth of new information that can be used in the design of experiments to examine how the protein alters its structure to achieve a stable binding interaction with lipid.

An important aspect of the three-dimensional structure of apoA-I reported by Ajees et al. [68] is that it provides an opportunity to develop coherent and rational hypotheses to examine how this protein alters its conformation upon association with a lipid surface. Others have employed three-dimensional structural information on insect apolipophorin III and human apolipoprotein E N-terminal domains in the design of experiments to decipher how these proteins alter their conformations in the presence of lipid. Examples of strategies include site-specific fluorescence resonance energy transfer [71–73] and disulfide bond engineering [74]. Given the availability of the apoA-I helix bundle structure, these approaches can be applied to test models of lipid-associated apoA-I, as well as differences in apoA-I conformation that may arise from binding to different sized lipid particles. In the meantime, however, Ajees and coworkers [68] have presented a model of potential lipid-induced structural conversions in apoA-I that ultimately result in an open extended conformation. These authors suggest that loop segments connecting helices in the bundle can function as hinges. Upon initiation of lipid binding, presumably through exposed hydrophobic regions on the exterior of the protein, a cooperative binding process is likely to ensue, in which conformational changes in the protein maximize accessibility of helix segments to lipid surface sites. The authors suggest two geometrically conceivable ways in which the four-helix bundle can open: one in which loop segments at one end of the protein function as hinges, and another in which the single loop at the opposite end serves this role. In either case the end result is an unfurling of the bundle to achieve an open, extended conformation. Given the resolution of this structure it should be possible to use structure-guided protein engineering to determine which end of the helix bundle initiates contact with lipid and which end serves as the hinge domain, permitting the adoption of an open conformation.

Comparison of the X-ray structure of apoA-I with other known apolipoprotein structures reveals important similarities [75]. As one example, the presence of a compact, globular N-terminal four-helix bundle is highly reminiscent of the apolipoprotein E N-terminal domain [76] and is similar to the five-helix bundle structure observed for invertebrate apolipophorins [77,78]. In addition, the presence of an independently folded C-terminal domain with strong lipid-binding activity is similar to that observed for the C-terminal domain of human apolipoprotein E [79]. Lipid-free apoA-I thus shares structural similarities with other members of the exchangeable apolipoprotein family and so may be reflective of a general strategy for proteins that are capable of a dual existence in alternate lipid-free and lipid-associated states.

1.5 Apolipoprotein A-I Structure – Lipid-Bound State

The origin of nascent HDL in plasma is generally thought to be either the association of apoA-I with remnants of phospholipid derived from the lipolysis of triglyceride-rich lipoproteins [80] or direct secretion from the liver [81]. The conformation of apoA-I is a regulator of the synthesis, secretion, and formation of nascent and mature HDL [6,7,82], so an understanding of apoA-I conformation is essential to our comprehension of nascent HDL biogenesis and function. For the past four decades, considerable effort has gone into determining apoA-I's lipid-bound structure. This work has been hampered by apoA-I's conformational plasticity and its ability to adapt to the dynamic lipid milieu of the HDL particle. Consequently, an array of lipid-dependent conformations have been observed, and the predominant tertiary and quaternary structures of lipid-bound apoA-I thus remain incompletely defined. Determining apoA-I's lipid-associated conformation is further complicated by the spectrum of HDL subclasses (Table 1.1) and the growing number of HDL-associated proteins that influence apoA-I conformation and HDL affinity. As an exchangeable apolipoprotein, when apoA-I interacts with certain lipids or lipid vesicles, such as DMPC, it rapidly reorganizes the lipid into discoidal structures, first observed by electron microscopy [83]. Discoidal HDL is composed of a lipid-bilayer circumscribed by apolipoprotein. The acyl chains of the phospholipid at the edge of the disc are

Tab. 1.1 [a]Particle size and compositional heterogeneity of select HDL subspecies from plasma.

HDL diameter (nm)	Cholesterol/apoA-I[b]	Phospholipid/apoA-I[b]
8.1–8.3	13	5
8.5	11	15
9.5	29	30
10.4	33	57
10.9	38	72
11.4	83	91

[a]Adapted from [6].
[b]Molar ratios of total cholesterol and phospholipid to apoA-I are from Cheung et al. [105].

shielded from solution by protein. These structures typically range in size from 7.4 nm to 17.5 nm in diameter with densities of 1.063 to 1.21 g mL^{-1}. Disc complexes with diameters <10.5 nm are considered to possess two apoA-I molecules per particle.

Traditionally, plasma HDL particles were characterized by density ultracentrifugation of human serum [84], in which HDL particles fractionated into density-based subclasses. Advances in electrophoretic methodologies have allowed for the characterization of HDL based on charge and size. When examined by agarose native PAGE electrophoresis, HDL migrate by charge into a fast migrating α-HDL and an intermediate migrating pre-β-HDL, which migrates slightly more quickly than β-migrating lipoprotein particles, typically identified as low density lipoproteins (LDL). Since pre-β-HDL are lipid-poor particles and α-HDL are lipid-associated particles, the origins of the charge differences between α-HDL and pre-β-HDL were attributed to conformational changes in apoA-I arising from lipid association and the charge contribution of lipid [85]. They further suggest that increases in net negative charge arise from conversion of discoidal HDL into spherical HDL containing cholesteryl ester cores.

α-HDL particles can be further differentiated by size. Native PAGE analysis of plasma HDL reveals an array of HDL sizes, summarized in Table 1.1. These different sized HDL reflect specific ratios of phospholipid and cholesterol, in relation to apoA-I, and many of these subspecies of HDL can be reconstituted by combining specific ratios of apoA-I, purified phospholipids, and cholesterol. Thanks to the relative uniformity and predictability of reconstituted nascent discoidal HDL particle preparations, analysis of apoA-I's lipid-bound structure has been focused on this early stage of HDL biogenesis. For the remainder of this chapter, "lipid-bound" states of apoA-I will be referring to the structure of apoA-I on discoidal nascent HDL particles.

1.6 Structural Studies of ApoA-I on Discoidal Nascent HDL

Interestingly, the structural organization of apoA-I associated with reconstituted HDL was initially suggested to be a "belt-like" conformation on the discoidal particle [87]. An alternative model, the "picket fence", was proposed by Nolte and Atkinson [87] and is composed of a series of 22 amino acid amphipathic α helices, separated by

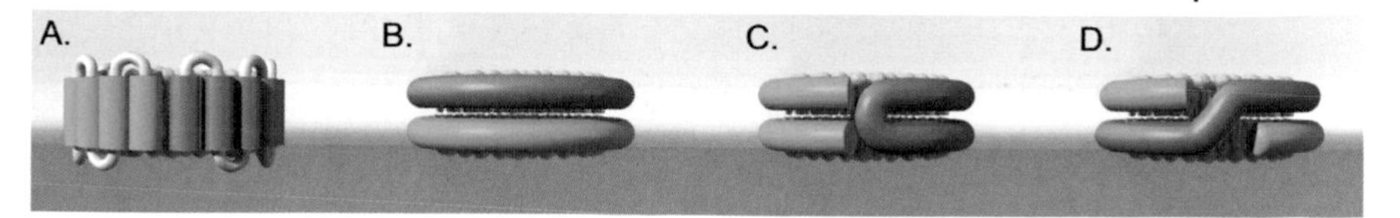

Fig. 1.3 *Models of apoA-I alignment on discoidal nascent HDL.* Proposed helical arrangement of apoA-I on the periphery of the HDL disc. ApoA-I has traditionally been presented as the "picket fence" (A) but recent evidence has eliminated this from consideration. Current data support the "belt" conformation on HDL (B) and other models in which apoA-I adopts extended helical structures. Direct evaluation suggests that the "tandem hairpin" (C) [101] and "z-scheme" (D) models [70,101,102] are unlikely conformational states of apoA-I.

reverse turns at proline residues. In the picket fence model the amphipathic helices align parallel to the fatty acyl chains of the phospholipids in the bilayer disc [25,35] (Fig. 1.3), where amphipathic α helix segments align perpendicular to the lipid plane of the disc. Because of the early absence of contradictory physical evidence, apoA-I on nascent HDL was traditionally presented in the form of the picket fence model (Fig. 1.3; for reviews see [7,82]).

While current evidence supports an extended belt-like conformation of apoA-I on nascent HDL, prior to the (Δ1–43)apoA-I crystal structure solution [15], significant contradictory evidence has supported both the belt and the picket fence models. Examination of the interaction between lipid and synthetic peptides that mimic apoA-I amphipathic helices, for instance, supported a picket fence orientation for apoA-I helices and showed that charged residues were involved in protein–protein interactions and did not appear to have a significant effect on lipid head group behavior [88]. This was corroborated by attenuated total reflection infrared spectroscopic analysis of apoA-I's association with lipid [89–91]. In contrast, Koppaka and colleagues had provided evidence to support the belt conformation on the basis of polarized internal reflection infrared spectroscopy [92], which was further substantiated by tryptophan fluorescence quenching studies with strategically labeled phospholipids [93].

In addition to the belt model, alternative extended helical structures (summarized in Fig. 1.3), including a "hairpin" model in which a given apoA-I molecule loops back on itself, have been proposed. The hairpin model was initially proposed by Tricerri et al. [94] on the basis of fluorescence resonance energy transfer (FRET) data, through which N- and C-terminal amino acids (residues 9 and 232) were found to be proximal. Interestingly, subsequent belt models are consistent with the proximal locale of residues 9 and 232. In fact, because of the similarities in alignment found in the hairpin and belt models, they have been difficult to distinguish; FRET experiments, for example, provided support for both the belt [95] and the hairpin [96] models. Davidson et al. [97] initially proposed that both belt and hairpin conformations may coexist on the same HDL disc particle. In contrast, Li et al. [98] suggested that apoA-I monomers adopt a belt conformation, yet can align in alternate helix registries with respect to its paired apoA-I counterpart on nascent HDL. Recent studies of apoA-I conformation on discoidal nascent HDL utilized a combination of chemical cross-linking and mass spectrometry [70,99]. Both studies support an extended belt model, similar to that described by Segrest et al. [100], in which apoA-I monomers are arranged in an antiparallel alignment with a paired apoA-I molecule, centered at residue 129.

In consideration of the extended conformation of apoA-I in the belt model, protein–protein contact on the edge of the bilayer disc must be predominantly intermolecular, whereas the conformation of the hairpin model necessitates that protein–protein contact be intramolecular. Martin et al. [101] utilized specifically engineered apoA-I to discriminate between intra- versus intermolecular association by FRET. In this study FRET was predominantly intermolecular, consistently with an extended belt conformation. Intramolecular FRET observed for probes located at the N and C termini (residues 50 and 230) are consistent with the observations of Tricerri et al [96], and hence with an antiparallel alignment of paired apoA-I molecules around the edge of the disc. The absence of intermolecular energy transfer in this case provides additional support for an antiparallel belt model.

1.7 Hinge Domain

Silva et al. [70] have suggested that apoA-I assumes a belt conformation possessing a hinge-like feature made up of amino acids 99–143, similar to that described earlier by Calabresi et al. [58], who based their conclusion on the results of monoclonal antibody binding studies. Increases in HDL particle phospholipid and cholesterol content each lead to specific conformational changes in this region [29]. Using nitroxide spin label quenching of tryptophan fluorescence, Maiorano et al. [102] identified a region of high flexibility on 7.8 nm HDL, whereas on 9.6 nm HDL these residues were relatively immobile. Curtiss and Banka [34] determined that these conformational changes are a result of tertiary and/or quaternary structural rearrangements and not the result of secondary structural rearrangement or rotation of α helices. Li et al. [103] extended the analysis of apoA-I conformations by proposing that the central region of apoA-I makes contact with the 10.5 nm nascent HDL disc edge but is lifted off the edge of the 9.5 nm particle. The proposal that specific segments of apoA-I make size-specific contact with the nascent HDL disc edge offers a partial explanation for the discontinuous population of nascent HDL sized particles often observable in plasma and reconstituted HDL preparations (Table 1.1).

Martin et al. [101] applied site-directed spin label electron paramagnetic resonance spectroscopy analysis to map the secondary structure of apoA-I on nascent HDL and to identify sites of intermolecular contact. Consistently with circular dichroism analysis, they observed predominantly α helical secondary structure but identified spin coupling at position 133 and position 146. Spin coupling in this instance indicates that these sequence positions reside close (<15 Å) to their counterparts in the 9.5 nm nascent HDL. In contrast, spin labels positioned adjacent to or between these two residues exhibited no spin coupling, suggesting that a loop-like structural element may reside between positions 133 and 146, such that residues 133 and 146 are proximal but the intervening residues are greater than 15 Å apart. Residues 134–145 may thus represent the hinge domain initially hypothesized by Calibresi et al. [58].

As described above, the X-ray crystal structure of full-length lipid-free apoA-I [68] reveals an N-terminal four-helix bundle and two C-terminal helices. Among the interesting structural features of the four-helix bundle conformation is the conspicuous presence of a solvent-exposed loop segment at one end of the bundle that connects helix C and helix D (Fig. 1.2). This loop segment corresponds to residues 137–146. It is thus conceivable that, following interaction with lipid surface binding sites, the helix bundle opens with elongation of the molecule into an extended helical conformation. At the same time, it is plausible that, depending on the lipid environment, the exposed loop segment between residues 137 and 146 may be retained in the lipid-bound state.

The presence of such a loop segment in lipid-associated apoA-I would have the effect of bringing residues 143–165 of one apoA-I into proximity with its paired counterpart on a 9.5 nm nascent HDL. Because residues in this region have been implicated in LCAT activation [18,37–42,44] it is conceivable that loop segment conformational adaptability could serve to modulate LCAT activity. The size of the loop is inversely related to nascent HDL diameter – the loop segment would be larger on a 7.4 nm particle than on a 9.4 nm nascent HDL and, possibly, nonexistent on larger nascent HDL. The central loop could be the primary source of the known conformational adaptability of apoA-I on nascent HDL. Furthermore, centering the loop segment over the nexus of apoA-I alignment allows for conformational adaptation through changes in loop size with minimal disruption of established intermolecular contacts.

1.8 Computer Models

In addition to direct experimental analysis of apoA-I structure, considerable effort has also been dedicated to molecular dynamic simulations of apoA-I on nascent HDL. A majority of computer simulations were applied to test the veracity of proposed models of apoA-I and led to molecular details unavailable from applied biophysical analyses. Recently, Catte et al. [104] presented a unique approach to the evaluation of lipid-bound models of apoA-I and a putative conformational transition process from lipid-free to 9.5 nm diameter nascent HDL particles. From defined conformational initiation states, Catte and colleagues eliminated lipid molecules and dynamically modeled the response of apoA-I to the gradual shrinking of the lipid bilayer component of the nascent HDL particle [104]. From this dynamic analysis, a surprising prediction was made, in which apoA-I's response to a shrinking particle circumference was not to extend segments of protein off the particle, as would be suggested by the hinge segment hypothesis, but the apoA-I ring structure and lipid bilayer twisted into either "figure eight" or "chair" conformations. This process presents an alternative means by which apoA-I adapts to the presence of lipid, potentially independent of the hinge hypothesis.

1.9
Spherical HDL

Little is known of the conformation of apoA-I on spherical HDL particles. Borhani et al. [15] presented one of the first views of apoA-I on spherical HDL. In their hypothetical model, they presented a particle containing four apoA-I units, in which it is proposed that paired apoA-I molecules orient perpendicularly to each other. While not supported by experimental evidence, this model presents a testable scenario for future analysis.

1.10
Future Directions

With higher resolution models of apoA-I structure both in the presence and in the absence of lipid, new avenues of investigation avail themselves: a) the effects of alternate nascent HDL compositions on apoA-I structure, b) the structural consequences of mutations with known biological consequences, c) the conformational adaptation process undergone by apoA-I during nascent HDL formation, and d) the effect of HDL-associated proteins on apoA-I structure. The past decade has been an exciting period for the field of apoA-I and HDL structure. During this time there have been significant and revolutionary advances in investigative methodologies and resultant knowledge from these new approaches. The challenges ahead are to expand our knowledge of other species of HDL, relating structural knowledge to HDL function, and to apply this information to enhance HDL's ability to attenuate and prevent atherosclerosis and its clinical sequelae.

1.11
Acknowledgments

The authors thank Jennifer Beckstead for technical assistance and Diane Kohlman for manuscript preparation. Research support from the National Institutes of Health (HL-77268 and HL-073061) is gratefully acknowledged.

References

1 Noma, A. Yokosuka, T. Kitamura, K. (1983) Plasma lipids and apolipoproteins as discriminators for presence and severity of angiographically defined coronary artery disease. *Atherosclerosis*, **49**, 1–7.

2 Fielding, C. J. Fielding, P. E. (1995) Molecular physiology of reverse cholesterol transport. *J. Lipid Res.*, **36**, 211–228.

3 Oram, J. F. Lawn, R. M. (2001) The gatekeeper for eliminating excess tissue cholesterol. *J. Lipid Res.*, **42**, 1173–1179.

4 Yokoyama, S. (1998) Apolipoprotein-mediated cellular cholesterol

efflux. *Biochim. Biophys. Acta*, **1392**, 1–15.

5 Brouillette, C. G. Anantharamaiah, G. M. (1995) Structural models of human apolipoprotein A-I *Biochim. Biophys. Acta*, **1256**, 103–129.

6 Brouillette, C. G. Anantharamaiah, G. M. Engler, J. A. Borhani, D. W. (2001) Structural models of human apolipoprotein A-I: a critical analysis and review *Biochim. Biophys. Acta*, **1531**, 4–46.

7 Marcel, Y. L. Kiss, R. S. (2003) Structure–function relationship of apolipoprotein A-I: a flexible protein with dynamic lipid association. *Curr. Opin. Lipidol.*, **14**, 151–157.

8 Ryan, R. O. Yokoyama, S. Liu, H. Czarnecka, H. Oikawa, K. Kay, C. M. (1992) Human apolipoprotein A-I liberated from high density lipoprotein without denaturation *Biochemistry*, **31**, 4509–4514.

9 Wang, G. (2002) How the lipid-free structure of the N-terminal truncated human apoA-I converts to the lipid-bound form: new insight from NMR and X-ray structural comparison *FEBS Lett.*, **529**, 157–161.

10 Segrest, J. P. Garber, D. W. Brouillette, C. G. Harvey, S. C. Anantharamaiah, G. M. (1994) The amphipathic α helix: a multifunctional structural motif in plasma apolipoproteins *Adv. Protein Chem.*, **45**, 303–369.

11 Brouillette, C. G. Dong, W. J. Yang, Z. W. Ray, M. J. Protasevich, I. I. Cheung, H. C. Engler, J. A. (2005) Forster resonance energy transfer measurements are consistent with a helical bundle model for lipid-free apolipoprotein A-I. *Biochemistry*, **44**, 16413–16425.

12 Kiss, R. S. Kay, C. M. Ryan, R. O. (1999) Amphipathic α-helix bundle organization of lipid-free chicken apolipoprotein A-I. *Biochemistry*, **38**, 4327–4334.

13 Segrest, J. P. Jones, M. K. DeLoof, H. Brouillette, C. G. Venkatachalapathi, Y. V. Anantharamaiah, G. M. (1992) The amphipathic helix in the exchangeable apolipoproteins: a review of secondary structure and function. *J. Lipid Res.*, **33**, 141–166.

14 Li, W. H. Tanimura, M. Luo, C. C. Datta, S. Chan, L. (1988) The apolipoprotein multigene family: biosynthesis, structure, structure–function relationships, and evolution. *J. Lipid Res.*, **29**, 245–271.

15 Borhani, D. W. Rogers, D. P. Engler, J. A. Brouillette, C. G. (1997) Crystal structure of truncated human apolipoprotein A-I suggests a lipid-bound conformation. *Proc. Natl. Acad. Sci. U.S.A.*, **94**, 12291–12296.

16 Rogers, D. P. Brouillette, C. G. Engler, J. A. Tendian, S. W. Roberts, L. Mishra, V. K. Anantharamaiah, G. M. Lund-Katz, S. Phillips, M. C. Ray, M. J. (1997) Truncation of the amino terminus of human apolipoprotein A-I substantially alters only the lipid-free conformation. *Biochemistry*, **36**, 288–300.

17 Rogers, D. P. Roberts, L. M. Lebowitz, J. Datta, G. Anantharamaiah, G. M. Engler, J. A. Brouillette, C. G. (1998) The lipid-free structure of apolipoprotein A-I: effects of amino-terminal deletions. *Biochemistry*, **37**, 11714–11725.

18 Roosbeek, S. Vanloo, B. Duverger, N. Caster, H. Breyne, J. DeBeun, I. Patel, H. Vandekerckhove, J. Shoulders, C. Rosseneu M., M. Peelman, F. (2001) Three arginine residues in apolipoprotein A-I are critical for activation of lecithin-cholesterol acyltransferase. *J. Lipid Res.*, **42**, 31–40.

19 Palgunachari, M. N. Mishra, V. K. Lund-Katz, S. Phillips, M. C. Adeyeye, S. O. Alluri, S. Anantharamaiah, G. M. Segrest, J. P. (1996) Only the two end helixes of eight tandem amphipathic helical domains of human apoA-I have a significant lipid affinity. Implications for HDL assembly. *Arterioscler. Thromb. Vasc. Biol.*, **16**, 328–338.

20 Wisniewski, T. Frangione, B. (1995) Amyloidosis in Alzheimer's disease. *Lancet*, **346**, 441.

21 Nichols, W. C. Gregg, R. E. Brewer, H. B., Jr. Benson, M. D. (1990) A

mutation in apolipoprotein A-I in the Iowa type of familial amyloidotic polyneuropathy. *Genomics*, **8**, 318–323.

22 Booth, D. R. Tan, S. Y. Booth, S. E. Hsuan, J. J. Totty, N. F. Nguyen, O. Hutton, T. Vigushin, D. M. Tennent, G. A. Hutchinson, W. L. (1995) A new apolipoprotein A-I variant, Trp50Arg, causes hereditary amyloidosis. *QJM*, **88**, 695–702.

23 Soutar, A. K. Hawkins, P. N. Vigushin, D. M. Tennent, G. A. Booth, S. E. Hutton, T. Nguyen, O. Totty, N. F. Feest, T. G. Hsuan, J. J. (1992) MB Pepys, Apolipoprotein AI mutation Arg-60 causes autosomal dominant amyloidosis. *Proc. Natl. Acad. Sci. U.S.A.*, **89**, 7389–7393.

24 Persey, M. R. Booth, D. R. Booth, S. E. vanZyl-Smit, R. Adams, B. K. Fattaar, A. B. Tennent, G. A. Hawkins, P. N. Pepys, M.B., M. B. (1998) Hereditary nephropathic systemic amyloidosis caused by a novel variant apolipoprotein A-I. *Kidney Int.*, **53**, 276–281.

25 Andreola, A. Bellotti, V. Giorgetti, S. Mangione, P. Obici, L. Stoppini, M. Torres, J. Monzani, E. Merlini, G. Sunde, M. (1993) Conformational switching and fibrillogenesis in the amyloidogenic fragment of apolipoprotein A-I. *J. Biol. Chem.*, **278**, 2444–2451.

26 Dobson, C. M. (2001) Protein folding and its links with human disease. *Biochem. Soc. Symp.*, 1–26

27 Scott, B. R. McManus, D. C. Franklin, V. McKenzie, A. G. Neville, T. Sparks, D. L. Marcel, Y. L. (2001) The N-terminal globular domain and the first class A amphipathic helix of apolipoprotein A-I are important for lecithin: cholesterol acyltransferase activation and the maturation of high density lipoprotein *in vivo*. *J. Biol. Chem.*, **276**, 48716–48724.

28 Jonas, A. Wald, J. H. Toohill, K. L. Krul, E. S. Kezdy, K. E. (1990) Apolipoprotein A-I structure and lipid properties in homogeneous, reconstituted spherical and discoidal high density lipoproteins. *J. Biol. Chem.*, **265**, 22123–22129.

29 Bergeron, J. Frank, P. G. Scales, D. Meng, Q. H. Castro, G. Marcel, Y. L. (1995) Apolipoprotein A-I conformation in reconstituted discoidal lipoproteins varying in phospholipid and cholesterol event. *J. Biol. Chem.*, **270**, 27429–27438.

30 Marcel, Y. L. Provost, P. R. Koa, H. Raffai, E. Dac, N. V. Fruchart, J. C. Rassart, E. (1991) The epitopes of apolipoprotein A-I define distinct structural domains including a mobile middle region. *J. Biol. Chem.*, **266**, 3644–3653.

31 Kosek, A. B. Durbin, D. Jonas, A. (1999) Binding affinity and reactivity of lecithin cholesterol acyltransferase with native lipoproteins. *Biochem. Biophys. Res. Commun.*, **258**, 548–551.

32 Sparks, D. L. Phillips, M. C. Lund-Katz, S. (1992) The conformation of apolipoprotein A-I in discoidal and spherical recombinant high density lipoprotein particles. ^{13}C NMR studies of lysine ionization behavior. *J. Biol. Chem.*, **267**, 25830–25838.

33 Curtiss, L. K. Bonnet, D. J. Rye, K. A. (2000) The conformation of apolipoprotein A-I in high-density lipoproteins is influenced by core lipid composition and particle size: a surface plasmon resonance study. *Biochemistry*, **39**, 5712–5721.

34 Curtiss, L. K. Banka, C. L. (1996) Selection of monoclonal antibodies for linear epitopes of an apolipoprotein yields antibodies with comparable affinity for lipid-free and lipid-associated apolipoprotein. *J. Lipid Res.*, **37**, 884–892.

35 Banka, C. L. Bonnet, D. J. Black, A. S. Smith, R. S. Curtiss, L. K. (1991) Localization of an apolipoprotein A-I epitope critical for activation of lecithin-cholesterol acyltransferase. *J. Biol. Chem.*, **266**, 23886–23892.

36 Uboldi, P. Spoladore, M. Fantappie, S. Marcovina, S. Catapano, A. L. (1996) Localization of apolipoprotein A-I epitopes involved in the activation of lecithin:cholesterol acyltransferase. *J. Lipid Res.*, **37**, 2557–2568.

37 Lindholm, E. M. Bielicki, J. K. Curtiss, L. K. Rubin, E. M. Forte, T. M. (1998) Deletion of amino acids

Glu146→Arg160 I in human apolipoprotein A-I (ApoA-ISeattle) alters lecithin:cholesterol acyltransferase activity and recruitment of cell phospholipid. *Biochemistry*, **37**, 4863–4868.

38 McManus, D. C. Scott, B. R. Frank, P. G. Franklin, V. Schultz, J. R. Marcel, Y. L. (2000) Distinct central amphipathic α-helices in apolipoprotein A-I contribute to the *in vivo* maturation of high density lipoprotein by either activating lecithin-cholesterol acyltransferase or binding lipids. *J. Biol. Chem.*, **275**, 5043–5051.

39 Minnich, A. Collet, X. Roghani, A. Cladaras, C. Hamilton, R. L. Fielding, C. J. Zannis, V. I. (1992) Site-directed mutagenesis and structure–function analysis of the human apolipoprotein A-I. Relation between lecithin-cholesterol acyltransferase activation and lipid binding. *J. Biol. Chem.*, **267**, 16553–16560.

40 Sorci-Thomas, M. Kearns, M. W. Lee, J. P. (1993) Apolipoprotein A-I domains involved in lecithin-cholesterol acyltransferase activation. Structure:function relationships. *J. Biol. Chem.*, **268**, 21403–21409.

41 Sorci-Thomas, M. G. Curtiss, L. Parks, J. S. Thomas, M. J. Kearns, M. W. Landrum, M. (1998) The hydrophobic face orientation of apolipoprotein A-I amphipathic helix domain 143–164 regulates lecithin:cholesterol acyltransferase activation. *J. Biol. Chem.*, **273**, 11776–11782.

42 Sorci-Thomas, M. G. Thomas, M., Curtiss, L., Landrum, M., (2000) Single repeat deletion in ApoA-I blocks cholesterol esterification and results in rapid catabolism of delta6 and wild-type ApoA-I in transgenic mice. *J. Biol. Chem.*, **275**, 12156–12163.

43 Sviridov, D., Hoang, A., Sawyer, W. H., Fidge, N. H., (2000) Identification of a sequence of apolipoprotein A-I associated with the activation of Lecithin:Cholesterol acyltransferase. *J. Biol. Chem.*, **275**, 19707–19712.

44 Cho, K. H., Durbin, D. M., Jonas, A., (2001) Role of individual amino acids of apolipoprotein A-I in the activation of lecithin:cholesterol acyltransferase and in HDL rearrangements. *J. Lipid Res.*, **42**, 379–389.

45 Miettinen, H. E., Jauhiainen, M., Gylling, H., Ehnholm, S., Palomaki, A., Miettinen, T. A., Kontula, K., (1997) Apolipoprotein AI-FIN (Leu159→Arg) mutation affects lecithin cholesterol acyltransferase activation and subclass distribution of HDL but not cholesterol efflux from fibroblasts. *Arterioscler. Thromb. Vasc. Biol.*, **17**, 3021–3032.

46 Rall, S. C., Jr., Weisgraber, K. H., Mahley, R. W., Ogawa, Y., Fielding, C. J., Utermann, G., Haas, J., Steinmetz, A., Menzel, H. J., Assmann, G., (1984) Abnormal lecithin:cholesterol acyltransferase activation by a human apolipoprotein A-I variant in which a single lysine residue is deleted. *J. Biol. Chem.*, **259**, 10063–10070.

47 Sorci-Thomas, M. G., Thomas, M. J., (2002) The effects of altered apolipoprotein A-I structure on plasma HDL concentration. *Trends Cardiovasc. Med.*, **12**, 121–128.

48 Han, H., Sasaki, J., Matsunaga, A., Hakamata, H., Huang, W., Ageta, M., Taguchi, T., Koga, T., Kugi, M., Horiuchi, S., Arakawa, K., (1999) A novel mutant, apoA-I nichinan (Glu235→0), is associated with low HDL cholesterol levels and decreased cholesterol efflux from cells. *Arterioscler. Thromb. Vasc. Biol.*, **19**, 1447–1455.

49 Huang, W., Sasaki, J., Matsunaga, A., Han, H., Li, W., Koga, T., Kugi, M., Ando, S., Arakawa, K., (2000) A single amino acid deletion in the carboxy terminal of apolipoprotein A-I impairs lipid binding and cellular interaction. *Arterioscler. Thromb. Vasc. Biol.*, **20**, 210–216.

50 Ji, Y., Jonas, A., (1995) Properties of an N-terminal proteolytic fragment of apolipoprotein AI in solution and in reconstituted high density lipoproteins. *J. Biol. Chem.*, **270**, 11290–11297.

51 Holvoet, P., Zhao, Z., Vanloo, B., Vos, R., Deridder, E., Dhoest, A., Taveirne, J., Brouwers, E., Demarsin, E., Engelborghs, Y., Rosseneu, M., Collen, D., Brasseur, R., (1995) Phospholipid binding and lecithin-cholesterol acyltransferase activation properties of apolipoprotein A-I mutants. *Biochemistry*, **34**, 13334–13342.

52 Vitello, L. B., Scanu, A. M., (1976) Studies on human serum high density lipoproteins. Self-association of apolipoprotein A-I in aqueous solutions. *J. Biol. Chem.*, **251**, 1131–1136.

53 Laccotripe, M., Makrides, S. C., Jonas, A., Zannis, V. I., (1997) The carboxyl-terminal hydrophobic residues of apolipoprotein A-I affect its rate of phospholipid binding and its association with high density lipoprotein. *J. Biol. Chem.*, **272**, 17511–17522.

54 Frank, P. G., N'Guyen, D., Franklin, V., Neville, T., Desforges, M., Rassart, E., Sparks, D. L., Marcel, Y. L., (1998) Importance of central α-helices of human apolipoprotein A-I in the maturation of high-density lipoproteins. *Biochemistry*, **37**, 13902–13909.

55 Ren, X., Zhao, L., Sivashanmugam, A., Miao, Y., Korando, L., Yang, Z., Reardon, C. A., Getz, G. S., Brouillette, C. G., Jerome, W. G., (2005) Engineering mouse apolipoprotein A-I into a monomeric, active protein useful for structural determination. *Biochemistry*, **44**, 14907–14919.

56 Panagotopulos, S. E., Witting, S. R., Horace, E. M., Hui, D. Y., Maiorano, J. N., Davidson, W. S., (2002) The role of apolipoprotein A-I helix 10 in apolipoprotein-mediated cholesterol efflux via the ATP-binding cassette transporter ABCA1. *J. Biol. Chem.*, **277**, 39477–39484.

57 Brubaker, G., Peng, D. Q., Somerlot, B., Abdollahian, D. J., Smith, J. D., (2006) ApolipoproteinA-I lysine modification effects on helical content, lipid binding and cholesterol acceptor activity. *Biochim. Biophys. Acta*, **1761**, 64–72.

58 Calabresi, L., Meng, Q. H., Castro, G. R., Marcel, Y. L., (1993) Apolipoprotein A-I conformation in discoidal particles: evidence for alternate structures. *Biochemistry*, **32**, 6477–6484.

59 Collet, X., Perret, B., Simard, G., Raffai, E., Marcel, Y. L., (1991) Differential effects of lecithin and cholesterol on the immunoreactivity and conformation of apolipoprotein A-I in high density lipoproteins. *J. Biol. Chem.*, **266**, 9145–9152.

60 Pio, F., De Loof, H., Vu Dac N., Clavey, V., Fruchart, J. C., Rosseneu, M., (1988) Immunochemical characterization of two antigenic sites on human apolipoprotein A-I: localization and lipid modulation of these epitopes. *Biochim. Biophys. Acta*, **959**, 160–168.

61 Jonas, A., Kezdy, K. E., Wald, J. H., (1989) Defined apolipoprotein A-I conformations in reconstituted high density lipoprotein discs. *J. Biol. Chem.*, **264**, 4818–4824.

62 Oda, M. N., Forte, T. M., Ryan, R. O., Voss, J. C., (2003) The C-terminal domain of apolipoprotein A-I contains a lipid-sensitive conformational trigger. *Nat. Struct. Biol.*, **10**, 455–460.

63 Carr, C. M., Kim, P. S., (1993) A spring-loaded mechanism for the conformational change of influenza hemagglutinin. *Cell*, **73**, 823–832.

64 Tanaka, M., Saito, H., Dhanasekaran, P., Wehrli, S., Handa, T., Lund-Katz, S., Phillips, M. C., (2005) Effects of the core lipid on the energetics of binding of ApoA-I to model lipoprotein particles of different sizes. *Biochemistry*, **44**, 10689–10695.

65 Behling Agree, A. K., Tricerri, M. A., Arnvig McGuire, K., Tian, S. M., Jonas, A., (2002) Folding and stability of the C-terminal half of apolipoprotein A-I examined with a Cys-specific fluorescence probe. *Biochim. Biophys. Acta*, **1594**, 286–296.

66 Fang, Y., Gursky, O., Atkinson, D., (2003) Structural studies of N- and C-terminally truncated human

apolipoprotein A-I. *Biochemistry*, **42**, 6881–6890.

67 Beckstead, J. A., Block, B. L., Bielicki, J. K., Kay, C. M., Oda, M. N., Ryan, R. O., (2005) Combined N- and C-terminal truncation of human apolipoprotein A-I yields a folded, functional central domain. *Biochemistry*, **44**, 4591–4599.

68 Ajees, A. A., Anantharamaiah, G. M., Mishra, V. K., Hussain, M. M., Murthy, H. M., (2006) Crystal structure of human apolipoprotein A-I: insights into its protective effect against cardiovascular diseases. *Proc. Natl. Acad. Sci. U.S.A.*, **103**, 2126–2131.

69 Saito, H., Lund-Katz, S., Phillips, M. C., (2004) Contributions of domain structure and lipid interaction to the functionality of exchangeable human apolipoproteins. *Prog. Lipid Res.*, **43**, 350–380.

70 Silva, R. A., Hilliard, G. M., Li, L., Segrest, J. P., Davidson, W. S., (2005) A mass spectrometric determination of the conformation of dimeric apolipoprotein A-I in discoidal high density lipoproteins. *Biochemistry*, **44**, 8600–8607.

71 Fisher, C. A., Narayanaswami, V., Ryan, R. O., (2000) The lipid-associated conformation of the low density lipoprotein receptor binding domain of human apolipoprotein E. *J. Biol. Chem.*, **275**, 33601–33606.

72 Fisher, C. A., Ryan, R. O., (1999) Lipid binding-induced conformational changes in the N-terminal domain of human apolipoprotein E. *J. Lipid Res.*, **40**, 93–99.

73 Narayanaswami, V., Szeto, S. S., Ryan, R. O., (2001) Lipid association-induced N- and C-terminal domain reorganization in human apolipoprotein E3. *J. Biol. Chem.*, **276**, 37853–37860.

74 Narayanaswami, V., Wang, J., Kay, C. M., Scraba, D. G., Ryan, R. O., (1996) Disulfide bond engineering to monitor conformational opening of apolipophorin III during lipid binding. *J. Biol. Chem.*, **271**, 26855–26862.

75 Saito, H., Dhanasekaran, P., Nguyen, D., Holvoet, P., Lund-Katz, S., Phillips, M. C., (2003) Domain structure and lipid interaction in human apolipoproteins A-I and E, a general model. *J. Biol. Chem.*, **278**, 23227–23232.

76 Wilson, C., Wardell, M. R., Weisgraber, K. H., Mahley, R. W., Agard, D. A., (1991) Three-dimensional structure of the LDL receptor-binding domain of human apolipoprotein E. *Science*, **252**, 1817–1822.

77 Breiter, D. R., Kanost, M. R., Benning, M. M., Wesenberg, G., Law, J. H., Wells, M. A., Rayment, I., Holden, H. M., (1991) Molecular structure of an apolipoprotein determined at 2.5 Å resolution. *Biochemistry*, **30**, 603–608.

78 Wang, J., Sykes, B. D., Ryan, R. O., (2002) Structural basis for the conformational adaptability of apolipophorin III, a helix-bundle exchangeable apolipoprotein. *Proc. Natl. Acad. Sci. U.S.A.*, **99**, 1188–1193.

79 Weisgraber, K. H., (1994) Apolipoprotein E: structure–function relationships. *Adv. Protein Chem.*, **45**, 249–302.

80 Eisenberg, S., (1984) High density lipoprotein metabolism. *J. Lipid Res.*, **25**, 1017–1058.

81 Hamilton, R. L., Williams, M. C., Fielding, C. J., Havel, R. J., (1976) Discoidal bilayer structure of nascent high density lipoproteins from perfused rat liver. *J. Clin. Invest.*, **58**, 667–680.

82 Klon, A. E., Segrest, J. P., Harvey, S. C., (2002) Molecular dynamics simulations on discoidal HDL particles suggest a mechanism for rotation in the apo A-I belt model. *J. Mol. Biol.*, **324**, 703–721.

83 Forte, T. M., Nichols, A. V., Gong, E. L., Levy, R. I., Lux, S., (1971) Electron microscopic study on reassembly of plasma high density apoprotein with various lipids. *Biochim. Biophys. Acta*, **248**, 381–386.

84 Anderson, D. W., Nichols, A. V., Forte, T. M., Lindgren, F. T., (1977)

Particle distribution of human serum high density lipoproteins. *Biochim. Biophys. Acta*, **493**, 55–68.

85 Davidson, W. S., Sparks, D. L., Lund-Katz, S., Phillips, M. C., (1994) The molecular basis for the difference in charge between pre-ß and α-migrating high density lipoproteins. *J. Biol. Chem.*, **269**, 8959–8965.

86 Tall, A. R., Small, D. M., Deckelbaum, R. J., Shipley, G. G., (1977) Structure and thermodynamic properties of high density lipoprotein recombinants. *J. Biol. Chem.*, **252**, 4701–4711.

87 Nolte, R. T., Atkinson, D., D., (1992) Conformational analysis of apolipoprotein A-I and E-3 based on primary sequence and circular dichroism. *Biophys. J.*, **63**, 1221–1239.

88 Epand, R. M., Surewicz, W. K., Hughes, D. W., Mantsch, H., Segrest, J. P., Allen, T. M., Anantharamaiah, G. M., (1989) Properties of lipid complexes with amphipathic helix-forming peptides. Role of distribution of peptide charges. *J. Biol. Chem.*, **264**, 4628–4635.

89 Brasseur, R., De Meutter, J., Vanloo, B., Goormaghtigh, E., Ruysschaert, J. M., Rosseneu, M., (1990) Mode of assembly of amphipatic helical segments in model high-density lipoproteins. *Biochim. Biophys. Acta*, **1043**, 245–252.

90 Corijn, J., Deleys, R., Labeur, C., Vanloo, B., Lins, L., Brasseur, R., Baert, J., Ruysschaert, J. M., Rosseneu, M., (1993) Synthetic model peptides for apolipoproteins. II. Characterization of the discoidal complexes generated between phospholipids and synthetic model peptides for apolipoproteins. *Biochim. Biophys. Acta*, **1170**, 8–16.

91 Wald, J. H., Coormaghtigh, E., DeMeutter, J., Ruysschaert, J. M., Jonas, A., (1990) Investigation of the lipid domains and apolipoprotein orientation in reconstituted high density lipoproteins by fluorescence and IR methods. *J. Biol. Chem.*, **265**, 20044–20050.

92 Koppaka, V., Silvestro, L., Engler, J. A., Brouillette, C. G., Axelsen, P. H., (1999) The structure of human lipoprotein A-I. Evidence for the "belt" model. *J. Biol. Chem.*, **274**, 14541–14544.

93 Panagotopulos, S. E., Horace, E. M., Maiorano, J. N., Davidson, W. S., (2001) Apolipoprotein A-I adopts a belt-like orientation in reconstituted high density lipoproteins. *J. Biol. Chem.*, **276**, 42965–42970.

94 Tricerri, M. A., Behling Agree, A. K., Sanchez, S. A., Jonas, A., (2000) Characterization of apolipoprotein A-I structure using a cysteine-specific fluorescence probe. *Biochemistry*, **39**, 14682–14691.

95 Li, H., Lyles, D. S., Thomas, M. J., Pan, W., Sorci-Thomas, M. G., (2000) Structural determination of lipid-bound apoA-I using fluorescence resonance energy transfer. *J. Biol. Chem.*, **275**, 37048–37054.

96 Tricerri, M. A., Behling Agree, A. K., Sanchez, S. A., Bronski, J., Jonas, A., (2001) Arrangement of apolipoprotein A-I in reconstituted high-density lipoprotein disks: an alternative model based on fluorescence resonance energy transfer experiments. *Biochemistry*, **40**, 5065–5074.

97 Davidson, W. S., Hilliard, G. M., (2003) The spatial organization of apolipoprotein A-I on the edge of discoidal high density lipoprotein particles: a mass spectrometry study. *J. Biol. Chem.*, **278**, 27199–27207.

98 Li, H. H., Lyles, D. S., Pan, W., Alexander, E., Thomas, M. J., Sorci-Thomas, M. G., (2002) ApoA-I structure on discs and spheres. Variable helix registry and conformational states. *J. Biol. Chem.*, **277**, 39093–39101.

99 Bhat, S., Sorci-Thomas, M. G., Alexander, E. T., Samuel, M. P., Thomas, M. J., (2005) Inter-molecular contact between globular N-terminal fold and C-terminal domain of apoA-I stabilizes its lipid-bound conformation: studies employing chemical cross-linking and mass

spectrometry. *J. Biol. Chem.*, **280**, 33015–33025.

100 Segrest, J. P., Jones, M. K., Klon, A. E., Sheldahl, C. J., Hellinger, M., De Loof, H., Harvey, S. C., (1999) A detailed molecular belt model for apolipoprotein A-I in discoidal high density lipoprotein. *J. Biol. Chem.*, **274**, 31755–31758.

101 Martin, D. D., Budamagunta, M. S., Ryan, R. O., Voss, J. C., Oda, M. N., (2006) Apolipoprotein A-I assumes a "looped belt" conformation on reconstituted high density lipoprotein. *J. Biol. Chem.*, **281**, 20418–20426.

102 Maiorano, J. N., Jandacek, R. J., Horace, E. M., Davidson, W. S., (2004) Identification and structural ramifications of a hinge domain in apolipoprotein A-I discoidal high-density lipoproteins of different size. *Biochemistry*, **43**, 11717–11726.

103 Li, L., Chen, J., Mishra, V. K., Kurtz, J. A., Cao, D., Klon, A. E., Harvey, S. C., Anantharamaiah, G. M., Segrest, J. P., (2004) Double belt structure of discoidal high density lipoproteins: molecular basis for size heterogeneity. *J. Mol. Biol.*, **343**, 1293–1311.

104 Catte, A., Patterson, J. C., Jones, M. K., Jerome, W. G., Bashtovyy, D., Su, Z., Gu, F., Chen, J., Aliste, M. P., Harvey, S. C., Harvey, S. C., Li, L., Weinstein, G., Segrest, J. P., (2006) Novel changes in discoidal high density lipoprotein morphology: a molecular dynamics study. *Biophys. J.*, **90**, 4345–4360.

105 Cheung, M. C., Segrest, J. P., Albers, J. J., Cone, J. T., Brouillette, C. G., Chung, B. H., Kashyap, M., Glasscock, M. A., Anantharamaiah, G. M., (1987) Characterization of high density lipoprotein subspecies: structural studies by single vertical spin ultracentrifugation and immunoaffinity chromatography. *J. Lipid Res.*, **28**, 913–929.

106 Segrest, J. P., De Loof, H., Dohlman, J. G., Brouillette, C. G., Anantharamaiah, G. M., (1990) Amphipathic helix motif: classes and properties. *Proteins*, **8**, 103–117.

2
Apolipoprotein A-II

Joan Carles Escolà-Gil, Jesús M. Martín-Campos, Josep Julve, Francisco Blanco-Vaca

2.1
Apolipoprotein A-II Structure and Plasma Level Regulation

2.1.1
Structural Organization and Transcriptional Regulation of the Gene

Apolipoprotein A-II (apoA-II) is a member of the apolipoprotein multigene superfamily, which includes genes encoding other apolipoproteins: apoAs (I, II, IV, and V), apoCs (I, II, and III), and apoE. The members of the family share a similar structural organization, which suggests that they evolved from a common ancestor by duplication/deletion of a 22 amino acid repeat [1]. The human apoA-II gene (*APOA2*) is located between two Alu sequences in chromosome 1q23.3, approximately 159.46 megabases (Mb) from the p end of the chromosome and spanning a region of 2.7 kb (NCBI 36 assembly of the human genome). *APOA2* has four exons and three introns (Fig. 2.1). The transcription of human *APOA2*, which takes place mainly in the liver, produces a 473-base pair (bp) mRNA. Transcription is controlled by a complex array of regulatory elements located in the promoter between nucleotides −903 to −33 with respect to the transcription start site. The promoter elements can be separated into four proximal (A to CD), four middle (D to H), and six distal (I to N) regulatory elements [2]. This 870 bp region is sufficient to restrict *APOA2* expression to the liver in transgenic mice [3]. Liver-specific expression of apoA-II is controlled by the interaction of transcription factors such as the heat-stable protein upstream stimulatory factors (USF) 1 and 2, the CCAAT enhancer-binding protein (C/EBP), and the heat-labile factor AIIABI [4]. USF1/USF2a heterodimers and USF2a homodimers bind to the proximal element B as well as to distal elements K and L. USF2a also induces apo A-II expression by cooperative binding with the orphan nuclear receptor hepatic nuclear factor 4 (HNF-4) to the hormone response element present in the distal regulatory element J. Phosphorylation is a well documented modulator of HNF-4 activity [5], and a recent study showed that a high plasma level of glucose could increase apoA-II transcription by inactivation of HNF-4 through phosphorylation [6]. Other orphan nuclear receptors, such as v-Erb-related receptor 2 (EAR-2), EAR-3, and

High-Density Lipoproteins: From Basic Biology to Clinical Aspects. Edited by Christopher J. Fielding

ISBN: 978-3-527-31717-2

apolipoprotein regulatory element 1 (ARP-1), also bind to distal element J and reduce apoA-II expression. A positive regulator of apoA-II transcription is the peroxisomal proliferator-activated receptor α (PPARα), which, forming heterodimers with retinoid X receptor α (RXRα), binds RXRα or PPARα ligands and agonists (see Section 2.1.4). Proximal element AB contains a functional thyroid hormone response element that strongly binds RXRα/T3Rβ heterodimer, which also binds to element J [7]. Sterol regulatory element-binding protein 1 (SREBP-1) and SREBP-2 bind to inverted palindromic or direct repeat motifs in proximal regulatory elements B and CD, middle regulatory element DE, and distal element K. SREBP-1 also binds to the middle regulatory element HI [8,9]. Binding motifs in B and K elements overlap with the USF binding site [10], though their role in the *in vivo* transcription of *APOA2* has not been established. These data point to mechanisms by which plasma levels of apoA-II may change through alteration of its gene transcription rate. This may be physiologically important, since plasma levels of apoA-II are more influenced, at population level, by its synthesis rate than by its catabolism [11]. A good illustration of this can be found in a report of low apoA-II, apoC-III, Lp(a), and triglyceride levels in individuals heterozygous for an HNF-4α mutation [12].

ApoA-II transcription is also modulated by some regulatory elements located in introns and exons. The first intron (169 bp long) is situated in the 5′ untranslated region of the gene, 24 bp upstream of the initiator methionine codon, and appears to contain two negative regulatory elements (NRE): NRE-I and NRE-II [13]. The second intron (293 bp) is positioned very close to the peptidase cleavage site and shows an acceptor site characterized by a polymorphic GT repeat instead of the consensus polypyrimidine tract. The effect of this unusual acceptor signal on intron splicing seems to be compensated for by an exonic splicing enhancer (ESE) located between

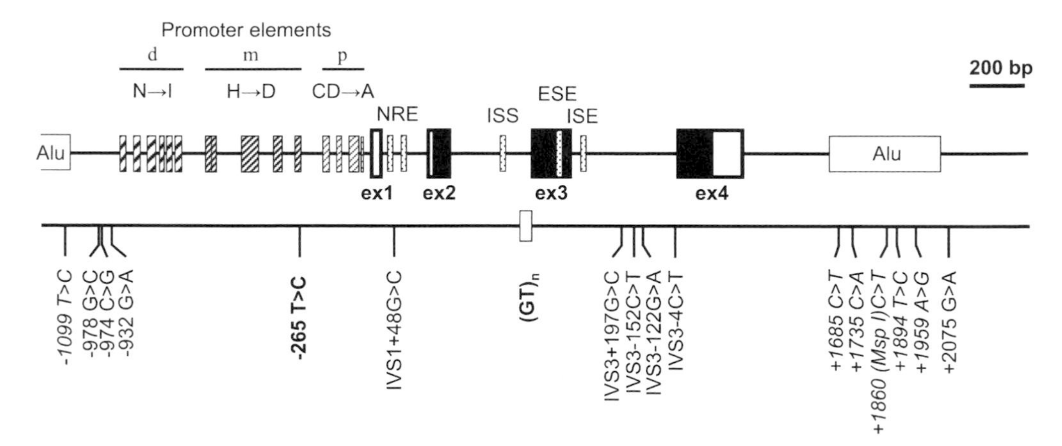

Fig. 2.1 Schematic representation of the *APOA2* gene showing described functional regions and the position of polymorphic sites identified in population studies. Regulatory elements located in the promoter are divided into distal (d), middle (m) and proximal (p) functional regions. NRE: negative regulatory elements I and II in intron 1. ISS: intronic splicing silencer. ESE: exonic splicing enhancer. ISE: intronic splicing enhancer. Polymorphic sites located within the Alu sequences are in italics, and putative functional variants in bold.

nucleotides 91 and 102 (+664 to +675) of exon 3 [14]. This ESE is a *cis*-acting element that binds serine-arginine-rich (SR) proteins, alternative splicing factor/splicing factor 2 (ASF/SF2), and splicing component 35 (SC35). Also located in the second intron are two inhibitors of exon 3 incorporation into the mature mRNA: an intronic splicing silencer (ISS) located between positions +448 and +477, which binds factor hnRNPH1, and the GT repeat which binds the splicing factor TDP-43 [15]. The cystic fibrosis transmembrane regulator gene (*CTFR*) also presents a polymorphic GT tract near the intron 8 acceptor site. Exon 9 of *CTFR* undergoes alternative splicing and its inclusion is inversely correlated with the length of the GT tract [16]. It is interesting to note that *in vitro* studies using the $(GT)_{16}$ allele have been conducted with apoA-II, while a significant association between the $(GT)_{19}$ allele and decreased apoA-II and LpA-I:A-II plasma levels has been reported in a healthy European population [17]. Finally, the third intron (395 bp) divides the region encoding the mature protein into two portions, possibly with different functions, and contains an intron splicing enhancer (ISE) located between positions +746 and +772 that binds SRp40 and SRp55. *In vitro* studies suggest that the incorporation of exon 3 to the mature RNA could be regulated by several *cis*- and *trans*-acting factors and, therefore, that regulation changes could cause alternative splicing *in vivo*. In fact, the existence of an alternative mRNA without the sequence of exon 3 has been reported in liver (EMBL accession number AB106565) in relation to late-onset Alzheimer's disease. However, no further details are currently available in sequence databases.

2.1.2
ApoA-II Mutations and Polymorphisms

Several population studies have shown the existence of genetic variability in the *APOA2* region [18–20], although none of the nucleotide variants found affects the amino acid sequence (Fig. 2.1). Alterations in apoA-II protein sequence have only been described in very rare cases. One was a Val to Met substitution at codon 68 (V68M), caused by a G→A transition, partially associated in heterozygosis with late-onset Alzheimer's disease found in one of 67 unrelated affected Caucasian families [21]. Another was a C-terminal extension of 21 amino acids caused by four different mutations at the stop codon: transitions T→C (X78R) [22] and T→A (X78R) [23] and transversions T→G (X78G) [24] and G→C (X78S), which are dominantly associated with non-neuropathic renal amyloidosis [25]. Another rare mutation was a G→A transition called $A\text{-}II_{\text{Hiroshima}}$ (IVS3 + 1G > A) found in a Japanese family [26]. This mutation caused loss of intron 3 normal splicing and, in homozygosis, was associated with apoA-II deficiency. ApoA-II deficiency was not associated with major effects in the lipid and lipoprotein profiles of affected family members.

Twenty-five polymorphic sites between the two flanking Alu sequences of the *APOA2* region are listed in the genetic variation databases, but only 10 have been published: nine single-nucleotide polymorphisms (SNP) and the $(GT)_n$ repeat mentioned previously, which has been used as a marker in several linkage studies. Four of these SNPs are located in the promoter region, but only one (rs5082; transition −265T > C) affects one of the regulatory elements described in the *APOA2* promoter (element D). Transient transfection studies in HepG2 cells showed a reduced transcription rate of the −265C allele, which is consistent with a significant association between decreased plasma apoA-II concentration and the −265C allele found in a healthy Swedish population [18]. It is thus noteworthy that −265T > C was the most common genetic variant found in population studies on *APOA2* variability, fluctuating between 0.23 in 1686 African-American and 0.39 in 1776 non-Hispanic European-American samples [20]. The other eight SNPs – three in the promoter (−978G > C, −974C > G, and −932G > A), one in intron 1 (IVS1 + 48G > C), and four in intron 3 (IVS3 + 197G > C, IVS3-152C > T, IVS3-122G > A, and IVS3-4C > T) – do not appear to affect functional regions, so −265T > C seems to be the only functional variant segregating in populations at a frequency higher than 0.01 %. Polymorphic positions in the two flanking Alu sequences have also been described; one of them (rs12 143 180), located 528 bp downstream of the end of the apoA-II transcript, alters a recognition sequence for the restriction enzyme *Msp* I and has often been used for association studies. These studies are discussed in Section 2.2.2.3, but the results seem to show that this variant could exist in linkage disequilibrium with another functional variant located in the adjacent regions.

2.1.3 Protein Structure, Synthesis, Secretion, and Metabolism

Structural Characteristics Most apoA-II is present in human plasma in the form of disulfide-linked homodimers of 17.4 kDa formed by two 77-residue-long monomers. Human apoA-II, unlike that of many other species, has a single cysteine (Cys6) that forms disulfide dimers with other cysteine-containing apolipoproteins such as apoE or apoD [27–33]. The two animal species in which apoA-II has been mostly studied are man and mouse (Table 2.1). In contrast with the human form, murine apoA-II exists exclusively as a monomer because of the absence of a Cys residue in position 6 [31,34]. Human and murine apoA-II amino acid sequences differ by approximately 40 %, which explains why monomeric human apoA-II expression in transgenic mice did not have effects similar to those of murine apoA-II [35]. These differences in primary structure may therefore account for some of the different effects of mouse and human apoA-II on lipoprotein metabolism (see Section 2.2).

Both the apoA-II dimers and the carboxymethylated monomer associate with lipids to form lipoprotein complexes. The structural bases of interaction with lipids of human apoA-II are common to other apolipoproteins and attributable to the amphipathic α helices present in each of them. They present helical 11-mer sequence repeats that are highly conserved among species [36]. The 11-mer periodicity in apoA-II has been shown to be the integral structural unit of the class A amphipathic α helices, the major apolipoprotein lipid motif [37]. Circular dichroism measurements suggest that lipid-bound plasma apoA-II is 70 % helical, corresponding to 54 α helical residues, which could be organized in three 18-residue class A amphipathic α helices [37]. Three such helices had been identified in apoA-II and shown to contribute to the lipid-binding properties of this apolipoprotein [36–38]. Such an apoA-II class A amphipathic helix is, therefore, characterized by a high mean hydrophobic moment

Tab. 2.1 Main comparative phenotype characteristics of apoA-II in humans and genetically modified mice.

Phenotype characteristics	Humans	Transgenic mice Human apoA-II	Mouse apoA-II	ApoA-II-deficient mice
HDL cholesterol	positive correlation with apoA-II [31,34]	↓ [60]	↑[63]	↓ [64]
CETP activity	↓ in LpA-I/A-II compared with LpA-I [66,67] = in reconstituted apoA-II-HDL [65]	a heat-labile inhibitory activity was found in HDL of transgenic mice [68] = in apoA-II/CETP transgenic mice [69,70]		
PLTP activity	↓ PLTP-mediated interconversion of LpA-I/A-II compared with LpA-I [75]			
LCAT activity	↓ in LpA-I/A-II compared with LpA-I and ↓ in reconstituted apoA-II-HDL [76–79,81]	↓ [60]		
HL activity	↑ or ↓ in apoA-II-containing HDL [59,70,84–89]	↓ [70,122] or ↑ [88,99]	↓ [92]	↑ [93]
EL activity		↓ in double apoA-I/apoA-II transgenic mice compared with apoA-I transgenic mice [94]		
cholesterol efflux	Usually ↓ in LpA-I/A-II compared with LpA-I [31,34]	↑ in plasma in presence of ABCA1-expressing macrophages [98] ↓ in plasma in presence of SR-BI-expressing cells [83,98,99]	= in HDL in presence of macrophages [97]	
SR-BI-mediated uptake	↓ in LpA-I/A-II compared with LpA-I and ↓ in apoA-II-containing HDL [104–107] ↑ in one study performed with reconstituted apoA-II-HDL in SR-BI transfected CHO cells [109].	↓ [106]	↓ [108]	↑ [108]

Tab. 2.1 *(Continued)*

Phenotype characteristics	Humans	Transgenic mice Human apoA-II	Mouse apoA-II	ApoA-II-deficient mice
macrophage RCT		↑ in in mice fed a chow diet [101] = in mice fed an atherogenic diet [101]		
PON1 activity	↓ in LpA-I/A-II compared with LpA-I and ↓ in reconstituted apoA-II-HDL [112,113]	↓ in plasma and HDL [111,115]	↓ in HDL [97]	
antioxidant properties of HDL		↑ in mice fed a chow diet [115] ↓ in mice fed an atherogenic diet [111]	↓ in mice fed a chow or an atherogenic diet [97]. HDL is proinflammatory [97]	
VLDL triglyceride	↑ in apoA-II-containing VLDL [46,118]	↑ [60,69,70,83,122,123]	↑ [63,92,121,125]	↓ in apoE knockout mice [64]
VLDL metabolism	Low apoA-II levels enhance postprandial metabolism of large VLDL [18]	↓ postprandial VLDL catabolism in mice fed a chow diet [122] ↑ VLDL production in fasted mice fed an atherogenic diet or cross-breeding with apoE knockout [123,124] ↑ VLDL production and ↓ VLDL catabolism in CETP transgenic mice [69]		↑ VLDL catabolism in apoE knockout mice [64]
LPL activity	↓ in apoA-II-containing VLDL [46,118]	↓ in fed mice [122] = in fasted mice [69,123,124]		
plasma FFA	positive correlation with apoA-II levels [119]	= in mice fed a chow diet [124]	↑ in mice fed a chow diet [125]	↓ [64]

		↑ in mice fed an atherogenic or Western diet and in apoE-knockout and CETP transgenic mice [55,69,123,124]	
adiposity	low apoA-II levels were associated with reduced waist circumference [18]	= in mice fed an atherogenic or Western diet [55,124]	↑ in mice fed a chow diet [125]
glucose tolerance test		= in mice fed an atherogenic or Western diet [55,124] ↓ in mice fed a carbohydrate-rich diet [6]	↓ in mice fed a chow diet [125]
atherogenesis	apoA-II levels are = or ↓ in CAD patients [31,34]	↑ in mice fed an atherogenic diet or cross-bred with apoE knockout mice [88,123] ↑ in CETP transgenic mice [69]	↑ in mice fed a chow or an atherogenic diet [121,171]

and by its single charge distribution. Positively charged residues are clustered at the polar/nonpolar interface, whereas negatively charged residues are found at the center of the polar face [36]. The apolar faces in apoA-II helices are nearly linear, contributing to its lipid-binding properties, since the resulting geometry is well suited to the binding of extended lipid surfaces. In solution, apoA-II adopts fluctuating helical conformations that are thought to facilitate protein–lipid association. Such a dynamic conformational situation has been demonstrated in the association of apoA-I and apoA-II or their peptide analogues with phospholipids, which consistently produce substantial increases in their α helicities. Experiments with synthetic peptides have suggested the presence of at least two lipid-associating domains in apoA-II, located at opposite ends of the molecule [36]. However, the C-terminal helix pair of apoA-II, which constitutes the most hydrophobic domain, is the main factor responsible for the lipid-binding properties of the entire protein [1,36].

Synthesis and Secretion The first translation product of apoA-II mRNA consists of a 100 amino acid precursor, preproapoA-II, containing an 18-residue-long signal peptide [1]. PreproapoA-II signal peptide is co-translationally cleaved to proapoA-II. The newly secreted apoA-II consists of 82 amino acid residues and is converted into the mature form by proteolytic cleavage of a five-residue-long prosegment (Ala-Leu-Val-Arg-Arg) attached to the $-NH_2$ terminus [1]. This prosegment, cleaved by a thiol protease to produce the mature protein form [1], may serve as a signal for apolipoprotein targeting along the secretory pathway, extracellular processing, or maturation of nascent lipoproteins [39]. ApoA-II has no N-linked glycosylation consensus sequence and is not significantly glycosylated in plasma. In lymph and in HepG2 culture media, however, approximately half of apoA-II is *O*-glycosylated. Although *O*-linked glycosylation has been shown not to be necessary for secretion, it reduced the association of the newly synthesized apoA-II with high-density lipoproteins (HDL) [40].

Plasma Concentration and Metabolism ApoA-II is the second major HDL protein, comprising 20 % of total HDL protein. Its concentration in normolipidemic humans is around 30–35 mg dL^{-1}[31,34], although these levels may be somewhat variable in different populations [41–43]. Interestingly, over 20 % of patients with coronary artery disease (CAD) could have concentrations higher than 40 mg dL^{-1}[44,45].

Most apoA-II is found in HDL particles containing both apoA-I and apoA-II (LpA-I/A-II), whereas a minor proportion of apoA-II exists in HDL particles that do not contain apoA-I (LpA-II) [31,34]. A small fraction of apoA-II has also been found associated with chylomicrons and very low density lipoproteins (VLDL) [46]. A relatively major part of LpA-I/A-II is found in HDL_3[31,34]. ApoA-II and LpA-I/A-II levels are mainly influenced by the apoA-II production rate [11,41,47]. *In vivo*, ApoA-II is rapidly incorporated into spherical LpA-I/A-II in plasma by a mechanism that seems to involve a lecithin:cholesterol acyltransferase-dependent (LCAT-dependent) LpA-I/LpA-II particle fusion process [48]. ApoA-II catabolism is complex and, probably, occurs at different rates in different pools [41]. The results of a recent report showed that scavenger receptor B-I (SR-BI) processing of HDL results in a more rapid clearance of apoA-II than of apoA-I and the segregation of apoA-I and apoA-II into HDL remnants [49]. However, the specific mechanism by which apoA-I and apoA-II catabolism are segregated is as yet unknown [49].

2.1.4 Other Determinants of ApoA-II Levels

Several dietary factors affect apoA-II levels. Alcohol consumption increases apoA-II levels, due to a higher apoA-II transport rate [50,51], although a reduced apoA-II fractional catabolic rate (FCR) has been observed after red wine consumption [51]. Different dietary fatty acid (FA) interventions have different marked effects on apoA-II. A palmolein-enriched diet increased plasma apoA-II levels in postmenopausal women [52], whereas a dietary switch from an olive oil and sunflower oil blend to extra virgin olive oil lowered apoA-II, as well as low density lipoprotein (LDL), levels and the estimated cardiovascular risk [53]. This is consistent with the reduction in apoA-II levels in men fed a high-monounsaturated fatty acid (MUFA) diet [54].

Dietary fats also modulate apoA-II metabolism in genetically engineered mice. As in human studies, human apoA-II transgenic mice fed a Western-type diet containing $200\,\mathrm{g\,kg^{-1}}$ fat with a polyunsaturated/saturated FA ratio of 0.07 also showed increased apoA-II levels [55]. Furthermore, the two isomers of linoleic acid (LA) – *cis*-9 *trans*-11- and *trans*-10-*cis*-12-LA – had opposite effects on apoA-II levels and atherogenesis in apoE-deficient mice [56]. Taken together, the results of these studies suggest that fatty acid composition modulates apoA-II levels and HDL particle composition and, thus, may alter the latter's atheroprotective properties. In fact, the human *APOA2* promoter is transactivated by fatty acid derivates, fibrates, and RXRα ligands and agonists through a PPARα/RXRα activation [45,57]. Fenofibrate, however, reduced apoA-II mRNA levels in rat hepatocytes, although these levels were not affected after bezafibrate, gemfibrozil, or Wy-14643 treatment [58], suggesting that apoA-II expression is differently regulated by fibrates in humans and rodents.

2.2 Pathophysiological Role of ApoA-II

2.2.1 On HDL Metabolism

2.2.1.1 HDL Remodeling

ApoA-II can both increase HDL particle stability and cause conformational changes in apoA-I on synthetic HDL particles [59]. These structural alterations influence HDL function by affecting the interfacial protein association with HDL, so HDL remodeling may in part be determined by apoA-II influence on HDL affinity towards lipid transfer proteins, enzymes, and the receptors involved in HDL metabolism, and also by its ability to displace apoA-I competitively from the HDL surface. Analysis of genetically modified mice has provided insight into the species-specific role of this apolipoprotein in HDL remodeling. Overexpression of human apoA-II in transgenic mice induced the formation of smaller HDL particles, which were heterogeneous in size and similar to those present in human plasma, in contrast to the monodisperse HDL population found in mice [60,61]. Interestingly, human apoA-I expression in transgenic mice also caused HDL size heterogeneity [62], suggesting that species-specific HDL particle size differences depend at least on apoA-I and apoA-II structure. In support of this interpretation, overexpression of murine apoA-II in transgenic mice induced the presence of HDL particles larger than those in control mice [63], whereas apoA-II-deficient mice presented decreased HDL particle size [64]. In all these studies, the presence of larger HDL particles in plasma was associated with increased concentrations of HDL cholesterol, whereas the presence of smaller HDL particles was associated with decreased HDL cholesterol [60,61,63,64].

Cholesteryl Ester Transfer Protein (CETP) ApoA-II has been shown to inhibit CETP lipid transfer activity in some, but not all, studies [65–67]. The effect on CETP activity of the apoA-II moiety of HDL has been studied *in vitro* by measurement of the rate of transfer of radiolabeled cholesteryl ester between LDL and HDL_3 [67]. In this study, progressive enrichment of HDL_3 with apoA-II significantly decreased the rate of cholesteryl ester exchange induced by purified human CETP both from apoA-II-enriched HDL_3 to LDL and vice versa [67]. As observed with HDL_3 artificially enriched with apoA-II, cholesteryl ester transfer rates to LpA-I/A-II were significantly decreased in relation to those to Lp-A-I [67]. However, this was not observed in reconstituted HDL [65].

The expression of human apoA-I and/or apoA-II in transgenic mice showed the presence of heat-labile lipid transfer inhibitory activity only in human apoA-II transgenic mice, and not in mice expressing human apoA-I or both human apoA-I and apoA-II [68]. However, this study was conducted *in vitro* by the addition of HDL isolated from plasma. Evidence of CETP activity inhibition was not apparent *in vivo* in another two studies in double human apoA-II/CETP transgenic mice [69,70].

Phospholipid Transfer Protein (PLTP) Plasma PLTP plays an important role in plasma HDL regulation and governs the distribution of HDL subpopulations in a time- and concentration-dependent manner [71]. PLTP has been shown to exist in both active and inactive forms. Active PLTP is present in HDL particles with and without apoA-II; its distribution between these two HDL subpopulations varies widely among individuals [72]. The ability of PLTP to bind both purified apoA-I and apoA-II has been directly shown [73]. A PTLP effect on HDL size conversion has been observed both in LpA-I and in LpA-I/A-II [74], but with lower efficiency for apoA-II-containing particles [75]. In addition, a decreased apoA-I/A-II molar ratio in the HDL particle inhibited PLTP-mediated HDL conversion, but not phospholipid transfer activity, which suggests that apoA-I is necessary for particle fusion [75].

Lecithin:cholesterol Acyltransferase (LCAT) ApoA-II is less efficient than apoA-I for LCAT activation [76]. The affinity of LCAT towars apoA-I-containing reconstituted HDL has been found to be higher than that towards those reconstituted with apoA-II

[77,78]. LCAT-mediated cholesterol esterification in discoidal, reconstituted HDL containing apoA-I and apoA-II has been shown to be lower than in reconstituted HDL containing only apoA-I [79]. Displacement (at high apoA-II/apoA-I ratios) or destabilization (at the physiological apoA-II/apoA-I ratio of 0.5:1) of apoA-I in HDL by apoA-II or its C-terminal end inhibit cholesterol esterification [79] and this mechanism may be important in modulating LCAT activity [77,79,80]. The effect on LCAT activity has been attributed, depending on the study, either to a decrease in K_m or to a decrease in V_{max} [77,79,80]. These observations are consistent with the fact that a larger proportion of LCAT is bound to LpA-I than to LpA-I/A-II in human plasma [81] and also with findings in human apoA-I and apoA-II transgenic mice [60,82,83]. Interestingly, in one of those studies there was a mild but significant corneal accumulation of free cholesterol that mimicked that found in Fish Eye disease [83].

Hepatic Lipase (HL) Studies both *in vitro* and *in vivo* and also with genetically modified animals have reported that apoA-II can both stimulate [84–88] and inhibit [70,89] HL activity. In contrast, they appear to agree in considering that apoA-II increases the affinity of HL towards HDL [90]. *In vitro*, HL displays greater affinity (K_m) (with regard either to phospholipid or to triglyceride hydrolysis) for LpA-II than for LpA-I [91]. The addition of apoA-I (regardless of whether it was present in the same particle as LpA-I/A-II or was added as LpA-I to the incubation of HL and LpA-II) increased phospholipid hydrolysis of LpA-II [89]. In this context, apoA-II appears to affect HL by increasing enzyme association with HDL and inhibiting lipid hydrolysis [59], indicating that apoA-II can inhibit HL-mediated lipid hydrolysis by decreasing the ability of the enzyme to shuttle between HDL substrates or particles. Consistently with this suggestion, studies on mouse apoA-II transgenic [92] and apoA-II-deficient mice crossed with HL-knockout mice [93] provided evidence that mouse apoA-II inhibits HL. However, it should be borne in mind that, unlike in humans, a significant part of mouse HL is associated with HDL.

Endothelial Lipase (EL) The expression of EL in human apoA-I/apoA-II doubly transgenic mice resulted in a lower decrease in HDL cholesterol, phospholipids, and apoA-I levels than in human apoA-I transgenic mice, suggesting that apoA-II inhibits EL activity *in vivo* [94].

2.2.1.2 Reverse Cholesterol Transport Function

Role in Cholesterol Efflux The quantitative *in vivo* importance of apoA-II in the first step of reverse cholesterol transport (RCT) is controversial [31,34,95,96], although there now seems to be some consensus in considering LpA-I to be more active than LpA-I/A-II in cholesterol efflux, at least in some cell types [31,34]. Plasma from murine apoA-II transgenic mice maintained cholesterol efflux from macrophages similar to that of control plasma [97], whereas that from human apoA-II transgenic mice increased cholesterol efflux from J774 macrophages [98] but reduced efflux from Fu5AH hepatoma cells [83,98,99]. This is consistent with the observation that human apoA-II transgenic mice showed an increase in pre-β HDL particles [100,101], which were capable of displaying greater cholesterol and phospholipid efflux capacity from ABCA1-expressing cells [102,103].

Effects on Cholesteryl Ester (CE) Uptake ApoA-II exerted inhibitory effects on selective cholesterol uptake in HepG2, adrenal cells, fibroblasts, and SR-BI-transfected cells [104–107]. Furthermore, the contrasting levels of apoA-II in HDLs from apoA-II transgenic and apoA-II knockout mice were inversely correlated with HDL binding and selective CE uptake to SR-BI and CD36 [108]. These results, however, contrast with those of another study that showed reduced association but increased CE uptake of reconstituted HDL containing apoA-II in SR-BI-transfected CHO cells [109]. Comparisons among these studies are difficult, since one used reconstituted, homogeneous HDL particles and cells overexpressing SR-BI, while the other used mouse HDL and cell systems that could express other HDL receptors beyond SR-BI.

Effects on Macrophage-Specific RCT Although the role of apoA-II on individual RCT steps has been studied in detail [31,34,95,96], little is known as to how apoA-II regulates the entire RCT rate. Furthermore, total RCT may not reflect specific RCT of macrophages, which are thought to be the most important cholesterol-accumulating cells in atherosclerosis [110]. Our group recently demonstrated that human apoA-II maintains effective RCT from macrophages to feces in transgenic mice *in vivo* despite an HDL and LCAT deficiency [101]. However, the potential influence of CETP on HDL remodeling was absent in these mice [66] and its impact on macrophage RCT should be the subject of future investigation.

2.2.1.3 Antioxidant and Antiinflammatory Effects

HDL isolated from murine apoA-II transgenic mice stimulated lipid hydroperoxide formation in artery wall cells and induced monocyte transmigration [97], indicating that mouse apoA-II-containing HDL were proinflammatory. The impaired ability of the HDL of these mice to protect LDL against cell-mediated oxidation was attributed to a decreased paraoxonase (PON1) activity [97]. Similar results were found in transgenic mice overexpressing human apoA-II, which presented impaired HDL protection against oxidative modification of apoB-containing lipoproteins [111]. In these transgenic mice, there was a decrease in apoA-I levels, as well as in PON1 and platelet-activating factor acetylhydrolase (PAF-AH) activities, possibly contributing both to the mentioned HDL impairment and to the increased atherosclerosis susceptibility [111]. The displacement of apoA-I and PON from HDL by human apoA-II could, at least in part, explain these findings [111], which are consistent with the observation in humans that PON1 is mostly found in HDL particles with apoA-I but without apoA-II [112,113]. They are also consistent with the finding that substitution of the central domain of apoA-I by a segment of human apoA-II in apoE-deficient mice enhanced oxidative stress and impaired HDL protection against LDL oxidative modification [114]. In contrast, an independently generated line of human apoA-II transgenic mice was reported to have HDL that displayed increased protection against VLDL oxidative modification despite decreased cholesterol, PON, and PAF-AH content [115]. The reasons for this discrepancy are unclear, but at least two factors may be of importance. Firstly, the latter study was performed in mice fed a chow diet, a situation in which human apoA-II transgenic mice do not display increased atherosclerosis (see Section 2.2.3), and secondly, their plasma human apoA-II concentration was lower than in transgenic mice fed a high-fat/high-cholesterol atherogenic diet [111,115]. It is well known that long-term feeding with an atherogenic diet decreases PON1 [116], whereas higher human plasma apoA-II concentrations may further impair the rest of the HDL antioxidant system formed by apoA-I, PAF-AH, and LCAT.

2.2.2 Role of ApoA-II in Triglyceride and Glucose Metabolism

2.2.2.1 Triglyceride and Free Fatty Acid (FFA) Metabolism

Evidence for a role of apoA-II in triglyceride and FFA metabolism was provided by several early reports that showed a positive correlation between apoA-II synthesis and VLDL-apoB in human beings [117]. Also, apoA-II-containing VLDL obtained from patients with Tangier disease and type V hyperlipoproteinemia were reported to be poor lipoprotein lipase (LPL) substrates [46,118]. The *APOA2* locus has been linked to a locus controlling plasma levels of apoA-II and FFA in both mice and humans [119]. Further evidence for a role of apoA-II in triglyceride and FFA metabolism was provided by the previously mentioned study that found an association of the −265C allele with decreased plasma apoA-II concentration, enhanced postprandial metabolism of large VLDL, and decreased waist circumference in healthy 50-year-old Swedish men [18]. In contrast, the −265C allele appeared to increase visceral adipose tissue accretion susceptibility in a study of white American premenopausal women [120].

A direct effect of apoA-II on triglyceride and FFA metabolism is supported by the studies conducted in genetically modified mice. Mice expressing mouse and human apoA-II presented clear increases in FFA, cholesterol, and/or triglycerides in apoB-containing lipoproteins in the majority of studies [55,60,63,69,70,83,92,121–125], although this was not found in one study [61]. Knocking out of mice apoA-II decreased plasma VLDL concentration and increased remnant catabolism [64]. However, it remains controversial whether apoA-II variability causes increased VLDL synthesis, decreases VLDL catabolism, or both [31,34,95,96,126]. One report showed very high postprandial hypertriglyceridemia in human apoA-II transgenic mice, which also presented a significant increase in plasma human apoA-II in fed mice relative to fasted mice [122]. In the former, their VLDL were a poor substrate for LPL and HL [122]. We have unpublished data showing a dramatic postprandial triglyceride peak at 2–3 hours after an oral fat test in our independently generated human apoA-II transgenic mice (Escolà-Gil et al; unpublished data). In our animals, however, plasma concentrations of human apoA-II did not differ according to fed/fasted status [124]. We speculate that the difference in the plasma apoA-II regulation mechanisms of these two independently generated transgenic mice could be due to the fact that the 3 kb genomic human *APOA2* injected into the mice were not the same: ours was obtained by *Msp* I digestion [60], whereas the other was obtained by *Hind* III digestion [122]. The upstream *Hind* III cleavage site is adjacent to regulatory element AIIN, and this could be significant in human apoA-II synthesis if the lack of sequences located upstream of the *Hind* III site influences binding of transcription factors to distal regulatory elements [31]. Our transgenic mice, either fed an atherogenic, cholate-containing diet or cross-bred with apoE knockout mice and fasted overnight, presented higher VLDL production and normal VLDL-triglyceride catabolism and postheparin lipolytic activities [123,124]. Double CETP/apoA-II transgenic mice were more hyperlipidemic than their parents, due to increased VLDL triglyceride production and impaired VLDL triglyceride catabolism [69]. This may imply that multiple mechanisms were involved in the origin of the hypertriglyceridemia in apoA-II transgenic mice. However, the changes detected in liver VLDL production during fasting could also have been a consequence of a marked delay in postprandial VLDL catabolism. Further, fenofibrate did not correct hypertriglyceridemia in our human apoA-II transgenic mice [127], as occurs in some patients with type I and type V hyperlipidemia.

2.2.2.2 Glucose Homeostasis

The linkage between apoA-II and FFA suggests a potential role of apoA-II in insulin resistance. Indeed, murine apoA-II transgenic mice showed insulin resistance and increased adiposity [125] and apoA-II-deficient mice presented insulin hypersensitivity [64]. HDL is a ligand for the scavenger receptor CD36 [108], which raises the possibility of an apoA-II-CD36 interaction that altered FFA metabolism. However, studies in combined murine apoA-II transgenic and CD36-deficient mice showed that the major effects of mouse apoA-II on FFA were CD36-independent [128]. This study did demonstrate that the β-oxidation decrease present in these apoA-II mice was, at least in part, due to a decrease in medium chain acyl-CoA dehydrogenase [128]. However, we did not find insulin resistance and increased adiposity in our human apoA-II transgenic mice fed either the atherogenic or a Western diet [55,124]. Since it is likely that FFA increase is critical in insulin resistance, these observations could be explained by a more powerful effect of murine apoA-II on FFA in relation to the human protein [55,124,125]. This could also explain the observation that refeeding the human apoA-II transgenic mice – acquired by injecting genomic human *APOA2* obtained by *Hind* III digestion [122] – with a carbohydrate-rich diet activated apoA-II transcription, increased blood glucose levels, and delayed plasma clearance of a glucose bolus [6].

2.2.2.3 Role of ApoA-II in Dyslipidemia and Type 2 Diabetes Mellitus

Several studies using the *APOA2 Msp* I polymorphism – located within the downstream Alu element – showed significant associations with plasma triglycerides [129–131], whereas this observation was not replicated in other studies [132–137]. Since the *Msp* I polymorphism does not affect the protein sequence or any regulatory signal, this lack of reproducibility was interpreted as the potential linkage disequilibrium between the *Msp* I polymorphic site and a functional variant located in the region. If this functional variant had been located in the *APOA2* locus, its potential effect should have been at expression level, since no alterations in the protein sequence have been described in population studies. However, only two of these studies found a significant association between *Msp* I polymorphism and apoA-II plasma level [130,138]. The effect of the rare allele (M^-) on apoA-II plasma level was the opposite in these two studies, although M^-M^- subjects presented a decreased apoA-I/apoA-II ratio in both.

It is noteworthy that the *Msp* I site is in strong linkage disequilibrium with the two variants reported in the *APOA2* locus that seem to be functional: $(GT)_{19}$ and −265C. The M^- allele is always associated with a haplotype characterized by the presence of both IVS3 + 197C and IVS3-4T alleles. They have always been found in perfect linkage disequilibrium, but almost never in the same haplotype as $(GT)_{19}$ or −265C alleles (Table 2.2). Since both alleles were associated with low apoA-II plasma levels [17,18], we would expect an association of M^- with a higher plasma level of apoA-II, as observed in a British sample [138]. However, the opposite was observed in a Chinese sample [130], although genetic variability between these two populations could be expected.

Tab. 2.2 APOA2 SNP haplotypes and frequencies in population samples from Spain (Martín-Campos et al., unpublished data), the US, and Finland [20].

	T -265	G IVS3+197G>C	C IVS3-4C>T	C +1860 (Msp I)C>T	Spain			US		Finland
					B (236)	B-HFC (206)	B-DM2 (172)	A-A (1672)	R (1770)	N (900)
H1	C	·	·	·	0.398	0.374	0.401	0.223	0.385	0.343
H2	·	·	·	·	0.301	0.345	0.308	0.661	0.316	0.409
H3	·	C	·	T	0.191	0.180	0.192	0.070	0.168	0.109
H4	·	·	T	·	0.102	0.102	0.099	0.022	0.115	0.081
H5	·	C	·	·	0.008	nd	nd	0.022	0.014	0.057
H6	C	·	T	·	nd	nd	nd	nd	0.001	nd
H7	·	·	·	T	nd	nd	nd	nd	nd	0.001
H8	C	C	·	·	nd	nd	nd	nd	0.001	nd
H9	C	C	·	T	nd	nd	nd	0.002	nd	nd
H10	C	C	T	T	nd	nd	nd	0.001	nd	nd

B: Healthy control subjects. B-FCHL: Familial combined hyperlipidemia subjects. B-DM2: Type 2 diabetes subjects, Barcelona, Spain. A-A: African-American subjects from Jackson, Mississippi. R: Non-Hispanic European-Americans from Rochester, Minnesota. N: Europeans from North Karelia, Finland. Sample size shown in parentheses. These four SNPs were selected because their variation reflected the major patterns of variation among the haplotypes identified by sequencing
Nd: not detected.

One study in human apoA-II transgenic mice supported a possible relationship between apoA-II, through its effect on triglycerides and FFA metabolism, and familial combined hyperlipidemia (FCHL) [123]. Several linkage studies in FCHL families found a positive signal at 1q21–24 with respect to triglycerides [139] and HDL-c [140,141]. A positive signal at 1q21–24 to FCHL was also obtained in Finnish families [142] and was replicated in affected families from Germany and China [143], the USA [144], and Mexico [145], but not in those from the UK [146]. On the other hand, another genome scan performed both in Dutch and Finnish FCHL families revealed differences between both populations with respect to the chromosome 1q21–q24 contribution to FCHL traits [147]. In one study of Dutch Caucasian FCHL families, differences in apoA-II plasma levels and apoA-II/HDL-C ratio were found between hyperlipidemic and normolipidemic subjects [148]. Furthermore, apoA-II showed a significant linear correlation with age and triglycerides, HDL-C, apoA-I, and apoB levels. These data support a modulatory, rather than causal, role of apoA-II in FCHL, however, since the chromosomal regions associated with apoA-II plasma levels [148,149] did not match those showing linkage with FCHL [150,151]. The ubiquitous transcription factor USF1, also located in 1q21–24, seems to be that responsible for this positive linkage signal in FCHL families. Several genetic variants within *USF1* were associated with FCHL and triglyceride levels in the same Finnish FCHL families in which previous evidence of 1q21–24 linkage was reported [152]. This result has been replicated in FCHL families from Utah [153].

Several studies in humans also linked the *APOA2* region to type 2 diabetes (DM2). Chromosome 1q21–q24, initially identified in Pima Indians [154] and Utah Caucasians [155], was also the best replicated positive signal with linkage to DM2 in Caucasians from France [156], the UK [157], and the USA (Framingham [158], Amish [159]), and also in three Chinese populations [160–162]. A meta-analysis of four European populations [163] showed that, despite the existing differences among samples from Sweden, Finland, France, and the UK, when the pooled genotype data were used, a region including the *APOA2* gene showed significant linkage ($P = 0.027$) with DM2. It is possible that 1q21–q24 contains several loci, rather than a major gene, involved in the susceptibility to DM2, since this region encompasses at least two or three linkage peaks [164]. However, a study of *APOA2* variability in DM2-affected Caucasian families in Utah failed to detect an association between polymorphisms at this locus and the disease, although some degree of association with parameters of glucose metabolism was observed [19].

In summary, it is unlikely that the functional variants identified in *APOA2*, which reduce rather than increase apoA-II concentration, were directly related to the expression of FCHL or DM2. A study in a Spanish population (Table 2.2) also showed no significant differences between gene or haplotype frequencies in healthy control subjects and FCHL or DM2 subjects (Martín-Campos et al., unpublished data). However, the results of several studies support a potentially modulatory role of apoA-II in FCHL and, perhaps, DM2.

2.2.3
ApoA-II in Atherogenesis

Similarly to those of apoA-I, a well known antiatherogenic molecule, plasma apoA-II levels have been inversely correlated with CAD risk in epidemiological studies [31,34]. Recent data from a large prospective epidemiological study on myocardial infarction revealed that both LpA-I and LpA-I/A-II were inversely related to CAD [165]. However, the only kindred known to have inherited apoA-II deficiency did not present clinical evidence of coronary atherosclerosis [26], whereas increased concentrations of apoAII in interstitial fluid were found in CAD patients [166]. Increased apoA-II containing VLDL was related to the progression of CAD in patients treated with lovastatin [167] and the ratio of apoA-II/HDL cholesterol was found to be increased in patients with this disease [168].

The development of transgenic mice expressing either human or murine apoA-II provided further insight into the role of apoA-II in atherosclerosis susceptibility [31,34]. An early report showed that human apoA-II coexpression with human apoA-I in doubly transgenic mice produced a 15-fold reduction in the antiatherogenic effects of human apoA-I in transgenic mice fed a high-fat, atherogenic diet [169]. However, the lesion area in these human apoA-II transgenic mice was not significantly different from that in control mice [170]. Overexpression of mouse apoA-II was found to be proatherogenic, particularly even when the mice were on a regular chow diet [121], an observation that was consistent with early results of a congenic strain that exhibited both increased apoA-II synthesis and aortic fatty streak development [171]. Developed human apoA-II transgenic mice have also generally been found to display increased atherosclerosis susceptibility, but when fed exclusively an atherogenic diet [69,88], when cross-bred with apoE-deficient mice and fed a regular chow diet [123], or as double apoA-II/CETP transgenics [69]. The proatherogenicity of apoA-II overexpression may be explained by the increased concentration of apoB-containing lipoprotein and the decreased protection towards LDL oxidation, two common characteristics of the most studied lines of both human and mouse apoA-II transgenic mice. Only one study using human apoA-II transgenic mice showed decreased atherosclerosis susceptibility after they had been fed an atherogenic diet [99]. In that case, however, the injected transgene construct contained the human antithrombin III promoter, and the apoA-II levels were lower than in other human apoA-II transgenic mice [99]. The main conclusion of these studies is therefore that the single overexpression of apoA-II is likely to be proatherogenic and that increasing its plasma concentration therefore does not seem a desirable target for treating or preventing atherosclerosis.

2.3
Acknowledgments

We are grateful to Christine O'Hara for editorial assistance. This work was funded by FIS grants 02/0373, 03/1058, and 05/1267. J.C.E.-G. is a Ramón y Cajal researcher, funded by the Ministerio de Educación y Ciencia.

References

1 Li, W. H., Tanimura, M., Luo, C. C., Datta, S., Chan, L. (1988) The apolipoprotein multigene family: biosynthesis, structure, structure-function relationships, and evolution. *J Lipid Res.*, **29**, 245–271.

2 Zannis, V. I., Kan, H. Y., Kritis, A., Zanni, E., Kardassis, D. (2001) Transcriptional regulation of the human apolipoprotein genes. *Front Biosci.*, **6**, D456–D504.

3 Le Beyec, J., Benetollo, C., Chauffeton, V., Domiar, A., Lafreniere, M.-J., Chambaz, J., Cardot, P., Kalopissis, A. (1998) The −911/+25 sequence of human apolipoprotein A-II promoter is sufficient to confer liver restricted expression, developmental changes and response to a high fat diet in transgenic mice. *Transgenics*, **2**, 211–220.

4 Cardot, P., Chambaz, J., Cladaras, C., Zannis, V. I. (1991) Regulation of the human ApoA-II gene by the synergistic action of factors binding to the proximal and distal regulatory elements. *J Biol Chem.*, **266**, 24460–24470.

5 Viollet, B., Kahn, A., Raymondjean, M. (1997) Protein kinase A-dependent phosphorylation modulates DNA-binding activity of hepatocite nuclear factor 4. *Mol Cell Biol.*, **17**, 4208–4219.

6 Sauvaget, D., Chauffeton, V., Dugue-Pujol, S., Kalopissis, A. D., Guillet-Deniau, I., Foufelle, F., Chambaz, J., Leturque, A., Cardot, P., Ribeiro, A. (2004) *In vitro* transcriptional induction of the human apolipoprotein A-II gene by glucose *Diabetes*, **53**, 672–678.

7 Hatzivassiliou, E., Koukos, G., Ribeiro, A., Zannis, V., Kardassis, D. (2003) Functional specificity of two hormone response elements present on the human apoA-II promoter that bind retinoid X receptor α/thyroid receptor β heterodimers for retinoids and thyroids: synergistic interactions between thyroid receptor β and upstream stimulatory factor 2a. *Biochem J.*, **376**, 423–431.

8 Kan, H. Y., Pissios, P., Chambaz, J., Zannis, V. I. (1999) DNA binding specificity and transactivation properties of SREBP-2 bound to multiple sites on the human apoA-II promoter. *Nucleic Acids Res.*, **27**, 1104–1117.

9 Pissios, P., Kan, H. Y., Nagaoka, S., Zannis, V. I. (1999) SREBP-1 binds to multiple sites and transactivates the human ApoA-II promoter *in vitro*: SREBP-1 mutants defective in DNA binding or transcriptional activation repress ApoA-II promoter activity. *Arterioscler Thromb Vasc Biol.*, **19**, 1456–1469.

10 Ogami, K., Kardassis, D., Cladaras, C., Zannis, V. I. (1991) Purification and characterization of a heat stable nuclear factor CIIIB1 involved in the regulation of the human ApoC-III gene. *J Biol Chem.*, **266**, 9640–9646.

11 Ikewaki, K., Zech, L. A., Kindt, M., Brewer, H. B., Jr., Rader, D. J. (1995) Apolipoprotein A-II production rate is a major factor regulating the distribution of apolipoprotein A-I among HDL subclasses LpA-I and LpA-I:A-II in normolipidemic humans. *Arterioscler Thromb Vasc Biol.*, **15**, 306–312.

12 Shih, D. Q., Dansky, H. M., Fleisher, M., Assmann, G., Fajans, S. S., Stoffel, M. (2000) Genotype/phenotype relationships in HNF-4α/MODY1: haploinsufficiency is associated with reduced apolipoprotein (AII), apolipoprotein (CIII), lipoprotein(a), and triglyceride levels. *Diabetes*, **49**, 832–837.

13 Bossu, J. P., Chartier, F. L., Vu-Dac, N., Fruchart, J. C., Laine, B. (1994) Transcription of the human apolipoprotein A-II is down-regulated by the first intron of its gene. *Biochem Biophys Res Commun.* **202**, 822–829.

14 Arrisi-Mercado, P., Romano, M., Muro, A. F., Baralle, F. E. (2004) An exonic splicing enhancer offsets the atypical GU-rich 3′ splice site of human apolipoprotein A-II exon 3. *J Biol Chem.*, **279**, 39331–39339.

15 Arrisi-Mercado, P., Ayala, Y. M., Romano, M., Buratti, E., Baralle, F. (2005) Depletion of TDP 43 overrides the need for exonic and intronic splicing enhancers in the human apoA-II gene. *Nucleic Acids Res.*, **33**, 6000–6010.

16 Chu, C. S., Trapnell, B. C., Curristin, S., Cutting, G. R., Crystal, R. G. (1993) Genetic basis of variable exon 9 skipping in cystic fibrosis transmembrane conductance regulator mRNA. *Nat Genet*, **3**, 151–156.

17 Brousseau, T., Dupuy-Gorce, A. M., Evans, A., Arveiler, D., Ruidavets, J. B., Haas, B., Cambou, J. P., Luc, G., Ducimetiere, P., Amouyel, P., Helbecque, N. (2002) Significant impact of the highly informative (CA)n repeat polymorphism of the APOA-II gene on the plasma APOA-II concentrations and HDL subfractions: The ECTIM study. *Am J Med Genet*, **110**, 19–24.

18 van't Hooft, F. M., Ruotolo, G., Boquist, S., deFaire, U., Eggertsen, G., Hamsten, A. (2001) Human evidence that the apolipoprotein A-II gene is implicated in visceral fat accumulation and metabolism of triglyceride-rich lipoproteins. *Circulation*, **104**, 1223–1228.

19 Elbein, S. C., Chu, W., Ren, Q., Wang, H., Hemphill, C., Hasstedt, S. J. (2002) Evaluation of apolipoprotein A-II as a positional candidate gene for familial Type II diabetes, altered lipid concentrations, and insulin resistance. *Diabetologia*, **45**, 1026–1033.

20 Fullerton, S. M., Clark, A. G., Weiss, K. M., Taylor, S. L., Stengard, J. H., Salomaa, V., Boerwinkle, E., Nickerson, D. A. (2002) Sequence polymorphism at the human apolipoprotein AII gene (APOA2): unexpected deficit of variation in an African-American sample. *Hum Genet*, **111**, 75–87.

21 Kamino, K., Wijsman, E., Anderson, L., Nemens, E., Yamagata, H., Ohta, S., Bird, T. H., Schellenberg, G. D., (1999) Iqbal, K. Swaab, D. F. Winblad, B. Wisniewski H. M. *Alzheimer's disease and related disorders* John Wiley & Sons, Indianapolis, 501–506.

22 Yazaki, M., Liepnieks, J. J., Barats, M. S., Cohen, A. H., Benson, M. D. (2003) Hereditary systemic amyloidosis associated with a new apolipoprotein AII stop codon mutation Stop78Arg. *Kidney Int.*, **64**, 11–16.

23 De Gracia, R., Fernandez, E. J., Riñon, C., Selgas, R., Garcia-Bustos, J. (2006) Hereditary renal amyloidosis associated with a novel mutation in the apolipoprotein AII gene. *QJ Med.*, **99**, 274.

24 Benson, M. D., Liepnieks, J. J., Yazaki, M., Yamashita, T., Hamidi Asl, K., Guenther, B., Kluve-Beckerman, B. (2001) A new human hereditary amyloidosis: the result of a stop-codon mutation in the apolipoprotein AII gene. *Genomics*, **72**, 272–277.

25 Yazaki, M., Liepnieks, J. J., Yamashita, T., Guenther, B., Skinner, M., Benson, M. D. (2001) Renal amyloidosis caused by a novel stop-codon mutation in the apolipoprotein A-II gene. *Kidney Int.*, **60**, 1658–1665.

26 Deeb, S. S., Takata, K., Peng, R. L., Kajiyama, G., Albers, J. J. (1990) A splice-junction mutation responsible for familial apolipoprotein A-II deficiency. *Am J Hum Genet*, **46**, 822–827.

27 Brewer, H. B., Jr., Lux, S. E., Ronan, R., John, K. M. (1972) Amino acid sequence of human apoLp-Gln-II (apoA-II), an apolipoprotein isolated from the high-density lipoprotein complex. *Proc Natl Acad Sci U S A.*, **69**, 1304–1308.

28 Weisgraber, K. H. and Mahley, R. W. (1978) Apoprotein (E--A-II) complex of human plasma lipoproteins. I. Characterization of this mixed

disulfide and its identification in a high density lipoprotein subfraction. *J Biol Chem.*, **253**, 6281–6288.
29 Borghini, I., James, R. W., Blatter, M. C., Pometta, D. (1991) Distribution of apolipoprotein E between free and A-II complexed forms in very-low- and high-density lipoproteins: functional implications. *Biochim Biophys Acta*, **1083**, 139–146.
30 Blanco-Vaca, F., Via, D. P., Yang, C. Y., Massey, J. B., Pownall, H. J. (1992) Characterization of disulfide-linked heterodimers containing apolipoprotein D in human plasma lipoproteins. *J Lipid Res.*, **33**, 1785–1796.
31 Blanco-Vaca, F., Escola-Gil, J. C., Martin-Campos, J. M., Julve, J. (2001) Role of apoA-II in lipid metabolism and atherosclerosis: advances in the study of an enigmatic protein. *J Lipid Res.*, **42**, 1727–1739.
32 Puppione, D. L., Whitelegge, J. P., Yam, L. M., Schumaker, V. N. (2005) Mass spectral analysis of pig (Sus scrofa) apo HDL: Identification of pig apoA-II, a dimeric apolipoprotein. *Comp Biochem Physiol B Biochem Mol Biol.*, **141**, 89–94.
33 Puppione, D. L., Whitelegge, J. P., Yam, L. M., Bassilian, S., Schumaker, V. N., MacDonald, M. H. (2005) Mass spectral analysis of domestic and wild equine apoA-I and A-II: detection of unique dimeric forms of apoA-II. *Comp Biochem Physiol B Biochem Mol Biol.*, **142**, 369–373.
34 Tailleux, A., Duriez, P., Fruchart, J. C., Clavey, V. (2002) Apolipoprotein A-II, HDL metabolism and atherosclerosis. *Atherosclerosis*, **164**, 1–13.
35 Gong, E. L., Stoltfus, L. J., Brion, C. M., Murugesh, D., Rubin, E. M. (1996) Contrasting *in vivo* effects of murine and human apolipoprotein A-II. Role of monomer versus dimer. *J Biol Chem.*, **271**, 5984–5987.
36 Segrest, J. P., Jones, M. K., De Loof, H., Brouillette, C. G., Venkatachalapathi, Y. V., Anantharamaiah, G. M. (1992) The amphipathic helix in the exchangeable apolipoproteins: a review of secondary structure and function. *J Lipid Res.*, **33**, 141–166.
37 Benetollo, C., Lambert, G., Talussot, C., Vanloo, E., Cauteren, T. V., Rouy, D., Dubois, H., Baert, J., Kalopissis, A., Denefle, P., Chambaz, J., Brasseur, R., Rosseneu, M. (1996) Lipid-binding properties of synthetic peptide fragments of human apolipoprotein A-II. *Eur J Biochem.*, **242**, 657–664.
38 Kumar, M. S., Carson, M., Hussain, M. M., Murthy, H. M. (2002) Structures of apolipoprotein A-II and a lipid-surrogate complex provide insights into apolipoprotein-lipid interactions. *Biochemistry*, **41**, 11681–11691.
39 Hussain, M. M. and Zannis, V. I. (1990) Intracellular modification of human apolipoprotein AII (apoAII) and sites of apoAII mRNA synthesis: comparison of apoAII with apoCII and apoCIII isoproteins. *Biochemistry*, **29**, 209–217.
40 Remaley, A. T., Wong, A. W., Schumacher, U. K., Meng, M. S., Brewer, H. B., Jr., Hoeg, J. M. (1993) O-linked glycosylation modifies the association of apolipoprotein A-II to high density lipoproteins. *J Biol Chem.*, **268**, 6785–6790.
41 Ikewaki, K., Zech, L. A., Brewer, H. B., Jr., Rader, D. J. (1996) ApoA-II kinetics in humans using endogenous labeling with stable isotopes: slower turnover of apoA-II compared with the exogenous radiotracer method. *J Lipid Res.*, **37**, 399–407.
42 Sakurabayashi, I. Saito, Y., Kita, T., Matsuzawa, Y., Goto, Y. (2001) Reference intervals for serum apolipoproteins A-I, A-II, B, C-II, C-III, and E in healthy Japanese determined with a commercial immunoturbidimetric assay and effects of sex, age, smoking, drinking, and Lp(a) level. *Clin Chim Acta*, **312**, 87–95.
43 Todo, Y., Kobayashi, J., Higashikata, T., Kawashiri, M., Nohara, A., Inazu, A., Koizumi, J., Mabuchi, H. (2004)

Detailed analysis of serum lipids and lipoproteins from Japanese type III hyperlipoproteinemia with apolipoprotein E2/2 phenotype. *Clin Chim Acta*, **348**, 35–40.

44 Bu, X., Warden, C. H., Xia, Y. R., De Meester, C., Puppione, D. L., Teruya, S., Lokensgard, B., Daneshmand, S., Brown, J., Gray, R. J. *et al.* (1994) Linkage analysis of the genetic determinants of high density lipoprotein concentrations and composition: evidence for involvement of the apolipoprotein A-II and cholesteryl ester transfer protein loci. *Hum Genet*, **93**, 639–648.

45 Vu-Dac, N., Schoonjans, K., Kosykh, V., Dallongeville, J., Fruchart, J. C., Staels, B., Auwerx, J. (1995) Fibrates increase human apolipoprotein A-II expression through activation of the peroxisome proliferator-activated receptor. *J Clin Invest.*, **96**, 741–750.

46 Alaupovic, P., Knight-Gibson, C., Wang, C. S., Downs, D., Koren, E., Brewer, H. B., Jr., Gregg, R. E. (1991) Isolation and characterization of an apoA-II-containing lipoprotein (LP-A-II:B complex) from plasma very low density lipoproteins of patients with Tangier disease and type V hyperlipoproteinemia. *J Lipid Res.*, **32**, 9–19.

47 Velez-Carrasco, W., Lichtenstein, A. H., Li, Z., Dolnikowski, G. G., Lamon-Fava, S., Welty, F. K., Schaefer, E. J. (2000) Apolipoprotein A-I and A-II kinetic parameters as assessed by endogenous labeling with [(2)H(3)]leucine in middle-aged and elderly men and women. *Arterioscler Thromb Vasc Biol.*, **20**, 801–806.

48 Hime, N. J., Drew, K. J., Wee, K., Barter, P. J., Rye, K. A. (2006) Formation of high density lipoproteins containing both apolipoprotein A-I and A-II in the rabbit. *J Lipid Res.*, **47**, 115–122.

49 de Beer, M. C., van der Westhuyzen, D. R., Whitaker, N. L., Webb, N. R., de Beer, F. C. (2005) SR-BI-mediated selective lipid uptake segregates apoA-I and apoA-II catabolism. *J Lipid Res.*, **46**, 2143–2150.

50 De Oliveira, E.S.E.R., Foster, D., McGee Harper M., Seidman, C. E., Smith, J. D., Breslow, J. L., Brinton, E. A. (2000) Alcohol consumption raises HDL cholesterol levels by increasing the transport rate of apolipoproteins A-I and A-II. *Circulation*, **102**, 2347–2352.

51 Gottrand, F., Beghin, L., Duhal, N., Lacroix, B., Bonte, J. P., Fruchart, J. C., Luc, G. (1999) Moderate red wine consumption in healthy volunteers reduced plasma clearance of apolipoprotein AII. *Eur J Clin Invest.*, **29**, 387–394.

52 Sanchez-Muniz, F. J., Merinero, M. C., Rodriguez-Gil, S., Ordovas, J. M., Rodenas, S., Cuesta, C. (2002) Dietary fat saturation affects apolipoprotein AII levels and HDL composition in postmenopausal women. *J Nutr.*, **132**, 50–54.

53 Rodenas, S., Rodriguez-Gil, S., Merinero, M. C., Sanchez-Muniz, F. J. (2005) Dietary exchange of an olive oil and sunflower oil blend for extra virgin olive oil decreases the estimate cardiovascular risk and LDL and apolipoprotein AII concentrations in postmenopausal women. *J Am Coll Nutr.*, **24**, 361–369.

54 Desroches, S., Paradis, M. E., Perusse, M., Archer, W. R., Bergeron, J., Couture, P., Bergeron, N., Lamarche, B. (2004) Apolipoprotein A-I, A-II, and VLDL-B-100 metabolism in men: comparison of a low-fat diet and a high-monounsaturated fatty acid diet. *J Lipid Res.*, **45**, 2331–2338.

55 Escola-Gil, J. C., Blanco-Vaca, F., Julve, J. (2002) Overexpression of human apolipoprotein A-II in transgenic mice does not increase their susceptibility to insulin resistance and obesity. *Diabetologia*, **45**, 600–601.

56 Arbones-Mainar, J. M., Navarro, M. A., Acin, S., Guzman, M. A., Arnal, C., Surra, J. C., Carnicer, R., Roche, H. M., Osada, J. (2006) Trans-10, cis-12- and cis-9, trans-11-conjugated

linoleic acid isomers selectively modify HDL-apolipoprotein composition in apolipoprotein E knockout mice. *J Nutr.*, **136**, 353–359.

57 Vu-Dac, N., Schoonjans, K., Kosykh, V., Dallongeville, J., Heyman, R. A., Staels, B., Auwerx, J. (1996) Retinoids increase human apolipoprotein A-11 expression through activation of the retinoid X receptor but not the retinoic acid receptor. *Mol Cell Biol.*, **16**, 3350–3360.

58 Berthou, L., Saladin, R., Yaqoob, P., Branellec, D., Calder, P., Fruchart, J. C., Denefle, P., Auwerx, J., Staels, B. (1995) Regulation of rat liver apolipoprotein A-I, apolipoprotein A-II and acyl-coenzyme A oxidase gene expression by fibrates and dietary fatty acids. *Eur J Biochem.*, **232**, 179–187.

59 Boucher, J., Ramsamy, T. A., Braschi, S., Sahoo, D., Neville, T. A., Sparks, D. L. (2004) Apolipoprotein A-II regulates HDL stability and affects hepatic lipase association and activity. *J Lipid Res.*, **45**, 849–858.

60 Marzal-Casacuberta, A., Blanco-Vaca, F., Ishida, B. Y., Julve-Gil, J., Shen, J., Calvet-Marquez, S., Gonzalez-Sastre, F., Chan, L. (1996) Functional lecithin:cholesterol acyltransferase deficiency and high density lipoprotein deficiency in transgenic mice overexpressing human apolipoprotein A-II. *J Biol Chem.*, **271**, 6720–6728.

61 Schultz, J. R., Gong, E. L., McCall, M. R., Nichols, A. V., Clift, S. M., Rubin, E. M. (1992) Expression of human apolipoprotein A-II and its effect on high density lipoproteins in transgenic mice. *J Biol Chem.*, **267**, 21630–21636.

62 Rubin, E. M., Ishida, B. Y., Clift, S. M., Krauss, R. M. (1991) Expression of human apolipoprotein A-I in transgenic mice results in reduced plasma levels of murine apolipoprotein A-I and the appearance of two new high density lipoprotein size subclasses. *Proc Natl Acad Sci U S A.*, **88**, 434–438.

63 Hedrick, C. C., Castellani, L. W., Warden, C. H., Puppione, D. L., Lusis, A. J. (1993) Influence of mouse apolipoprotein A-II on plasma lipoproteins in transgenic mice. *J Biol Chem.*, **268**, 20676–20682.

64 Weng, W. and Breslow, J. L. (1996) Dramatically decreased high density lipoprotein cholesterol, increased remnant clearance, and insulin hypersensitivity in apolipoprotein A-II knockout mice suggest a complex role for apolipoprotein A-II in atherosclerosis susceptibility. *Proc Natl Acad Sci U S A*, **93**, 14788–14794.

65 Rye, K. A. and Barter, P. J. (1994) The influence of apolipoproteins on the structure and function of spheroidal, reconstituted high density lipoproteins. *J Biol Chem.*, **269**, 10298–10303.

66 Rye, K. A., Wee, K., Curtiss, L. K., Bonnet, D. J., Barter, P. J. (2003) Apolipoprotein A-II inhibits high density lipoprotein remodeling and lipid-poor apolipoprotein A-I formation. *J Biol Chem.*, **278**, 22530–22536.

67 Lagrost, L., Persegol, L., Lallemant, C., Gambert, P. (1994) Influence of apolipoprotein composition of high density lipoprotein particles on cholesteryl ester transfer protein activity. Particles containing various proportions of apolipoproteins AI and AII. *J Biol Chem.*, **269**, 3189–3197.

68 Masson, D., Duverger, N., Emmanuel, F., Lagrost, L. (1997) Differential interaction of the human cholesteryl ester transfer protein with plasma high density lipoproteins (HDLs) from humans, control mice, and transgenic mice to human HDL apolipoproteins. Lack of lipid transfer inhibitory activity in transgenic mice expressing human apoA-I. *J Biol Chem.*, **272**, 24287–24293.

69 Escola-Gil, J. C., Julve, J., Marzal-Casacuberta, A., Ordonez-Llanos, J., Gonzalez-Sastre, F., Blanco-Vaca, F. (2001) ApoA-II expression in CETP transgenic mice increases VLDL

production and impairs VLDL clearance. *J Lipid Res.*, **42**, 241–248.

70 Zhong, S., Goldberg, I. J., Bruce, C., Rubin, E., Breslow, J. L., Tall, A. (1994) Human ApoA-II inhibits the hydrolysis of HDL triglyceride and the decrease of HDL size induced by hypertriglyceridemia and cholesteryl ester transfer protein in transgenic mice. *J Clin Invest.*, **94**, 2457–2467.

71 Rye, K. A., Jauhiainen, M., Barter, P. J., Ehnholm, C. (1998) Triglyceride-enrichment of high density lipoproteins enhances their remodelling by phospholipid transfer protein. *J Lipid Res.*, **39**, 613–622.

72 Cheung, M. C. and Albers, J. J. (2006) Active plasma phospholipid transfer protein is associated with apoA-I- but not apoE-containing lipoproteins. *J Lipid Res.*, **47**, 1315–1321.

73 Pussinen, P. J., Jauhiainen, M., Metso, J., Pyle, L. E., Marcel, Y. L., Fidge, N. H., Ehnholm, C. (1998) Binding of phospholipid transfer protein (PLTP) to apolipoproteins A-I and A-II: location of a PLTP binding domain in the amino terminal region of apoA-I. *J Lipid Res.*, **39**, 152–161.

74 Albers, J. J., Wolfbauer, G., Cheung, M. C., Day, J. R., Ching, A. F., Lok, S., Tu, A. Y. (1995) Functional expression of human and mouse plasma phospholipid transfer protein: effect of recombinant and plasma PLTP on HDL subspecies. *Biochim Biophys Acta*, **1258**, 27–34.

75 Pussinen, P. J., Jauhiainen, M., Ehnholm, C. (1997) ApoA-II/apoA-I molar ratio in the HDL particle influences phospholipid transfer protein-mediated HDL interconversion. *J Lipid Res.*, **38**, 12–21.

76 Jonas, A., Sweeny, S. A., Herbert, P. N. (1984) Discoidal complexes of A and C apolipoproteins with lipids and their reactions with lecithin: cholesterol acyltransferase. *J Biol Chem.*, **259**, 6369–6375.

77 Durbin, D. M. and Jonas, A. (1999) Lipid-free apolipoproteins A-I and A-II promote remodeling of reconstituted high density lipoproteins and alter their reactivity with lecithin:cholesterol acyltransferase. *J Lipid Res.*, **40**, 2293–2302.

78 Forte, T. M., Bielicki, J. K., Goth-Goldstein, R., Selmek, J., McCall, M. R. (1995) Recruitment of cell phospholipids and cholesterol by apolipoproteins A-II and A-I: formation of nascent apolipoprotein-specific HDL that differ in size, phospholipid composition, and reactivity with LCAT. *J Lipid Res.*, **36**, 148–157.

79 Durbin, D. M. and Jonas, A. (1997) The effect of apolipoprotein A-II on the structure and function of apolipoprotein A-I in a homogeneous reconstituted high density lipoprotein particle. *J Biol Chem.*, **272**, 31333–31339.

80 Labeur, C., Lambert, G., Van Cauteren, T., Duverger, N., Vanloo, B., Chambaz, J., Vandekerckhove, J., Castro, G., Rosseneu, M. (1998) Displacement of apo A-I from HDL by apo A-II or its C-terminal helix promotes the formation of pre-β1 migrating particles and decreases LCAT activation. *Atherosclerosis*, **139**, 351–362.

81 Leroy, A., Dallongeville, J., Fruchart, J. C. (1995) Apolipoprotein A-I-containing lipoproteins and atherosclerosis. *Curr Opin Lipidol.*, **6**, 281–285.

82 Francone, O. L., Gong, E. L., Ng, D. S., Fielding, C. J., Rubin, E. M. (1995) Expression of human lecithin-cholesterol acyltransferase in transgenic mice. Effect of human apolipoprotein AI and human apolipoprotein All on plasma lipoprotein cholesterol metabolism. *J Clin Invest.*, **96**, 1440–1448.

83 Julve-Gil, J., Ruiz-Perez, E., Casaroli-Marano, R. P., Marzal-Casacuberta, A., Escola-Gil, J. C., Gonzalez-Sastre, F., Blanco-Vaca, F. (1999) Free cholesterol deposition in the cornea of human apolipoprotein A-II transgenic mice with functional lecithin: cholesterol acyltransferase deficiency. *Metabolism*, **48**, 415–421.

84 Mowri, H. O., Patsch, W., Smith, L. C., Gotto, A. M., Jr., Patsch, J. R. (1992) Different reactivities of high density lipoprotein2 subfractions with hepatic lipase. *J Lipid Res.*, **33**, 1269–1279.

85 Mowri, H. O., Patsch, J. R., Gotto, A. M., Jr., Patsch, W. (1996) Apolipoprotein A-II influences the substrate properties of human HDL2 and HDL3 for hepatic lipase. *Arterioscler Thromb Vasc Biol.*, **16**, 755–762.

86 Jahn, C. E., Osborne, J. C., Jr., Schaefer, E. J., Brewer, H. B., Jr. (1981) *In vitro* activation of the enzymic activity of hepatic lipase by apoA-II*FEBS Lett.*, **131**, 366–368.

87 Jahn, C. E., Osborne, J. C., Jr., Schaefer, E. J., Brewer, H. B., Jr. (1983) Activation of the enzymic activity of hepatic lipase by apolipoprotein A-II. Characterization of a major component of high density lipoprotein as the activating plasma component *in vitro*. *Eur J Biochem.*, **131**, 25–29.

88 Escola-Gil, J. C., Marzal-Casacuberta, A., Julve-Gil, J., Ishida, B. Y., Ordonez-Llanos, J., Chan, L., Gonzalez-Sastre, F., Blanco-Vaca, F. (1998) Human apolipoprotein A-II is a pro-atherogenic molecule when it is expressed in transgenic mice at a level similar to that in humans: evidence of a potentially relevant species-specific interaction with diet. *J Lipid Res.*, **39**, 457–462.

89 Hime, N. J., Barter, P. J., Rye, K. A. (2001) Evidence that apolipoprotein A-I facilitates hepatic lipase-mediated phospholipid hydrolysis in reconstituted HDL containing apolipoprotein A-II. *Biochemistry*, **40**, 5496–5505.

90 Thuren, T. (2000) Hepatic lipase and HDL metabolism. *Curr Opin Lipidol.*, **11**, 277–283.

91 Hime, N. J., Barter, P. J., Rye, K. A. (1998) The influence of apolipoproteins on the hepatic lipase-mediated hydrolysis of high density lipoprotein phospholipid and triacylglycerol. *J Biol Chem.*, **273**, 27191–27198.

92 Hedrick, C. C., Castellani, L. W., Wong, H., Lusis, A. J. (2001) *In vivo* interactions of apoA-II, apoA-I, and hepatic lipase contributing to HDL structure and antiatherogenic functions. *J Lipid Res.*, **42**, 563–570.

93 Weng, W., Brandenburg, N. A., Zhong, S., Halkias, J., Wu, L., Jiang, X. C., Tall, A., Breslow, J. L. (1999) ApoA-II maintains HDL levels in part by inhibition of hepatic lipase. Studies In apoA-II and hepatic lipase double knockout mice. *J Lipid Res.*, **40**, 1064–1070.

94 Broedl, U. C., Jin, W., Fuki, I. V., Millar, J. S., Rader, D. J. (2006.) Endothelial lipase is less effective in influencing HDL metabolism *in vivo* in mice expressing apoA-II. *J Lipid Res.*, **47**, 2191–2197

95 Kalopissis, A. D., Pastier, D., Chambaz, J. (2003) Apolipoprotein A-II: beyond genetic associations with lipid disorders and insulin resistance. *Curr Opin Lipidol.*, **14**, 165–172.

96 Martin-Campos, J. M., Escola-Gil, J. C., Ribas, V., Blanco-Vaca, F. (2004) Apolipoprotein A-II, genetic variation on chromosome 1q21–q24, and disease susceptibility. *Curr Opin Lipidol.*, **15**, 247–253.

97 Castellani, L. W., Navab, M., Van Lenten, B. J., Hedrick, C. C., Hama, S. Y., Goto, A. M., Fogelman, A. M., Lusis, A. J. (1997) Overexpression of apolipoprotein AII in transgenic mice converts high density lipoproteins to proinflammatory particles. *J Clin Invest.*, **100**, 464–474.

98 Fournier, N., Cogny, A., Atger, V., Pastier, D., Goudouneche, D., Nicoletti, A., Moatti, N., Chambaz, J., Paul, J. L., Kalopissis, A. D. (2002) Opposite effects of plasma from human apolipoprotein A-II transgenic mice on cholesterol efflux from J774 macrophages and Fu5AH hepatoma cells. *Arterioscler Thromb Vasc Biol.*, **22**, 638–643.

99 Tailleux, A., Bouly, M., Luc, G., Castro, G., Caillaud, J. M., Hennuyer, N., Poulain, P., Fruchart, J. C., Duverger, N., Fievet, C. (2000) Decreased susceptibility to diet-induced atherosclerosis in human apolipoprotein A-II transgenic mice. *Arterioscler Thromb Vasc Biol.*, **20**, 2453–2458.

100 Pastier, D., Dugue, S., Boisfer, E., Atger, V., Tran, N. Q., van Tol, A., Chapman, M. J., Chambaz, J., Laplaud, P. M., Kalopissis, A. D. (2001) Apolipoprotein A-II/A-I ratio is a key determinant *in vivo* of HDL concentration and formation of pre-β HDL containing apolipoprotein A-II. *Biochemistry*, **40**, 12243–12253.

101 Rotllan, N., Ribas, V., Calpe-Berdiel, L., Martin-Campos, J. M., Blanco-Vaca, F., Escola-Gil, J. C. (2005) Overexpression of human apolipoprotein A-II in transgenic mice does not impair macrophage-specific reverse cholesterol transport *in vivo*. *Arterioscler Thromb Vasc Biol.*, e128–132.

102 Remaley, A. T., Stonik, J. A., Demosky, S. J., Neufeld, E. B., Bocharov, A. V., Vishnyakova, T. G., Eggerman, T. L., Patterson, A. P., Duverger, N. J., Santamarina-Fojo, S., Brewer, H. B., Jr. (2001) Apolipoprotein specificity for lipid efflux by the human ABCAI transporter. *Biochem Biophys Res Commun.*, **280**, 818–823.

103 Fitzgerald, M. L., Morris, A. L., Chroni, A., Mendez, A. J., Zannis, V. I., Freeman, M. W. (2004) ABCA1 and amphipathic apolipoproteins form high-affinity molecular complexes required for cholesterol efflux. *J Lipid Res.*, **45**, 287–294.

104 Rinninger, F., Kaiser, T., Windler, E., Greten, H., Fruchart, J. C., Castro, G. (1998) Selective uptake of cholesteryl esters from high-density lipoprotein-derived LpA-I and LpA-I:A-II particles by hepatic cells in culture. *Biochim Biophys Acta*, **1393**, 277–291.

105 Pilon, A., Briand, O., Lestavel, S., Copin, C., Majd, Z., Fruchart, J. C., Castro, G., Clavey, V. (2000) Apolipoprotein AII enrichment of HDL enhances their affinity for class B type I scavenger receptor but inhibits specific cholesteryl ester uptake. *Arterioscler Thromb Vasc Biol.*, **20**, 1074–1081.

106 Julve, J., Escola-Gil, J. C., Ribas, V., Gonzalez-Sastre, F., Ordonez-Llanos, J., Sanchez-Quesada, J. L., Blanco-Vaca, F. (2002) Mechanisms of HDL deficiency in mice overexpressing human apoA-II. *J Lipid Res.*, **43**, 1734–1742.

107 Rinninger, F., Brundert, M., Budzinski, R. M., Fruchart, J. C., Greten, H., Castro, G. R. (2003) Scavenger receptor BI (SR-BI) mediates a higher selective cholesteryl ester uptake from LpA-I compared with LpA-I:A-II lipoprotein particles. *Atherosclerosis*, **166**, 31–40.

108 de Beer, M. C., Castellani, L. W., Cai, L., Stromberg, A. J., de Beer, F. C., van der Westhuyzen, D. R. (2004) ApoA-II modulates the association of HDL with class B scavenger receptors SR-BI and CD36. *J Lipid Res.*, **45**, 706–715.

109 de Beer, M. C., Durbin, D. M., Cai, L., Mirocha, N., Jonas, A., Webb, N. R., de Beer, F. C., van Der Westhuyzen, D. R. (2001) Apolipoprotein A-II modulates the binding and selective lipid uptake of reconstituted high density lipoprotein by scavenger receptor BI. *J Biol Chem.*, **276**, 15832–15839.

110 Escola-Gil, J. C., Calpe-Berdiel, L., Palomer, X., Ribas, V., Ordonez-Llanos, J., Blanco-Vaca, F. (2006) Antiatherogenic role of high-density lipoproteins: insights from genetically engineered-mice. *Front Biosci.*, **11**, 1328–1348.

111 Ribas, V., Sanchez-Quesada, J. L., Anton, R., Camacho, M., Julve, J., Escola-Gil, J. C., Vila, L., Ordonez-Llanos, J., Blanco-Vaca, F. (2004) Human apolipoprotein A-II enrichment displaces paraoxonase from HDL and impairs its antioxidant properties: a new mechanism linking HDL protein

composition and antiatherogenic potential. *Circ Res.*, **95**, 789–797.

112 Bergmeier, C., Siekmeier, R., Gross, W. (2004) Distribution Spectrum of Paraoxonase Activity in HDL Fractions. *Clin Chem.*, **50**, 2309–2315.

113 Gaidukov, L. and Tawfik, D. S. (2005) High affinity, stability, and lactonase activity of serum paraoxonase PON1 anchored on HDL with ApoA-I. *Biochemistry*, **44**, 11843–11854.

114 Holvoet, P., Peeters, K., Lund-Katz, S., Mertens, A., Verhamme, P., Quarck, R., Stengel, D., Lox, M., Deridder, E., Bernar, H., Nickel, M., Theilmeier, G., Ninio, E., Phillips, M. C. (2001) Arg123-Tyr166 domain of human ApoA-I is critical for HDL-mediated inhibition of macrophage homing and early atherosclerosis in mice. *Arterioscler Thromb Vasc Biol.*, **21**, 1977–1983.

115 Boisfer, E., Stengel, D., Pastier, D., Laplaud, P. M., Dousset, N., Ninio, E., Kalopissis, A. D. (2002) Antioxidant properties of HDL in transgenic mice overexpressing human apolipoprotein A-II. *J Lipid Res.*, **43**, 732–741.

116 Shih, D. M., Gu, L., Hama, S., Xia, Y. R., Navab, M., Fogelman, A. M., Lusis, A. J. (1996) Genetic-dietary regulation of serum paraoxonase expression and its role in atherogenesis in a mouse model. *J Clin Invest.*, **97**, 1630–1639.

117 Magill, P., Rao, S. N., Miller, N. E., Nicoll, A., Brunzell, J., St Hilaire, J., Lewis, B. (1982) Relationships between the metabolism of high-density and very-low-density lipoproteins in man: studies of apolipoprotein kinetics and adipose tissue lipoprotein lipase activity. *Eur J Clin Invest.*, **12**, 113–120.

118 Wang, C. S., Alaupovic, P., Gregg, R. E., Brewer, H. B., Jr. (1987) Studies on the mechanism of hypertriglyceridemia in Tangier disease. Determination of plasma lipolytic activities, k1 values and apolipoprotein composition of the major lipoprotein density classes. *Biochim Biophys Acta*, **920**, 9–19.

119 Warden, C. H., Daluiski, A., Bu, X., Purcell-Huynh, D. A., De Meester, C., Shieh, B. H., Puppione, D. L., Gray, R. M., Reaven, G. M., Chen, Y. D. *et al.* (1993) Evidence for linkage of the apolipoprotein A-II locus to plasma apolipoprotein A-II and free fatty acid levels in mice and humans. *Proc Natl Acad Sci U S A*, **90**, 10886–10890.

120 Lara-Castro, C., Hunter, G. R., Lovejoy, J. C., Gower, B. A., Fernandez, J. R. (2005) Apolipoprotein A-II polymorphism and visceral adiposity in African-American and white women. *Obes Res.*, **13**, 507–512.

121 Warden, C. H., Hedrick, C. C., Qiao, J. H., Castellani, L. W., Lusis, A. J. (1993) Atherosclerosis in transgenic mice overexpressing apolipoprotein A-II. *Science*, **261**, 469–472.

122 Boisfer, E., Lambert, G., Atger, V., Tran, N. Q., Pastier, D., Benetollo, C., Trottier, J. F., Beaucamps, I., Antonucci, M., Laplaud, M., Griglio, S., Chambaz, J., Kalopissis, A. D. (1999) Overexpression of human apolipoprotein A-II in mice induces hypertriglyceridemia due to defective very low density lipoprotein hydrolysis. *J Biol Chem.*, **274**, 11564–11572.

123 Escola-Gil, J. C., Julve, J., Marzal-Casacuberta, A., Ordonez-Llanos, J., Gonzalez-Sastre, F., Blanco-Vaca, F. (2000) Expression of human apolipoprotein A-II in apolipoprotein E-deficient mice induces features of familial combined hyperlipidemia. *J Lipid Res.*, **41**, 1328–1338.

124 Julve, J., Escola-Gil, J. C., Marzal-Casacuberta, A., Ordonez-Llanos, J., Gonzalez-Sastre, F., Blanco-Vaca, F. (2000) Increased production of very-low-density lipoproteins in transgenic mice overexpressing human apolipoprotein A-II and fed with a high-fat diet. *Biochim Biophys Acta*, **1488**, 233–244.

125 Castellani, L. W., Goto, A. M., Lusis, A. J. (2001) Studies with apolipoprotein A-II transgenic mice indicate a role for HDLs in adiposity

and insulin resistance. *Diabetes*, **50**, 643–651.

126 Julve, J., Blanco-Vaca, F., Carles Escola-Gil, J. (2002) On the mechanisms by which human apolipoprotein A-II gene variability relates to hypertriglyceridemia. *Circulation*, **105** e129; author reply e129.

127 Ribas, V., Palomer, X., Roglans, N., Rotllan, N., Fievet, C., Tailleux, A., Julve, J., Laguna, J. C., Blanco-Vaca, F., Escola-Gil, J. C. (2005) Paradoxical exacerbation of combined hyperlipidemia in human apolipoprotein A-II transgenic mice treated with fenofibrate. *Biochim Biophys Acta*, **1737**, 130–137.

128 Castellani, L. W., Gargalovic, P., Febbraio, M., Charugundla, S., Jien, M. L., Lusis, A. J. (2004) Mechanisms mediating insulin resistance in transgenic mice overexpressing mouse apolipoprotein A-II. *J Lipid Res.*, **45**, 2377–2387.

129 Ferns, G. A., Shelley, C. S., Stocks, J., Rees, A., Paul, H., Baralle, F., Galton, D. J. (1986) A DNA polymorphism of the apoprotein AII gene in hypertriglyceridaemia. *Hum Genet*, **74**, 302–306.

130 Saha, N., Tan, J. A., Tay, J. S. (1992) MspI polymorphism of the apolipoprotein A-II gene, serum lipids and apolipoproteins in Chinese from Singapore. *Hum Hered.*, **42**, 293–297.

131 Hong, S. H., Kang, B. Y., Park, W. H., Kim, J. Q., Lee, C. C. (1998) Association between apolipoprotein A2 MspI polymorphism and hypertriglyceridemia in Koreans. *Hum Biol.*, **70**, 41–46.

132 Dupuy-Gorce, A. M., Desmarais, E., Vigneron, S., Buresi, C., Nicaud, V., Evans, A., Luc, G., Arveiler, D., Marques-Vidal, P., Cambien, F., Tiret, L., Crastes de Paulet, A., Roizes, G. (1996) DNA polymorphisms in linkage disequilibrium at the 3′ end of the human APO AII gene: relationships with lipids, apolipoproteins and coronary heart disease. *Clin Genet*, **50**, 191–198.

133 Rajput-Williams, J., Eyre, J., Nanjee, M. N., Crook, D., Scott, J., Miller, N. E. (1989) Plasma lipoprotein lipids in relation to the MspI polymorphism of the apolipoprotein AII gene in Caucasian men. Lack of association with plasma triglyceride concentration. *Atherosclerosis*, **77**, 31–36.

134 Civeira, F., Genest, J., Pocovi, M., Salem, D. N., Herbert, P. N., Wilson, P. W., Schaefer, E. J., Ordovas, J. M. (1992) The MspI restriction fragment length polymorphism 3′ to the apolipoprotein A-II gene: relationships with lipids, apolipoproteins, and premature coronary artery disease. *Atherosclerosis*, **92**, 165–176.

135 Vohl, M. C., Lamarche, B., Bergeron, J., Moorjani, S., Prud'homme, D., Nadeau, A., Tremblay, A., Lupien, P. J., Bouchard, C., Despres, J. P. (1997) The MspI polymorphism of the apolipoprotein A-II gene as a modulator of the dyslipidemic state found in visceral obesity. *Atherosclerosis*, **128**, 183–190.

136 Myklebost, O., Rogne, S., Hjermann, I., Olaisen, B., Prydz, H. (1990) Association analysis of lipid levels and apolipoprotein restriction fragment length polymorphisms. *Hum Genet*, **86**, 209–214.

137 Thorn, J. A., Stocks, J., Reichl, D., Alcolado, J. C., Chamberlain, J. C., Galton, D. (1993) Variability of plasma apolipoprotein (apo) A-II levels associated with an apo A-II gene polymorphism in monozygotic twin pairs. *Biochim Biophys Acta*, **1180**, 299–303.

138 Scott, J., Knott, T. J., Priestley, L. M., Robertson, M. E., Mann, D. V., Kostner, G., Miller, G. J., Miller, N. E. (1985) High-density lipoprotein composition is altered by a common DNA polymorphism adjacent to apoprotein AII gene in man. *Lancet*, **1**, 771–773.

139 Han, Z., Heath, S. C., Shmulewitz, D., Li, W., Auerbach, S. B., Blundell, M. L., Lehner, T., Ott, J., Stoffel, M., Friedman, J. M., Breslow, J. L. (2002)

Candidate genes involved in cardiovascular risk factors by a family-based association study on the island of Kosrae, Federated States of Micronesia. *Am J Med Genet*, **110**, 234–242.

140 Elbein, S. C. Hasstedt, S. J. (2002) Quantitative trait linkage analysis of lipid-related traits in familial type 2 diabetes: evidence for linkage of triglyceride levels to chromosome 19q. *Diabetes*, **51**, 528–535.

141 Lilja, H. E., Soro, A., Ylitalo, K., Nuotio, I., Viikari, J. S., Salomaa, V., Vartiainen, E., Taskinen, M. R., Peltonen, L., Pajukanta, P. (2002) A candidate gene study in low HDL-cholesterol families provides evidence for the involvement of the APOA2 gene and the APOA1C3A4 gene cluster. *Atherosclerosis*, **164**, 103–111.

142 Pajukanta, P., Nuotio, I., Terwilliger, J. D., Porkka, K. V., Ylitalo, K., Pihlajamaki, J., Suomalainen, A. J., Syvanen, A. C., Lehtimaki, T., Viikari, J. S., Laakso, M., Taskinen, M. R., Ehnholm, C., Peltonen, L. (1998) Linkage of familial combined hyperlipidaemia to chromosome 1q21–q23. *Nat Genet*, **18**, 369–373.

143 Pei, W., Baron, H., Muller-Myhsok, B., Knoblauch, H., Al-Yahyaee, S. A., Hui, R., Wu, X., Liu, L., Busjahn, A., Luft, F. C., Schuster, H. (2000) Support for linkage of familial combined hyperlipidemia to chromosome 1q21–q23 in Chinese and German families. *Clin Genet*, **57**, 29–34.

144 Coon, H., Myers, R. H., Borecki, I. B., Arnett, D. K., Hunt, S. C., Province, M. A., Djousse, L., Leppert, M. F. (2000) Replication of linkage of familial combined hyperlipidemia to chromosome 1q with additional heterogeneous effect of apolipoprotein A-I/C-III/A-IV locus. The NHLBI Family Heart Study. *Arterioscler Thromb Vasc Biol.*, **20**, 2275–2280.

145 Huertas-Vazquez, A., del Rincon, J. P., Canizales-Quinteros, S., Riba, L., Vega-Hernandez, G., Ramirez-Jimenez, S., Auron-Gomez, M., Gomez-Perez, F. J., Aguilar-Salinas, C. A., Tusie-Luna, M. T. (2004) Contribution of chromosome 1q21–q23 to familial combined hyperlipidemia in Mexican families. *Ann Hum Genet*, **68**, 419–427.

146 Naoumova, R. P., Bonney, S. A., Eichenbaum-Voline, S., Patel, H. N., Jones, B., Jones, E. L., Amey, J., Colilla, S., Neuwirth, C. K., Allotey, R., Seed, M., Betteridge, D. J., Galton, D. J., Cox, N. J., Bell, G. I., Scott, J., Shoulders, C. C. (2003) Confirmed locus on chromosome 11p and candidate loci on 6q and 8p for the triglyceride and cholesterol traits of combined hyperlipidemia. *Arterioscler Thromb Vasc Biol.*, **23**, 2070–2077.

147 Pajukanta, P., Allayee, H., Krass, K. L., Kuraishy, A., Soro, A., Lilja, H. E., Mar, R., Taskinen, M. R., Nuotio, I., Laakso, M., Rotter, J. I., de Bruin, T. W., Cantor, R. M., Lusis, A. J., Peltonen, L. (2003) Combined analysis of genome scans of Dutch and Finnish families reveals a susceptibility locus for high-density lipoprotein cholesterol on chromosome 16q. *Am J Hum Genet*, **72**, 903–917.

148 Allayee, H., Castellani, L. W., Cantor, R. M., de Bruin, T. W., Lusis, A. J. (2003) Biochemical and genetic association of plasma apolipoprotein A-II levels with familial combined hyperlipidemia. *Circ Res.*, **92**, 1262–1267.

149 Klos, K. L., Kardia, S. L., Ferrell, R. E., Turner, S. T., Boerwinkle, E., Sing, C. F. (2001) Genome-wide linkage analysis reveals evidence of multiple regions that influence variation in plasma lipid and apolipoprotein levels associated with risk of coronary heart disease. *Arterioscler Thromb Vasc Biol.*, **21**, 971–978.

150 Aouizerat, B. E., Allayee, H., Cantor, R. M., Davis, R. C., Lanning, C. D., Wen, P. Z., Dallinga-Thie, G. M., de Bruin, T. W., Rotter, J. I., Lusis, A. J. (1999) A genome scan for familial combined hyperlipidemia reveals

evidence of linkage with locus on chromosome 11. *Am J Hum Genet*, **65**, 397–410.

151 Pajukanta, P., Terwilliger, J. D., Perola, M., Hiekkalinna, T., Nuotio, I., Ellonen, P., Parkkonen, M., Hartiala, J., Ylitalo, K., Pihlajamaki, J., Porkka, K., Laakso, M., Viikari, J., Ehnholm, C., Taskinen, M. R., Peltonen, L. (1999) Genomewide scan for familial combined hyperlipidemia genes in Finnish families, suggesting multiple susceptibility loci influencing triglyceride, cholesterol, and apolipoprotein B levels. *Am J Hum Genet*, **64**, 1453–1463.

152 Pajukanta, P., Lilja, H. E., Sinsheimer, J. S., Cantor, R. M., Lusis, A. J., Gentile, M., Duan, X. J., Soro-Paavonen, A., Naukkarinen, J., Saarela, J., Laakso, M., Ehnholm, C., Taskinen, M. R., Peltonen, L. (2004) Familial combined hyperlipidemia is associated with upstream transcription factor 1 (USF1). *Nat Genet*, **36**, 371–376.

153 Coon, H., Xin, Y., Hopkins, P. N., Cawthon, R. M., Hasstedt, S. J., Hunt, S. C. (2005) Upstream stimulatory factor 1 associated with familial combined hyperlipidemia, LDL cholesterol, and triglycerides. *Hum Genet*, **117**, 444–451.

154 Hanson, R. L., Ehm, M. G., Pettitt, D. J., Prochazka, M., Thompson, D. B., Timberlake, D., Foroud, T., Kobes, S., Baier, L., Burns, D. K., Almasy, L., Blangero, J., Garvey, W. T., Bennett, P. H., Knowler, W. C. (1998) An autosomal genomic scan for loci linked to type II diabetes mellitus and body-mass index in Pima Indians. *Am J Hum Genet*, **63**, 1130–1138.

155 Elbein, S. C., Hoffman, M. D., Teng, K., Leppert, M. F., Hasstedt, S. J. (1999) A genome-wide search for type 2 diabetes susceptibility genes in Utah Caucasians. *Diabetes*, **48**, 1175–1182.

156 Vionnet, N., Hani El, H., Dupont, S., Gallina, S., Francke, S., Dotte, S., De Matos, F., Durand, E., Lepretre, F., Lecoeur, C., Gallina, P., Zekiri, L., Dina, C., Froguel, P. (2000) Genomewide search for type 2 diabetes-susceptibility genes in French whites: evidence for a novel susceptibility locus for early-onset diabetes on chromosome 3q27–qter and independent replication of a type 2-diabetes locus on chromosome 1q21–q24. *Am J Hum Genet*, **67**, 1470–1480.

157 Wiltshire, S., Hattersley, A. T., Hitman, G. A., Walker, M., Levy, J. C., Sampson, M., O'Rahilly, S., Frayling, T. M., Bell, J. I., Lathrop, G. M., Bennett, A., Dhillon, R., Fletcher, C., Groves, C. J., Jones, E., Prestwich, P., Simecek, N., Rao, P. V., Wishart, M., Bottazzo, G. F., Foxon, R., Howell, S., Smedley, D., Cardon, L. R., Menzel, S., McCarthy, M. I. (2001) A genomewide scan for loci predisposing to type 2 diabetes in a UK population (the Diabetes UK Warren 2 Repository): analysis of 573 pedigrees provides independent replication of a susceptibility locus on chromosome 1q. *Am J Hum Genet*, **69**, 553–569.

158 Meigs, J. B., Panhuysen, C. I., Myers, R. H., Wilson, P. W., Cupples, L. A. (2002) A genome-wide scan for loci linked to plasma levels of glucose and HbA(1c) in a community-based sample of Caucasian pedigrees: The Framingham Offspring Study. *Diabetes*, **51**, 833–840.

159 Hsueh, W. C., St Jean, P. L., Mitchell, B. D., Pollin, T. I., Knowler, W. C., Ehm, M. G., Bell, C. J., Sakul, H., Wagner, M. J., Burns, D. K., Shuldiner, A. R. (2003) Genome-wide and fine-mapping linkage studies of type 2 diabetes and glucose traits in the Old Order Amish: evidence for a new diabetes locus on chromosome 14q11 and confirmation of a locus on chromosome 1q21–q24. *Diabetes*, **52**, 550–557.

160 Xiang, K., Wang, Y., Zheng, T., Jia, W., Li, J., Chen, L., Shen, K., Wu, S., Lin, X., Zhang, G., Wang, C., Wang, S., Lu, H., Fang, Q., Shi, Y., Zhang, R., Xu, J., Weng, Q. (2004) Genome-wide search for type 2 diabetes/

impaired glucose homeostasis susceptibility genes in the Chinese: significant linkage to chromosome 6q21–q23 and chromosome 1q21–q24. *Diabetes*, **53**, 228–234.

161 Ng, M. C., So, W. Y., Cox, N. J., Lam, V. K., Cockram, C. S., Critchley, J. A., Bell, G. I., Chan, J. C. (2004) Genome-wide scan for type 2 diabetes loci in Hong Kong Chinese and confirmation of a susceptibility locus on chromosome 1q21–q25. *Diabetes*, **53**, 1609–1613.

162 Zhao, J. Y., Xiong, M. M., Huang, W., Wang, H., Zuo, J., Wu, G. D., Chen, Z., Qiang, B. Q., Zhang, M. L., Chen, J. L., Ding, W., Yuan, W. T., Xu, H. Y., Jin, L., Li, Y. X., Sun, Q., Liu, Q. Y., Boerwinkle, E., Fang, F. D. (2005) An autosomal genomic scan for loci linked to type 2 diabetes in northern Han Chinese. *J Mol Med.*, **83**, 209–215.

163 Demenais, F., Kanninen, T., Lindgren, C. M., Wiltshire, S., Gaget, S., Dandrieux, C., Almgren, P., Sjogren, M., Hattersley, A., Dina, C., Tuomi, T., McCarthy, M. I., Froguel, P., Groop, L. C. (2003) A meta-analysis of four European genome screens (GIFT Consortium) shows evidence for a novel region on chromosome 17p11.2–q22 linked to type 2 diabetes. *Hum Mol Genet*, **12**, 1865–1873.

164 Das, S. K., Hasstedt, S. J., Zhang, Z., Elbein, S. C. (2004) Linkage and association mapping of a chromosome 1q21–q24 type 2 diabetes susceptibility locus in northern European Caucasians. *Diabetes*, **53**, 492–499.

165 Luc, G., Bard, J. M., Ferrieres, J., Evans, A., Amouyel, P., Arveiler, D., Fruchart, J. C., Ducimetiere, P. (2002) Value of HDL cholesterol, apolipoprotein A-I, lipoprotein A-I, and lipoprotein A-I/A-II in prediction of coronary heart disease: the PRIME Study. Prospective Epidemiological Study of Myocardial Infarction. *Arterioscler Thromb Vasc Biol.*, **22**, 1155–1161.

166 Luc, G., Majd, Z., Poulain, P., Elkhalil, L., Fruchart, J. C. (1996) Interstitial fluid apolipoprotein A-II: an association with the occurrence of myocardial infarction. *Atherosclerosis*, **127**, 131–137.

167 Alaupovic, P., Mack, W. J., Knight-Gibson, C., Hodis, H. N. (1997) The role of triglyceride-rich lipoprotein families in the progression of atherosclerotic lesions as determined by sequential coronary angiography from a controlled clinical trial. *Arterioscler Thromb Vasc Biol.*, **17**, 715–722.

168 Kawaguchi, A., Miyao, Y., Noguchi, T., Nonogi, H., Yamagishi, M., Miyatake, K., Kamikubo, Y., Kumeda, K., Tsushima, M., Yamamoto, A., Kato, H. (2000) Intravascular free tissue factor pathway inhibitor is inversely correlated with HDL cholesterol and postheparin lipoprotein lipase but proportional to apolipoprotein A-II. *Arterioscler Thromb Vasc Biol.*, **20**, 251–258.

169 Schultz, J. R., Verstuyft, J. G., Gong, E. L., Nichols, A. V., Rubin, E. M. (1993) Protein composition determines the anti-atherogenic properties of HDL in transgenic mice. *Nature.*, **365**, 762–764.

170 Schultz, J. R. and Rubin, E. M. (1994) The properties of HDL in genetically engineered mice. *Curr Opin Lipidol.*, **5**, 126–137.

171 Mehrabian, M., Qiao, J. H., Hyman, R., Ruddle, D., Laughton, C., Lusis, A. J. (1993) Influence of the apoA-II gene locus on HDL levels and fatty streak development in mice. *Arterioscler Thromb.*, **13**, 1–10.

3
Apolipoprotein E and Reverse Cholesterol Transport: The Role of Recycling in Plasma and Intracellular Lipid Trafficking

Larry L. Swift, Alyssa H. Hasty, MacRae F. Linton, Sergio Fazio

3.1
Introduction

Apolipoprotein E (apoE) is unique among the apoproteins because of its many different functions in biology, ranging from modulation of plasma lipoprotein metabolism, through control of cellular cholesterol homeostasis, to regulation of neuronal function. Although apoE is synthesized by several tissues and cell types, its circulating levels are influenced primarily by the liver. Within the plasma compartment apoE readily redistributes among chylomicrons, remnant particles, and high-density lipoproteins (HDL), where it has functional roles both in directing triglyceride-rich particles to sites of catabolism and clearance and in determining size expansion of the HDL. Its role as a ligand for the receptor-mediated endocytosis of lipoproteins is well established [1–3]. It is recognized by the low-density lipoprotein (LDL) receptor (LDLR) [2,3], the LDLR-related protein 1 (LRP1) [1,4,5], and heparan sulfate proteoglycans (HSPG), either alone [6,7] or in concert with the LRP1 [8]. Additionally, within the periphery apoE interacts with lipoprotein lipase (LPL) and hepatic lipase (HL) to modulate triglyceride hydrolysis [9,10]. Within the cell apoE is known to serve many biologically relevant roles. Studies have suggested that apoE directs the intracellular routing of internalized remnant lipoproteins [11,12] and modulates intracellular lipid metabolism, in particular the hydrolysis and utilization of triglyceride [13]. In hepatocytes, it plays a critical role in the assembly and secretion of VLDL, augmenting the incorporation of triglycerides into newly forming particles [14–16]. Finally, apoE is a critical component of the reverse cholesterol transport (RCT) pathway. It promotes cholesterol efflux from macrophages [17–19] and plays major roles in HDL formation, maturation, and hepatic uptake, as well as in the selective uptake of HDL cholesteryl ester by the liver [20–23]. ApoE is also expressed by macrophage-derived foam cells, while a number of studies support an important antiatherogenic role for macrophage apoE *in vivo* [24–27].

The unique nature of apoE is also demonstrated by the fact that, after internalization, this apoprotein follows specialized cellular pathways allowing it to exert maximum impact on cellular lipid homeostasis. In particular, studies both in our

High-Density Lipoproteins: From Basic Biology to Clinical Aspects. Edited by Christopher J. Fielding

ISBN: 978-3-527-31717-2

laboratory and in others have shown that a portion of internalized apoE escapes degradation and is resecreted. The pathophysiologic relevance of apoE recycling is being elucidated, and it seems clear that the recycling process is linked with cellular cholesterol trafficking and efflux. Ultimately, the recycling pathway impacts HDL metabolism, and as such has significant repercussions on vascular health and the process of atherogenesis. In this chapter we summarize the literature on the role of apoE in lipoprotein and cellular lipid metabolism, with specific emphasis on RCTand HDL metabolism, and review the studies that link apoE recycling and cholesterol homeostasis.

3.2
Role of ApoE in Triglyceride-Rich Lipoprotein Metabolism

Although this chapter focuses on the role of apoE in HDL metabolism and its effects on the reverse cholesterol transport system, a summary of the canonical role played by this protein in VLDL metabolism is necessary for full understanding of the complexity of apoE's involvement in lipid metabolism. Throughout this chapter reference is made to studies of both human and murine apoE. However, it must be clarified that human apoE is a polymorphic protein whereas murine apoE is not. The basis for the polymorphism in human populations resides in common arginine-for-cysteine ($R > C$) substitutions at positions 112 and 158 in the apoE molecule [28]. There are three isoforms of apoE: E2, E3, and E4. ApoE3 (C112, R158) is considered the wild-type allele product because it is expressed in highest frequency in all populations studied [29]. ApoE2 (C112, C158) is the rarest isoform, binds poorly to the LDLR, and in homozygosity can cause a severe lipoprotein abnormality called type III hyperlipoproteinemia (HLP) [30]. ApoE4 (R112, R158) is the isoform with intermediate population frequency, binds normally to receptors, and is linked to development of Alzheimer's disease through direct effects on the neuronal cytoskeleton [31]. Murine apoE has no cysteines and is therefore comparable to human apoE4.

3.2.1
ApoE and Lipoprotein Assembly and Secretion

ApoE is a physiologic component of the VLDL particle. It is synthesized and secreted by the hepatocyte at a rate that is basically unmodified by the lipogenic status of the cell [32]. However, the majority of apoE is secreted as a lipoprotein-free protein when lipoproteins are not being produced, and as part of the VLDL particle at times of increased lipid packaging and lipoprotein synthesis [32,33]. VLDL remnants are very rapidly and efficiently cleared from the circulation because they are enriched in apoE, which acts as a strong ligand for specific cell surface receptors such as HSPG, LDLR, and LRP1 [29,34]. Chylomicrons and their remnants are also enriched in apoE, even though this enrichment does not result from inclusion of apoE at the time of lipoprotein assembly but rather from acquisition of apoE in the extracellular space after secretion of the lipoprotein from the intestinal epithelial cells into the lymph [35].

The fact that chylomicrons isolated from the lymph of the thoracic duct are relatively apoE-free, yet those isolated from the general circulation are very apoE-rich, indicates that apoE does not need to become part of the lipoprotein at the time of assembly in order to express normal functionality. Experimental validation of this concept was produced by experiments performed by us and others showing that in mice lacking apoE the introduction of apoE-producing macrophages through bone marrow transplantation resulted in apoE being present in the plasma compartment. Furthermore, the association of apoE with the circulating lipoproteins (in this case an extracellular event only) produced rapid clearance of plasma lipoproteins, leading to normalization of cholesterol levels [27,36,37]. Even though apoE is a physiologic component of VLDL and chylomicrons and is essential for their processing and clearance, it is not a necessary structural component for the lipoprotein to be normally assembled and secreted. This is seen in both humans and mice without apoE, where VLDL and chylomicron assembly and secretion are normal whereas their processing and clearance are clearly defective. The fact that apoE is produced by the liver cell at a nearly constant rate and placed either on the VLDL during lipogenesis or in proximity of the cell surface in periods of decreased lipoprotein production suggests a wide set of biologic functions for apoE, either as a lipoprotein-associated protein or as a membrane-associated protein.

Another interesting observation regarding the association of apoE with lipoproteins is the fact that the different apoE isoforms display relative preference for particles of different curvatures. It is known that apoE can associate with any lipoprotein, but *in vitro* experiments have shown that in the presence of equivalent amounts of large (VLDL and chylomicrons) versus small (HDL) lipoproteins, significantly more apoE will associate with lipoproteins of larger curvature rather than with the smaller particles [38]. This is particularly true for the apoE4 isoform and its natural and artificial mutants. Heterozygous subjects with an E2/4 or E3/4 genotype have a preponderance of the E4 allele product in the VLDL and a relative preponderance of the other allele product in the HDL particle population [39]. This phenomenon may have important repercussions in lipid metabolism and may influence the genetic penetrance and clinical severity of disease phenotype in subjects born with dysfunctional mutations of apoE. As an example, $apoE_{Leiden}$ – a duplication of seven amino acids in the middle of the molecule, leading to reduced receptor binding – produces severe dyslipidemia with an apparently dominant mode of inheritance and nearly complete penetrance [40]. The mutation occurs on an apoE4 substrate and is associated with predominant distribution of the mutant apoE in the VLDL fraction of heterozygous carriers [41]. This is in contrast with the recessive mode of inheritance and incomplete penetrance of the same disease phenotype in the most common presentations of type III HLP due to homozygosity for the apoE2 isoform [29]. Likewise, another mutation characterized by the presence of a cysteine at position 142 (and referred to as $apoE_{cys142}$) on an otherwise apoE4 allele shows a pattern of complete clinical penetrance and dominant genetic transmission of the hyperlipidemic phenotype through a mechanism that may in part be due to its preponderant accumulation on the triglyceride-rich particles [42]. It is proposed that, in heterozygous carriers, a receptor mutant form of apoE is more likely to produce dyslipidemia if it accumulates on the biologic target (in this case the triglyceride-rich lipoproteins) in high concentrations, overwhelming the wild-type allele product and producing a condition of *de facto* homozygosity.

The presence of apoE increases the amount of triglycerides packaged in newly formed VLDL. In the absence of apoE, secreted VLDL are relatively deficient in triglycerides. This phenomenon is observed not only in the animal models but also in humans, as type III HLP patients carrying a mutant form of apoE [43] have higher triglyceride levels than patients with apoE deficiency [44], while transgenic mice expressing isoform or mutant apoE [45] have higher triglyceride levels than apoE-deficient mice [46,47]. This is an intriguing observation because it applies exclusively to the secretion of hepatic lipoproteins and not to the chylomicrons, the largest triglyceride-rich lipoproteins, which are made by the intestine in the absence of apoE. Therefore, even though the presence of apoE is not mandatory or even required for packaging of large amounts of triglycerides in the newly formed lipoprotein (such as the intestinal chylomicron), the availability of apoE will nonetheless influence triglyceride content at least in the liver and modify the structure of the lipoprotein.

3.2.2
ApoE and Lipoprotein Processing and Catabolism

Once in the circulation, intestinal and hepatic triglyceride-rich lipoproteins use apoE for functions that are critical for the destiny of the lipoprotein particle: namely, engaging LPL and HL [48] and activating the mechanisms of receptor-mediated clearance mostly from the liver [1,6,49]. A large body of literature on the interaction between apoE and the intravascular lipases exists [50–56]. The lipoprotein characteristic of type III HLP, an abnormal pre-β VLDL enriched in cholesterol and triglycerides, is the most obvious evidence of influence of apoE mutations on processing of the triglyceride-rich lipoproteins by LPL. In addition, the disruption of the interaction between apoE and HL may be at the heart of why remnant lipoproteins deficient in apoE or carrying dysfunctional apoE are unable to be taken up by specific mechanisms in the liver. It appears that the final step of hydrolysis in the hepatic sinusoid, mediated by HL, is crucial for the unmasking of the apoE regions that engage HSPG and other membrane receptors involved in binding and uptake of remnant lipoproteins [22]. It is also believed that apoE-free LDL is indeed the end-product of HL action on the VLDL remnant [20].

3.2.3
ApoE and Lipoprotein Uptake

The final step in the extracellular processing of triglyceride-rich lipoproteins also relies heavily on the presence of apoE. Through the heparin binding domain of apoE, remnants bind to cell surface HSPG and eventually interact with specific receptors such as the LDLR and the LRP1, mediating the internalization and subsequent degradation of the lipoprotein and its components [6,8,57,58]. The presence of apoE is essential for several crucial functions of this final stage of the remnant's life. The interaction of apoE with HL described above also clears the way for the attachment of apoE to the cell surface. It is known that humans with HL mutations accumulate large amounts of LDL-like particles, which, unlike normal LDL, contain apoE and are enriched in triglyceride [59]. This suggests that the VLDL remnant in the space of Disse faces a crossroad leading to opposite destinies: either fast clearance from plasma or termination as LDL, a particle with notoriously slow plasma clearance. These alternative destinies may be decided by the efficiency of the interaction between apoE and HL, as the abnormal lipoprotein that accumulates in HL deficiency indicates that without the lipid hydrolysis step in the space of Disse the remnant will not be captured by the liver and will end up within the LDL density range but with a full complement of apoE.

The relevance of apoE in remnant uptake is also confirmed by the presence of an additional mechanism that appears to be in place to increase the available concentration of apoE for remnant uptake. This pathway, defined as "apoE secretion-capture", is present in hepatocytes [60] but may also be operational in macrophages [63] and in adipocytes [62] and may have physiologic effects in those cell types as well [63]. The basic workings of the secretion-capture mechanisms include enrichment of the pericellular space with newly secreted lipoprotein-free apoE to increase the ligand concentration of incoming lipoproteins. The already apoE-rich remnants reaching the space of Disse will acquire significantly larger amounts of apoE through the secretion-capture mechanism, and the resulting higher density of apoE molecules on the remnant surface will lead to more efficient uptake and internalization. Our laboratory provided powerful support for the *in vivo* relevance of the apoE secretion-capture mechanism through studies in mice lacking both LDLR and apoE. After reconstitution with apoE-producing bone marrow, these mice showed progressive accumulation of apoE in plasma and in the liver. Even though the plasma apoE levels reached 17 times those of normal mouse plasma, there was no uptake by the liver and no resolution of the dyslipidemia [64].

Rapid remnant uptake and normalization of the lipid profile was instead quickly achieved by adenoviral expression of the LDLR in the liver. Therefore, the secretion-capture mechanism seems to be in place to ensure clearance of remnants from a non-LDLR route, possibly LRP1 [65]. The secretion capture mechanism may also be a way for the cell to direct incoming lipoproteins to a specific receptor pathway, which may be important in situations in which a lipoprotein can be taken up by different receptors, such as in the liver (LDLR vs. LRP1) or in the macrophages (scavenger receptors vs. saturable receptors). As an example, the uptake of modified lipoproteins through scavenger receptors is the classic route for accelerated formation of foam cells, the hallmark of the atherosclerotic plaque. Physiologic enrichment of the oxidized or partially modified lipoprotein with apoE may direct the incoming particle away from the atherogenic pathway and into one (possibly linked to LRP1) leading to a more efficient processing and routing of the lipoprotein-derived cholesterol.

The apoE secretion-capture mechanism is an important example of the value of apoE in biologic cycles encompassing the intracellular and pericellular environments. The other important example, discussed later in this chapter, is the phenomenon of apoE recycling and resecretion that has been established for hepatocytes [66], macrophages [61], and CHO cells [67]. Although no connection between the mechanisms of apoE secretion-capture and apoE recycling has been established, it can be hypothesized that one system may potentiate the other to achieve higher efficiency in both lipoprotein clearance and cholesterol efflux processes.

3.3 Role of ApoE in HDL Metabolism

3.3.1 HDL Formation and Maturation

Blood levels of HDL cholesterol are inversely related to the risk of coronary heart disease [68]. One of the major antiatherogenic functions of HDL cholesterol is the transport of excess cholesterol from peripheral cells to the liver for disposal in bile, a process known as reverse cholesterol transport (RCT) [69]. Roughly half of the apoE in fasting plasma is found on HDL cholesterol [29], and apoE is required for HDL maturation [23,70]. In the artery wall, HDL cholesterol serves as an acceptor for cholesterol efflux from macrophage-derived foam cells, and apoE has been shown to promote cholesterol efflux [71]. Thus, apoE plays critical roles in HDL formation, RCT, and macrophage cholesterol efflux.

HDL cholesterol particles are heterogeneous, and qualitative and quantitative differences in their lipid and protein compositions give rise to HDL subclasses [72]. The main structural apoprotein of both nascent and mature HDL particles is apoAI, and deficiency of apoAI in humans [73,74] and in mice [75] results in low to undetectable levels of HDL cholesterol. Most of the apoAI in plasma is found on particles with α-electrophoretic mobility on agarose gel electrophoresis [76]. These spherical HDL particles also contain apoE and can be further separated by density into HDL_2 and HDL_3, which are less and more dense, respectively [76]. Around 5–15 % of apoAI in plasma is found to migrate with pre-β HDL mobility on agarose gels [72]. Nascent pre-β HDL particles contain lipid-poor apoAI with phospholipid and cholesterol [77]. Lipid-poor particles containing only apoE (γ-LpE) are also found in plasma [78]. The concentration of lipid-poor apoproteins is increased relative to α-HDL in the intravascular space and lymph, indicating local enrichment of the cholesterol acceptors in the environment for peripheral cell efflux [79]. Lipid rich α-HDL particles are derived from lipid-poor or lipid-free apoproteins.

The discovery of ATP-binding cassette A1 (ABCA1) as the mutant gene responsible for Tangier disease, an autosomal recessive disorder with low HDL, undetectable apoAI, xanthomatosis, and early atherosclerosis [80–82], has led to rapid advances in our understanding of the biogenesis of HDL cholesterol. ABCA1 plays a critical role in formation of nascent HDL by promoting efflux of cellular phospholipids and unesterified cholesterol (UC) to lipid-free or lipid-poor apoAI particles, forming pre-β HDL particles [83]. Although the primary sites of apoAI synthesis are in the liver and the intestine, the primary site for biogenesis of HDL cholesterol was a matter of debate, given the critical role of nascent HDL in accepting cholesterol from peripheral cells, such as macrophages. This issue was recently resolved by the development of tissue-specific knockouts of the ABCA1 gene in mice; these studies have clearly demonstrated that the liver is responsible for the biogenesis of the majority of nascent HDL [84], and that the intestine contributes approximately 30 % of plasma HDL formation [85]. In contrast, elimination of ABCA1 expression in bone marrow-derived cells has shown that macrophages contribute relatively little to the total amount of circulating HDL cholesterol [86]. However, the ability of cholesterol-loaded macrophages to efflux UC to lipid-poor apoproteins and pre-β HDL is a critical first step in RCT and an important determinant of the development of atherosclerosis.

The development of apoE-deficient (apo$E^{-/-}$) mice through gene targeting [46,47] has contributed substantially to our understanding of the role of apoE in HDL metabolism. The HDL in apo$E^{-/-}$ mice is characterized by markedly abnormal lipid composition, apoprotein content, and associated antioxidative enzymes, supporting the critical role of apoE in HDL formation. The amount of cholesterol carried by the HDL-sized lipoproteins is about 45 % of normal, and the majority of cholesterol is found in VLDL-sized particles containing increased amounts of apoAI and apoAIV [46,47]. The HDL cholesterol particles have reduced apoAI, but also have increased apoAIV content [87] and reduced capacity to accept cholesterol [88]. In addition, the HDL cholesterol in apo$E^{-/-}$ mice is enriched in sphingomyelin, due to combined defects in its synthesis and clearance [89].

The antioxidative functions of HDL are maintained by a number of enzymes, including lecithin:cholesterol acyltransferase (LCAT), paraoxonase 1 (PON1), and platelet-activating factor acetylhydrolase (PAF-AH). LCAT mediates the maturation of HDL by esterifying UC and facilitates RCT by maintaining an UC gradient from interstitial fluid to plasma [90,91]. LCAT also protects against oxidation by hydrolyzing long-chain oxidized PL [92]. PON1 is a lipoprotein-associated esterase that decreases the oxidative stress, possibly by hydrolyzing lipid hydroperoxides [93]. Decreased HDL PON1 activity is associated with increased oxidation of LDL and risk of atherosclerosis [94]. PAF-AH (also called Lp-PLA2) is a lipoprotein-associated enzyme that hydrolyzes proinflammatory PAF and short-chain PL hydroperoxides. PAF-AH activity modulates inflammation and LDL oxidation [92,95]. Forte et al. reported that apo$E^{-/-}$ mice on a chow diet have impaired maturation of HDL, due at least in part to reduced (28 %) LCAT activity in relation to C57BL/6 mice and $LDLR^{-/-}$ mice [96]. In response to a Western diet, the apo$E^{-/-}$ mice showed a further reduction in LCAT activity and reduced PAF-AH, while PON activity was reduced by 38 %. Altered activities of these antiatherogenic enzymes are likely to contribute to the formation of dysfunctional HDL and accelerated atherogenesis in apo$E^{-/-}$ mice.

Insights into the role of apoE in HDL metabolism have been gleaned from bone marrow transplantation experiments in which apo$E^{-/-}$ mice were transplanted with wild-type bone marrow, resulting in the reconstitution of macrophage apoE expression. In addition to promoting the clearance of cholesterol-rich lipoproteins in the VLDL-IDL size range, macrophage apoE promotes the near normalization of HDL cholesterol levels [27,36]. Spangenberg and Curtiss found that hepatic apoAI mRNA levels and total plasma apoAI levels are not affected in the apo$E^{+/+}$→apo$E^{-/-}$ mice, but HDL cholesterol was increased twofold [87]. Data from our laboratory provided evidence that macrophage apoE in the apoE$^{+/+}$→apoE$^{-/-}$ mice fully displaces apoAIV from the surface of remnant lipoproteins, but is not sufficient to redistribute apoAI back to the HDL [97]. Extrahepatic apoE is thus able to correct many of the abnormalities due to apoE deficiency in HDL formation.

3.3.2
Cholesterol Efflux and RCT

Accumulation of macrophages overloaded with cholesteryl esters (CE) in the artery wall plays a central role in the pathogenesis of atherosclerosis, so macrophage cholesterol homeostasis is believed to play a critical role in the initiation and progression of atherosclerotic lesions. Incubation of macrophages with modified lipoproteins increases their CE content and leads to increased apoE secretion [98], due to enhanced apoE synthesis [99] and decreased degradation [100]. Macrophages get rid of excess cholesterol through the efflux of UC to acceptors, such as HDL cholesterol or lipid-poor apoproteins, including apoAI and apoE. Inhibition of ACAT, the enzyme that catalyzes the addition of free fatty acids to UC to form CE, has been considered as a possible strategy to reduce foam cell formation and atherosclerosis. However, studies in mice lacking expression of macrophage ACAT1 have shown that excess accumulation of UC by macrophages is toxic and promotes atherosclerosis [101,102]. Recent studies with ACAT inhibitors in humans gave similar results, with ACAT inhibition promoting progression of atherosclerosis [103,104]. It seems likely that either the availability of acceptors or the improper trafficking of UC for efflux from macrophages becomes limiting in this scenario, causing a breakdown in the sequence of events that prevent accumulation of UC to toxic levels in the macrophage.

A number of studies in murine models of atherosclerosis have shown that macrophage expression of apoE is antiatherogenic. In bone marrow transplantation studies, elimination of macrophage apoE expression promotes atherogenesis in the absence of significant changes in plasma lipoproteins in several murine models fed on atherogenic diets, including C57BL/6 mice [24,105], LDLR$^{-/-}$ mice [26], and apoAI$^{-/-}$ mice [106]. The strong local effect of apoE in reducing cholesterol in arterial macrophages is evidenced by the fact that expression of apoE at levels too low to affect plasma lipoprotein levels significantly delays atherogenesis [25,107,108]. Although apoE may impact atherogenesis through a number of mechanisms, including anti-inflammatory and antioxidant effects [109], its role in cholesterol efflux is likely to be a major mechanism responsible for the antiatherogenic effects.

The absence of endogenous apoE expression by the macrophage leads to a significant reduction in cholesterol elimination, whereas transfection of apoE into apoE-negative macrophage lines increases efflux [71,110]. HDL particles containing only apoE and no apoAI are efficient acceptors of cholesterol efflux [74,111], while incubation of HDL with macrophage foam cells results in secretion of apoE-enriched HDL [112,113]. ApoE can also be found in lipid-poor particles in the plasma known as (γ-LpE), and these particles have been shown to be acceptors for cellular cholesterol efflux [78]. The upregulation of apoE synthesis from cholesterol loading is linked to an LXR responsive element (ME1) in the macrophage apoE promoter [114]. The fact that both ABCA1 and apoE are under the control of LXR suggests a coordinated mechanism that controls the expression of cholesterol efflux proteins when the cell is exposed to oxysterols or LXRα agonists [114].

It is important to remember that apoAI is not expressed by macrophages, so the local concentration of apoE in the developing plaque is likely to be significantly higher than the concentration of apoAI. We examined the ability of transgenic expression of human apoAI by macrophages to counter the proatherogenic effects of macrophage apoE deficiency. Transgenic expression of apoAI from apoE$^{-/-}$ mouse macrophages *in vivo* is able to reduce the proatherogenic effects of macrophage apoE deficiency dramatically [106,115]. Furthermore, the apoE$^{-/-}$ macrophages expressing transgenic apoAI show increased cholesterol efflux in relation to control apoE$^{-/-}$ macrophages [116]. Human monocyte-derived macrophages secrete apoE associated with lipids independently of the presence of exogenous apoAI, and the efficiency of this process is affected by the apoE genotype, with apoE4 being least efficient [117]. In the setting of excess exogenous acceptor apoproteins, cholesterol efflux was greater from cells with endogenous expression of apoE (e.g., wild-type murine macrophages or J774 cells expressing human apoE) than apoE$^{-/-}$ control cells [18,19]. It is interesting to note that apoE deficiency in macrophages promotes foam cell formation despite the presence of functional ABCA1 and abundant circulating apoAI, and that efflux is decreased in apoE$^{-/-}$ macrophages even though the ABCA1 transporter is functional and probably upregulated [118]. These observations raise the question of whether endogenous apoE secretion may be a more efficient mediator of cholesterol efflux via ABCA1 than lipid-free apoAI.

3.3.3
Roles of apoE in RCT and HDL Metabolism

RCT is the process by which excess cholesterol is delivered from peripheral cells to the liver for excretion in the bile (Figure 3.1). ApoE promotes RCT both by enhancing efflux of UC from peripheral cells to maturing HDL cholesterol particles and by promoting the uptake of CE-enriched lipoproteins by the liver [119]. Lipid-poor apoAI particles, or pre-β HDL, are secreted by the liver and small intestine or can be produced from chylomicron remnants or HDL_2 through the action of cholesterol ester transfer protein (CETP) [120], phospholipid transfer protein [121], or HL [122]. The nascent pre-β HDL particles act as acceptors for phospholipids and UC transported from peripheral cells via ABCA1 [123]. The UC in these disc-shaped lipoproteins is subsequently esterified through the action of LCAT, and the resulting CE forms the hydrophobic core, converting the nascent disc to a spherical HDL_3 particle. Scavenger receptor class B type I (SR-BI) and ABCG1/4 have been implicated in promoting efflux of UC to HDL_3, and LCAT-mediated conversion of the UC to CE results in further expansion of the hydrophobic core and formation of HDL_2. Matsuura et al. recently reported that ABCG1/4-mediated macrophage cholesterol efflux is apoE-dependent [124]. In these studies, apoE-containing HDL_3 particles obtained from CETP-deficient patients were able to accept ABCG1/4-mediated efflux

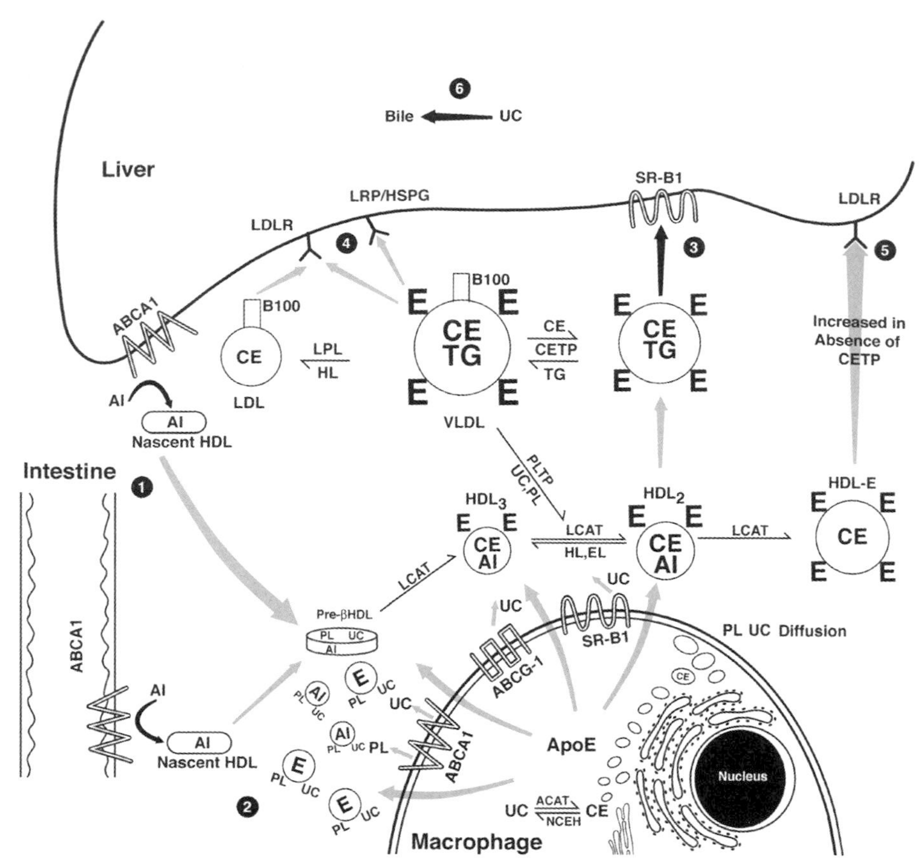

Fig. 3.1 Role of apoE in macrophage cholesterol efflux and reverse cholesterol transport. (*1*) The liver and the intestine are the primary sites for synthesis of apoAI and ABCA1-dependent formation of nascent HDL particles. (*2*) Efflux of unesterified cholesterol (UC) by macrophage-derived foam cells in the artery wall is a critical determinant of foam cell formation and progression of atherosclerosis. Acceptors of ABCA1-mediated efflux of UC and PL include lipid-poor apoAI, apoE, nascent or pre-β HDL, as well as α-HDL particles. In addition to promoting cholesterol efflux, apoE plays an important structural role in allowing expansion of discoidal pre-β HDL particles into spherical α-HDL. ABCG1 and SR-BI have been implicated in promoting UC efflux to maturing α-HDL particles. (*3*) Mature α-HDL particles may deliver CE directly to the liver through SR-BI mediated selective uptake, or (*4*) indirectly, when CETP is present (e.g., in humans) by transferring CE to VLDL/LDL cholesterol. ApoE plays a crucial role in mediating uptake of VLDL remnant lipoproteins by the LDLR or LRP/HSPG remnant pathway, whereas apoB100 mediates uptake of LDL cholesterol uptake by LDLR. (*5*) In the absence of CETP (e.g., in mice), large HDL-E particles play an increased role in reverse cholesterol transport through apoE-mediated clearance by LDLR. (*6*) Cholesterol can be excreted by the liver either directly into the bile or after conversion into bile acids. Abbreviations: UC: unesterified cholesterol. EC: esterified cholesterol. PL: phospholipid. ACAT: Acyl-CoA:cholesterol acyltransferase. NCEH: neutral cholesteryl ester hydrolase. LCAT: lecithin:cholesterol acyltransferase. CETP: cholesteryl ester transfer protein. PLTP: phospholipid transfer protein. HL: hepatic lipase. EL: endothelial lipase. ABCA1: ATP-binding cassette transporter A1. ABCG1: ATP-binding cassette transporter G1. SR-B1: scavenger receptor-type B1. HDL: high-density lipoprotein. VLDL: very low density lipoprotein. LDL: low-density lipoprotein. A–I apolipoprotein A–I. E: apolipoprotein E. B100: apolipoprotein B100. LRP: LDL receptor-related protein. HSPG: heparan sulfate proteoglycans.

of UC from macrophages *in vitro* [124]. Surprisingly, recent bone marrow transplantation studies in LDLR$^{-/-}$ recipient mice have shown that deficiency of ABCG1/4 in bone marrow-derived cells results in reduced atherosclerosis, calling into question the physiological relevance of ABCG1/4-mediated cholesterol efflux in reducing foam cell formation and atherosclerosis *in vivo* [125,126]. Interestingly, the authors attribute this paradoxical effect either to increased expression of ABCA1 associated with a dramatic increase in macrophage apoE secretion by the ABCG1/4$^{-/-}$ macrophages [125] or to increased apoptosis [126]. ApoAI is the main structural component of pre-β HDL, HDL_3, and HDL_2; however, apoE is present on HDL_3 and HDL_2 and plays a critical role in maturation of the HDL particles [119]. It is believed that apoAI adopts a belt-like conformation on HDL particles and that apoAI-containing HDL can hold only a limited amount of CE in its core. In contrast, the addition of apoE to the HDL particle allows for significant expansion of the amount of CE in the core and increased size of the HDL particle. Weisgraber and coworkers have reported a model based on low-resolution X-ray crystallography of apoE bound to phospholipid, which facilitates visualization of how apoE performs this critical role in HDL maturation [127]. In this model, apoE wraps around a spherical phospholipid HDL particle in two circular horseshoe-shaped bands associating with the polar heads of the phospholipid groups, allowing easy expansion of the CE-rich core under the influence of LCAT [119,127].

The next step in RCT is the delivery of CE from HDL_2 particles to the liver. HDL_2 may deliver CE directly to the liver by interacting with SR-BI [128] or, alternatively, apoE may facilitate this step in RCT through its role as a ligand for hepatic uptake of CE-enriched lipoproteins. In humans, CETP promotes the transfer of CE from HDL_2 to VLDL cholesterol [129]. The VLDL cholesterol particles undergo lipolysis and the resulting IDL cholesterol can either be cleared directly by the liver or be further lipolyzed to form LDL cholesterol. These apoB-containing lipoproteins deliver their CE to the liver through LDLR-mediated uptake. ApoE binds to the LDLR with high affinity and mediates the uptake of VLDL and IDL remnant lipoproteins, whereas apoB100 serves as the ligand for uptake of LDL cholesterol by the LDLR. In species that lack CETP, such as mice, apoE also provides an important alternative RCT pathway for the delivery of the CE to the liver directly through HDL. In the absence of CETP, the HDL contains a subclass of large apoE-enriched particles referred to as HDL_1, HDL_c, or HDL-with apoE [119]. These apoE-enriched HDL particles can be taken up directly by the hepatic LDL receptors, providing a direct pathway for apoE-mediated clearance of HDL cholesterol that is independent of SR-BI.

3.3.4
Critical Roles for ApoE and SR-BI in RCT

Mice null for apoE and SR-BI (DKO) develop severe hypercholesterolemia, occlusive coronary atherosclerosis, myocardial infarction, and premature death at 6–8 weeks of age [130]. The dyslipidemia is characterized by the accumulation of cholesterol in VLDL and large UC-enriched HDL [131]. However, the lethality of the phenotype is disproportionate to the degree of hypercholesterolemia or elevation in apoB-containing lipoproteins, given that apoE$^{-/-}$ and LDLR$^{-/-}$ mice develop more severe hypercholesterolemia on an atherogenic diet without developing the rapidly lethal phenotype [132]. In mice, deletion of either SR-BI or apoE results in formation of HDL particles with dysfunctional properties and accelerated atherosclerosis. Studies have shown that deletion of apoE reduces the selective uptake of HDL-CE by the liver despite increased hepatic expression of SR-BI, and the reduced selective uptake was attributed to the absence of apoE on HDL and in the liver [133]. Also consistent with a role of apoE in HDL clearance is the marked accumulation of apoE on HDL in SR-BI deficient mice [134]. Deletion of SR-BI increases HDL-C and results in the formation of large, UC-enriched HDL [134]. The combined deficiency of SR-BI and apoE results in a greater increase in UC content of HDL than is seen with single deletion of SR-BI [134], and with the HDL particles being abnormally large and similar to lipoprotein X [135,136]. Interestingly, recent studies by Krieger and colleagues demonstrated reduced accumulation of abnormally large HDL particles when low levels of hepatic apoE were expressed in DKO mice [137]. Analysis of the apoAI-containing lipoprotein particles from DKO mice by 2-D gel electrophoresis revealed the presence of markedly enlarged apoAI-containing particles that migrate with electrophoretic mobility between pre-β and α [138]. As the combined deficiency of SR-BI and apoE results in formation of abnormally enlarged and UC-enriched HDL, it is probable that these particles are severely dysfunctional in relation to the HDL from single knockout mice and that these particles also contribute significantly to the accelerated atherosclerosis in DKO mice. Consistent with this concept, we have recently demonstrated that expression of macrophage apoE in DKO mice markedly reduces atherosclerosis while reducing the accumulation of the enlarged, UC-enriched, apoAI-containing HDL particles and producing an HDL composition and subpopulation profile similar to that in SR-BI$^{-/-}$ mice [138]. Combined deficiency of apoE and SR-BI thus leads to a disruption in RCT with rapidly lethal consequences, demonstrating the critical importance of having alternative pathways for HDL-mediated delivery of cholesterol to the liver.

3.4 ApoE and Recycling

3.4.1 Evidence for apoE Recycling

We hypothesized, on the basis of the multiple and critical functions of apoE in the metabolism of lipids and lipoproteins, that apoE follows unique pathways of secretion and internalization, allowing it to have maximal effects on cellular cholesterol homeostasis. Specifically, we hypothesized that a fraction of apoE on internalized lipoproteins escapes degradation and is recycled. Wong had previously reported that ~60 % of rat intrahepatic apoE, compared with 7–10 % of the apoB, was derived from the plasma compartment, suggesting that internalized apoE did not undergo complete degradation and was available for resecretion [139]. In addition, Takahashi and Smith [140] found that a substantial amount of lipoprotein-associated apoE, internalized by the murine macrophage RAW 264.7 cells, was retained inside the cell for periods extending far beyond the time for degradation of apoB. They estimated that as much as 30 % of internalized apoE was resecreted. Consequently, a series of experiments in our laboratory, using a variety of models and approaches, have provided overwhelming evidence for resecretion of internalized apoE:

1. We recovered intact radiolabeled apoE with hepatic Golgi lipoproteins after injecting radioiodinated mouse $d < 1.019\,g\,mL^{-1}$ lipoproteins into wild-type mice. This finding suggests that a portion of internalized apoE escaped degradation and was trafficked to the Golgi complex, where it associated with newly forming lipoproteins [66].
2. We found apoE associated with nascent Golgi lipoproteins recovered from the livers of apoE-deficient mice that had been reconstituted with bone marrow from C57BL/6 mice [141]. Furthermore, primary cultures of hepatocytes from these transplanted apoE-deficient mice secreted apoE, evidence that endocytosed apoE can be retained by the liver cell, processed through the Golgi apparatus, and eventually resecreted [141].
3. As further corroboration of recycling, we demonstrated resecretion of apoE from apoE-deficient hepatocytes after incubation either with apoE-containing mouse VLDL or with HDL [142]. The discovery that apoE recycles even when associated with HDL adds significantly to our understanding of this process, as it shows that the recycling pathway can be reached by multiple routes, and indicates that this phenomenon may be physiologically correlated to RCT.

ApoE mediates endocytosis of lipoprotein particles via the LDLR [2,3] and LRP1 [6,7]. It may also be internalized by binding to heparan sulfate proteoglycans, with [6,7] or without [8,57] binding to LRP1. ApoE-containing HDL has also been shown to be internalized by the LDLR [20,23], LRP1 [21], and HSPGs [22]. Studies by Rensen et al. [143] suggested that the LDLR is critical for apoE recycling in hepatocytes, while studies by Heeren et al. [144] pointed to LRP1 as the predominant pathway for apoE recycling from triglyceride-rich lipoproteins in human skin fibroblasts and hepatoma cell lines. However, our studies with hepatocytes demonstrated that apoE recycling is not altered in the absence of the LDLR [66] or under conditions in which LRP1 expression is downregulated [145]. The ability of apoE to recycle is thus not limited to a specific entry point into the cell; however, it is not clear if the specific intracellular recycling pathway is influenced by the point of entry.

Demonstration of apoE resecretion in hepatocytes led to questions about recycling as a general cellular process. We initially speculated that apoE recycling provides alternate routes for lipid trafficking or lipoprotein secretion in a cell that faces a tremendous lipid burden. The discovery that recycling occurs in macrophages [61,142] proved consistent with this hypothesis. On the other hand, it was unclear whether intracellular lipid content or endogenous apoE production in both cell types impacted apoE recycling, so we studied apoE recycling in Chinese hamster ovary (CHO) cells. These cells do not produce apoE, nor do they process large amounts of intracellular lipid; however, they do express many of the critical agents in cholesterol transport, such as LDLR, LRP1, and ABCA1 and ABCG1. Our studies revealed that apoE does recycle in CHO cells [67], although the process is not identical to that in hepatocytes and macrophages (discussed below). Recycling has also been shown to occur in human skin fibroblasts [144,146–148], so recycling is not specific to those cells that produce the apoprotein or process large amounts of lipid. ApoE recycling appears to be a general cellular process that functions to conserve the apoprotein, allowing it to exert maximal impact on intra- and extracellular functions. The differences in recycling among different cell types may be a reflection of the relative ability of the cell to process lipid, which in turn may depend on whether the cell is equipped to secrete its own apoE.

The determinants of apoE recycling are unknown. For any endocytosed molecule, the fate of the ligand and the receptor is determined within the endosome (Figure. 3.2). Ligands such as asialoglycoprotein and LDL apparently dissociate from their receptors in the acidic environment and are routed to the lysosomal compartment [149,150]. While there is some evidence for retroendocytosis of LDL in monkey skin fibroblasts [151], most of the apoB100 is trafficked to the lysosome and degraded. On the other hand, a large fraction of apoE escapes the lysosomal pathway and is recycled. This difference between apoB100 and apoE may be explained by the relative abilities of the two proteins to remain associated with the receptor in the acidic environment: apoE has a 23-fold higher affinity for the LDLR than that of apoB100 [3]. Upon a change in pH in the endosome, apoE may remain attached to the receptor while the lipoprotein particle dissociates and traffics on to the lysosome. ApoE could then transit to other intracellular destinations or back to the surface of the cell. Consistent with this hypothesis is the observation that apoE4 exhibits impaired HDL-induced recycling in relation to apoE3 [147], while apoE4 has been shown to have a greater affinity than apoE3 for lipid [152]. In addition, apoE4 has a greater propensity than apoE3 to form molten globules at low pH [153]. The drop in pH in the early endosomes may thus produce conformational changes in apoE4 that lead to dissociation from the receptor, and coupled with the greater affinity of this isoform for lipid, apoE4 may not be ushered into the recycling pathway as apoE3. However, despite the fact that apoE4 is not recycled to the same extent as apoE3, apoE4 degradation is apparently not increased in HuH7 cells [147]. This suggests that intact apoE4 is retained within the cell, unable to be resecreted and is not trafficked to the lysosome for degradation.

It is important to note that binding of apoE to a lipoprotein is not essential for recycling to occur. Our studies have shown that the N-terminal 22 kD thrombin cleavage fragment of apoE recycles in mouse hepatocytes [145]. The C-terminal 10 kD fragment contains the major lipid-binding domain for apoE [154–156], and although the N-terminal fragment has some lipid binding affinity [2,157], only a

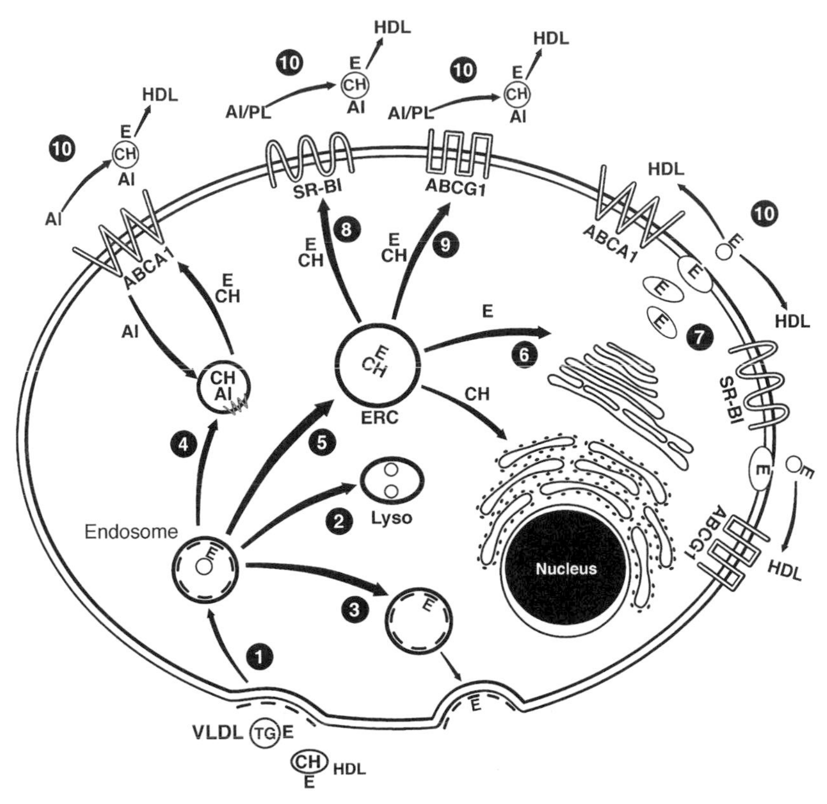

Fig. 3.2 General model for apoE recycling. (*1*) ApoE enters the cell on VLDL or HDL through receptor-mediated pathways. (*2*) Within the endosome apoE dissociates from the lipoprotein, which moves on to the lysosome (lyso). From the endosome, apoE may traverse a number of pathways. (*3*) It may recycle back to the cell surface. (*4*) The apoE-containing endosome may fuse with apoAI/ABCA1-containing vesicles, leading to formation of cholesterol/apoE/apoAI complexes that are resecreted. (*5*) ApoE may move to the endocytic recycling compartment (ERC). (*6*) From the ERC, apoE may traffic to the Golgi where it becomes lipidated and (*7*) secreted from the cell. ApoE resecretion may be linked to cellular cholesterol removal via phospholipid-containing acceptors and SR-BI (*8*) and ABCG1 (*9*). Both SR-BI and ABCG1 have been localized to the ERC, a major intracellular cholesterol storage organelle. (*10*) Regardless of the pathway of resecretion, recycling apoE promotes cholesterol efflux and plays a major role in HDL formation and maturation. Abbreviations: TG: triglyceride. CH: cholesterol. PL: phospholipid. E: apoE. AI: apoAI. –: lipoprotein receptor.

small fraction of the 22 kD fragment actually associates with VLDL [145]. These results suggest that, regardless of the fact that the 22 kD fragment has decreased affinity for lipid and lipoprotein surfaces, this domain of apoE contains sufficient information to enter and navigate the recycling pathway. This also indicates that association of apoE with lipids is not essential for recycling to occur, which may be of significance in considering recycling of apoE from HDL. For example, one could postulate that when HDL docks to the membrane via SR-BI [128], apoE diffuses away from the particle and becomes membrane-bound and internalized either as such or as part of an incoming remnant lipoprotein. Thus, irrespective of entry point or association with lipid, a fraction of apoE internalized by the cell is spared degradation and, in the case of apoE3, ultimately resecreted. The regulation of this process and the impact on cellular lipid trafficking are critical to understanding its physiologic relevance.

3.4.2
Regulation of ApoE recycling

We have reported that human apoAI increases apoE resecretion from mouse hepatocytes nearly fourfold over control [142]. Concomitant with the increased resecretion were decreased amounts of cellular apoE relative to cells not incubated with apoAI. Heeren et al. reported that HDL serves as an extracellular acceptor for apoE internalized on triglyceride-rich lipoproteins by human skin fibroblasts and human Hep3b hepatoma cells [144]. Later, this concept was expanded with the finding that HDL_3 acts as an extracellular acceptor for recycled apoE and stimulates the recycling of apoE internalized by Hep3b cells [148]. It is possible that apoE recycling is linked to RCT both by serving as a signaling mechanism for HDL cholesterol entry into the cell and by increasing intracellular cholesterol efflux in response to HDL signaling [142]. In addition, apoE recycling in macrophages may have important implications for atherogenesis, as even small amounts of apoE produced in macrophages reduce plaque formation [25,158], and HDL and apoAI stimulate apoE recycling in mouse peritoneal macrophages [61]. Furthermore, we reported that macrophages from mice harboring a macrophage-specific human apoAI expression construct exhibited increased apoE recycling in relation to macrophages from wild-type mice [61]. This increase was noted even though the total accumulation of apoAI in the media was $<1\%$ of the amount added in the experiments with exogenous apoAI. Further evidence supporting a relationship between recycling and HDL metabolism was the finding that treatment of macrophages with the liver X receptor (LXR) agonist TO-901317, which stimulates expression of ABCA1, potentiated apoE recycling.

The mechanism by which apoAI enhances the recycling of apoE is unknown, but it is our hypothesis that the effect is related to cellular cholesterol efflux, as both apoE and apoAI have been shown to play crucial roles in this process. ApoE promotes cholesterol efflux through intracellular and pericellular effects [110], and apoAI serves as an acceptor for cholesterol efflux [159,160] through the ABCA1 pathway [161–163]. In addition, there is evidence that binding of HDL/apoAI to the cell surface mobilizes cholesterol from intracellular pools to the cell surface [164,165] where the lipid can be transported out of the cell through ABCA1. We speculate that the addition of apoAI to the medium mobilizes intracellular cholesterol and activates the ABCA1 pathway, which in turn stimulates apoE resecretion. Huang et al. [118] have presented evidence that macrophage cholesterol efflux mediated by endogenous apoE is not dependent on ABCA1, but this does not preclude a role for recycling apoE in cholesterol efflux via the ABCA1 pathway. On the other hand, Remaley et al. [166] reported that HeLa cells transfected to express ABCA1 had increased efflux of cholesterol and phospholipid to apoE, suggesting a role for ABCA1 in apoE-mediated cholesterol efflux. Rees et al. demonstrated that apoAI stimulates the secretion of apoE by macrophages, but concluded that apoAI primarily stimulated the secretion of newly synthesized apoE with little effect on a cell surface pool [167]. However, it should be noted that stimulation of apoE secretion was seen only after 8 h of incubation with apoAI. In contrast, we found a fourfold increase in apoE resecretion within 30 min, suggesting that recycling apoE may respond to extracellular signals more rapidly than newly synthesized apoE.

Taken together, the data suggest that apoE recycling is linked to RCT both by serving as a biosensor of cholesterol balance and by regulating cholesterol efflux by ABCA1-dependent and independent mechanisms.

3.4.3
ApoE Recycling and Cholesterol Efflux

The observations that the apoE recycling pathway in hepatocytes and macrophages was upregulated by two fundamental components of the RCT pathway (apoAI and HDL) [61,142] and that apoE recycling was potentiated in macrophages when ABCA1 expression was stimulated [61] are highly suggestive that apoE recycling modulates intracellular cholesterol trafficking and functions in RCT. Studies by Heeren and colleagues [148] provided direct evidence for a link between apoE recycling and cholesterol efflux. These investigators found apoE recycling to be accompanied by a reduction in filipin staining in peripheral compartments of the cell. In addition, they showed that HDL_3-derived apoAI is internalized to preexisting apoE/cholesterol-containing endosomes. Time-lapse fluorescence microscopy confirmed the intracellular formation of apoE/cholesterol/apoAI-containing complexes, leading to the conclusion that these complexes are involved in recycling. It has been suggested that apoAI and ABCA1 may be endocytosed to intracellular lipid depots where ABCA1 facilitates the movement of lipid to the vesicle lumen so that it associates with apoAI before resecretion [168]. It is thus possible that the apoAI/ABCA1-containing vesicles fuse with the apoE/cholesterol-containing endosomes, leading to the formation of cholesterol/apoE/apoAI complexes, which are subsequently secreted (Fig. 3.2). The effect of apoAI on apoE recycling [61,142] and the stimulation of recycling through upregulation of ABCA1 by LXR agonists [61] could both be explained through this mechanism and support a role for ABCA1 in recycling. However, Heeren and coworkers reported that apoE recycling was not altered in fibroblasts from patients with Tangier's disease [147], suggesting that ABCA1 expression and activity are not required for apoE recycling in fibroblasts.

It is important to note that while apoE recycling may be linked with intracellular cholesterol trafficking and efflux, the mechanisms may be quite different in different cell types. For example, studies in our laboratory have shown that apoE recycling in CHO cells is not stimulated by apoAI, nor by LXR upregulation of ABCA1 [67]. These results are in marked contrast to those obtained in macrophages and hepatocytes, and suggest that apoE recycling in CHO cells is not linked to the ABCA1-mediated mobilization of cholesterol. Consistent with this concept are the observations that apoE recycling in CHO cells was minimally stimulated with acceptors that promote ABCA1 lipid efflux, including lipid-free apoAI and lipid-poor apoproteins. In addition, apoE recycling was not affected by brefeldin A (BFA) or monensin, known inhibitors of ABCA1-mediated lipid efflux [169,170]. However, β-cyclodextrin, which preferentially extracts cholesterol from plasma membrane lipid rafts [171,172] by a mechanism independent of ABCA1 [169], markedly increased apoE recycling in CHO cells. Our studies therefore suggest that apoE recycling in CHO cells may be linked to cellular cholesterol removal through phospholipid-containing acceptors and hence possibly through either SR-BI or ABCG1 [173–176] (Fig. 3.2). The extraction of membrane cholesterol by β-cyclodextrin would lead to a redistribution of newly synthesized and endosomal cholesterol [177] to the plasma membrane. Furthermore, both SR-BI [178,179] and ABCG1 [180,181] have been localized to the endosomal recycling compartment (ERC), a major intracellular cholesterol storage organelle [182], and SR-BI has been shown to mediate the endocytosis and resecretion of HDL [179,183]. Because cholesterol is transported from the ERC to the plasma membrane through the same vesicles that carry recycling proteins [182], cholesterol efflux by β-cyclodextrin or phospholipid-containing acceptors may enhance apoE recycling by stimulating cholesterol trafficking to the plasma membrane through the cholesterol-rich ERC.

Thus, whereas apoE recycling in general is clearly associated with cholesterol trafficking and efflux, the specific links differ depending on the cell type. In macrophages and hepatocytes, apoE recycling appears to be linked to cholesterol efflux through apoAI and ABCA1, whereas apoE recycling in non-macrophage peripheral cells is linked to cellular cholesterol removal through phospholipid-containing acceptors and hence possibly through either SR-BI or ABCG1 (Fig. 3.2).

3.4.4 Form of Recycled ApoE

Our first studies on recycling suggested that apoE internalized by the liver reassociated with lipoproteins within the cell. We found that recycled apoE was associated with lipoproteins of $d = 1.019–1.210\ \mathrm{g\ mL^{-1}}$ recovered from the Golgi apparatus [66]. However, the liver may represent a unique situation for the exit of recycling apoE since hepatocytes secrete lipoproteins of different compositions and sizes, and apoE would probably readily associate with the lipid surfaces of lipoprotein particles. Heeren et al. [148] reported that apoE resecreted from HuH7 cells in the presence of human serum was recovered both with the HDL fraction and with a lipid-free fraction. They speculated that lipid-free apoAI associated with recycled apoE. Interestingly, when HuH7 cells were chased in serum- and HDL-free media, nearly all of the recycled apoprotein was recovered with the lipid-free fraction. Even though HuH7 cells would be expected to secrete lipoproteins, the time of the chase in these experiments (60 min) would not be sufficiently long to accumulate any mass of lipoproteins with which apoE could associate.

Recycled apoE also appears to exit macrophages on a lipidated HDL-like particle when the cells are chased in the presence of apoAI [61]. In contrast, the bulk of apoE recycled from CHO cells is either lipid-free or associated with limited amounts of lipid [67]. Furthermore, apoAI does not affect the form in which apoE recycles. To explain these differences, we speculate that recycled apoE is resecreted from all cells in a cholesterol-containing complex, with the level of cholesterol enrichment dependent upon the cholesterol status of the cell. In addition, we speculate that the effects of apoAI on apoE recycling and the recycling complexes depend in large part on the degree of lipidation of apoAI by ABCA1. It has been shown, for example, that macrophages form larger, UC-enriched nascent HDLs, and lipidate more apoAI through ABCA1 than fibroblasts [184]. Stimulation of cholesterol efflux in macrophages by apoAI would thus enhance the resecretion of apoE/cholesterol complexes that could subsequently fuse with nascent apoAI-containing HDL. In contrast, fibroblasts (and perhaps CHO cells) inefficiently lipidate apoAI through ABCA1, thus leading to resecretion of minimally lipidated apoE complexes. In this regard a physiologic function of the apoE recycling pathway may be to produce poorly lipidated apoE particles for use as cholesterol acceptors and as seeds for size expansion of apoAI/lipid particles (Fig. 3.2). The efficiency with which recycled apoE is used as an acceptor would depend on cellular cholesterol status and existing efflux mechanisms of the particular cell type. Another intriguing hypothesis is that recycled apoE associates with high activity phospholipid transfer protein (PLTP), thereby facilitating HDL formation and remodeling. High activity PLTP has been shown to interact with apoE [185,186], but at present there are no data to link PLTP with recycled apoE.

3.4.5
ApoE Recycling: Relevance to Atherosclerosis

Whereas apoE recycling may serve different functions depending on the cell type, we hypothesize that in macrophages this process serves a protective role against the development of foam cells and atherogenesis. A number of studies have shown that apoE secreted by macrophages delays plaque formation [24–27,107,108,187]. Reconstitution of apoE-deficient mice with marrow from wild-type mice leads to a normalization of plasma cholesterol levels and >50-fold reduction in atherosclerotic lesion area [24]. Reconstitution of C57BL/6 mice with marrow from apoE-deficient mice resulted in a tenfold increase in lesion area [24]. In addition, expression of apoE in murine apoE$^{-/-}$ macrophages through retroviral vectors led to a 60 % reduction in atherosclerotic lesion area even though plasma lipid levels were unaffected [25,158]. It is our hypothesis that apoE recycling in macrophages enhances cellular cholesterol efflux and delays atherosclerotic lesion formation. The antiatherogenic effects of HDL and apoAI may be mediated by a number of mechanisms including the ability of apoAI both to upregulate the secretion of apoE [167,188] and to stimulate apoE recycling in macrophages trapped within atherosclerotic lesions. We thus propose that apoE recycling is an additional protective mechanism by which macrophages can eliminate excess cholesterol, making it available for RCT.

3.5
Summary

ApoE influences HDL structure, function, and metabolism through intracellular, pericellular, and intravascular effects. At the cellular level, apoE synergizes cholesterol entry and exit mechanisms and is the strongest regulator of cholesterol efflux from arterial macrophages. At the pericellular level, apoE-containing particles extract additional cholesterol and fuse with apoAI particles to create larger HDL. At the intravascular level, apoE coordinates the final disposal of the HDL particle through direct and indirect modulation of CETP, SR-BI, and LDLR delivery pathways. ApoE recycling may significantly contribute to some of these effects, depending on cell type: within hepatocytes, for example, apoE recycling may represent a mechanism to ensure increased availability of apoE in the hepatic sinusoid, thus potentiating the secretion-capture process and accelerating the uptake of remnant particles. ApoE recycling in hepatocytes might also be important in enhancing the selective uptake of HDL cholesteryl esters through SR-BI [133], thereby playing a role in the final steps of RCT in the liver. Additionally, recycling apoE may also serve as a chaperone for proper targeting and repositioning of recycling LRP1 or other receptors to the cell surface. In macrophages, apoE recycling may enhance cholesterol efflux and contribute to the mechanisms countering foam cell formation. In non-apoE-producing peripheral cells, apoE recycling may act as a biosensor of cholesterol entry and exit pathways to help the cell avoid the dangers of cholesterol accumulation or depletion.

References

1 Beisiegel, U., Weber, W., Ihrke, G., Herz, J., Stanley, K.K. (1989) The LDL-receptor-related protein, LRP, is an apolipoprotein E-binding protein. *Nature*, **341**, 162–164.

2 Innerarity, T. L., Friedlander, E. J., Rall, S. C., Jr.,Weisgraber, K. H., Mahley, R. W. (1983) The receptor-binding domain of human apolipoprotein E. Binding of apolipoprotein E fragments. *J. Biol. Chem.*, **258**, 12341–12347.

3 Pitas, R. E., Innerarity, T. L., Arnold, K. S., Mahley, R. W. (1979) Rate and equilibrium constants for binding of apo-E HDL_c (a cholesterol-induced lipoprotein) and low density lipoproteins to human fibroblasts: evidence for multiple receptor binding of apo-E HDL_c. *Proc. Natl. Acad. Sci. USA*, **76**, 2311–2315.

4 Herz, J., Hamann, U., Rogne, S., Myklebost, O., Gausepohl, H., Stanley, K. K. (1988) Surface location and high affinity for calcium of a 500-kd liver membrane protein closely related to the LDL-receptor suggest a physiological role as lipoprotein receptor. *EMBO J.*, **7**, 4119–4127.

5 Willnow, T. E., Sheng, Z., Ishibashi, S., Herz, J. (1994) Inhibition of hepatic chylomicron remnant uptake by gene transfer of a receptor antagonist. *Science*, **264**, 1471–1474.

6 Ji, Z. S., Pitas, R. E., Mahley, R. W. (1998) Differential cellular accumulation/retention of

apolipoprotein E mediated by cell surface heparan sulfate proteoglycans. *J. Biol. Chem.*, **273**, 13452–13460.

7 Al-Haideri, M., Goldberg, I. J., Galeano, N. F., Gleeson, A., Vogel, T., Gorecki, M., Turley, S. L., Deckelbaum, R. J. (1997) Heparan sulfate proteoglycan-mediated uptake of apolipoprotein E-triglyceride-rich lipoprotein particles: a major pathway at physiological particle concentrations.*Biochemistry*, **36**, 12766–12772.

8 Ji, Z. S., Brecht, W. J., Miranda, R. D., Hussain, M. M., Innerarity, T. L., Mahley, R. W. (1993) Role of heparan sulfate proteoglycans in the binding and uptake of apolipoprotein E-enriched remnant lipoproteins by cultured cells. *J. Biol. Chem.*, **268**, 10160–10167.

9 Rensen, P.C.N. and van Berkel, T.J.C. (1996) Apolipoprotein E effectively inhibits lipoprotein lipase-mediated lipolysis of chylomicron-like triglyceride-rich lipid emulsions in vitro and in vivo. *J. Biol. Chem.*, **271**, 14791–14799.

10 Jong, M. C., Dahlmans, V. E. H., van Gorp, P. J. J., Breuer, M. L., Mol, M. J. T. M., van der Zee, A., Frants, R. R., Hofker, M. H., Havekes, L. M. (1996) Both lipolysis and hepatic uptake of VLDL are impaired in transgenic mice coexpressing human apolipoprotein E*3Leiden and human apolipoproteinC1. *Arterioscler. Thromb. Vasc. Biol.*, **16**, 934–940.

11 Tabas, I., Myers, J. N., Innerarity, T. L., Xu, X. X., Arnold, K., Boyles, J., Maxfield, F. R. (1991) The influence of particle size and multiple apoprotein E-receptor interactions on the endocytic targeting of beta-VLDL in mouse peritoneal macrophages.*J. Cell Biol.*, **115**, 1547–1560.

12 Tabas, I., Lim, S., Xu, X. X., Maxfield, F. R. (1990) Endocytosed beta-VLDL and LDL are delivered to different intracellular vesicles in mouse peritoneal macrophages. *J. Cell Biol.*, **111**, 929–940.

13 Schwiegelshohn, B., Presley, J. F., Gorecki, M., Vogel, T., Carpentier, Y. A., Maxfield, F. R., Deckelbaum, R. J. (1995) Effects of apoprotein E on intracellular metabolism of model triglyceride-rich particles are distinct from effects on cell particle uptake. *J. Biol. Chem.*, **270**, 1761–1769.

14 Huang, Y., Ji, Z.-S., Brecht, W. J., Rall, S. C., Jr.,Taylor, J. M., Mahley, R. W. (1999) Overexpression of apolipoprotein E3 in transgenic rabbits causes combined hyperlipidemia by stimulating hepatic VLDL production and impairing VLDL lipolysis. *Arterioscler. Thromb. Vasc. Biol.*, **19**, 2952–2959.

15 Kuipers, F., Jong, M. C., Lin, Y., van Eck, M., Havinga, R., Bloks, V., Verkade, H. J., Hofker, M. H., Moshage, H., van Berkel, T. J. C., Vonk, R. J., Havekes, L. M. (1997) Impaired secretion of very low density lipoprotein-triglycerides by apolipoprotein E-deficient mouse hepatocytes. *J. Clin. Invest.*, **100**, 2915–2922.

16 Mensenkamp, A. R., Jong, M. C., van Goor, H., van Luyn, M. J. A., Bloks, V., Havinga, R., Voshol, P. J., Hofker, M. H., van Dijk, K. W., Havekes, L. M., Kuipers, F. (1999) Apolipoprotein E participates in the regulation of very low density lipoprotein-triglyceride secretion by the liver. *J. Biol. Chem.*, **274**, 35711–35718.

17 Kinoshita, M., Kawamura, M., Maeda, T., Fujimaki, Y., Fujita, M., Kojima, K., Teramoto, T. (2000) Apolipoprotein E accelerates the efflux of cholesterol from macrophages: mechanism of xanthoma formation in apolipoprotein deficiency. *J. Atheroscler. Thromb.*, **6**, 22–27.

18 Langer, C., Yadong, H., Cullen, P., Wiesenhutter, B., Mahley, R. W., Assmann, G., von Eckardstein, A. (2000) Endogenous apolipoprotein E modulates cholesterol efflux and cholesteryl ester hydrolysis mediated by high-density lipoprotein-3 and lipid-free apoproteins in mouse

peritoneal macrophages. *J. Mol. Med.*, **78**, 217–222.

19 Lin, C. Y., Duan, H., Mazzone, T. (1999) Apolipoprotein E-dependent cholesterol efflux from macrophages: kinetic study and divergent mechanisms for endogenous versus exogenous apolipoprotein E. *J. Lipid Res.*, **40**, 1618–1627.

20 Mahley, R. W. (1988) Apolipoprotein E: cholesterol transport protein with expanding role in cell biology. *Science*, **240**, 622–630.

21 Fagan, A. M., Bu, G., Sun, Y., Daugherty, A., Holtzman, D. M. (1996) Apolipoprotein E-containing high density lipoprotein promotes neurite outgrowth and is a ligand for the low density lipoprotein receptor-related protein. *J. Biol. Chem.*, **271**, 30121–30125.

22 Ji, Z. S., Dichek, H. L., Miranda, R. D., Mahley, R. W. (1997) Heparan sulfate proteoglycans participate in hepatic lipase and apolipoprotein E-mediated binding and uptake of plasma lipoproteins, including high density lipoproteins. *J. Biol. Chem.*, **272**, 31285–31292.

23 Koo, C., Innerarity, T. L., Mahley, R. W. (1985) Obligatory role of cholesterol and apolipoprotein E in the formation of large cholesterol-enriched and receptor-active high density lipoproteins. *J. Biol. Chem.*, **260**, 11934–11943.

24 Fazio, S., Babaev, V. R., Murray, A. B., Hasty, A. H., Carter, K. J., Gleaves, L. A., Atkinson, J. B., Linton, M. F. (1997) Increased atherosclerosis in C57BL/6 mice reconstituted with apolipoprotein E null macrophages. *Proc. Natl. Acad. Sci. USA*, **94**, 4647–4652.

25 Hasty, A. H., Linton, M. F., Brandt, S. J., Babaev, V. R., Gleaves, L. A., Fazio, S. (1999) Retroviral gene therapy in apoE-deficient mice: apoE expression in the artery wall reduces early foam cell lesion formation. *Circulation*, **99**, 2571–2576.

26 Fazio, S., Babaev, V. R., Burleigh, M. E., Major, A. S., Hasty, A. H., Linton, M. F. (2002) Physiological expression of macrophage apoE in the artery wall reduces atherosclerosis in severely hyperlipidemic mice. *J. Lipid Res.*, **43**, 1602–1609.

27 Linton, M. F., Atkinson, J. B., Fazio, S. (1995) Prevention of atherosclerosis in apolipoprotein E deficient mice by bone marrow transplantation. *Science*, **267**, 1034–1037.

28 Weisgraber, K. H., Troxler, R. F., Rall, S. C., Mahley, R. W. (1980) Comparison of the human, canine, and swine E apoproteins. *Biochem. Biophys. Res. Commun.*, **95**, 374–380.

29 Mahley, R. W. and Rall, S. C., Jr. (1995) Scriver, C. R., Beaudet, A. L. Sly, W. S., Valle, D., *The Metabolic Basis of Inherited Disease*, McGraw-Hill, Inc., New York 1953–1980.

30 Weisgraber, K. H., Rall, S. C., Jr., Innerarity, T. L., Mahley, R. W. (1983) Schettler, F., Gotto, A., Middelhoff, G., Habenicht, A., Jurutka, K. *Atherosclerosis VI*, Springer-Verlag, Berlin, 537–542.

31 Mahley, R. W. and Huang, Y. (1999) Apolipoprotein E: from atherosclerosis to Alzheimer's disease and beyond. *Curr. Opin. Lipidol.*, **10**, 207–217.

32 Fazio, S., Yao, Z., McCarthy, B. J., Rall, S. C., Jr. (1992) Synthesis and secretion of apolipoprotein E occur independently of synthesis and secretion of apolipoprotein B-containing lipoproteins in HepG2 cells. *J. Biol. Chem.*, **267**, 6941–6945.

33 Davis, R. A., Dluz, S. M., Leighton, J. K., Brengaze, V. A. (1989) Increased translatable mRNA and decreased lipogenesis are responsible for the augmented secretion of lipid-deficient apolipoprotein E by hepatocytes from fasted rats. *J. Biol. Chem.*, **264**, 8970–8977.

34 Herz, J. and Willnow, T. E. (1995) Lipoprotein and receptor interactions *in vivo*. *Curr. Opin. Lipidol.*, **6**, 97–103.

35 Mahley, R. W. and Hussain, M. M. (1991) Chylomicron and chylomicron remnant catabolism. *Curr. Opin. Lipidol.*, **2**, 170–176.

36 Boisvert, W. A., Spangerberg, J., Curtiss, L. K. (1995) Treatment of severe hypercholesterolemia in apolipoprotein E-deficient mice by bone marrow transplantation. *J. Clin. Invest.*, **96**, 1118–1124.

37 van Eck, M., Herijgers, N., Yates, J., Pearce, N. J., Hoogerbrugge, P. M., Groot, P. H., van Berkel, T. J. (1997) Bone marrow transplantation in apolipoprotein E-deficient mice. Effect of ApoE gene dosage on serum lipid concentrations, (beta)VLDL catabolism, and atherosclerosis. *Arterioscler. Thromb. Vasc. Biol.*, **17**, 3117–3126.

38 Weisgraber, K. H. (1990) Apolipoprotein E distribution among human plasma lipoproteins: role of the cysteine-arginine interchange at residue 112. *J. Lipid Res.*, **31**, 1503–1511.

39 Steinmetz, A., Jakobs, C., Motzny, S., Kaffarnik, H. (1989) Differential distribution of apolipoprotein E isoforms in human plasma lipoproteins. *Arteriosclerosis*, **9**, 405–411.

40 Wardell, M. R., Weisgraber, K. H., Havekes, L. M., Rall, S. C., Jr. (1989) Apolipoprotein E3-Leiden contains a seven-amino acid insertion that is a tandem repeat of residues 121 to 127. *J. Biol. Chem.*, **264**, 21205–21210.

41 Fazio, S., Horie, Y., Weisgraber, K. H., Havekes, L. M., Rall, S. C., Jr. (1993) Preferential association of apolipoprotein E Leiden with very low density lipoproteins of human plasma. *J. Lipid Res.* **34**, 447–453.

42 Horie, Y., Fazio, S., Westerlund, J. R., Weisgraber, K. H., Rall, S. C., Jr. (1992) The functional characteristics of a human apolipoprotein E variant (cysteine at residue 142) may explain its association with dominant expression of type III hyperlipoproteinemia. *J. Biol. Chem.*, **267**, 1962–1968.

43 Mahley, R. W., Huang, Y., Rall, S. C., Jr. (1999) Pathogenesis of type III hyperlipoproteinemia (dysbetalipoproteinemia). Questions, quandaries, and paradoxes. *J. Lipid Res.*, **40**, 1933–1949.

44 Schaefer, E. J., Gregg, R. E., Ghiselli, G., Forte, T. M., Ordovas, J. M., Zech, L. A., Brewer, H. B., Jr. (1986) Familial apolipoprotein E deficiency. *J. Clin. Invest.*, **78**, 1206–1219.

45 Huang, Y., Liu, X. Q., Rall, S. C., Jr.,Taylor, J. M., von Eckardstein, A., Assmann, G., Mahley, R. W. (1998) Overexpression and accumulation of apolipoprotein E as a cause of hypertriglyceridemia. *J. Biol. Chem.*, **273**, 26388–26393.

46 Plump, A. S., Smith, J. D., Hayek, T., Aalto-Setälä, K., Walsh, A., Verstuyft, J. G., Rubin, E. M., Breslow, J. L. (1992) Severe hypercholesterolemia and atherosclerosis in apolipoprotein E-deficient mice created by homologous recombination in ES cells. *Cell*, **71**, 343–353.

47 Zhang, S., Reddick, R., Piedrahita, J., Maeda, N. (1992) Spontaneous hypercholesterolemia and arterial lesions in mice lacking apolipoprotein E. *Science*, **258**, 468–471.

48 Medh, J. D., Fry, G. L., Bowen, S. L., Ruben, S., Wong, H., Chappell, D. A. (2000) Lipoprotein lipase- and hepatic triglyceride lipase-promoted very low density lipoprotein degradation proceeds via an apolipoprotein E-dependent mechanism. *J. Lipid Res.*, **41**, 1858–1871.

49 Ehnholm, C., Mahley, R. W., Chappell, D. A., Weisgraber, K. H., Ludwig, E., Witztum, J. L. (1984) Role of apolipoprotein E in the lipolytic conversion of β-very low density lipoproteins to low density lipoproteins in type III hyperlipoproteinemia. *Proc. Natl. Acad. Sci. USA*, **81**, 5566–5570.

50 Steinberg, F. M., Tsai, E. C., Brunzell, J. D., Chait, A. (1996) ApoE enhances lipid uptake by macrophages in lipoprotein lipase deficiency during pregnancy. *J. Lipid Res.*, **37**, 972–984.

51 Takahashi, S., Suzuki, J., Kohno, M., Oida, K., Tamai, T., Miyabo, S., Yamamoto, T., Nakai, T. (1995)

Enhancement of the binding of triglyceride-rich lipoproteins to the very low density lipoprotein receptor by apolipoprotein E and lipoprotein lipase. *J. Biol. Chem.*, **270**, 15747–15754.

52 Brasaemle, D. L., Cornely-Moss, K., Bensadoun, A. (1993) Hepatic lipase treatment of chylomicron remnants increases exposure of apolipoprotein E. *J. Lipid Res.*, **34**, 455–465.,

53 Breckenridge, W. C., Little, J. A., Alaupovic, P., Wang, C. S., Kuksis, A., Kakis, G., Lindgren, F., Gardiner, G. (1982) Lipoprotein abnormalities associated with a familial deficiency of hepatic lipase. *Atherosclerosis*, **45**, 161–179.

54 Landis, B. A., Rotolo, F. S., Meyers, W. C., Clark, A. B., Quarfordt, S. H. (1987) Influence of apolipoprotein E on soluble and heparin-immobilized hepatic lipase. *Am. J. Physiol.*, **252**, G805–G810.

55 Shafi, S., Brady, S. F., Bensadoun, A., Havel, R. J. (1994) Role of hepatic lipase in the uptake and processing of chylomicron remnants in rat liver. *J. Lipid Res.*, **35**, 709–720.

56 Thuren, T., Weisgraber, K. H., Sisson, P., Waite, M. (1992) Role of apolipoprotein E in hepatic lipase catalyzed hydrolysis of phospholipid in high-density lipoproteins. *Biochemistry*, **31**, 2332–2338.

57 Ji, Z. S., Fazio, S., Mahley, R. W. (1994) Variable heparan sulfate proteoglycan binding of apolipoprotein E variants may modulate the expression of type III hyperlipoproteinemia. *J. Biol. Chem.*, **269**, 13421–13428.

58 Ji, Z. S., Sanan, D. A., Mahley, R. W. (1995) Intravenous heparinase inhibits remnant lipoprotein clearance from the plasma and uptake by the liver: in vivo role of heparan sulfate proteoglycans. *J. Lipid Res.*, **36**, 583–592.

59 Connelly, P. W., Maguire, G. F., Lee, M., Little, J. A. (1990) Plasma lipoproteins in familial hepatic lipase deficiency. *Arteriosclerosis*, **10**, 40–48.

60 Ji, Z. S., Fazio, S., Lee, Y. L., Mahley, R. W. (1994) Secretion-capture role for apolipoprotein E in remnant lipoprotein metabolism involving cell surface heparan sulfate proteoglycans. *J. Biol. Chem.*, **269**, 2764–2772.

61 Hasty, A. H., Plummer, M. R., Weisgraber, K. H., Linton, M. F., Fazio, S., Swift, L. L. (2005) The recycling of apolipoprotein E in macrophages: influence of HDL and apolipoprotein A-I. *J. Lipid Res.*, **46**, 1433–1439.

62 Vassiliou, G. and McPherson, R. (2004) A novel efflux-recapture process underlies the mechanism of high-density lipoprotein cholesteryl ester-selective uptake mediated by the low-density lipoprotein receptor-related protein. *Arterioscler. Thromb. Vasc. Biol.*, **24**, 1538–1539.

63 Fazio, S. and Linton, M. F. (2004) Unique pathway for cholesterol uptake in fat cells. *Arterioscler. Thromb. Vasc. Biol.*, **24**, 1538–1539.

64 Linton, M. F., Hasty, A. H., Babaev, V. R., Fazio, S. (1998) Hepatic apoE expression is required for remnant lipoprotein clearance in the absence of the low density lipoprotein receptor. *J. Clin. Invest.*, **101**, 1726–1736.

65 Raffai, R. L., Hasty, A. H., Wang, Y., Mettler, S. E., Sanan, D. A., Linton, M. F., Fazio, S., Weisgraber, K. H. (2003) Hepatocyte-derived ApoE is more effective than non-hepatocyte-derived ApoE in remnant lipoprotein clearance. *J. Biol. Chem.*, **278**, 11670–11675.

66 Fazio, S., Linton, M. F., Hasty, A. H., Swift, L. L. (1999) Recycling of apolipoprotein E in mouse liver. *J. Biol. Chem.*, **274**, 8247–8253.

67 Braun, N. A., Mohler, P. J., Weisgraber, K. H., Hasty, A. H., Linton, M. F., Yancey, P. G., Su, Y. R., Fazio, S., Swift, L. L. (2006) Intracellular trafficking of recycling apolipoprotein E in Chinese hamster ovary cells. *J. Lipid Res.*, **47**, 1176–1186.

68 Castelli, W. P., Doyle, J. T., Gordon, T., Hames, C. G., Hjortland, M. C., Hulley, S. B., Kagan, A., Zukel, W. J. (1977) HDL cholesterol and other lipids in coronary heart disease. The Cooperative Lipoprotein Phenotyping Study. *Circulation*, **55**, 767–772.

69 Glomset, J. A. (1968) The plasma lecithin:cholesterol acyltransferase reaction. *J. Lipid Res.*, **9**, 155–167.

70 Gordon, V., Innerarity, T. L., Mahley, R. W. (1983) Formation of cholesterol- and apoprotein E-enriched high density lipoproteins in vitro. *J. Biol. Chem.*, **258**, 6202–6212.

71 Mazzone, T. (1996) Apolipoprotein E secretion by macrophages: its potential physiological functions. *Curr. Opin. Lipidol.*, **7**, 303–307.

72 von Eckardstein, A., Nofer, J. R., Assmann, G. (2001) High density lipoproteins and arteriosclerosis. Role of cholesterol efflux and reverse cholesterol transport. *Arterioscler. Thromb. Vasc. Biol.*, **21**, 13–27.

73 Funke, H., von Eckardstein, A., Pritchard, P. H., Karas, M., Albers, J. J., Assmann, G. (1991) A frameshift mutation in the human apolipoprotein A-I gene causes high density lipoprotein deficiency, partial lecithin: cholesterol-acyltransferase deficiency, and corneal opacities. *J. Clin. Invest.*, **87**, 371–376.

74 Genest, J., Jr.,Marcil, M., Denis, M., Yu, L. (1999) High density lipoproteins in health and in disease. *J. Invest. Med.*, **47**, 31–42.

75 Li, H., Reddick, R. L., Maeda, N. (1993) Lack of apoA-I is not associated with increased susceptibility to atherosclerosis in mice. *Arterioscler. Thromb.*, **13**, 1814–1821.

76 von Eckardstein, A., Huang, Y., Assmann, G. (1994) Physiological role and clinical relevance of high-density lipoprotein subclasses. *Curr. Opin. Lipidol.*, **5**, 404–416.

77 Fielding, C. J. (1991) Reverse cholesterol transport. *Curr. Opin. Lipidol.*, **2**, 376–378.

78 Huang, Y., von Eckardstein, A., Wu, S., Maeda, N., Assmann, G. (1994) A plasma lipoprotein containing only apolipoprotein E and with gamma mobility on electrophoresis releases cholesterol from cells. *Proc. Natl. Acad. Sci. USA*, **91**, 1834–1838.

79 Asztalos, B. F., Sloop, C. H., Wong, L., Roheim, P. S. (1993) Comparison of apo A-I-containing subpopulations of dog plasma and prenodal peripheral lymph: evidence for alteration in subpopulations in the interstitial space. *Biochim. Biophys. Acta*, **1169**, 301–304.

80 Bodzioch, M., Orso, E., Klucken, J., Langmann, T., Bottcher, A., Diederich, W., Drobnik, W., Barlage, S., Buchler, C., Porsch-Ozcurumez, M., Kaminski, W. E., Hahmann, H. W., Oette, K., Rothe, G., Aslanidis, C., Lackner, K. J., Schmitz, G. (1999) The gene encoding ATP-binding cassette transporter 1 is mutated in Tangier disease. *Nat. Genet*, **22**, 347–351.

81 Brooks-Wilson, A., Marcil, M., Clee, S. M., Zhang, L. H., Roomp, K., van Dam, M., Yu, L., Brewer, C., Collins, J. A., Molhuizen, H. O., Loubser, O., Ouelette, B. F., Fichter, K., Ashbourne-Excoffon, K. J., Sensen, C. W., Scherer, S., Mott, S., Denis, M., Martindale, D., Frohlich, J., Morgan, K., Koop, B., Pimstone, S., Kastelein, J. J., Hayden, M. R. (1999) Mutations in ABC1 in Tangier disease and familial high-density lipoprotein deficiency. *Nat. Genet*, **22**, 336–345.

82 Rust, S., Rosier, M., Funke, H., Real, J., Amoura, Z., Piette, J. C., Deleuze, J. F., Brewer, H. B., Duverger, N., Denefle, P., Assmann, G. (1999) Tangier disease is caused by mutations in the gene encoding ATP-binding cassette transporter 1. *Nat. Genet*, **22**, 352–355.

83 Lee, J. Y. and Parks, J. S. (2005) ATP-binding cassette transporter AI and its role in HDL formation. *Curr. Opin. Lipidol.*, **16**, 19–25.

84 Timmins, J. M., Lee, J. Y., Boudyguina, E., Kluckman, K. D., Brunham, L. R., Mulya, A., Gebre, A. K., Coutinho, J. M., Colvin, P. L.,

Smith, T. L., Hayden, M. R., Maeda, N., Parks, J. S. (2005) Targeted inactivation of hepatic Abca1 causes profound hypoalphalipoproteinemia and kidney hypercatabolism of apoA-I. *J. Clin. Invest.*, **115**, 1333–1342.

85 Brunham, L. R., Kruit, J. K., Iqbal, J., Fievet, C., Timmins, J. M., Pape, T. D., Coburn, B. A., Bissada, N., Staels, B., Groen, A. K., Hussain, M. M., Parks, J. S., Kuipers, F., Hayden, M. R. (2006) Intestinal ABCA1 directly contributes to HDL biogenesis in vivo. *J. Clin. Invest.*, **116**, 1052–1062.

86 Aiello, R. J., Brees, D., Bourassa, P. A., Royer, L., Lindsey, S., Coskran, T., Haghpassand, M., Francone, O. L. (2002) Increased atherosclerosis in hyperlipidemic mice with inactivation of ABCA1 in macrophages. *Arterioscler. Thromb. Vasc. Biol.*, **22**, 630–637.

87 Spangenberg, J. and Curtiss, L. K. (1997) Influence of macrophage-derived apolipoprotein E on plasma lipoprotein distribution of apolipoprotein A-I in apolipoprotein E-deficient mice. *Biochim. Biophys. Acta*, **1349**, 109–121.

88 Zhu, Y., Bellosta, S., Langer, C., Bernini, F., Pitas, R. E., Mahley, R. W., Assmann, G., von Eckardstein, A. (1998) Low-dose expression of a human apolipoprotein E transgene in macrophages restores cholesterol efflux capacity of apolipoprotein E-deficient mouse plasma. *Proc. Natl. Acad. Sci. USA*, **95**, 7585–7590.

89 Jeong, T., Schissel, S. L., Tabas, I., Pownall, H. J., Tall, A. R., Jiang, X. (1998) Increased sphingomyelin content of plasma lipoproteins in apolipoprotein E knockout mice reflects combined production and catabolic defects and enhances reactivity with mammalian sphingomyelinase. *J. Clin. Invest.*, **101**, 905–912.

90 Wong, L., Curtiss, L., Huang, J., Mann, C., Maldonado, B., Roheim, P. (1992) Altered epitope expression of human interstitial fluid apolipoprotein A-I reduces its ability to activate lecithin cholesterol acyl transferase. *J. Clin. Invest.*, **90**, 2370–2375.

91 Glomset, J. A., Assmann, G., Ghone, E., Norum, K. R. (1995) Scriver, C. R., Beaudet, A. L., Sly, W. S. Valle D. *The Metabolic Basis of Inherited Diseases*, McGraw-Hill, New York, 1933–1952.

92 Subramanian, V., Goyal, J., Miwa, M., Sugatami, J., Akiyama, M., Liu, M., Subbaiah, P. (1999) Role of lecithin-cholesterol acyltransferase in the metabolism of oxidized phospholipids in plasma: studies with platelet activating factor-acetyl hydrolase-deficient plasma. *Biochim. Biophys. Acta*, **1439**, 95–109.

93 Aviram, M. and Rosenblat, M. (2004) Paraoxonases 1, 2, and 3, oxidative stress, and macrophage foam cell formation during atherosclerosis development. *Free Rad. Biol. Med.*, **37**, 1304–1316.

94 Rozenberg, O., Rosenblat, M., Coleman, R., Shih, D., Aviram, M. (2003) Paraoxonase (PON1) Deficiency is Associated with Increased Macrophage Oxidative Stress: Studies in PON-1 Knockout Mice. *Free Rad. Biol. Med.*, **34**, 774–784.

95 Tselepis, A. and Chapman, M. (2002) Inflammation bioactive lipids and atherosclerosis: potential roles of a lipoprotein-associated phospholipase A2, platelet activating factor-acetylhydrolase. *Atheroscler. Suppl.*, **3**, 57–68.

96 Forte, T. M., Subbanagounder, G., Berliner, J. A., Blanche, P. J., Clermont, A. O., Jia, Z., Oda, M. N., Krauss, R. M., Bielicki, J. K. (2002) Altered activities of anti-atherogenic enzymes LCAT, paraoxonase, and platelet-activating factor acetylhydrolase in atherosclerosis-susceptible mice. *J. Lipid Res.*, **43**, 477–485.

97 Linton, M. F. and Fazio, S. (1999) Macrophages, lipoprotein metabolism, and atherosclerosis: insights from murine bone marrow transplantation studies. *Curr. Opin. Lipidol.*, **10**, 97–105.

98 Basu, S. K., Brown, M. S., Ho, Y. K., Havel, R. J., Goldstein, J. L. (1981) Mouse macrophages synthesize and secrete a protein resembling apolipoprotein E. *Proc. Natl. Acad. Sci. USA*, **78**, 7545–7549.

99 Mazzone, T., Gump, H., Diller, P., Getz, G. (1987) Macrophage free cholesterol content regulates apolipoprotein E synthesis. *J. Biol. Chem.*, **262**, 11657–11662.

100 Duan, H., Lin, C. Y., Mazzone, T. (1997) Degradation of macrophage ApoE in a nonlysosomal compartment. Regulation by sterols. *J. Biol. Chem.*, **272**, 31156–31162.

101 Fazio, S., Major, A. S., Swift, L. L., Gleaves, L. A., Accad, M., Linton, M. F., Farese, R. V., Jr. (2001) Increased atherosclerosis in LDL receptor-null mice lacking ACAT1 in macrophages. *J. Clin. Invest.*, **107**, 163–171.

102 Su, Y. R., Dove, D. E., Major, A. S., Hasty, A. H., Boone, B., Linton, M. F., Fazio, S. (2005) Reduced ABCA1-mediated cholesterol efflux and accelerated atherosclerosis in apolipoprotein E-deficient mice lacking macrophage-derived AC AT1. *Circulation*, **111**, 2373–2381.

103 Nissen, S. E., Tuzcu, E. M., Brewer, H. B., Sipahi, I., Nicholls, S. J., Ganz, P., Schoenhagen, P., Waters, D. D., Pepine, C. J., Crowe, T. D., Davidson, M. H., Deanfield, J. E., Wisniewski, L. M., Hanyok, J. J., Kassalow, L. M. (2006) ACAT Intravascular Atherosclerosis Treatment Evaluation (ACTIVATE) Investigators. Effect of ACAT inhibition on the progression of coronary atherosclerosis. *N. Engl. J. Med.*, **354**, 1253–1263.

104 Fazio, S. and Linton, M. F. (2006) Failure of ACAT inhibition to retard atherosclerosis. *N. Engl. J. Med.*, **354**, 1307–1309.

105 van Eck, M., Herijgers, N., Vidgeon-Hart, M., Pearce, N. J., Hoogerbrugge, P. M., Groot, P. H., van Berkel, T. J. (2000) Accelerated atherosclerosis in C57Bl/6 mice transplanted with ApoE-deficient bone marrow. *Atherosclerosis.*, **150**, 71–80.

106 Ishiguro, H., Yoshida, H., Major, A. S., Zhu, T., Babaev, V. R., Linton, M. F., Fazio, S. (2001) Retrovirus-mediated expression of apolipoprotein A-I in the macrophage protects against atherosclerosis in vivo. *J. Biol. Chem.*, **276**, 36742–36748.

107 Bellosta, S., Mahley, R. W., Sanan, D. A., Murata, J., Newland, D. L., Taylor, J. M., Pitas, R. E. (1995) Macrophage-specific expression of human apolipoprotein E reduces atherosclerosis in hypercholesterolemic apolipoprotein E-null mice. *J. Clin. Invest.*, **96**, 2170–2179.

108 Shimano, H., Ohsuga, J., Shimada, M., Namba, Y., Gotoda, T., Harada, K., Katsuki, M., Yazaki, Y., Yamada, N. (1995) Inhibition of diet-induced atheroma formation in transgenic mice expressing apolipoprotein E in the arterial wall. *J. Clin. Invest.*, **95**, 469–476.

109 Curtiss, L. K. and Boisvert, W. A. (2000) Apolipoprotein E and atherosclerosis. *Curr. Opin. Lipidol.*, **11**, 243–251.

110 Mazzone, T. and Reardon, C. (1994) Expression of heterologous human apolipoprotein E by J774 macrophages enhances cholesterol efflux to HDL3. *J. Lipid Res.*, **35**, 1345–1353.

111 Krimbou, L., Marcil, M., Chiba, H., Genest, J., Jr. (2003) Structural and functional properties of human plasma high density-sized lipoprotein containing only apoE particles. *J. Lipid Res.*, **44**, 884–892.

112 Schmitz, G., Robenek, H., Lohmann, U., Assmann, G. (1985) Interaction of high density lipoproteins with cholesteryl ester-laden macrophages: biochemical and morphological characterization of cell surface receptor binding, endocytosis and resecretion of high density lipoproteins by macrophages. *EMBO J.*, **4**, 613–622.

113 Dory, L. (1989) Synthesis and secretion of apo E in thioglycolate-

elicited mouse peritoneal macrophages: effect of cholesterol efflux. *J. Lipid Res.*, **30**, 809–816.

114 Laffitte, B. A., Repa, J. J., Joseph, S. B., Wilpitz, D. C., Kast, H. R., Mangelsdorf, D. J., Tontonoz, P. (2001) LXRs control lipid-inducible expression of the apolipoprotein E gene in macrophages and adipocytes. *Proc. Natl. Acad. Sci. USA*, **98**, 507–512.

115 Su, Y. R., Ishiguro, H., Major, A. S., Dove, D. E., Zhang, W., Hasty, A. H., Babaev, V. R., Linton, M. F., Fazio, S. (2003) Macrophage apolipoprotein A-I expression protects against atherosclerosis in ApoE-deficient mice and up-regulates ABC transporters. *Mol. Therap.*, **8**, 576–583.

116 Major, A. S., Dove, D. E., Ishiguro, H., Su, Y. R., Brown, A. M., Liu, L., Carter, K. J., Linton, M. F., Fazio, S. (2001) Increased cholesterol efflux in apolipoprotein AI (ApoAI)-producing macrophages as a mechanism for reduced atherosclerosis in ApoAI((−/−)) mice. *Arterioscler. Thromb. Vasc. Biol.*, **21**, 1790–1795.

117 Cullen, P., Cignarella, A., Brennhausen, B., Mohr, S., Assmann, G., von Eckardstein, A. (1998) Phenotype-dependent differences in apolipoprotein E metabolism and in cholesterol homeostasis in human monocyte-derived macrophages. *J. Clin. Invest.*, **101**, 1670–1677.

118 Huang, Z. H., Lin, C. Y., Oram, J. F., Mazzone, T. (2001) Sterol efflux mediated by endogenous macrophage apoE expression is independent of ABCA1. *Arterioscler. Thromb. Vasc. Biol.*, **21**, 2019–2025.

119 Mahley, R. W., Huang, Y., Weisgraber, K. H. (2006) Putting cholesterol in its place: apoE and reverse cholesterol transport. *J. Clin. Invest.*, **116**, 1226–1229.

120 Francone, O. L., Royer, L., Haghpassand, M. (1996) Increased prebeta-HDL levels, cholesterol efflux, and LCAT-mediated esterification in mice expressing the human cholesteryl ester transfer protein (CETP) and human apolipoprotein A-I (apoA-I) transgenes. *J. Lipid Res.*, **37**, 1268–1277.

121 von Eckardstein, A., Jauhiainen, M., Huang, Y., Metso, J., Langer, C., Pussinen, P., Wu, S., Ehnholm, C., Assmann, G. (1996) Phospholipid transfer protein mediated conversion of high density lipoproteins generates pre beta 1-HDL. *Biochim. Biophys. Acta*, **1301**, 255–262.

122 Barrans, A., Collet, X., Barbaras, R., Jaspard, B., Manent, J., Vieu, C., Chap, H., Perret, B. (1994) Hepatic lipase induces the formation of pre-beta 1 high density lipoprotein (HDL) from triacylglycerol-rich H DL2.A study comparing liver perfusion to in vitro incubation with lipases. *J. Biol. Chem.*, **269**, 11572–11577.

123 Yokoyama, S. (2006) ABCA1 and biogenesis of HDL. *J. Atheroscler. Thromb.*, **13**, 1–15.

124 Matsuura, F., Wang, N., Chen, W., Jiang, X.-C., Tall, A. R. (2006) HDL from CETP-deficient subjects shows enhanced ability to promote cholesterol efflux from macrophages in an apoE- and ABCG1-dependent pathway. *J. Clin. Invest.*, **116**, 1435–1442.

125 Ranalletta, M., Wang, N., Han, S., Yvan-Charvet, L., Welch, C., Tall, A. R. (2006) Decreased atherosclerosis in low-density lipoprotein receptor knockout mice transplanted with Abcg1$^{-/-}$ bone marrow. *Arterioscler. Thromb. Vasc. Biol.*, **26**, 2308–2315.

126 Baldán, Á., Pei, L., Lee, R., Tarr, P., Tangirala, R. K., Weinstein, M. M., Frank, J., Li, A. C., Tontonoz, P., Edwards, P. A. (2006) Impaired development of atherosclerosis in hyperlipidemic Ldlr$^{-/-}$ and apoE$^{-/-}$ mice transplanted with Abcg1$^{-/-}$ bone marrow. *Arterioscler. Thromb. Vasc. Biol.*, **26**, 2301–2307.

127 Peters-Libeu, C. A., Newhouse, Y., Hatters, D. M., Weisgraber, K. H. (2006) Model of biologically active apolipoprotein E bound to dipalmitoylphosphatidylcholine. *J. Biol. Chem.*, **281**, 1073–1079.

128 Acton, S., Rigotti, A., Landschulz, K. T., Xu, S., Hobbs, H. H., Krieger, M. (1996) Identification of scavenger receptor SR-BI as a high density lipoprotein receptor. *Science*, **271**, 518–520.

129 Tall, A. R., Jiang, X. C., Luo, Y., Silver, D. (2000) 1999 George Lyman Duff memorial lecture: Lipid transfer proteins, HDL metabolism, and atherogenesis. *Arterioscler. Thromb. Vasc. Biol.*, **20**, 1185–1188.

130 Braun, A., Trigatti, B. L., Post, M. J., Sato, K., Simons, M., Edelberg, J. M., Rosenberg, R. D., Schrenzel, M., Krieger, M. (2002) Loss of SR-BI expression leads to the early onset of occlusive atherosclerotic coronary artery disease, spontaneous myocardial infarctions, severe cardiac dysfunction, and premature death in apolipoprotein E-deficient mice. *Circ. Res.*, **90**, 270–276.

131 Trigatti, B., Rayburn, H., Vinals, M., Braun, A., Miettinen, H., Penman, M., Hertz, M., Schrenzel, M., Amigo, L., Rigotti, A., Krieger, M. (1999) Influence of the high density lipoprotein receptor SR-BI on reproductive and cardiovascular pathophysiology. *Proc. Natl. Acad. Sci. USA*, **96**, 9322–9327.

132 Fazio, S. and Linton, M. F. (2001) Mouse models of hyperlipidemia and atherosclerosis. *Front. Biosci.*, **6**, D515–525.

133 Arai, T., Rinninger, F., Varban, L., Fairchild-Huntress, V., Liang, C. P., Chen, W., Seo, T., Deckelbaum, R., Huszar, D., Tall, A. R. (1999) Decreased selective uptake of high density lipoprotein cholesteryl esters in apolipoprotein E knock-out mice. *Proc. Natl. Acad. Sci. USA*, **96**, 12050–12055.

134 Rigotti, A., Trigatti, B. L., Penman, M., Rayburn, H., Herz, J., Krieger, M. (1997) A targeted mutation in the murine gene encoding the high density lipoprotein (HDL) receptor scavenger receptor class B type I reveals its key role in HDL metabolism. *Proc. Natl. Acad. Sci. USA*, **94**, 12610–12615.

135 Davit-Spraul, A., Pourci, M., Atger, V., Cambillau, M., Hadchouel, M., Moatti, N., Legrand, A. (1996) Abnormal lipoprotein pattern in patients with Alagille syndrome depends on icterus severity. *Gastroenterology*, **111**, 1023–1032.

136 Braun, A., Zhang, S., Miettinen, H. E., Ebrahim, S., Holm, T. M., Vasile, E., Post, M. J., Yoerger, D. M., Picard, M. H., Krieger, J. L., Andrews, N. C., Simons, M., Krieger, M. (2003) Probucol prevents early coronary heart disease and death in the high-density lipoprotein receptor SR-BI/apolipoprotein E double knockout mouse. *Proc. Natl. Acad. Sci. USA*, **100**, 7283–7288.

137 Zhang, S., Picard, M. H., Vasile, E., Zhu, Y., Raffai, R. L., Weisgraber, K. H., Krieger, M. (2005) Diet-induced occlusive coronary atherosclerosis, myocardial infarction, cardiac dysfunction, and premature death in scavenger receptor class B type I-deficient, hypomorphic apolipoprotein ER61 mice. *Circulation*, **111**, 3457–3464.

138 Yu, H., Zhang, W., Yancey, P. G., Koury, M. J., Zhang, Y., Fazio, S., Linton, M. F. (2006) Macrophage apolipoprotein E reduces atherosclerosis and prevents premature death in apolipoprotein E and scavenger receptor-class BI double-knockout mice. *Arterioscler. Thromb. Vasc. Biol.*, **26**, 150–156.

139 Wong, L. (1989) Contribution of endosomes to intrahepatic distribution of apolipoprotein B and apolipoprotein E. *J. Cell. Physiol.*, **141**, 441–452.

140 Takahashi, Y. and Smith, J. D. (1999) Cholesterol efflux to apolipoprotein AI involves endocytosis and resecretion in a calcium-dependent pathway. *Proc. Natl. Acad. Sci. USA*, **96**, 11358–11363.

141 Swift, L. L., Farkas, M. H., Major, A. S., Valyi-Nagy, K., Linton, M. F., Fazio, S. (2001) A recycling pathway for resecretion of internalized

apolipoprotein E in liver cells. *J. Biol. Chem.*, **276**, 22965–22970.

142 Farkas, M. H., Swift, L. L., Hasty, A. H., Linton, M. F., Fazio, S. (2003) The recycling of apolipoprotein E in primary cultures of mouse hepatocytes. *J. Biol. Chem.*, **278**, 9412–9417.

143 Rensen, P. C. N., Jong, M. C., van Vark, L. C., van der Boom, H., Hendriks, W. L., van Berkel, T. J. C., Biessen, E. A. L., Havekes, L. M. (2000) Apolipoprotein E is resistant to intracellular degradation in vitro and in vivo. *J. Biol. Chem.*, **275**, 8564–8571.

144 Heeren, J., Weber, W., Beisiegel, U. (1999) Intracellular processing of endocytosed triglyceride-rich lipoproteins comprises both recycling and degradation. *J. Cell Sci.*, **112**, 349–359.

145 Farkas, M. H., Weisgraber, K. W., Shepherd, V. L., Linton, M. F., Fazio, S., Swift, L. L. (2004) The recycling of apolipoprotein E and its amino-terminal 22 kDa fragment: evidence for multiple redundant pathways. *J. Lipid. Res.*, **45**, 1546–1554.

146 Heeren, J., Grewal, T., Jackle, S., Beisiegel, U. (2001) Recycling of apolipoprotein E and lipoprotein lipase through endosomal compartments in vivo. *J. Biol. Chem.*, **276**, 42333–42338.

147 Heeren, J., Grewal, T., Laatsch, A., Becker, N., Rinninger, F., Rye, K. A., Beisiegel, U. (2004) Impaired recycling of apolipoprotein E4 is associated with intracellular cholesterol accumulation. *J. Biol. Chem.*, **279**, 55483–55492.

148 Heeren, J., Grewal, T., Laatsch, A., Rottke, D., Rinninger, F., Enrich, C., Beisiegel, U. (2003) Recycling of apoprotein E is associated with cholesterol efflux and high density lipoprotein internalization. *J. Biol. Chem.*, **278**, 14370–14378.

149 Geuze, H. J., Slot, J. W., Strous, G.J. A.M. (1983) Intracellular site of asialoglycoprotein receptor-ligand uncoupling: double label imunoelectron microscopy during receptor-mediated endocytosis. *Cell*, **32**, 277–287.

150 Brown, M. S., Anderson, R. G. W., Basu, S. K., Goldstein, J. L. (1982) Recycling of cell-surface receptors: observations from the LDL receptor system. *Cold Spring Harbor Symp. Quant. Biol.*, **46**, 713–721.

151 Greenspan, P. and St. Clair, R. W. (1984) Retroendocytosis of low density lipoprotein. *J. Biol. Chem.*, **259**, 1703–1713.

152 Saito, H., Dhanasekaran, P., Baldwin, F., Weisgraber, K. H., Phillips, M. C., Lund-Katz, S. (2003) Effects of polymorphism on the lipid interaction of human apolipoprotein E. *J. Biol. Chem.*, **278**, 40723–40729.

153 Morrow, J. A., Hatters, D. M., Lu, B., Hochtl, P., Oberg, K. A., Rupp, B., Weisgraber, K. H. (2002) Apolipoprotein E4 forms a molten globule. A potential basis for its association with disease. *J. Biol. Chem.*, **277**, 50380–50385.

154 Wetterau, J. R., Aggerbeck, L. P., Rall, S. C., Jr., Weisgraber, K. H. (1988) Human apolipoprotein E3 in aqueous solution. I. Evidence for two structural domains. *J. Biol. Chem.*, **263**, 6240–6248.

155 Aggerbeck, L. P., Wetterau, J. R., Weisgraber, K. H., Wu, C.-S.C., Lindgren, F. T. (1988) Human apolipoprotein E3 in aqueous solution. II. Properties of the amino- and carboxyl-terminal domains. *J. Biol. Chem.*, **263**, 6249–6258.

156 Bradley, W. A., Hwang, S.-L.C., Karlin, J. B., Lin, A. H. Y., Prasad, S. C., Gotto, A. M., Jr., Gianturco, S. H. (1984) Low-density lipoprotein receptor binding determinants switch from apolipoprotein E to apolipoprotein B during conversion of hypertriglyceridemic very-low-density lipoprotein to low-density lipoproteins. *J. Biol. Chem.*, **259**, 14728–14735.

157 Gianturco, S. H., Gotto, A. M., Jr.,Hwang, S. L. C., Karlin, J. B., Lin, A. H. Y., Prasad, S. C., Bradley, W. A. (1983) Apolipoprotein E mediates uptake of S_f 100–400

hypertriglyceridemic very low density lipoproteins by the low density lipoprotein receptor pathway in normal human fibroblasts. *J. Biol. Chem.*, **258**, 4526–4533.

158 Yoshida, H., Hasty, A., Major, A. S., Ishiguro, H., Su, Y., Gleaves, L. A., Babaev, V. R., Linton, M. F., Fazio, S. (2001) Isoform-specific effects of apolipoprotein E on atherogenesis: gene transduction studies in apoE-null mice. *Circulation*, **104**, 2820–2825.

159 Hara, H. and Yokayama, S. (1991) Interaction of free apolipoproteins with macrophages. Formation of high density lipoprotein-like lipoproteins and reduction of cellular cholesterol. *J. Biol. Chem.*, **266**, 3080–3086.

160 Kritharides, L., Jessup, W., Mander, E. L., Dean, R. T. (1995) Apolipoprotein A-I-mediated efflux of sterols from oxidized LDL-loaded macrophages. *Arterioscler. Thromb. Vasc. Biol.*, **15**, 276–289.

161 Oram, J. F., Lawn, R. M., Garvin, M. R., Wade, D. P. (2000) ABCA1 is the cAMP-inducible apolipoprotein receptor that mediates cholesterol secretion from macrophages. *J. Biol. Chem.*, **275**, 34508–34511.

162 Wang, N., Silver, D. L., Costet, P., Tall, A. R. (2000) Specific binding of apoA-I, enhanced cholesterol efflux, and altered plasma membrane morphology in cells expressing ABC1. *J. Biol. Chem.*, **275**, 33053–33058.

163 Bortnick, A. E., Rothblat, G. H., Stoudt, G., Hoppe, K. L., Royer, L. J., McNeish, J., Francone, O. L. (2000) The correlation of ATP-binding cassette 1 mRNA levels with cholesterol efflux from various cell lines. *J. Biol. Chem.*, **275**, 28634–28640.

164 Slotte, J. P., Oram, J. F., Bierman, E. L. (1987) Binding of high density lipoproteins to cell receptors promotes translocation of cholesterol from intracellular membranes to the cell surface. *J. Biol. Chem.*, **262**, 12904–12907.

165 Walter, M., Gerdes, U., Seedorf, U., Assman, G. (1994) The high density lipoprotein- and apolipoprotein A-I-induced mobilization of cellular cholesterol is impaired in fibroblasts from Tangier disease subjects. *Biochem. Biophys. Res. Commun.*, **205**, 850–856.

166 Remaley, A. T., Stonik, J. A., Demosky, S. J., Neufeld, E. B., Bocharov, A. V., Vishnyakova, T. G., Eggerman, T. L., Patterson, A. P., Duverger, N. J., Santamarina-Fojo, S., Brewer, H. B., Jr. (2001) Apolipoprotein specificity for lipid efflux by the human ABCA1 transporter. *Biochem. Biophys. Res. Commun.*, **280**, 818–823.

167 Rees, D., Sloane, T., Jessup, W., Dean, R. T., Kritharides, L. (1999) Apolipoprotein A-I stimulates secretion of apolipoprotein E by foam cell macrophages. *J. Biol. Chem.*, **274**, 27925–27933.

168 Oram, J. F. (2002) ATP-binding cassette transporter A1 and cholesterol trafficking. *Curr. Opin. Lipidol.*, **13**, 373–381.

169 Remaley, A. T., Schumacher, U. K., Stonik, J. A., Farsi, B. D., Nazih, H., Brewer, H. B., Jr. (1997) Decreased reverse cholesterol transport from Tangier Disease fibroblasts. Acceptor specificity and effect of brefeldin on lipid efflux. *Arterioscler. Thromb. Vasc. Biol.*, **17**, 1813–1821.

170 Neufeld, E. B., Remaley, A. T., Demosky, S. J., Stonik, J. A., Cooney, A. M., Comly, M., Dwyer, N. K., Zhang, M., Blanchette-Mackey, E., Santamarina-Fojo, S., Brewer, H. B., Jr. (2001) Cellular localization and trafficking of the human ABCA1 transporter. *J. Biol. Chem.*, **276**, 27584–27590.

171 Thomsen, P., Roepstorff, K., Stahlhut, M., van Deurs, B. (2002) Caveolae are highly immobile plasma membrane microdomains, which are not involved in constitutive endocytic trafficking. *Mol. Biol. Cell*, **13**, 238–250.

172 Haynes, M. P., Phillips, M. C., Rothblat, G. H. (2000) Efflux of cholesterol from different cellular pools. *Biochemistry*, **39**, 4508–4517.

173 Kennedy, M. A., Barrera, G. C., Nakamura, K., Baldan, A., Tarr, P., Fishbein, M. C., Frank, J., Francone, O. L., Edwards, P. A. (2005) ABCG1 has a critical role in mediating cholesterol efflux to HDL and preventing cellular lipid accumulation. *Cell Metab.*, **1**, 121–131.

174 Yancey, P. G., de la Llera-Moya, M., Swarnakar, S., Monza, P., Klein, S. M., Connelly, M. A., Johnson, W. J., Williams, D. L., Rothblat, G. H. (2000) High density lipoprotein phospholipid composition is a major determinant of the bi-directional flux and net movement of cellular free cholesterol mediated by scavenger receptor BI. *J. Biol. Chem.*, **275**, 36596–36604.

175 Yancey, P. G., Kawashiri, M. A., Moore, R., Glick, J. M., Williams, D. L., Connelly, M. A., Rader, D. J., Rothblat, G. H. (2004) In vivo modulation of HDL phospholipid has opposing effects on SR-BI- and ABCA1-mediated cholesterol efflux. *J. Lipid Res.*, **45**, 337–346.

176 Wang, N., Lan, D., Chen, W., Matsuura, F., Tall, A. R. (2004) ATP-binding cassette transporters G1 and G4 mediate cellular cholesterol efflux to high-density lipoproteins. *Proc. Natl. Acad. Sci. USA*, **101**, 9774–9779.

177 Hao, M., Mukherjee, S., Sun, Y., Maxfield, F. R. (2004) Effects of cholesterol depletion and increased lipid unsaturation on the properties of endocytic membranes. *J. Biol. Chem.*, **279**, 14171–14178.

178 Wustner, D., Mondal, M., Huang, A., Maxfield, F. R. (2004) Different transport routes for high density lipoprotein and its associated free sterol in polarized hepatic cells. *J. Lipid Res.*, **45**, 427–437.

179 Silver, D. L., Wang, N., Xiao, X., Tall, A. R. (2001) High density lipoprotein (HDL) particle uptake mediated by scavenger receptor class B type 1 results in selective sorting of HDL cholesterol from protein and polarized cholesterol secretion. *J. Biol. Chem.*, **276**, 25287–25293.

180 Lorkowski, S., Kratz, M., Wenner, C., Schmidt, R., Weitkamp, B., Fobker, M., Reinhardt, J., Rauterberg, J., Galinski, E. A., Cullen, P. (2001) Expression of the ATP-binding cassette transporter gene ABCG1 (ABC8) in Tangier Disease. *Biochem. Biophys. Res. Commun.*, **283**, 821–830.

181 Klucken, J., Buchler, C., Kaminski, W. E., Porsch-Ozcurumez, M., Liebisch, G., Kapinsky, M., Diederich, W., Dean, M., Allikmets, R., Schmitz, G. (2000) ABCG1 (ABC8), the human homolog of the Drosophila white gene, is a regulator of macrophage cholesterol and phospholipid transport. *Proc. Natl. Acad. Sci. USA*, **97**, 817–822.

182 Hao, M., Lin, S. X., Karylowski, O. J., Wustner, D., McGraw, T. E., Maxfield, F. R. (2002) Vesicular and non-vesicular sterol transport in living cells. The endocytic recycling compartment is a major sterol storage organelle. *J. Biol. Chem.*, **277**, 609–617.

183 Silver, D. L., Wang, N., Tall, A. R. (2000) Defective HDL particle uptake in ob/ob hepatocytes causes decreased recycling, degradation, and selective lipid uptake. *J. Clin. Invest.*, **105**, 151–159.

184 Duong, P. T., Collins, H. L., Nickel, M., Lund-Katz, S., Rothblat, G. H., Phillips, M. C. (2006) Characterization of nascent HDL particles and microparticles formed by ABCA1-mediated efflux of cellular lipids to apoA-I. *J. Lipid Res.*, **47**, 832–843.

185 Kärkkäinen, M., Oka, T., Olkkonen, V., Metso, J., Hattori, H., Hauhiainen, M., Ehnholm, C. (2002) Isolation and partial characterization of the inactive and active forms of human plasma phospolipid transfer protein (PLTP). *J. Biol. Chem.*, **277**, 15413–15418.

186 Siggins, S., Jauhiainen, M., Olkkonen, V. M., Tenhunen, J., Ehnholm, C. (2003) PLTP secreted by HepG2 cells resembles the high-

activity PLTP form in human plasma. *J. Lipid Res.*, **44**, 1698–1704.

187 van Eck, M., Herijgers, N., van Dijk, K. W., Havekes, L. M., Hofker, M. H., Groot, P. H., van Berkel, T. J. (2000) Effect of macrophage-derived mouse ApoE, human ApoE3-Leiden, and human ApoE2 (Arg158 -> Cys) on cholesterol levels and atherosclerosis in ApoE-deficient mice. *Arterioscler. Thromb. Vasc. Biol.*, **20**, 119–127.

188 Bielicki, J. K., McCall, M. R., Forte, T. M. (1999) Apolipoprotein A-I promotes cholesterol release and apolipoprotein E recruitment from THP-1 macrophage-like foam cells. *J. Lipid Res.*, **40**, 85–92.

4
ApoM – A Novel Apolipoprotein with Antiatherogenic Properties

Ning Xu, Peter Nilsson-Ehle

4.1
Introduction

The antiatherogenic function of HDL is well established, and its ability to promote cholesterol efflux from foam cells in atherosclerotic lesions is generally regarded as one of the key mechanisms behind the protective function. However, HDL also displays a variety of properties that may affect the complex atherosclerotic process by other mechanisms, thus also being involved in processes related to defense against oxidants, the immune system, and systemic effects in septicemia. ApoM, a recently discovered HDL apolipoprotein with antiatherosclerotic properties, may provide a link between these diverse effects.

Apolipoprotein M (apoM), first identified and characterized in 1999, is the latest addition to the apolipoprotein family. The apoM gene is located in a highly conserved segment in the major histocompatibility complex (MHC) class III locus on chromosome 6, and codes for a 22 kDa protein that structurally belongs to the lipocalin superfamily. In plasma, apoM is mainly confined to HDL, but it may also occur in other lipoprotein classes, such as in triglyceride-rich particles after fat intake.

ApoM is selectively expressed in hepatocytes and in the tubular epithelium of kidney. Data from knockout mice show that apoM, in this species, is critical for the formation of HDL, notably pre-β HDL_1. In transgenic mouse models, apoM has a strong protective effect against atherosclerosis. The functions and physiological roles of apoM in the kidney are still speculative; it has been proposed that epithelium-derived apoM might serve to bind vital substances in the tubular space to prevent their loss in the urine.

Although observations in experimental animals have given distinctive clues as to the functions of apoM, the physiological and pathophysiological role(s) of apoM in man remain to be clarified. Available data suggest that, besides dyslipoproteinemia and atherosclerosis, apoM may be linked to important clinical entities such as diabetes, obesity, and inflammation.

High-Density Lipoproteins: From Basic Biology to Clinical Aspects. Edited by Christopher J. Fielding

ISBN: 978-3-527-31717-2

4.2
Cloning and Characterization of Human ApoM

Human apolipoprotein (apoM) was first identified and isolated by Xu and Dahlbäck in 1999, from triglyceride-rich lipoproteins (TGRLP) in postprandial plasma [1]. When they performed SDS-PAGE of delipidated human TGRLP and sequenced protein bands ranging from 6 to 45 kDa, the N-terminal sequence of one of the sequences was characterized as MFHQIWAALLYFYGI. No homologous protein was identified in public databases, but several human expressed sequence tags (EST) showed similarities to this N-terminal amino acid sequence. From these sequences, a full-length cDNA coding for this novel 188 amino acid protein was obtained [1].

Rabbit antibodies were raised against five synthetic peptides based on the protein sequence, and the pooled antisera were used to analyze the distribution of the protein among various lipoprotein subclasses by Western blotting. Under reducing conditions, a 26 kDa band was particularly abundant in high-density lipoprotein (HDL), but minor amounts were also observed in low-density lipoprotein (LDL) and TGRLP. In addition, a less pronounced band (approximately 23 kDa), corresponding in size to a nonglycosylated variant of the protein [1] (Fig. 4.1), was also observed. As the

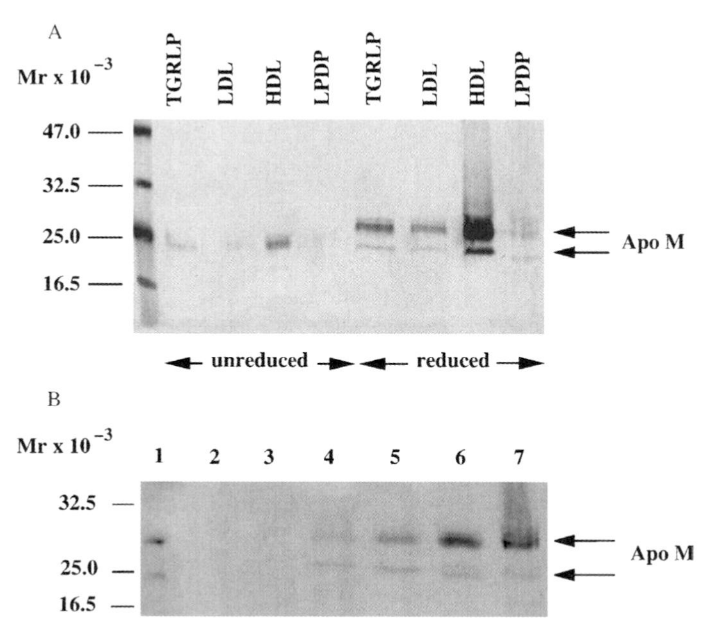

Fig. 4.1 ApoM distribution in different lipoprotein fractions. A) Apolipoproteins from TGRLP, LDL, and HDL (5 μg in each lane) and lipoprotein-deficient plasma (LPDP; 5 μg of plasma protein) were applied to 4–15 % gradient SDS-PAGE under reducing and nonreducing conditions and detected by Western blotting with pooled anti-peptide apoM antisera. B) Increasing amounts of normal plasma proteins were applied to 8–18 % gradient SDS-PAGE and analyzed by Western blotting. Lanes 2–7 contain 0.75, 1.25, 2.5, 5, 10 and 20 μg of plasma proteins, respectively.

protein is extensively associated with lipoproteins in plasma, it fulfills the criteria for classification as an apolipoprotein. The novel protein was named apolipoprotein M (apoM) [1], following the last previously identified apolipoprotein, called apoL [2].

4.3
Gene Location and Amino Acid Sequence of ApoM

The human apoM gene is located in the major histocompatibility complex (MHC) class III locus (chromosome 6, p21.31) and contains six exons [3,4] (chromosome 17 in mouse). The genomic sequence of this region was determined and the human apoM gene identified (GenBank accession number AF118393). In the human genome, the apoM gene is surrounded by BAT4 and NG34 on one side and BAT3 on the other (Fig. 4.2). Both in mouse and man, the apoM gene is predicted to contain six exons enclosed in a 1.6 kb genomic region, which is consistent with the results of Southern blotting. Southern blot analysis of different species gave positive signals in all mammalian genomes, but not in DNA from chicken and yeast [1] (Fig. 4.3).

The human apoM cDNA (734 base pairs) encodes for a 188 amino acid residue protein. The 5′-untranslated region was 33 nucleotides and the 3′-untranslated region 120 nucleotides, excluding the poly(A) tail. The calculated molecular mass of the protein was 21 256. The amino acid sequences of human and mouse apoM are 79 % identical (human and rat apoM: 82 %) (Fig. 4.4). The gene structure of apoM is also preserved in other species, such as the dog, cow, chimpanzee, and orangutan [5].

Several genes in the MHC locus (e.g., TNF, lymphotoxin B, and BAT3) are associated with the immune response. A recent report concludes that TNF, LTA (or its putative teleost homologue TNF-N), apoM, and BAT3 have remained together for over 450 million years, predating the divergence of mammals from fish [6]. The observed enrichment in conserved sequences within the inflammatory region suggests conservation at the transcriptional regulatory level [6].

In man, mouse, and rat, the apoM gene sequences predict the presence in the protein of a signal peptide sequence. Generally, such sequences are split from extracellular proteins prior to secretion from the cell of origin. ApoM, however, retains this signal peptide sequence (20 aa) in the mature protein, as the signal peptide sequence is not followed by a signal peptidase cleavage site [1]. Interestingly,

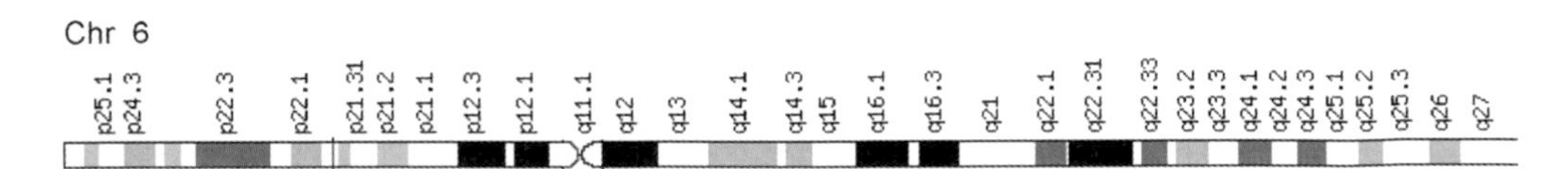

Fig. 4.2 ApoM gene location in chromosome 6. ApoM gene is located in chromosome 6 p21.31 (http://bioinfo.weizmann.ac.il/cards-bin/carddisp?APOM).

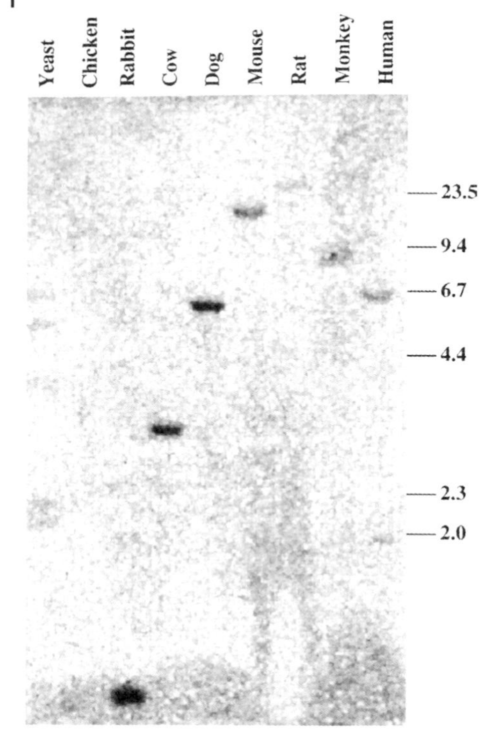

Fig. 4.3 Southern blot of apoM. The zoo blot contained nine different species. The positions and sizes of marker DNA (HindIII-digested λDNA) are indicated at the right.

```
ApoM-Rat    MFHQVWAALL YLYGLLFNSM NQCPEHSQLM TLGMDDKETP EPHLGLWYFI AGAAPTMEEL
ApoM-Human  ....I..... .F..IIL..I Y........T ...V.G..F. .V...Q.... ......K...
ApoM-Mouse  .......... S......... .........T A....T.... .......... ....S.T...

ApoM-Rat    ATFDQVDNIV FNMAAGSAPR QLQLRATIRT KNGVCVPRKW TYHLTEGKGN TELRTEGRPD
ApoM-Human  ....P..... .........M ..H......M .D.L...... I.....--.S .D........
ApoM-Mouse  ....P..... .......... .......... .S........ ..R....... M.........

ApoM-Rat    MKTDLFSISC PGGIMLKETG QGYQRFLLYN RSPHPPEECV EEFQSLTSCL DFKAFLVTPR
                             •
ApoM-Human  ...E...S.. ......N... .......... .......K.. ...K...... .S....L...
ApoM-Mouse  .......S.. .......... .......... .......K.. .......... ..........

ApoM-Rat    NQEACPLSSK
ApoM-Human  .....E..NN
ApoM-Mouse  ..........
```

Fig. 4.4 Comparison of apoM amino acid sequence of rat, human, and mouse. Dots indicate residues that are identical to the top line (rat). One potential site for N-linked glycosylation site (Asn-Glu-Thr) is indicated by a large dot above the sequence of human apoM (•). The two underlined sequences indicate the typical lipocalin motifs.

apoM shares this feature with two other proteins that are also transported bound to HDL particles in plasma: haptoglobin-related protein (HRP) and paraoxonase 1 (PON 1).

The amino acid sequence of apoM contains six cysteines, which may be involved in the formation of three disulfide bridges. There is one potential site for N-linked glycosylation at Asn135 (Asn-Glu-Thr), whereas Asn148 (Asn-Arg-Ser-Pro) is less likely to be glycosylated because Pro-151 follows Ser-150.

4.4 Protein Structure of ApoM

Through sensitive sequence searches it was proposed that apoM is related, like apoD, to the lipocalin protein superfamily [1]. Lipocalins are involved in numerous biological functions: some are enzymatically active, others bind signal substances such as pheromones, while still others are have regulatory functions in cellular metabolism and immunological responses [7,8]. Using computer protein modeling of two lipocalins – mouse major urinary protein (MUP) and human retinol binding protein (RBP) – as initial templates to build the apoM protein structure, Duan et al. [9] demonstrated that apoM has a structure similar to that of other members of the lipocalin protein superfamily. As mentioned above, apoM retains an uncleaved N-terminal signal peptide; this hydrophobic sequence most probably serves to anchor the molecule into the single layer of amphipathic lipids on the surface of the lipoprotein particle [9]. The predominant phospholipid in HDL is phosphatidyl choline, which has a positively charged choline group exposed on the surface of the lipoprotein particle. This property may mediate the binding to electronegative regions of apoM. Two such sequences are present in the apoM model, located around the N terminus and the opening of the binding pocket, respectively.

The three-dimensional model of apoM [9] is characterized by an eight-stranded antiparallel β-barrel. This harbors a segment including Asn135, which can adopt a closed or open conformation. In analogy with other lipocalins, this arrangement reflects the presence of a hydrophobic binding pocket, the specificity and properties of which remain to be clarified.

ApoM presents three disulfide bridges, which would make it a member of the lipocalin subgroup with three S–S bonds [9]. These connect Cys23 and Cys167, Cys 95 and Cys183, and Cys128 and Cys157. The overall shape of apoM resembles a coffee filter holder, while on the outside an α-helix runs like a handle (Fig. 4.5).

Several isoforms of apoM have been identified in plasma. These most probably represent various degrees of glycosylation (there is a glycosylation site at residue Asn135), sialylation, or phosporylation. Karlsson et al., using two-dimensional gel electrophoresis and mass spectrometry, demonstrated that two isoforms of apoM are present in human HDL and three isoforms in LDL particles, probably due to differences in glycosylation or sialylation [10,11]. However, there is only one form of apoM found in VLDL [12]. The physiological implications of these isoforms remain unclear.

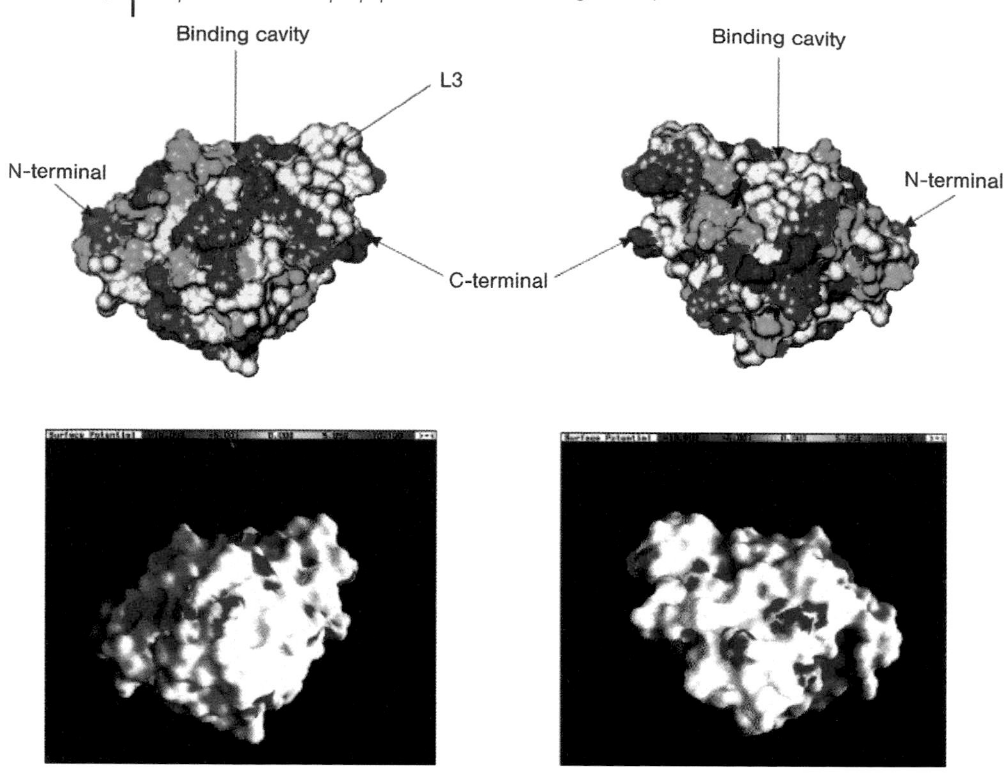

Fig. 4.5 ApoM structure. The apoM protein structure was modulated by computer.

4.5 Distribution of ApoM in Plasma Lipoproteins

Because of its hydrophobic features, including the nonpolar character of the retained signal peptide, ApoM is almost exclusively transported in plasma as a component of plasma lipoproteins, preferentially HDL.

The plasma concentration of apoM in man has been reported at 20–150 mg L^{-1}; although these estimates are uncertain for methodological reasons, it has been estimated that apoM constitutes a minor proportion of HDL apolipoproteins in man (less than 10 % of the concentrations of apoAI) [13]. Like most other small apolipoproteins, it seems that apoM can transfer between lipoprotein particles (e.g., between HDL subfractions as well as between lipoprotein classes). During the postprandial phase, for example, the concentrations of apoM in triglyceride-rich lipoproteins increase, probably as a result of transfer from HDL particles [1].

Although originally identified in TGRLP, human apoM is mainly transported with HDL. It is present both in HDL_2 and in HDL_3 particles (unpublished data), to a large extent in HDL particles that also contain apoAI [14]. Using monoclonal antibodies and immunoaffinity chromatography, Christoffersson et al. [13] recently demonstrated that about 5 % of all HDL particles in human plasma contain apoM; these were defined as HDL-apoM+. Such apoM+ particles contained more cholesterol than apoM− particles. ApoM+ HDL is quite heterogenous with regard to protein composition; besides apoAI and AII, it also contains several other apolipoproteins such as apoCI, CII, and CIII. In mice, apoM is an important component of pre-β-HDL.

Observations in genetically modified mice have added to our understanding of the transport of apoM in plasma. ApoM is thus associated with HDL-sized particles in wild-type and apoA-I deficient mice, whereas in LDL receptor-deficient (hypercholesterolemic) mice it is found in HDL- and LDL-sized particles [15]. In apoE-deficient mice fed a high-fat, high-cholesterol diet, apoM is found mainly in VLDL-sized particles [15]. It thus associates primarily with HDL under normal conditions, but it may also occur in pathologically increased lipoprotein fractions regardless of the nature of the lipoprotein particles.

To investigate the impact on plasma lipoprotein metabolism of primary derangements in apoM processing, Wolfrum et al. [16] modulated hepatic apoM expression in mice through the use of apoM-silencing RNA or apoM-adenovirus. Decreased apoM expression was accompanied by the accumulation of large HDL_1 particles in plasma, while pre-β-HDL disappeared. In analogy, HNF-1-α knockout mice (which also have low apoM expression; see below) exhibit a lipoprotein pattern similar to that induced by apoM-silencing RNA; in this model, the aberrations in HDL fractions could be reversed by injection of apoM-adenovirus [16]. Taken together, these observations demonstrate that apoM is critically involved in the formation of HDL, notably pre-β-HDL_1.

4.6 Tissue Distribution and Cellular Expression of ApoM

Northern blot analyses of multiple tissues (including spleen, thymus, prostate, testis, ovary, small intestine, colon, leukocytes, heart, brain, placenta, lung, liver, skeletal muscle, kidney, pancreas, stomach, thyroid, spinal cord, lymph node, trachea, adrenal gland, and bone marrow) showed that apoM was expressed mainly in kidney and liver [1]. Furthermore, human tissue expression array studies indicated that apoM is only expressed in liver and in kidney, while small amounts were also found in fetal liver and in fetal kidney [17] (Fig. 4.6).

To elucidate whether and when apoM is expressed, Zhang et al. investigated apoM expression patterns during mouse and human embryogenesis [18]. ApoM transcripts were detectable in mouse embryos from day 7.5 to day 18.5, and apoM was expressed at low levels at day 7.5 and then increased up to day 18.5 (i.e., almost to parturition; Figure 4.7a). ApoM-positive cells appeared mainly in the livers of day 12 embryos as detected by *in situ* hybridization. In day 15 embryos, apoM was expressed in both liver and kidney.

During human embryogenesis, apoM was strongly expressed in livers of 3–5-month-old embryos and continued to be so throughout embryogenesis. In the kidney, apoM expression was highest in 5–9-month-old embryos. There was some

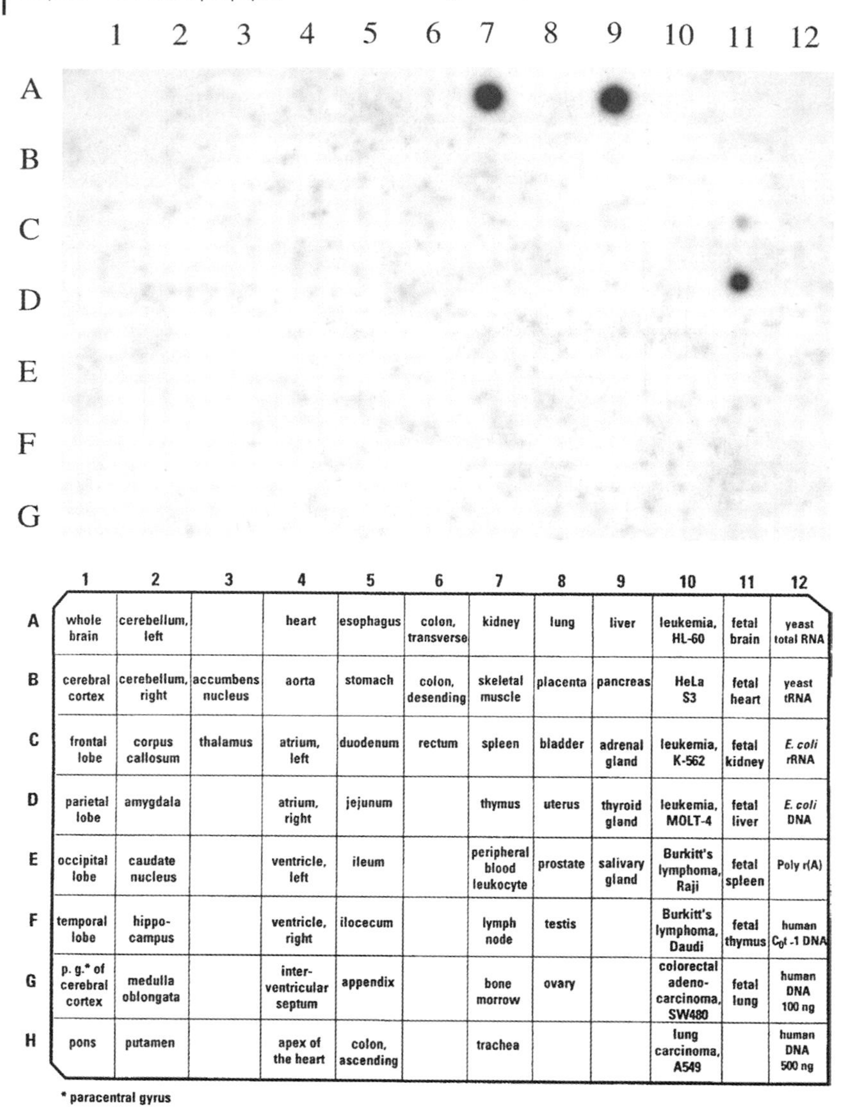

	1	2	3	4	5	6	7	8	9	10	11	12
A	whole brain	cerebellum, left		heart	esophagus	colon, transverse	kidney	lung	liver	leukemia, HL-60	fetal brain	yeast total RNA
B	cerebral cortex	cerebellum, right	accumbens nucleus	aorta	stomach	colon, desending	skeletal muscle	placenta	pancreas	HeLa S3	fetal heart	yeast tRNA
C	frontal lobe	corpus callosum	thalamus	atrium, left	duodenum	rectum	spleen	bladder	adrenal gland	leukemia, K-562	fetal kidney	*E. coli* rRNA
D	parietal lobe	amygdala		atrium, right	jejunum		thymus	uterus	thyroid gland	leukemia, MOLT-4	fetal liver	*E. coli* DNA
E	occipital lobe	caudate nucleus		ventricle, left	ileum		peripheral blood leukocyte	prostate	salivary gland	Burkitt's lymphoma, Raji	fetal spleen	Poly r(A)
F	temporal lobe	hippo-campus		ventricle, right	ilocecum		lymph node	testis		Burkitt's lymphoma, Daudi	fetal thymus	human C_0t-1 DNA
G	p. g.* of cerebral cortex	medulla oblongata		inter-ventricular septum	appendix		bone morrow	ovary		colorectal adeno-carcinoma, SW480	fetal lung	human DNA 100 ng
H	pons	putamen		apex of the heart	colon, ascending		trachea			lung carcinoma, A549		human DNA 500 ng

* paracentral gyrus

Fig. 4.6 Tissue array of apoM expression. A dot blot containing poly(A)+ RNA from multiple human tissue expression array was hybridized to a randomly primed cDNA probe of apoM. Random priming was achieved with radiolabeled ^{32}P-dCTP, and a final specific activity greater than 10^9 cpm g^{-1} of DNA was obtained.

expression of apoM in small intestine, particularly in the later stages of embryogenesis. In skeletal muscle, minute apoM expression was found in 3–5-month-old embryos, and some apoM expression was found in the stomach during middle stages of embryogenesis (Fig. 4.7b) [18].

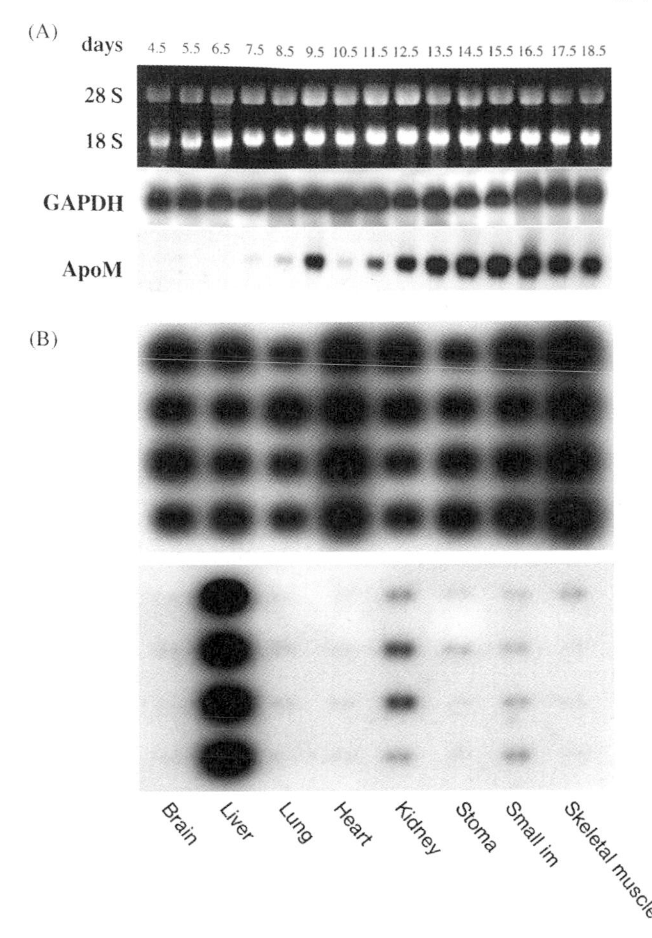

Fig. 4.7 *A*) Mouse embryogenesis of apoM. Northern blots containing total RNA from 7.5–18.5 day mouse embryos were hybridized to a randomly primed cDNA probe of apoM. Embryonic stages by dpc are indicated. Each lane contained 20 ug total RNA from whole embryo. *B*) Human embryogenesis of apoM. mRNA array containing different fetal tissues with different embryo stages was hybridized to a randomly primed cDNA probe of apoM (Panel B). Embryonic stages by months are indicated. Random priming was achieved with radiolabeled [^{32}P]dCTP, and a final specific activity greater than 10^9 cpm g^{-1} of DNA was obtained. Human GAPDH cDNA (Panel A) was used as control.

4.7 Putative Functions of ApoM

ApoM is therefore selectively expressed in hepatocytes and in tubular epithelium in all species studied so far. The high tissue specificity of apoM expression strongly suggests that the physiological role of apoM is related to hepatic lipid metabolism, lipid transport in the circulation, and/or renal functions.

Lipoprotein Metabolism and Atherosclerosis The complex synthetic machinery for lipoprotein formation in the liver includes the synthesis of apolipoproteins, formation of nascent lipoprotein particles, and their secretion into the circulation. The circulating lipoproteins are, in turn, closely bound up with cardiovascular status and the development of atherosclerotic lesions in the arterial vessel wall. HDL, which carries the predominant portion of apoM in plasma, is generally regarded as antiatherogenic, an attribute mainly ascribed to its role in "reverse cholesterol transport".

In mice, apoM is essential for the formation of HDL in the liver and its metabolism in the circulation. Wolfrum et al. [16] demonstrated that treatment with apoM-silencing RNA led to the accumulation of large HDL_1 particles in plasma at the expense of normal pre-β HDL particles, while overexpression of apoM in HNF-1α knockout mice (which have low apoM expression; see below) by treatment with apoM adenovirus increased the formation of such pre-β HDL particles.

Modulation of apoM levels also affects the development of atherosclerosis. Wolfrum et al. [16] used LDL-receptor knockout mice fed a cholesterol-rich diet for 12 weeks, and then administered apoM-adenovirus, which increased apoM levels about twofold. After three weeks, atherosclerotic lesions were reduced by about 50 % in animals with elevated apoM (and pre-β HDL) levels.

The remarkable antiatherogenic effect of elevated apoM levels may reflect several mechanisms. As demonstrated by *in vitro* experiments, HDL without pre-β HDL were less efficient in promoting the efflux of cholesterol from cultured cells [16], so the concomitant rise in pre-β HDL after administration of apoM adenovirus may conceivably increase the efficacy of "reverse cholesterol transport". To test whether this mechanism was also relevant for man, Christoffersson et al. [13] compared the properties of human HDL particles that contained apoM with those that did not. ApoM+ particles were significantly more efficient in promoting the efflux of cholesterol from prelabeled THP-1 cells, lending support to the notion that one mechanism behind the antiatherogenic effect of apoM reflects a role in reverse cholesterol transport.

However, it is also possible that apoM interacts with other steps in the complex formation of atherosclerotic lesions. Recent data [13] indicate that apoM may also affect the oxidative processes that increase the atherogenicity of LDL. Oxidized LDL particles, which have reduced affinity for LDL receptors, are instead removed by scavenger receptors in, for example, macrophages; this is a critical step for the generation of foam cells, and thus of atherosclerotic lesions. ApoM+ HDL was more efficient than apoM- HDL in preventing Cu^{2+}-induced oxidation of LDL *in vitro* [13], indicating that an antioxidative function of apoM may also contribute to its antiatherogenic effect. It is not known whether this is due to the presence of an (as yet unknown) ligand to apoM, or whether it reflects other properties of apoM as such.

ApoM and Kidney Function The high levels of expression of apoM in proximal tubular epithelium of the kidney suggest a physiological role of apoM in excretion or reabsorption of metabolites in the urine [19]. Megalin is a receptor located in tubular epithelial membranes that strongly binds to various substances in urine, including lipocalins, thereby mediating their reabsorption and preservation in the body. Megalin-deficient mice consequently excrete lipocalins (e.g., RBP, MUP-6, and vitamin D-binding protein) in urine.

It has been demonstrated that megalin has high affinity for apoM [19], suggesting that tubulus-derived apoM may also be a ligand for megalin. It is therefore interesting that megalin-deficient mice (unlike normal mice) excrete apoM in urine [19]. It has been proposed that the function of apoM in the kidney might be related to its lipocalin structure. Thus, if apoM in the urine binds lipophilic vitamins or other lipids it may, through subsequent binding and reabsorption by the megalin receptor pathway, prevent the loss of vital substances in the urine [19].

4.8 Regulation of ApoM

The expression of ApoM and its concentration in plasma are dependent upon a number of nuclear transcription factors, and also subject to hormonal and metabolic regulation. Information collected from *in vitro* studies, cell cultures, genetically modified mice, and from clinical observations yields a pattern so far neither comprehensive nor consistent, but some features are emerging. Available data thus demonstrate that several different regulatory pathways are involved in the regulation of apoM. Also, it appears that alterations in apoM metabolism are linked to clinically important entities such as inflammation, diabetes, and obesity.

Nuclear Transcription Factors and Receptors Hepatocyte nuclear factor 1α (HNF-1α) belongs to the helix loop/helix homeodomain transcription factor family and was first identified by its interaction with regulatory sequences of liver-specific gene promoters [20]. It has important roles in development, cell differentiation, and metabolism, primarily in the liver, intestine, kidney, and the exocrine pancreas.

HNF-1α protein can bind to the HNF-1 binding site of apoM promoter *in vitro* [21], while HNF-1α *in vivo* is a potent activator of apoM gene promoter [21]. Mutant HNF-1$\alpha^{-/-}$ mice thus completely lack expression of apoM in liver and kidney, and apoM is absent from plasma. In heterozygous HNF-1$\alpha^{+/-}$ mice, serum levels of apoM are reduced by 50 % in relation to wild-type animals. The HNF-binding site of the apoM promoter, which is highly preserved, has been identified, and specific mutations to this binding site abolished transcriptional activation of the apoM gene [21]. As described in more detail below, mutations in the HNF-1α gene are closely related to diabetes, notably to the MODY3 (maturity onset diabetes in the young) type. These patients have low plasma concentrations of apoM [21].

Peroxisome proliferator activated receptors (PPARs) are nuclear transcription factors that regulate lipid and lipoprotein metabolism, glucose homeostasis, and the inflammatory response [22]. The PPAR family consists of three proteins $-\alpha$, β/δ, and $\gamma-$ that all display tissue-specific expression patterns reflecting their biological functions. PPARα is principally expressed in tissues exhibiting high rates of β-oxidation, such as liver, kidney, heart, and muscle, while PPARγ is expressed at high levels in adipose tissue. PPARβ/δ, however, is ubiquitously expressed [23]. The molecular actions of fibrates and statins, two of the conventional hypolipidemic agents, involve the functions of hepatic PPAR(α) [24–26].

Exposure of HepG2 and Hep3B cells to the PPAR-α activator gemfibrozil resulted in a twofold induction of apoAI mRNA and a one-third reduction in apoB mRNA, but had no significant effect on apoE mRNA levels [27]. Ciprofibrate treatment decreases hepatic apoB mRNA editing and alters the pattern of hepatic lipoprotein secretion [28]. Linden et al. reported that the PPAR-α agonist WY 14 643 decreased the secretion of apoB-100 in VLDL, but not that of apoB-48, and decreased triglyceride biosynthesis and secretion from primary rat hepatocytes [29].

While the effect of PPAR-α on the production of apoB-containing lipoproteins is well established, data on the impact of PPARs on apoM regulation are still scarce. Recently, Anderson et al., in a microarray study, demonstrated that PPAR-α agonists did not affect apoM expression *in vivo* (while downregulating hepatic apoB expression as expected) [30]. In HepG2 cell cultures, we found that neither PPAR-α nor PPAR-γ agonists affected apoM expression, while downregulating apoB expression [31], though the PPAR-β/δ agonist GW 501 516 inhibited both apoM and apoB expression in such cells [31]. Available data thus suggest that the PPAR pathways affect the regulation of apoB and apoM differently; the significance or the inhibitory effect of PPAR-β/δ on apoM expression remains speculative.

Liver X Receptors and Liver Retinoid X Receptors Liver X receptors (LXR) are key regulators of cholesterol and bile acid metabolism in hepatocytes, and also target genes involved in steroid hormone synthesis, growth hormone signaling, and inflammation. The retinoid X receptors (RXR) bind the biologically active vitamin A, 9-*cis*-retinoic acid, and are involved in a variety of cellular functions including cell differentiation and fatty acid metabolism. To integrate the cellular responses to various stimuli there is excessive "cross-talk" not only between LXRs and RXR, but also with the PPARs [24].

As part of a microarray study on the interaction between these receptors, Auboeuf et al. found that LXR agonists inhibited apoM expression *in vivo* [32]. Recent studies in HepG2 cells demonstrate that both LXR and RXR agonists regulate apoM expression *in vitro*. Both T0 901 317 (a LXR agonist) and 9-*cis*-retinoic acid (a RXR ligand) significantly inhibited apoM expression, but not apoB expression, in HepG2 cell cultures (Fig. 4.8), indicating that apoM expression may also be modulated by the LXR-RXR pathway.

Growth Factors Several growth factors influence the transcription and secretion of apolipoproteins in HepG2 cells. ApoB expression, for example, is markedly downregulated by transforming growth factor-β (TGF-β) [33,34]. In the case of apoM, we have reported that TGF-β was also able to downregulate apoM expression and secretion from HepG2 cells [35].

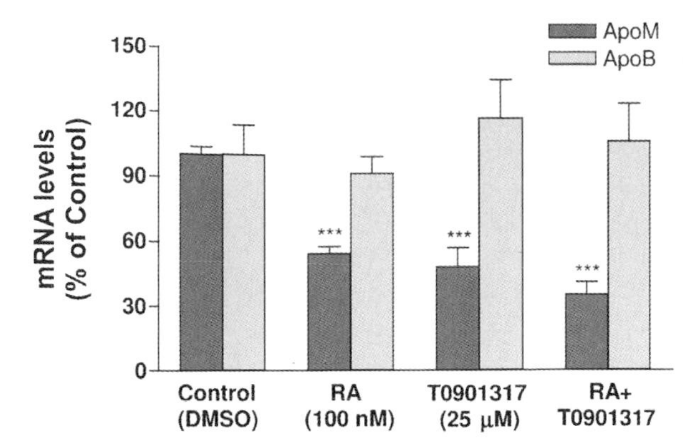

Fig. 4.8 Regulation of apoM expression by the LXR-RXR pathway. HepG2 cells were incubated with T 0 901 317 (a LXR agonist) and/or retinoic acid (RA, a RXR ligand) for 24 hrs. ApoM mRNA levels were determined by real-time RT-PCR analyses. Each experimental group contains 6 replicates and real-time RT-PCP was run in triplicate. Data are means ±SD ($n = 6$ for each sample group). Control group represented 100 %. $^{***}P < 0.001$ vs. control group.

Inflammation and Immune Response Since the apoM gene is located in a highly conserved region (histocompatibility complex III (HMC-III) region on chromosome 6) in close proximity to genes related to the immune response (i.e., TNF, lymphotoxin B, and BAT3), it is reasonable to hypothesize that apoM may also be related to the immune response system, or regulated by cytokines or other inflammatory factors. *In vitro* studies lend some support to this notion.

Thus, in HepG2 cells, platelet-activating factor (PAF) upregulates apoM expression, whereas lexipafant (a PAF receptor antagonist) significantly suppressed mRNA levels and secretion of apoM in a dose-dependent manner [36]. However, neither tumor necrosis factor-α (TNF-α) nor interleukin-1α (IL-1α) influenced apoM expression or secretion in HepG2 cell cultures [36].

A couple of observations in more complex models are compatible with the idea that apoM may be involved in tissue defense mechanisms. During local ischemia-reperfusion injury of livers in rats, hepatic apoM mRNA levels increased significantly during 1 hr ischemia followed by 0.5 to 3 hrs reperfusion, which was similar to what has been observed for the heat shock protein HSP70 [37]. However, the plasma concentrations of apoM were not affected by ischemia-reperfusion injury. Using cDNA microarrays to compare apoM expression in renal biopsy specimens from kidney recipients, Hauser et al. [38] found that apoM expression was higher in allograft kidneys from living donors than in cadaveric donor kidney recipients. This, however, may also reflect differences in renal function in the two groups.

Hepatitis and Carcinoma Chronic infection with hepatitis B virus (HBV) greatly increases the risks of developing liver cirrhosis and hepatocellular carcinoma (HCC), which also influences lipid, lipoprotein, and apolipoprotein metabolism *in vivo*[39]. It has been demonstrated that the expression of most apolipoproteins, including apoM, is downregulated in HepG2 cells infected with HBV [40].

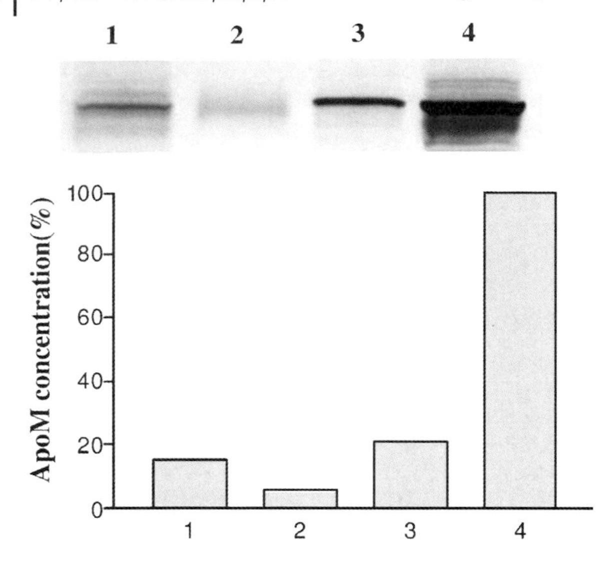

Fig. 4.9 ApoM protein levels in liver cancer and normal tissues. 30 μg tissue homogenates were applied to the SDS-PAGE and apoM was determined by Western blotting. ApoM protein level in normal liver tissue was considered as 100 %.

Although information is still scarce, it seems that apoB expression is also reduced in malignant tumors. Thus, apoM expression is downregulated in human mammary carcinomas [41], and treatment with tanshinone IIA, a potential anticancer agent, can upregulate apoM expression in such tumors [42]. Similarly, we have found that apoM protein levels are significantly lower in hepatocellular carcinoma tissue than in the normal liver (Fig. 4.9).

4.9 Hormonal Regulation of ApoM

Leptin Leptin has multiple functions in appetite regulation and energy metabolism, not least in the liver. In a comprehensive investigation of its role in key metabolic pathways in the liver, Liang and Tall found that the mRNA level of apoM was decreased in leptin-deficient *ob/ob* mice [43] relative to wild-type mice. Furthermore, leptin administration to such rats increased the expression of apoM, demonstrating that the leptin pathway is involved in the regulation of apoM *in vivo*.

Along the same lines, we found the expression of apoM to be significantly lower in both liver and kidney of leptin-deficient *ob/ob* mice than in control mice; in addition, apoM expression in liver and kidney was decreased in leptin receptor-deficient *db/db* mice. Leptin administration increased plasma apoM levels and apoM mRNA levels in liver and in kidney in *ob/ob* mice [44]. Thus, both leptin and the leptin-receptor are essential for the apoM expression.

In vitro, however, leptin inhibits apoM expression in HepG2 cell cultures [45]. The contradictory effects of leptin on apoM expression *in vivo* and *in vitro* might indicate that the effects of leptin are not direct, but are exerted through some other biological signal.

Insulin Plasma apoM concentrations in alloxan-induced diabetic mice were significantly lower than those in control mice, while the expression of apoM both in liver and kidney was less [46]. Administration of insulin increased plasma apoM levels as well as apoM mRNA levels in liver and kidney. These effects could be explained by insulin as such, but may also reflect other metabolic defects of the diabetic state, such as hyperglycemia.

In vitro data point in favor of the latter explanation. Thus, insulin, insulin-like growth factor (IGF), and IGF-potential peptide all inhibited apoM expression in HepG2 cells [31]. The inhibitory effect of insulin seems to be mediated through the activation of phosphatidylinositol 3-kinase (PI3K) [31]. However, glucose also strongly inhibited apoM expression in HepG2 cells, and the inhibitory effects of insulin and glucose were additive (Fig. 4.10).

Other Hormonal Effects Other than insulin and leptin, as described above, there are only a few reports on hormonal regulation of apoM. Epinephrine has been reported to upregulate apoM in human skeletal muscle *in vivo,* as determined by microarray techniques [47]. This does not seem to be a selective effect on apoM, however, since most other apolipoprotein genes were also upregulated.

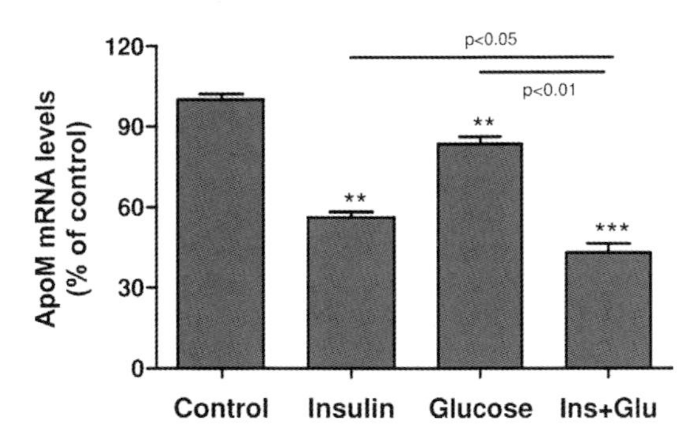

Fig. 4.10 Effects of glucose and insulin on apoM expression in HepG2 cells. HepG2 cells were incubated with glucose (30 mM) and/or insulin (1 μg mL^{-1}) for 24 hrs. ApoM mRNA levels were determined by real-time RT-PCR analyses. Each experimental group contains 6 replicates and real-time RT-PCT was run in triplicate. Data are means ±SD ($n = 6$ for each sample group). Control group represented 100 %. ** $P < 0.01$, *** $P < 0.001$ vs. control group.

Administration of adrenocorticotropic hormone (ACTH) has rapid, beneficial effects on plasma lipoproteins [48–52], with decreases in plasma total cholesterol and LDL cholesterol by 20–40 % within one week [51–54]. The hypolipidemic effect of ACTH seems, at least in part, to be due to a downregulation of apoB synthesis [55]. ACTH did not, however, influence apoM expression or secretion *in vivo* or *in vitro* [55,56]. This is consistent with clinical data, demonstrating that the hypolipidemic effect is restricted to the apoB-containing lipoproteins, while HDL increases slightly during ACTH administration [51–54].

4.10 Clinical Observations

Diabetes As mentioned above, there is a strong relationship between mutations in the HNF-1α gene and specific types of maturity onset diabetes in the young (MODY3) [21]. Mutations in the HNF-1α gene lead to impaired pancreatic β-cell function and impaired insulin secretion. Because of the multiple actions of HNF-1α, however, it is not surprising that such mutations also affect other critical metabolic functions.

As described above, HNF-1α is a potent activator of the apoM promoter. Richter et al. [21], following up observations in HNF-1α -deficient rats with low apoM levels, measured apoM concentrations in the sera of nine HNF-1α/MODY3 patients, nine normal matched control subjects (HNF-1α+/+), and nine HNF-4α/MODY1 subjects (carrying a mutation in HNF-4α). Serum levels of apoM were significantly decreased in HNF-1α/MODY3 subjects, in relation both to control subjects and to HNF-4α/MODY1 subjects, indicating that HNF-1α haploinsufficiency rather than hyperglycemia is the primary cause of decreased serum apoM protein concentrations. Serum levels of apoM may therefore be a useful serum marker for the identification of MODY3 patients [21].

More recently, it was reported that a single-nucleotide polymorphism of the apoM proximal promoter region of the apoM gene (SNP T-778C) is associated with type 2 diabetes in a Chinese population [57]. Although it is well established that such patients develop atherogenic disturbances in lipoprotein metabolism, including low HDL concentrations, hypertriglyceridemia, and small dense LDL, it was not possible to evaluate the impact of this polymorphism on plasma lipoprotein concentrations since the patients were all under treatment.

Obesity The relationships between plasma apoM, insulin, and leptin levels and lipoprotein concentrations were studied in normal and overweight females (BMI 18.9–57.1 kg m^{-2})[58]. ApoM concentrations were positively correlated to leptin, BMI, and fasting insulin, and negatively correlated to total cholesterol and LDL-cholesterol. The correlations between apoM and cholesterol and between apoM and leptin remained significant after adjustment for the influence of BMI. Forward stepwise multiple regressions showed that cholesterol and leptin were independent predictors of circulating apoM. Together, these two parameters explained about 30 % of the variance in apoM. Hence, apoM is positively correlated to leptin and negatively correlated to cholesterol levels in humans [58].

Alzheimer's Disease Alzheimer's disease (AD) is a complex, multifactor disorder, probably the result of interaction between environmental and genetic factors [59–61]. Increasing evidence points to a link between cholesterol turnover and AD [62,63]. It is well known that the apoE genotype and the apoE receptor are related to AD [63,64]. In this context, Kabbara et al. examined the association of apoM with the risk of developing AD in a case-control study [65]. The results excluded apoM as a genetic determinant of AD.

4.11 Conclusions and Perspectives

Since its identification seven years ago, apoM has been extensively characterized with regard to gene and protein structure, while fundamental regulatory mechanisms have also been identified. Its highly selective expression, in hepatocytes and in renal tubular epithelium, indicates that apoM has its principal roles in hepatic lipoprotein metabolism and renal function. In addition, the lipocalin structure of apoM suggests that it may be engaged in binding and transport of specific (as yet unidentified) ligands.

In genetically modified mice, apoM is critical for the formation of HDL particles, notably pre-β HDL. In LDL-receptor-deficient cholesterol-fed rats, over-expression of apoM protects against the development of atherosclerosis. ApoM/pre-β HDL also promotes the efflux of cholesterol from macrophages, suggesting a critical role in reverse cholesterol transport, while apoM (or some ligand) also seems to be involved in the antioxidative functions of HDL.

Clinical studies of apoM are still scarce, but indicate links between apoM and cardiovascular risk conditions such as diabetes and obesity. It is a challenge for the future to define the roles of apoM in man, with regard to both lipoprotein metabolism and renal function, and to outline its potential role in pathological processes, especially in atherosclerosis.

References

1 Xu, N. and Dahlback, B. (1999) A novel human apolipoprotein (apoM). *J Biol Chem*, **274**, 31286–31290.

2 Duchateau, P. N., Pullinger, C. R., Orellana, R. E., Kunitake, S. T., Naya-Vigne, J., O'Connor, P. M., Malloy, M. J. and Kane, J. P. (1997) Apolipoprotein L, a new human high density lipoprotein apolipoprotein expressed by the pancreas. Identification, cloning, characterization, and plasma distribution of apolipoprotein L. *J Biol Chem*, **272**, 25576–25582.

3 Xie, T., Rowen, L., Aguado, B., Ahearn, M. E., Madan, A., Qin, S., Campbell, R. D. and Hood, L. (2003) Analysis of the gene-dense major histocompatibility complex class III region and its comparison to mouse. *Genome Res*, **13**, 2621–2636.

4 Sultmann, H., Sato, A., Murray, B. W., Takezaki, N., Geisler, R., Rauch, G. J. and Klein, J. (2000) Conservation of Mhc class III region synteny between zebrafish and human as determined by radiation hybrid mapping. *J Immunol*, **165**, 6984–6993.

5 Dahlback, B. and Nielsen, L. B. (2006) Apolipoprotein M – a novel player in high-density lipoprotein metabolism and atherosclerosis. *Curr Opin Lipidol*, **17**, 291–295.

6 Deakin, J. E., Papenfuss, A. T., Belov, K., Cross, J. G., Coggill, P., Palmer, S., Sims, S., Speed, T. P., Beck, S. and Graves, J. A. (2006) Evolution and comparative analysis of the MHC Class III inflammatory region. *BMC Genomics*, **7**, 281.

7 Flower, D. R., North, A. C. and Sansom, C. E. (2000) The lipocalin protein family: structural and sequence overview. *Biochim Biophys Acta*, **1482**, 9–24.

8 Xu, S. and Venge, P. (2000) Lipocalins as biochemical markers of disease. *Biochim Biophys Acta*, **1482**, 298–307.

9 Duan, J., Dahlback, B. and Villoutreix, B. O. (2001) Proposed lipocalin fold for apolipoprotein M based on bioinformatics and site-directed mutagenesis. *FEBS Lett*, **499**, 127–132.

10 Karlsson, H., Leanderson, P., Tagesson, C. and Lindahl, M. (2005) Lipoproteomics I: mapping of proteins in low-density lipoprotein using two-dimensional gel electrophoresis and mass spectrometry. *Proteomics*, **5**, 551–565.

11 Karlsson, H., Leanderson, P., Tagesson, C. and Lindahl, M. (2005) Lipoproteomics II: mapping of proteins in high-density lipoprotein using two-dimensional gel electrophoresis and mass spectrometry. *Proteomics*, **5**, 1431–1445.

12 Mancone, C., Amicone, L., Fimia, G. M., Bravo, E., Piacentini, M., Tripodi, M. and Alonzi, T. (2007) Proteomic analysis of human very low-density lipoprotein by two-dimensional gel electrophoresis and MALDI-TOF/TOF. *Proteomics*, **7**, 143–54.

13 Christoffersen, C., Nielsen, L. B., Axler, O., Andersson, A., Johnsen, A. H. and Dahlback, B. (2006) Isolation and characterization of human apolipoprotein M-containing lipoproteins. *J Lipid Res*, **47**, 1833–43.

14 Ogorzalek Loo, R. R., Yam, L., Loo, J. A. and Schumaker, V. N. (2004) Virtual two-dimensional gel electrophoresis of high-density lipoproteins. *Electrophoresis*, **25**, 2384–2391.

15 Faber, K., Axler, O., Dahlback, B. and Nielsen, L. B. (2004) Characterization of apoM in normal and genetically modified mice. *J Lipid Res*, **45**, 1272–1278.

16 Wolfrum, C., Poy, M. N. and Stoffel, M. (2005) Apolipoprotein M is required for prebeta-HDL formation and cholesterol efflux to HDL and protects against atherosclerosis. *Nat Med*, **11**, 418–422.

17 Zhang, X. Y., Dong, X., Zheng, L., Luo, G. H., Liu, Y. H., Ekstrom, U., Nilsson-Ehle, P., Ye, Q., and Xu, N. (2003) Specific tissue expression and cellular localization of human apolipoprotein M as determined by in situ hybridization. *Acta Histochem*, **105**, 67–72.

18 Zhang, X. Y., Jiao, G. Q., Hurtig, M., Dong, X., Zheng, L., Luo, G. H., Nilsson-Ehle, P., Ye, Q. and Xu, N. (2004) Expression pattern of apolipoprotein M during mouse and human embryogenesis. *Acta Histochem*, **106**, 123–128.

19 Faber, K., Hvidberg, V., Moestrup, S. K., Dahlback, B. and Nielsen, L. B. (2006) Megalin is a receptor for apolipoprotein M, and kidney-specific megalin-deficiency confers urinary excretion of apolipoprotein M. *Mol Endocrinol*, **20**, 212–218.

20 Mendel, D. B. and Crabtree, G. R. (1991) HNF-1, a member of a novel class of dimerizing homeodomain proteins. *J Biol Chem*, **266**, 677–680.

21 Richter, S., Shih, D. Q., Pearson, E. R., Wolfrum, C., Fajans, S. S., Hattersley, A. T. and Stoffel, M. (2003) Regulation of Apolipoprotein M Gene Expression by MODY3 Gene Hepatocyte Nuclear Factor-1alpha, Haploinsufficiency Is Associated With Reduced Serum Apolipoprotein M Levels. *Diabetes*, **52**, 2989–2995.

22 Duval, C., Chinetti, G., Trottein, F., Fruchart, J. and Staels, B. (2002) The

role of PPARs in atherosclerosis. *Trends Mol Med*, **8**, 422.

23 Gervois, P., Torra, I. P., Fruchart, J. C. and Staels, B. (2000) Regulation of lipid and lipoprotein metabolism by PPAR activators. *Clin Chem Lab Med*, **38**, 3–11.

24 Krey, G., Braissant, O., L'Horset, F., Kalkhoven, E., Perroud, M., Parker, M. G. and Wahli, W. (1997) Fatty acids, eicosanoids, and hypolipidemic agents identified as ligands of peroxisome proliferator-activated receptors by coactivator-dependent receptor ligand assay. *Mol Endocrinol*, **11**, 779–791.

25 Inoue, I., Itoh, F., Aoyagi, S., Tazawa, S., Kusama, H., Akahane, M., Mastunaga, T., Hayashi, K., Awata, T., Komoda, T. *et al.* (2002) Fibrate and statin synergistically increase the transcriptional activities of PPARalpha/RXRalpha and decrease the transactivation of NFkappaB. *Biochem Biophys Res Commun*, **290**, 131–139.

26 Post, S. M., Duez, H., Gervois, P. P., Staels, B., Kuipers, F. and Princen, H. M. (2001) Fibrates suppress bile acid synthesis via peroxisome proliferator-activated receptor-alpha-mediated downregulation of cholesterol 7alpha-hydroxylase and sterol 27-hydroxylase expression. *Arterioscler Thromb Vasc Biol*, **21**, 1840–1845.

27 Tam, S. P. (1991) Effects of gemfibrozil and ketoconazole on human apolipoprotein AI, B and E levels in two hepatoma cell lines, HepG2 and Hep3B. *Atherosclerosis*, **91**, 51–61.

28 Fu, T., Mukhopadhyay, D., Davidson, N. O. and Borensztajn, J. (2004) The peroxisome proliferator-activated receptor alpha (PPARalpha) agonist ciprofibrate inhibits apolipoprotein B mRNA editing in low density lipoprotein receptor-deficient mice: effects on plasma lipoproteins and the development of atherosclerotic lesions. *J Biol Chem*, **279**, 28662–28669.

29 Linden, D., Lindberg, K., Oscarsson, J., Claesson, C., Asp, L., Li, L., Gustafsson, M., Boren, J. and Olofsson, S. O. (2002) Influence of peroxisome proliferator-activated receptor alpha agonists on the intracellular turnover and secretion of apolipoprotein (Apo) B-100 and ApoB-48. *J Biol Chem*, **277**, 23044–23053.

30 Anderson, S. P., Howroyd, P., Liu, J., Qian, X., Bahnemann, R., Swanson, C., Kwak, M. K., Kensler, T. W. and Corton, J. C. (2004) The transcriptional response to a peroxisome proliferator-activated receptor alpha (PPAR alpha) agonist includes increased expression of proteome maintenance genes.*J Biol Chem.*, **279**, 52390–8.

31 Xu, N., Ahren, B., Jiang, J. and Nilsson-Ehle, P. (2006) Down-regulation of apolipoprotein M expression is mediated by phosphatidylinositol 3-kinase in HepG2 cells. *Biochim Biophys Acta*, **1761**, 256–260.

32 Auboeuf, D., Rieusset, J., Fajas, L., Vallier, P., Frering, V., Riou, J. P., Staels, B., Auwerx, J., Laville, M. and Vidal, H. (1997) Tissue distribution and quantification of the expression of mRNAs of peroxisome proliferator-activated receptors and liver X receptor-alpha in humans: no alteration in adipose tissue of obese and NIDDM patients. *Diabetes*, **46**, 1319–1327.

33 Singh, K., Batuman, O. A., Akman, H. O., Kedees, M. H., Vakil, V. and Hussain, M. M. (2002) Differential, tissue-specific, transcriptional regulation of apolipoprotein B secretion by transforming growth factor beta. *J Biol Chem*, **277**, 39515–39524.

34 Zannis, V. I., Kan, H. Y., Kritis, A., Zanni, E. and Kardassis, D. (2001) Transcriptional regulation of the human apolipoprotein genes. *Front Biosci*, **6**, D456–504.

35 Xu, N., Hurtig, M., Zhang, X. Y., Ye, Q. and Nilsson-Ehle, P. (2004) Transforming growth factor-beta down-regulates apolipoprotein M in HepG2 cells. *Biochim Biophys Acta*, **1683**, 33–37.

36 Xu, N., Zhang, X. Y., Dong, X., Ekstrom, U., Ye, Q. and Nilsson-Ehle,

P. (2002) Effects of Platelet-Activating Factor, Tumor Necrosis Factor, and Interleukin-1alpha on the Expression of Apolipoprotein M in HepG2 Cells. *Biochem Biophys Res Commun*, **292**, 944–950.

37 Xu, X., Ye, Q., Xu, N., He, X., Luo, G., Zhang, X., Zhu, J., Zhang, Y. and Nilsson-Ehle, P. (2006) Effects of ischemia-reperfusion injury on apolipoprotein M expression in the liver. *Transplant Proc*, **38**, 2769–2773.

38 Hauser, P., Schwarz, C., Mitterbauer, C., Regele, H. M., Muhlbacher, F., Mayer, G., Perco, P., Mayer, B., Meyer, T. W. and Oberbauer, R. (2004) Genome-wide gene-expression patterns of donor kidney biopsies distinguish primary allograft function. *Lab Invest*, **84**, 353–361.

39 Jiang, J., Nilsson-Ehle, P. and Xu, N. (2006) Influence of liver cancer on lipid and lipoprotein metabolism. *Lipids Health Dis*, **5**, 4.

40 Norton, P. A., Gong, Q., Mehta, A. S., Lu, X. and Block, T. M. (2003) Hepatitis B virus-mediated changes of apolipoprotein mRNA abundance in cultured hepatoma cells. *J Virol*, **77**, 5503–5506.

41 Xu, X. Q., Emerald, B. S., Goh, E. L., Kannan, N., Miller, and L. D., Gluckman, P. D., Liu, E. T. and Lobie, P. E. (2005) Gene expression profiling to identify oncogenic determinants of autocrine human growth hormone in human mammary carcinoma. *J Biol Chem*, **280**, 23987–24003.

42 Wang, X., Wei, Y., Yuan, S., Liu, G., Lu, Y., Zhang, J. and Wang, W. (2005) Potential anticancer activity of tanshinone IIA against human breast cancer. *Int J Cancer*, **116**, 799–807.

43 Liang, C. P. and Tall, A. R. (2001) Transcriptional profiling reveals global defects in energy metabolism, lipoprotein, and bile acid synthesis and transport with reversal by leptin treatment in ob/ob mouse liver. *J Biol Chem*, **276**, 49066–49076.

44 Xu, N., Nilsson-Ehle, P., Hurtig, M. and Ahren, B. (2004) Both leptin and leptin-receptor are essential for apolipoprotein M expression in vivo. *Biochem Biophys Res Commun*, **321**, 916–921.

45 Luo, G., Hurtig, M., Zhang, X., Nilsson-Ehle, P. and Xu, N. (2005) Leptin inhibits apolipoprotein M transcription and secretion in human hepatoma cell line, HepG2 cells. *Biochim Biophys Acta*, **1734**, 198–202.

46 Xu, N., Nilsson-Ehle, P. and Ahren, B. (2006) Suppression of apolipoprotein M expression and secretion in alloxan-diabetic mouse: Partial reversal by insulin. *Biochem Biophys Res Commun*, **342**, 1174–1177.

47 Viguerie, N., Clement, K., Barbe, P., Courtine, M., Benis, A., Larrouy, D., Hanczar, B., Pelloux, V., Poitou, C., Khalfallah, Y. *et al.* (2004) In vivo epinephrine-mediated regulation of gene expression in human skeletal muscle. *J Clin Endocrinol Metab*, **89**, 2000–2014.

48 Berg, A. L. and Nilsson-Ehle, P. (1996) ACTH lowers serum lipids in steroid-treated hyperlipemic patients with kidney disease. *Kidney Int*, **50**, 538–542.

49 Berg, A. L., Hansson, P. and Nilsson-Ehle, P. (1991) ACTH 1-24decreases hepatic lipase activities and low density lipoprotein concentrations in healthy men. *J Intern Med*, **229**, 201–203.

50 Berg, A. L. and Nilsson-Ehle, P. (1994) Direct effects of corticotropin on plasma lipoprotein metabolism in man – studies *in vivo* and *in vitro*. *Metabolism*, **43**, 90–97.

51 Berg, A. L. and Arnadottir, M. (2000) ACTH revisited – potential implications for patients with renal disease [In Process Citation]. *Nephrol Dial Transplant*, **15**, 940–942.

52 Berg, A. L., Nilsson-Ehle, P. and Arnadottir, M. (1999) Beneficial effects of ACTH on the serum lipoprotein profile and glomerular function in patients with membranous nephropathy. *Kidney Int*, **56**, 1534–1543.

53 Arnadottir, M., Berg, A. L., Dallongeville, J., Fruchart, J. C. and

Nilsson-Ehle, P. (1997) Adrenocorticotrophic hormone lowers serum Lp(a) and LDL cholesterol concentrations in hemodialysis patients. *Kidney Int*, **52**, 1651–1655.

54 Arnadottir, M., Dallongeville, J., Nilsson-Ehle, P. and Berg, A. L. (2001) Effects of short-term treatment with corticotropin on the serum apolipoprotein pattern. *Scand J Clin Lab Invest*, **61**, 301–306.

55 Xu, N., Ekstrom, U. and Nilsson-Ehle, P. (2001) Acth decreases the expression and secretion of apolipoprotein b in hepg2 cell cultures. *J Biol Chem*, **276**, 38680–38684.

56 Xu, N., Hurtig, M., Ekstrom, U. and Nilsson-Ehle, P. (2004) Adrenocorticotrophic hormone retarded metabolism of low-density lipoprotein in rats. *Scand J Clin Lab Invest*, **64**, 217–222.

57 Niu, N., Zhu, X., Liu, Y., Du, T., Wang, X., Chen, D., Sun, B., Gu, H. F. and Liu, Y. (2006) Single nucleotide polymorphisms in the proximal promoter region of apolipoprotein M gene (apoM) confer the susceptibility to development of type 2 diabetes in Han Chinese.*Diabetes Metab Res Rev.*, **23**, 21–5.

58 Xu, N., Nilsson-Ehle, P., and Ahren, B. (2004) Correlation of apolipoprotein M with leptin and cholesterol in normal and obese subjects. *J Nutr Biochem*, **15**, 579–582.

59 Bertram, L. and Tanzi, R. E. (2004) Alzheimer's disease: one disorder, too many genes?. *Hum Mol Genet*, **13** Spec No 1,, R135–141.

60 Poirier, J. (2000) Apolipoprotein E and Alzheimer's disease. A role in amyloid catabolism. *Ann N Y Acad Sci*, **924**, 81–90.

61 Strittmatter, W. J. (2001) Apolipoprotein E and Alzheimer's disease: signal transduction mechanisms. *Biochem Soc Symp*, 101–109.

62 Yanagisawa, K. (2002) Cholesterol and pathological processes in Alzheimer's disease. *J Neurosci Res*, **70**, 361–366.

63 Austen, B., Christodoulou, G. and Terry, J. E. (2002) Relation between cholesterol levels, statins and Alzheimer's disease in the human population. *J Nutr Health Aging*, **6**, 377–382.

64 Holtzman, D. M. and Fagan, A. M. (1998) Potential Role of apoE in Structural Plasticity in the Nervous System; Implications for Disorders of the Central Nervous System. *Trends Cardiovasc Med*, **8**, 250–255.

65 Kabbara, A., Payet, N., Cottel, D., Frigard, B., Amouyel, P. and Lambert, J. C. (2004) Exclusion of CYP46 and APOM as candidate genes for Alzheimer's disease in a French population. *Neurosci Lett*, **363**, 139–143.

II
HDL Plasma Metabolic Factors

High-Density Lipoproteins: From Basic Biology to Clinical Aspects. Edited by Christopher J. Fielding

ISBN: 978-3-527-31717-2

5
ATP Binding Cassette Transporters

Laetitia Dufort, Giovanna Chimini

5.1
The Family of ABC Transporters – General Features

ATP binding cassette transporters (ABC) constitute one of the largest families of proteins, extremely highly conserved across evolution [1,2]. ABC genes are abundant in all vertebrate and invertebrate eukaryotic genomes, indicating that their arising predates eukaryotic evolution. These proteins drive ATP-dependent vectorial transport of solutes across cellular membranes [3], and the extremely varied nature of substrates transported implicates individual members of the family in a large variety of physiological processes. Historically, one of the most relevant pathophenotypes associated with ABC proteins in humans has been resistance to multiple drugs developed both *in vivo* and *in vitro* by tumor cells upon chemotherapy [4]. Amplified expression of ABC transporters is often present among the factors underlying the phenotype, and this effectively decreases the intracellular level of the drug by active outward pumping [5].

The complexity of ABC transporters in bacteria and lower organisms had been known previously, but only in the early 1990s did we highlight the fact that large numbers of ABC genes were also to be found in mammals [6]. To date, 49 human ABC proteins have been identified [1]. On the basis of their sequence homology and domain organization, they have been classified into seven families designated A to G. (Table 5.1). Interestingly, ABC genes organization in subfamilies is conserved not only in all vertebrates but also in plants and insects, although extensive expansion/reduction in the number of genes within subfamilies has occurred [7]. Notably, the A class is absent from yeast [8] and has undergone the highest rate of gene duplication and loss during evolution. Only seven of the human genes are found in fish, while rodents show evidence of recent duplications absent in other species. Conversely, the G family has undergone a huge expansion in insect genomes [1].

The evolutionary history of ABC transporters has been extensively investigated and the alignment of individual ATP-binding domains has allowed the building of a phylogenetic tree organizing individual genes within the subfamilies [9]. This

High-Density Lipoproteins: From Basic Biology to Clinical Aspects. Edited by Christopher J. Fielding

ISBN: 978-3-527-31717-2

Tab. 5.1 ABC genes subfamilies in characterized eukaryotes [1].

Subfamily	Homo sapiens	Mus musculus	Zebrafish Daniorerio	Drosophila melanogaster	Dictyostelium discoideum	Caenorhabditis elegans	Arabidopsis thaliana	Saccharomyces cerevisiae
A	13	16	7	10	12	7	16	0
B	11	12	9	10	11	21	27	4
C	12	11	11	12	12	9	14	7
D	4	4	4	2	4	5	2	2
E	1	1	1	1	1	1	2	0
F	3	4	3	3	3	3	5	5
G	5	6	5	15	4	9	40	10
H	0	0	1	3	1	0	0	0
Total	49	54	41	56	48	55	106	31

analysis provides compelling evidence for extensive duplication of ATP-binding domains within the major subfamilies in ABC transporters, and repeated duplications of isolated ancestral modules are regarded as the most likely event generating the diversity of the family. On the basis of the syntenic distribution of ABC genes in mouse and human genomes it is likely that duplications occurred before speciation.

In accordance with the variety of substrate that they can transport, ABC proteins decorate all cellular membranes; they have been localized at the plasma membrane and on intracellular organelles such as the peroxisome, mitochondria, endoplasmic reticulum, Golgi and lysosomes.

ABC transporters are symmetrical proteins; to function they need four core domains, two transmembrane domains (TMDs), and two nucleotide or ATP-binding domains (NBDs) facing the cytosolic side of the membrane [10]. The TMDs provide membrane anchoring through a tight succession of six transmembrane helices. In addition to the four core domains, common to all ABC transporters, accessory domains can also be found. These perform different functions and can variously be separate from, attached to, or integrated into the different parts of the core ABC transporter [11]. They may be classified into different groups on the basis of their location (extrinsic or intrinsic to the membrane) and function (regulatory or catalytic) within the cognate transporter. Extracytoplasmic domains are exemplified by the large extracellular loops in the A subfamily [12] or by the substrate binding proteins in periplasmic permeases [13], while membrane-embedded domains consist of extra transmembrane helices or of a whole multispanner anchoring domain [14–16]. Finally, regulatory and catalytic domains face the cytosol and can either control the transcription of the transporter itself or regulate its activity (e.g., the R domain of CFTR) [17]. A schematic representation of the various arrangements found in mammalian ABC proteins is provided in Fig. 5.1.

ABC transporters may be classified as either half-transporters or full transporters on the basis of the number of domains encoded as a single polypeptide [3]. Full transporters consist of the typical two pairs of TMDs and NBDs encoded by a single gene, while half-transporters consist of only one TMD and NBD and must combine with other half-transporters to gain functionality. Half-transporters can form homodimers if two identical ABC transporters join, but frequently assemble into heterodimers. In some cases, several pairs are possible within a single family of half-transporters; this fine-tunes substrate specificity and physiological attributes. A paradigmatic example is that of the *Drosophila white* gene, belonging to the G family, a major controller of eye pigmentation in the fly [18]. This hemi-transporter can dimerize with either *scarlet* or *brown*, two close molecules sharing similar structural arrangements. Interestingly, while *white/brown* dimers transport drosopterin precursors, namely guanine, heterodimerization with *scarlet* shifts the specificity of transport toward tryptophan, a precursor of xanthommatin.

In eukaryotic ABC transporters, two topological arrangements exist: TMD-NBD-TMD-NBD [or (TMD-NBD) $\times$ 2], or the reverse one, with the NBD domain preceding the TMD. The latter is, in mammals, the exclusive prerogative of the G subfamily [19,20].

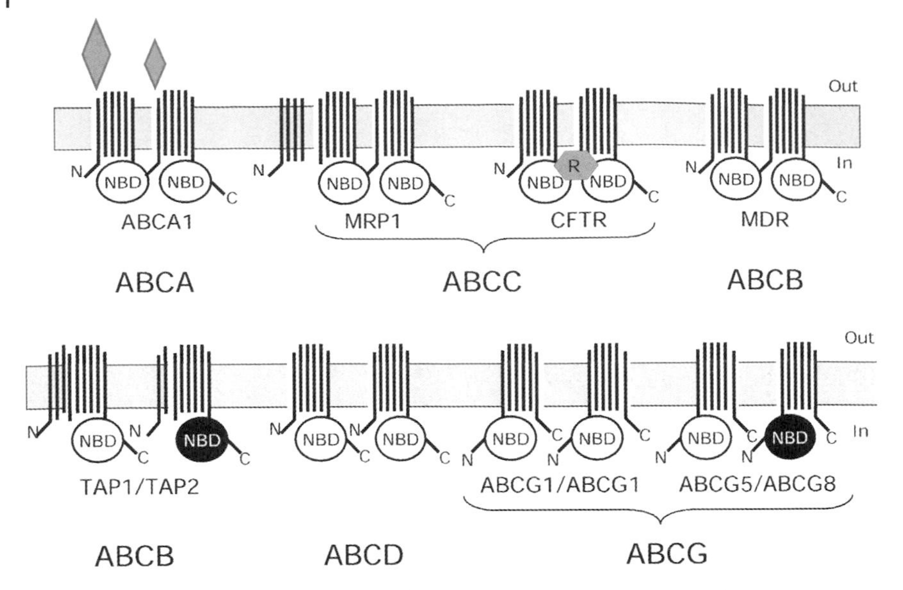

Fig. 5.1 *Schematic representation of the modular arrangements of ABC transporters in mammals.* The symmetrical assembly of four core domains required for function is shown. In the case of TAP (B class) and for all ABCD and ABCG proteins, it is achieved through the association of two half-transporters. The ABCG proteins have an inverted orientation with the NBD preceding the TMD. Additional domains, such as the extracellular loops or the regulatory domain of CFTR, are indicated in gray. Extra transmembrane spanners are also indicated in the case of ABCC proteins and TAP proteins. The NH_2 and COOH termini are indicated for clarity. Hetero- or homodimerization is schematized by the shading of the NBDs.

While the sequence variability of TMD is extensive and therefore hampers any significant evolutionary or comparative analysis, conservation at the NBDs is quite high. There are contained the "ABC signature" motifs [21], known as Walker A, Walker B, and LSGGQ motif (alias ABC signature or C motif). The A and B sites are shared by most nucleotide binding proteins; in the case of ABC transporters the consensus has been defined as follows: (G-X(4)-G-K-[ST]) for the Walker A and ([RK]-X(3)-G-X(3)-L-[hydrophobic] for Walker B(3)). The C motif is exclusively found in ABC transporters approximately 20 amino acids upstream of the Walker B site. Other signatures are also present and have been designated as Q and H loop located in close proximity to Walker A and B, respectively. Recently an additional aromatic A loop upstream of Walker A has been identified [22]. All these motifs are thought to interact cooperatively with bound ATP molecules in a complex pattern implicating both NBDs. In fact there is reasonable evidence, derived from crystal structure studies on NBDs from several transporters [23–27], that these undergo dimerization during the transport cycle, and consequently that the ATP molecules interact in a sandwich lined by residues from both NDBs as schematized in Fig. 5.2.

In order to understand the translocator function of ABC transporters precisely a complete definition of the intermediate conformational states during the transport

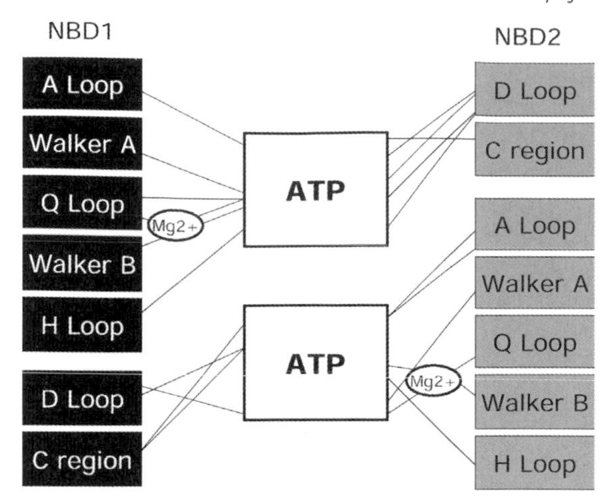

Fig. 5.2 *Molecular contacts between ABC signatures and ATP molecules.* Sandwich-like interaction between the ATP molecules and the two NDBs is predicted according to their observed dimerization. Nomenclature of the motifs follows Ambudkar and coworkers [22].

pathway would be required, but this is so far beyond actual knowledge [27,28]. These states occur during a catalytic cycle and include, schematically, i) binding of substrate and nucleotide, ii) reorientation and nucleotide hydrolysis, and iii) release of phosphate and ADP. Though they are independent and take place at different sites on the molecule, the events of substrate and nucleotide binding are thought to be inextricably coupled. In fact, the high-affinity conformation of the substrate binding sites, required for initial binding, undergoes conformational changes concomitant with the binding of nucleotide. This reorientation, called nucleotide fit, generates huge modifications of conformation, both at the NBDs and at the TMD, and includes dimerization of the NBDs. There is now considerable evidence that the transition to a dimeric juxtaposition of NBDs occurs upon binding of the nucleotide in a variety of transporters [23–27]. Once the transporter is fully loaded the process of reorientation of the molecule and of substrate release begins. Nucleotide hydrolysis should in principle provide the final switch of substrate binding site to low affinity and start the resetting process, resulting in reversion, through the subsequent release of phosphate and dissociation of ADP, to the initial molecular conformation. What these transitions entail at the molecular level and how they take place is mostly beyond our current knowledge, though it is likely that differences exist from one transporter to another and that these intermediate states may also lead to transient molecular associations with specific partners. In this context an analysis of the quaternary structure of ABC transporters by approaches alternative to the classical structural ones could turn out to be extremely useful. Indeed, by applying native gel electrophoresis techniques and biophysical approaches, such as fluorescent resonance energy transfer, to the study of the ABCA1 transporter we were recently able to find evidence that catalytic intermediates may be instrumental for the assembly in molecular complexes [29].

5.2 ABC Transporters and Lipid Transport

Cell membranes are essentially phospholipid bilayers. They are not assembled *de novo*, but are rather created by fusion and fission from preexisting membranes with continuous traffic of material. Theoretically, this should result in a homogenous composition, but in fact single cell compartments maintain distinctive membrane compositions. A detailed discussion on variations in lipid composition and on the mechanisms of its maintenance is beyond the scope of this chapter. As an example it is worth recalling that while the membrane in the endoplasmic reticulum is highly fluid and the composition of the two leaflets is equivalent, due to rapid equilibration of phospholipids, this is not the case at the plasma membrane, where an asymmetric arrangement of phospholipids is present and actively maintained. Of the major phospholipids, phosphatidylserine and phosphatidylethanolamine are enriched in the internal leaflet, while choline-containing phospholipids – sphingomyelin and phosphatidylcholine – are enriched in the outer layer. This asymmetry is achieved both by active transport and by spontaneous exchange rates. The first involves protein translocators, such as the amino phospholipid translocase, which actively displaces negatively charged lipids inward. Conversely, spontaneous flip is dependent on the structure of the lipid itself and the physical packing at the membrane. Highly polar head groups and large-sized backbones reduce the rate of spontaneous movements, whereas loose packing or membrane bending increase that rate [30–33]. Lipids can also spontaneously desorb from membrane and partition into the aqueous phase in spite of their low solubility in water. In the case of fatty acids and lysophospholipids, and even of cholesterol, spontaneous desorption occurs at half-times (seconds to hours) relevant for biological processes [34–36]. Similarly to the spontaneous transbilayer translocation, the partitioning into water is positively modulated by membrane curvature and fluidity. Other factors such as vesicular collision or protein-mediated lipid transfer can also dramatically increase the off-rate of membrane desorption [37].

Shedding of lipid from cell membranes is a ubiquitous event and involves highly diversified physiological processes. Many human diseases have been related to defective desorption of lipids, and members of the ABC transporter family appear at the origin of the defect in nearly all cases. A relationship between ABC transporters and lipid transport was first found during the analysis of mice harbouring invalidation of the mdr 2 gene, a close relative to the human P-glycoprotein multidrug transporter [38]. These mice showed jaundice, due to lack of biliary excretion of phosphatidylcholine. At the same time the accumulation of very long fatty acids in adrenoleukodystrophy was directly linked to the loss of function of an ABC transporter of the D class: ABCD1 [39]. Afterwards, defects in ABC transporters genes were identified as the causal trait underlying more and more lipid-related and

Tab. 5.2 ABCD subfamily members and their mapping in mouse and human genome.

Gene	Mus musculus location	Homo sapiens location	Reference
ABCD1	XB	Xq28	[39]
ABCD2	15. E–F	12q1–q12	[167,168]
ABCD3	3G–H1	1p22–p21	[169]
ABCD4	12D1	14q24.3	[170]

inherited diseases. At present, nearly half of the 49 mammalian transporters are thought to be involved in different aspects of lipid transport. In their vast majority, they belong to the A, D, and G classes.

The D class includes four half-transporters localized at the membrane of the peroxisome and involved in peroxisome biogenesis and/or in the transport of fatty acids across the membrane of these organelles (Table 5.2) [40]. Two of them have been directly linked to the onset of genetically transmitted fatal disorders (i.e., adrenoleukodystrophy and Zellweger syndrome). Though their importance in the cellular handling of lipids and their clinical relevance are beyond doubt, they are not discussed here since their function is unrelated to HDL biogenesis.

Most of the A class members are believed to transport lipid substrates from the cytosol to the exofacial leaflet of the membrane, to the extracellular space, or onto specialized membranes in intracellular organelles [41–43]. Many of them have been linked to the onset of pathological traits related to handling of lipids in specialized cell types. We will discuss in detail two of them – ABCA1 and ABCA7 – whose implication in HDL biogenesis has generated extensive studies

Four out of the ABCG hemi-transporters have been implicated in the transport of sterols, either in the intestine or in liver cells and macrophages [44]. In the latter ABCG1 controls the cellular efflux of cholesterol to mature HDL synergistically to the ABCA1 transporter [45].

5.3 The ABCA Subfamily and the Handling of Excess Lipids

This subfamily was identified in our laboratory in 1994, with the cloning of the prototype ABCA1 [6]. To date it comprises 12 full transporters in man and 16 in the mouse (Table 5.3). Phylogenetic analysis highlighted two major gene clusters [1,46]. The first includes eight genes present in both species and dispersed over different chromosomes: ABCA1, ABCA2, ABCA3, ABCA4, and ABCA7, together with the more recently cloned ABCA5, ABCA12, [47], and ABCA13 [48]. A group of ABCA3-related genes exists and is functional in mouse but unproductive in man (ABCA14, ABCA15, ABCA 16, and ABCA 17). These are clustered in the mouse 7F3 region; in man in the syntenic 16p12 region only scattered exons unable to code for functional ABC transporters are present [49,50]. All these genes in rodents appear to be transcribed in the testes, though no precise definition of their cellular expression or function has been provided.

Tab. 5.3 ABCA subfamily members and their mapping in mouse and human genome.

Gene	Mus musculus location	Homo sapiens location	Reference
ABCA1	4A5B3	9q31.1	[59]
ABCA2	2A2B	9q34.3	[6,171]
ABCA3	17A3.3	16p13.3	[170,172]
ABCA4	3G1	1p22.1–1p21.3	[170]
ABCA5	11E1	17q24.3	[170,173]
ABCA6	11E1	17q24	[173]
ABCA7	10B4C1	19p13.3	[174]
ABCA8	11E1	17q24	[173]
ABCA9	11E1	17q24	[173]
ABCA10		17q24	[173]
ABCA11	p[1]	P	
ABCA12	1C1.3	2q34	[47]
ABCA13	11A2	7p12.3	[48]
ABCA14	7F3		[49]
ABCA15	7F3		[49]
ABCA16	7F3		[49]
ABCA17	17B	16p13.3	[50,175]

[1]p = pseudogene.

The second set of ABCA genes is clustered, both in mouse and in man, on a single chromosome (ABCA6, ABCA8, ABCA9, ABCA11 on human chromosome 17q24 and mouse 11), where ABCA5 also maps [9]. Since their sequence similarity exceeds 70 %, it is assumed that they have arisen from recent duplications of a protypical transporter [46].

The ABCA genes encode some of the largest ABC proteins, several of them >2100 amino acids long and more than 200 kDa in predicted molecular weight. The general topological arrangement of ABC transporters of this class has been addressed in the cases of ABCA1 and ABCA4 by classical approaches such as epitope mapping or glycosylation patterns [51–54]; it can be confidently transposed to the other members of the class since all of them share similar hydrophobicity profiles [55]. The most salient feature of the A class is the presence of two additional domains in the form of symmetrical extracellular loops between the first and second transmembrane helices of each TMD. These domains are extremely variable in length and sequence in individual transporters. Though their function remains elusive, it is likely that they participate in the functional specification of individual transporters and might allow interaction with other proteins.

Most ABCA members show non-overlapping patterns of preferential expression, suggesting that their basic lipid transport function has been adjusted to specific tissue environments [56].

Four members of the ABCA family have been identified as defective in inherited diseases. Loss of ABCA1 function leads to the dyslipidaemia known as Tangier disease [57–59]. The extremely low levels of circulating HDL characteristic of this disorder are due to defective handling of cellular lipids [60]. ABCA4 is exclusively expressed in photoreceptors, where it controls disposal of toxic derivates of retinol

(N-retinylidene PE) generated during the visual cycle [61]. Its dysfunction leads to a broad spectrum of degenerative pathologies of the retina, the graveness of which is directly correlated to the degree of functional impairment induced by the genotypic mutations [62–64]. ABCA3 is involved in surfactant secretion in the lung lining epithelium and its dysfunction leads to fatal surfactant deficiency in newborns [41,65–67]. ABCA12 mutations have recently been found in harlequin ichthyosis, a frequently fatal disease of keratinization leading to the coverage of the body with an armor-like scale layer [43,68–70].

All the ABCA members so far studied decorate the plasma membrane and/or lysosomal-like compartments. From the phenotypes that their dysfunction generates, they are thought to facilitate outward movements of lipids, and the evidence gathered so far is indeed consistent with this view. However, since in most cases no formal evidence on the vectoriality of transport has yet been provided, it may still be worth bearing in mind that it has been deduced from sequence analysis that the A class of transporters should behave as importers rather than exporters [20] and that ABCA4 indeed actually generates an inward flux of lipids (i.e., from the disc lumen in the photoreceptor outer segment, equivalent to the extracellular space, into the cytosol) [61].

In the following paragraphs we review current knowledge not only of the physiology of ABCA1, whose role in HDL biogenesis is beyond any doubt, but also of its closest relative, ABCA7. In fact, in spite of its tenuous link to HDL biogenesis, this transporter has been extensively studied in the context of lipid handling.

5.3.1 ABCA1

The ABCA1 gene has been mapped to human chromosome 9q22–31 and to the syntenic region on mouse chromosome 4 [6]. In both species the gene spans over 150 kb and is composed of 50 exons, encoding a polypeptide of 2261 amino acids in length [71,72]. Like all members of the A family in mammals, the ABCA1 protein is a full-size ABC transporter containing two transmembrane domains, each composed of six α helices and two intracellular nucleotide-binding domains. ABCA1 was originally identified as an engulfment receptor on macrophages, scavenging cells dying by apoptosis [73]. In 1999 ABCA1 mutations were detected in patients suffering from Tangier disease, thus unambiguously linking cellular cholesterol effluxes to this transporter [57–59]. Since then, a number of phenotypes associated with the loss of ABCA1 have been identified [74–77]. All of them are consistent with the function of ABCA1 as a gatekeeper of membrane lipid organization.

5.3.1.1 ABCA1 Expression and its Control

ABCA1 transcript is widely expressed at low levels. The highest mRNA expression levels are detected in placenta and pregnant uterus, liver, lung, adrenal glands, heart, small intestine, and fetal tissues; the lowest expression is found in pancreas, skeletal muscle, ovary, colon, prostate, mammary glands, and bone marrow. At the cellular level, tissue macrophages and macrophage-like cell lines of mouse or human origin consistently express high levels of the transporter [78 and reference therein].

The transcriptional regulation of ABCA1 appears to be exceptionally complex and has been extensively reviewed by Schmitz and coworkers [79]. ABCA1's identification as a sterol-sensitive gene is well characterized, both in human and mouse systems. Most of the inductive transcriptional effect of natural or synthetic lipid appears to be mediated through the nuclear orphan receptor LXR, which has emerged as a key regulator of lipid metabolism. LXRs form obligate heterodimers with RXR, and in this configuration they recognize specific DNA response elements consisting of two direct hexanucleotide repeats separated by four nucleotides (DR4 elements) present in ABCA1 promoter [72,80]. Two LXR isoforms exist: LXR α and β, encoded by different genes. Unlike LXR β, which is ubiquitously expressed, the α isoform shows prominent activity in macrophages and liver. LXR target genes are involved in the whole spectrum of steps crucial in lipid metabolic pathways, from absorption of dietary cholesterol to cellular cholesterol effluxes and reverse cholesterol transport, to the metabolism of lipoproteins and the synthesis and esterification of fatty acids [81]. The regulation of cholesterol homeostasis is particularly relevant in macrophages, since these cells accumulate massive amounts of cholesterol during the development of atherosclerosis [82]. The PPAR class of nuclear hormone receptors α and γ also participates in the upregulation of ABCA1 expression indirectly through enhanced transcription of LXR α [83–85].

Cytokines have been shown to exert pleiotropic and antinomic effects on ABCA1 transcription. As a general rule, proinflammatory cytokines, tumor necrosis factor α (TNFα), interleukin-1β (IL-1β), and interferon-γ downregulate the LXR-mediated enhancement of ABCA1 transcription and protein expression [86,87], whereas transforming growth factor β (TGF β) has the reverse effect and induces ABCA1 expression [88]. These results reinforce the recent findings of multiple crossroads between cellular handling of cholesterol and inflammatory responses, mostly mediated at the LXR level. Indeed, the activation of LXR exerts a global antiinflammatory effect and promotes macrophage survival [89,90]. Though this is likely to depend on the coordinate activation of multiple sets of genes, it may be worthwhile to investigate whether ABCA1, as a lipid transporter, may directly modulate the transduction of signals instrumental for macrophage activation responses [91–93]. This could in fact provide an explanation for the phenotypes unrelated to HDL metabolism that have been observed in the mouse models of ABCA1 inactivation.

The expression of the ABCA1 protein has been reported not to follow strictly the levels of transcript implying, as a first interpretation, a high level of posttranscriptional regulation of the transporter [94]. It may, however, be worth recalling that the detection of the transporter is rather difficult, and misleading underestimations might have been reached. At any rate, it seems reasonably well established that ABCA1 posttranscriptional regulation acts mainly through modulation of protein stability [95,96]. Several factors including the interaction with apoproteins and inhibition of degradative pathways have been reported to increase the stability of the transporter above the basal half life of 1–2 hrs. Conversely, metabolites associated with metabolic disorders such as diabetes may destabilize ABCA1 through activation of phosphorylation pathways and thus contribute to the accumulation of cellular cholesterol in these diseases [96–99].

5.3.1.2 Structural Considerations

The structural arrangement of ABCA1 follows the rules of the A family: ABCA1 is a full-sized transporter possessing the canonical four domains encoded by a single gene. Two additional and symmetrical domains are present and consist of extracellular loops of large size interrupting the TMD after the first and seventh transmembrane helices, respectively (Fig. 5.3) [53,54]. Through analogy with ABCA4 it is accepted that the two halves are linked together by extracellularly located disulfide bridges involving conserved cysteines [100]. The ABCA1 transporter decorates the plasma membrane in naturally expressing or transfected cells. In addition, as deduced from analysis of the intracellular traffic of ABCA1/GFP chimeras, the transporter decorates endolysosomes and the Golgi apparatus [101]. While its presence at the membrane is sufficient to justify most of the cellular phenotypes associated with ABCA1, it has been proposed that lysosomal ABCA1 may also be functional and actually participate in the disposal of cellular cholesterol through retroendocytosis of HDL particles formed in the lumen of these vesicles [102,103].

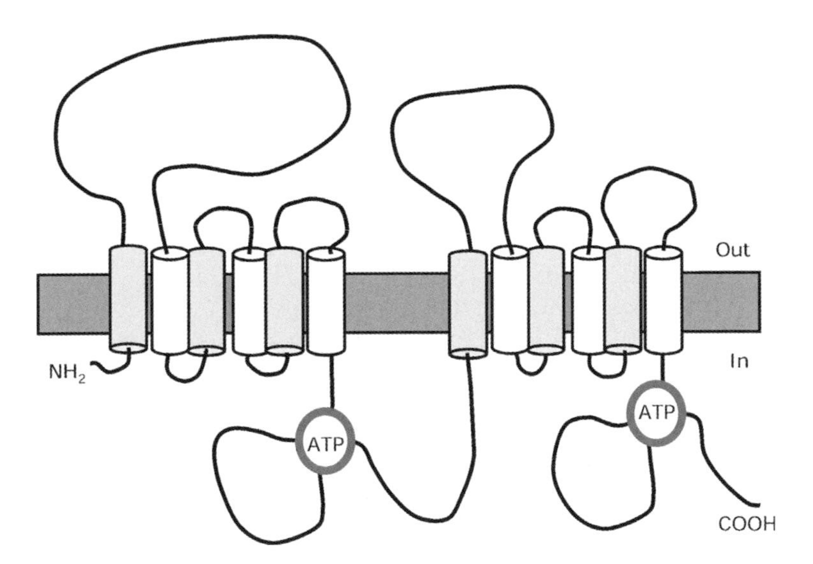

Fig. 5.3 *Predicted topology of ABCA1 monomer as a prototype of the A class of transporters.* Six transmembrane spanners are symmetrically present; they are interrupted by large extracellular loops between the first and second spanners in each TMD. The size of the loops varies largely between individual members of the A class. The suggested disulfide bridge between conserved cysteines in the loops [100] is not indicated in the figure. In view of the high conservation of hydrophobicity plots, a similar topology can be extended to all ABCA proteins [55].

A large body of studies has addressed structure/function relationships of individual residues and structural motifs in the transporter molecules [54,104–106]. Collectively, these results indicate that the folding of the transporter is complex and that retention in the endoplasmic reticulum results from minimal mutations dispersed over the whole molecule. The assembly of the transporter at the membrane also deserves interest since, in the field of ABC transporters, defining the "real" size of the functional molecule has long been a concern. In this light we recently analyzed the quaternary structure of ABCA1 and found evidence that it is predominantly associated, both in naturally expressing cells and upon transfection, in dimers [29]. This implies a basal conformation consisting of four nucleotide-binding domains, as has also been suggested in the past for CFTR and MDR and more recently for ABCG2 [107–109]. Through the study of several ABCA1 mutants we also found evidence that ATP binding induces not only major conformational changes, but also associations in higher-order structures [29]. These structures should in principle include additional partners. Notably, a similar analysis carried out on other ABCA members did not provide evidence of oligomerization, suggesting that these complexes are specifically related to the ABCA1 transport function (Trompier, unpublished). Their relationship to lipid effluxes is still elusive, but Denis et al. [110] have reported that multimerization of ABCA1 is required to achieve binding of apolipoproteins at the cell surface.

5.3.1.3 **ABCA1 and Cellular Effluxes of Phospholipid and Cholesterol – Facts and Speculation**

As mentioned previously, mutations in the ABCA1 gene were found in patients affected by Tangier disease [57–59]. A quite similar pathology can be observed in mice bearing a genetic invalidation of ABCA1, thus unambiguously indicating that it is solely the loss of ABCA1 function that determines reduced effluxes of cellular cholesterol to apoproteins, the seminal defect in Tangier disease [101,111,112]. Indeed, ABCA1 controls the crucial step in the biogenesis of mature HDL particles, levels of which are extremely low both in Tangier disease patients and in animal models of ABCA1 invalidation. In view of the wide distribution of the transporters in the tissues classically concerned with HDL biogenesis, such as gut, liver, and macrophages, sophisticated animal models bearing tissue-specific loss of ABCA1 were generated to provide insight into the relative contributions to maintenance of plasma HDL levels. These studies unambiguously clarified the pivotal contributions of liver and gut to the biogenesis of circulating HDL [113–116], while that of macrophages appeared irrelevant. Conversely, macrophage ABCA1 expression turned out to be protective towards the development of atherosclerotic lesions, irrespective of HDL levels [117,118]. This is in line with the accumulation of sterol-laden foam macrophages in the artery wall as an early causal event in the development of atheroma.

Several ABCA1 polymorphisms have been identified from family studies and population screens [104,106,119]. Many of the missense mutations were detected in subjects with modified HDL levels and were therefore assumed to be causally linked to the clinical sign [120]. Similarly, and in line with the previous findings in mice, a significant inverse correlation was found between ABCA1 activity *in vitro* and severity of cardiovascular disease, at least in subjects bearing heterozygous mutations [106,121].

A large body of *in vitro* evidence has substantiated the direct link between ABCA1 surface expression and cellular effluxes of phospholipid and cholesterol to a broad range of lipid-poor HDL apoproteins including Apo-AI [122], -AII, -E, -CI, -CII, -CIII, and -AIV [123]. Similarly, cells from knockout animals confirmed the absolute requirement for ABCA1 in efficient handling of cellular lipids [101,111,112]. Many laboratories have also analyzed the effects of naturally occurring or *ad hoc* introduced mutation on the activity of the transporter [54,104,105]. As expected, a whole spectrum of functional impairments can be identified, with distinct influences on binding of apoproteins or the effluxes of lipids.

At this point, however, the answers to a number of key questions remain elusive; these concern the formal elucidation of how ABCA1 orchestrates the lipid effluxes, how the effluxes of phospholipids and cholesterol are linked, and how the interaction between ABCA1 and acceptor apoproteins takes place. Basically, ABCA1 transport of cholesterol as a direct substrate has been excluded on the basis of several experiments [124,125]. No direct interaction with photoactivable derivates of cholesterol could be shown [125], while Ueda and co-authors did not find any evidence of cholesterol transport in ABCA1 reconstitutions in proteolipisome [100]. These approaches suggested rather that phospholipids, and among them choline-containing ones, could be the substrates; however, in ABCA1 knockout animals the transbilayer movement of phosphatidylcholine seemed normal or even increased [101]. We and others have also found evidence of a direct correlation between the outward exposure of phosphatidylserine and ABCA1 activity [101,126]; this does not indicate formally that phosphatidylserine is a substrate but supports the notion that ABCA1 activity at the membrane results in redistribution of lipids between the two membrane leaflets, with obvious consequences on their lateral distribution. This implies that the ABCA1 ATPase function destabilizes the proximal membrane microenvironment and is in agreement with the finding that ABCA1 reduces the size of membrane rafts [127,128]. Collectively, these data support a model in which ABCA1, by increasing the flip-flop of one or more phospholipids, modifies (at least locally) the packing of lipid at the membrane. In the context of lipid effluxes this may be influential in that it favors at least two seminal events. Firstly, the docking of apoproteins and the ABCA1-dependent maturation is improved in the modified microenvironment [129], and secondly, the bioavailability of cholesterol for effluxes is augmented [130] through positive displacement into membrane pools accessible for effluxes.

Mechanistically, the transport of lipids out of the cell should occur according to the classical flippase model [131], since, in our view, the binding of apoproteins follows lipid translocation across leaflets. The destabilization of membrane lipid domains, predicted here as the primary action of ABCA1, accommodates its ability to promote cellular interaction with various apopoproteins, but does not easily fit with the sensitivity of Apo A-I docking to structural mutations in ABCA1 [54]. Among the possible explanations we favor a two-step model of interaction between acceptors and transporter. Surface docking is the initial step: it is ABCA1 dependent and does not imply physical interaction but rather membrane insertion. The maturation of Apo A-I follows and only at this point does its physical binding to the transporter occur. This might require appropriate configurations (e.g., multimers of both Apo A-I and transporter), exquisitely sensitive to subtle structural mutations in either partner [110]. This model theoretically accommodates most of the literature but is unfortunately not easily testable, since no experimental system has so far provided a clear temporal or physical dissociation of lipid flipping from the surface docking of apoproteins.

5.3.2 ABCA7

ABCA7 is a full-size transporter, its physiological function being unknown to date [132]. In view of its close structural relationship to ABCA1, the function of ABCA7 has mainly been investigated in the context of cellular lipid metabolism and, more recently, engulfment of dying cells. The results of studies on *in vitro* reconstituted systems, which suggested a functional role in lipid handling, have not been corroborated by study of animal models of genetic invalidation of ABCA7. This, together with a global lack of information, hampers in our opinion any speculations on the physiological function of this transporter.

5.3.2.1 From the Gene to the Protein

Human [133], mouse [134], and rat [135] ABCA7 cDNAs have been cloned; they encode proteins of 2146, 2159, and 2170 amino acids, respectively. The human ABCA7 gene has been mapped to chromosome 19p13 and is composed of 48 exons. ABCA7 shows the highest degree of homology to ABCA1 (54 % between the human proteins), but a relatively low degree of interspecies identity of protein sequences (79 % between mouse and human, against 95 % in the case of ABCA1) [134]. The ABCA7 gene is widely expressed in both adult and embryonic tissue in human, mouse, and rat. Unlike that of ABCA1, the highest expression of ABCA7 is found in the thymus and other immune and hematopoietic tissues (such as spleen, lymph node, peripheral leukocytes, and bone marrow) [133,134,136]. A splicing variant mRNA of human ABCA7 has been detected [136]; it involves the alternative use of a new initiation codon located in exon 5b and its usage leads to a shorter protein (type II) possessing 28 novel amino acids at the N terminus rather than the 1–166 residues of the full-length product (type I). Tissue-specific alternative splicing was detected by RT-PCR. Among the tissues examined, expression of type II mRNA was higher in spleen, thymus, lymph node, and trachea, while type I was equal to or more than type II mRNA in bone marrow and brain. While both ABCA1 and ABCA7 are expressed in macrophages, their relative expression levels are substantially different, ABCA1 being clearly predominant by several orders of magnitude. Available data on the tissue distribution of the ABCA7 protein are so far limited, due to the lack of sensitive tools [136]. In agreement with its induction during adipocyte differentiation in *in vitro* systems (Dufort, unpublished), high levels of ABCA7 have been found in white adipose tissue.

The transcriptional regulation of the ABCA7 gene appears to differ substantially from that of ABCA1, in spite of initial reports suggesting a similar sterol dependency [133]. The predicted promoter regions of both human and mouse ABCA7 show conservation of two modules of putative binding sites [134]. These are targets either for ubiquitous transcription factors or for liver- or lymphoid-specific transcription factors. Unlike that of ABCA1, the transcription of ABCA7 in mouse peritoneal or bone marrow-derived macrophages is unaffected by pharmacological or naturally occurring LXR agonists [137].

Recent results have shown that sterol cellular content is inversely correlated to induction of the ABCA7 transcript; this is apparently mediated by the SREBP system [138]. These observations suggest that the function of ABCA1 and ABCA7, though possibly similar in essence, serves distinct metabolic needs. The subcellular localization of ABCA7 remains a mystery. Indeed, several laboratories including our own have found that ABCA7 localizes at the plasma membrane in several cellular systems upon transfection [132]. No evidence of plasma membrane localization has been found in physiological systems, in which conversely an intracellular localization has been claimed, although a detailed description of the compartment concerned is missing [139,140]. The lack of information may be due to the combined difficulty of low levels of expression and low sensitivity of detection tools. Several laboratories have associated the forced cellular expression of ABCA7 with an increased ability to efflux lipids and to generate mature HDL upon contact with the acceptor ApoAI [139,141,142]. Collectively, and in spite of a number of inconsistencies, these data indicate that surface ABCA7 can function as a phospholipid transporter, though with lower efficiency than ABCA1 [143]. The ability to promote cholesterol effluxes is still a matter of debate. Yokoyama et al. reported that the compositions of HDL particles formed upon the influence of ABCA7 differ substantially from those generated by ABCA1 [144], reinforcing the idea that the transport functions of these two proteins do not necessarily follow the same rules.

5.3.2.2 The ABCA7 Knockout Mouse – A Model with some Surprises

Two sources of ABCA7 knockout animals have been reported [137,140]; in spite of the use of similar strategies for the genomic deletion and equivalent genetic backgrounds, the homozygous deletion of ABCA7 led to embryonic lethality in one case and to viable animals in the second. No explanation for the discrepancy has yet been provided, and nor have any detailed investigation clarifying the early causes of death.

The viable animals, which have been validated for lack of ABCA7 expression, do not show either evidence of lipid-associated disorder or alterations in lipid effluxes from isolated macrophages [137]. These data reinforce the idea that under physiological conditions only ABCA1 is in charge of the maintenance of HDL levels. It can, however, be conceived that ABCA7 takes over after relocalization at the plasma membrane, so investigation of which physiological conditions control the intracellular trafficking of ABCA7 appears determinant to provision of insight into its as yet elusive physiology. Recently, a link between ABCA7 and the engulfment of dying cells has been reported [140]. Unfortunately the evidence is still tenuous and no definitive conclusion can be reached in the absence of a detailed analysis of death-related phenotypes in the available mice.

5.4
The ABCG Subfamily and Sterol Trafficking

In 1910 Morgan published the identification of the first genetic mutation in *Drosophila melanogaster,* which he called "*white*", due to the fact that the flies lacked the characteristic red eyes [145]. Seventy years later it was shown that the *Drosophila white* gene is a member of the ABC transporter superfamily, along with two other genes controlling eye color in the fly: *brown* and *scarlet* [146]. When mammalian ABCG1 was originally cloned it was called murine or human white because of its similarity to the *Drosophila white* gene [19,147].

The mammalian ABCG family consists of five half-transporters (Table 5.4): ABCG1, ABCG2, ABCG4, ABCG5, and ABCG8. In rodents a sixth gene – Abcg3 – has been identified, though it is unlikely to give rise to a functional transporter, due to the acquisition of a number of mutations in the ATP-binding domains.

The ABCG members have a unique architecture, since the NBD is located at the N terminus and precedes the TMD [44] (Fig. 5.4). Homo- or heterodimerization between members is required to function; with the exception of ABCG5 and ABCG8 heterodimers, which control intestinal absorption of sitosterols, the remaining ABCG proteins are thought to homodimerize. ABCG2 was originally identified as overexpressed in multidrug-resistant cell lines [148], whereas ABCG5 and ABC8 were identified by reverse genetics approaches [149].

Although ABCG5 and ABCG8 are not involved in HDL biogenesis, they are required for the transport of plant- and fish-derived sterols in the gut [149]. Indeed their loss of function in humans generates sitosterolemia, a rare autosomal recessive disorder characterized by a decreased back efflux of absorbed food sterols. In spite of the clear indication of their function as cholesterol transporters, no experimental evidence on the mechanistic of transport has yet been provided. Their obligate heterodimerization has been demonstrated by several approaches, including inefficient maturation to a functional transporter in the presence of a single subunit [150] and the development of signs of disease in animals carrying genetic invalidation of either ABC5 or ABCG8 [151,152].

Tab. 5.4 ABCG subfamily members and their mapping in mouse and human genome.

Gene	Mus musculus location	Homo sapiens location	Reference
ABCG1	17 A2-B	21q22.3	[153]
ABCG2	6B3	4q22	[170]
ABCG3	5E5		[176]
ABCG4	9A5.3	11q23	[177]
ABCG5	17E4	2p21	[178]
ABCG8	17E4	2p21	[149]

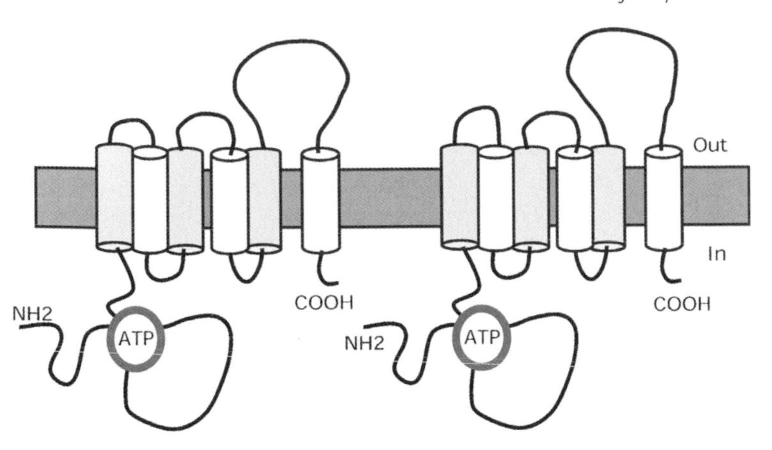

Fig. 5.4 *Predicted topology of a functional ABCG transporter.* Notably, the NBD is located at the N terminus of the transporter, preceding the whole set of transmembrane spanners. A similar arrangement is predicted for ABCG5 and ABCG8. While ABCG1 or ABCG2 associate in homodimers, ABCG5 and ABCG8 heterodimerization is obligate to achieve function.

5.4.1 **ABCG1**

Mouse and human ABCG1 cDNAs were first identified in 1996 and 1997 and were each shown to encode a protein of 74 kDa with 33 % identity and 58 % similarity to the drosophila white gene [19,147,153]. The ABCG1 gene has been mapped to chromosome 21q22.3 in the human genome and to the syntenic region in chromosome 17 in mouse [19]. It is composed of 23 exons and has multiple transcripts, differing in fact only in the use of alternative promoters [154]. These transcripts are predicted to give rise to proteins that differ at their N termini. Recent evidence, however, indicates that a predominant transcript exists in both man and mouse [155].

5.4.1.1 **Expression and its Regulation**

mRNA analysis and *in situ* hybridization demonstrated that ABCG1 mRNAs are expressed in numerous murine and human tissues, with the highest levels in spleen, brain, and lung, and in general in tissues with high macrophage contents. Evidence was also provided quite early that its expression is under the control of the LXR pathway in a fashion superimposable with that of ABCA1 [154,156–158]. In addition to an LXR-mediated axis of activation, ABCG1 transcription can also be activated directly through Peroxisome proliferator activated receptor y (PPAR y) [159]. This LXR-independent activation involves exclusively ABCG1. Like ABCA1, ABCG1 is upregulated during the differentiation process of monocytes into mature macrophages and is strongly induced by foam cell conversion of these macrophages under sterol-loading conditions. In the liver, the expression is highest in Kupffer cells and lowest in parenchymal cells. The sterol-sensitive induction seen in myeloid cells appears independent of proinflammatory stimuli and the oxidative state of the cell after treatment with tumor necrosis factor or lipopolysaccharide has not impact on ABCG1 expression. Treatment with cAMP analogues of murine cell lines also strongly induced ABCA1 expression but was devoid of effect on ABCG1 [95].

The localization of the ABCG1 protein is predominantly intracellular in macrophages, though a precise identification of the compartment is lacking, while in transfected cells ABCG1 has been found at the plasma membrane [160]. It has been reported that LXR agonists stimulate ABCG1 traffic to the membrane, thus performing the dual task of increasing both its transcription and its availability for cellular effluxes [161]. No formal evidence of heterodimerization with similar transporters has been provided so far and it is accepted that homodimers have to form to achieve function. This is mostly based on transfection-based studies in which the exclusive presence of ABCG1 was able to induce function [160].

5.4.1.2 Physiopathology and Functional Predictions

Studies on knockout animals have demonstrated no effect on plasma lipids, but a massive accumulation of lipids in macrophages in different tissues under a high-fat diet [162]. In spite of the wide expression of ABCG1, only selected tissues, such as lung, display this pronounced phenotype. Currently it is not known what governs the reported selectivity.

Studies both in overexpressing cells and in macrophages issued from knockout animals have contributed to the definition of ABCG1 function as a promoter of cholesterol effluxes to already formed HDL or composite lipidated particles [160,161]. In no instances has any ABCG1 activity on the effluxes of phospholipids been reported.

The activity in controlling cholesterol effluxes from macrophages predicts that ABCG1 should be atheroprotective, but so far the results obtained from transplantations of ABCG1$^{-/-}$ bone marrow cells in atherogenic models indicate only mildly increased or even decreased incidences of atherosclerotic lesions [163–165]. This unexpected *in vivo* result may be due to the beneficial effect of the enhanced ABCA1 expression overriding the inactivation of ABCG1 function.

On the basis of *in vitro* reconstitution systems suggesting a specific cholesterol-related efflux induced by ABCG1, a model has been proposed [160,166]. This, represented schematically in Fig. 5.5, envisions that ABCA1 and ABCG1 work in succession to promote the generation of fully mature cholesterol-rich HDL particles. Only plasma membrane ABCA1 can orchestrate the lipidation of nascent apoproteins and the initial formation of mature HDL. These ABCA1-primed particles may well serve as acceptors for the ABCG1-mediated cholesterol effluxes. Whether ABCG1 serves primarily as an enhancement of the LXR-controlled ABCA1-mediated pathway in overall body turnover of cholesterol or acts independently to promote cell surface redistribution of intracellular cholesterol is at this point unclear. Only sophisticated tools enabling the *in vivo* dissociation of the ABCA1- and the ABCG1-mediated effects would allow further understanding of the complex mechanisms collectively called "cellular cholesterol homeostasis".

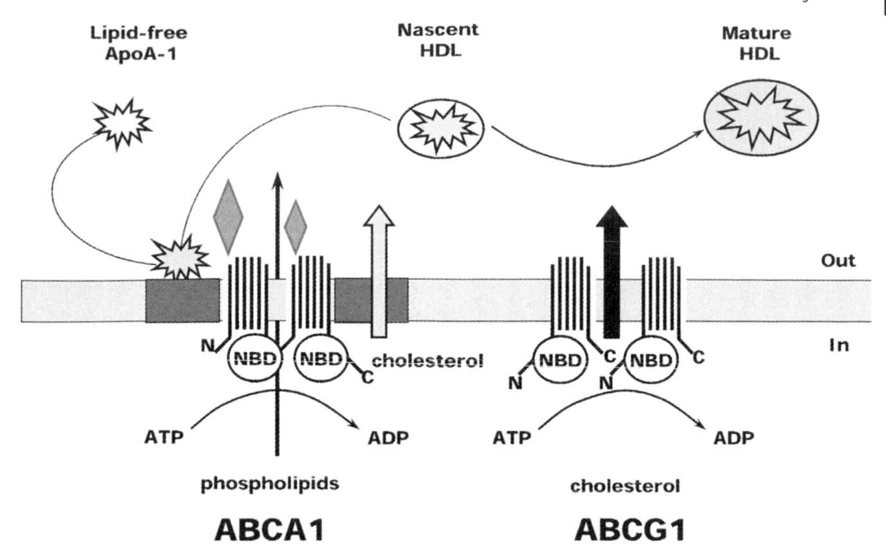

Fig. 5.5 *Schematic representation of ABCA1 and ABCG1 cooperation in the biogenesis of mature cholesterol-rich HDL particles.* The ABCA1 modification of lipid architecture at the membrane is indicated by the dark gray rectangles. It allows docking of lipid-free apoproteins and their ABCA1-dependent maturation. Successive loading with phospholipids and secondarily with cholesterol takes place. The details of the interaction between ABCA1 and ApoA-I are yet unclear. HDL formed upon the influence of ABCA1 can undergo further enrichment in cholesterol by an ABCG1-mediated pathway.

5.5 Acknowledgments

The authors wish to thank the members of the laboratory and all investigators working in the field whose work could not be quoted for reasons of space. The laboratory at the CIML is funded by institutional grants and specific grants from the European Community (FLIPPASERTN and MPCM STREP).

References

1 Dean, M. and Annilo, T. (2005) Evolution of the ATP-binding cassette (ABC) transporter superfamily in vertebrates. *Annu Rev Genomics Hum Genet.*, **6**, 123–142.

2 Higgins, C. F., Hiles, I. D., Salmond, G. P. C., Gill, D. R., Downie, J. A. *et al.* (1986) A family of related ATP-binding subunits coupled to many distinct biological processes in bacteria. *Nature*, **323**, 448–450.

3 Higgins, C. F. (1992) ABC transporters: from microorganisms

to man. *Annu.Rev.Cell Biol.*, **8**, 67–113.

4 Leonard, G. D., Fojo, T., Bates, S. E. (2003) The role of ABC transporters in clinical practice. *Oncologist*, **8**, 411–424.

5 Gottesman, M. M. and Ling, V. (2006) The molecular basis of multidrug resistance in cancer: the early years of P-glycoprotein research. *FEBS Lett*, **580**, 998–1009.

6 Luciani, M. F., Denizot, F., Savary, S., Mattei, M. G., Chimini, G. (1994) Cloning of two novel ABC transporters mapping on human chromosome 9. *Genomics*, **21**, 150–159.

7 Theodoulou, F. L. (2000) Plant ABC transporters. *Biochim Biophys Acta*, **1465**, 79–103.

8 Decottignies, A. and Goffeau, A. (1997) Complete inventory of the yeast ABC proteins. *Nature Genet.*, **15**, 137–145.

9 Annilo, T., Chen, Z. Q., Shulenin, S., Dean, M. (2003) Evolutionary analysis of a cluster of ATP-binding cassette (ABC) genes. *Mamm Genome*, **14**, 7–20.

10 Ames, G.F.-L. (1986) Bacterial periplasmic transport system: structure, mechanism and evolution. *Annu. Rev. Biochem.*, **55**, 377–425.

11 Biemans-Oldehinkel, E., Doeven, M. K., Poolman, B. (2006) ABC transporter architecture and regulatory roles of accessory domains. *FEBS Lett*, **580**, 1023–1035.

12 Peelman, F., Labeur, C., Vanloo, B., Roosbeek, S., Devaud, C. *et al.* (2003) Characterization of the ABCA transporter subfamily: identification of prokaryotic and eukaryotic members, phylogeny and topology. *J Mol Biol*, **325**, 259–274.

13 Biemans-Oldehinkel, E. and Poolman, B. (2003) On the role of the two extracytoplasmic substrate-binding domains in the ABC transporter OpuA. *Embo J*, **22**, 5983–5993.

14 Koch, J., Guntrum, R., Heintke, S., Kyritsis, C., Tampe, R. (2004) Functional dissection of the transmembrane domains of the transporter associated with antigen processing (TAP). *J Biol Chem*, **279**, 10142–10147.

15 Westlake, C. J., Cole, S. P., Deeley, R. G. (2005) Role of the NH2-terminal membrane spanning domain of multidrug resistance protein 1/ABCC1 in protein processing and trafficking. *Mol Biol Cell*, **16**, 2483–2492.

16 Chan, K. W., Zhang, H., Logothetis, D. E. (2003) N-terminal transmembrane domain of the SUR controls trafficking and gating of Kir6 channel subunits. *Embo J*, **22**, 3833–3843.

17 Rich, D. P., Gregory, R. J., Anderson, M. P., Manavalan, P., Smith, A. E. *et al.* (1991) Effect of deleting the R domain on CFTR-generated chloride channels. *Science*, **253**, 205–207.

18 Ewart, G. D. and Howells, A. J. (1998) ABC transporters involved in transport of eye pigment precursors in Drosophila melanogaster. *Methods Enzymol.*, **292**, 213–224.

19 Savary, S., Denizot, F., Luciani, M. F., Mattei, M. G., Chimini, G. (1996) Molecular cloning of a mammalian ABC transporter homologous to Drosophila white gene. *Mammalian Genome*, **7**, 673–676.

20 Dassa, E. and Bouige, P. (2001) The ABC of ABCS: a phylogenetic and functional classification of ABC systems in living organisms. *Res Microbiol*, **152**, 211–229.

21 Gottesman, M. M., Pastan, I., Ambudkar, S. V. (1996) P-glycoprotein and multidrug resistance. *Curr Opin Genet Dev*, **6**, 610–617.

22 Ambudkar, S. V., Kim, I. W., Xia, D., Sauna, Z. E. (2006) The A-loop, a novel conserved aromatic acid subdomain upstream of the Walker A motif in ABC transporters, is

critical for ATP binding. *FEBS Lett*, **580**, 1049–1055.

23 Hung, L. W., Wang, I., Nikaido, K., Liu, P. Q., Ferro-Luzzi Ames, G. *et al.* (1998) Crystal structure of the ATP- binding subunit of an ABC transporter. *Nature*, **396**, 703–707.

24 Chang, G. and Roth, C. B. (2001) Structure of MsbA from E. coli: a homolog of the multidrug resistance ATP binding cassette (ABC) transporters. *Science*, **293**, 1793–1800.

25 Chen, J., Lu, G., Lin, J., Davidson, A. L., Quiocho, F. A. (2003) A tweezers-like motion of the ATP-binding cassette dimer in an ABC transport cycle. *Mol Cell*, **12**, 651–661.

26 Locher, K. P., Lee, A. T., Rees, D. C. (2002) The *E. coli* BtuCD structure: a framework for ABC transporter architecture and mechanism. *Science*, **296**, 1091–1098.

27 Zaitseva, J., Jenewein, S., Wiedenmann, A., Benabdelhak, H., Holland, I. B. *et al.* (2005) Functional characterization and ATP-induced dimerization of the isolated ABC-domain of the haemolysin B transporter. *Biochemistry*, **44**, 9680–9690.

28 Callaghan, R., Ford, R. C., Kerr, I. D. (2006) The translocation mechanism of P-glycoprotein. *FEBS Lett*, **580**, 1056–1063.

29 Trompier, D., Alibert, M., Davanture, S., Hamon, Y., Pierres, M. *et al.* (2006) Transition from dimers to higher oligomeric forms occurs during the ATPase cycle of the ABCA1 transporter. *J Biol Chem*, **281**, 20283–20290.

30 Pohl, E. E., Peterson, U., Sun, J., Pohl, P. (2000) Changes of intrinsic membrane potentials induced by flip-flop of long-chain fatty acids. *Biochemistry*, **39**, 1834–1839.

31 Steck, T. L., Ye, J., Lange, Y. (2002) Probing red cell membrane cholesterol movement with cyclodextrin. *Biophys J*, **83**, 2118–2125.

32 Hamilton, J. A. (2003) Fast flip-flop of cholesterol and fatty acids in membranes: implications for membrane transport proteins. *Curr Opin Lipidol*, **14**, 263–271.

33 Bai, J. and Pagano, R. E. (1997) Measurement of spontaneous transfer and transbilayer movement of BODIPY-labeled lipids in lipid vesicles. *Biochemistry*, **36**, 8840–8848.

34 Sampaio, J. L., Moreno, M. J., Vaz, W. L. (2005) Kinetics and thermodynamics of association of a fluorescent lysophospholipid derivative with lipid bilayers in liquid-ordered and liquid-disordered phases. *Biophys J*, **88**, 4064–4071.

35 Abreu, M. S., Moreno, M. J., Vaz, W. L. (2004) Kinetics and thermodynamics of association of a phospholipid derivative with lipid bilayers in liquid-disordered and liquid-ordered phases. *Biophys J*, **87**, 353–365.

36 Phillips, M. C., Johnson, W. J., Rothblat, G. H. (1987) Mechanisms and consequences of cellular cholesterol exchange and transfer. *Biochim Biophys Acta*, **906**, 223–276.

37 Jones, J. D. and Thompson, T. E. (1990) Mechanism of spontaneous, concentration-dependent phospholipid transfer between bilayers. *Biochemistry*, **29**, 1593–1600.

38 Smit, J. J. M., Schinkel, A. H., Oude Elferink, R. P. J., Groen, A. K., Wagenaar, E. *et al.* (1993) Homozygous disruption of the murine mdr2 P-glycoprotein gene leads to a complete absence of phospholipid from bile and to liver disease. *Cell*, **75**, 451–462.

39 Mosser, J., Douar, A.-M., Sarde, C. O., Kioschis, P., Feil, R. (1993) *et al.* Putative X-linked adrenoleucodystrophy gene shares unexpected homology with ABC transporters. *Nature*, **361**, 726–730.

40 Theodoulou, F. L., Holdsworth, M., Baker, A. (2006) Peroxisomal ABC transporters. *FEBS Lett*, **580**, 1139–1155.

41 Shulenin, S., Nogee, L. M., Annilo, T., Wert, S. E., Whitsett, J. A. *et al.* (2004) ABCA3 gene mutations in newborns with fatal surfactant deficiency. *N Engl J Med*, **350**, 1296–1303.
42 Uitto, J. (2005) The gene family of ABC transporters--novel mutations, new phenotypes. *Trends Mol Med*, **11**, 341–343.
43 Kelsell, D. P., Norgett, E. E., Unsworth, H., Teh, M. T., Cullup, T. *et al.* (2005) Mutations in ABCA12 underlie the severe congenital skin disease harlequin ichthyosis. *Am J Hum Genet*, **76**, 794–803.
44 Kusuhara, H., and Sugiyama, Y. (2007) ATP-binding cassette, subfamily G (ABCG family). Pflugers Arch. **453**, 735–744.
45 Schmitz, G., Langmann, T., Heimerl, S. (2001) Role of ABCG1 and other ABCG family members in lipid metabolism. *J Lipid Res*, **42**, 1513–1520.
46 Kaminski, W. E., Piehler, A., Wenzel, J. J. (2006) ABC A-subfamily transporters: structure, function and disease. *Biochim Biophys Acta*, **1762**, 510–524.
47 Annilo, T., Shulenin, S., Chen, Z. Q., Arnould, I., Prades, C. *et al.* (2002) Identification and characterization of a novel ABCA subfamily member, ABCA12, located in the lamellar ichthyosis region on 2q34. *Cytogenet Genome Res*, **98**, 169–176.
48 Prades, C., Arnould, I., Annilo, T., Shulenin, S., Chen, Z. Q. *et al.* (2002) The human ATP binding cassette gene ABCA13, located on chromosome 7p12.3, encodes a 5058 amino acid protein with an extracellular domain encoded in part by a 4.8-kb conserved exon. *Cytogenet Genome Res*, **98**, 160–168.
49 Chen, Z. Q., Annilo, T., Shulenin, S., Dean, M. (2004) Three ATP-binding cassette transporter genes, Abca14, Abca15, and Abca16, form a cluster on mouse Chromosome 7 F3. *Mamm Genome*, **15**, 335–343.
50 Piehler, A. P., Wenzel, J. J., Olstad, O. K., Haug, K. B., Kierulf, P. *et al.* (2006) The human ortholog of the rodent testis-specific ABC transporter Abca17 is a ubiquitously expressed pseudogene (ABCA17P) and shares a common 5′ end with AB CA3. *BMC Mol Biol*, **7**, 28.
51 Fitzgerald, M. L., Mendez, A. J., Moore, K. J., Andersson, L. P., Panjeton, H. A. *et al.* (2001) ABCA1 contains an N-terminal signal-anchor sequence that translocates the protein's first hydrophilic domain to the exoplasmic space. *J Biol Chem*, **276**, 15137–15145.
52 Fitzgerald, M. L., Morris, A. L., Rhee, J. S., Andersson, L. P., Mendez, A. J., *et al.* (2002) Naturally Occurring Mutations in the Largest Extracellular Loops of ABCA1 Can Disrupt Its Direct Interaction with Apolipoprotein A-I. *J Biol Chem*, **277**, 33178–33187.
53 Bungert, S., Molday, L. L., Molday, R. S. (2001) Membrane Topology of the ATP Binding Cassette Transporter ABCR and Its Relationship to ABC1 and Related ABCA Transporters. *J Biol Chem*, **276**, 23539–23546.
54 Rigot, V., Hamon, Y., Chambenoit, O., Alibert, M., Duverger, N. *et al.* (2002) Distinct sites on ABCA1 control distinct steps required for cellular release of phospholipids. *J Lipid Res*, **43**, 2077–2086.
55 Tusnady, G. E., Sarkadi, B., Simon, I., Varadi, A. (2006) Membrane topology of human ABC proteins. *FEBS Lett*, **580**, 1017–1022.
56 Dean, M., Hamon, Y., Chimini, G. (2001) The human ATP-binding cassette (ABC) transporter superfamily. *J Lipid Res*, **42**, 1007–1017.
57 Rust, S., Rosier, M., Funke, H., Real, J., Amoura, Z. *et al.* (1999) Tangier disease is caused by mutations in the gene encoding ATP-binding cassette transporter 1. *Nature Genet.*, **22**, 352–355.

58 Brooks-Wilson, A., Marcil, M., Clee, S. M., Zhang, L. H., Roomp, K., *et al.* (1999) Mutations in ABC1 in Tangier disease and familial high-density lipoprotein deficiency. *Nature Genet.*, **22**, 336–345.

59 Bodzioch, M., Orso, E., Klucken, J., Langmann, T., Böttcher, A. *et al.* (1999) The gene encoding ATP-binding cassette transporter 1 is mutated in Tangier disease. *Nature Genet.*, **22**, 347–351.

60 Assmann, G., vonEckardstein, A., Brewer, H. B. (2001) Familial Analphalipoproteinemia: Tangier Disease. Scriver, C. R.Beaudet, A. L.Sly, W. S.ValleD. *In The Metabolic Basis of Inherited Disease*, 8th edn McGraw-Hill, New York, 2937–2960.

61 Beharry, S., Zhong, M., Molday, R. S. (2004) N-retinylidene-phosphatidylethanolamine is the preferred retinoid substrate for the photoreceptor-specific ABC transporter ABCA4 (ABCR). *J Biol Chem*, **279**, 53972–53979.

62 Sun, H., Smallwood, P. M., Nathans, J. (2000) Biochemical defects in ABCR protein variants associated with human retinopathies. *Nat Genet*, **26**, 242–246.

63 Aldred, M. A. (2000) The ABCR of visual impairment. *Mol Med Today*, **6**, 417.

64 Shroyer, N. F., Lewis, R. A., Lupski, J. R. (2000) Complex inheritance of ABCR mutations in Stargardt disease: linkage disequilibrium, complex alleles, and pseudodominance. *Hum Genet*, **106**, 244–248.

65 Yamano, G., Funahashi, H., Kawanami, O., Zhao, L. X., Ban, N. *et al.* (2001) ABCA3 is a lamellar body membrane protein in human lung alveolar type II cells. *FEBS Lett*, **508**, 221–225.

66 Fitzgerald, M. L., Xavier, R., Haley, K. J., Welti, R., Goss, J. L. *et al.* (2007) ABCA3 inactivation in mice causes respiratory failure, loss of pulmonary surfactant and depletion of lung phosphatidylglycerol. J Lipid Res. **48**, 621–632.

67 Bullard, J. E., Wert, S. E., Nogee, L. M. (2006) ABCA3 Deficiency: Neonatal Respiratory Failure and Interstitial Lung Disease. *Semin Perinatol*, **30**, 327–334.

68 Lefevre, C., Audebert, S., Jobard, F., Bouadjar, B., Lakhdar, H. *et al.* (2003) Mutations in the transporter ABCA12 are associated with lamellar ichthyosis type 2. *Hum Mol Genet*, **12**, 2369–2378.

69 Akiyama, M., Sugiyama-Nakagiri, Y., Sakai, K., McMillan, J. R., Goto, M. *et al.* (2005) Mutations in lipid transporter ABCA12 in harlequin ichthyosis and functional recovery by corrective gene transfer. *J Clin Invest*, **115**, 1777–1784.

70 Hovnanian, A. (2005) Harlequin ichthyosis unmasked: a defect of lipid transport. *J Clin Invest*, **115**, 1708–1710.

71 Langmann, T., Klucken, J., Reil, M., Liebisch, G., Luciani, M. F. *et al.* (1999) Molecular Cloning of the Human ATP-Binding Cassette Transporter 1 (hABC1): Evidence for Sterol-Dependent Regulation in Macrophages. *Biochem. Biophys. Res. Commun.*, **257**, 29–33.

72 Santamarina-Fojo, S., Peterson, K., Knapper, C., Qiu, Y., Freeman, L. *et al.* (2000) Complete genomic sequence of the human ABCA1 gene: Analysis of the human and mouse ATP-binding cassette A promoter. *Proc Natl Acad Sci USA*, **97**, 7987–7992.

73 Luciani, M. F. and Chimini, G. (1996) The ATP binding cassette transporter ABC1, is required for the engulfment of corpses generated by apoptoic cell death. *EMBO J.*, **15**, 226–235.

74 Combes, V., Coltel, N., Alibert, M., van Eck, M., Raymond, C. *et al.* (2005) ABCA1 gene deletion protects against cerebral malaria:

potential pathogenic role of microparticles in neuropathology. *Am J Pathol*, **166**, 295–302.

75 Koldamova, R., Staufenbiel, M., Lefterov, I. (2005) Lack of ABCA1 considerably decreases brain ApoE level and increases amyloid deposition in APP23 mice. *J Biol Chem*, **280**, 43224–43235.

76 Wahrle, S. E., Jiang, H., Parsadanian, M., Hartman, R. E., Bales, K. R. *et al.* (2005) Deletion of Abca1 increases Abeta deposition in the PDAPP transgenic mouse model of Alzheimer disease. *J Biol Chem*, **280**, 43236–43242.

77 Hirsch-Reinshagen, V., Maia, L. F., Burgess, B. L., Blain, J. F., Naus, K. E. *et al.* (2005) The absence of ABCA1 decreases soluble ApoE levels but does not diminish amyloid deposition in two murine models of Alzheimer disease. *J Biol Chem*, **280**, 43243–43256.

78 Trompier, D., and Chimini, G. (2005) ABCA1 AfCS/ Nature-Molecular pages doi 0.1038/ mp.a002593.01

79 Schmitz, G. and Langmann, T. (2005) Transcriptional regulatory networks in lipid metabolism control ABCA1 expression. *Biochim Biophys Acta*, **1735**, 1–19.

80 Fitzgerald, M. L., Moore, K. J., Freeman, M. W. (2002) Nuclear hormone receptors and cholesterol trafficking: the orphans find a new home. *J Mol Med*, **80**, 271–281.

81 Li, A. C. and Glass, C. K. (2004) PPAR- and LXR-dependent pathways controlling lipid metabolism and the development of atherosclerosis. *J Lipid Res*, **45**, 2161–2173.

82 Li, A. C. and Glass, C. K. (2002) The macrophage foam cell as a target for therapeutic intervention. *Nat Med*, **8**, 1235–1242.

83 Chinetti, G., Lestavel, S., Bocher, V., Remaley, A. T., Neve, B. *et al.* (2001) PPAR-alpha and PPAR-gamma activators induce cholesterol removal from human macrophage foam cells through stimulation of the ABCA1 pathway. *Nat Med*, **7**, 53–58.

84 Chawla, A., Boisvert, W. A., Lee, C. H., Laffitte, B. A., Barak, Y. *et al.* (2001) A PPAR gamma-LXR-ABCA1 pathway in macrophages is involved in cholesterol efflux and atherogenesis. *Mol Cell*, **7**, 161–171.

85 Chinetti-Gbaguidi, G., Rigamonti, E., Helin, L., Mutka, A. L., Lepore, M., *et al.* (2005) Peroxisome proliferator-activated receptor alpha controls cellular cholesterol trafficking in macrophages. *J Lipid Res*, **46**, 2717–2725.

86 Lusis, A. J. (2000) Atherosclerosis. *Nature*, **407**, 233–241.

87 Panousis, C. G. and Zuckerman, S. H. (2000) Interferon-gamma induces downregulation of Tangier disease gene (ATP-binding-cassette transporter 1) in macrophage-derived foam cells. *Arterioscler Thromb Vasc Biol*, **20**, 1565–1571.

88 Panousis, C. G., Evans, G., Zuckerman, S. H. (2001) TGF-beta increases cholesterol efflux and ABC-1 expression in macrophage-derived foam cells: opposing the effects of IFN-gamma. *J Lipid Res*, **42**, 856–863.

89 Castrillo, A., Joseph, S. B., Vaidya, S. A., Haberland, M., Fogelman, A. M. *et al.* (2003) Crosstalk between LXR and toll-like receptor signaling mediates bacterial and viral antagonism of cholesterol metabolism. *Mol Cell*, **12**, 805–816.

90 Joseph, S. B., Bradley, M. N., Castrillo, A., Bruhn, K. W., Mak, P. A., *et al.* (2004) LXR-dependent gene expression is important for macrophage survival and the innate immune response. *Cell*, **119**, 299–309.

91 Joseph, S. B., Castrillo, A., Laffitte, B. A., Mangelsdorf, D. J., Tontonoz, P. (2003) Reciprocal regulation of inflammation and lipid metabolism by liver X receptors. *Nat Med*, **9**, 213–219.

92 Valledor, A. F. (2005) The innate immune response under the

control of the LXR pathway. *Immunobiology*, **210**, 127–132.

93 Freeman, M. W. and Moore, K. J. (2003) eLiXiRs for restraining inflammation. *Nat Med*, **9**, 168–169.

94 Wellington, C. L., Walker, E. K., Suarez, A., Kwok, A., Bissada, N. *et al.* (2002) ABCA1 mRNA and protein distribution patterns predict multiple different roles and levels of regulation. *Lab Invest*, **82**, 273–283.

95 Oram, J. F., Lawn, R. M., Garvin, M. R., Wade, D. P. (2000) ABCA1 is the cAMP-inducible apolipoprotein receptor that mediates cholesterol secretion from macrophages. *J Biol Chem*, **275**, 34508–34511.

96 Wang, Y. and Oram, J. F. (2002) Unsaturated fatty acids inhibit cholesterol efflux from macrophages by increasing degradation of ATP-binding cassette transporter A1. *J Biol Chem*, **277**, 5692–5697.

97 Oram, J. F. and Vaughan, A. M. (2006) ATP-Binding cassette cholesterol transporters and cardiovascular disease. *Circ Res*, **99**, 1031–1043.

98 Wang, N., Chen, W., Linsel-Nitschke, P., Martinez, L. O., Agerholm-Larsen, B. *et al.* (2003) A PEST sequence in ABCA1 regulates degradation by calpain protease and stabilization of ABCA1 by apoA-I. *J Clin Invest*, **111**, 99–107.

99 Wang, C. S., Alaupovic, P., Gregg, R. E., Brewer, H. B., Jr. (1987) Studies on the mechanism of hypertriglyceridemia in Tangier disease. Determination of plasma lipolytic activities, k1 values and apolipoprotein composition of the major lipoprotein density classes. *Biochim Biophys Acta*, **920**, 9–19.

100 Takahashi, K., Kimura, Y., Kioka, N., Matsuo, M., Ueda, K. (2006) Purification and ATPase Activity of Human AB CA1. *J Biol Chem*, **281**, 10760–10768.

101 Hamon, Y., Broccardo, C., Chambenoit, O., Luciani, M. F., Toti, F. *et al.* (2000) ABC1 promotes engulfment of apoptotic cells and transbilayer redistribution of phosphatidylserine. *Nat Cell Biol*, **2**, 399–406.

102 Boadu, E. and Francis, G. A. (2006) The role of vesicular transport in ABCA1-dependent lipid efflux and its connection with NPC pathways. *J Mol Med*, **84**, 266–275.

103 Takahashi, Y. and Smith, J. D. (1999) Cholesterol efflux to apolipoprotein AI involves endocytosis and resecretion in a calcium-dependent pathway [see comments]. *Proc Natl Acad Sci USA*, **96**, 11358–11363.

104 Singaraja, R. R., Visscher, H., James, E. R., Chroni, A., Coutinho, J. M. *et al.* (2006) Specific mutations in ABCA1 have discrete effects on ABCA1 function and lipid phenotypes both in vivo and in vitro. *Circ Res*, **99**, 389–397.

105 Attie, A. D., Hamon, Y., Brooks-Wilson, A. R., Gray-Keller, M. P., MacDonald, M. L. *et al.* (2002) Identification and functional analysis of a naturally occurring E89K mutation in the ABCA1 gene of the WHAM chicken. *J Lipid Res*, **43**, 1610–1617.

106 Singaraja, R. R., Brunham, L. R., Visscher, H., Kastelein, J. J., Hayden, M. R. (2003) Efflux and atherosclerosis: the clinical and biochemical impact of variations in the ABCA1 gene. *Arterioscler Thromb Vasc Biol*, **23**, 1322–1332.

107 Schillers, H., Shahin, V., Albermann, L., Schafer, C., Oberleithner, H. (2004) Imaging CFTR: a tail to tail dimer with a central pore. *Cell Physiol Biochem*, **14**, 1–10.

108 Ramjeesingh, M., Kidd, J. F., Huan, L. J., Wang, Y., Bear, C. E. (2003) Dimeric cystic fibrosis transmembrane conductance regulator exists in the plasma membrane. *Biochem J*, **374**, 793–797.

109 Xu, J., Liu, Y., Yang, Y., Bates, S., Zhang, J. T. (2004) Characterization of oligomeric human half-ABC transporter ATP-binding cassette G2. *J Biol Chem*, **279**, 19781–19789.

110 Denis, M., Haidar, B., Marcil, M., Bouvier, M., Krimbou, L., *et al.* (2004) Characterization of oligomeric human ATP binding cassette transporter A1.Potential implications for determining the structure of nascent high density lipoprotein particles. *J Biol Chem*, **279**, 41529–41536.

111 McNeish, J., Aiello, R. J., Guyot, D., Turi, T., Gabel, C. *et al.* (2000) High density lipoprotein deficiency and foam cell accumulation in mice with targeted disruption of ATP-binding cassette transporter-1. *Proc Natl Acad Sci USA*, **97**, 4245–4250.

112 Orso, E., Broccardo, C., Kaminski, W. E., Bottcher, A., Liebisch, G. *et al.* (2000) Transport of lipids from golgi to plasma membrane is defective in tangier disease patients and Abc1-deficient mice. *Nat Genet*, **24**, 192–196.

113 Singaraja, R. R., vanEck, M., Bissada, N., Zimetti, F., Collins, H. L. *et al.* (2006) Both hepatic and extrahepatic ABCA1 have discrete and essential functions in the maintenance of plasma high-density lipoprotein cholesterol levels in vivo. *Circulation*, **114**, 1301–1309.

114 Singaraja, R. R., Stahmer, B., Brundert, M., Merkel, M., Heeren, J. *et al.* (2006) Hepatic ATP-binding cassette transporter A1 is a key molecule in high-density lipoprotein cholesteryl ester metabolism in mice. *Arterioscler Thromb Vasc Biol*, **26**, 1821–1827.

115 Brunham, L. R., Kruit, J. K., Iqbal, J., Fievet, C., Timmins, J. M. *et al.* (2006) Intestinal ABCA1 directly contributes to HDL biogenesis *in vivo*. *J Clin Invest*, **116**, *1052–1062.*

116 Timmins, J. M., Lee, J. Y., Boudyguina, E., Kluckman, K. D., Brunham, L. R. *et al.* (2005) Targeted inactivation of hepatic Abca1 causes profound hypoalphalipoproteinemia and kidney hypercatabolism of apoA-I. *J Clin Invest*, **115**, 1333–1342.

117 Haghpassand, M., Bourassa, P. A., Francone, O. L., Aiello, R. J. (2001) Monocyte/macrophage expression of ABCA1 has minimal contribution to plasma HDL levels. *J Clin Invest*, **108**, 1315–1320.

118 Lee, J. Y. and Parks, J. S. (2005) ATP-binding cassette transporter AI and its role in HDL formation. *Curr Opin Lipidol*, **16**, 19–25.

119 Cohen, J. C., Kiss, R. S., Pertsemlidis, A., Marcel, Y. L., McPherson, R. *et al.* (2004) Multiple rare alleles contribute to low plasma levels of HDL cholesterol. *Science*, **305**, 869–872.

120 Frikke-Schmidt, R., Nordestgaard, B. G., Jensen, G. B., Tybjaerg-Hansen, A. (2004) Genetic variation in ABC transporter A1 contributes to HDL cholesterol in the general population. *J Clin Invest*, **114**, 1343–1353.

121 Clee, S. M., Zwinderman, A. H., Engert, J. C., Zwarts, K. Y., Molhuizen, H. O. *et al.* (2001) Common genetic variation in ABCA1 is associated with altered lipoprotein levels and a modified risk for coronary artery disease. *Circulation*, **103**, 1198–1205.

122 Chambenoit, O., Hamon, Y., Marguet, D., Rigneault, H., Rosseneu, M. *et al.* (2001) Specific docking of apolipoprotein A-I at the cell surface requires a functional ABCA1 transporter. *J Biol Chem*, **276**, 9955–9960.

123 Remaley, A. T., Stonik, J. A., Demosky, S. J., Neufeld, E. B., Bocharov, A. V. *et al.* (2001) Apolipoprotein specificity for lipid efflux by the human ABCAI transporter. *Biochem Biophys Res Commun*, **280**, 818–823.

124 Kiss, R. S., Maric, J., Marcel, Y. L. (2005) Lipid efflux in human and mouse macrophagic cells: evidence for differential regulation of phospholipid and cholesterol efflux. *J Lipid Res*, **46**, 1877–1887.

125 Wang, N., Silver, D. L., Thiele, C., Tall, A. R. (2001) ABCA1 functions as a cholesterol efflux regulatory protein. *J Biol Chem*, **276**, 23742–23747.

126 Alder-Baerens, N., Muller, P., Pohl, A., Korte, T., Hamon, Y. *et al.* (2005) Headgroup-specific exposure of phospholipids in ABCA1-expressing cells. *J Biol Chem*, **280**, 26321–26329.

127 Koseki, M., Hirano, K. I., Masuda, D., Ikegami, C., Tanaka, M. *et al.* (2007) Increased lipid rafts and accelerated lipopolysaccharide-induced tumor necrosis factor alpha secretion in Abca1-deficient macrophages. *J Lipid Res.*, **48**, 299–306.

128 Landry, Y. D., Denis, M., Nandi, S., Bell, S., Vaughan, A. M. *et al.* (2006) ATP-binding cassette transporter A1 expression disrupts raft membrane microdomains through its ATPase-related functions. *J Biol Chem*, **281**, 36091–36101.

129 Chau, P., Nakamura, Y., Fielding, C. J., Fielding, P. E. (2006) Mechanism of prebeta-HDL formation and activation. *Biochemistry*, **45**, 3981–3987.

130 Vaughan, A. M. and Oram, J. F. (2003) ABCA1 redistributes membrane cholesterol independent of apolipoprotein interactions. *J Lipid Res.*, **44**, 1373–1380.

131 vanMeer, G., Halter, D., Sprong, H., Somerharju, P., Egmond, M. R. (2006) ABC lipid transporters: Extruders, flippases, or flopless activators?. *FEBS Lett*, **580**, 1171–1177.

132 Trompier, D. and Chimini, G. (2005) ABCA7 AfCS-Nature Molecular Page doi doi:10.1038/mp.a002599.01.

133 Kaminski, W. E., Orso, E., Diederich, W., Klucken, J., Drobnik, W. *et al.* (2000) Identification of a novel human sterol-sensitive ATP-binding cassette transporter (ABCA7). *Biochem Biophys Res Commun*, **273**, 532–538.

134 Broccardo, C., Osorio, J., Luciani, M. F., Schriml, L. M., Prades, C. *et al.* (2001) Comparative analysis of the promoter structure and genomic organization of the human and mouse ABCA7 gene encoding a novel ABCA transporter. *Cytogenet Cell Genet*, **92**, 264–270.

135 Sasaki, M., Shoji, A., Kubo, Y., Nada, S., Yamaguchi, A. (2003) Cloning of rat ABCA7 and its preferential expression in platelets. *Biochem Biophys Res Commun*, **304**, 777–782.

136 Ikeda, Y., Abe-Dohmae, S., Munehira, Y., Aoki, R., Kawamoto, S. *et al.* (2003) Posttranscriptional regulation of human ABCA7 and its function for the apoA-I-dependent lipid release. *Biochem Biophys Res Commun*, **311**, 313–318.

137 Kim, W. S., Fitzgerald, M. L., Kang, K., Okuhira, K., Bell, S. A. *et al.* (2005) Abca7 null mice retain normal macrophage phosphatidylcholine and cholesterol efflux activity despite alterations in adipose mass and serum cholesterol levels. *J Biol Chem*, **280**, 3989–3995.

138 Iwamoto, N., Abe-Dohmae, S., Sato, R., Yokoyama, S. (2006) ABCA7 expression is regulated by cellular cholesterol through the SREBP2 pathway and associated with phagocytosis. *J Lipid Res*, **47**, 1915–1927.

139 Linsel-Nitschke, P., Jehle, A. W., Shan, J., Cao, G., Bacic, D. *et al.* (2005) Potential role of ABCA7 in cellular lipid efflux to apoA-I. *J Lipid Res*, **46**, 86–92.

140 Jehle, A. W., Gardai, S. J., Li, S., Linsel-Nitschke, P., Morimoto, K. *et al.* (2006) ATP-binding cassette transporter A7 enhances phagocytosis of apoptotic cells and associated ERK signaling in macrophages. *J Cell Biol*, **174**, 547–556.

141 Abe-Dohmae, S., Ikeda, Y., Matsuo, M., Hayashi, M., Okuhira, K. *et al.*

(2004) Human ABCA7 supports apolipoprotein-mediated release of cellular cholesterol and phospholipid to generate high density lipoprotein. *J Biol Chem*, **279**, 604–611.

142 Wang, N., Lan, D., Gerbod-Giannone, M., Linsel-Nitschke, P., Jehle, A. W. *et al.* (2003) ATP-binding cassette transporter A7 (ABCA7) binds apolipoprotein A-I and mediates cellular phospholipid but not cholesterol efflux. *J Biol Chem*, **278**, 42906–42912.

143 Abe-Dohmae, S., Ueda, K., Yokoyama, S. (2006) ABCA7, a molecule with unknown function. *FEBS Lett*, **580**, 1178–1182.

144 Hayashi, M., Abe-Dohmae, S., Okazaki, M., Ueda, K., Yokoyama, S. (2005) Heterogeneity of high density lipoprotein generated by ABCA1 and AB CA7. *J Lipid Res*, **46**, 1703–1711.

145 Morgan, T. H. (1910) Sex limited inheritance in Drosophila. *Science*, **32**, 120–122.

146 Dreesen, T. D., Johnson, D. H., Henikoff, S. (1988) The brown protein of Drosophila melanogaster is similar to the white protein and to components of active transport complexes. *Mol Cell Biol*, **8**, 5206–5215.

147 Croop, J. M., Tiller, G. E., Fletcher, J. A., Lux, M. L., Raab, E. *et al.* (1997) Isolation and characterization of a mammalian homolog of the Drosophila white gene. *Gene*, **185**, 77–85.

148 Doyle, L. A., Yang, W., Abruzzo, L. V., Krogmann, T., Gao, Y. *et al.* (1998) A multidrug resistance transporter from human MCF-7 breast cancer cells. *Proc Natl Acad Sci USA*, **95**, 15665–15670.

149 Berge, K. E., Tian, H., Graf, G. A., Yu, L., Grishin, N. V. *et al.* (2000) Accumulation of dietary cholesterol in sitosterolemia caused by mutations in adjacent ABC transporters. *Science*, **290**, 1771–1775.

150 Graf, G. A., Li, W. P., Gerard, R. D., Gelissen, I., White, A. *et al.* (2002) Coexpression of ATP-binding cassette proteins ABCG5 and ABCG8 permits their transport to the apical surface. *J Clin Invest*, **110**, 659–669.

151 Plosch, T., Bloks, V. W., Terasawa, Y., Berdy, S., Siegler, K. *et al.* (2004) Sitosterolemia in ABC-transporter G5-deficient mice is aggravated on activation of the liver-X receptor. *Gastroenterology*, **126**, 290–300.

152 Klett, E. L., Lu, K., Kosters, A., Vink, E., Lee, M. H. *et al.* (2004) A mouse model of sitosterolemia: absence of Abcg8/sterolin-2 results in failure to secrete biliary cholesterol. *BMC Med*, **2**, 5.

153 Chen, H., Rossier, C., Lalioti, M. D., Lynn, A., Chakravarti, A. *et al.* (1996) Cloning of the cDNA for a human homolog of the Drosophila white gene and mapping to chromosome 21q22.3. *Am J Hum Genet*, **59**, 66–75.

154 Kennedy, M. A., Venkateswaran, A., Tarr, P. T., Xenarios, I., Kudoh, J. *et al.* (2001) Characterization of the human ABCG1 gene: liver X receptor activates an internal promoter that produces a novel transcript encoding an alternative form of the protein. *J Biol Chem*, **276**, 39438–39447.

155 Nakamura, K., Kennedy, M. A., Baldan, A., Bojanic, D. D., Lyons, K. *et al.* (2004) Expression and regulation of multiple murine ATP-binding cassette transporter G1 mRNAs/isoforms that stimulate cellular cholesterol efflux to high density lipoprotein. *J Biol Chem*, **279**, 45980–45989.

156 Sabol, S. L., Brewer, H. B., Jr., Santamarina-Fojo, S. (2005) The human ABCG1 gene: identification of LXR response elements that modulate expression in macrophages and liver. *J Lipid Res*, **46**, 2151–2167.

157 Venkateswaran, A., Laffitte, B. A., Joseph, S. B., Mak, P. A., Wilpitz, D. C. *et al.* (2000) Control of cellular cholesterol efflux by the nuclear oxysterol receptor LXR alpha. *Proc Natl Acad Sci USA*, **97**, 12097–12102.

158 Venkateswaran, A., Repa, J. J., Lobaccaro, J. M., Bronson, A., Mangelsdorf, D. J. *et al.* (2000) Human white/murine ABC8 mRNA levels are highly induced in lipid-loaded macrophages. A transcriptional role for specific oxysterols. *J Biol Chem*, **275**, 14700–14707.

159 Li, A. C., Binder, C. J., Gutierrez, A., Brown, K. K., Plotkin, C. R. *et al.* (2004) Differential inhibition of macrophage foam-cell formation and atherosclerosis in mice by PPARalpha, beta/delta, and gamma. *J Clin Invest*, **114**, 1564–1576.

160 Jessup, W., Gelissen, I. C., Gaus, K., Kritharides, L. (2006) Roles of ATP binding cassette transporters A1 and G1, scavenger receptor BI and membrane lipid domains in cholesterol export from macrophages. *Curr Opin Lipidol*, **17**, 247–257.

161 Vaughan, A. M. and Oram, J. F. (2005) ABCG1 redistributes cell cholesterol to domains removable by high density lipoprotein but not by lipid-depleted apolipoproteins. *J Biol Chem*, **280**, 30150–30157.

162 Kennedy, M. A., Barrera, G. C., Nakamura, K., Baldan, A., Tarr, P. *et al.* (2005) ABCG1 has a critical role in mediating cholesterol efflux to HDL and preventing cellular lipid accumulation. *Cell Metab*, **1**, 121–131.

163 Baldan, A., Pei, L., Lee, R., Tarr, P., Tangirala, R. K. *et al.* (2006) Impaired development of atherosclerosis in hyperlipidemic Ldlr−/− and ApoE−/− mice transplanted with Abcg1−/− bone marrow. *Arterioscler Thromb Vasc Biol*, **26**, 2301–2307.

164 Ranalletta, M., Wang, N., Han, S., Yvan-Charvet, L., Welch, C. *et al.* (2006) Decreased atherosclerosis in low-density lipoprotein receptor knockout mice transplanted with Abcg1−/− bone marrow. *Arterioscler Thromb Vasc Biol*, **26**, 2308–2315.

165 Out, R., Hoekstra, M., Hildebrand, R. B., Kruit, J. K., Meurs, I. *et al.* (2006) Macrophage ABCG1 deletion disrupts lipid homeostasis in alveolar macrophages and moderately influences atherosclerotic lesion development in LDL receptor-deficient mice. *Arterioscler Thromb Vasc Biol*, **26**, 2295–2300.

166 Gelissen, I. C., Harris, M., Rye, K. A., Quinn, C., Brown, A. J. *et al.* (2005). ABCA1 and ABCG1 Synergize to Mediate Cholesterol Export to ApoA-I. *Arterioscler Thromb Vasc Biol*,

167 Lombard-Platet, G., Savary, S., Sarde, C. O., Mandel, J. L., Chimini, G. (1996) A close relative of the adrenoleukodystrophy (ALD) gene codes for a peroxisomal protein with a specific expression pattern. *Proc Natl Acad Sci USA*, **93**, 1265–1269.

168 Savary, S., Allikmets, R., Denizot, F., Luciani, M. F., Mattei, M. G. *et al.* (1997) Isolation and chromosomal mapping of a novel ATP-binding cassette transporter conserved in mouse and human. *Genomics*, **41**, 275–278.

169 Gartner, J., Moser, H., Valle, D. (1992) Mutations in the 70 K peroxisomal membrane protein gene in Zellweger syndrome. *Nature Genet.*, **1**, 16–22.

170 Allikmets, R., Gerrard, B., Hutchinson, A., Dean, M. (1996) Characterization of the human ABC superfamily: Isolation and mapping of 21 new genes using the expressed sequence tags database. *Hum Mol Genet*, **5**, 1649–1655.

171 Allikmets, R., Gerrard, B., Ravnic-Glavac, M., Jenkins, N. A., Gilbert, D. J. *et al.* (1995) Characterization and mapping of

three new mammalian ATP-binding transporter genes from EST database. *Mamm Genome,* **6**, 114–117.

172 Klugbauer, N. and Hofmann, F. (1996) Primary structure of a novel ABC transporter with a chromosomal localization on the band encoding the multidrug resistance-associated protein. *FEBS Lett.,* **391**, 61–65.

173 ArnouldI.,S. L., Prades C., Lachtermacher-TriunfolM., Schneider T., Maintoux C., Lemoine C., Debono D., Devaud C., Naudin L. *et al.* (2001) Identification and characterization of a cluster of five new ATP-binding cassette transporter genes on human chromosome 17q24: a new sub-group within the ABCA sub-family. *GeneScreen,* **1**, 157–164.

174 Kaminski, W. E., Piehler, A., Schmitz, G. (2000) Genomic organization of the human cholesterol-responsive ABC transporter ABC A7:tandem linkage with the minor histocompatibility antigen HA-1 gene. *Biochem Biophys Res Commun,* **278**, 782–789.

175 Ban, N., Sasaki, M., Sakai, H., Ueda, K., Inagaki, N. (2005) Cloning of ABCA17, a novel rodent sperm-specific ABC (ATP-binding cassette) transporter that regulates intracellular lipid metabolism. *Biochem J,* **389**, 577–585.

176 Mickley, L., Jain, P., Miyake, K., Schriml, L. M., Rao, K. *et al.* (2001) An ATP-binding cassette gene (ABCG3) closely related to the multidrug transporter ABCG2 (MXR/ABCP) has an unusual ATP-binding domain. *Mamm Genome,* **12**, 86–88.

177 Annilo, T., Tammur, J., Hutchinson, A., Rzhetsky, A., Dean, M. *et al.* (2001) Human and mouse orthologs of a new ATP-binding cassette gene, AB CG4. *Cytogenet Cell Genet,* **94**, 196–201.

178 Shulenin, S., Schriml, L. M., Remaley, A. T., Fojo, S., Brewer, B. *et al.* (2001) An ATP-binding cassette gene (ABCG5) from the ABCG (White) gene subfamily maps to human chromosome 2p21 in the region of the Sitosterolemia locus. *Cytogenet Cell Genet,* **92**, 204–208.

6
Reverse Cholesterol Transport – New Roles for Preβ_1-HDL and Lecithin:Cholesterol Acyltransferase

Christopher J. Fielding, Phoebe E. Fielding

6.1
Introduction

Reverse Cholesterol Transport (RCT) is the net transfer of free cholesterol (FC) from peripheral tissues via plasma high density lipoprotein (HDL). After esterification by lecithin:cholesterol acyltransferase (LCAT), the cholesteryl ester (CE) formed is cleared by the liver, hydrolyzed intracellularly to FC, and then catabolized to bile acid [1]. In rodents, CE in the RCT pathway is cleared as part of HDL, by the combined activities of scavenger receptor BI (SR-BI) (Chapter 7) and hepatic HDL holoreceptors (Chapter 12). In human plasma, much of the HDL-CE is first transferred to very low and low density lipoprotein (VLDL, LDL) by a cholesteryl ester transfer protein (CETP) (Chapter 7) and cleared by the LDL receptor pathway (Fig. 6.1) with apoB and/or apoE as ligand [2]. While plasma HDL total cholesterol levels (HDL-C) are an important predictor of cardiovascular health, they represent only one intermediate in a complex pathway, and do not directly measure or predict rates of RCT. For example, in cardiac transplant recipients, HDL-C is increased but RCT is inhibited [3]. In mice overexpressing SR-BI, HDL-C is decreased while RCT (measured as bile acid production) is stimulated [4].

Apolipoprotein A1 (apoA-1), the major protein of HDL (Chapter 1), plays a key role in RCT. In the absence of apoA-1, normal HDL is not formed [5]. ApoA-1 has an unusually low free energy of folding and unfolding [6], which helps it cycle between unfolded (lipid-rich) and contracted (lipid-poor) states during RCT. By recycling, each apoA-1 can carry multiple quanta of cholesterol from peripheral cells to the liver (Fig. 6.1). Other apolipoproteins may partially replace apoA-1 as lipid carriers when apoA-1 is reduced or absent, but in normal plasma successful RCT depends critically on the unique properties of apoA-1 in HDL [7].

The recycling role for apoA-1 was described over a decade ago [8] and there have been several excellent reviews since that time [9–12]. One important development has been a better understanding that major parts of the RCT pathway involve apoA-1 in lipid-poor complexes falling outside the classical density range of HDL ($1.063 < d < 1.21$ g mL^{-1}) [8,9,13]. Centrifugally isolated HDL can be a misleading surrogate for native HDL. Like a synthetic lipid vesicle, its transfers of FC are largely

High-Density Lipoproteins: From Basic Biology to Clinical Aspects. Edited by Christopher J. Fielding

ISBN: 978-3-527-31717-2

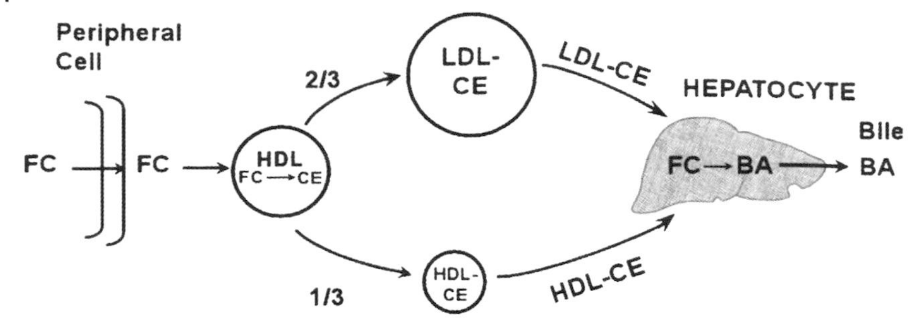

Fig. 6.1 Outline model of RCT, showing the major fluxes in human plasma that link peripheral cell membrane FC and its delivery, as CE, to the liver. FC: free cholesterol. CE: cholesteryl ester. BA: bile acids.

nonspecific and represent nonproductive exchange, or diffusion down a concentration gradient. Centrifuged HDL almost completely lacks the lipid-poor HDL fraction that is the most effective as an initial acceptor of FC for RCT [14]. In comparison with native HDL, the centrifuged isolate is enriched in apoE transferred from VLDL during isolation [15]. Finally, centrifugation leads to significant changes in the distribution of apoA-1 between HDL subfractions.

There has also been a growing recognition that isolated lipid-free serum-derived apoA-1 – centrifuged, delipidated, denatured with urea, dialyzed, and then refrozen or lyophilized – has physical properties in solution that differ in significant ways from those of native lipid-poor HDL. This is particularly true in its interaction with cells. Unlike native lipid-poor HDL, serum-derived lipid-free apoA-1 self-associates [16] and forms large discoidal particles with cell-derived PL and FC [17].

With the development of milder techniques for the isolation and characterization of HDL from biological fluids and cell culture media, a significantly revised picture of RCT has emerged.

6.2
Structure and Properties of Preβ$_1$-HDL

Intercellular fluid, lymph and plasma all contain a lipid-poor HDL whose protein moiety consists only of apoA-1. This HDL contains in addition small amounts of PL and FC (<10 % of HDL mass). By agarose gel electrophoresis, lipid-poor HDL shows preβ-electrophoretic migration, separating it from most other HDL, which has α-migration. By nondenaturing PAGE this lipid-poor HDL (now called preβ$_1$-HDL) has a MW_{app} of ~65 kDa [15,18]. From this value, based on globular protein standards, it had originally been concluded that each preβ$_1$-HDL contained two molecules of apoA-1. However, no apoA-1 dimers were detected in preβ$_1$-HDL treated with dimethyl suberimidate, a crosslinking agent, even though these were clearly identified in other HDL particles under the same conditions [19]. More recently, sedimentation velocity and equilibrium data found a molecular mass for preβ_1-HDL of ~33 kDa [18]. This difference is almost certainly the result of a high degree of asymmetry (an axial ratio of ~6.5:1) in preβ_1-HDL. A similar phenomenon has been widely recognized for other rod-shaped proteins [20,21].

Preβ_1-HDL becomes fully lipid-loaded when it contains two to three molecules of PL, mainly phosphatidylcholine (PC) [19,22]. Further information on the mechanism of lipid-binding by preβ_1-HDL has been provided by studies of the apoA-1 secreted by human liver- and intestine-derived cells. The major initial secretion product from these cells was a globular apoA-1 monomer with a Stokes radius of 2.6 nm (Stokes diameter 5.2 nm) that was unable to bind PL, and had pre-α electrophoretic migration. Secretion of apoA-1 in this PL-resistant form may be a mechanism to promote distribution of lipid-free apoA-1 to the extracellular space of peripheral tissues, prior to the inception of RCT. Conversion of the precursor particle into a lipid-poor HDL required the presence of the ABCA1 transporter protein (Chapter 5). Most cells express ABCA1 both at their surfaces and on internal membranes, including those of Golgi vacuoles [23]. In the presence of ABCA1, 2.6 nm apoA-1 became refolded into an elongate form with a Stokes radius of 3.6 nm (Stokes diameter 7.2 nm) that was similar to that found for native plasma preβ_1-HDL [18]. Glyburide, which blocked labeled PL transfer by ABCA1 from ^{3}H-choline-labeled cells, did not inhibit conversion of 5.2 nm apoA-1 into 7.2 nm apoA-1, which shows that protein folding and PL transfer are sequential activities of this transporter with apoA-1. The refolded apoA-1 is a substrate for secondary ABCA1-dependent PL transfer (Fig. 6.2). Like plasma preβ_1-HDL, the newly formed lipid-poor HDL particle accelerated the transfer of FC from peripheral cell membranes, both in cell culture medium and also in native human plasma.

In summary, lipid-poor apoA-1-particles, similar or identical to plasma preβ_1-HDL in structure and biological properties, are a major component of HDL newly produced by human liver- and intestine-derived cells. In some cultured liver cell

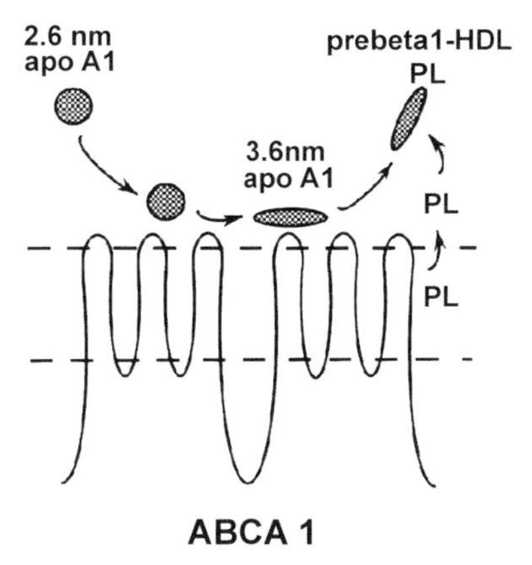

Fig. 6.2 Processing of newly secreted, PL-resistant apoA1 into PL-receptive preβ_1-HDL. ApoA1 species are defined by Stokes Radius (SR). PL: phospholipid. From Ref. 18.

models, mixtures of both lipid-free and lipid-poor apoA-1 particles were found at early time points in the culture medium [24,25]. This possibly reflects the relative rates of apoA-1 production and ABCA1 activity under different experimental conditions. Recent studies agree that lipid-poor HDL, and not lipid-rich discoidal or spheroidal HDL particles, are the main HDL product of apoA-1-secreting cells.

6.3
Cell Membrane Origin of FC for RCT

There has been controversy both about the membrane origin of FC destined for RCT and about the mechanism of its transfer out of the cell [review: 26]. It is now broadly accepted that the plasma membranes of living cells are made up of a mosaic of FC-rich and FC-poor microdomains stabilized by membrane proteins. FC-rich microdomains may be planar ("lipid rafts") or invaginated ("caveolae"). Much of the earlier research on FC efflux to apoA-1 was based on a working hypothesis that the properties of sterol domains in natural membranes would be similar to those in synthetic membrane bilayers, and that most or all of FC efflux was nonspecific, driven by simple (Fickian) diffusion, or possibly collisional energy. More recently, significant differences have been described in the properties of native and synthetic surface bilayers, especially in terms of FC transfer.

FC condenses and stiffens PL acyl chains and reduces FC exchange in synthetic bilayers [27]. From this, it had been predicted that cell surface microdomains with low FC content would transfer FC more effectively to extracellular acceptors than those rich in FC [28]. Recent evidence, however, suggests the opposite to be the case. In biological membranes, FC depletion increased membrane stiffness [29]. FC-rich microdomains of these membranes were more water-permeable (not less) than noncaveolar domains [30], reflecting looser, not tighter, lipid packing. FC transferred more effectively out of FC-rich fibroblast caveolae than from the noncaveolar membrane fraction in the same cells [31]. Even nonspecific FC acceptors such as cyclodextrin removed FC from caveolae more effectively than from FC-poor domains [32,33]. These findings suggested that physical interaction between cell surface FC-rich microdomains and specific subfractions of HDL might be important in driving RCT.

The interaction of cell membranes with native HDL was initially studied by following ^{3}H-FC from equilibrium-labeled fibroblast monolayers incubated in human plasma. Preβ_1-HDL was an unexpectedly effective early acceptor of FC from the plasma membranes of peripheral cells [34,35]. When cells equilibrated with ^{3}H-FC were briefly incubated with native plasma, 30–40 % of label was recovered after nondenaturing 2D-electrophoresis in preβ_1-HDL, though this fraction contained only 3–5 % of total plasma apoA-1, and $<$0.5 % of total HDL-FC. Little if any PL was transferred to preβ_1-HDL under these conditions. In contrast, PL transfer to serum-derived lipid-free apoA-1 was much faster than that of FC [26]. Analysis of FC transfers from FC-poor and FC-rich domains to preβ_1-HDL during FC efflux found the latter to be the major contributors to preβ_1-HDL in native plasma (Fig. 6.2). Preβ_1-HDL contains a single molecule of bound FC [18].

RCT was initiated by transfer of FC from the surfaces of peripheral cells to lipid-poor apoA-1. FC in caveolae was the preferred donor to preβ_1-HDL in human vascular smooth muscle cells [36,37]. Though the molecular basis for these effects is not fully understood, HDL of small diameter may have better access to the membrane surface than larger HDL by penetrating the cell's unstirred water layer [28]. Small HDL may bind directly to FC-rich lipid rafts or caveolae at the cell surface [38]. Finally, FC appears to bind to apoA-1 in preβ_1-HDL with higher affinity than to other HDL, thereby reducing the "off-rate" redistributing FC on preβ_1-HDL to other HDL [13]. As predictable, when HDL was centrifuged, removing its preβ_1-HDL fraction, it no longer showed the FC-sequestering properties of preβ_1-HDL [39].

In summary, efficient selective transfer of FC from caveolae and lipid rafts to preβ_1-HDL seems to depend both on the properties of caveolae/lipid rafts and on the conformation of apoA-1 in preβ_1-HDL [40].

6.4
Role of ABCA1 in FC Transfer

Since preβ_1-HDL is an ineffective acceptor of cell-derived PL, experiments designed to compare PL and FC transfer in RCT have mostly relied on the interaction of lipid-free apoA-1 with radiolabeled washed cell monolayers.

From these studies, there is broad agreement that PL-transfer to newly secreted apoA-1 is dependent on the activity of ABCA1. Plasma phospholipid transfer protein (PLTP) (Chapter 8) was recently shown to be an acceptor (competitive with apoA-1) of PL from ABCA1 [41,42]. Since most of the cells used in studies of PL transfer secrete PLTP [43], some of the labeled PL transferred to apoA-1 may have passed through PLTP. Further research will be needed to estimate the contribution of this "indirect" transfer.

Other points of contention have been whether FC transfer in RCT is ABCA1-dependent, and if PL and FC transfer to apoA-1 is sequential or simultaneous.

Two main models have been proposed (Fig. 6.3):

1. FC and PL transfer driven simultaneously by ABCA1 from the same microdomain ("microsolubilization") [44]. In this model, the rates of FC and PL transfer would depend on their relative concentrations in the vicinity of plasma membrane ABCA1.
2. Energy-dependent PL transport driven by ABCA1 from a PL-rich, FC-poor microdomain, followed by FC transfer from an adjacent, FC-rich microdomain ("two-step" transfer) [45,46]. In this model, partition of the membrane into FC-poor/PL-rich and FC-rich/PL-poor microdomains would help to drive RCT.

When FC and PL binding were assayed over a time course of 5–60 min, parallel transfers of FC and PL transfer were reported [47]. However, cell surface microdomains are very small (typically 100 nm in diameter) while the movement of apoA-1

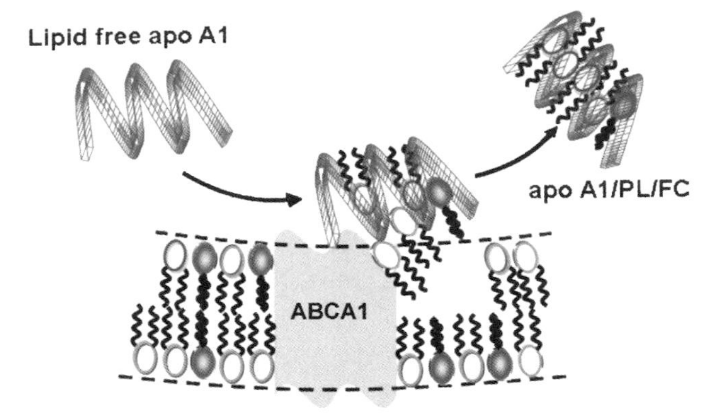

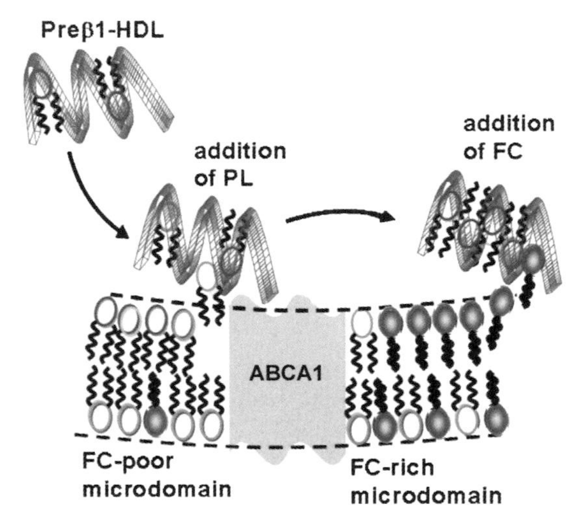

Fig. 6.3 Alternative models for PL and FC transfer to apoA-1. *A*) Simultaneous transfer of PL and FC ("microsolubilization") to serum-derived lipid-free apoA1. *B*) Transfer of PL and FC ("two-step") to preβ_1-HDL.

between domains is relatively rapid [48]. To establish unambiguously that PL and FC had been transferred from the same domain would require measurement over a very short time course (~10 millisec) [49]. This information is not yet available.

There is broad agreement that co-transport of FC is not obligatory for maximal rates of ABCA1-dependent PL transfer. HEK cells were transfected with murine ABCA1 cDNA and treated with 2-hydroxypropyl-cyclodextrin [46]. FC transfer was inhibited in relation to control conditions, but PL transfer was unchanged. In a second study, PL without FC was transferred to apoA-1 by vanadate-treated SMC. FC without PL was then transferred to the apoA-1/PL complexes formed from the caveolae of endothelial cells that lacked functional ABCA1 [45]. Finally, Yamauchi et al. found that L929 and CHO-KI cells transferred PL but not FC to apoA-1, under the influence of ABCA1 [50].

ABCA1 plays an essential role in the PL lipidation of newly secreted lipid-free apoA-1 to form preβ_1-HDL. It does not seem to be required for transfer of cell-derived FC.

6.5 Mechanism of Lecithin:Cholesterol Acyltransferase (LCAT)

This is the only enzyme reaction in the RCT pathway. LCAT is a 416 amino acid glycoprotein secreted by hepatocytes and it catalyzes the transesterification of rapidly exchangeable FC with poorly diffusible phosphatidylcholine (PC) to generate cholesteryl ester (CE), which is almost insoluble. Most of the second product, lyso-PC, becomes bound to serum albumin. The effect of LCAT is to sequester FC in HDL particles as CE (Fig. 6.4). While the transfer of FC for LCAT is rapid, PL transfers more slowly, mainly from VLDL and LDL, in a reaction dependent on PLTP. A significant role for the PLTP in the LCAT reaction *in vivo* is confirmed by decreased levels of HDL-C found in PLTP$^{-/-}$ mice [51].

LCAT is a serine-dependent hydrolase, the catalytic triad of which consists of ser-182, asp-345, and his-377 (codon numbers are those of the mature protein) [52]. These data, and some sequence homology to other lipases, support a model of the 3-D structure of isolated LCAT that includes a hydrophobic "lid" (residues 50–74) covering a lipid-binding cavity. There is no information on the conformation of the LCAT bound to HDL.

The LCAT reaction is strongly stimulated by the presence of apoA-1. Other apolipoproteins, though showing structural similarity (Chapter 1), are significantly less effective. A small amount of LCAT is bound to LDL. If FC (relative to PL) in HDL is reduced, LCAT becomes a phospholipase A-2, generating free fatty acid (FFA) instead of CE [55]. This PLase activity, like the acyltransferase activity of LCAT, is

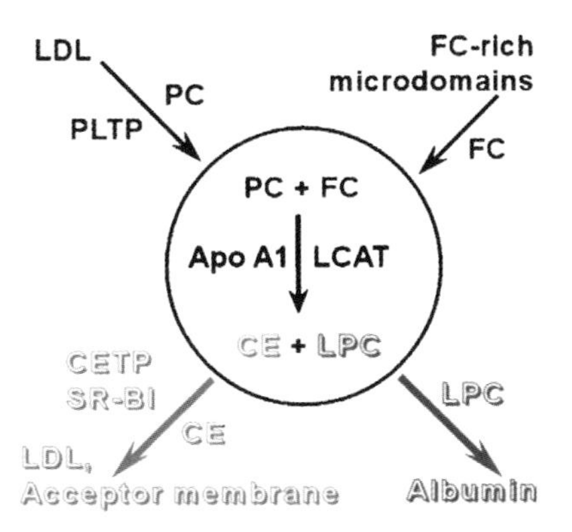

Fig. 6.4 Lipid transfers by LCAT in complexation with HDL-apoA-1.

apoA-1-dependent. It may represent a mechanism self-regulating CE synthesis in response to the supply of FC from cells or plasma lipoproteins to HDL.

Further information on the biological role of LCAT has been gained from the structures and compositions of circulating lipoproteins in human LCAT deficiency, while confirmatory data have been obtained from the plasma of LCAT$^{-/-}$ mice. LCAT deficiency is characterized by an almost complete absence of the major (α-migrating) fraction of HDL. The circulating lipoproteins in LCAT-deficient plasma include lipid-poor particles similar or identical to preβ_1-HDL, discoidal particles rich in apoE, and several classes of triglyceride-rich lipoproteins that contain apolipoprotein B and float in the LDL and VLDL density ranges. Much, though not all, of the CE found in normal LDL is generated by the LCAT reaction, as a result of CETP-mediated transfer from HDL. CE levels in both HDL and LDL are significantly reduced in LCAT deficiency, and both "forward" (liver-to-periphery) and "reverse" (periphery-to-liver) sterol fluxes are diminished. As a result, LCAT deficiency has a smaller effect on FC homeostasis than would be predicted from the central role of this enzyme in RCT [56].

A second rare genetic disorder, fish-eye disease (FED), is also the result of mutation in the LCAT gene sequence. FED is associated with corneal opacity, secondary to sterol deposition in the eye. In FED, LCAT can esterify FC on LDL (albeit inefficiently) but not on HDL [53]. The distribution of mutations in the LCAT gene sequence in FED has been used to map domains in the LCAT protein likely to interact with apoA-1. LCAT mutations V46E, G71R, Y83A, S91P, T123I, and R140H are among those leading to FED. LCAT mutant E149A has significantly increased phospholipase (relative to acyltransferase) activity, in relation to the wild-type enzyme [57]. This may identify a domain important in FC binding by LCAT. The sequence immediately preceding E149 (–V_{139}RAAP*Y*DW*R*L–) contains a well marked cholesterol recognition amino acid consensus (CRAC) motif (V/L x_{1-4} Y x_{1-4} R/K) recently identified in a number of FC-binding proteins [58], though no direct evidence that this sequence of LCAT binds FC has yet been obtained.

A number of spontaneous mutants of apoA-1 are associated with low plasma HDL-C *in vivo*, while many of them activate LCAT poorly *in vitro* [59]. They are clustered in repeat 6 (residues 143–165) of the apoA-1 amino acid sequence, which is also adjacent to a CRAC motif (–L_{163}AP*Y*SDEL*R*–). A monoclonal antibody recognizing residues 137–144 of apoA-1 blocked LCAT-mediated CE synthesis in human plasma [60], while haptoglobin binding to residues 141–164 of apoA-1 displaced LCAT from HDL [61]. These data are all consistent with protein–protein binding between specific domains of apoA-1 and LCAT that is needed for effective CE synthesis. It is possible that the FC-binding motifs of each protein contribute to a common FC-binding cavity within the enzyme–substrate complex.

6.6
Origins of FC and PL for LCAT Activity

Until recently, a model of RCT has prevailed in which preβ_1-HDL containing apoA-1, by continuing to incorporate FC and PL, is converted into a discoidal particle structurally similar or identical to the synthetic recombinants, containing two to three molecules of apoA-1, that are formed from serum-derived lipid-free apoA-1, PL, and FC at high pressure or by cholate dialysis. The biophysical structures of these recombinants have been extensively investigated [62]. Synthetic discoidal HDL can react with LCAT to generate a spheroidal, α-migrating HDL particle, rich in CE. The presence of discoidal HDL of similar appearance in the plasma of humans and rodents with primary or secondary LCAT deficiency made this model attractive [53], but a number of difficulties have since become apparent.

1. Discoidal HDL was absent from normal biological fluids, including plasma, lymph, or blister fluid (a model for lymph), even when LCAT activity (in lymph) was low (<10 % that of plasma) [65]. In addition, the proportion of preβ_1-HDL in lymph was similar to that in plasma [66] which seemed to argue against a precursor–product relationship between preβ_1-HDL and discoidal HDL.
2. In primary or acquired LCAT deficiency, discoidal HDL contain little if any apoA-1. They are rich in apoE, but the apoA-1 present in plasma is in the form of preβ_1-HDL [69].
3. ABCA1 and PLTP are the suggested mechanisms that could drive PL transfer to preβ_1-HDL. However overexpression of ABCA1 in mice had little or no effect on HDL particle distribution [63], while overexpression of PLTP in mice increased plasma levels of preβ_1-HDL, but did not lead to the appearance of discoidal HDL [64].
4. Preβ_1-HDL was recently shown to be a direct substrate for LCAT.

Sparks et al. [67] found that synthetic preβ-migrating apoA-1/PL/FC complexes with compositions equivalent to that of preβ_1-HDL were transesterified by LCAT. Nakamura et al. [19] found that when native human plasma was incubated with ^{3}H-labeled fibroblast monolayers to label preβ_1-HDL, LCAT bound directly to this particle to generate ^{3}H-CE with high efficiency. Consistently with this, a preβ-migrating, small HDL particle that contained CE and lysolecithin was found in ovarian follicular fluid (which contains LCAT) by Jaspart et al. [68]. This may be identical with the small CE-containing product of preβ-HDL and LCAT identified in native plasma [19].

Together, these data argue against a necessary role for discoidal HDL in the conversion of preβ_1-HDL into α-HDL. In plasma and also probably in lymph, LCAT can act directly on preβ_1-HDL, to esterify cell-derived FC. Cellular FC drawn into these complexes to replace that esterified limits dilution of the preβ_1-HDL pool of FC by other lipoproteins, increasing the efficiency of RCT. The apoE-rich discoidal HDL present in LCAT-deficient plasma may be the result of microsolubilization, or a similar mechanism, of plasma membrane domains from macrophages and other cells secreting apoE. If this view is correct, discoidal HDL would represent not an intermediate in LCAT activity, but a response to LCAT deficiency and the failure of normal RCT.

6.7
Further Metabolism of LCAT/Preβ_1-HDL Complexes

When native plasma interacted with ^{3}H-FC equilibrated cells to generate labeled preβ_1-HDL, a "ladder" of CE-labeled HDL fractions of increasing MW_{app} was generated (Fig. 6.5A). The sizes of these products differed by roughly that of an apoA-1 subunit (~30 kDa). A precursor–product relationship between the smaller and larger particles suggested fusion of apoA-1 in preβ_1-HDL subunits into larger particles (Fig. 6.5B). The rate of FC esterification by LCAT in these complexes exceeded that of FC in the fraction of large HDL by a factor of 30 (MW $>$ 220 kDa) [19]. Little cell-derived ^{3}H-FC was transferred to the MW $>$ 220 kDa fraction, and less was esterified there. The mass of LCAT in plasma ($\sim$5 μg mL^{-1}) is only enough to bind about 1 % of circulating HDL particles. As a result, LCAT must transfer between HDL particles, or else a small subset of HDL must act as generators of CE that are distributed by CETP to the others. The data in this study showed that a disproportionate amount of LCAT (relative to the distribution of apoA-1) esterified FC newly derived from the cell surface present on preβ_1-HDL.

Little is known about what might drive fusion of preβ_1-HDL containing one apoA-1 to form larger HDL containing two to three apoA-1 units. LysoPC and FFA generated by LCAT are potential fusogens. PLTP was reported to promote fusion of centrifugally isolated HDL [70], but this is absent from preβ_1-HDL and from the other small HDL that accumulate ^{3}H-CE from cell-derived ^{3}H-FC [19]. Hepatic lipase has been implicated in HDL fusion [71] but the self-association of preβ_1-HDL precursors occurs *in vitro* in native plasma containing little triglyceride lipase activity.

CETP promotes the movement of CE between plasma lipoprotein particles (Chapter 7). Though its main metabolic significance is considered to be the promotion of net exchange between CE on HDL and TG on VLDL and LDL [72], CETP also promotes an uncompensated (unidirectional) transfer of CE from HDL to the SR-BI cell surface acceptor [73]. It also transfers CE from small, preβ-derived "generators" to other HDL particles [19], as was shown by the retention of label there when CETP was inhibited (Fig. 6.5C). Little of the ^{3}H-CE formed by LCAT from cell-derived ^{3}H-FC was transferred to VLDL or LDL; most was transferred from small HDL to larger HDL. The greatest effect of the inhibitor was seen on transfers from the smallest HDL, where ^{3}H-CE was first generated. These data indicate that fusion of small HDL and CETP-mediated transfers both contribute to the accumulation of ^{3}H-CE in plasma HDL.

Pathways in the metabolism of large HDL are incompletely understood. FC transfer to these particles may be driven by spontaneous diffusion from the surface of VLDL and LDL, or possibly by a gradient dependent on ABCG1 on peripheral cell surfaces (Chapter 5) and driven by LCAT antigen associated with these particles.

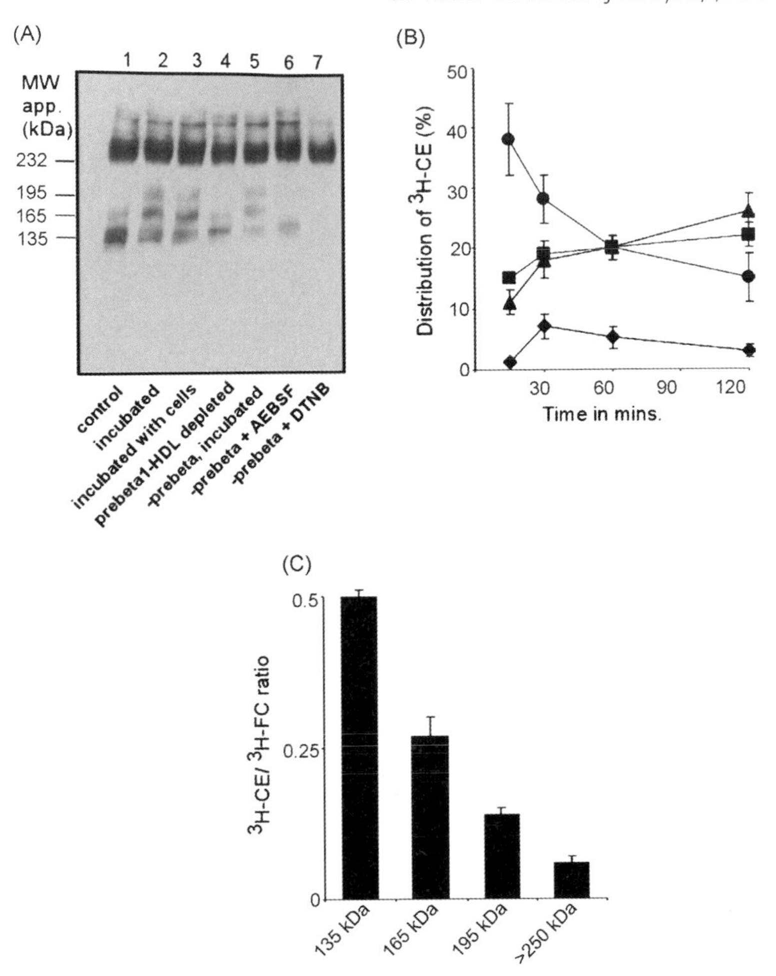

Fig. 6.5 CE synthesis and transfer on complexes of LCAT and preβ_1-HDL. *A)* LCAT in "small" (MW < 165 kDa) and "large" (MW > 220 kDa) HDL, detected by Western blot after nondenaturing PAGE. The effects on LCAT distribution of incubation+/− fibroblasts, prior removal of preβ_1-HDL, and inhibition of LCAT activity by AEBSF or DTDB are shown. *B)* Time course of formation and distribution of ^{3}H-CE in native human plasma incubation with ^{3}H-FC-labeled human skin fibroblasts. After incubation (15 min, 37 °C) of native plasma with ^{3}H-FC labeled fibroblasts to label preβ_1-HDL, the medium was transferred to unlabeled cells, and incubation was continued over the time course shown. Circles: ^{3}H-CE in preβ_1-HDL containing one APOA1 + LCAT. Squares: the same in two-apoA1 particles. Triangles: the same in 3-apoA1 particles. Diamonds: one-apoA-1 particles from which LCAT is dissociated. The time course shows the recovery of ^{3}H-CE in larger HDL as a function of time.C) Effect of the CETP inhibitor CFTP on transfers of ^{3}H-CE between HDL. Data from Ref. 19.

6.8
ApoA-1 Recycling and the Production of Preβ_1-HDL

The recycling of apoA-1 subunits from mature HDL constitutes an integral part of RCT, because a major part of apoA-1 in preβ_1-HDL is recycled from α-HDL [74]. One force driving the dissociation of lipid-depleted apoA-1 from large HDL could be loss of core lipid, through the activity of SR-BI. Consistently with this, the level of preβ_1-HDL in native plasma was increased by the presence of cells, an effect lost when the cell surfaces had been pretreated with protease [75]. However preβ-HDL is still formed in plasma, even when there is no change in the surface/volume ratio of HDL. For example, though SR-BI is exclusively cell-associated, preβ_1-HDL accumulates in isolated native plasma when LCAT is inhibited [75].

Particle disproportion is a second potential mechanism by which preβ_1-HDL may be generated. Incubation of two medium-sized centrifuged HDL particles (each with two apoA-1 units) in the presence of PLTP generated one large HDL (with three apoA-1 units) with a volume double that of its precursors, together with a preβ_1-HDL particle [76]. However, in native plasma, generation of preβ_1-HDL is not accompanied by a change in the MW distribution of HDL particles. When serum amyloid A (SAA) is added to native human plasma, it binds to large HDL. Preβ_1-HDL is formed, but the size distribution of α-HDL again remains the same [77]. A similar result was obtained in mouse plasma both *in vivo* and *in vitro*: apoA-1 mimetic peptides increased serum preβ-HDL levels, without changing size distribution in the α-HDL fraction [78,79]. These data suggest that disproportion is particularly a property of centrifuged, but not of native, HDL particles.

A third mechanism to displace apoA-1 (as preβ_1-HDL) from the surface of large HDL could be to replace it with a lipid or with another apolipoprotein. There are two obvious candidates. Several studies have linked apoA-1 recycling and preβ_1-HDL production to the activity of PLTP [80,81].

Overexpression of PLTP in mice led to increased levels of preβ-HDL, HDL-FC, and HDL-PL in plasma, without any change in the size distribution of α-HDL. When LCAT was inhibited by AEBSF in human plasma *in vitro*, PLTP activity and preβ_1-HDL levels both increased [19]. Replacement of apoA-1 by PL at the cell surface could be a significant factor driving preβ_1-HDL production.

A second factor closely linked to preβ-HDL levels is apolipoprotein M (apoM). This protein is mainly bound to HDL, while the balance is in a lipid-free/lipid-poor particle. Its agarose migration is slower than that of apoA-1 in preβ_1-HDL [82]. In apoM$^{-/-}$ mice, preβ_1-HDL was undetectable. In apoM$^{-/-}$ mice, α-HDL became enlarged. ApoM deficiency was associated with increased atherogenesis in a mouse model, suggesting this factor could play a role in promoting RCT. In apoM-overexpressing mice, preβ_1-HDL levels were increased, while atherogenesis was diminished.

These data suggest that apoM, which is probably a PL-binding protein (Chapter 4), may be able to displace apoA-1 (in the form of preβ_1-HDL) from the surface of mature HDL. However the relative significance of PLTP and apoM in preβ_1-HDL formation have not been established.

6.9 Revised Model of RCT

From the data above, some modifications to the earlier model of human RCT now appear to be justified.

1. One major influence of ABCA1 appears to be on the initial formation of preβ_1-HDL. Preβ_1-HDL contains a single apoA-1 unit. Transfer of PL to recycling preβ_1-HDL may be mediated by PLTP.
2. LCAT can bind directly to preβ_1-HDL and generate CE. The balance of evidence no longer supports the hypothesis that a discoidal apoA-1 particle is an obligatory intermediate in RCT. The effect of direct LCAT activity with preβ_1-HDL is to increase

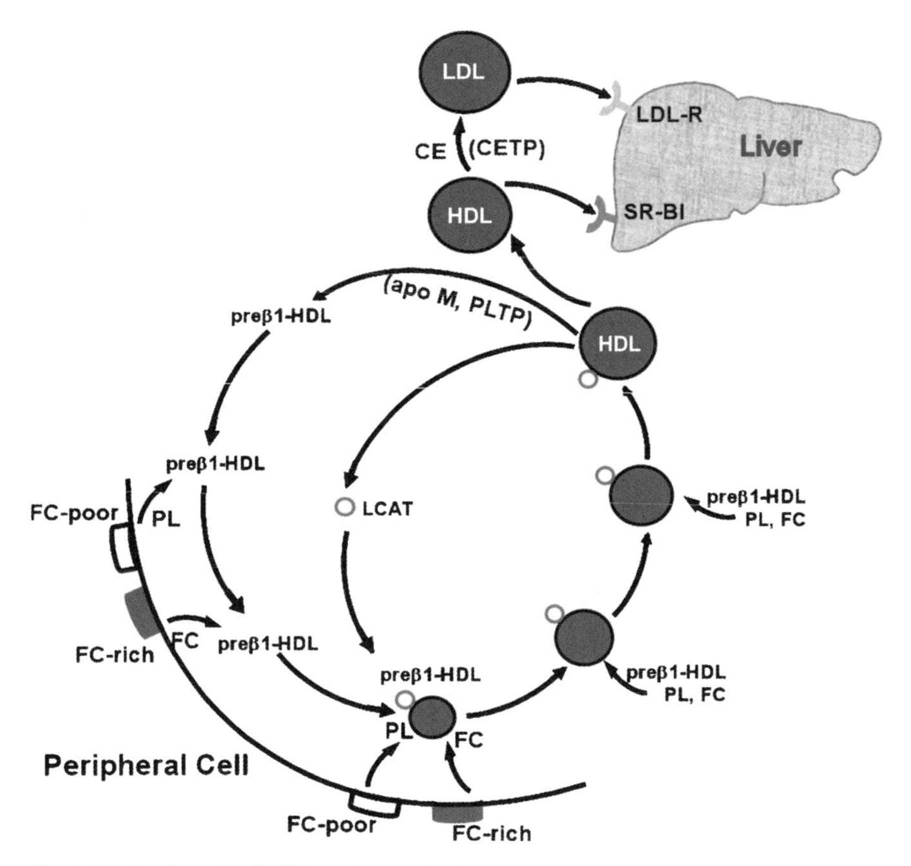

Fig. 6.6 Revised model of RCT, based on preβ_1-HDL recycling, two-step sequential transfer of PL and FC to HDL, and direct esterification of cell-derived FC on preβ_1-HDL by LCAT. Clearance of cell-derived sterol as CE by the LDL receptor (LDL-R) and SR-BI pathways is shown.

the efficiency by which FC newly derived from the surface of peripheral cells can be sequestered as CE.

3. CETP can transfer CE out of preβ_1-HDL/LCAT complexes to larger HDL, as well as to VLDL and LDL. Increase in HDL size in plasma takes place through incorporation of preβ_1-HDL, as well as CETP-mediated transfer.
4. ApoA-1 recycling is an important source of preβ_1-HDL. Important roles for PLTP and apoM in displacing apoA-1 (as preβ_1-HDL) from the surface of HDL are now indicated.

These steps are included in the revised model of RCT shown in Fig. 6.6. Nevertheless, more studies are undoubtedly needed to understand the molecular physiology of this complex but important cycle.

References

1 Lewis, G. F. (2006) Determinants of plasma HDL concentrations and reverse cholesterol transport. *Curr. Opin. Cardiol.*, **21**, 345–352.

2 Mardones, P., Quinones, V., Amigo, L., Moreno, M., Miguel, J. F., Schwartz, R., Miettenen, H. E., Trigatti, B., Krieger, M., Van Patten, S., Choen, D. E., Rigotti, A. (2001) Hepatic cholesterol and bile acid metabolism and intestinal cholesterol absorption in scavenger receptor class B type-I deficient mice. *J. Lipid Res.*, **42**, 170–180.

3 Sviridov D., Chin-Dusting, J., Nestel, P., Kingwell, B., Hoang, A., Olchawa, B., Starr, J., Dart, A. (2006) Elevated HDL cholesterol is functionally ineffective in cardiac transplant recipients: evidence for impaired reverse cholesterol transport. *Transplantation*, **81**, 361–366.

4 Kozarsky, K. F., Donahee, M. H., Rigotti, A., Iqbal, S. N., Edelman, E. R., Krieger, M. (1997) Overexpression of the HDL receptor SR-BI alters plasma HDL and bile cholesterol levels. *Nature*, **387**, 414–417.

5 Marcel, Y. L. and Kiss, R. S. (2003) Structure-function relationships of apolipoprotein A-I: a flexible protein with dynamic lipid associations. *Curr. Opin. Lipidol.*, **14**, 151–157.

6 Gursky, O. and Atkinson, D. (1996) Thermal unfolding of human high-density apolipoprotein A-1: implications for a lipid-free molten globular state. *Proc. Natl. Acad. Sci. USA*, **93**, 2991–2995.

7 Fielding, C. J. and Fielding, P. E. (1981) Evidence for a lipoprotein carrier in human plasma catalyzing sterol efflux from cultured fibroblasts and its relationship to lecithin:cholesterol acyltransferase. *Proc. Natl. Acad. Sci. USA*, **78**, 3911–3914.

8 Fielding, C. J. and Fielding, P. E. (1995) Molecular physiology of reverse cholesterol transport. *J. Lipid Res.*

9 von Eckardstein, A., Nofer, J. R., Assmann, G. (2001) High density lipoproteins and arteriosclerosis. Role of cholesterol efflux and reverse cholesterol transport. *Arterioscler. Thromb. Vasc. Biol.*, **21**, 13–27.

10 Lewis, G. F. and Rader, D. J. (2005) New insights into the regulation of HDL metabolism and reverse cholesterol transport. *Circ. Res.*, **96**, 1221–1232.

11 Curtiss, L. K., Valenta, D. T., Hime, N. J., Rye, K. A. (2006) What is so special about apolipoprotein A1 in reverse cholesterol transport. *Arterio. Thromb. Vasc. Biol.*, **26**, 12–19.

12 Zannis, V. I., Chroni, A., Krieger, M. (2006) Role of ABCA1, LCAT and SR-BI in the biogenesis of HDL. *J. Mol. Med.*, **84**, 276–294.
13 Sviridov, D., Miazaki, O., Theodore, K., Hoang, A., Fukamachi, I., Nestel, P. (2002) Delinieation of the role of preβ-1-HDL in cholesterol efflux using isolated preβ-1-HDL. *Arterioscler. Thromb. Vasc. Biol.*, **22**, 1482–1488.
14 Asztalos, B. F., Sloop, C. H., Wong, L., Roheim, P. S. (1993) Two-dimensional electrophoresis of plasma lipoproteins: recognition of new apoA-I-containing sub-populations. *Biochim. Biophys. Acta*, **1169**, 291–300.
15 Castro, G. R. and Fielding, C. J. (1984) Evidence for the distribution of apolipoprotein E between lipoprotein classes in human normocholesterolemic plasma and the origin of unassociated apolipoprotein E (Lp-E). *J. Lipid Res.*, **25**, 58–67.
16 Gianazza, E., Calabresi, L., Santi, O., Sirtori, C. R., Franceschini, G. (1997) Denaturation and self-association of apolipoprotein A-I investigated by electrophoretic techniques. *Biochemistry*, **36**, 7898–7905.
17 Duong, P. T., Collins, H. L., Nickel, M., Lund-Katz, S., Rothblat, G. H., Phillips, M. C. (2006) Characterization of nascent HDL particles and microparticles formed by ABCA1-mediated efflux of cellular lipids to apoA1. *J. Lipid Res.*, **47**, 832–843.
18 Chau, P., Nakamura, Y., Fielding, C. J., Fielding, P. E. (2006) Mechanism of preβ-HDL formation and activation. *Biochemistry*, **45**, 3981–3987.
19 Nakamura, Y., Kotite, L., Gan, Y., Spencer, T. A., Fielding, C. J., Fielding, P. E. (2004) Molecular mechanism of reverse cholesterol transport: reaction of preβ-migrating high-density lipoprotein with plasma lecithin/cholesterol acyltransferase. *Biochemistry*, **43**, 14811–14820.
20 Colfen, H., Harding, S. E., Boulter, J. M., Watts, A. (1996) Hydrodynamic examination of the dimeric cytoplasmic domain of the human erythrocyte anion transporter, band 3. *Biophys. J.*, **71**, 1611–1615.
21 Marshall, C. B., Chakrabartty, A., Davies, P. L. (2005) Hyperactive antifreeze protein from winter flounder is a very long rod-like dimer of α-helices. *J. Biol. Chem.*, **280**, 17920–17929.
22 Lee, J. Y., Lanningham-Foster, L., Boudyguina, E. Y., Smith, T. L., Young, E. R., Colvin, P. L., Thomas, M. J., Parks, J. S. (2004) Preβ high density lipoprotein has two metabolic fates in human apolipoprotein A-I transgenic mice. *J. Lipid Res.*, **45**, 716–728.
23 Neufeld, E. A., Stonik, J. A., Demosky, S. J., Knapper, C. L., Combs, C. A., Coney, A., Comly, M., Dwyer, N., Blanchette-Mackie, J., Remaley, A. T., Santamarina-Fojo, S., and Brewer, H. B. (2004) The ABCA1 transporter modulates late endocytic trafficking: insights from the correction of the genetic defect in Tangier Disease. *J. Biol. Chem.*, **279**, 15571–15578.
24 Chisholm, J. W., Burleson, E. R., Shelness, G. S., Parks, J. S. (2002) ApoA-I secretion from HepG2 cells: evidence for the secretion of both lipid-poor apoA-I and intracellularly assembled nascent HDL. *J. Lipid Res.*, **43**, 36–44.
25 Kiss, R. S., McManus, D. C., Franklin, V., Tan, W. L., McKenzie, A., Chimini, G., Marcel, Y. L. (2003) The lipidation by hepatocytes of human apolipoprotein A-I occurs by both ABCA1-dependent and ABCA1-independent pathways. *J. Biol. Chem.*, **278**, 10119–10127.
26 Fielding, C. J. and Fielding, P. E. (2001) Cellular cholesterol efflux. *Biochim. Biophys. Acta*, **1533**, 175–189.
27 Smaby, J. M., Brockman, H. L., Brown, R. E. (1994) Cholesterol's interfacial interactions with sphingomyelins and phosphatidylcholines: hydrocarbon chain structure determines the magnitude of condensation. *Biochemistry*, **33**, 9135–9142.
28 Rothblat, G. H., Mahlberg, F. H., Johnson, W. J., Phillips, M. C. (1992)

Apolipoproteins, membrane cholesterol domains, and the regulation of cholesterol efflux. *J. Lipid Res.*, **33**, 1091–1097.

29 Byfield, F. J., Aranda-Espinoza, H., Romanenko, V. G., Rothblat, G. H., Levitan, I. (2004) Cholesterol depletion increases membrane stiffness of aortic endothelial cells. *Biophys. J.*, **87**, 3336–3343.

30 Hill, W. G., Almasri, E., Ruiz, W. G., Apodaca, G., Zeidel, M. L. (2005) Water and solute permeability of rat lung caveolae: high permeabilities explained by acyl chain unsaturation. *Am. J. Physiol. Cell Physiol.*, **289**, C33–C41.

31 Gallegos, A. M., McIntosh, A. L., Atshaves, B. P., Schroeder, F. (2004) Structure and cholesterol domain dynamics of an enriched caveolae/lipid raft isolate. *Biochem. J.*, **382**, 451–461.

32 Zeidan, A., Broman, J., Hellstrand, P., Sward, K. (2003) Cholesterol dependence of vascular ERK1/2 activation and growth in response to stretch: role of endothelin-1. *Arterioscler. Thromb. Vasc. Biol.*, **23**, 1528–1534.

33 Graziani, A., Bricko, V., Carmignani, M., Graier, W. F., Groschner, K. (2004) Cholesterol- and caveolin-rich membrane domains are essential for phospholipase A2-dependent EDHF formation. *Cardiovascular. Res.*, **64**, 234–242.

34 Castro, G. R. and Fielding, C. J. (1988) Early incorporation of cell-derived cholesterol into preβ-migrating high density lipoprotein. *Biochemistry*, **27**, 25–29.

35 Huang, Y., von Eckardstein, A., Assmann, G. (1993) Cell-derived unesterified cholesterol cycles between different HDLs and LDL for its effective esterification in plasma. *Arterioscler. Thromb.*, **13**, 445–458.

36 Fielding, P. E. and Fielding, C. J. (1995) Plasma membrane caveolae mediate the efflux of cellular free cholesterol. *Biochemistry*, **34**, 14288–14292.

37 Fielding, P. E. and Fielding, C. J. (1996) Intracellular transport of low density lipoprotein derived free cholesterol begins at clathrin-coated pits and terminates at cell surface caveolae. *Biochemistry*, **35**, 14932–14938.

38 Nion, S., Briand, O., Lestavel, S., Torpier, G., Nazih, F., Delbart, C., Fruchart, J. C., Clavey, V. (1997) High-density lipoprotein subfraction–3 interaction with glycophosphatidylinositol-anchored proteins. *Biochem. J.*, **328**, 415–423.

39 Frank, P. G., Cheung, M. W., Pavlides, S., Llaverias, G., Park, D. S., Lisanti, M. P. (2006) Caveolin-1 and regulation of cellular cholesterol homeostasis. *Am. J. Physiol. Heart Circ. Physiol.*, **291**H677–H686.

40 Li, Q., Tsujita, M., Yokoyama, S. (1997) Selective down-regulation by protein kinase C inhibitors of apolipoprotein-mediated cellular cholesterol efflux in macrophages. *Biochemistry*, **36**, 12045–12052.

41 Oram, J. F., Wolfbauer, G., Vaughan, A. M., Tang, C., Albers, J. J. (2003) Phospholipid transfer protein interacts with and stabilizes ATP-binding cassette transporter A1 and enhances cholesterol efflux from cells. *J. Biol. Chem.*, **278**, 52379–52385.

42 Lee-Rueckert, M., Vikstedt, R., Metso, J., Ehnholm, C., Kovanen, P. T., Jauhiainen, M. (2006) Absence of endogeneous phospholipids transfer protein impairs ABCA1-dependent efflux of cholesterol from macrophage foam cells. *J. Lipid Res.*, **47**, 1725–1732.

43 Day, J. R., Albers, J. J., Lofton-Day, C. E., Gilbert, T. L., Ching, A. F., Grant, F. J., O'Hara, P. J., Marcovina, S. M., Adolphson, J. L. (1994) Complete cDNA encoding human phospholipid transfer protein from human endothelial cells. *J. Biol. Chem.*, **269**, 9388–9391.

44 Gillotte, K. L., Davidson, W. S., Lund-Katz, S., Rothblat, G. H., Phillips, M. C. (1998) Removal of cellular cholesterol by preβ-HDL involves plasma membrane microsolubilization. *J. Lipid Res.*, **39**, 1918–1928.

45 Fielding, P. E., Russel, J. S., Spencer, T. A., Hakamata, H., Nagao, K.,

Fielding, C. J. (2002) Sterol efflux to apolipoprotein A1 originates from caveolin-rich microdomains and potentiates PDGF-dependent protein kinase activity. *Biochemistry*, **41**, 4929–4937.

46 Wang, N., Silver, D. L., Thiele, C., Tall, A. R. (2001) ATP-binding cassette transporter A1 (ABCA1) functions as a cholesterol efflux regulatory protein. *J. Biol. Chem.*, **276**, 23742–23747.

47 Smith, J. D., LeGoff, W., Settle, M., Brubaker, G., Waelde, C., Horwitz, A., Oda, M. N. (2004) ABCA1 mediates concurrent cholesterol and phospholipid efflux to apolipoprotein A-I. *J. Lipid Res.*, **45**, 635–644.

48 Chambenoit, O., Hamon, Y., Marguet, D., Rigneault, H., Rosseneu, M., Chimini, G. (2001) Specific docking of apolipoprotein A-I at the cell surface requires a functional ABCA1 transporter. *J. Biol. Chem.*, **276**, 9955–9960.

49 Fielding, C. J. and Fielding, P. E. (2001) Cellular cholesterol efflux. *Biochim. Biophys. Acta.*, **1533**, 175–189.

50 Yamauchi, Y., Hayashi, M., Abe-Dohmae, S., Yokoyama, S. (2003) Apolipoprotein A-I activates protein kinase C α signaling to phosphorylate and stabilize ATP binding cassette transporter A1 for the high density lipoprotein assembly. *J. Biol. Chem.*, **278**, 47890–47897.

51 Jiang, X. C., Bruce, C., Mar, J., Lin, M., Francone, O. L., Tall, A. R. (1999) Targeted mutation of plasma phospholipids transfer protein gene markedly reduces high density lipoprotein levels. *J. Clin. Invest.*, **103**, 907–914.

52 Jonas, A. (2000) Lecithin cholesterol acyltransferase. *Biochim. Biophys. Acta.*, **1529**, 245–256.

53 Kuivenhoven, J. A., Pritchard, H., Hill, J., Frohlich, J., Assmann, G., Kastelein, J. (1997) The molecular pathology of lecithin:cholesterol acyltransferase (LCAT) deficiency syndromes. *J. Lipid Res.*, **38**, 191–205.

54 Ng, D. S., Francone, O. L., Forte, T. M., Zhang, J., Haghpassand, M., Rubin, E. M. (1997) Disruption of the murine lecithin:cholesterol acyltransferase gene causes impairment of adrenal lipid delivery and up-regulation of scavenger receptor class B type I. *J. Biol. Chem.*, **272**, 15777–15781.

55 Aron, L., Jones, S., Fielding, C. J. (1978) Human plasma lecithin-cholesterol acyltransferase. Characterization of cofactor-dependent phospholipase activity. *J. Biol. Chem.*, **253**, 7220–7226.

56 Ayyobi, A. F., McGladdery, S. H., Chan, S., Mancini, G. B., Hill, J. S., Frohlich, J. J. (2004) Lecithin:cholesterol acyltransferase (LCAT) deficiency and risk of vascular disease: 25 year follow-up. *Atherosclerosis*, **177**, 361–366.

57 Zhao, Y., Gebre, A. K., Parks, J. S. (2004) Amino acids 149 and 294 of human lecithin:cholesterol acyltransferase affect fatty acyl specificity. *J. Lipid Res.*, **45**, 2310–2316.

58 Fielding, C. J. and Fielding, P. E. (2006) Role of cholesterol in signal transduction from caveolae. *Lipid Rafts and Caveolae*, Wiley-VCH, pp. 91–113.

59 Sorci-Thomas, M. and Thomas, M. J. (2002) The effects of altered apolipoprotein A-I structure on plasma HDL concentration. *Trends Cardiovasc. Med.*, **12**, 121–128.

60 Meng, Q. H., Calabresi, L., Fruchart, J. C., Marcel, Y. L. (1993) Apolipoprotein A-I domains involved in the activation of lecithin:cholesterol acyltransferase. Importance of the central domain. *J. Biol. Chem.*, **268**, 16966–16973.

61 Spagnuolo, M. S., Cigliano, L., D'Andrea, L. D., Pedone, C., Abrescia, P. (2005) Assignment of the binding site for haptoglobin on apolipoprotein A-I. *J. Biol. Chem.*, **280**, 1193–1198.

62 Li, L., Chen, J., Mishra, V. K., Kurtz, J. A., Cao, D., Klon, A. E., Harvey, S. C., Anantharamaiah, G. M., Segrest, J. P. (2004) Double belt structure of discoidal high density lipoproteins: molecular basis for size heterogeneity. *J. Mol. Biol.*, **343**, 1293–1311.

63 Vaisman, B. L., Lambert, G., Amar, M., Joyce, C., Ito, T., Shamburek, R.

D., Cain, W. J., Fruchart-Najib, J., Neufeld, E. D., Remaley, A. T., Brewer, H. B., Santamarina-Fojo, S. (2001) ABCA1 overexpression leads to hyperαlipoproteinemia and increased biliary cholesterol in transgenic mice. *J. Clin. Invest.*, **108**, 303–309.

64 Foger, B., Santamarina-Fojo, S., Shamburek, R. D., Parrot, C. L., Talley, G. D., Brewer, H. B. (1997) Plasma phospholipid transfer protein. Adenovirus-mediated overexpression in mice leads to decreased plasma high density lipoprotein (HDL) and enhanced hepatic uptake of phospholipids and cholesteryl esters from HDL. *J. Biol. Chem.*, **272**, 27393–27400.

65 Wong, L., Curtiss, L. K., Huang, J., Mann, C. J., Maldonado, B., Roheim, P. S. (1992) Altered epitope expression of human interstitial fluid apolipoprotein A-I reduces its ability to activate lecithin:cholesterol acyltransferase. *J. Clin. Invest.*, **90**, 2370–2375.

66 Nanjee, M. N., Cooke, C. J., Olszewski, W. L., Miller, N. E. (2000) Concentrations of electrophoretic and size subclasses of apolipoprotein A-I-containing particles in human peripheral lymph. *Arterioscler. Thromb. Vasc. Biol.*, **20**, 2148–2155.

67 Sparks, D. L., Frank, P. G., Braschi, S., Neville, T. A., Marcel, Y. L. (1999) Effect of apolipoprotein A-I lipidation on the formation and function of preβ- and α-migrating LpA-I particles. *Biochemistry*, **38**, 1727–1735.

68 Jaspard, B., Collet, X., Barbaras, R., Manent, J., Vieu, C., Parinaud, J., Chap, H., Perret, B. (1996) Biochemical characterization of preβ-1 high density lipoprotein from human ovarian follicular fluid: evidence for the presence of a lipid core. *Biochemistry*, **35**, 1352–1357.

69 Mitchell, C. D., King, W. C., Applegate, K. R., Forte, T., Glomset, J. A., Norum, K. R., Gjone, E. (1980) Characterization of apolipoprotein E-rich high density lipoproteins in familial lecithin: cholesterol acyltransferase deficiency. *J. Lipid Res.*, **21**, 625–634.

70 Settasatian, N., Duong, M., Curtiss, L. K., Ehnholm, C., Jauhiainen, M., Huuskonen, J., Rye, K. A. (2001) The mechanism of the remodeling of high density lipoproteins by phospholipid transfer protein. *J. Biol. Chem.*, **276**, 26898–26905.

71 Clay, M. A., Newnham, H. H., Barter, P. J. (1991) Hepatic lipase promotes a loss of apolipoprotein A-I from triglyceride-enriched human high density lipoproteins during incubation in vitro. *Arterioscler. Thromb.*, **11**, 415–422.

72 Bruce, C., Chouinard, R. A., Tall, A. R. (1998) Plasma lipid transfer proteins, high-density lipoproteins, and reverse cholesterol transport. *Ann Rev. Nutr.*, **18**, 297–330.

73 Gauthier, A., Lau, P., Zha, X., Milne, R., McPherson, R. (2005) Cholesteryl ester transfer protein directly mediates selective uptake of high density lipoprotein cholesteryl esters by the liver. *Arterioscler. Thromb. Vasc. Biol.*, **25**, 2177–2184.

74 Chetiveaux, M., Ouguerram, K., Zair, Y., Maugere, P., Falconi, I., Nazih, H., Krempf, M. (2004) New model for kinetic studies of HDL metabolism in humans. *Eur. J. Clin. Invest.*, **34**, 262–267.

75 Miida, T., Kawano, M., Fielding, C. J., Fielding, P. E. (1992) Regulation of the concentration of preβ high density lipoprotein in normal plasma by cell membranes and lecithin:cholesterol acyltransferase. *Biochemistry*, **31**, 11112–11117.

76 Rye, K. A., Hime, N. J., Barter, B. J. (1997) Evidence that cholesteryl ester transfer protein-mediated reductions in reconstituted high density lipoprotein size involve particle fusion. *J. Biol. Chem.*, **272**, 3953–3960.

77 Miida, T., Yamada, T., Yamadera, T., Ozaki, K., Inano, K., Okada, M. (1999) Serum amyloid A protein generates preβ-1 high density lipoprotein from α-migrating high density lipoprotein. *Biochemistry*, **38**, 16958–16962.

78 Navab, M., Anantharamaiah, G. M., Reddy, S. T., Hama, S., Hough, G., Grijalva, V. R., Wagner, A. C., Frank, J. S., Datta, G., Garber, D., Fogelman, A. M. (2004) Oral D-4F causes formation of preβ high density lipoprotein and improves high density lipoprotein-mediated cholesterol efflux and reverse cholesterol transport from macrophages in apolipoprotein E-null mice. *Circulation*, **109**, 3215–3220.

79 Dashti, N., Datta, G., Manchekar, M., Chaddha, M., Anantharamaiah, G. M. (2004) Model class A and class L peptides increase the production of apoA1-containing lipoproteins in HepG2 cells. *J. Lipid Res.*, **45**, 1919–1928.

80 Lie, J., de Crom, R., Jauhiainen, M., van Gent, T., van Haperen, R., Scheek, L., Jansen, H., Ehnholm, C., van Tol, A. (2001) Evaluation of phospholipid transfer protein and cholesteryl ester transfer protein as contributors in the generation of preβ-high-density lipoproteins. *Biochem. J.*, **360**, 379–385.

81 Siggins, S., Bykov, I., Hermansson, M., Somerharju, P., Lindros, K., Miettenen, T. A., Jauhiai, M., Oikkonen, V. M., Ehnholm, C. (2007) Altered hepatic lipid status and apolipoprotein A1 metabolism in mice lacking phospholipid transfer protein. *Atherosclerosis*, **190**, 114–123.

82 Wolfrum, C., Poy, M. N., Stoffel, M. (2005) Apolipoprotein M is required for preβ-HDL formation and cholesterol efflux to HDL and protects against atherosclerosis. *Nat. Med.*, **11**, 418–422.

7
HDL Remodeling by CETP and SR-BI

Christopher J. Harder, Ruth McPherson

7.1
Introduction

Cholesteryl ester transfer protein (CETP) and Scavenger Receptor class BI (SR-BI) play important and partially overlapping roles in HDL metabolism. CETP has a dual function in this process. To date, attention has primarily been focused on the function of CETP in lipoprotein remodeling within the plasma compartment. However, as is discussed here, CETP also mediates the selective uptake of HDL-cholesteryl ester (CE) by hepatocytes [1] and adipocytes, analogously to the well described role of SR-BI. Despite the widely accepted view that plasma HDL-cholesterol (HDL-C) concentrations are a strong negative predictor of coronary artery disease risk, SR-BI deficiency is associated with elevated HDL-C levels but increased atherosclerosis risk whereas CETP decreases HDL-C levels, but under conditions of SR-BI deficiency, is atheroprotective.

7.2
Atheroprotective Function of HDL

Reduced plasma concentrations of HDL-C are a significant risk factor for cardiovascular disease, including both coronary artery disease (CAD) and stroke. This relationship has been epidemiologically documented in numerous population studies and further supported through interventional randomized controlled trials with a variety of pharmacological agents [2,3]. HDL mitigates against atherosclerosis through multiple mechanisms, including enhancement of reverse cholesterol transport [3,4], regulation of eNOS and endothelial function [5,6], suppression of the induction of cell-adhesion molecules [7], antiinflammatory effects [8], and antioxidative properties 9.

One of the protective actions of HDL involves its ability to act as an acceptor of cholesterol from peripheral cells, including arterial macrophage foam cells. The ABCA1 transporter mediates phospholipid and cholesterol efflux to lipid-free or lipid-poor apoA-I, whereas other proteins including SR-BI, ABCG1, and ABCG4 enhance cholesterol efflux to larger lipidated HDL particles. Cholesterol extracted

High-Density Lipoproteins: From Basic Biology to Clinical Aspects. Edited by Christopher J. Fielding

ISBN: 978-3-527-31717-2

by efflux from peripheral tissues is esterified within HDL by lecithin:cholesterol acyltransferase (LCAT), thereby contributing to HDL maturation. HDL particles highly enriched in apoE can be cleared by holoparticle uptake through members of the LDL receptor family, such as the LDL receptor related protein 1 (LRP1). ApoA-I binding to an ecto-F(1)-ATPase has also been reported to stimulate the production of extracellular ADP that activates a P2Y(13)-mediated HDL endocytosis pathway [10,11].

7.3 Cholesteryl Ester Transfer Protein (CETP)

CETP is a 74 kD hydrophobic glycoprotein that mediates the hetero- and homo-exchange of neutral lipid between apoA-I- and apoB-containing lipoproteins. It thus mediates a net transfer of CE from HDL to apolipoprotein B-containing (apoB-containing) lipoproteins, resulting in the cholesterol depletion and TG enrichment of HDL [12]. These changes in HDL composition make it a better substrate for human hepatic lipase (LIPC) [13], a phospholipid and triglyceride lipase that plays an important role in HDL remodeling and catabolism. Under conditions of efficient hepatic apoB lipoprotein clearance, CETP may promote cholesterol transport from HDL to the liver for subsequent biliary secretion by this pathway (Fig. 7.1). However, high plasma concentrations of CETP are associated with low levels of HDL-C, factors that are commonly associated with increased atherosclerosis risk.

7.4 Selective Uptake

HDL-CE is also directly returned to the liver by a process known as selective uptake. Selective uptake of HDL-derived CE is characterized by its entry into reversible and irreversible compartments [15]. In particular, the reversible incorporation of HDL-derived CE occurs at the plasma membrane and is followed by transfer of the lipid to an irreversible pool by mechanisms that do not involve coated pit-mediated endocytosis [14] (Fig. 7.2). It is proposed that the reversible compartment includes CE that has entered the plasma membrane and remains accessible to extraction by extracellular, unlabeled HDL [16]. This reversible pool attains a plateau value after 2 h, possibly because the plasma membrane has a limited capacity to store CE or because the incorporation of CE into the plasma membrane reaches an equilibrium with the transfer of CE out of this compartment. The plasma membrane CE is subsequently internalized to an irreversible compartment and becomes inaccessible to extraction by extracellular unlabeled HDL. Selective uptake of HDL-derived CE results in the production of small, lipid-depleted apoA-1 particles that are returned to the circulation, where they may acquire more cholesterol and phospholipids from peripheral cells or be cleared directly by the kidney.

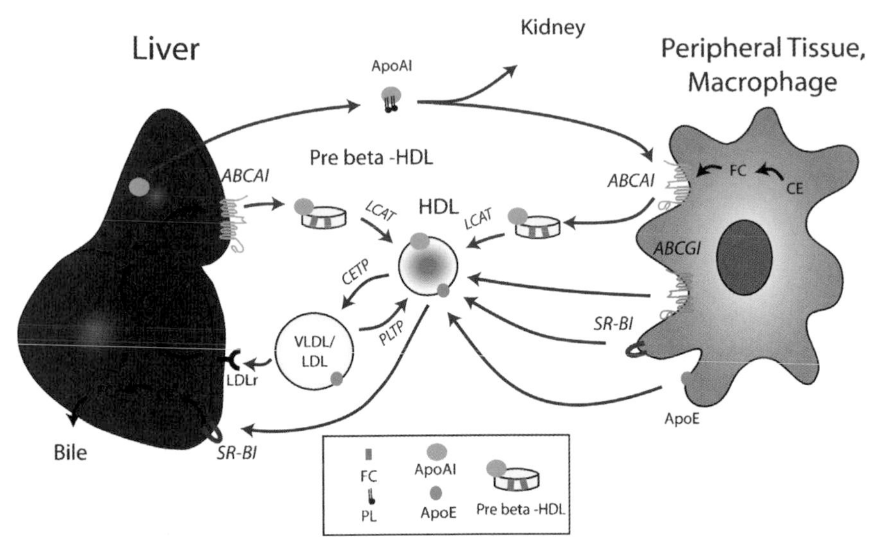

Fig. 7.1 *HDL-mediated reverse cholesterol transport*. Lipid-poor apoA-1 particles are secreted by the liver and the intestine and readily acquire cholesterol and phospholipids from the liver and peripheral tissues through ABCA1-mediated efflux, thereby forming nascent HDL. ABCG1, ABCG4, and SR-BI efflux cellular cholesterol to mature HDL particles to be esterified by LCAT. Additional apolipoproteins (A-II, A-IV, E, and Cs) are transferred to HDL during the lipolysis of chylomicrons and VLDL. HDL undergoes further remodeling in the plasma compartment. PLTP mediates the exchange of phospholipids between lipoproteins and participates in the formation of smaller HDL particles. Endothelial lipase (LIPG) plays an important role as an HDL phospholipase and facilitates HDL catabolism. CETP mediates the exchange of CE for TG between HDL and apoB-containing lipoproteins, respectively, resulting in the cholesterol depletion and TG enrichment of HDL (making it a better substrate for hepatic lipase (LIPC)). HDL particles enriched in apoE can be cleared directly through members of the LDLr family. HDL-CE is also directly returned to the liver by a process known as selective uptake, mediated by SR-BI (SCARB1 gene product) and CETP. Thus, apoA-I, ABCA1, ABCG1, ABCG4, and LCAT all play important roles in the genesis of HDL, whereas CETP, PLTP, LIPG, LIPC, and SR-BI function in the remodeling and catabolism of HDL.

7.5 SR-BI Function and Trafficking in Hepatocytes

SR-BI is the primary receptor responsible for hepatic HDL-Cholesterol clearance in the mouse [17] and is a cell surface glycoprotein with two transmembrane domains, a large extracellular loop, and short intracellular N- and C-terminal domains (Fig. 7.3) [18]. The receptor is expressed ubiquitously, with the highest levels of expression being in steroidogenic, intestinal, and hepatic cells [19]. SR-BI binds a large array of ligands including HDL and native or modified low-density lipoproteins (LDL). In contrast with the holo-particle uptake of the LDL receptor pathway, SR-BI mediates cholesterol, CE, and phospholipid uptake by a selective pathway whereby lipids are transferred down their concentration gradient through a hydrophobic channel into

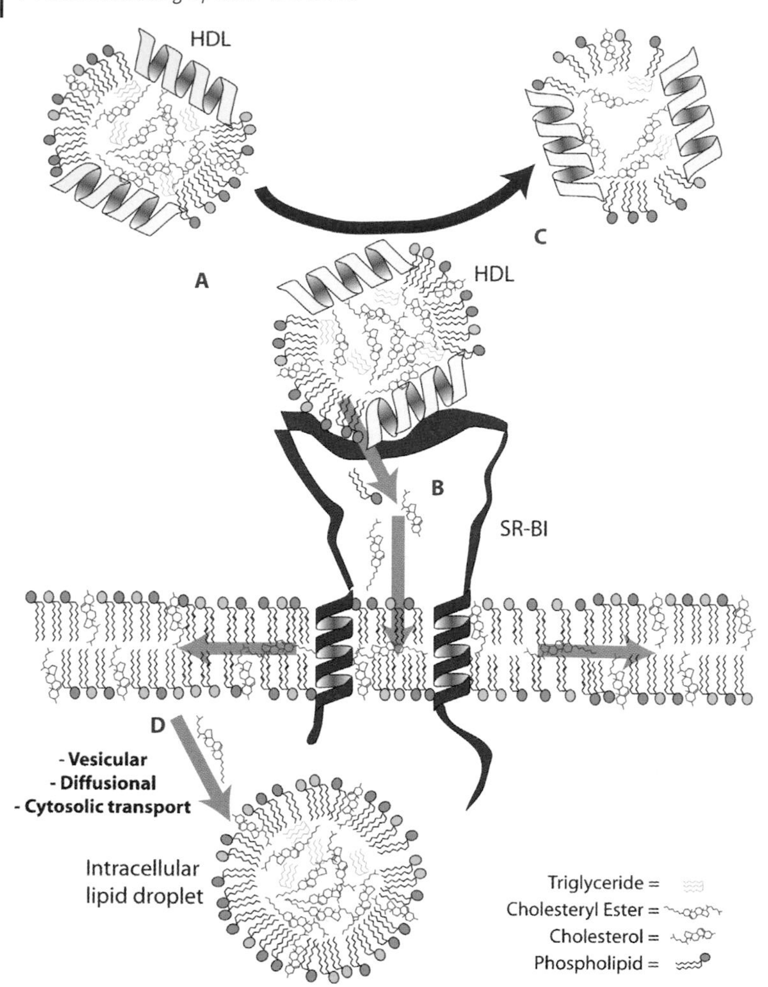

Fig. 7.2 *SR-BI-mediated selective uptake.* This process is characterized by HDL binding to SR-BI (*A*) and the selective transfer of lipids down a concentration gradient (*B*). For continual uptake, the dissociation of the HDL particle from the receptor (*C*) and the irreversible internalization of SR-BI appear to be required (*D*). Step (*B*) is a reversible process, such that transferred lipids can be extracted by an excess of poorly lipidated HDL. The site of intracellular accumulation of non-hydrolyzed CE is the lipid droplet. The irreversible transport of FC and CE to intracellular sites can occur through vesicular transport, diffusion, or by cytosolic carrier proteins.

the membrane [20]. If the concentration gradient is reversed, SR-BI can also mediate cellular cholesterol efflux [21]. Hepatic SR-BI protects against atherosclerosis by promoting the final stages of macrophage reverse cholesterol transport [22].

SR-BI localizes to the apical domain of isolated primary mouse hepatocytes [23], primary mouse heptocyte couplets [24], and hepatic tissues sections [25], but has also been detected on the bile canaliculus [26] and has been shown to undergo regulated

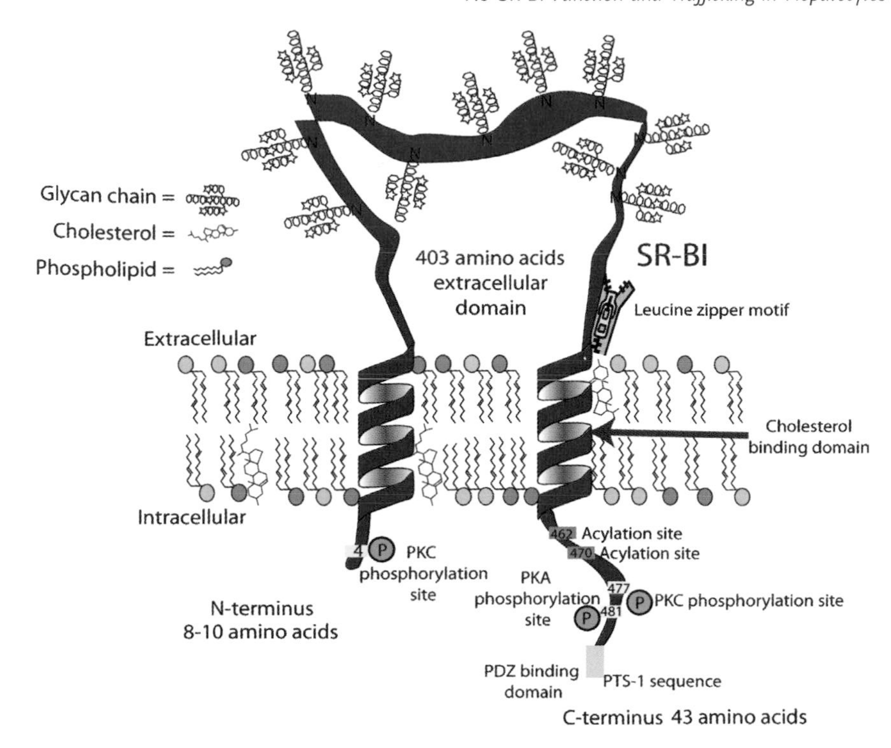

Fig. 7.3 *SR-BI protein – distinguishing features.* The putative structure of SR-BI includes a large, heavily glycosylated extracellular domain, two cytosolic N- and C-terminal tails, and two transmembrane domains. Putative phosphorylation, acylation, dimerization (leucine zipper motif), cholesterol-binding, and regulatory protein binding recognition sites are shown. The structures of the intracellular and extracellular domains are still unsolved.

transcytotic movement in polarized Madin–Darby Canine Kidney (MDCK) cells [27]. Although lipid delivery to and from lipoprotein particles occurs mainly on the plasma membrane, both SR-BI and HDL have been shown to internalize into and recycle from endocytic compartments [28–30]. Our laboratory has recently demonstrated that in primary mouse hepatocytes, SR-BI achieves efficient selective uptake of CE in the absence of endocytosis [31]. These studies, confirmed by others [32–34], indicate that endocytosis and recycling of HDL are not required for selective uptake by SR-BI.

The functional implications of the hepatic trafficking of SR-BI thus remain unclear. Given its polarized distribution and role in lipid transport, it is plausible that SR-BI could mediate selective sorting and secretion of HDL-derived cholesterol into the bile. This is an attractive hypothesis, given that both gene deletion [35] and overexpression of SR-BI [36] reciprocally affect biliary cholesterol secretion *in vivo*. However, there is currently no link between the basolateral selective uptake function of SR-BI in polarized hepatocytes and the *in vivo* data demonstrating that hepatocyte SR-BI expression correlates with biliary cholesterol secretion. The ATP binding cassette (ABC) half-transporters ABCG5 and ABCG8, which localize exclusively to the bile canaliculus, have been shown to be rate-limiting for cholesterol secretion into

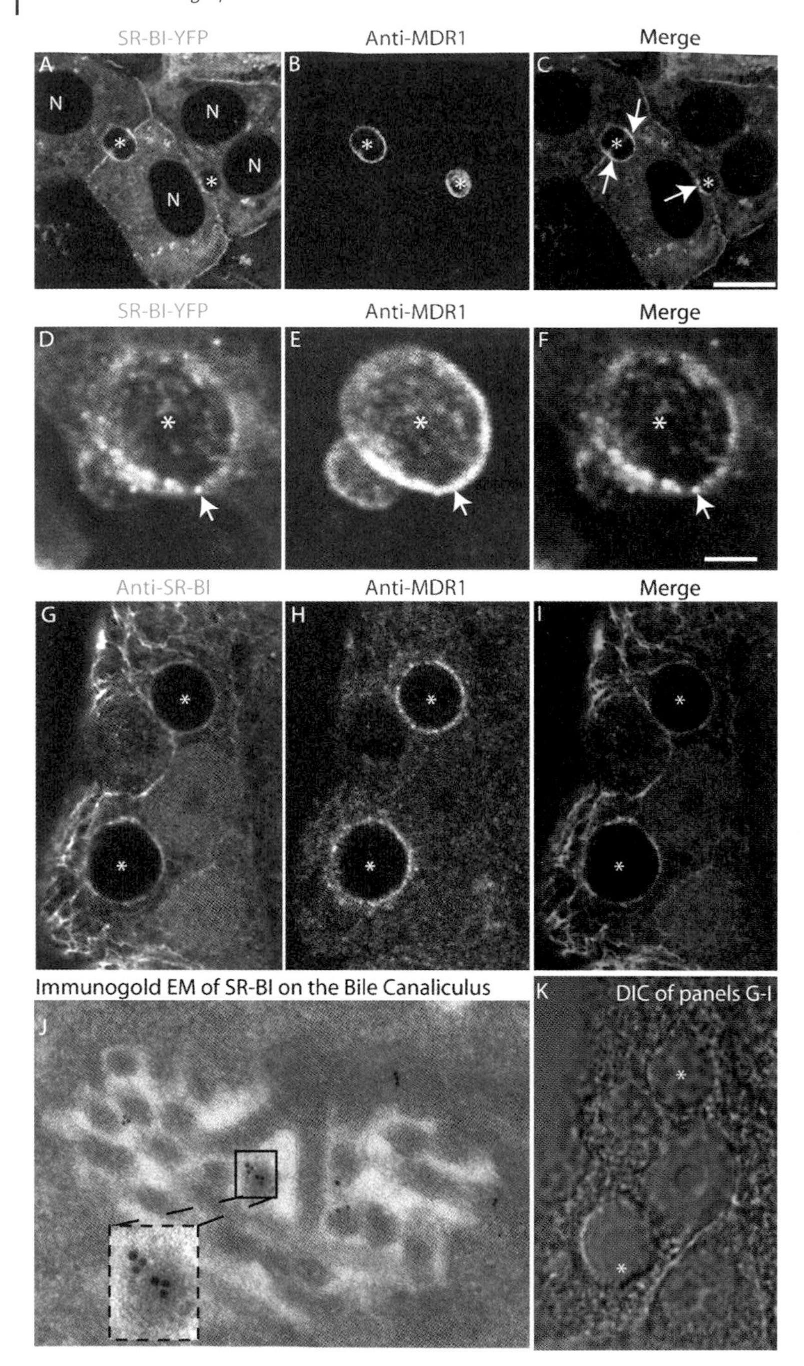
SR-BI-YFP
Anti-MDR1
Merge
A
B
C
N
N
N
N
SR-BI-YFP
Anti-MDR1
Merge
D
E
F
Anti-SR-BI
Anti-MDR1
Merge
G
H
I
Immunogold EM of SR-BI on the Bile Canaliculus
J
K
DIC of panels G-I

bile [37]. However, ABCG5/G8 expression does not always correlate with biliary cholesterol secretion and there are data indicating that other proteins including NPC1 [38], SCP-2 [39], and caveolin-1 [40] may be involved [41]. SR-BI may therefore participate with ABCG5/G8 at the bile canaliculus in the regulation of cholesterol secretion.

To gain further insight into the hepatic function and trafficking of SR-BI, we recently studied the movement of fluorescently tagged SR-BI-YFP in a well characterized polarized hepatocyte cell model (WIF-B) [42,43]. WIF-B cells grow in monolayers and, like functional hepatic cells, form bile canalicular-like spaces between adjacent cells. Each bile canaiculus is spatially restricted by tight junctions, rendering it inaccessible to surrounding media [44,45]. Although some earlier experiments failed to confirm the localization of SR-BI on the bile canaliculus [46,47], these studies clearly demonstrated the presence of SR-BI on the canalicular membrane [48] (Fig. 7.4).

Using both fluorescence and electron microscopy, we demonstrated that SR-BI localizes to the bile canaliculus as well as the basolateral membrane. With this system, we established a link between cellular sterol availability and hepatic SR-BI trafficking. Cholesterol loading was found to elicit SR-BI transcytosis from the basolateral membrane to the bile canaliculus. Following cholesterol loading, SR-BI translocation was delayed by approximately 60 minutes and then preceded with a half time of 34 minutes. Notably, transcytosis occurred with both methyl β-cyclodextrin (MβCD) and lipoprotein (LDL or HDL) cholesterol loading, demonstrating that the direct plasma membrane binding of HDL to SR-BI and SR-BI-mediated signaling [49] are not required. SR-BI transcytosis thus appears to be tightly tied to plasma membrane free cholesterol levels and possibly to SR-BI raft cholesterol levels. The slow time course noted for SR-BI translocation resembles that of another transcytotic raft protein, 5′nucleotidase, a GPI-anchored protein that associates with and uses membrane rafts for its slow transcytotic itinerary [50]. Future studies will inevitably focus on fully understanding SR-BI transcytosis, includingthe contribution of cellular signaling pathways, the receptor's intracellular trafficking itinerary, and the functional significance of the translocation in terms of biliary cholesterol secretion.

Fig. 7.4 *SR-BI localizes on the bile canaliculus in WIF-B cells.* WIF-B cells infected with SR-BI-YFP (*A*) were labeled with an antibody against MDR1 (*B*). Bile canalicular SR-BI-YFP are highlighted with white arrows in the merge image (*C*). SR-BI-YFP localized in microdomains on the BC is seen after immunolabeling with anti-MDR1/Rhodamine Red-X (*D–F*). Fixing and immunostaining of uninfected WIF-B cells with anti-SR-BI/Alexa 488 and anti-MDR1/Alexa 647 demonstrates that endogenous SR-BI also localizes on the bile canaliculus (*G–I*). The DIC image of (*G–I*) is seen in panel (*K*). Canalicular SR-BI-YFP in polarized WIF-B cells was detected by EM with an Anti-YFP antibody visualized with 15 nm gold coupled to an anti-mouse secondary antibody (*J*). Clusters of SR-BI-YFP on bile canaliculi are highlighted in the insets. Asterisks denote BC. A–C, G–I and K: Bars = 10 μm. D–E: Bars = 5 μm. J: Bars = 100 nm.

7.6 Cellular Mechanisms of SR-BI-Mediated Biliary Cholesterol Secretion

By facilitating selective uptake, hepatic SR-BI provides a pool of cholesterol available for secretion and, as demonstrated by Schwartz et al., HDL-derived cholesterol is the preferential substrate for biliary secretion [51]. Since there is only a partial decrease in biliary cholesterol secretion in SR-BI (−/−) mice [52], hepatic SR-BI is not an absolute requirement for this process. Silver et al. [53] and Wustner et al. [54] have proposed that selective depletion of HDL cholesterol destined for transcytotic delivery to the bile canaliculus occurs as HDL undergoes recycling in hepatocytes. This process requires active transport of cholesterol in lipoproteins or cholesterol-enriched domains. However, in contrast to the rapid transcytosis seen by other recycling cargo such as transferrin, our studies revealed a lag phase of about 60 min for SR-BI transcytosis in response to cholesterol loading. Furthermore, SR-BI transcytosis was independent of lipoprotein binding and was not specific to HDL.

SR-BI has been shown to exist in detergent-resistant microdomains or lipid rafts [55–58]. Given that rafts have long been studied as transcytotic vehicles [59], it is possible that cholesterol is actively transported across the cell in rafts, albeit – from the recently reported recycling dynamics of SR-BI to and from the bile canaliculus [48] – such a trafficking process is unlikely to account for significant amounts of selective cholesterol transport. Sehayek et al. also detected SR-BI on the bile canaliculus and suggested that it may facilitate cholesterol efflux to phospholipid micelles [60], a hypothesis consistent with the known function of SR-BI as a bidirectional channel [61]. SR-BI has been shown to localize in phosphatidylcholine-enriched membranes [62] of type similar to canalicular membranes. It is possible that SR-BI creates a unique lipid environment on the bile canaliculus conducive to cholesterol and/or phospholipid efflux. Relevantly to this hypothesis, we have demonstrated that SR-BI dynamically associates with microdomains on the bile canaliculus. SR-BI has been shown to bind cholesterol [63] and may provide a pool of cholesterol on the bile canaliculus available for secretion. Additionally, SR-BI may promote the formation of canalicular microvilli as seen in other cell types [64], thereby enhancing the curvature of the canalicular membrane to facilitate cholesterol desorption into bile acid micelles.

SR-BI clearly has an important role in HDL remodeling and the final stages of reverse cholesterol transport. As we have shown, SR-BI-mediated selective uptake does not require HDL endocytosis. However, many questions remain unanswered, including the teleological significance of SR-BI-mediated HDL endocytosis and recycling, the function of the apparently stable pool of SR-BI on the late endosome, and, in hepatocytes, the molecular regulation of cholesterol-mediated transcytosis to the bile canaliculus and its possible role in biliary cholesterol secretion (Fig. 7.5).

7.7 Role of CETP in Direct Clearance of HDL-derived CE (Selective Uptake)

CETP mRNA is predominantly expressed in the liver, spleen, and adipose tissue in humans and is upregulated by cholesterol by a mechanism that involves both a

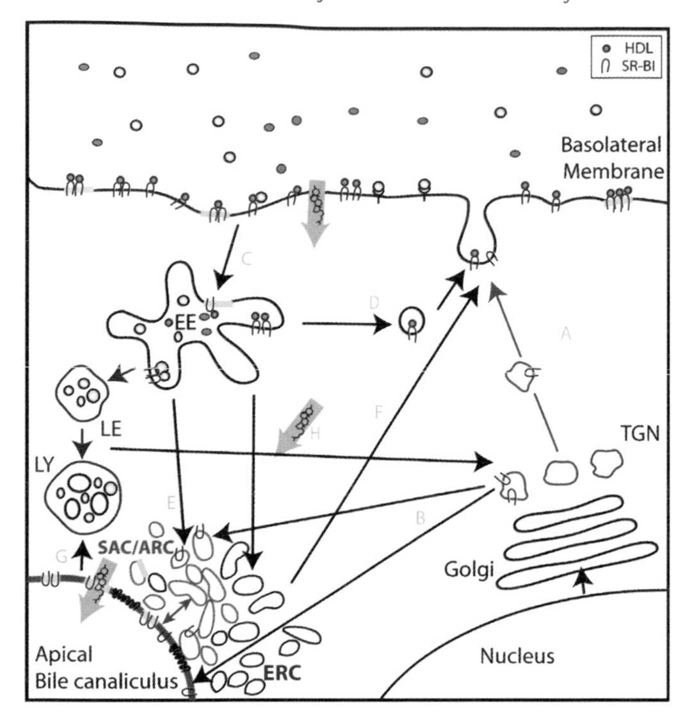

Fig. 7.5 *Intracellular trafficking of SR-BI and cholesterol in a polarized hepatocyte.* In normal hepatocytes, SR-BI is transported directly from the Golgi to the basolateral plasma membrane (*A*). Some SR-BI may traffic from the Golgi to the bile canaliculus bypassing the plasma membrane (*B*). This trafficking would either occur directly or through a subapical compartment (SAC/ARC). SR-BI endocytosed from the basolateral surface is probably internalized into an early endosome (*C*), recycles to the basolateral surface (*D*), or undergoes transcytosis to the bile canaliculus (*E*). On the bile canaliculus, SR-BI dynamically associates with membrane microdomains (green/yellow membrane bars), from where it either recycles to the basolateral surface (*F*) or traffics to the lysosomes for degradation (*G*). Cholesterol is separated from HDL or LDL at the basolateral surface and undergoes transcytosis in a non-vesicular fashion (*H*). On the bile canaliculus, the combined actions of ABCG5/G8, and possibly SR-BI, mediate the secretion of this cholesterol into bile.

sterol response element (CRE) between nucleotides –361 and –138, which binds YY1 and SREBP-1 [65], and by a DR-4 element, which binds LXR/RXR [66]. Notably, plasma concentrations of CETP are increased in various hyperlipidemic states and in response to cholesterol intake [67,68]. Dietary or pharmacological treatments that lower plasma lipids, such as statin therapy, also reduce circulating CETP levels [68,69].

The overall role of CETP in atherosclerosis remains unclear. High plasma concentrations of CETP are associated with low HDL cholesterol levels, which are a risk factor for coronary artery disease (CAD). This observation has led to the development of CETP inhibitors as a potential therapy to reduce atherosclerosis. On the other hand, under conditions of efficient hepatic apoB lipoprotein clearance, CETP may promote cholesterol transport from HDL to the liver.

Although SR-BI has been shown to be the primary receptor responsible for hepatic HDL-cholesterol clearance in the mouse [70], a species that intrinsically lacks CETP

[5], the role of the human homologue of SR-BI – CLA-1 – in lipoprotein metabolism is less clear. Genetic deficiency of CETP in humans results in a marked increase in plasma concentrations of HDL-CE [72,73], which suggests that CETP is required for normal HDL cholesterol clearance in humans. Interestingly, the CE-rich, low-density HDL particles characteristic of SR-BI deficiency in mice are very similar to the HDL particles seen in CETP-deficient patients [74]. These particles contain increased amounts of apoE and can be cleared by members of the LDL receptor family.

CETP and SR-BI each function to reduce HDL size and cholesterol content. We have demonstrated in adipocytes [75,76] and more recently in hepatocytes [77] that CETP mediates the selective acquisition of CE from HDL in a fashion similar to SR-BI. Furthermore, experiments using primary hepatocytes from either SR-BI- or LDL-deficient mice demonstrated that CETP mediates the selective uptake of HDL-derived CE by a mechanism that does not involve SR-BI or the transfer of CE to apo B-containing lipoproteins and their subsequent uptake by a member of the LDL receptor gene family [78]. For these studies, the receptor associated protein (RAP) was used to block uptake by the members of the LDLr family through competitive binding and the presence of heparin was used to disrupt binding of HL to the heparin sulfate proteoglycan (HSPG) matrix. When adenovirally expressed *in vivo*, hepatic expression of CETP resulted in significant remodeling of plasma HDL, consistently with a major physiological role for hepatocyte CETP as a receptor mediating selective uptake.

In other studies, we used the CETP inhibitor torcetrapib [79,80] as a tool to unravel the distinct mechanisms by which exogenous and cell-associated CETP mediate selective uptake. In order to determine the effects of cell-associated CETP on selective uptake, we first used low titers of adenovirus to express CETP in isolated primary murine hepatocytes. Under these conditions, there was no detectable CETP protein or CETP transfer activity in the cell lysates. Low levels of endogenously expressed CETP increased selective uptake of HDL-CE by 50 %, and this effect was not inhibited by torcetrapib. Addition of physiological concentrations (0.48 $\mu g\ mL^{-1}$) of CETP to the media of primary hepatocytes increased selective uptake fivefold and torcetrapib was found to reduce selective uptake of HDL-derived cholesterol mediated by exogenous CETP by 80 %. However, torcetrapib did not completely prevent selective uptake and even in the presence of 25 μm torcetrapib, exogenous CETP increased ^{3}H-CE incorporation twofold. Exogenous CETP thus appears to increase selective uptake of HDL-CE by hepatocytes through two distinct pathways. Only one of these is sensitive to torcetrapib but, as we have shown, it is not contingent on transfer of CE to apoB-containing lipoproteins and does not require the participation of SR-BI. Torcetrapib has been shown to increase the association of CETP with its lipoprotein substrates to create a nonfunctional complex [81], which would be expected to impair the ability of CETP to shuttle CE directly to the plasma membrane. This may account for the torcetrapib-sensitive component of CETP-mediated selective uptake. In contrast, the torcetrapib-insensitive pathway may represent cell-associated CETP. Neither the reversible nor the irreversible phases of selective uptake mediated by cell-associated CETP were affected by torcetrapib.

From studies in hepatocytes [82] and in adipocytes [83,84], we have proposed two pathways to explain CETP-mediated selective uptake of CE. Firstly, CETP may mediate selective uptake by shuttling CE directly from HDL to the plasma membrane. This process may or may not require the participation of another protein on the cell surface, but clearly does not require SR-BI, members of the LDL receptor family, or an intact HSPG matrix. It is possible that CETP transfers lipids to membrane protrusions that have a high curvature and may, therefore, be comparable to a lipoprotein at a molecular level [85]. Secondly, cell-associated CETP appears to perform selective uptake by a mechanism that does not involve CETP-mediated shuttling of CE from HDL to the plasma membrane but, rather, represents direct interaction of HDL with CETP on the cell surface. One hypothesis is that plasma membrane associated CETP may mediate the transient fusion of the HDL amphipathic coat with the membrane outer leaflet, allowing CE transfer without HDL particle uptake. CETP contains a C-terminal peptide that has a tilted orientation relative to the lipid–water interface, and this peptide has fusogenic properties similar to those described for viral fusion peptides [86]. Indeed, CETP has been shown to mediate lipoprotein fusion under certain circumstances [87]. Evidently, further studies will be required to dissect the molecular mechanisms of CETP-mediated selective uptake and its possible role in intracellular cholesterol trafficking (Fig. 7.6).

Mice deficient in SR-BI have been shown to develop atherosclerosis on a high-fat, high-cholesterol diet [88]. Mice do not express CETP, and the introduction of the human CETP transgene results in a marked reduction in HDL-Cholesterol concentrations. The effect on atherosclerosis susceptibility is highly contingent on the level of CETP expression and on the genetic and metabolic context. Thus, the atherosclerosis in apoE and LDL receptor knockout mice is compounded when they are crossed with CETP transgenic mice [89,90]. In contrast, introduction of the CETP transgene results in decreased atherosclerosis in mice deficient in either apo CIII [91] or LCAT [92]. Plasma CETP concentrations in cholesterol-fed rabbits are 10 times those of humans, while inhibition of CETP expression or function by use of antisense oligonucleotides [93], a CETP vaccine [94], or a CETP inhibitor [95] reduced atherosclerosis in this model. In more recent studies, we have demonstrated that the human CETP transgene protects against diet-induced atherosclerosis in SR-BI deficient mice, thereby providing new and important insight into the possible roles of CETP and SR-BI as alternative routes for HDL-CE clearance by the liver [100]. These findings have significant implications for the role of CETP in the final steps of reverse cholesterol transport.

7.8 Summary

CETP and SR-BI play important roles in HDL remodeling and hepatic clearance of HDL-derived cholesterol. CETP participates in lipoprotein remodeling in the plasma compartment as well as in the selective uptake of HDL-CE by hepatocytes [96] and adipocytes. Mice do not express CETP and the introduction of the human transgene results in a marked reduction in HDL-cholesterol concentrations. However, the effect on atherosclerosis susceptibility is highly contingent on the level of CETP expression and on the genetic and metabolic context. Important to note is the fact that steady state

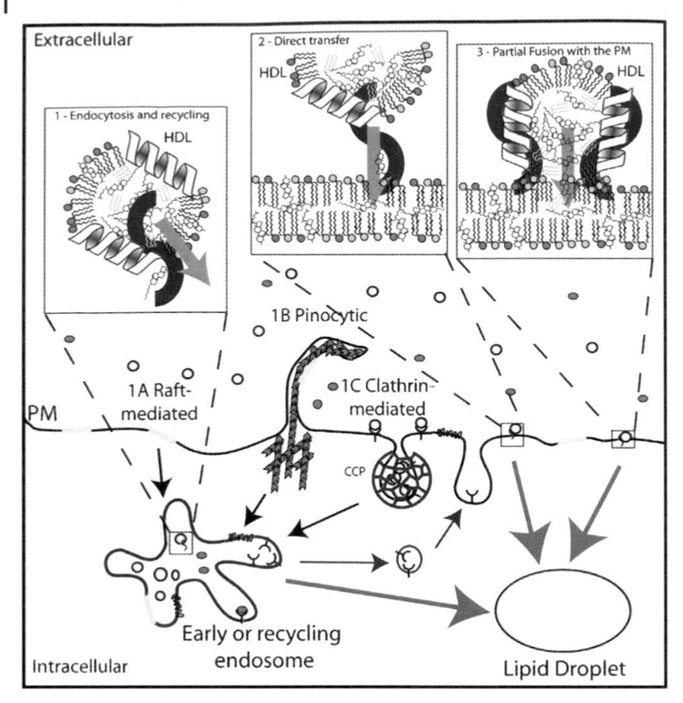

Fig. 7.6 *Three putative mechanisms of CETP-mediated CE selective uptake.* Three hypothetical underlying mechanisms of CETP-mediated selective uptake of CE to hepatocytes and adipocytes exist. The first involves selective depletion of CE during a process of endocytosis and recycling. CETP endocytosis could be raft-mediated (1A), pinocytotic (1B), or through clathrin-coated pit (CCP) internalization (1C). The second mechanism may involve CETP tethering HDL to the cell surface and mediating a direct hydrophobic transfer through its neutral lipid-binding pocket. The third mechanism may involve CETP mediating the partial-fusion of HDL with the plasma membrane. Importantly, CETP endocytosis and recycling most probably coincides with either direct transfer or partial-fusion mediated selective uptake. Following selective uptake, the CE is probably mostly shuttled to a lipid droplet or to the bile canaliculus in polarized hepatocytes (not shown).

plasma concentrations of HDL-C do not necessarily reflect HDL functionality in terms of reverse cholesterol transport. Overall, low plasma levels of HDL-C are an important risk factor for atherosclerosis. However, human alleles that modify HDL-C concentrations may not all have synchronous effects on atherosclerosis. Notably, in a recent study, the CETP polymorphisms associated with HDL-C were distinct from the CETP SNP predicting myocardial infarction [97]. Single SNPs (single nucleotide polymerphism) in CETP predicted HDL-C levels in the Intermountain Heart Collaborative Study but did not predict angiographic coronary artery disease [98]. Furthermore, in a recent large clinical study, the CETP inhibitor torcetrapib increased rather than decreased clinical cardiovascular events. Although active debate on the role of CETP in modifying CHD risk continues [99], we propose that very high plasma concentrations of CETP have unfavorable effects on plasma lipids and possibly on atherogenesis. On the other hand, by mediating the hepatic clearance of HDL-derived cholesterol, CETP may have an important atheroprotective function.

SR-BI has a well established role as an HDL receptor that mediates selective uptake and promotes HDL remodeling. Again, contrary to the widely accepted view that plasma HDL-cholesterol concentrations are a strong negative predictor of coronary artery disease risk, SR-BI deficiency is associated with elevated HDL-C but increased atherosclerosis susceptibility. The cellular functions and trafficking of SR-BI are only beginning to be unraveled. Future studies will focus on the molecular mechanisms by which SR-BI promotes the final steps of reverse cholesterol transport and may provide important new information on the regulation of biliary cholesterol secretion.

References

1 Gauthier, A., Lau, P., Zha, X. H., Milne, R., McPherson, R. (2005) Cholesteryl ester transfer protein directly mediates selective uptake of high density lipoprotein cholesteryl esters by the liver. *Arteriosclerosis Thrombosis and Vascular Biology,* **25**(10), 2177–2184.

2 Ashen, M. D. and Blumenthal, R. S. (22-9-2005) Clinical practice: Low HDL cholesterol levels. *N Engl J Med.,* **353**(12), 1252–1260.

3 Genest, J., Jr.,Bard, J.-M., Fruchart, J.-C., Ordovas, J. M., Schaefer, E. J. (1993) Familial hypoalphalipoproteinemia in premature coronary artery disease. *Arterioscler Thromb.,* **13**, 1728–1737.

4 Brewer, H. B., Remaley, A. T., Neufeld, E. B., Basso, F., Joyce, C. (2004) Regulation of plasma high-density lipoprotein levels by the ABCA1 transporter and the emerging role of high-density lipoprotein in the treatment of cardiovascular disease. *Arteriosclerosis Thrombosis and Vascular Biology,* **24**(10), 1755–1760.

5 Uittenbogaard, A., Shaul, P. W., Yuhanna, I. S., Blair, A., Smart, E. J. (14-4-2000) High density lipoprotein prevents oxidized low density lipoprotein-induced inhibition of endothelial nitric-oxide synthase localization and activation in caveolae. *J Biol Chem.,* **275**(15), 11278–11283.

6 Yuhanna, I. S., Zhu, Y., Cox, B. E., Hahner, L. D., Osborne-Lawrence, S., Marcel, Y. L., Anderson, R. G. W., Mendelsohn, M. E., Hobbs, H. H., Shaul, P. W. (2001) High-density lipoprotein binding to scavenger receptor-BI activates endothelial nitric oxide synthase. *Nature Med.,* **7**(7), 855–859.

7 Cockerill, G. W., Rye, K. A., Gamble, J. R., Vadas, M. A., Barter, P. J. (1995) High-density lipoproteins inhibit cytokine-induced expression of endothelial cell adhesion molecules. *Arterioscler Thromb Vasc Biol.,* **15**, 1987–1994.

8 Barter, P. J., Nicholls, S., Rye, K. A., Anantharamaiah, G. M., Navab, M., Fogelman, A. M. (15-10-2004) Antiinflammatory properties of HDL. *Circ Res.,* **95**(8), 764–772.

9 Navab, M., Hama, S. Y., Cooke, C. J., Anantharamaiah, G. M., Chaddha, M., Jin, L., Subbanagounder, G., Faull, K. F., Reddy, S. T., Miller, N. E., Fogelman, A. M. (2000) Normal high density lipoprotein inhibits three steps in the formation of mildly oxidized low density lipoprotein: step 1. *J Lipid Res.,* **41**(9), 1481–1494.

10 Jacquet, S., Malaval, C., Martinez, L. O., Sak, K., Rolland, C., Perez, C., Nauze, M., Champagne, E., Terce, F., Gachet, C., Perret, B., Collet, X., Boeynaems, J. M., Barbaras, R. (2005) The nucleotide receptor P2Y(13) is a key regulator of hepatic high-density lipoprotein (HDL) endocytosis. *Cellular and Molecular Life Sciences,* **62**(21), 2508–2515.

11 Martinez, L. O., Jacquet, S., Esteve, J. P., Rolland, C., Cabezon, E., Champagne, E., Pineau, T., Georgeaud, V., Walker, J. E., Terce, F., Collet, X., Perret, B., Barbaras, R. (2-1-2003) Ectopic beta-chain of ATP

synthase is an apolipoprotein A-I receptor in hepatic HDL endocytosis. *Nature*, **421**(6918), 75–79.

12 McPherson, R. (15-6-1993) Function of cholesteryl ester transfer protein in reverse cholesterol transport. *CMAJ*, **148**(12), 2165–2166.

13 Rashid, S., Watanabe, T., Sakaue, T., Lewis, G. F. (2003) Mechanisms of HDL lowering in insulin resistant, hypertriglyceridemic states: the combined effect of HDL triglyceride enrichment and elevated hepatic lipase activity. *Clin Biochem.*, **36**(6), 421–429.

14 Acton, S., Rigotti, A., Landschulz, K. T., Xu, S., Hobbs, H. H., Krieger, M. (26-1-1996) Identification of scavenger receptor SR-BI as a high density lipoprotein receptor. *Science*, **271**(5248), 518–520.

15 Vassiliou, G., Benoist, F., Lau, P., Kavaslar, G. N., McPherson, R. (28-12-2001) The low density lipoprotein receptor-related protein contributes to selective uptake of high density lipoprotein cholesteryl esters by SW872 liposarcoma cells and primary human adipocytes. *J Biol Chem.*, **276**(52), 48823–48830.

16 Knecht, T. P. and Pittman, R. C. (1989) A plasma membrane pool of cholesteryl esters that may mediate the selective uptake of cholesteryl esters from high-density lipoproteins. *Biochim Biophys Acta.*, **1002**, 365–375.

17 Brundert, M., Ewert, A., Heeren, J., Behrendt, B., Ramakrishnan, R., Greten, H., Merkel, M., Rinninger, F. (2005) Scavenger receptor class B type I mediates the selective uptake of high-density lipoprotein-associated cholesteryl ester by the liver in mice. *Arterioscler Thromb Vasc Biol.*, **25**(1), 143–148.

18 Babitt, J., Trigatti, B., Rigotti, A., Smart, E. J., Anderson, R. G., Xu, S., Krieger, M. (1997) Murine SR-BI, a high density lipoprotein receptor that mediates selective lipid uptake, is N-glycosylated and fatty acylated and colocalizes with plasma membrane caveolae. *J Biol Chem.*, **272**(20), 13242–13249.

19 Acton, S., Rigotti, A., Landschulz, K. T., Xu, S., Hobbs, H. H., Krieger, M. (26-1-1996) Identification of scavenger receptor SR-BI as a high density lipoprotein receptor. *Science*, **271**(5248), 518–520.

20 Rodrigueza, W. V., Thuahnai, S. T., Temel, R. E., Lund-Katz, S., Phillips, M. C., Williams, D. L. (16-7-1999) Mechanism of scavenger receptor class B type I-mediated selective uptake of cholesteryl esters from high density lipoprotein to adrenal cells. *J Biol Chem.*, **274**(29), 20344–20350.

21 Ji, Y., Jian, B., Wang, N., Sun, Y., Moya, M. D., Phillips, M. C., Rothblat, G. H., Swaney, J. B., Tall, A. R. (22-8-1997) Scavenger receptor BI promotes high density lipoprotein-mediated cellular cholesterol efflux. *J Biol Chem.*, **272**(34), 20982–20985.

22 Zhang, Y. Z., Da Silva, J. R., Reilly, M., Billheimer, J. T., Rothblat, G. H., Rader, D. J. (2005) Hepatic expression of scavenger receptor class B type I (SR-BI) is a positive regulator of macrophage reverse cholesterol transport in vivo. *J Clin Invest*, **115**(10), 2870–2874.

23 Sehayek, E., Wang, R., Ono, J. G., Zinchuk, V. S., Duncan, E. M., Shefer, S., Vance, D. E., Ananthanarayanan, M., Chait, B. T., Breslow, J. L. (2003) Localization of the PE methylation pathway and SR-BI to the canalicular membrane: evidence for apical PC biosynthesis that may promote biliary excretion of phospholipid and cholesterol. *J Lipid Res.*, **44**(9), 1605–1613.

24 Silver, D. L., Wang, N., Xiao, X., Tall, A. R. (6-7-2001) High density lipoprotein (HDL) particle uptake mediated by scavenger receptor class B type 1 results in selective sorting of HDL cholesterol from protein and polarized cholesterol secretion. *J Biol Chem.*, **276**(27 PT 2,), 25287–25293.

25 Kozarsky, K. F., Donahee, M. H., Rigotti, A., Iqbal, S. N., Edelman, E. R., Krieger, M. (1997) Overexpression of the HDL receptor SR-BI alters plasma HDL and bile

cholesterol levels. *Nature,* **387**(6631), 414–417.

26 Sehayek, E., Wang, R., Ono, J. G., Zinchuk, V. S., Duncan, E. M., Shefer, S., Vance, D. E., Ananthanarayanan, M., Chait, B. T., Breslow, J. L. (2003) Localization of the PE methylation pathway and SR-BI to the canalicular membrane: evidence for apical PC biosynthesis that may promote biliary excretion of phospholipid and cholesterol. *J Lipid Res.,* **44**(9), 1605–1613.

27 Burgos, P. V., Klattenhoff, C., De la Fuente, E., Rigotti, A., Gonzalez, A. (16-3-2004) Cholesterol depletion induces PKA-mediated basolateral-to-apical transcytosis of the scavenger receptor class B type I in MDCK cells. *Proc Natl Acad Sci USA,* **101**(11), 3845–3850.

28 Silver, D. L., Wang, N., Xiao, X., Tall, A. R. (6-7-2001) High density lipoprotein (HDL) particle uptake mediated by scavenger receptor class B type 1 results in selective sorting of HDL cholesterol from protein and polarized cholesterol secretion. *J Biol Chem.,* **276**(27 PT 2), 25287–25293.

29 Rhainds, D., Bourgeois, P., Bourret, G., Huard, K., Falstrault, L., Brissette, L. (1-7-2004) Localization and regulation of SR-BI in membrane rafts of HepG2 cells. *J Cell Sci.,* **117**(Pt 15), 3095–3105.

30 Wustner, D. (25-2-2005) Mathematical analysis of hepatic high density lipoprotein transport based on quantitative imaging data. *J Biol Chem.,* **280**(8), 6766–6779.

31 Harder, C. J., Vassiliou, G., McBride, H. M., McPherson, R. (2006) Hepatic SR-BI-mediated cholesteryl ester selective uptake occurs with unaltered efficiency in the absence of cellular energy. *J Lipid Res.,* **47**(3), 492–503.

32 Eckhardt, E. R., Cai, L., Sun, B., Webb, N. R., Vander Westhuyzen, D. R. (2-4-2004) High density lipoprotein uptake by scavenger receptor SR-BII. *J Biol Chem.,* **279**(14), 14372–14381.

33 Nieland, T. J. F., Ehrlich, M., Krieger, M., Kirchhausen, T. (1-5-2005) Endocytosis is not required for the selective lipid uptake mediated by murine SR-BI. *Biochimica et Biophysica Acta-Molecular and Cell Biology of Lipids,* **1734**(1), 44–51.

34 Pittman, R. C., Knecht, T. P., Rosenbaum, M. S., Taylor, C. A., Jr. (1987) A nonendocytotic mechanism for the selective uptake of high density lipoprotein-associated cholesterol esters. *J Biol Chem.,* **262**, 2443–2450.

35 Mardones, P., Quiñones, V., Amigo, L., Moreno, M., Miquel, J. F., Schwarz, M., Miettinen, H. E., Trigatti, B., Krieger, M., VanPatten, S., Cohen, D. E., Rigotti, A. (2001) Hepatic cholesterol and bile acid metabolism and intestinal cholesterol absorption in scavenger receptor class B type I-deficient mice. *J Lipid Res.,* **42**(2), 170–180.

36 Kozarsky, K. F., Donahee, M. H., Rigotti, A., Iqbal, S. N., Edelman, E. R., Krieger, M. (1997) Overexpression of the HDL receptor SR-BI alters plasma HDL and bile cholesterol levels. *Nature,* **387**(6631), 414–417.

37 Yu, L. Q., Li-Hawkins, J., Hammer, R. E., Berge, K. E., Horton, J. D., Cohen, J. C., Hobbs, H. H. (2002) Overexpression of ABCG5 and ABCG8 promotes biliary cholesterol secretion and reduces fractional absorption of dietary cholesterol. *J Clin Invest,* **110**(5), 671–680.

38 Amigo, L., Mendoza, H., Castro, J., Quinones, V., Miquel, J. F., Zanlungo, S. (2002) Relevance of Niemann-Pick type C1 protein expression in controlling plasma cholesterol and biliary lipid secretion in mice. *Hepatology,* **36**(4), 819–828.

39 Amigo, L., Zanlungo, S., Miquel, J. F., Glick, J. M., Hyogo, H., Cohen, D. E., Rigotti, A., Nervi, F. (2003) Hepatic overexpression of sterol carrier protein-2 inhibits VLDL production and reciprocally enhances biliary lipid secretion. *J Lipid Res.,* **44**(2), 399–407.

40 Moreno, M., Molina, H., Amigo, L., Zanlungo, S., Arrese, M., Rigotti, A., Miquel, J. F. (2003) Hepatic

overexpression of caveolins increases bile salt secretion in mice. *Hepatology*, **38**(6), 1477–1488.

41 Geuken, E., Visser, D. S., Leuvenink, H. G. D., de Jong, K. P., Peeters, P. M. J. G., Slooff, M. J. H., Kuipers, F., Porte, R. J. (2005) Hepatic expression of ABC transporters G5 and G8 does not correlate with biliary cholesterol secretion in liver transplant patients. *Hepatology*, **42**(5), 1166–1174.

42 Ihrke, G., Neufeld, E. B., Meads, T., Shanks, M. R., Cassio, D., Laurent, M., Schroer, T. A., Pagano, R. E., Hubbard, A. L. (1993) Wif-B Cells – An In-Vitro Model for Studies of Hepatocyte Polarity. *J Cell Biol.*, **123**(6), 1761–1775.

43 Shanks, M. R., Cassio, D., Lecoq, O., Hubbard, A. L. (1994) An improved polarized rat hepatoma hybrid cell line. Generation and comparison with its hepatoma relatives and hepatocytes in vivo. *J Cell Sci.*, **107**(Pt 4), 813–825.

44 Ihrke, G., Neufeld, E. B., Meads, T., Shanks, M. R., Cassio, D., Laurent, M., Schroer, T. A., Pagano, R. E., Hubbard, A. L. (1993) Wif-B Cells – An In-Vitro Model for Studies of Hepatocyte Polarity. *J Cell Biol.*, **123**(6), 1761–1775.

45 Shanks, M. R., Cassio, D., Lecoq, O., Hubbard, A. L. (1994) An improved polarized rat hepatoma hybrid cell line. Generation and comparison with its hepatoma relatives and hepatocytes in vivo. *J Cell Sci.*, **107**(Pt 4), 813–825.

46 Mardones, P., Quiñones, V., Amigo, L., Moreno, M., Miquel, J. F., Schwarz, M., Miettinen, H. E., Trigatti, B., Krieger, M., VanPatten, S., Cohen, D. E., Rigotti, A. (2001) Hepatic cholesterol and bile acid metabolism and intestinal cholesterol absorption in scavenger receptor class B type I-deficient mice. *J Lipid Res.*, **42**(2), 170–180.

47 Stangl, H., Graf, G. A., Yu, L., Cao, G., Wyne, K. (2002) Effect of estrogen on scavenger receptor BI expression in the rat. *Journal of Endocrinology*, **175**(3), 663–672.

48 Harder, C. J., Meng, A., Rippstein, P., McBride, H., McPherson, R., (2007) SR-BI undergoes cholesterol-stimulated transcytosis to the bile canaliculus in polarized WIF-B cells. *J Biol Chem.* **282**, 1445–1555.

49 Assanasen, C., Mineo, C., Seetharam, D., Yuhanna, I. S., Marcel, Y. L., Connelly, M. A., Williams, D. L., Llera-Moya, M., Shaul, P. W., Silver, D. L. (2005) Cholesterol binding, efflux, and a PDZ-interacting domain of scavenger receptor-BI mediate HDL-initiated signaling. *J Clin Invest.*, **115**(4), 969–977.

50 Schell, M. J., Maurice, M., Stieger, B., Hubbard, A. L. (1992) 5′ Nucleotidase Is Sorted to the Apical Domain of Hepatocytes Via An Indirect Route. *Journal of Cell Biology*, **119**(5), 1173–1182.

51 Schwartz, C. C., Halloran, L. G., Vlahcevic, Z. R., Gregory, D. H., Swell, L. (1978) Preferential Utilization of Free Cholesterol from High-Density Lipoproteins for Biliary Cholesterol Secretion in Man. *Science*, **200**(4337), 62–64.

52 Trigatti, B., Rayburn, H., Viñals, M., Braun, A., Miettinen, H., Penman, M., Hertz, M., Schrenzel, M., Amigo, L., Rigotti, A., Krieger, M. (3-8-1999) Influence of the high density lipoprotein receptor SR-BI on reproductive and cardiovascular pathophysiology. *Proc Natl Acad Sci USA*, **96**(16), 9322–9327.

53 Silver, D. L., Wang, N., Xiao, X., Tall, A. R. (6-7-2001) High density lipoprotein (HDL) particle uptake mediated by scavenger receptor class B type 1 results in selective sorting of HDL cholesterol from protein and polarized cholesterol secretion. *J Biol Chem.*, **276**(27 PT 2), 25287–25293.

54 Wustner, D., Mondal, M., Huang, A., Maxfield, F. R. (2004) Different transport routes for high density lipoprotein and its associated free sterol in polarized hepatic cells. *J Lipid Res.*, **45**(3), 427–437.

55 Burgos, P. V., Klattenhoff, C., De la Fuente, E., Rigotti, A., Gonzalez, A. (16-3-2004) Cholesterol depletion induces PKA-mediated basolateral-to-apical transcytosis of the scavenger receptor class B type I in MDCK cells. *Proc Natl Acad Sci USA*, **101**(11), 3845–3850.

56 Camarota, L. M., Chapman, J. M., Hui, D. Y., Howles, P. N. (25-6-2004) Carboxyl ester lipase cofractionates with scavenger receptor BI in hepatocyte lipid rafts and enhances selective uptake and hydrolysis of cholesteryl esters from H DL3. *J Biol Chem.*, **279**(26), 27599–27606.

57 Peng, Y., Akmentin, W., Connelly, M. A., Lund-Katz, S., Phillips, M. C., Williams, D. L. (2004) Scavenger receptor BI (SR-BI) clustered on microvillar extensions suggests that this plasma membrane domain is a way station for cholesterol trafficking between cells and high-density lipoprotein. *Mol Biol Cell.*, **15**(1), 384–396.

58 Rhainds, D., Bourgeois, P., Bourret, G., Huard, K., Falstrault, L., Brissette, L. (1-7-2004) Localization and regulation of SR-BI in membrane rafts of HepG2 cells. *J Cell Sci.*, **117**(Pt 15), 3095–3105.

59 Nyasae, L. K., Hubbard, A. L., Tuma, P. L. (2003) Transcytotic efflux from early endosomes is dependent on cholesterol and glycosphingolipids in polarized hepatic cells. *Mol Biol Cell.*, **14**(7), 2689–2705.

60 Sehayek, E., Wang, R., Ono, J. G., Zinchuk, V. S., Duncan, E. M., Shefer, S., Vance, D. E., Ananthanarayanan, M., Chait, B. T., Breslow, J. L. (2003) Localization of the PE methylation pathway and SR-BI to the canalicular membrane: evidence for apical PC biosynthesis that may promote biliary excretion of phospholipid and cholesterol. *J Lipid Res.*, **44**(9), 1605–1613.

61 Rodrigueza, W. V., Thuahnai, S. T., Temel, R. E., Lund-Katz, S., Phillips, M. C., Williams, D. L. (16-7-1999) Mechanism of scavenger receptor class B type I-mediated selective uptake of cholesteryl esters from high density lipoprotein to adrenal cells. *J Biol Chem.*, **274**(29), 20344–20350.

62 Parathath, S., Connelly, M. A., Rieger, R. A., Klein, S. M., Abumrad, N. A., Llera-Moya, M., Iden, C. R., Rothblat, G. H., Williams, D. L. (1-10-2004) Changes in plasma membrane properties and phosphatidylcholine subspecies of insect Sf9 cells due to expression of scavenger receptor class B, type I, and CD36. *J Biol Chem.* **279**(40), 41310–41318.

63 Assanasen, C., Mineo, C., Seetharam, D., Yuhanna, I. S., Marcel, Y. L., Connelly, M. A., Williams, D. L., Llera-Moya, M., Shaul, P. W., Silver, D. L. (2005) Cholesterol binding, efflux, and a PDZ-interacting domain of scavenger receptor-BI mediate HDL-initiated signaling. *J Clin Invest.*, **115**(4), 969–977.

64 Williams, D. L., Wong, J. S., Hamilton, R. L. (2002) SR-BI is required for microvillar channel formation and the localization of HDL particles to the surface of adrenocortical cells in vivo. *J Lipid Res.*, **43**(4), 544–549.

65 Gauthier, B., Robb, M., Gaudet, F., Ginsburg, G. S., McPherson, R. (1999) Characterization of a cholesterol response element (CRE) in the promoter of the cholesteryl ester transfer protein gene: functional role of the transcription factors SREBP-1a, -2, and YY1. *J Lipid Res.*, **40**(7), 1284–1293.

66 Luo, Y. and Tall, A. R. (2000) Sterol upregulation of human CETP expression in vitro and in transgenic mice by an LXR element. *J Clin Invest.*, **105**(4), 513–520.

67 Martin, L. J., Connelly, P. W., Nancoo, D., Wood, N., Zhang, Z. J., Maguire, G., Quinet, E., Tall, A. R., Marcel, Y. L., McPherson, R. (1993) Cholesteryl ester transfer protein and high density lipoprotein responses to

cholesterol feeding in men: relationship to apolipoprotein E genotype. *J Lipid Res.*, **34**(3), 437–446.

68 McPherson, R., Mann, C. J., Tall, A. R., Hogue, M., Martin, L., Milne, R. W., Marcel, Y. L. (1991) Plasma concentrations of cholesteryl ester transfer protein in hyperlipoproteinemia: Relation to cholesteryl ester transfer protein activity and other lipoprotein variables. *Arterioscler Thromb.*, **11**, 797–804.

69 McPherson, R. (1999) Comparative effects of simvastatin and cholestyramine on plasma lipoproteins and CETP in humans. *Can J Clin Pharmacol.*, **6**(2), 85–90.

70 Brundert, M., Ewert, A., Heeren, J., Behrendt, B., Ramakrishnan, R., Greten, H., Merkel, M., Rinninger, F. (2005) Scavenger receptor class B type I mediates the selective uptake of high-density lipoprotein-associated cholesteryl ester by the liver in mice. *Arterioscler Thromb Vasc Biol.*, **25**(1), 143–148.

71 Hogarth, C. A., Roy, A., Ebert, D. L. (2003) Genomic evidence for the absence of a functional cholesteryl ester transfer protein gene in mice and rats. *Comp Biochem Physiol B Biochem Mol Biol.*, **135**(2), 219–229.

72 Inazu, A., Brown, M. L., Hesler, C. B., Agellon, L. B., Koizumi, J., Takata, K., Maruhama, Y., Mabuchi, H., Tall, A. R. (1990) Increased high-density lipoprotein levels caused by a common cholesteryl-ester transfer protein gene mutation. *N Engl J Med.*, **323**, 1234–1238.

73 Inazu, A., Jiang, X-C., Haraki, T., Yagi, K., Kamon, N., Koizumi, J., Mabuchi, H., Takeda, R., Takata, K., Moriyama, Y., Doi, M., Tall, A. (1994) Genetic cholesteryl ester transfer protein deficiency caused by two prevalent mutations as a major determinant of increased levels of high density lipoprotein cholesterol. *J Clin Invest.*, **94**, 1872–1882.

74 Chiba, H., Akita, H., Tsuchihashi, K., Hui, S. P., Takahashi, Y., Fuda, H., Suzuki, H., Shibuya, H., Tsuji, M., Kobayashi, K. (1997) Quantitative and compositional changes in high density lipoprotein subclasses in patients with various genotypes of cholesteryl ester transfer protein deficiency. *J Lipid Res.*, **38**(6), 1204–1216.

75 Benoist, F., Lau, P., McDonnell, M., Doelle, H., Milne, R., McPherson, R. (19-9-1997) Cholesteryl ester transfer protein mediates selective uptake of high density lipoprotein cholesteryl esters by human adipose tissue. *J Biol Chem.* **272**(38), 23572–23577.

76 Vassiliou, G. and McPherson, R. (2004) Role of cholesteryl ester transfer protein in selective uptake of high density lipoprotein cholesteryl esters by adipocytes. *J Lipid Res.*, **45**(9), 1683–1693.

77 Gauthier, A., Lau, P., Zha, X. H., Milne, R., McPherson, R. (2005) Cholesteryl ester transfer protein directly mediates selective uptake of high density lipoprotein cholesteryl esters by the liver. *Arteriosclerosis Thrombosis and Vascular Biology*, **25**(10), 2177–2184.

78 Gauthier, A., Lau, P., Zha, X. H., Milne, R., McPherson, R. (2005) Cholesteryl ester transfer protein directly mediates selective uptake of high density lipoprotein cholesteryl esters by the liver. *Arteriosclerosis Thrombosis and Vascular Biology*, **25**(10), 2177–2184.

79 Clark, R. W., Sutfin, T. A., Ruggeri, R. B., Willauer, A. T., Sugarman, E. D., Magnus-Aryitey, G., Cosgrove, P. G., Sand, T. M., Wester, R. T., Williams, J. A., Perlman, M. E., Bamberger, M. J. (2004) Raising high-density lipoprotein in humans through inhibition of cholesteryl ester transfer protein: an initial multidose study of torcetrapib. *Arterioscler Thromb Vasc Biol.*, **24**(3), 490–497.

80 Lloyd, D. B., Lira, M. E., Wood, L. S., Durham, L. K., Freeman, T. B., Preston, G., Qiu, X., Sugarman, E., Bonnette, P., Lanzetti, A., Milos, P. M., Thompson, J. F. (2005) Cholesteryl ester transfer protein

variants have differential stability but uniform inhibition by torcetrapib. *J Biol Chem.* **280**, 14918–14922.

81 Clark, R. W., Sutfin, T. A., Ruggeri, R. B., Willauer, A. T., Sugarman, E. D., Magnus-Aryitey, G., Cosgrove, P. G., Sand, T. M., Wester, R. T., Williams, J. A., Perlman, M. E., Bamberger, M. J. (2004) Raising high-density lipoprotein in humans through inhibition of cholesteryl ester transfer protein: an initial multidose study of torcetrapib. *Arterioscler Thromb Vasc Biol.*, **24**(3), 490–497.

82 Gauthier, A., Lau, P., Zha, X. H., Milne, R., McPherson, R. (2005) Cholesteryl ester transfer protein directly mediates selective uptake of high density lipoprotein cholesteryl esters by the liver. *Arteriosclerosis Thrombosis and Vascular Biology*, **25**(10), 2177–2184.

83 Benoist, F., Lau, P., McDonnell, M., Doelle, H., Milne, R., McPherson, R. (19-9-1997) Cholesteryl ester transfer protein mediates selective uptake of high density lipoprotein cholesteryl esters by human adipose tissue. *J Biol Chem.*, **272**(38), 23572–23577.

84 Vassiliou, G. and McPherson, R. (2004) Role of cholesteryl ester transfer protein in selective uptake of high density lipoprotein cholesteryl esters by adipocytes. *J Lipid Res.*, **45**(9), 1683–1693.

85 Vassiliou, G. and McPherson, R. (2004) Role of cholesteryl ester transfer protein in selective uptake of high density lipoprotein cholesteryl esters by adipocytes. *J Lipid Res.*, **45**(9), 1683–1693.

86 Brasseur, R., Pillot, T., Lins, L., Vandekerckhove, J., Rosseneu, M. (1997) Peptides in membranes: tipping the balance of membrane stability. *Trends Biochem Sci.*, **22**(5), 167–171.

87 Rye, K. A., Hime, N. J., Barter, P. J. (1997) Evidence that cholesteryl ester transfer protein-mediated reductions in reconstituted high density lipoprotein size involve particle fusion. *J Biol Chem.*, **272**(7), 3953–3960.

88 Van Eck, M., Twisk, J., Hoekstra, M., Van Rij, B. T., Van der Lans, C. A., Bos, I. S., Kruijt, J. K., Kuipers, F., Van Berkel, T. J. (27-6-2003) Differential effects of scavenger receptor BI deficiency on lipid metabolism in cells of the arterial wall and in the liver. *J Biol Chem.*, **278**(26), 23699–23705.

89 Plump, A. S., Masucci-Magoulas, L., Bruce, C., Bisgaier, C. L., Breslow, J. L., Tall, A. R. (1999) Increased atherosclerosis in apoE and LDL receptor gene knock-out mice as a result of human cholesteryl ester transfer protein transgene expression. *Arteriosclerosis Thrombosis and Vascular Biology*, **19**(4), 1105–1110.

90 Westerterp, M., van der Hoogt, C. C., de Haan, W., Offerman, E. H., Dallinga-Thie G. M., Jukema, J. W., Havekes, L. M., Rensen, P. C. (2006) Cholesteryl ester transfer protein decreases high-density lipoprotein and severely aggravates atherosclerosis in APOE*3-Leiden mice. *Arterioscler Thromb Vasc Biol.*, **26**(11), 2552–2559.

91 Hayek, T., Masucci-Magoulas, L., Jiang, X., Walsh, A., Rubin, E., Breslow, J. L., Tall, A. R. (1995) Decreased early atherosclerotic lesions in hypertriglyceridemic mice expressing cholesteryl ester transfer protein transgene. *J Clin Invest.*, **96**(4), 2071–2074.

92 Foger, B., Chase, M., Amar, M. J., Vaisman, B. L., Shamburek, R. D., Paigen, B., Fruchart-Najib, J., Paiz, J. A., Koch, C. A., Hoyt, R. F., Brewer, H. B., Jr., Santamarina-Fojo, S. (24-12-1999) Cholesteryl ester transfer protein corrects dysfunctional high density lipoproteins and reduces aortic atherosclerosis in lecithin cholesterol acyltransferase transgenic mice. *J Biol Chem.*, **274**(52), 36912–36920.

93 Sugano, M., Makino, N., Sawada, S., Otsuka, S., Watanabe, M., Okamoto, H., Kamada, M., Mizushima, A.

(27-2-1998) Effect of antisense oligonucleotides against cholesteryl ester transfer protein on the development of atherosclerosis in cholesterol-fed rabbits. *J Biol Chem.*, **273**(9), 5033–5036.

94 Rittershaus, C. W., Miller, D. P., Thomas, L. J., Picard, M. D., Honan, C. M., Emmett, C. D., Pettey, C. L., Adari, H., Hammond, R. A., Beattie, D. T., Callow, A. D., Marsh, H. C., Ryan, U. S. (2000) Vaccine-induced antibodies inhibit CETP activity in vivo and reduce aortic lesions in a rabbit model of atherosclerosis. *Arterioscler Thromb Vasc Biol.*, **20**(9), 2106–2112.

95 Okamoto, H., Yonemori, F., Wakitani, K., Minowa, T., Maeda, K., Shinkai, H. (13-7-2000) A cholesteryl ester transfer protein inhibitor attenuates atherosclerosis in rabbits. *Nature*, **406**(6792), 203–207.

96 Gauthier, A., Lau, P., Zha, X. H., Milne, R., McPherson, R. (2005) Cholesteryl ester transfer protein directly mediates selective uptake of high density lipoprotein cholesteryl esters by the liver. *Arteriosclerosis Thrombosis and Vascular Biology*, **25**(10), 2177–2184.

97 Thompson, J. F., Durham, L. K., Lira, M. E., Shear, C., Milos, P. M. (2005) CETP polymorphisms associated with HDL cholesterol may differ from those associated with cardiovascular disease. *Atherosclerosis*, **181**(1), 45–53.

98 Whiting, B. M., Anderson, J. L., Muhlestein, J. B., Horne, B. D., Bair, T. L., Pearson, R. R., Carlquist, J. F. (2005) Candidate gene susceptibility variants predict intermediate end points but not angiographic coronary artery disease. *Am Heart J.*, **150**(2), 243–250.

99 Stein, O. and Stein, Y. (2005) Lipid transfer proteins (LTP) and atherosclerosis. *Atherosclerosis*, **178**(2), 217–230.

100 Harder, C., Lau, P., Meng, A., Whitman, S., McPherson, R. (2007) Cholesteryl ester transfer protein (CETP) expression protects against diet induced atherosclerosis in SR-BI mice. *Arterioscler Thrombos Vasc Biol.* **26**, 2552–2559.

8
Human Plasma Phospholipid Transfer Protein (PLTP) – Structural and Functional Features

Sarah Siggins, Kerry-Ann Rye, Vesa M. Olkkonen, Matti Jauhiainen, Christian Ehnholm

8.1
Introduction

Atherosclerosis is the leading cause of mortality in industrialized western societies, while lipid abnormalities are key factors involved in the development of cardiovascular disease (CVD). Indeed, lipid-modifying drugs – in particular statins – that decrease plasma low-density lipoprotein (LDL) cholesterol levels considerably reduce the risk of cardiovascular events. Nevertheless, there seem to be limits to the degree of benefit that can be achieved by lowering LDL-cholesterol levels alone, and this has led to increased interest in targeting other lipid-related risk factors for cardiovascular disease, such as low levels of high-density lipoproteins (HDL). Epidemiological studies, which demonstrate that a 10 mg per liter increase in HDL cholesterol concentration is associated with a 2–3 % decrease in cardiovascular risk, suggest that HDL-raising therapies may reduce the risk of cardiovascular disease [1], though this association has not yet been directly proven. Both epidemiological and clinical studies provide strong evidence that a low HDL level is a major risk factor for development of coronary heart disease. Although the exact mechanism(s) behind this observation is/are still unsolved, one of the reasons that this lipoprotein class protects against CVD is thought to be due to its participation in the pathway of reverse cholesterol transport (RCT) [2]. The HDL in human plasma consists of several subpopulations of particles of distinct structure, composition, and function; this heterogeneity is a result of continuous remodeling of HDL by a number of cellular and plasma factors, including the cellular scavenger receptor class B type 1 (SR-BI), the ATP-binding cassette transporters A1 (ABCA1) and G1 (ABCG1), as well as plasma proteins such as cholesteryl ester transfer protein (CETP), phospholipid transfer protein (PLTP), lecithin-cholesterol acyltransferase (LCAT), and the endothelial-bound enzymes lipoprotein lipase (LPL), hepatic lipase (HL), and endothelial lipase (EL) [1]. As indicated by the inverse relationship between HDL-cholesterol and coronary artery disease, plasma HDL levels have a major impact on the progression of atherosclerosis. Among the many HDL-modifying factors, PLTP is a crucial one that affects both the distribution of HDL subpopulations and the levels of circulating HDL.

High-Density Lipoproteins: From Basic Biology to Clinical Aspects. Edited by Christopher J. Fielding

ISBN: 978-3-527-31717-2

8.2 Lipid Transfer Proteins

In human plasma, the transfer of lipids between lipoprotein particles occurs through the action of two distinct transfer proteins: cholesteryl ester transfer protein (CETP) and phospholipid transfer protein (PLTP). CETP is a hydrophobic glycoprotein that is secreted primarily by the liver and circulates in plasma bound to HDL [3] and promotes the redistribution of cholesteryl esters from HDL to triglyceride-rich lipoproteins and of triglycerides from triglyceride-rich lipoproteins to LDL and HDL. PLTP mediates the transfer of phospholipids from triglyceride-rich lipoproteins to HDL and also plays a key role in HDL remodeling [142]. Together with CETP, PLTP is important in RCT because it modifies the metabolic properties and size distribution of HDL [4,5].

CETP was originally designated Lipid Transfer Protein-I (LTP-I), while PLTP was designated LTP-II [6]. Sequence analysis of CETP and PLTP has revealed that they share a 21.7 % amino acid identity, and they belong to the same gene family: the lipid transfer/lipopolysaccharide binding proteins [7]. Other family members include the bactericidal/permeability-increasing protein (BPI) and lipopolysaccharide-binding protein (LBP). BPI and LBP have antibacterial and proinflammatory activities as a consequence of their binding to lipopolysaccharide (LPS) [8,9]. PLTP activity has been detected in all species studied thus far, while CETP activity is also present in a wide range of species [10,11]. Importantly, though, CETP is not present in the mouse, the most commonly used experimental animal model of atherosclerosis [12]. Computational models of CETP and PLTP based on the crystal structure of BPI [13] have been constructed [14,15], and these models predict an elongated, boomerang-like structure containing two highly conserved, hydrophobic lipid-binding pockets. The amino terminal domains in CETP and PLTP are highly homologous, whereas the carboxy terminal domains vary in hydrophobicity and, hence, functional specificity.

The possibility that CETP and PLTP might act as independent risk factors in the development of atherosclerosis has raised the question of whether they protect against or are detrimental to disease progression. This review focuses on the structural and functional features of PLTP.

8.3 Phospholipid Transfer Protein (PLTP)

8.3.1 Characteristics of PLTP

The human PLTP gene is located in chromosome 20q12–q13.1, in close proximity to the genes for LBP and BPI [16,17]. Evidence that PLTP, BPI, and LBP all originate from a common ancestral gene comes from the findings that their intron positionings and exon sizes and distributions are almost identical, and that the structural organization of the entire LPS-binding/lipid transfer protein family is highly conserved [18,19].

The PLTP gene is approximately 13.3 kilobases (kb) in length and comprises 16 exons that range in size from 43 base pairs (bp) (exon 13) to 304 bp (exon 16), along with 15 intervening introns [20,21]. The first exon contains the 5′-untranslated region with an interruption by intron 1 occurring 11 bp upstream of the translation initiation site, a feature characteristic of most apolipoprotein genes [22]. The second exon encodes the entire signal peptide, which is cleaved off upon secretion, together with 16 amino acids of the mature protein. Exons 3 to 15 encode the mature protein; of these, exon 12 encodes a segment of 21 amino acids (residues 372–392) including seven basic amino acids that form a heparin binding site on PLTP. Exon 16 partly contains coding, together with the entire 3′-untranslated region [20,21].

The mouse PLTP gene [143,146], which is localized to chromosome 2, corresponding to human chromosome 20, exhibits intron–exon junctions that are highly conserved and share an 81.1 % nucleotide sequence identity in the promoter region. PLTP complementary DNA (cDNA) is 1750 bp in length and encodes a 17 amino acid hydrophobic signal peptide and a 476 amino acid mature protein [7]. The messenger RNA (mRNA) of PLTP is ubiquitously expressed, with the single 1.8 kb transcript occurring at high levels in the ovary, thymus, and placenta, and at moderate levels in the pancreas, small intestine, testes, lung, and prostate. Relatively low levels of the transcript have been detected in the kidney, liver, and spleen, as well as very low levels in the heart, colon, skeletal muscle, leukocytes, and brain [7,23], indicating that PLTP gene expression has a tissue-specific regulation. It is apparent that this expression differs between human and mouse tissues [23]. Although only moderate levels of PLTP mRNA have been found in the liver, the relatively large mass of this organ may still make it a major contributor to plasma PLTP levels.

Mature PLTP has a predicted molecular mass of 55 kDa; the 81 kDa observed by sodium dodecylsulfate-polyacrylamide gel electrophoresis (SDS-PAGE) of purified plasma PLTP under reducing conditions is due to extensive glycosylation [7,24]. The major secreted forms of PLTP carry complex N-glycans, and N-glycosylation is required for secretion of human PLTP [24,25]. When PLTP is expressed in a baculovirus/insect cell system capable of performing only high-mannose-type glycosylation, efficient secretion and full activity of the protein is still observed [26]. PLTP also has numerous potential O-glycosylation sites that may also contribute to the increased mass estimation by SDS-PAGE. PLTP contains four cysteine residues with the potential to form two intrachain disulfide bonds, and the disulfide bridge that forms between cysteine residues 146 and 185 is important for the structural integrity of the protein [15]. According to homology modeling, PLTP displays a two-domain architecture composed of two symmetrical barrels with hydrophobic pockets (Fig. 8.1). Site-directed mutagenesis of conserved amino acid residues has revealed that the N-terminal pocket is critical for PLTP transfer activity and that the C-terminal pocket is required for HDL binding [15]. The final 30 C-terminal amino acid residues of PLTP are not required for secretion or activity [15]. From secondary structure predictions, PLTP is also thought to contain two potential transmembrane regions spanning from residues 169 through 181 and residues 288 through to 304 [27], although there is no evidence that these would anchor PLTP integrally to membranes.

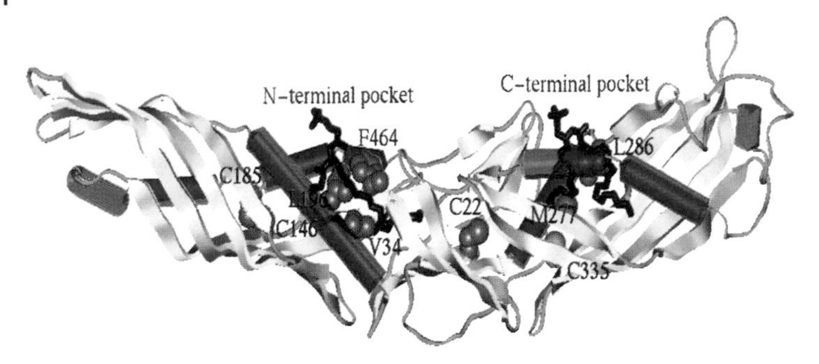

Fig. 8.1 Structural model of human phospholipid transfer protein (adapted from [15]).

8.3.2
Regulation of PLTP Gene Expression

Transcriptional regulation of many genes that are involved in bile acid and cholesterol metabolism occurs through the pivotal role of various members of the orphan nuclear hormone receptor superfamily. The PLTP gene promoter contains response elements to nonsteroidal receptors such as farnesoid X-activated receptor (FXR), peroxisome proliferator-activated receptorα (PPARα), and the liver X-activated receptor (LXR) [28,29]. These nuclear receptors bind DNA as heterodimers with their obligate partner, retinoid X receptor (RXR) [30]. In addition, the PLTP promoter contains binding sites for sterol regulatory element binding protein (SREBP) and CCAT/enhancer binding protein (C/EBP). Within the PLTP promoter region there is an inverted repeat-1 response element (IR-1), to which FXR-RXRα heterodimers bind. Chenodeoxycholic acid (CDCA), a bile acid and potent activator of FXR, is capable of inducing human PLTP promoter eight- to tenfold in the presence of FXR-RXRα [31] and can also induce mouse PLTP gene expression [28]. It is also noteworthy that this effect of CDCA is lost in FXR-KO animals [32].

In addition to FXR response elements, the PLTP promoter region also contains PPARα-responsive element-like sequences, to which PPAR-RXRα heterodimers bind after fatty acid- or fibrate-induced activation. Fibrates constitute a class of drugs that efficiently decrease plasma triglycerides and increase HDL levels in humans [33]. Fenofibrate profoundly affects the promoter activity of the human PLTP gene by lowering its expression [34]. Furthermore, as C/EBP and PPARα share the same binding region on the PLTP gene, C/EBP may compete for PPARα binding, subsequently modifying promoter function in response to fibrates. It has been documented that the ratio of PPARα and C/EBP proteins within cells seems to dictate PLTP gene transcription in response to fibrates [34]. Interestingly, when fenofibrate is administered to the mouse, PLTP gene expression significantly increases, which is the opposite of what occurs in humans [28]. This indicates that, despite the mechanisms for basic transcription of the PLTP gene in human and mouse being highly conserved, the regulatory pathways of PLTP gene expression in response to environmental stimuli are species-specific.

LXRs, which also belong to the orphan nuclear receptor superfamily, act as master transcription factors for the regulation of cholesterol absorption in the intestine, cholesterol catabolism in the liver, and cholesterol efflux from peripheral tissues [35]. The LXR subfamily of nuclear receptors are bound and transcriptionally activated by oxysterols [36]. LXR target genes include apoE [37], LPL [38], ABCG1 [39], ABCG5 and ABCG8 [40], FAS [41], ABCA1, CETP, rodent Cyp7A, and SREBP1c [42]. As the PLTP promoter region contains a high-affinity LXR response element bound by LXR/RXR heterodimers, PLTP is also a direct target gene for LXR [29]. LXR/RXR heterodimers are known to bind to direct repeats of a hexanucleotide repeat spaced by 4 bp (DR-4). The functional LXR-response elements in the PLTP gene that are necessary for maximal induction correspond to DR-4A and DR-4B sequences, and show significant similarity to other known LXR-response elements [29]. Oxysterols and synthetic LXR ligands strongly regulate the expression of murine and human PLTP genes in the liver and in macrophages [29,43,44]. When specific LXR agonists are administered to mice, hepatic PLTP mRNA and plasma PLTP activity increase. This occurs alongside HDL cholesterol and phospholipid elevation, which generates enlarged HDL particles that are enriched in cholesterol, apoA-I, apoE, and phospholipid [43]. Such HDL enlargement may also involve the induction of ABCA1 and apoE: genes that are also involved in HDL metabolism and regulated by LXRs. Short-term administration of synthetic LXR ligands in animals induces ABCA1 and increases plasma HDL cholesterol and phospholipid levels, yet plasma triglycerides also rise [41,45]. With the goal of overcoming the hypertriglyceridemia mediated by LXR activation, but without losing the advantageous effects of LXR-mediated HDL enlargement and hepatic PLTP mRNA induction, a simultaneous activation of LXRs and PPARα by synthetic ligands has been shown to reduce the synthesis of triglycerides in the liver and plasma without hindering the favorable formation of enlarged HDL and increased PLTP [46].

8.4 Functions of PLTP

8.4.1 Phospholipid Transfer Activity

PLTP transfers phospholipids from triglyceride-rich lipoproteins that are being lipolyzed by lipoprotein lipase to HDL (Fig. 8.2) [47]. PLTP also facilitates the transfer of radiolabeled phosphatidylcholine from unilamellar egg phosphatidylcholine vesicles to HDL [47]. The substrate specificity of plasma PLTP has been determined by use of pyrene-labeled phospholipids [48,49]. These studies demonstrate that PLTP is capable of transferring all common phospholipid classes nonspecifically, although phosphatidylethanolamine is transferred more slowly than the other classes. As no PLTP-phospholipid intermediates were detected in these studies, it appears unlikely

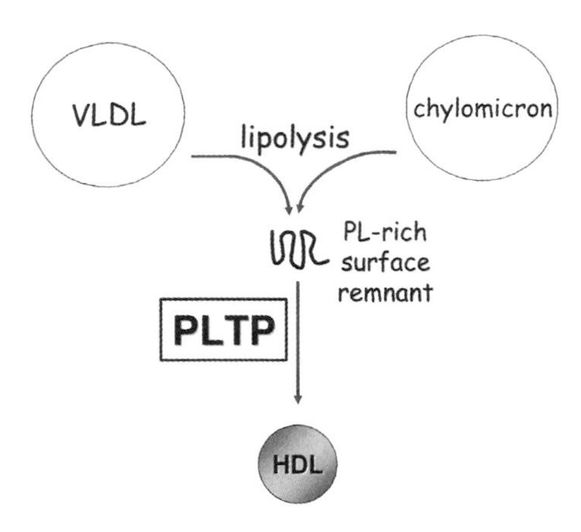

Fig. 8.2 Schematic representation of the function of PLTP in PL transfer to HDL. PL-rich surface remnants are generated during lipolysis of triglyceride-rich chylomicrons and VLDL particles, through the action of lipoprotein lipase. These remnants are transported by PLTP to HDL.

that PLTP forms a tight complex with its lipid substrate. Accordingly, rather than being a true lipid carrier it is more probable that PLTP function involves a ternary complex between donor and acceptor particles to facilitate the transfer of phospholipids. The exact mechanism underlying PLTP-facilitated phospholipid transfers, which in humans is responsible for approximately half of the overall exchange activity between lipoproteins, has not been fully resolved [50]. However, the physiological importance of PLTP *in vivo* is evident in PLTP-deficient mice, which show a total absence of transfer of phospholipids from VLDL to HDL in plasma, along with markedly decreased HDL levels and a subsequent accumulation of apoA-IV-rich lamellar lipoproteins [51,52].

The nonspecificity of PLTP-mediated phospholipid transfer also extends to its ability to transfer diacylglycerol [49,53] and other lipophilic substances such as α-tocopherol (vitamin E) [54], and LPS [55,56]. It also transfers cholesterol between cells and lipoproteins [57]. PLTP plays an important role in determining the distribution of α-tocopherol between lipoproteins and cell membranes [58,59]. α-Tocopherol acts as an antioxidant in plasma lipoproteins and cell membranes, inhibiting LDL oxidation and protecting cellular membranes from oxidative damage, so the deleterious loss of α-tocopherol from apoB-containing lipoproteins due to PLTP activity can be verified by the concomitant increase in the oxidizability of these lipoproteins [59]. PLTP also can modulate phosphatidylserine externalization in erythrocytes, thereby affecting coagulation [60]. On the other hand, PLTP deficiency in the brain, a rich source of α-tocopherol in the body, results in decreased levels of brain α-tocopherol and increases potential for oxidative brain injury [61]. Such findings suggest a novel function for PLTP as a local transporter of α-tocopherol, representing a new route of transfer of α-tocopherol. Furthermore, PLTP deficiency in mice has been shown to impair sperm motility and male fertility; this was due to significantly reduced motility of vitamin E-deficient spermatozoa [62].

8.4.2 HDL Remodeling

In addition to transferring lipophilic substances, PLTP remodels plasma HDL_3 into large HDL_2 particles, together with a concomitant release of lipid-free or lipid-poor apoA-I (termed preβ_1-HDL) [63,64]. The formation of large HDL and preβ_1-HDL (Fig. 8.3) involves particle fusion rather than a net lipid transfer or particle aggregation and relies on the interaction of PLTP with apoA-I [4,65,66,67] Settasatian et al. 2001. It has been suggested that the fusion of unstable particles is a response to the phospholipid transfer process that increases the surface pressure of HDL, thus releasing apoA-I from the surface [65]. HDL conversion is dependent both on PLTP concentration and on time and proceeds as a continuum of intermediate particle formation with the transfer of phospholipids preceding the change in particle size [49]. In addition to human PLTP, recombinant mouse PLTP and pig PLTP can also facilitate HDL conversion [23,68]. Further studies have shown that enrichment of HDL with triglycerides enhances the rate of remodeling, an interesting observation in view of the fact that plasma from hypertriglyceridemic patients contains higher

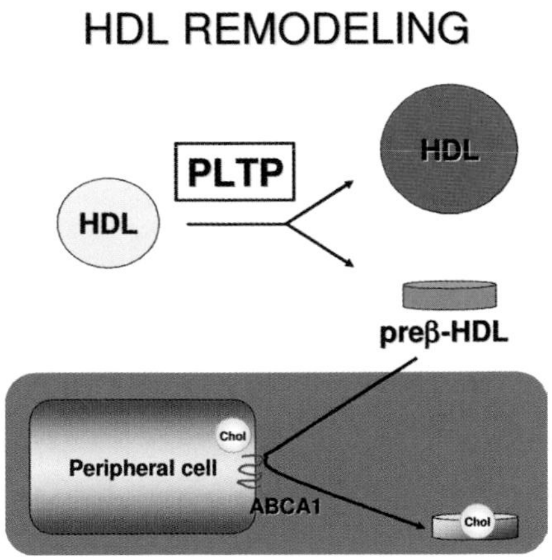

Fig. 8.3 Schematic representation of HDL remodeling facilitated by PLTP. PLTP increases HDL surface pressure, leading to the release of preβ-HDL and fusion of the destabilized HDL particles. The formed preβ-HDL acts as an acceptor for cellular cholesterol/phospholipid efflux mediated by ATP-binding cassette transporter A1 (ABCA1).

levels of small HDL relative to plasma from normolipidemic controls [69,70]. Remodeling is important for the formation of small preβ_1-HDL rich in apoA-I and devoid of triglycerides, having an approximately equimolar PC/SM ratio. These particles are the preferred acceptors of unesterified cholesterol from cell membranes [2,71]. HDL_2 can also act as a substrate for PLTP in generating preβ-HDL [72].

8.4.3 Cellular Cholesterol Efflux

In addition to its role in lipoprotein metabolism in the plasma compartment, increasing evidence points to a function for PLTP in hepatic and extrahepatic tissues. A positive correlation between human serum PLTP activity and cellular cholesterol efflux to HDL has been demonstrated [73]. Furthermore, exposure of cholesterol-enriched human skin fibroblasts to PLTP enhanced cholesterol efflux to HDL, but not to albumin. This was not apparent in fibroblasts from Tangier patients, who have mutations in the ABCA1 transporter [74]. This observation raises the possibility that the enhanced lipid efflux mediated by PLTP occurs through interplay with ABCA1. In the presence of HDL acceptor particles, PLTP efficiently stimulates phospholipid and cholesterol efflux, protects ABCA1 from calpain-mediated degradation, and, together with apoA-I, binds to the same – or closely related – sites on ABCA1 at the cell surface [75]. In the plasma of ABCA1-KO mice, PLTP and LCAT activities are decreased by more than 80 %. This has a significant impact on the maturation of HDL and provides a plausible explanation for the low HDL levels that are observed when ABCA1 is absent [76]. Studies in macrophages obtained from mice deficient in PLTP demonstrate that endogenous PLTP contributes to the optimal function of the ABCA1-mediated cholesterol efflux-promoting machinery in these cells (Lee-Rueckert et al. 2006).

PLTP also promotes cellular lipid efflux by increasing the concentration of plasma preβ_1-HDL *in vitro* [77], and *in vivo* [78]. Additionally, chymase, a neutral protease secreted by mast cells in human atherosclerotic lesions, can proteolytically degrade PLTP and irreversibly inhibit the PLTP-dependent formation of preβ_1-HDL. This is associated with impaired cholesterol efflux from macrophages [79]. Bone marrow transplantation experiments assessing the impact of macrophage-derived PLTP on lesion development have also suggested an atheroprotective role for macrophage PLTP in LDL receptor-deficient mice [80].

8.4.4 PLTP in Human Plasma and Tissues

Genetic PLTP deficiency has not been unequivocally described in humans thus far, apart from one case reported in an abstract form [81]. Studies on the effects of PLTP gene polymorphisms on lipoprotein metabolism have just started [82]. Plasma PLTP activity does not significantly differ between males and females, increases with age, and correlates positively with body mass index, serum cholesterol, and triglycerides [83]. In premenopausal women, PLTP activity correlates positively with IDL and buoyant LDL particles, but not with the dense LDL fractions, whilst PLTP activity in non-obese subjects of both sexes correlates positively with HDL [84–86]. Infusion of apoA-I/PC discs into healthy male subjects increases plasma PLTP activity and small preβ-HDL [87]. Furthermore, when an increase in HDL triglyceride content coincides with high plasma triglycerides, PLTP-mediated preβ-HDL formation is enhanced [70,88,89], so high plasma triglycerides and PLTP activity levels possibly act together in promoting HDL remodeling. Likewise, various studies have reported relationships between plasma PLTP activity, elevated plasma triglycerides, and indexes of obesity [84,90–93]. In a clinical setting, plasma PLTP activity rises following a 24 h intravenous fat load and conversely, it decreases with diet-induced weight loss accompanied by reductions in abdominal subcutaneous fat in males and females [94–96].

Plasma PLTP activity has also been associated with diabetes mellitus [145]. In type 1 diabetes PLTP activity is increased and is related to HDL subclass distribution [97,98]. Furthermore, PLTP activity has been reported to be elevated in type 2 diabetic patients in association with high plasma triglycerides and obesity, and is a positive determinant of carotid intima-media thickness [91,99]; However, plasma PLTP activity is not elevated in type 2 diabetic subjects with relatively normal body mass indexes [91,94]. In comparisons of insulin-resistant with insulin-sensitive subjects without diabetes mellitus, higher plasma PLTP activity is associated with insulin resistance and appears in the context of altered triglyceride metabolism [91,100]. The suppressive effect of insulin on PLTP activity has been demonstrated *in vitro* by use of HepG2 cells, while suppression of PLTP by leptin, the adipocyte-derived hormone that increases with body fat mass, has also been demonstrated [101].

Recently, plasma PLTP activity was found to be related to cardiovascular disease (CVD). Patients within the highest quintile of PLTP activity were shown to have a 1.9-fold increased risk for CVD in relation to patients within the lowest quintile [102], indicating that plasma PLTP activity is an independent predictive value for CVD. Furthermore, PLTP activity has also been shown to increase in conjunction with other independent risk factors for CVD, such as inflammatory markers [103,141] cigarette smoking [104], excessive alcohol intake [105,106], and overconsumption of the diterpenes present in filtered coffee [107,108].

Extracellular and cellular PLTP immunostaining is widespread in atherosclerotic lesions, especially in the macrophage foam cells and SMCs, which suggests that cellular cholesterol accumulation might increase PLTP expression [109,110]. However, the notion that PLTP may be a novel risk factor for CVD and a therapeutic target should be treated with extreme caution on the basis that it may also be beneficial in certain tissues and organs. PLTP mRNA, for example, is widely distributed throughout all regions of the central nervous system at levels comparable to those in other organs. PLTP activity is also present in cerebrospinal fluid (CSF), strongly suggesting that it is synthesized in the brain. Interestingly, a significant increase in PLTP levels in brain tissue homogenates from patients with Alzheimer's disease (AD) has been documented [111], yet when PLTP activity and mass from the CSF of AD patients are measured, an overall reduction in PLTP levels in relation to the control subjects is observed. PLTP activity in the CSF of patients with multiple sclerosis is also lower than in control subjects [112], while in the brains of patients with Down syndrome, there is a downregulation of PLTP mRNA [113]. This indicates that alterations in brain PLTP activity and lipid metabolism are closely linked with the pathogenesis of neurological diseases. Human seminal plasma and human tear fluid also contain high levels of PLTP activity [114,115]. Notably, lung tissue displays very high levels of PLTP mRNA expression relative to other tissues [7], suggesting that PLTP may serve an important role in maintaining the normal function of this organ, such as in the transport of surfactant components, primarily phospholipid. This is consistent with the PLTP gene being highly expressed in alveolar type II epithelial cells and its induction during hypoxia and in emphysema [116].

8.4.5
Low-Activity (LA) and High-Activity (HA) PLTP

A limiting factor in evaluation of the physiological significance of PLTP has been the absence of a suitable method for the quantification of PLTP concentration in biological samples. In order to measure PLTP mass and transfer activity from human samples simultaneously, three groups have developed specific ELISA methods [4,5,117–120]. As each assay employed a different combination of polyclonal and/or monoclonal anti-PLTP antibodies, discrepancies in the correlation of PLTP mass and activity have been observed [144]. An explanation for the lack of association between PLTP mass and transfer activity is that two forms of PLTP exist in human plasma, one being catalytically active (HA-PLTP) and the other inactive (LA-PLTP) [118,119,144]. Adenovirus-mediated overexpression of human PLTP in mice has also revealed the presence of two forms of PLTP [121]. In human plasma, these two forms are associated with macromolecular complexes of different size: the apparent size of LA-PLTP is 520 kDa and that of HA-PLTP is 160 kDa [122]. Partial characterization of the two forms revealed that LA-PLTP is complexed with apoA-I, while HA-PLTP co-purifies with apoE [122]. PLTP secreted by HepG2 cells resembles the HA form and is associated with apoE [123], while PLTP was shown by surface plasmon resonance analysis to bind apoA-I, apoE, and apoA-IV. Furthermore, the activation of LA-PLTP into an active form was achieved with proteoliposomes containing either apoE or apoA-IV, but not with those containing apoA-I [124]. The association of apoE as an important determinant of PLTP activity has also been demonstrated in two clinical trials [125,126]. A model based on these findings, in which nascent PLTP enters the circulation in a high-specific-activity form that is not associated with apoA-I, has been suggested. During or after the transfer of lipolytic surface remnants to HDL, PLTP is transferred to apoA-I-containing HDL particles and thereby rendered inactive. A PLTP ELISA method reported by Siggins et al. [120] allowed accurate measurement of both LA- and HA-PLTP from human plasma [127]. The mean plasma concentrations of LA- and HA-PLTP reported in this study were 3.5 mg L^{-1} and 3.0 mg L^{-1}, respectively, so approximately 45 % of plasma PLTP appears to display high activity and 55 % low activity.

8.5 PLTP and Atherosclerosis

8.5.1 Insights from Mouse Models

The mouse has become a widely accepted model for human diseases such as atherosclerosis, cardiovascular disease, and hyperlipidemia [12,128]. Human PLTP transgenic mice, first created in 1996 [129], display a complicated phenotype. Transgenic mice that express moderate levels (~30 % increase) of human PLTP do not exhibit marked changes in lipoprotein metabolism unless they are crossed into a human apoA-I background, after which increases in α-HDL and preβ-HDL are observed [129,130]. Overexpression of huPLTP by adenovirus-mediated infection increases preβ-HDL levels. It also decreases α-HDL and enhances the hepatic uptake of HDL-CE, thus increasing the fractional catabolic rate of these particles [131,132]. Transgenic mice expressing high PLTP levels (2.5–4.5-fold increases in activity) display 30–40 % decreases in plasma HDL cholesterol and concomitant rises in preβ-HDL formation, in relation to wild-type controls [78]. Similar findings were also made upon transient adenovirus-mediated overexpression of human PLTP [121].

Further analysis of PLTP overexpression in LDL-$R^{+/-}$ mice revealed an increased potential for atherosclerosis due to a moderate (1.5-fold) elevation in VLDL secretion [133]. In a study of the contributions of CETP and PLTP in the metabolism of apoB-100-containing lipoproteins, mice transgenic for CETP and PLTP displayed a similar moderate 1.5-fold elevation of VLDL secretion [134]. Conversely, when CETP and PLTP were overexpressed in the LDL-$R^{+/-}$ background, a strong PLTP dose-dependent decrease in VLDL and LDL cholesterol was observed and, additionally, the activities of the antiatherogenic enzymes PON-1 and PAF-AH were reduced [135]. As HDL cholesterol is also diminished in these animals, the authors concluded that these mice had an increased risk of atherosclerosis.

In PLTP-KO mice the *in vivo* phospholipid transfer from triglyceride-rich lipoproteins to HDL is completely abolished and when on a chow diet, a marked decrease in HDL phospholipid (60 %), cholesterol (65 %), and apoA-I (85 %) is observed. On a high-fat diet, the HDL levels in these mice are similarly decreased, but there is also an increase in VLDL and LDL phospholipids (210 %), free cholesterol (60 %), and cholesteryl ester (40 %), without changes in apoB-100 levels, thus suggesting an accumulation of surface components of triglyceride-rich lipoproteins [51]. In addition, the surface material deriving from the triglyceride-rich lipoproteins represents a fraction enriched in lamellar structures composed of primarily apoA-IV (55 %) and apoE (25 %) and is enriched in phospholipid and free cholesterol [52]. The small HDL pool in these animals has been suggested to be the result of hypercatabolism of HDL protein and CE, an impairment that may possibly lead to the condition of hypoalphalipoproteinemia. When the CETP transgene is expressed in the PLTP-KO background it is unable to compensate for the PLTP deficiency and causes a further lowering of plasma HDL concentrations [136]. Kawano and colleagues later showed that HL and SR-B1 play a major role in the clearance of free cholesterol and phospholipid surface remnants in the plasma of PLTP-deficient mice [137]. To address the involvement of PLTP in the turnover of apoB-100-containing lipoproteins and dyslipidemia, PLTP-KO mice were bred with different hyperlipidemic mouse strains. In this study, PLTP deficiency in apoB-100 transgenic and apoE-deficient backgrounds resulted in reduced production and levels of apoB-100-containing lipoproteins, as well as reduced atherosclerosis. Diminished secretion of apoB-100 from the hepatocytes of these animals was also observed and could be corrected when PLTP was reintroduced by adenovirus [138]. The apparent defect in apoB-100 secretion appeared to be not only a result of the lack of PLTP protein *per se* but also a product of the substantial reduction in the α-tocopherol content and elevated lipid peroxides in the livers of these mice. Replenishment of the isolated primary hepatocytes with α-tocopherol reduced the cellular ROS-dependent destruction of newly synthesized apoB-100 through a post-ER process. As the defect in apoB-100 secretion was fully restored by α-tocopherol alone, this suggested that PLTP inhibitors might be beneficial to increase the α-tocopherol content in plasma LDL [139]. Previous reports by the same authors demonstrated that the α-tocopherol content of VLDL and LDL was significantly increased in PLTP-KO mice [58]. Thus, in mice, the bioavailability of α-tocopherol in atherogenic lipoproteins appears to be downregulated by PLTP, indicating that PLTP deficiency could reduce their atherogenicity. When PLTP removes α-tocopherol from atherogenic lipoproteins, the propensity for atherosclerotic lesions to form is shown to increase [59]. PLTP deficiency was also found to improve the antiinflammatory properties of HDL, as well as to reduce the ability of LDL to induce monocyte chemotactic activity in human artery cell wall co-cultures exposed to human LDL [140]. However, recent studies using the bone marrow transplantation technique demonstrated that macrophage PLTP deficiency increased atherosclerotic lesion formation in LDL receptor-deficient mice, thus suggesting an atheroprotective role for macrophage-derived PLTP [80].

Genetic mouse models have played a crucial role in elucidating the role of PLTP in lipoprotein metabolism and atherosclerosis, whereby PLTP-transgenic and PLTP-KO mice have provided the first *in vivo* evidence for the vital role of PLTP in the maintenance of circulating HDL levels, in phospholipid transfer from post-lipolytic triglyceride-rich lipoproteins to HDL, and in the development of atherosclerosis. However, it still remains to be shown whether the same mechanisms operate in humans.

Data from mouse studies suggest that the antiatherogenic or proatherogenic effect of PLTP is dependent on its site of action. The actions of PLTP in the circulation are connected to the elevated production of apoB-containing lipoproteins, as well as to reduced levels of vitamin E in circulating lipoproteins. Locally produced PLTP within the subendothelial space is an important contributor to atherosclerosis, and a balance of PLTP activity in these compartments seems to determine the impact it has on the development of atherosclerosis.

References

1 Linsel-Nitschke, P. and Tall, A. R. (2005) HDL as a target in the treatment of atherosclerotic cardiovascular disease. *Nature Rev*, **4**, 193–205.

2 Fielding, C. J. and Fielding, P. E. (1995) Molecular physiology of reverse cholesterol transport. *J. Lipid Res*, **36**, 211–228.

3 Tall, A. R., Abreu, E., Shuman, J. (1983) Separation of a plasma phospholipid transfer protein from cholesterol ester/phospholipid exchange protein. *J Biol Chem*, **258** (4), 2174–80.

4 Huuskonen, J., Ekstrom, M., Tahvanainen, E., Vainio, A., Metso, J., Pussinen, P., Ehnholm, C., Olkkonen, V. M., Jauhiainen, M. (2000b) Quantification of human plasma phospholipid transfer protein (PLTP): relationship between PLTP mass and phospholipid transfer activity. *Atherosclerosis*, **151** (2), 451–61.

5 Huuskonen, J., Olkkonen, V. M., Ehnholm, C., Metso, J., Julkunen, I., Jauhiainen, M. (2000c) Phospholipid transfer is a prerequisite for PLTP-mediated HDL conversion. *Biochemistry*, **39** (51), 16092–16098.

6 Tollefson, J. H., Ravnik, S., Albers, J. J. (1988) Isolation and characterization of a phospholipid transfer protein (LTP-II) from human plasma. *J Lipid Res*, **29** (12), 1593–602.

7 Day, J. R., Albers, J. J., Lofton-Day, C. E., Gilbert, T. L., Ching, A. F., Grant, F. J., O'Hara, P. J., Marcovina, S. M., Adolphson, J. L. (1994) Complete cDNA encoding human phospholipid transfer protein from human endothelial cells. *J Biol Chem*, **269** (12), 9388–91.

8 Gray, P. W., Flaggs, G., Leong, S. R., Gumina, R. J., Weiss, J., Ooi, C. E., Elsbach, P. (1989) Cloning of the cDNA of a human neutrophil bactericidal protein. Structural and functional correlations. *J Biol Chem*, **264** (16), 9505–9.

9 Schumann, R. R., Leong, S. R., Flaggs, G. W., Gray, P. W., Wright, S. D., Mathison, J. C., Tobias, P. S., Ulevitch, R. J. (1990) Structure and function of lipopolysaccharide binding protein. *Science*, **249** (4975), 1429–31.

10 Ha, Y. L. and Barter, P. J. (1982) Differences in plasma cholesteryl ester transfer activity in sixteen vertebrate species. *Comp. Biochemistry and Physiology*, **71B**, 265–72.

11 Guyard-Dangremont, V., Desrumaux, C., Gambert, P., Lallemant, C., Lagrost, L. (1998) Phospholipid and cholesteryl ester transfer activities in plasma from 14 vertebrate species. Relation to atherogenesis susceptibility. *Comp Biochem Physiol B Biochem Mol Biol*, **120** (3), 517–25.

12 Breslow, J. L. (1996) Mouse models of atherosclerosis. *Science*, **272** (5262), 685–8.

13 Beamer, L. J., Carroll, S. F., Eisenberg, D. (1997) Crystal structure of human BPI and two bound phospholipids at 2.4 angstrom resolution. *Protein Sci*, **7**, 906–914.

14 Bruce, C., Beamer, L. J., Tall, A. R. (1998) The implications of the structure of the bactericidal/permeability-increasing protein on the lipid-transfer function of the cholesteryl ester transfer protein. *Curr Opin Struct Biol*, **8** (4), 426–34.

15 Huuskonen, J., Wohlfahrt, G., Jauhiainen, M., Ehnholm, C., Teleman, O., Olkkonen, V. M. (1999) Structure and phospholipid transfer activity of human PLTP: analysis by molecular modeling and site-directed mutagenesis. *J Lipid Res*, **40** (6), 1123–30.

16 Gray, P. W., Corcorran, A. E., Eddy, R. L., Jr.,Byers, M. G., Shows, T. B. (1993) The genes for the lipopolysaccharide binding protein (LBP) and the bactericidal permeability increasing protein (BPI) are encoded in the same region of

human chromosome 20. *Genomics*, **15** (1), 188–90.

17 Whitmore, T. E., Day, J. R., Albers, J. J. (1995) Localization of the human phospholipid transfer protein gene to chromosome 20q12–q13.1. *Genomics*, **28** (3), 599–600.

18 Hubacek, J. A., Buchler, C., Aslanidis, C., Schmitz, G. (1997) The genomic organization of the genes for human lipopolysaccharide binding protein (LBP) and bactericidal permeability increasing protein (BPI) is highly conserved. *Biochem Biophys Res Commun*, **236** (2), 427–30.

19 Kirschning, C. J., Au-Young, J., Lamping, N., Reuter, D., Pfeil, D., Seilhamer, J. J., Schumann, R. R. (1997) Similar organization of the lipopolysaccharide-binding protein (LBP) and phospholipid transfer protein (PLTP) genes suggests a common gene family of lipid-binding proteins. *Genomics*, **46** (3), 416–25.

20 Tu, A. Y., Deeb, S. S., Iwasaki, L., Day, J. R., Albers, J. J. (1995) Organization of human phospholipid transfer protein gene. *Biochem Biophys Res Commun*, **207** (2), 552–8.

21 Tu, A. Y., Wolfbauer, G., Albers, J. J. (1995) Functional characterization of the promoter region of the human phospholipid transfer protein gene. *Biochem Biophys Res Commun*, **217** (3), 705–11.

22 Li, W. H., Tanimura, M., Luo, C. C., Datta, S., Chan, L. (1988) The apolipoprotein multigene family: biosynthesis, structure, structure-function relationships, and evolution. *J Lipid Res*, **29** (3), 245–71.

23 Albers, J. J., Wolfbauer, G., Cheung, M. C., Day, J. R., Ching, A. F. T., Lok, S., Tu, A.-Y. (1995) Functional expression of human and mouse plasma phospholipid transfer protein: effect of recombinant and plama PLTP on HDL subspecies. *Biochimica et biophysica acta*, **1258**, 27–34.

24 Huuskonen, J., Jauhiainen, M., Ehnholm, C., Olkkonen, V. M. (1998) Biosynthesis and secretion of human plasma phospholipid transfer protein. *J Lipid Res*, **39** (10), 2021–2030.

25 Qu, S. J., Fan, H. Z., Gillard, B. K., Pownall, H. J. (2006) N-glycosylation is required for secretion-competent human plasma phospholipid transfer protein. *Protein J*, **25**, 167–173.

26 Huuskonen, J., Olkkonen, V. M., Jauhiainen, M., Sareneva, T., Somerharju, P., Ehnholm, C. (1998b) Oxidative modification of HDL3 in vitro and its effect on PLTP-mediated phospholipid transfer. *Biochim Biophys Acta*, **1391** (2), 181–92.

27 Albers, J. J., Tu, A. Y., Wolfbauer, G., Cheung, M. C., Marcovina, S. M. (1996b) Molecular biology of phospholipid transfer protein. *Curr Opin Lipidol*, **7** (2), 88–93.

28 Tu, A. Y. and Albers, J. J. (2001) Functional analysis of the transcriptional activity of the mouse phospholipid transfer protein gene. *Biochem Biophys Res Commun*, **287** (4), 921–6.

29 Laffitte, B. A., Joseph, S. B., Chen, M., Castrillo, A., Repa, J., Wilpitz, D., Mangelsdorf, D., Tontonoz, P. (2003) The phospholipid transfer protein gene is a liver X receptor target expressed by macrophages in atherosclerotic lesions. *Mol Cell Biol*, **23** (6), 2182–91.

30 Chawla, A., Repa, J. J., Evans, R. M., Mangelsdorf, D. J. (2001) Nuclear receptors and lipid physiology: opening the X-files. *Science*, **294** (5548), 1866–70.

31 Urizar, N. L., Dowhan, D. H., Moore, D. D. (2000) The farnesoid X-activated receptor mediates bile acid activation of phospholipid transfer protein gene expression. *J Biol Chem*, **275** (50), 39313–7.

32 Kast, H. R., Nguyen, C. M., Sinal, C. J., Jones, S. A., Laffitte, B. A., Reue, K., Gonzalez, F. J., Willson, T. M., Edwards, P. A. (2001) Farnesoid X-activated receptor induces apolipoprotein C-II transcription: a molecular mechanism linking plasma triglyceride levels to bile acids. *Mol Endocrinol*, **15** (10), 1720–8.

33 Brown, W. V. (1987) Potential use of fenofibrate and other fibric acid derivatives in the clinic. *Am J Med,* **83** (5B), 85–9.

34 Tu, A. Y. and Albers, J. J. (1999) DNA sequences responsible for reduced promoter activity of human phospholipid transfer protein by fibrate. *Biochem Biophys Res Commun,* **264** (3), 802–7.

35 Tontonoz, P. and Mangelsdorf, D. J. (2003) Liver X receptor signaling pathways in cardiovascular disease. *Mol Endocrinol,* **17**, 985–993.

36 Lehmann, J. M., Kliewer, S. A., Moore, L. B., Smith-Oliver, T. A., Oliver, B. B., Su, J. L., Sundseth, S. S., Winegar, D. A., Blanchard, D. E., Spencer, T. A., Willson, T. M. (1997) Activation of the nuclear receptor LXR by oxysterols defines a new hormone response pathway. *J Biol Chem,* **272** (6), 3137–40.

37 Laffitte, B. A., Repa, J. J., Joseph, S. B., Wilpitz, D. C., Kast, H. R., Mangelsdorf, D. J., Tontonoz, P. (2001) LXRs control lipid-inducible expression of the apolipoprotein E gene in macrophages and adipocytes. *Proc Natl Acad Sci USA,* **98** (2), 507–12.

38 Zhang, Y., Repa, J. J., Gauthier, K., Mangelsdorf, D. J. (2001) Regulation of lipoprotein lipase by the oxysterol receptors, LXRalpha and LXRbeta. *J Biol Chem,* **276** (46), 43018–24.

39 Kennedy, M. A., Venkateswaran, A., Tarr, P. T., Xenarios, I., Kudoh, J., Shimizu, N., Edwards, P. A. (2001) Characterization of the human ABCG1 gene: liver X receptor activates an internal promoter that produces a novel transcript encoding an alternative form of the protein. *J Biol Chem,* **276** (42), 39438–47.

40 Repa, J. J., Berge, K. E., Pomajzl, C., Richardson, J. A., Hobbs, H., Mangelsdorf, D. J. (2002) Regulation of ATP-binding cassette sterol transporters ABCG5 and ABCG8 by the liver X receptors alpha and beta. *J Biol Chem,* **277** (21), 18793–800.

41 Joseph, S. B., Laffitte, B. A., Patel, P. H., Watson, M. A., Matsukuma, K. E., Walczak, R., Collins, J. L., Osborne, T. F., Tontonoz, P. (2002) Direct and indirect mechanisms for regulation of fatty acid synthase gene expression by liver X receptors. *J Biol Chem,* **277** (13), 11019–25.

42 Edwards, P. A., Kast, H. R., Anisfeld, A. M. (2002) BAREing it all: the adoption of LXR and FXR and their roles in lipid homeostasis. *J Lipid Res,* **43** (1), 2–12.

43 Cao, G., Beyer, T. P., Yang, X. P., Schmidt, R. J., Zhang, Y., Bensch, W. R., Kauffman, R. F., Gao, H., Ryan, T. P., Liang, Y., Eacho, P. I., Jiang, X. C. (2002) Phospholipid transfer protein is regulated by liver X receptors in vivo. *J Biol Chem,* **277** (42), 39561–5.

44 Mak, P. A., Kast-Woelbern, H. R., Anisfeld, A. M., Edwards, P. A. (2002) Identification of PLTP as an LXR target gene and apoE as an FXR target gene reveals overlapping targets for the two nuclear receptors. *J Lipid Res,* **43** (12), 2037–41.

45 Schultz, J. R., Tu, H., Luk, A., Repa, J. J., Medina, J. C., Li, L., Schwendner, S., Wang, S., Thoolen, M., Mangelsdorf, D. J., Lustig, K. D., Shan, B. (2000) Role of LXRs in control of lipogenesis. *Genes Dev,* **14** (22), 2831–8.

46 Beyer, T. P., Schmidt, R. J., Foxworthy, P., Zhang, Y., Dai, J., Bensch, W. R., Kauffman, R. F., Gao, H., Ryan, T. P., Jiang, X. C., Karathanasis, S. K., Eacho, P. I., Cao, G. (2004) Coadministration of a liver X receptor agonist and a peroxisome proliferator activator receptor-alpha agonist in mice: effects of nuclear receptor interplay on high-density lipoprotein and triglyceride metabolism in vivo. *J Pharmacol Exp Ther,* **309** (3), 861–8.

47 Tall, A., Krumholz, S., Olivecrona, T., Deckelbaum, R. (1985) Plasma phospholipid transfer protein enhances transfer and exchange of phospholipids between very low density lipoproteins and high density lipoproteins during lipolysis. *Journal of Lipid Research,* **26**, 842–851.

48 Huuskonen, J., Olkkonen, V. M., Jauhiainen, M., Metso, J., Somerharju, P., Ehnholm, C. (1996) Acyl chain and headgroup specificity of human plasma phospholipid transfer protein. *Biochim Biophys Acta*, **1303** (3), 207–14.

49 Rao, R., Albers, J. J., Wolfbauer, G., Pownall, H. J. (1997) Molecular and macromolecular specificity of human plasma phospholipid transfer protein. *Biochemistry*, **36** (12), 3645–53.

50 Brown, M. L., Hesler, C., Tall, A. R. (1990) Plasma enzymes and transfer proteins in cholesterol metabolism. *Curr Opin Lipidology*, **1**, 122–127.

51 Jiang, X. C., Bruce, C., Mar, J., Lin, M., Ji, Y., Francone, O. L., Tall, A. R. (1999) Targeted mutation of plasma phospholipid transfer protein gene markedly reduces high-density lipoprotein levels. *J Clin Invest*, **103** (6), 907–914.

52 Qin, S., Kawano, K., Bruce, C., Lin, M., Bisgaier, C., Tall, A. R., Jiang, X. (2000) Phospholipid transfer protein gene knock-out mice have low high density lipoprotein levels, due to hypercatabolism, and accumulate apoA-IV-rich lamellar lipoproteins. *J Lipid Res*, **41** (2), 269–276.

53 Vieu, C., Jaspard, B., Barbaras, R., Manent, J., Chap, H., Perret, B., Collet, X. (1996) Identification and quantification of diacylglycerols in HDL and accessibility to lipase. *J Lipid Res.*, **37**, 1153–1161.

54 Kostner, G. M., Oettl, K., Jauhiainen, M., Ehnholm, C., Esterbauer, H., Dieplinger, H. (1995) Human plasma phospholipid transfer protein accelerates exchange/transfer of alpha-tocopherol between lipoproteins and cells. *Biochem J*, **305** (Pt 2), 659–67.

55 Hailman, E., Albers, J. J., Wolfbauer, G., Tu, A. Y., Wright, S. D. (1996) Neutralization and transfer of lipopolysaccharide by phospholipid transfer protein. *J Biol Chem*, **271** (21), 12172–8.

56 Levels, J. H., Marquart, J. A., Abraham, P. R., vanden Ende, A. E., Molhuizen, H. O., vanDeventer, S. J., Meijers, J. C. (2005) Lipopolysaccharide is transferred from high-density to low-density lipoproteins by lipopolysaccharide-binding protein and phospholipid transfer protein. *Infect Immun*, **73** (4), 2321–6.

57 Nishida, H. I. and Nishida, T. (1997) Phospholipid transfer protein mediates transfer of not only phosphatidylcholine but also cholesterol from phosphatidylcholine-cholesterol vesicles to high density lipoproteins. *J Biol Chem*, **272** (11), 6959–64.

58 Jiang, X. C., Tall, A. R., Qin, S., Lin, M., Schneider, M., Lalanne, F., Deckert, V., Desrumaux, C., Athias, A., Witztum, J. L., Lagrost, L. (2002) Phospholipid transfer protein deficiency protects circulating lipoproteins from oxidation due to the enhanced accumulation of vitamin E. *J Biol Chem*, **277** (35), 31850–6.

59 Yang, X. P., Yan, D., Qiao, C., Liu, R. J., Chen, J. G., Li, J., Schneider, M., Lagrost, L., Xiao, X., Jiang, X. C. (2003) Increased atherosclerotic lesions in apoE mice with plasma phospholipid transfer protein overexpression. *Arterioscler Thromb Vasc Biol*, **23** (9), 1601–7.

60 Klein, A., Deckert, V., Schneider, M., Dutrillaux, F., Hammann, A., Athias, A., LeGuern, N., deBarros, J.-P.P., Desrumaux, C., Masson, D., Jiang, X-C., Lagrost, L. (2006) α-Tocopherol modulates phosphatidylserine externalization in erythrocytes. Relevance in phospholipid transfer protein-deficient mice. *Arteriscler Thromb Vasc Biol*, **26**, 2160–2167.

61 Desrumaux, C., Risold, P. Y., Schroeder, H., Deckert, V., Masson, D., Athias, A., Laplanche, H., LeGuern, N., Blache, D., Jiang, X. C., Tall, A. R., Desor, D., Lagrost, L. (2005) Phospholipid transfer protein (PLTP) deficiency reduces brain vitamin E content and increases anxiety in mice. *Faseb J*, **19** (2), 296–7.

62 Drouineaud, V., Lagrost, L., Klein, A., Desrumaux, C., LeGuern, N., Athias,

A., Menetrier, F., Moiroux, P., Sagot, P., Jimenez, C., Masson, D., Deckert, V. (2006) Phospholipid transfer protein (PLTP) deficiency reduces sperm motility and impairs fertility of mouse males. *FASEB J*, **20**, 794–806.

63 Jauhiainen, M., Metso, J., Pahlman, R., Blomqvist, S., vanTol, A., Ehnholm, C. (1993) Human plasma phospholipid transfer protein causes high density lipoprotein conversion. *J Biol Chem*, **268** (6), 4032–6.

64 Tu, A. Y., Nishida, H. I., Nishida, T. (1993) High density lipoprotein conversion mediated by human plasma phospholipid transfer protein. *J Biol Chem*, **268** (31), 23098–105.

65 Lusa, S., Jauhiainen, M., Metso, J., Somerharju, P., Ehnholm, C. (1996) The mechanism of human plasma phospholipid transfer protein-induced enlargement of high-density lipoprotein particles: evidence for particle fusion. *Biochem J*, **313** (Pt 1), 275–82.

66 Korhonen, A., Jauhiainen, M., Ehnholm, C., Kovanen, P. T., Ala-Korpela, M. (1998) Remodeling of HDL by phospholipid transfer protein: demonstration of particle fusion by 1H NMR spectroscopy. *Biochem Biophys Res Commun*, **249** (3), 910–6.

67 Pownall, H. J. and Ehnholm, C. (2006) The unique role of apolipoprotein A-I in HDL remodeling and metabolism. *Curr Opin Lipidol*, **17**, 209–213.

68 Pussinen, P., Jauhiainen, M., Metso, J., Tyynela, J., Ehnholm, C. (1995) Pig plasma phospholipid transfer protein facilitates HDL interconversion. *J Lipid Res*, **36** (5), 975–85.

69 Murakami, T., Michelagnoli, S., Longhi, R., Gianfranceschi, G., Pazzucconi, F., Calabresi, L., Sirtori, C. R., Franceschini, G. (1995) Triglycerides are major determinants of cholesterol esterification/transfer and HDL remodeling in human plasma. *Arterioscler Thromb Vasc Biol*, **15** (11), 1819–28.

70 Rye, K. A., Jauhiainen, M., Barter, P. J., Ehnholm, C. (1998) Triglyceride-enrichment of high density lipoproteins enhances their remodelling by phospholipid transfer protein. *J Lipid Res*, **39** (3), 613–22.

71 Sviridov, D., Miyazaki, O., Theodore, K., Hoang, A., Fukamachi, I., Nestel, P. (2002) Delineation of the role of pre-beta 1-HDL in cholesterol efflux using isolated pre-beta 1-HDL. *Arterioscler Thromb Vasc Biol*, **22** (9), 1482–8.

72 Marques-Vidal, P., Jauhiainen, M., Metso, J., Ehnholm, C. (1997) Transformation of high density lipoprotein 2 particles by hepatic lipase and phospholipid transfer protein. *Atherosclerosis*, **133** (1), 87–95.

73 Syvanne, M., Castro, G., Dengremont, C., DeGeitere, C., Jauhiainen, M., Ehnholm, C., Michelagnoli, S., Franceschini, G., Kahri, J., Taskinen, M. R. (1996) Cholesterol efflux from Fu5AH hepatoma cells induced by plasma of subjects with or without coronary artery disease and non-insulin-dependent diabetes: importance of LpA-I:A-II particles and phospholipid transfer protein. *Atherosclerosis*, **127** (2), 245–53.

74 Wolfbauer, G., Albers, J. J., Oram, J. F. (1999) Phospholipid transfer protein enhances removal of cellular cholesterol and phospholipids by high-density lipoprotein apolipoproteins. *Biochim Biophys Acta*, **1439** (1), 65–76.

75 Oram, J. F., Wolfbauer, G., Vaughan, A. M., Tang, C., Albers, J. J. (2003) Phospholipid transfer protein interacts with and stabilizes ATP-binding cassette transporter A1 and enhances cholesterol efflux from cells. *J Biol Chem*, **278** (52), 52379–85.

76 Francone, O. L., Subbaiah, P. V., vanTol, A., Royer, L., Haghpassand, M. (2003) Abnormal phospholipid composition impairs HDL biogenesis and maturation in mice lacking Abca1. *Biochemistry*, **42** (28), 8569–78.

77 vonEckardstein, A., Jauhiainen, M., Huang, Y., Metso, J., Langer, C., Pussinen, P., Wu, S., Ehnholm, C., Assmann, G. (1996) Phospholipid transfer protein mediated conversion of high density lipoproteins generates pre beta 1-HDL. *Biochim Biophys Acta*, **1301** (3), 255–62.

78 vanHaperen, R., vanTol, A., Vermeulen, P., Jauhiainen, M., vanGent, T., vandenBerg, P., Ehnholm, S., Grosveld, F., vanderKamp, A., deCrom, R. (2000) Human plasma phospholipid transfer protein increases the antiatherogenic potential of high density lipoproteins in transgenic mice. *Arterioscler Thromb Vasc Biol*, **20** (4), 1082–8.

79 Lee, M., Metso, J., Jauhiainen, M., Kovanen, P. T. (2003) Degradation of phospholipid transfer protein (PLTP) and PLTP-generated pre-beta-high density lipoprotein by mast cell chymase impairs high affinity efflux of cholesterol from macrophage foam cells. *J Biol Chem*, **278** (15), 13539–45.

80 Valenta, D. T., Ogier, L., Bradshaw, G., Black, A. S., Bonnet, D. J., Lagrost, L., Curtiss, L. K., Desrumaux, C. M. (2006) Atheroprotective potential of macrophage-derived phospholipid transfer protein in low-density lipoprotein receptor-deficient mice is overcome by apolipoprotein A-I overproduction. *Arterioscler Thromb Vasc Biol*, **26**, 1572–1580.

81 Mallow, M. J., Zoppo, A., Tu, A.-Y., O'Connor, P., Kunitake, S. T., Hamilton, R. L., Robbins, E., Fielding, C., Kane, J. P. (1994) A new metabolic disorder: phospholipid transfer protein deficiency. *Clinical Research*, **42**:85 Abstract.

82 Aquizerat, B. E., Engler, M. B., Natnzon, Y., Kulkarni, M., Song, J., Eng, C., Huuskonen, J., Rivera, C., Poon, A., Bensley, M., Schnert, A., Zellner, C., Malloy, M., Kane, J., Pullinger, C. R. (2006) Genetic variation of PLTP modulated lipoprotein profiles in hypoalphalipoproteinemia. *J Lipid Res*, **47**, 787–793.

83 Tahvanainen, E., Jauhiainen, M., Funke, H., Vartiainen, E., Sundvall, J., Ehnholm, C. (1999) Serum phospholipid transfer protein activity and genetic variation of the PLTP gene. *Atherosclerosis*, **146** (1), 107–15.

84 Murdoch, S. J., Carr, M. C., Hokanson, J. E., Brunzell, J. D., Albers, J. J. (2000) PLTP activity in premenopausal women. Relationship with lipoprotein lipase, HDL, LDL, body fat, and insulin resistance. *J Lipid Res*, **41** (2), 237–44.

85 Cheung, M. C., Knopp, R. H., Retzlaff, B., Kennedy, H., Wolfbauer, G., Albers, J. J. (2002) Association of plasma phospholipid transfer protein activity with IDL and buoyant LDL: impact of gender and adiposity. *Biochim Biophys Acta*, **1587** (1), 53–9.

86 Murdoch, S. J., Carr, M. C., Kennedy, H., Brunzell, J. D., Albers, J. J. (2002b.) Selective and independent associations of phospholipid transfer protein and hepatic lipase with the LDL subfraction distribution. *J Lipid Res*, **43** (8), 1256–63.

87 Kujiraoka, T., Nanjee, M. N., Oka, T., Ito, M., Nagano, M., Cooke, C. J., Takahashi, S., Olszewski, W. L., Wong, J. S., Stepanova, I. P., Hamilton, R. L., Egashira, T., Hattori, H., Miller, N. E. (2003) Effects of intravenous apolipoprotein A-I/ phosphatidylcholine discs on LCAT, PLTP, and CETP in plasma and peripheral lymph in humans. *Arterioscler Thromb Vasc Biol*, **23** (9), 1653–9.

88 Dullaart, R. P. and vanTol, A. (2001) Role of phospholipid transfer protein and prebeta-high density lipoproteins in maintaining cholesterol efflux from Fu5AH cells to plasma from insulin-resistant subjects. *Scand J Clin Lab Invest*, **61** (1), 69–74.

89 Dullaart, R. P. and vanTol, A. (2001b) Short-term Acipimox decreases the ability of plasma from Type 2 diabetic patients and healthy subjects to stimulate cellular cholesterol efflux: a potentially adverse effect on reverse cholesterol transport. *Diabet Med*, **18** (6), 509–13.

90 Dullaart, R. P., Sluiter, W. J., Dikkeschei, L. D., Hoogenberg, K., VanTol, A. (1994b) Effect of adiposity on plasma lipid transfer protein activities: a possible link between insulin resistance and high density lipoprotein metabolism. *Eur J Clin Invest*, **24** (3), 188–94.

91 Riemens, S. C., vanTol, A., Sluiter, W. J., Dullaart, R. P. (1998b) Plasma phospholipid transfer protein activity is related to insulin resistance: impaired acute lowering by insulin in obese Type II diabetic patients. *Diabetologia*, **41** (8), 929–34.

92 Kaser, S., Sandhofer, A., Foger, B., Ebenbichler, C. F., Igelseder, B., Malaimare, L., Paulweber, B., Patsch, J. R. (2001b) Influence of obesity and insulin sensitivity on phospholipid transfer protein activity. *Diabetologia*, **44** (9), 1111–7.

93 Tzotzas, T., Dumont, L., Triantos, A., Karamouzis, M., Constantinidis, T., Lagrost, L. (2006) Early decreases in plasma lipid transfer proteins during weight reduction. *Obesity*, **14**, 1038–1045.

94 Riemens, S. C., VanTol, A., Sluiter, W. J., Dullaart, R. P. (1999) Acute and chronic effects of a 24-hour intravenous triglyceride emulsion challenge on plasma lecithin: cholesterol acyltransferase, phospholipid transfer protein, and cholesteryl ester transfer protein activities. *J Lipid Res*, **40** (8), 1459–66.

95 Murdoch, S. J., Kahn, S. E., Albers, J. J., Brunzell, J. D., Purnell, J. Q. (2003) PLTP activity decreases with weight loss: changes in PLTP are associated with changes in subcutaneous fat and FFA but not IAF or insulin sensitivity. *J Lipid Res*, **44** (9), 1705–12.

96 Kaser, S., Laimer, M., Sandhofer, A., Salzmann, K., Ebenbichler, C. F., Patsch, J. R. (2004) Effects of weight loss on PLTP activity and HDL particle size. *Int J Obes Relat Metab Disord*, **28** (10), 1280–2.

97 Colhoun, H. M., Scheek, L. M., Rubens, M. B., VanGent, T., Underwood, S. R., Fuller, J. H., VanTol, A. (2001) Lipid transfer protein activities in type 1 diabetic patients without renal failure and nondiabetic control subjects and their association with coronary artery calcification. *Diabetes*, **50** (3), 652–9.

98 Colhoun, H. M., Taskinen, M. R., Otvos, J. D., Van DenBerg, P., O'Connor, J., VanTol, A. (2002) Relationship of phospholipid transfer protein activity to HDL and apolipoprotein B-containing lipoproteins in subjects with and without type 1 diabetes. *Diabetes*, **51** (11), 3300–5.

99 Riemens, S., van Tol, A., Sluiter, W., Dullaart, R. (1998) Elevated plasma cholesteryl ester transfer in NIDDM: relationships with apolipoprotein B-containing lipoproteins and phospholipid transfer protein. *Atherosclerosis*, **140** (1), 71–9.

100 Jonkers, I. J., Smelt, A. H., Hattori, H., Scheek, L. M., van Gent, T., de Man, F. H., vander Laarse, A., van Tol, A. (2003) Decreased PLTP mass but elevated PLTP activity linked to insulin resistance in HTG: effects of bezafibrate therapy. *J Lipid Res*, **44** (8), 1462–9.

101 Kaser, S., Foger, B., Ebenbichler, C. F., Kirchmair, R., Gander, R., Ritsch, A., Sandhofer, A., Patsch, J. R. (2001) Influence of leptin and insulin on lipid transfer proteins in human hepatoma cell line, HepG2. *Int J Obes Relat Metab Disord*, **25** (11), 1633–1639.

102 Schlitt, A., Bickel, C., Thumma, P., Blankenberg, S., Rupprecht, H. J., Meyer, J., Jiang, X. C. (2003) High plasma phospholipid transfer protein levels as a risk factor for coronary artery disease. *Arterioscler Thromb Vasc Biol*, **23** (10), 1857–62.

103 Cheung, M. C., Brown, B. G., Marino Larsen, E. K., Frutkin, A. D., O'Brien, K. D., Albers, J. J. (2006) Phospholipid transfer protein activity is associated with inflammatory markers in patients with cardiovascular disease. *Biochim Biophys Acta*, **1762**, 131–137.

104 Dullaart, R. P., Hoogenberg, K., Dikkeschei, B. D., van Tol, A. (1994) Higher plasma lipid transfer protein activities and unfavorable lipoprotein changes in cigarette-smoking men. *Arterioscler Thromb*, **14** (10), 1581–5.

105 Lagrost, L., Athias, A., Herbeth, B., Guyard-Dangremont, V., Artur, Y., Paille, F., Gambert, P., Lallemant, C. (1996) Opposite effects of cholesteryl ester transfer protein and phospholipid transfer protein on the size distribution of plasma high density lipoproteins. Physiological relevance in alcoholic patients. *J Biol Chem*, **271** (32), 19058–65.

106 Liinamaa, M. J., Hannuksela, M. L., Kesaniemi, Y. A., Savolainen, M. J. (1997) Altered transfer of cholesteryl esters and phospholipids in plasma from alcohol abusers. *Arterioscler Thromb Vasc Biol*, **17** (11), 2940–7.

107 van Tol, A., Urgert, R., deJong-Caesar, R., van Gent, T., Scheek, L. M., de Roos, B., Katan, M. B. (1997) The cholesterol-raising diterpenes from coffee beans increase serum lipid transfer protein activity levels in humans. *Atherosclerosis*, **132** (2), 251–4.

108 De Roos, B., Van Tol, A., Urgert, R., Scheek, L. M., Van Gent, T., Buytenhek, R., Princen, H. M., Katan, M. B., (2000) Consumption of French-press coffee raises cholesteryl ester transfer protein activity levels before LDL cholesterol in normolipidaemic subjects. *J Intern Med*, **248** (3), 211–6.

109 Desrumaux, C. M., Mak, P. A., Boisvert, W. A., Masson, D., Stupack, D., Jauhiainen, M., Ehnholm, C., Curtiss, L. K. (2003) Phospholipid transfer protein is present in human atherosclerotic lesions and is expressed by macrophages and foam cells. *J Lipid Res*, **44** (8), 1453–61.

110 O'Brien, K. D., Vuletic, S., McDonald, T. O., Wolfbauer, G., Lewis, K., Tu, A. Y., Marcovina, S., Wight, T. N., Chait, A., Albers, J. J. (2003) Cell-associated and extracellular phospholipid transfer protein in human coronary atherosclerosis. *Circulation*, **108** (3), 270–4.

111 Vuletic, S., Jin, L. W., Marcovina, S. M., Peskind, E. R., Moller, T., Albers, J. J. (2003) Widespread distribution of PLTP in human CNS: evidence for PLTP synthesis by glia and neurons, and increased levels in Alzheimer's disease. *J Lipid Res*, **44** (6), 1113–23.

112 Vuletic, S., Peskind, E. R., Marcovina, S. M., Quinn, J. F., Cheung, M. C., Kennedy, H., Kaye, J. A., Jin, L. W., Albers, J. J. (2005) Reduced CSF PLTP activity in Alzheimer's disease and other neurologic diseases; PLTP induces ApoE secretion in primary human astrocytes *in vitro*. *J Neurosci Res*, **80** (3), 406–413.

113 Krapfenbauer, K., Yoo, B. C., Kim, S. H., Cairns, N., Lubec, G. (2001) Differential display reveals downregulation of the phospholipid transfer protein (PLTP) at the mRNA level in brains of patients with Down syndrome. *Life Sci*, **68** (18), 2169–79.

114 Masson, D., Drouineaud, V., Moiroux, P., Gautier, T., Dautin, G., Schneider, M., Fruchart-Najib, J., Jauhiainen, M., Ehnholm, C., Sagot, P., Gambert, P., Jimenez, C., Lagrost, L. (2003) Human seminal plasma displays significant phospholipid transfer activity due to the presence of active phospholipid transfer protein. *Mol Hum Reprod*, **9** (8), 457–64.

115 Jauhiainen, M., Setala, N. L., Ehnholm, C., Metso, J., Tervo, T. M., Eriksson, O., Holopainen, J. M. (2005) Phospholipid Transfer Protein Is Present in Human Tear Fluid. *Biochemistry*, **44** (22), 8111–8116.

116 Jiang, X. C., D'Armiento, J., Mallampalli, R. K., Mar, J., Yan, S. F., Lin, M. (1998) Expression of plasma phospholipid transfer protein mRNA in normal and emphysematous lungs and regulation by hypoxia. *J Biol Chem*, **273** (25), 15714–8.

117 Desrumaux, C., Athias, A., Bessede, G., Verges, B., Farnier, M., Persegol, L., Gambert, P., Lagrost, L. (1999) Mass concentration of plasma phospholipid transfer protein in

normolipidemic, type IIa hyperlipidemic, type IIb hyperlipidemic, and non-insulin-dependent diabetic subjects as measured by a specific ELISA. *Arterioscler Thromb Vasc Biol*, **19** (2), 266–75.

118 Oka, T., Kujiraoka, T., Ito, M., Egashira, T., Takahashi, S., Nanjee, M. N., Miller, N. E., Metso, J., Olkkonen, V. M., Ehnholm, C., Jauhiainen, M., Hattori, H. (2000) Distribution of phospholipid transfer protein in human plasma: presence of two forms of phospholipid transfer protein, one catalytically active and the other inactive. *J Lipid Res*, **41** (10), 1651–1657.

119 Oka, T., Kujiraoka, T., Ito, M., Nagano, M., Ishihara, M., Iwasaki, T., Egashira, T., Miller, N. E., Hattori, H. (2000) Measurement of human plasma phospholipid transfer protein by sandwich ELISA. *Clin Chem*, **46** (9), 1357–64.

120 Siggins, S., Kärkkäinen, M., Tenhunen, J., Metso, J., Tahvanainen, E., Olkkonen, V. M., Jauhiainen, M., Ehnholm, C. (2004) Quantitation of the active and low-active forms of human plasma phospholipid transfer protein by ELISA. *J Lipid Res*, **45**, 387–395.

121 Jaari, S., van Dijk, K. W., Olkkonen, V. M., van der Zee, A., Metso, J., Havekes, L., Jauhiainen, M., Ehnholm, C. (2001) Dynamic changes in mouse lipoproteins induced by transiently expressed human phospholipid transfer protein (PLTP): importance of PLTP in prebeta-HDL generation. *Comp Biochem Physiol B Biochem Mol Biol*, **128** (4), 781–92.

122 Kärkkäinen, M., Oka, T., Olkkonen, V. M., Metso, J., Hattori, H., Jauhiainen, M., Ehnholm, C. (2002) Isolation and partial characterization of the inactive and active forms of human plasma phospholipid transfer protein (PLTP). *J Biol Chem*, **277** (18), 15413–8.

123 Siggins, S., Jauhiainen, M., Olkkonen, V. M., Tenhunen, J., Ehnholm, C. (2003) PLTP secreted by HepG2 cells resembles the high-activity PLTP form in human plasma. *J Lipid Res*, **44**, 1698–1704.

124 Jänis, M. T., Metso, J., Lankinen, H., Strandin, T., Olkkonen, V. M., Rye, K. A., Jauhiainen, M., Ehnholm, C. (2005) Apolipoprotein E activates the low-activity form of human phospholipid transfer protein. *Biochem Biophys Res Commun*, **331** (1), 333–40.

125 Tan, K. C. B., Shiu, S. W. M., Wong, Y., Wong, W. K., Tam, S. (2006) Plasma apolipoprotein E concentration is an important determinant of phospholipid transfer protein activity in type 2 diabetes mellitus. *Diabetes Metab Res Rev*, **22**, 307–312.

126 Dallinga-Thie, G. M., van Tol, A., Hattori, H., Rensen, P. C., Sijbrands, E. J. (2006) Plasma phospholipid transfer protein activity is decreased in type 2 diabetes during treatment with atorvastatin: a role for apolipoprotein E? *Diabetes*, **55**, 1491–1496.

127 Jänis, M. T., Siggins, S., Tahvanainen, E., Vikstedt, R., Silander, K., Metso, J., Aromaa, A., Taskinen, M-R., Olkkonen, V. M., Jauhiainen, M., Ehnholm, C. (2004) Active and low-active forms of serum phospholipid transfer protein in a normal Finnish population sample. *J Lipid Res*, **45**, 2303–2309.

128 Paigen, B., Plump, A. S., Rubin, E. M. (1994) The mouse as a model for human cardiovascular disease and hyperlipidemia. *Curr Opin Lipidol*, **5** (4), 258–64.

129 Jiang, X., Francone, O. L., Bruce, C., Milne, R., Mar, J., Walsh, A., Breslow, J. L., Tall, A. R. (1996) Increased prebeta-high density lipoprotein, apolipoprotein AI, and phospholipid in mice expressing the human phospholipid transfer protein and human apolipoprotein AI transgenes. *J Clin Invest*, **98** (10), 2373–80.

130 Albers, J. J., Tu, A. Y., Paigen, B., Chen, H., Cheung, M. C., Marcovina,

S. M. (1996) Transgenic mice expressing human phospholipid transfer protein have increased HDL/non-HDL cholesterol ratio. *Int J Clin Lab Res*, **26** (4), 262–7.

131 Foger, B., Santamarina-Fojo, S., Shamburek, R. D., Parrot, C. L., Talley, G. D., Brewer, H. B., Jr. (1997) Plasma phospholipid transfer protein. Adenovirus-mediated overexpression in mice leads to decreased plasma high density lipoprotein (HDL) and enhanced hepatic uptake of phospholipids and cholesteryl esters from HDL. *J Biol Chem*, **272** (43), 27393–400.

132 Ehnholm, S., vanDijk, K. W., van't Hof, B., vanderZee, A., Olkkonen, V. M., Jauhiainen, M., Hofker, M., Havekes, L., Ehnholm, C. (1998) Adenovirus mediated overexpression of human phospholipid transfer protein alters plasma HDL levels in mice. *J Lipid Res*, **39** (6), 1248–53.

133 Van Haperen, R., Van Tol, A., Van Gent, T., Scheek, L., Visser, P., Van Der Kamp, A., Grosveld, F., De Crom, R. (2002) Increased risk of atherosclerosis by elevated plasma levels of phospholipid transfer protein. *J Biol Chem*, **277** (50), 48938–48943.

134 Lie, J., d., C. R., van Gent, T., van Haperen, R., Scheek, L., Lankhuizen, I., van Tol, A. (2002) Elevation of plasma phospholipid transfer protein in transgenic mice increases VLDL secretion. *J Lipid Res*, **43** (11), 1875–80.

135 Lie, J., de Crom, R., van Gent, T., van Haperen, R., Scheek, L., Sadeghi-Niaraki, F., van Tol, A. (2004) Elevation of plasma phospholipid transfer protein increases the risk of atherosclerosis despite lower apolipoprotein B-containing lipoproteins. *J Lipid Res*, **45** (5), 805–11.

136 Kawano, K., Qin, S. C., Lin, M., Tall, A. R., Jiang, X. C. (2000) Cholesteryl ester transfer protein and phospholipid transfer protein have nonoverlapping functions *in vivo*. *J Biol Chem*, **275** (38), 29477–81.

137 Kawano, K., Qin, S., Vieu, C., Collet, X., Jiang, X. C. (2002) Role of hepatic lipase and scavenger receptor BI in clearing phospholipid/free cholesterol-rich lipoproteins in PLTP-deficient mice. *Biochim Biophys Acta*, **1583** (2), 133–40.

138 Jiang, X. C., Qin, S., Qiao, C., Kawano, K., Lin, M., Skold, A., Xiao, X., Tall, A. R. (2001) Apolipoprotein B secretion and atherosclerosis are decreased in mice with phospholipid-transfer protein deficiency. *Nat Med*, **7** (7), 847–52.

139 Jiang, X. C., Li, Z., Liu, R., Yang, X. P., Pan, M., Lagrost, L., Fisher, E. A., Williams, K. J. (2005) Phospholipid transfer protein deficiency impairs apolipoprotein-B secretion from hepatocytes by stimulating a proteolytic pathway through a relative deficiency of vitamin E and an increase in intracellular oxidants. *J Biol Chem*, **280** (18), 18336–40.

140 Yan, D., Navab, M., Bruce, C., Fogelman, A. M., Jiang, X. C. (2004) PLTP deficiency improves the anti-inflammatory properties of HDL and reduces the ability of LDL to induce monocyte chemotactic activity. *J Lipid Res*, **45** (10), 1852–8.

141 Barlage, S., Frohlich, D., Bottcher, A., Jauhiainen, M., Muller, H. P., Noetzel, F., Rothe, G., Schutt, C., Linke, R. P., Lackner, K. J., Ehnholm, C., Schmitz, G. (2001) ApoE-containing high density lipoproteins and phospholipid transfer protein activity increase in patients with a systemic inflammatory response. *J Lipid Res*, **42** (2), 281–90.

142 Huuskonen, J. and Ehnholm, C. (2000) Phospholipid transfer protein in lipid metabolism. *Curr Opin Lipidol*, **11** (3), 285–289.

143 LeBoeuf, R. C., Caldwell, M., Tu, A., Albers, J. J. (1996) Phospholipid transfer protein maps to distal mouse chromosome 2. *Genomics*, **34** (2), 259–60.

144 Murdoch, S. J., Wolfbauer, G., Kennedy, H., Marcovina, S. M., Carr, M. C., Albers, J. J. (2002) Differences in reactivity of antibodies to active

versus inactive PLTP significantly impacts PLTP measurement. *J Lipid Res*, **43** (2), 281–9.

145 Riemens, S. C., van Tol, A., Sluiter, W. J., Dullaart, R. P. (1999b) Plasma phospholipid transfer protein activity is lowered by 24-h insulin and acipimox administration: blunted response to insulin in type 2 diabetic patients. *Diabetes*, **48** (8), 1631–7.

146 Tu, A. Y., Chen, H., Johnson, K. A., Paigen, B., Albers, J. J. (1997) Characterization of the mouse gene encoding phospholipid transfer protein. *Gene*, **188** (1), 115–8.

9
Use of Gene Transfer to Study HDL Remodeling

Kazuhiro Oka, Lawrence Chan

9.1
Introduction

Coronary artery disease (CAD) is a major cause of morbidity and mortality in industrial countries and our understanding of its pathogenesis has undergone a remarkable transformation during recent decades. Atherosclerosis, which underlies CAD, is a chronic inflammatory disease of arteries [1,2] and is a complex process in which multiple genetic and environmental factors are involved. Epidemiological studies, however, have repeatedly shown an inverse relationship between the risk of atherosclerotic cardiovascular disease and high-density lipoprotein (HDL) cholesterol (HDL-c) levels [3–6]. Apolipoprotein A-I (apoA-I), a major protein component of HDL, appears to be an even better indicator of CAD risk than HDL-cholesterol [7]. Recent population studies and statin trials [8,9] have lent further support for the protective role of HDL and apoA-I, an old notion that first caught the eye of lipoprotein pioneers over a half century ago [10].

HDL-c and apoA-I are not mere markers of CAD risk but are active players in the atherogenic process [11,12]. The most commonly cited mechanism for the relationship between HDL-c and CAD risk has been that HDL facilitates reverse cholesterol transport (RCT). RCT is a concept proposed by Glomset [13], in which cholesterol from peripheral tissues is transported to the liver for disposal into bile and feces. RCT was later suggested to be one of the mechanisms by which HDL protects against atherosclerosis [14]. Other antiatherogenic properties of HDL include antiinflammatory, antioxidant, antithrombotic and nitric oxide-inducing actions [15,16]. Although some clinical trials have shown that raising HDL-c can reduce CAD [17], HDL can also accumulate oxidized lipids under inflammatory conditions. Such modification of HDL lipids, as well as protein components, results in an impairment of HDL's ability to promote cholesterol efflux [18,19]. Therefore, raising HDL-c and improving its function would be one goal for the prevention/treatment of atherosclerosis. How to achieve these goals, however, is still problematic, as illustrated in the recently failed clinical trial of a torcetrapib combination therapy [20–23].

Somatic gene therapy was originally developed to correct or compensate for hereditary disorders through the use of a form of replacement therapy. This therapy

High-Density Lipoproteins: From Basic Biology to Clinical Aspects. Edited by Christopher J. Fielding

ISBN: 978-3-527-31717-2

is simple in concept but its implementation into clinical practice has been challenging. Nevertheless, the development of efficient gene delivery systems has transformed our approach in investigating the pathogenesis of CAD. This review highlights recent advances in gene delivery vectors relevant to research in HDL function and metabolism. We discuss the pros and cons of nonviral vs. viral gene transfer vectors and then review our current understanding of the various components of HDL remodeling, with an emphasis on the role of lipases and other enzymes that affect lipid oxidation in HDL.

9.2
Vectors for Gene Transfer

The underlying concept of gene therapy is that transfer of genetic informational molecules will modify disease phenotypes, yielding therapeutic results. However, the hurdle to clear in order to achieve efficient and safe gene transfer is high. The therapeutic gene must reach the target cell within the desired organ(s). Even as it gains access to the cell, it must travel through the plasma membrane, traffic through cytoplasm and reach the nucleus, where it can highjack the endogenous transcriptional machinery to unleash its encrypted information [24]. Gene transfer vectors can be divided into two categories: nonviral and viral. Direct transfection of naked DNA directly into cells or tissues is relatively inefficient. To effect efficient gene transfer, the gene (DNA) is usually coupled with a liposome or nanoparticle or incorporated in a recombinant viral vector. For effects on lipoproteins such as HDL the major target organ is the liver, since this is the only organ equipped with the necessary machinery for secreting molecules that modify lipoproteins and for excreting cholesterol in bile acids [25]. The efficiency of liver targeting by naked or liposome-complexed DNA is low and the resulting transgene expression tends to be short-lived.

9.2.1
Physical Delivery of Nonviral DNA

The use of naked plasmid DNA has obvious advantages, including the relative ease of DNA preparation and storage. It can be handled as a chemical instead of a biological. Furthermore, viral vectors are generally more immunogenic, which often precludes repeat administration of the same vector. The major limitation for transfection of the liver by naked DNA is poor transduction efficiency by systemic administration; it requires local (portal or hepatic arterial) administration or *ex vivo* transfection. To overcome this problem, a hydrodynamic delivery method has been developed [26]. The procedure involves the rapid injection of a large volume – of approximately 8–12 % of the body weight of the animal – within 5–8 seconds. This results in a transient increase in intravascular pressure, leading to an increase of more than a thousandfold in transgene expression, mainly in the liver. A single tail vein injection of plasmid DNA was able to produce ~40 % of hepatocyte transduction [27].

Increased hepatic transgene expression has also been reported for a hydrodynamic injection of lentiviral [28] and adenoviral vectors [29]. Other physical methods for plasmid DNA delivery include electroporation and ballistic delivery [30].

9.2.2 Nonviral Vectors

The expression of therapeutic genes delivered by nonviral vectors is mostly transient and cannot be maintained over an extended time period. To prolong transgene expression, recently developed nonviral vectors have integration systems incorporated into them [30]. Insertional mutagenesis is the primary concern in viral vectors; however, the integration of a transgene into a predetermined safe site in the host chromosome could avoid this problem.

9.2.2.1 Phage Integrase

Phage integrases are enzymes that mediate unidirectional site-specific recombination between two DNA recognition sequences: the phage attachment site (*attP*) and the bacterial attachment site (*attB*) [31]. Some integrases such as ΦC31 and R4 integrase do not require host cofactors, while others act with the help of phage cofactors and/or bacterial host factors. ΦC31 integrase is functional in mammalian cells [32] and it can integrate plasmid DNA containing an *attB* site into pseudo-*attP* sites in the mammalian genome [33]. ΦC31 integrase acts efficiently on a circular plasmid containing *attB*, while linear DNA serves as its substrate, albeit with low efficiency [34]. This ΦC31 integrase-mediated gene transfer has been used in small animal experiments [35–38]. The integrase system has no apparent upper size limit and, with its size of 11 kb, readily integrates the plasmid [36]. Repeated injections of plasmid DNA can be made without inducing a host immune response to the vector; Chen and Woo have treated a mouse model of phenylketonuria multiple times with multiple co-injections of a plasmid expressing phiBT1 integrase and a therapeutic gene to correct the phenotype [39]. Plasmid DNA is degraded rapidly if it is not integrated, which might be ideal for a plasmid coding for integrase since it should be eliminated from cells when recombination is ended. Although the *attP* recognition sequence is 39 bp in length, 101 integration sites have been identified in human cell lines [40]. Despite the presence of a more restricted pattern of integration sites than seen in Sleeping Beauty transposon [41,42], chromosomal abnormalities including chromosome translocation have been reported with the ΦC31 integrase-mediated transgene integration [43,44]. It is likely that the high level expression of integrase may be responsible for the observed abnormality.

9.2.2.2 Adeno-Associated Virus (AAV) Rep Protein-Mediated Integration

AAVs are the only known mammalian virus capable of site-specific integration in a human chromosome. During latency in humans, AAV2 preferentially integrates at a site on chromosome 19q13.3-qtr [45,46], the so-called "AAVS1" site, by targeting a sequence composed of a *rep* binding element and a terminal resolution site identical to the viral terminal repeats [47,48]. Similar sites have been identified in an African green monkey kidney cell line (CV-1 [49]) and on mouse chromosome 7 [50]. Despite its proximity to the muscle-specific myosin-binding subunit 85 gene, integration into the AAVS1 site has not been associated with any diseases. Alternative splicing of *rep* gene generates multiple Rep proteins (Rep68, Rep78, Rep52, and Rep40). Either Rep68 or Rep78 is sufficient in mediating site-specific integration, while the absence of Rep function allows the transgene to integrate randomly at chromosome breaks [51,52]. Integration of nonviral vector containing a transgene cassette flanked by AAV inverted terminal repeats (ITRs) has no apparent strict size limitation when *Rep* function is supplied *in trans* by a different plasmid. The 138 bp p5 promoter known as p5 integration efficiency element (p5IEE) enhances integration efficiency and is sufficient to mediate AAVS1 integration without ITRs [53]. Rep protein is cytotoxic through arresting of the cell cycle [54]. Another nonviral integrating system is based on DNA transposable elements. The Sleeping Beauty transposon has been used for gene delivery in animal models [38,55–57].

9.2.3
Viral Vectors

Viruses are highly evolved biological machines, the sole purpose of which is to gain access to the host cell and to replicate using the host's cellular machinery. Therefore, viruses have highly efficient gene transfer mechanisms fully tested in nature. There are two major groups of viruses, according to the natures of their genetic materials: DNA viruses and RNA viruses. Genetic materials of RNA viruses are reverse transcribed into the host chromosome before the viral gene is expressed. Although attempts to identify new viral species for vector development are in progress, the best characterized vectors in gene therapy are AAV, adenovirus (Ad), retrovirus, lentivirus, foamy virus, and herpes simplex virus (HSV). Since HSV is effective mainly in the nervous system, we will review only the first five vector systems here (Table 9.1).

9.2.3.1 DNA Viral Vectors

Adeno-associated Virus (AAV) AAVs are single-stranded DNA viruses that were first described as a contaminant of a tissue culture-grown simian adenovirus [58–60]. AAV is a member of the *Parvoviridae* family and has a capsid with icosahedral symmetry approximately 20 nm in diameter [61,62]. Taking advantage of the natural diversity and evolution of AAV in primate species, Gao et al. have identified more than 100 unique AAV capsid sequences [63]. Several AAV receptors and co-receptors have been identified; they include heparan sulfate proteoglycan [64], $\alpha_V\beta_5$ integrins [65,66], fibroblast growth factor receptor-1 [67], platelet-derived growth factor receptor [68], and sialic acid [69], while the 37/67 kDa laminin receptor has recently been identified as a receptor for AAV serotypes 2, 3, 8, and 9 [70]. The presence of these receptors or co-receptors in tissues determines cell type- or tissue-specific transduction, which must be considered for targeting tissues [71,72]. To date, eleven AAV serotypes have

Tab. 9.1 Main features of viral vectors.

Vector	Genetic material	Packaging capacity (kb)	Transduction	Integration	Immune problems	Special features
AAV	ssDNA	∼4.7(‘∼10)*	dividing cells nondividing cells	mainly episomal	low	nonpathogenic
HDAd	dsDNA	∼37	dividing cells nondividing cells	episomal	moderate	
retrovirus	RNA	7–10	dividing cells	integrated	low	
lentivirus	RNA	9–10	dividing cells nondividing cells	integrated	low	
foamy virus	RNA	∼8.5	dividing cells nondividing cells	integrated	low	nonpathogenic

AAV: adeno-associated virus. HDAd: helper-dependent adenovirus. The major advantage of integrating vectors is a persistent gene transfer. However, this feature can be disadvantageous as it may induce oncogenesis in some applications.

been characterized for gene therapy. In contrast to the cellular pathogenic response induced by an Ad vector infection, AAV vectors induce a nonpathogenic response affecting antiproliferative genes [73]. There is no known human disease associated with an AAV infection, making this vector a popular gene therapy vehicle. AAV vectors express the transgene mostly in an episomal form (>90 %), but also in an integrated form (<10 %) [74]. The relatively poor transduction efficiency in the liver has been a limitation to the use of AAV serotype 2 (AAV2), the best characterized AAV vector. However, AAV8 effectively transduces hepatocytes following intravenous administration [75–77].

One disadvantage of an AAV vector has been its cloning capacity of only ∼4.7 kb, although this size limitation can be partially overcome by an intermolecular joining method using two AAV vectors [78–80]. Chao et al. used this approach to transfer the 7 kb Factor VIII gene by intraportal injection of two split AAV vectors [81]. In comparison with an Ad vector, an AAV vector has a relatively long latent period between vector delivery and initiation of transgene expression; this lag period occurs because AAV vectors package a single-stranded genome and require host cell synthesis of the complementary strand for transduction. Use of a self-complementary AAV vector can overcome this problem but further limits the AAV cloning capacity [82,83].

There have been reports of a potential role of AAV2 in abortion [84] and hepatocellular carcinoma in mice [85], but studies involving large numbers of mice have failed to confirm an association between AAV vector treatment and tumor formation [86,87]. Upon infection with a helper virus such as Ad or herpes simplex virus, AAV enters the lytic cycle and undergoes replication and productive infection. Although most recombinant AAV vectors do not contain *rep* gene and express the transgene as an episome [52], the potential for mutation and oncogenesis may exist. In the absence of *rep* gene, AAV genome has been shown to integrate randomly into the mouse genome with propensity towards gene regulatory sequences, albeit at low frequency [88,89]. Despite some limitations, AAV has become a popular viral gene delivery system. AAV2/2, AAV2/7 (AAV pseudotyped with serotype 7 capsids), and AAV2/8 have been used to transfer LDLR and VLDLR genes into *Ldlr*$^{-/-}$ mice [90,91]. Of these, AAV8 was the most effective in inhibiting atherosclerotic lesion progression. ApoE gene transfer by AAV2/7 or AAV2/8 completely prevented the development of atherosclerosis in *Apoe*$^{-/-}$ mice for a year [92], while AAV2/1-mediated apoA-I expression was stable for up to a year in *Apoa1*$^{-/-}$ mice, raising the possibility of induced HDL-c as a therapy [93]. Sharifi et al. expressed apoA-I-Milano using AAV and found that AAV2/1 was most effective in intramuscular injections while intravenous injections of AAV2/1, AAV2/2, and AAV2/5 were all effective [94].

Adenoviral Vector Adenoviruses (Ad) are nonenveloped, double-stranded DNA viruses ~70–100 nm in diameter, each virus possessing an icosahedral protein shell surrounding a protein core that contains the linear viral genome of ~35 kb. Ad infects cells independently of the cell cycle. The primary attachment sites of the fiber protein of Ad subgroup C to cell surface receptors are coxsackievirus and adenovirus receptor (CAR) [95]. The internalization of Ad vector is mediated by interaction of the penton base protein with integrins $\alpha_V\beta_3$ or $\alpha_V\beta_5$ as the secondary internalization receptors [96,97]. Heparan sulfate proteoglycans and LDLR-related proteins have also been reported as cellular attachment sites for this subgroup of Ads [98,99]. In contrast, subgroup B, such as serotypes 11 and 35, uses CD46 as the entry site [100,101]. Of the over 50 different Ad serotypes, subgroup B serotypes 2 and 5 are the best characterized. First-generation Ad vectors (FGAds) are based on these serotypes. FGAds contain all of the essential Ad genome except E1A, which encodes the key regulators of early gene expression. FGAd is replication defective; however, leaky viral gene expression occurring in the absence of the E1A element accounts for some of the toxicity resulting from FGAd vector-mediated gene transfer *in vivo*. In addition to its toxicity and immunogenicity, transgene expression mediated by FGAd is short-lived. In order to attenuate its toxicity further, second- and third-generation Ad vectors have been developed, but there is only marginal improvement in the *in vivo* toxicity or duration of transgene expression in these vectors. Nevertheless, these early-generation Ad vectors have provided useful information on the role of gene products important in lipoprotein metabolism and the pathophysiology of atherosclerosis [102].

Helper-dependent Adenoviral Vector (HDAd) HDAds are the latest-generation Ad vectors, possessing a greatly improved safety profile and allowing for sustained transgene expression *in vivo* [103]. The HDAd lacks all viral protein-coding genes and encompasses only the inverted terminal repeats (ITR) for replication and the packaging signal. Helper virus, a first-generation Ad, provides the necessary viral proteins *in trans* for packaging in cultured cells, so the chronic hepatotoxicity seen in early-generation Ads has been eliminated in HDAd vectors, though some minimal acute toxicity still exists. The innate immune response to HDAd vectors underlies this residual toxicity [104–106], which is the result of the direct interaction of viral particles with innate immune effector cells upon systemic vector administration, an efficient route for global hepatocyte transduction. Transient depletion of Kuppfer cells is effective in reducing toxicity as well as in increasing hepatic transduction [107]. HDAds have a large cloning capacity, accommodating inserts of up to 37 kb. The Ad vector has a natural tropism for the liver after intravenous administration, so many studies have used Ad vectors for proof-of-principle experiments [25]. Transfer of apoE, apoA-I, and LDLR genes has been shown to lead to lifetime correction of phenotypes in mice lacking the corresponding genes [108–110], and the introduction of HDAd vectors has enabled us to study the long-term effects of induced hepatic transgene expression on atherosclerosis [111–114]. Importantly, the effects of long-term stable transgene expression on atherosclerotic lesions were different from those of short-term expression [112]; this is to be expected as atherosclerosis is a chronic disease and only long-term studies address the pathophysiology of true atherosclerotic lesions rather than fatty streak lesions of dubious significance.

9.2.3.2 RNA Viral Vectors

The Moloney murine leukemia retrovirus was the first gene transfer vector developed for gene therapy, while other RNA viral vectors include lentiviruses and foamy viruses. These RNA viral vectors integrate the transgene into the host chromosome, resulting in sustained transgene expression.

Oncoretroviral Vectors Retroviral vectors are single-stranded RNA viruses. The current vector is pseudotyped with the vesicular Stomatitis virus glycoprotein (VSV-G), which confers a broad host range and stabilization of the vector particles, allowing the vector stock to be concentrated to high titers by ultracentrifugation. The packaging capacity of this vector is about 8 kb. Transgene expression persists due to integration into the host chromosome, but the vector genome is integrated randomly, making insertional mutagenesis and oncogenesis rare but definite possibilities. Another drawback is that retroviral vectors work only in cells that are actively dividing because they can access the nucleus only if the nuclear membrane has been broken down. Retroviral vectors are mainly used in *ex vivo* gene therapy and are effective in transducing hematopoietic stem cells [24,74,115].

Lentiviral Vectors Lentiviral vectors were derived from human immunodeficiency virus (a member of the retrovirus family) and are mostly pseudotyped with VSV-G. The deletion of most of the parental genome in packaging constructs minimizes the possibility of generating a replication-competent virus. As in retroviral vectors, most vectors now in use are self-inactivated vectors that lack the regulatory elements in the downstream long-terminal repeat (LTR), eliminating transcription of the packaging signal. Lentiviral vectors can penetrate an intact nuclear membrane and transduce nondividing cells, so lentiviral vectors can be used both in *ex vivo* and in *in vivo* applications. Few immunological problems have been reported for both retroviral and lentiviral vectors. Retroviral vectors favor transgene integration near transcription start sites and show little preference for active transcription units, while lentiviral vectors preferentially integrate into active genes [116–118].

Foamy Viral Vector (FV) Foamy viral vectors are the most recently developed gene transfer vectors [119,120]. Foamy viruses comprise a genus of retroviruses but have distinct properties from oncoretroviruses and lentiviruses. They have the largest genome size of all retroviruses and are considered nonpathogenic [121]. The FV genome contains canonical retroviral *gag, pol,* and *env* genes, but its replication is dependent on a transcriptional activator, Tas, encoded by the *bel*1 gene [122]. FV has a broad tropism and infects both dividing and nondividing cells. FV vectors transduce G_0 fibroblasts, due to the persistence of a stable transduction intermediate in quiescent cells [123] and have a relatively large packaging capacity (up to 9.2 kb) [124] and a distinct integration profile in comparison with other types of retroviruses. Although they show a preference for integration near transcription sites and for CpG islands, they do not integrate preferentially within genes [125]. They have been found to be effective in transducing hematopoietic stem cells of mice and humans [126–128].

9.3 Roles of Lipases and HDL-Associated Enzymes in HDL Remodeling

Endothelial lipase (EL), hepatic lipase (HL), and lipoprotein lipase (LPL) directly interact with circulating lipoproteins and affect lipoprotein metabolism through their catalytic and noncatalytic activities. An overview of lipoprotein remodeling and metabolism in humans is shown in Figure 9.1. Many studies dissecting molecular mechanisms of lipoprotein remodeling have utilized genetically modified mice. There are two major differences between humans and mice in terms of lipoprotein metabolism. In humans, all apoB generated in the liver is apoB100, while the major species in mice is apoB48, because ~70 % of the apoB mRNA is edited in the liver in mice. This results in very-low-density lipoproteins (VLDLs) containing apoB48, which is more rapidly cleared by receptors recognizing apoE as a ligand. Another major difference is the lack of cholesteryl ester transfer protein in mice.

9.3.1 Endothelial Lipase (EL)

EL is a recently identified member of the vascular lipase gene family, which also encompasses LPL and HL, and plays a major role in HDL metabolism [129–131]. EL is expressed in the placenta, thyroid, liver, lung, kidney, testis, and ovary. However, EL expression has been reported not only in endothelial cells but also in hepatocytes and

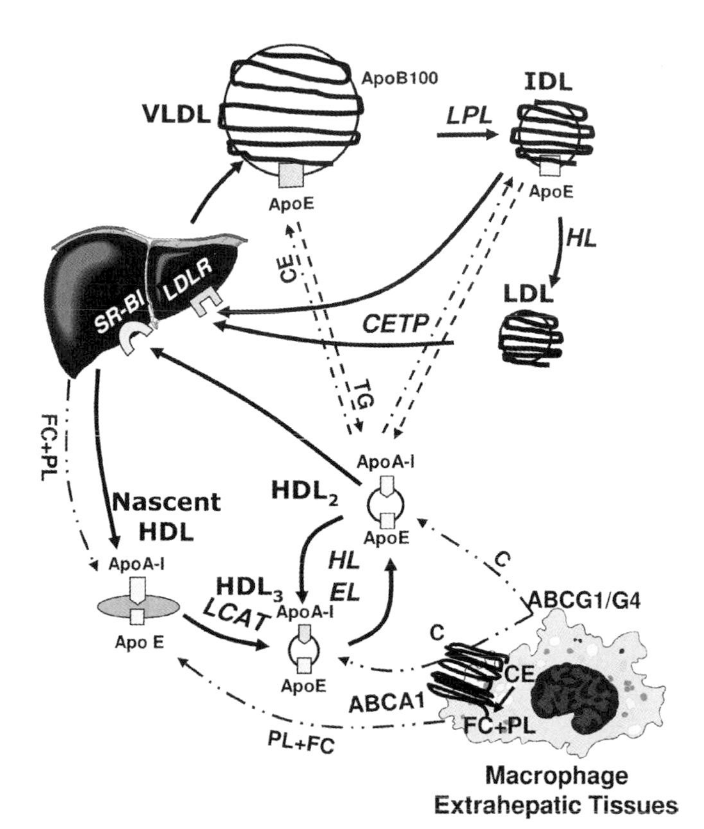

Fig. 9.1 *Overview of metabolism of liver-secreted lipoproteins in humans.* ApoB100, containing triglyceride-rich (TG-rich) very-low-density lipoprotein (VLDL), is secreted from the liver. TGs on VLDL are hydrolyzed to free fatty acids and glycerol by the action of lipoprotein lipase (LPL). This process remodels VLDL to a smaller, denser intermediated-density lipoprotein (IDL). IDL is taken up by the low-density lipoprotein receptor (LDLR) or other apoE receptors through an apoE ligand and further remodeled by hepatic lipase (HL). ApoB100 is the sole apolipoprotein (apo) in LDL particles taken up by LDLR.

Lipid-free or lipid-poor apoA-I is secreted from the liver and serves as an acceptor for ABC transporter-1-mediated (ABCA1-mediated) lipid efflux from hepatocytes as well as macrophages in extrahepatic tissues. ABCA1-mediated efflux of free cholesterol (FC) and phospholipids (PL) to apoA-I forms nascent or pre β-high-density lipoprotein (HDL), which is further modified by lecithin-cholesterol acyltransferase (LCAT). The resulting large, less dense HDL_2 and smaller, denser HDL_3 can serve as acceptors for ABCG1/ABCG4-mediated cholesterol (C) efflux. HL also hydrolyzes PL in HDL_2 and is therefore involved in the conversion of HDL_2 into HDL_3. Endothelial lipase (EL) is a recently identified lipase expressed in endothelial cells, hepatocytes, and macrophages. EL is considered to act on HDL-PL. Cholesteryl ester (CE) from HDL can be transferred to apoB-containing lipoproteins by the action of cholesteryl ester transfer protein (CETP) in exchange for TG. HDL cholesterol is taken up by the liver and is secreted into the bile for disposal through the scavenger receptor BI (SR-BI). There are substantial differences in lipoprotein metabolism between humans and mice. In mice, liver-secreted VLDL contains both apoB100 and apoB48, due to the presence of apoB's mRNA editing activity. TGs on apoB48-VLDL are hydrolyzed by LPL and are remodeled to smaller, denser apoB48-VLDL remnants. These particles are taken up through an interaction with apoE receptors. Moreover, the CETP activity is absent in mice. As a result, normal mice have fewer apoB100-containing lipoproteins and high levels of HDL.

macrophages [129,132]. Although both HL and EL have phospholipase activity, HL appears to act primarily on the triglyceride-rich (TG-rich) HDL particle, while EL may act primarily on HDL phospholipids (PLs) with relatively low TG lipase activity [133,134]. Hepatic overexpression of EL in mice resulted in markedly reduced HDL-c and apoA-I levels and was associated with an increase in post-heparin plasma phospholipase activity, HDL-apolipoprotein catabolism, and apoA-I uptake in kidney and liver [129,134]. The EL transgenic mice had reduced HDL-c [131], while $Hl^{-/-}$ mice had increased HDL-c [130,131]. Ad-mediated overexpression of catalytically inactive EL led to modest reductions in total plasma cholesterol, HDL-c, and phospholipid in HL-deficient mice, suggesting that EL may possess both a lipolytic and a nonlipolytic function in HDL metabolism [135]. Reduction of the maturation of nascent HDL was capable of inhibiting reverse cholesterol transport, whilst an increase in HDL catabolism may accelerate cholesterol disposal, so hepatic overexpression of EL may have both proatherogenic and antiatherogenic consequences. Ishida et al. have reported that a deficiency of EL in $apoE^{-/-}$ mice reduced atherosclerosis [136], while Ko et al. found no effects on atherosclerosis development in either $apoE^{-/-}$ mice or $Ldlr^{-/-}$ mice despite increased LDL-C [137]. HDL is known to inhibit the expression of endothelial adhesion molecules, which is thought to be an important mechanism for the antiinflammatory action of HDL. Ahmed et al. found that the lipase inhibitor tetrahydrolipstatin blunted HDL-mediated inhibition of VCAM-1 expression in TNF-α-stimulated human endothelial cells (ECs). It is likely that EL hydrolysis of HDL generates peroxisome proliferator-activated receptor (PPAR) ligands, which in turn activates PPARs. In support of this mechanism, HDL inhibited leukocyte adhesion to TNF-α-stimulated ECs isolated from wild-type, but not PPAR-α-deficient, mice [138]. LPL has also been reported to generate PPAR-α ligands, but mainly through VLDL hydrolysis [139,140].

9.3.2 Hepatic Lipase (HL)

HL hydrolyzes TG and PL in chylomicron remnants, intermediate-density lipoproteins (IDL), and HDL and plays an important role in lipoprotein metabolism [141]. HL is also a determinant of HDL levels through conversion of the phospholipid-rich (PL-rich) HDL_2 to HDL_3[142]. As in the case of LPL, a noncatalytic function of HL has been reported to be the enhancement of the binding or uptake of lipoproteins through interaction with cell surface receptors [141]. Ad-mediated hepatic overexpression of both wild-type and catalytically inactive HL in $Hl^{-/-}$ mice resulted in reductions of plasma cholesterol, PL, HDL-C, and apoA-II without affecting apoA-I levels. However, clearance of apoA-I-HDL was enhanced in mice treated with wild-type but not mutant HL, demonstrating an important role for catalytic HL in HDL catabolism [143]. Patients with HL deficiency have premature atherosclerosis and elevated plasma cholesterol, TG, VLDL, TG-rich LDL, and HDL subfractions, which are characteristic of type III hyperlipoproteinemia [144,145]. $Hl^{-/-}$ mice have mildly elevated plasma cholesterol (~30 %), PL, and HDL cholesterol. The less dense, large HDL_1 is increased in these mice, demonstrating the importance of HL in HDL metabolism [146]. In a hypertriglyceridemic state, HDL is enriched by TG and has a relative depletion of CE [147]. Rabbits are naturally deficient in HL, and Rashid et al. showed with this model that Ad-mediated overexpression of human HL in rabbits preferentially clears TG-rich HDL [148].

HL is a complex protein with multiple functions and may play both proatherogenic and antiatherogenic roles: overexpression of HL, for example, resulted in an anti-atherogenic lipid profile and reduced the aortic cholesterol content in cholesterol-fed transgenic mice [149–151]. However, HL deficiency in *Apoe*$^{-/-}$ mice also reduced atherosclerosis despite an increase in plasma cholesterol [152]. Similarly, HL deficiency in lecithin cholesterol acyltransferase transgenic mice increased proatherogenic apoB-containing lipoproteins but reduced atherosclerosis [153].

9.3.3
Lipoprotein Lipase (LPL)

LPL is a key enzyme that hydrolyzes TG in circulating TG-rich lipoproteins such as VLDL and chylomicrons. LPL also has noncatalytic activity through anchoring of atherogenic lipoproteins to matrix molecules within the arterial wall and plays a role in the selective uptake of lipoprotein cholesteryl ester [154], so it may modulate atherogenesis [154,155]. The importance of LPL in lipoprotein metabolism is well documented in patients with LPL deficiency, a rare, autosomal recessive disorder characterized by high TG and low HDL, a characteristic of type I hyperlipoproteinemia (hyperchylomicronemia). The low HDL is caused by an insufficient transfer of surface components from VLDL to HDL [156]. Functional LPL deficiency can also be caused by a deficiency of apoC-II, a cofactor of LPL. Many mutations in the LPL gene have been identified, with heterozygous mutations being common. When these mutations are compounded by other risk factors, a significantly increased risk of cardiovascular disease results [154,157].

LPL deficiency in mice is lethal [158,159]. At birth, *Lpl*$^{-/-}$ mice have elevated TG and VLDL cholesterol levels, and when permitted to suckle, *Lpl*$^{-/-}$ mice become cyanotic and finally die with severe hypertriglyceridemia. Heterozygotes had higher TG than wild-type mice. Muscle-specific expression of LPL rescues *Lpl*$^{-/-}$ mice [159].

Although LPL is mainly synthesized in adipose tissue, heart, and muscle, Ad-mediated ectopic hepatic expression reduced hyperlipidemias associated with apoE, LDLR, and LPL deficiencies in mice [160,161] and corrected the phenotype in a feline model of LPL deficiency [162]. Three percent of *Lpl*$^{-/-}$ mice survived during the weaning period after intraperitoneal injection of Ad expressing LPL. However, they lacked apoA-I-preβ-HDL and mature HDL and had undetectable HDL-c and HDL-apoA-I, suggesting a pivotal role of LPL in HDL maturation [163].

LPL^{S447X} (truncated LPL) is a naturally occurring variant [164] with increased enzyme activity and an improved lipoprotein profile [165–167]. Intramuscular (IM) injection of pseudotyped AAV1 expressing LPL^{S447X} corrected hypertriglyceridemia in *Lpl*$^{-/-}$ mice for a year [168], while use of the same vector to infect skeletal muscle isolated from patients with an LPL deficiency was effective in inducing LPL expression, suggesting that the approach may be feasible in humans with LPL deficiency [169]. However, phenotypic correction of an LPL deficiency in a feline model by an IM injection of AAV1-LPL^{S447X} was transient and immunosuppression was needed to sustain the effects, which underscores the importance of the host immune response against therapeutic gene products [170].

9.3.4
Structural and Functional Alterations of HDL by Enzymes

A methionine residue of apoA-I has been reported to play a role in the reduction of HDL phospholipid hydroperoxides to the corresponding hydroxides [171]. It has also been shown that the oxidation of methionine increases the potential antiatherogenic properties of apoA-I [172,173]. However, HDL's antioxidant effects may come from HDL-associated proteins.

9.3.5
Paraoxonases (PONs)

The genes for paraoxonase-1, -2, and -3 are closely located on human chromosome 7. They share considerable structural similarities and are presumably generated by gene duplication of a common ancestral gene [174,175]. PONs hydrolyze organophosphates and have peroxidase, lactonase, and arylesterase activities. Their names are derived from paraoxon, the most commonly used substrate for *in vitro* enzyme assays [176]. PONs have attracted attention for their ability to hydrolyze oxidized lipids in LDL and because of evidence showing their possible involvement in atherosclerosis [19,177].

PON1, a resident on HDL, metabolizes LDL lipid peroxides and thereby protects against oxidative modification of LDL [178]. PON1 also inhibits phospholipid oxidation in HDL [179]. The possible association of PON1 common polymorphisms with CAD has not been clear. PON1 activity is apparently a better predictor of a risk of CAD than gene polymorphisms [180,181]. However, PON1 serum levels vary considerably between individuals and are influenced by environmental factors [182].

Direct evidence for the connection between PON1 activity and the development of atherosclerosis was provided by a study of mice lacking PON1 [183]. *PON1*$^{-/-}$ mice were extremely sensitive to the toxic effects of organophosphate, while HDL and LDL isolated from *PON1*$^{-/-}$ mice were more susceptible to oxidation. When *PON1*$^{-/-}$ mice were fed a high-fat high-cholesterol diet, they developed more atherosclerosis, while PON1/apoE double knockout mice displayed further increases in lipoprotein oxidation and exhibited significantly more atherosclerosis. In addition, mRNA levels of heme oxygenase-1, PPAR-γ, scavenger receptor type A, CD36, and macrosialin were increased, which suggests the involvement of macrophages [184], while macrophages isolated from *PON1*$^{-/-}$ mice contained more peroxidized lipids and exhibited increased oxidative stress [185]. However, mouse macrophages express PON2 and PON3, but not PON1 [186], so the observed macrophage phenotype may be an indirect consequence of reduced exposure to plasma paraoxonase or to the increase of oxidized LDL and lipid peroxide. Consistent with the antiatherogenic effects of PON1, HDL isolated from PON1 transgenic mice provided protection against LDL oxidation, while atherosclerotic lesions were significantly reduced in mice fed an atherogenic diet or in *Apoe*$^{-/-}$ background [187].

Bradshaw et al. took an indirect approach to test the effects of hepatic PON1 expression. The increase in hepatic PON1 activity was sufficient to reduce atherosclerosis in *Ldlr*$^{-/-}$ mice, and so the authors generated macrophage-specific PON1 transgenic mice and transplanted bone marrow cells into either Kupffer cell-depleted or untreated *Ldlr*$^{-/-}$ mice. Only recipient mice depleted of Kupffer cells had less atherosclerosis, so the expression of PON1 by liver macrophages was necessary and sufficient to decrease atherosclerosis [188]. Adenovirus-mediated hepatic overexpression of PON1 also inhibited the development of atherosclerosis six weeks after treatment in mice with a combined deficiency of leptin and LDLR [189].

PON-2 does not associate with HDL but rather remains associated with the membrane and protects LDL against lipid peroxidation [190]. PON-2-deficient mice fed a high-fat and high-cholesterol diet developed increased atherosclerotic lesions despite having lower VLDL/LDL cholesterol levels than wild-type mice. Enhanced inflammatory properties of LDL, attenuated antiatherogenic capacity of HDL, and oxidative stress caused by an exacerbated inflammatory response of PON-2-deficient macrophages are likely explanations [191]. In support of PON-2's antiatherogenic action, Ad-mediated PON-2 gene transfer protected against development of atherosclerosis in *Apoe*$^{-/-}$ mice [192].

PON3 is also an HDL-associated esterase that possesses antioxidant properties. However, PON3 is present only at low levels in relation to PON1, and PON3 expression is detected only in mouse macrophages and not in human macrophages [186]. PON3 is not highly regulated by oxidative stress [193], so it probably does not play a significant role in LDL oxidation in humans.

9.3.6 Myeloperoxidase (MPO)

Myeloperoxidase, a heme protein expressed in neutrophils, monocytes, and macrophages, is another enzyme involved in HDL oxidation. It is synthesized as inactive apoproMPO, which forms a complex in the endoplasmic reticulum with calreticulin and calnexin. Following acquisition of heme, it is converted into proMPO, which undergoes proteolytic cleavages to generate heavy and light subunits before becoming a mature form of MPO. The primary function of MPO may be to act against invading microorganisms as a component of the innate immune system by reacting with H_2O_2 [194]. Large proportions of lipids in lipoproteins are oxidized in lesions [195], which might impair the protective effect of HDL. MPO is a potential candidate enzyme that participates in phagocyte-dependent HDL-oxidation. Hypochlorous acid/hypochlorite (HOCl) is an oxidant produced from H_2O_2 and chloride through the action of MPO, which co-localizes with HOCl-modified LDL in human atherosclerotic lesions [196]. MPO-catalyzed oxidation of HDL and apoA-I selectively inhibits ABCA1-dependent cholesterol efflux from macrophages [197,198]. Elevated levels of leukocyte- and blood-MPOs are associated with an increased risk of CAD [199,200]. In support of the pro-atherogenic action of MPO, macrophage-specific MPO expression increased atherosclerosis in *Ldlr*$^{-/-}$ mice [201]. Furthermore, male *Ldlr*$^{-/-}$ mice expressing MPO-463G/A, which is associated with an increased risk of

CAD in humans [202,203], developed hyperlipidemia and obesity and exhibited more atherosclerosis [204]. Paradoxically, however, *Ldlr*$^{-/-}$ mice transplanted with bone marrow cells derived from *Mpo*$^{-/-}$ mice also developed more atherosclerosis [205]. The primary role of MPO may be to inhibit overreaction of the innate immune system by producing oxidants, and a lack of protective oxidants from MPO may result in more inflammation, leading to an increased cholesterol load in macrophages. Further studies are needed to clarify the mechanisms involved.

9.4 Conclusion

It is generally accepted that HDL protects against atherosclerosis by promoting RCT. However, simply increasing HDL levels is not necessarily beneficial. The protective effects of HDL are influenced by its integrity, which is modulated by circulating and heparan sulfate-bound lipases, as well as by HDL-associated enzymes. A large body of evidence suggests that these enzymes play complex roles in the production of protective HDL and in the maintenance of HDL's integrity. Mouse genetics and genetically modified mice have contributed to our understanding of lipoprotein physiology and the pathogenesis of atherosclerosis. Somatic gene transfer *in vivo* is a powerful tool in proof-of-principle experiments. The technology was developed as a new frontier in gene medicine. Despite slow progress towards its intended goals, new strategies of gene delivery have shed much light on the mechanisms of HDL remodeling and metabolism *in vivo*. Continued advancement in gene therapy technology, together with an increasing understanding of HDL structure, function, and modification, could see clinical application in generation of protective HDL as a reality in the near future.

9.5 Perspective

Genetically modified animal models are a powerful tool with which to dissect lipoprotein metabolism and physiology and they have contributed in a major way to our current understanding of HDL metabolism and remodeling. The action of various lipases and HDL-associated peroxidases is complex. Furthermore, a large body of evidence suggests that the sites of action (i.e., the tissues or cell types where it happens) determine whether particular reactions of interactions are beneficial or harmful to the individual in terms of antiatherogenic effects. For *in vivo* experiments, the genetic approach in theory has an advantage over the pharmacological approach that aims to change levels of enzyme activities for therapeutic purposes, since the routes of gene delivery and specific tissue targeting are easily manipulated. Much of the recent progress notwithstanding, however, gene manipulation still faces considerable hurdles in achieving its therapeutic goals. Nevertheless, genetically modified animal models and progress in gene transfer technology should be a major driving force to move the field of HDL metabolism forward; knowledge gained from the use of these tools should go a long way toward the prevention and treatment of CAD.

9.6 Acknowledgments

We than Dr. H. Chao, S. Cormier, and Leslie Wu for their assistance in the preparation of the manuscript. This work was supported in part by NIH HL59314, HL51586, and HL73144.

References

1 Libby, P. (2002) Inflammation in atherosclerosis. *Nature*, **420**, 868–874.
2 Libby, P. and Theroux, P. (2005) Pathophysiology of coronary artery disease. *Circulation*, **111**, 3481–3488.
3 Miller, G. J. and Miller, N. E. (1975) Plasma-high-density-lipoprotein concentration and development of ischaemic heart-disease. *Lancet*, **1**, 16–19.
4 Castelli, W. P., Garrison, R. J., Wilson, P. W., Abbott, R. D., Kalousdian, S., Kannel, W. B. (1986) Incidence of coronary heart disease and lipoprotein cholesterol levels. The Framingham Study. *Jama*, **256**, 2835–2838.
5 Goldbourt, U., Yaari, S., Medalie, J. H. (1997) Isolated low HDL cholesterol as a risk factor for coronary heart disease mortality. A 21-year follow-up of 8000 men. *Arterioscler Thromb Vasc Biol.*, **17**, 107–113.
6 Boden, W. E. (2000) High-density lipoprotein cholesterol as an independent risk factor in cardiovascular disease: assessing the data from Framingham to the Veterans Affairs High-Density Lipoprotein Intervention Trial. *Am J Cardiol.*, **86**, 19L–22L.
7 Maciejko, J. J., Holmes, D. R., Kottke, B. A., Zinsmeister, A. R., Dinh, D. M., Mao, S. J. (1983) Apolipoprotein A-I as a marker of angiographically assessed coronary-artery disease. *N Engl J Med.*, **309**, 385–389.
8 Sniderman, A. D., Furberg, C. D., Keech, A., Roeters van Lennep, J. E., Frohlich, J., Jungner, I., Walldius, G. (2003) Apolipoproteins versus lipids as indices of coronary risk and as targets for statin treatment. *Lancet*, **361**, 777–780.
9 Walldius, G. and Jungner, I. (2004) Apolipoprotein B and apolipoprotein A-I: risk indicators of coronary heart disease and targets for lipid-modifying therapy. *J Intern Med.*, **255**, 188–205.
10 Barr, D. P., Russ, E. M., Eder, H. A. (1951) Protein-lipid relationships in human plasma. II. In atherosclerosis and related conditions. *Am J Med.*, **11**, 480–493.
11 Lewis, G. F. and Rader, D. J. (2005) New insights into the regulation of HDL metabolism and reverse cholesterol transport. *Circ Res.*, **96**, 1221–1232.
12 Rader, D. J. (2006) Molecular regulation of HDL metabolism and function: implications for novel therapies. *J Clin Invest.*, **116**, 3090–3100.
13 Glomset, J. A. (1968) The plasma lecithins:cholesterol acyltransferase reaction. *J Lipid Res.*, **9**, 155–167.
14 Ross, R. and Glomset, J. A. (1973) Atherosclerosis and the arterial smooth muscle cell: Proliferation of smooth muscle is a key event in the genesis of the lesions of

atherosclerosis. *Science*, **180**, 1332–1339.
15 Barter, P. J., Nicholls, S., Rye, K. A., Anantharamaiah, G. M., Navab, M., Fogelman, A. M. (2004) Antiinflammatory properties of HDL. *Circ Res.*, **95**, 764–772.
16 Mineo, C., Deguchi, H., Griffin, J. H., Shaul, P. W. (2006) Endothelial and antithrombotic actions of HDL. *Circ Res.*, **98**, 1352–1364.
17 (2001) Executive Summary of The Third Report of The National Cholesterol Education Program (NCEP) Expert Panel on Detection, Evaluation, Treatment of High Blood Cholesterol In Adults (Adult Treatment Panel III). *Jama*, **285**, 2486–2497.
18 Navab, M., Anantharamaiah, G. M., Reddy, S. T., VanLenten, B. J., Ansell, B. J., Fogelman, A. M. (2006) Mechanisms of disease: proatherogenic HDL–an evolving field. *Nat Clin Pract Endocrinol Metab.*, **2**, 504–511.
19 Ferretti, G., Bacchetti, T., Negre-Salvayre, A., Salvayre, R., Dousset, N., Curatola, G. (2006) Structural modifications of HDL and functional consequences. *Atherosclerosis*, **184**, 1–7.
20 Pearson, H. (2006) When good cholesterol turns bad. *Nature*, **444**, 794–795.
21 (2006) Cholesterol: the good, the bad, and the stopped trials. *Lancet*, **368**, 2034.
22 Davidson, M. H., McKenney, J. M., Shear, C. L., Revkin, J. H. (2006) Efficacy and safety of torcetrapib, a novel cholesteryl ester transfer protein inhibitor, in individuals with below-average high-density lipoprotein cholesterol levels. *J Am Coll Cardiol.*, **48**, 1774–1781.
23 McKenney, J. M., Davidson, M. H., Shear, C. L., Revkin, J. H. (2006) Efficacy and safety of torcetrapib, a novel cholesteryl ester transfer protein inhibitor, in individuals with below-average high-density lipoprotein cholesterol levels on a background of atorvastatin. *J Am Coll Cardiol.*, **48**, 1782–1790.
24 O'Connor, T. P. and Crystal, R. G. (2006) Genetic medicines: treatment strategies for hereditary disorders. *Nat Rev Genet.*, **7**, 261–276.
25 Oka, K. and Chan, L. (2004) Liver-directed gene therapy for dyslipidemia and diabetes. *Curr Atheroscler Rep.*, **6**, 203–209.
26 Herweijer, H. and Wolff, J. A. (2007) Gene therapy progress and prospects: Hydrodynamic gene delivery. *Gene Ther.*, **14**, 99–107.
27 Liu, F., Song, Y., Liu, D. (1999) Hydrodynamics-based transfection in animals by systemic administration of plasmid DNA. *Gene Ther.*, **6**, 1258–1266.
28 Condiotti, R., Curran, M. A., Nolan, G. P., Giladi, H., Ketzinel-Gilad, M., Gross, E., Galun, E. (2004) Prolonged liver-specific transgene expression by a non-primate lentiviral vector. *Biochem Biophys Res Commun.*, **320**, 998–1006.
29 Brunetti-Pierri, N., Palmer, D. J., Mane, V., Finegold, M., Beaudet, A. L., Ng, P. (2005) Increased hepatic transduction with reduced systemic dissemination and proinflammatory cytokines following hydrodynamic injection of helper-dependent adenoviral vectors. *Mol Ther.*, **12**, 99–106.
30 Glover, D. J., Lipps, H. J., Jans, D. A. (2005) Towards safe, non-viral therapeutic gene expression in humans. *Nat Rev Genet.*, **6**, 299–310.
31 Groth, A. C. and Calos, M. P. (2004) Phage integrases: biology and applications. *J Mol Biol.*, **335**, 667–678.
32 Groth, A. C., Olivares, E. C., Thyagarajan, B., Calos, M. P. (2000) A phage integrase directs efficient site-specific integration in human cells. *Proc Natl Acad Sci USA*, **97**, 5995–6000.
33 Thyagarajan, B., Olivares, E. C., Hollis, R. P., Ginsburg, D. S., Calos, M. P. (2001) Site-specific genomic integration in mammalian cells mediated by phage phiC31 integrase. *Mol Cell Biol.*, **21**, 3926–3934.
34 Belteki, G., Gertsenstein, M., Ow, D. W., Nagy, A. (2003) Site-specific

cassette exchange and germline transmission with mouse ES cells expressing phiC31 integrase. *Nat Biotechnol.*, **21**, 321–324.

35 Ortiz-Urda, S., Thyagarajan, B., Keene, D. R., Lin, Q., Fang, M., Calos, M. P., Khavari, P. A. (2002) Stable nonviral genetic correction of inherited human skin disease. *Nat Med.*, **8**, 1166–1170.

36 Quenneville, S. P., Chapdelaine, P., Rousseau, J., Beaulieu, J., Caron, N. J., Skuk, D., Mills, P., Olivares, E. C., Calos, M. P., Tremblay, J. P. (2004) Nucleofection of muscle-derived stem cells and myoblasts with phiC31 integrase: stable expression of a full-length-dystrophin fusion gene by human myoblasts. *Mol Ther.*, **10**, 679–687.

37 Held, P. K., Olivares, E. C., Aguilar, C. P., Finegold, M., Calos, M. P., Grompe, M. (2005) In vivo correction of murine hereditary tyrosinemia type I by phiC31 integrase-mediated gene delivery. *Mol Ther.*, **11**, 399–408.

38 Ehrhardt, A., Xu, H., Huang, Z., Engler, J. A., Kay, M. A. (2005) A direct comparison of two nonviral gene therapy vectors for somatic integration: in vivo evaluation of the bacteriophage integrase phiC31 and the Sleeping Beauty transposase. *Mol Ther.*, **11**, 695–706.

39 Chen, L. and Woo, S. L. (2005) Complete and persistent phenotypic correction of phenylketonuria in mice by site-specific genome integration of murine phenylalanine hydroxylase cDNA. *Proc Natl Acad Sci USA*, **102**, 15581–15586.

40 Chalberg, T. W., Portlock, J. L., Olivares, E. C., Thyagarajan, B., Kirby, P. J., Hillman, R. T., Hoelters, J., Calos, M. P. (2006) Integration specificity of phage phiC31 integrase in the human genome. *J Mol Biol.*, **357**, 28–48.

41 Vigdal, T. J., Kaufman, C. D., Izsvak, Z., Voytas, D. F., Ivics, Z. (2002) Common physical properties of DNA affecting target site selection of sleeping beauty and other Tc1/mariner transposable elements. *J Mol Biol.*, **323**, 441–452.

42 Yant, S. R., Wu, X., Huang, Y., Garrison, B., Burgess, S. M., Kay, M. A. (2005) High-resolution genome-wide mapping of transposon integration in mammals. *Mol Cell Biol.*, **25**, 2085–2094.

43 Liu, J., Jeppesen, I., Nielsen, K., Jensen, T. G., (2006) Phi c31 integrase induces chromosomal aberrations in primary human fibroblasts. *Gene Ther.*, **13**, 1188–1190.

44 Ehrhardt, A., Engler, J. A., Xu, H., Cherry, A. M., Kay, M. A. (2006) Molecular analysis of chromosomal rearrangements in mammalian cells after phiC31-mediated integration. *Hum Gene Ther.*, **17**, 1077–1094.

45 Kotin, R. M., Siniscalco, M., Samulski, R. J., Zhu, X. D., Hunter, L., Laughlin, C. A., McLaughlin, S., Muzyczka, N., Rocchi, M., Berns, K. I. (1990) Site-specific integration by adeno-associated virus. *Proc Natl Acad Sci USA*, **87**, 2211–2215.

46 Samulski, R. J., Zhu, X., Xiao, X., Brook, J. D., Housman, D. E., Epstein, N., Hunter, L. A. (1991) Targeted integration of adeno-associated virus (AAV) into human chromosome 19. *Embo J.*, **10**, 3941–3950.

47 Weitzman, M. D., Kyostio, S. R., Kotin, R. M., Owens, R. A. (1994) Adeno-associated virus (AAV) Rep proteins mediate complex formation between AAV DNA and its integration site in human DNA. *Proc Natl Acad Sci USA*, **91**, 5808–5812.

48 Linden, R. M., Ward, P., Giraud, C., Winocour, E., Berns, K. I. (1996) Site-specific integration by adeno-associated virus. *Proc Natl Acad Sci USA*, **93**, 11288–11294.

49 Amiss, T. J., McCarty, D. M., Skulimowski, A., Samulski, R. J. (2003) Identification and characterization of an adeno-associated virus integration site in CV-1 cells from the African green monkey. *J Virol.*, **77**, 1904–1915.

50 Dutheil, N., Yoon-Robarts, M., Ward, P., Henckaerts, E., Skrabanek, L., Berns, K. I., Campagne, F., Linden, R. M. (2004) Characterization of the mouse adeno-associated virus AAVS1 ortholog. *J Virol.*, **78**, 8917–8921.

51 Miller, D. G., Petek, L. M., Russell, D. W. (2004) Adeno-associated virus vectors integrate at chromosome breakage sites. *Nat Genet.*, **36**, 767–773.

52 McCarty, D. M., Young, S. M., Jr., Samulski, R. J. (2004) Integration of adeno-associated virus (AAV) and recombinant AAV vectors. *Annu Rev Genet.*, **38**, 819–845.

53 Philpott, N. J., Gomos, J., Falck-Pedersen, E. (2004) Transgene expression after rep-mediated site-specific integration into chromosome 19. *Hum Gene Ther.*, **15**, 47–61.

54 Berthet, C., Raj, K., Saudan, P., Beard, P. (2005) How adeno-associated virus Rep78 protein arrests cells completely in S phase. *Proc Natl Acad Sci USA*, **102**, 13634–13639.

55 Yant, S. R., Meuse, L., Chiu, W., Ivics, Z., Izsvak, Z., Kay, M. A. (2000) Somatic integration and long-term transgene expression in normal and haemophilic mice using a DNA transposon system. *Nat Genet.*, **25**, 35–41.

56 Ohlfest, J. R., Frandsen, J. L., Fritz, S., Lobitz, P. D., Perkinson, S. G., Clark, K. J., Nelsestuen, G., Key, N. S., McIvor, R. S., Hackett, P. B., Largaespada, D. A. (2005) Phenotypic correction and long-term expression of factor VIII in hemophilic mice by immunotolerization and nonviral gene transfer using the Sleeping Beauty transposon system. *Blood*, **105**, 2691–2698.

57 Liu, L., Mah, C., Fletcher, B. S. (2006) Sustained FVIII expression and phenotypic correction of hemophilia A in neonatal mice using an endothelial-targeted sleeping beauty transposon. *Mol Ther.*, **13**, 1006–1015.

58 Atchison, R. W., Casto, B. C., Hammon, W. M. (1965) Adenovirus-Associated Defective Virus Particles. *Science*, **149**, 754–756.

59 Hoggan, M. D., Blacklow, N. R., Rowe, W. P. (1966) Studies of small DNA viruses found in various adenovirus preparations: physical, biological, and immunological characteristics. *Proc Natl Acad Sci USA*, **55**, 1467–1474.

60 Hoggan, M. D. (1970) Adenovirus associated viruses. *Prog Med Virol.*, **12**, 211–239.

61 Berns, K. I. (1990) Parvovirus replication. *Microbiol Rev.*, **54**, 316–329.

62 Carter, B. J. (2004) Adeno-associated virus and the development of adeno-associated virus vectors: a historical perspective. *Mol Ther.*, **10**, 981–989.

63 Gao, G., Alvira, M. R., Somanathan, S., Lu, Y., Vandenberghe, L. H., Rux, J. J., Calcedo, R., Sanmiguel, J., Abbas, Z., Wilson, J. M. (2003) Adeno-associated viruses undergo substantial evolution in primates during natural infections. *Proc Natl Acad Sci USA*, **100**, 6081–6086.

64 Summerford, C. and Samulski, R. J. (1998) Membrane-associated heparan sulfate proteoglycan is a receptor for adeno-associated virus type 2 virions. *J Virol.*, **72**, 1438–1445.

65 Summerford, C., Bartlett, J. S., Samulski, R. J. (1999) AlphaVbeta5 integrin: a co-receptor for adeno-associated virus type 2 infection. *Nat Med.*, **5**, 78–82.

66 Sanlioglu, S., Benson, P. K., Yang, J., Atkinson, E. M., Reynolds, T., Engelhardt, J. F. (2000) Endocytosis and nuclear trafficking of adeno-associated virus type 2 are controlled by rac1 and phosphatidylinositol-3 kinase activation. *J Virol.*, **74**, 9184–9196.

67 Qing, K., Mah, C., Hansen, J., Zhou, S., Dwarki, V., Srivastava, A. (1999) Human fibroblast growth factor receptor 1 is a co-receptor for infection by adeno-associated virus 2. *Nat Med.*, **5**, 71–77.

68 Di Pasquale, G., Davidson, B. L., Stein, C. S., Martins, I., Scudiero, D., Monks, A., Chiorini, J. A. (2003)

Identification of PDGFR as a receptor for AAV-5 transduction. *Nat Med.*, **9**, 1306–1312.

69 Walters, R. W., Yi, S. M., Keshavjee, S., Brown, K. E., Welsh, M. J., Chiorini, J. A., Zabner, J. (2001) Binding of adeno-associated virus type 5 to 2,3-linked sialic acid is required for gene transfer. *J Biol Chem.*, **276**, 20610–20616.

70 Akache, B., Grimm, D., Pandey, K., Yant, S. R., Xu, H., Kay, M. A. (2006) The 37/67-kilodalton laminin receptor is a receptor for adeno-associated virus serotypes 8, 2, 3, and 9. *J Virol.*, **80**, 9831–9836.

71 Grimm, D., Kay, M. A., Kleinschmidt, J. A. (2003) Helper virus-free, optically controllable, and two-plasmid-based production of adeno-associated virus vectors of serotypes 1 to 6. *Mol Ther.*, **7**, 839–850.

72 Burger, C., Gorbatyuk, O. S., Velardo, M. J., Peden, C. S., Williams, P., Zolotukhin, S., Reier, P. J., Mandel, R. J., Muzyczka, N. (2004) Recombinant AAV viral vectors pseudotyped with viral capsids from serotypes 1, 2, and 5 display differential efficiency and cell tropism after delivery to different regions of the central nervous system. *Mol Ther.*, **10**, 302–317.

73 Stilwell, J. L. and Samulski, R. J. (2004) Role of viral vectors and virion shells in cellular gene expression. *Mol Ther.*, **9**, 337–346.

74 Thomas, C. E., Ehrhardt, A., Kay, M. A. (2003) Progress and problems with the use of viral vectors for gene therapy. *Nat Rev Genet.*, **4**, 346–358.

75 Gao, G. P., Alvira, M. R., Wang, L., Calcedo, R., Johnston, J., Wilson, J. M. (2002) Novel adeno-associated viruses from rhesus monkeys as vectors for human gene therapy. *Proc Natl Acad Sci USA*, **99**, 11854–11859.

76 Sarkar, R., Tetreault, R., Gao, G., Wang, L., Bell, P., Chandler, R., Wilson, J. M., Kazazian, H. H., Jr. (2004) Total correction of hemophilia A mice with canine FVIII using an AAV 8 serotype. *Blood*, **103**, 1253–1260.

77 Nakai, H., Fuess, S., Storm, T. A., Muramatsu, S., Nara, Y., Kay, M. A. (2005) Unrestricted hepatocyte transduction with adeno-associated virus serotype 8 vectors in mice. *J Virol.*, **79**, 214–224.

78 Duan, D., Yue, Y., Yan, Z., Engelhardt, J. F. (2000) A new dual-vector approach to enhance recombinant adeno-associated virus-mediated gene expression through intermolecular cis activation. *Nat Med.*, **6**, 595–598.

79 Nakai, H., Storm, T. A., Kay, M. A. (2000) Increasing the size of rAAV-mediated expression cassettes in vivo by intermolecular joining of two complementary vectors. *Nat Biotechnol.*, **18**, 527–532.

80 Sun, L., Li, J., Xiao, X. (2000) Overcoming adeno-associated virus vector size limitation through viral DNA heterodimerization. *Nat Med.*, **6**, 599–602.

81 Chao, H., Sun, L., Bruce, A., Xiao, X., Walsh, C. E. (2002) Expression of human factor VIII by splicing between dimerized AAV vectors. *Mol Ther.*, **5**, 716–722.

82 McCarty, D. M., Monahan, P. E., Samulski, R. J. (2001) Self-complementary recombinant adeno-associated virus (scAAV) vectors promote efficient transduction independently of DNA synthesis. *Gene Ther.*, **8**, 1248–1254.

83 McCarty, D. M., Fu, H., Monahan, P. E., Toulson, C. E., Naik, P., Samulski, R. J. (2003) Adeno-associated virus terminal repeat (TR) mutant generates self-complementary vectors to overcome the rate-limiting step to transduction in vivo. *Gene Ther.*, **10**, 2112–2118.

84 Tobiasch, E., Rabreau, M., Geletneky, K., Larue-Charlus, S., Severin, F., Becker, N., Schlehofer, J. R. (1994) Detection of adeno-associated virus DNA in human genital tissue and in material from spontaneous abortion. *J Med Virol.*, **44**, 215–222.

85 Donsante, A., Vogler, C., Muzyczka, N., Crawford, J. M., Barker, J., Flotte, T., Campbell-Thompson, M., Daly, T.,

Sands, M. S. (2001) Observed incidence of tumorigenesis in long-term rodent studies of rAAV vectors. *Gene Ther.*, **8**, 1343–1346.

86 Bell, P., Wang, L., Lebherz, C., Flieder, D. B., Bove, M. S., Wu, D., Gao, G. P., Wilson, J. M., Wivel, N. A. (2005) No Evidence for Tumorigenesis of AAV Vectors in a Large-Scale Study in Mice. *Mol Ther.*, **12**, 299–306.

87 Bell, P., Moscioni, A. D., McCarter, R. J., Wu, D., Gao, G., Hoang, A., Sanmiguel, J. C., Sun, X., Wivel, N. A., Raper, S. E., Furth, E. E., Batshaw, M. L., Wilson, J. M. (2006) Analysis of tumors arising in male B6C3F1 mice with and without AAV vector delivery to liver. *Mol Ther.*, **14**, 34–44.

88 Nakai, H., Montini, E., Fuess, S., Storm, T. A., Grompe, M., Kay, M. A. (2003) AAV serotype 2 vectors preferentially integrate into active genes in mice. *Nat Genet.*, **34**, 297–302.

89 Nakai, H., Wu, X., Fuess, S., Storm, T. A., Munroe, D., Montini, E., Burgess, S. M., Grompe, M., Kay, M. A. (2005) Large-scale molecular characterization of adeno-associated virus vector integration in mouse liver. *J Virol.*, **79**, 3606–3614.

90 Chen, S. J., Rader, D. J., Tazelaar, J., Kawashiri, M., Gao, G., Wilson, J. M. (2000) Prolonged correction of hyperlipidemia in mice with familial hypercholesterolemia using an adeno-associated viral vector expressing very-low-density lipoprotein receptor. *Mol Ther.*, **2**, 256–261.

91 Lebherz, C., Gao, G., Louboutin, J. P., Millar, J., Rader, D., Wilson, J. M. (2004) Gene therapy with novel adeno-associated virus vectors substantially diminishes atherosclerosis in a murine model of familial hypercholesterolemia. *J Gene Med.*, **6**, 663–672.

92 Kitajima, K., Marchadier, D. H., Miller, G. C., Gao, G. P., Wilson, J. M., Rader, D. J. (2006) Complete prevention of atherosclerosis in apoE-deficient mice by hepatic human apoE gene transfer with adeno-associated virus serotypes 7 and 8. *Arterioscler Thromb Vasc Biol.*, **26**, 1852–1857.

93 Kitajima, K., Marchadier, D. H., Burstein, H., Rader, D. J. (2006) Persistent liver expression of murine apoA-l using vectors based on adeno-associated viral vectors serotypes 5 and 1. *Atherosclerosis.*, **186**, 65–73.

94 Sharifi, B. G., Wu, K., Wang, L., Ong, J. M., Zhou, X., Shah, P. K. (2005) AAV serotype-dependent apolipoprotein A-I Milano gene expression. *Atherosclerosis.*, **181**, 261–269.

95 Bergelson, J. M., Cunningham, J. A., Droguett, G., Kurt-Jones, E. A., Krithivas, A., Hong, J. S., Horwitz, M. S., Crowell, R. L., Finberg, R. W. (1997). Isolation of a common receptor for Coxsackie B viruses and adenoviruses 2 and 5. *Science*, **275**, 1320–1323.

96 Wickham, T. J., Mathias, P., Cheresh, D. A., Nemerow, G. R. (1993) Integrins alpha v beta 3 and alpha v beta 5 promote adenovirus internalization but not virus attachment. *Cell*, **73**, 309–319.

97 Nemerow, G. R. (2000) Cell receptors involved in adenovirus entry. *Virology*, **274**, 1–4.

98 Dechecchi, M. C., Melotti, P., Bonizzato, A., Santacatterina, M., Chilosi, M., Cabrini, G. (2001) Heparan sulfate glycosaminoglycans are receptors sufficient to mediate the initial binding of adenovirus types 2 and 5. *J Virol.*, **75**, 8772–8780.

99 Shayakhmetov, D. M., Gaggar, A., Ni, S., Li, Z. Y., Lieber, A. (2005) Adenovirus binding to blood factors results in liver cell infection and hepatotoxicity. *J Virol.*, **79**, 7478–7491.

100 Gaggar, A., Shayakhmetov, D. M., Lieber, A. (2003) CD46 is a cellular receptor for group B adenoviruses. *Nat Med.*, **9**, 1408–1412.

101 Segerman, A., Atkinson, J. P., Marttila, M., Dennerquist, V., Wadell,

G., Arnberg, N. (2003) Adenovirus type 11 uses CD46 as a cellular receptor. *J Virol.*, **77**, 9183–9191.

102 Oka, K. and Chan, L. (2002) Recent advances in liver-directed gene therapy for dyslipidemia. *Curr Atheroscler Rep.*, **4**, 199–207.

103 Kochanek, S. (1999) High-capacity adenoviral vectors for gene transfer and somatic gene therapy. *Hum Gene Ther.*, **10**, 2451–2459.

104 Schnell, M. A., Zhang, Y., Tazelaar, J., Gao, G. P., Yu, Q. C., Qian, R., Chen, S. J., Varnavski, A. N., LeClair, C., Raper, S. E., Wilson, J. M. (2001) Activation of innate immunity in nonhuman primates following intraportal administration of adenoviral vectors. *Mol Ther.*, **3**, 708–722.

105 Brunetti-Pierri, N., Palmer, D. J., Beaudet, A. L., Carey, K. D., Finegold, M., Ng, P. (2004) Acute toxicity after high-dose systemic injection of helper-dependent adenoviral vectors into nonhuman primates. *Hum Gene Ther.*, **15**, 35–46.

106 Muruve, D. A., Cotter, M. J., Zaiss, A. K., White, L. R., Liu, Q., Chan, T., Clark, S. A., Ross, P. J., Meulenbroek, R. A., Maelandsmo, G. M., Parks, R. J. (2004) Helper-dependent adenovirus vectors elicit intact innate but attenuated adaptive host immune responses in vivo. *J Virol.*, **78**, 5966–5972.

107 Schiedner, G., Hertel, S., Johnston, M., Dries, V., van Rooijen, N., Kochanek, S. (2003). Selective depletion or blockade of Kupffer cells leads to enhanced and prolonged hepatic transgene expression using high-capacity adenoviral vectors. *Mol Ther.*, **7**, 35–43.

108 Kim, I. H., Jozkowicz, A., Piedra, P. A., Oka, K., Chan, L. (2001) Lifetime correction of genetic deficiency in mice with a single injection of helper-dependent adenoviral vector. *Proc Natl Acad Sci USA*, **98**, 13282–13287.

109 Nomura, S., Merched, A., Nour, E., Dieker, C., Oka, K., Chan, L. (2004) Low-density lipoprotein receptor gene therapy using helper-dependent adenovirus produces long-term protection against atherosclerosis in a mouse model of familial hypercholesterolemia. *Gene Ther.*, **11**, 1540–1548.

110 Oka, K., Belalcazar, L. M., Dieker, C., Nour, E. A., Nuno-Gonzalez, P., Paul, A., Cormier, S., Shin, J. K., Finegold, M., Chan, L. (2007) Sustained phenotypic correction in a mouse model of hypoalphalipoproteinemia with a helper-dependent adenovirus vector. *Gene Ther.*, **14**, 191–202.

111 Oka, K., Pastore, L., Kim, I. H., Merched, A., Nomura, S., Lee, H. J., Merched-Sauvage, M., Arden-Riley, C., Lee, B., Finegold, M., Beaudet, A., Chan, L. (2001) Long-term stable correction of low-density lipoprotein receptor-deficient mice with a helper-dependent adenoviral vector expressing the very low-density lipoprotein receptor. *Circulation*, **103**, 1274–1281.

112 Belalcazar, L. M., Merched, A., Carr, B., Oka, K., Chen, K. H., Pastore, L., Beaudet, A., Chan, L. (2003) Long-term stable expression of human apolipoprotein A-I mediated by helper-dependent adenovirus gene transfer inhibits atherosclerosis progression and remodels atherosclerotic plaques in a mouse model of familial hypercholesterolemia. *Circulation*, **107**, 2726–2732.

113 MacDougall, E. D., Kramer, F., Polinsky, P., Barnhart, S., Askari, B., Johansson, F., Varon, R., Rosenfeld, M. E., Oka, K., Chan, L., Schwartz, S. M., Bornfeldt, K. E. (2006) Aggressive very low-density lipoprotein (VLDL) and LDL lowering by gene transfer of the VLDL receptor combined with a low-fat diet regimen induces regression and reduces macrophage content in advanced atherosclerotic lesions in LDL receptor-deficient mice. *Am J Pathol.*, **168**, 2064–2073.

114 Johnson, J. L., Baker, A. H., Oka, K., Chan, L., Newby, A. C., Jackson, C.

L., George, S. J. (2006) Suppression of atherosclerotic plaque progression and instability by tissue inhibitor of metalloproteinase-2: involvement of macrophage migration and apoptosis. *Circulation*, **113**, 2435–2444.

115 Sinn, P. L., Sauter, S. L., McCray, P. B., Jr. (2005) Gene therapy progress and prospects: development of improved lentiviral and retroviral vectors--design, biosafety, and production. *Gene Ther.*, **12**, 1089–1098.

116 Schroder, A. R., Shinn, P., Chen, H., Berry, C., Ecker, J. R., Bushman, F. (2002) HIV-1 integration in the human genome favors active genes and local hotspots. *Cell*, **110**, 521–529.

117 Wu, X., Li, Y., Crise, B., Burgess, S. M. (2003) Transcription start regions in the human genome are favored targets for MLV integration. *Science*, **300**, 1749–1751.

118 Mitchell, R. S., Beitzel, B. F., Schroder, A. R., Shinn, P., Chen, H., Berry, C. C., Ecker, J. R., Bushman, F. D. (2004) Retroviral DNA integration: ASLV, HIV, and MLV show distinct target site preferences. *PLoS Biol.*, **2**, E234.

119 Hill, C. L., Bieniasz, P. D., McClure, M. O. (1999) Properties of human foamy virus relevant to its development as a vector for gene therapy. *J Gen Virol.*, **80** (Pt 8), 2003–2009.

120 Heinkelein, M., Dressler, M., Jarmy, G., Rammling, M., Imrich, H., Thurow, J., Lindemann, D., Rethwilm, A. (2002) Improved primate foamy virus vectors and packaging constructs. *J Virol.*, **76**, 3774–3783.

121 Schweizer, M., Turek, R., Hahn, H., Schliephake, A., Netzer, K. O., Eder, G., Reinhardt, M., Rethwilm, A., Neumann-Haefelin, D. (1995) Markers of foamy virus infections in monkeys, apes, and accidentally infected humans: appropriate testing fails to confirm suspected foamy virus prevalence in humans. *AIDS Res Hum Retroviruses.*, **11**, 161–170.

122 Verma, I. M. and Weitzman, M. D. (2005) Gene therapy: twenty-first century medicine. *Annu Rev Biochem.*, **74**, 711–738.

123 Trobridge, G. and Russell, D. W. (2004) Cell cycle requirements for transduction by foamy virus vectors compared to those of oncovirus and lentivirus vectors. *J Virol.*, **78**, 2327–2335.

124 Trobridge, G., Josephson, N., Vassilopoulos, G., Mac, J., Russell, D. W. (2002) Improved foamy virus vectors with minimal viral sequences. *Mol Ther.*, **6**, 321–328.

125 Trobridge, G. D., Miller, D. G., Jacobs, M. A., Allen, J. M., Kiem, H. P., Kaul, R., Russell, D. W. (2006) Foamy virus vector integration sites in normal human cells. *Proc Natl Acad Sci USA*, **103**, 1498–1503.

126 Vassilopoulos, G., Trobridge, G., Josephson, N. C., Russell, D. W. (2001) Gene transfer into murine hematopoietic stem cells with helper-free foamy virus vectors. *Blood*, **98**, 604–609.

127 Josephson, N. C., Trobridge, G., Russell, D. W. (2004) Transduction of long-term and mobilized peripheral blood-derived NOD/SCID repopulating cells by foamy virus vectors. *Hum Gene Ther.*, **15**, 87–92.

128 Leurs, C., Jansen, M., Pollok, K. E., Heinkelein, M., Schmidt, M., Wissler, M., Lindemann, D., VonKalle, C., Rethwilm, A., Williams, D. A., Hanenberg, H. (2003) Comparison of three retroviral vector systems for transduction of nonobese diabetic/severe combined immunodeficiency mice repopulating human CD34+ cord blood cells. *Hum Gene Ther.*, **14**, 509–519.

129 Jaye, M., Lynch, K. J., Krawiec, J., Marchadier, D., Maugeais, C., Doan, K., South, V., Amin, D., Perrone, M., Rader, D. J. (1999) A novel endothelial-derived lipase that modulates HDL metabolism. *Nat Genet.*, **21**, 424–428.

130 Ma, K., Cilingiroglu, M., Otvos, J. D., Ballantyne, C. M., Marian, A. J., Chan, L. (2003) Endothelial lipase is

a major genetic determinant for high-density lipoprotein concentration, structure, and metabolism. *Proc Natl Acad Sci USA*, **100**, 2748–2753.

131 Ishida, T., Choi, S., Kundu, R. K., Hirata, K., Rubin, E. M., Cooper, A. D., Quertermous, T. (2003) Endothelial lipase is a major determinant of HDL level. *J Clin Invest.*, **111**, 347–355.

132 Hirata, K., Dichek, H. L., Cioffi, J. A., Choi, S. Y., Leeper, N. J., Quintana, L., Kronmal, G. S., Cooper, A. D., Quertermous, T. (1999) Cloning of a unique lipase from endothelial cells extends the lipase gene family. *J Biol Chem.*, **274**, 14170–14175.

133 McCoy, M. G., Sun, G. S., Marchadier, D., Maugeais, C., Glick, J. M., Rader, D. J. (2002) Characterization of the lipolytic activity of endothelial lipase. *J Lipid Res.*, **43**, 921–929.

134 Maugeais, C., Tietge, U. J., Broedl, U. C., Marchadier, D., Cain, W., McCoy, M. G., Lund-Katz, S., Glick, J. M., Rader, D. J. (2003) Dose-dependent acceleration of high-density lipoprotein catabolism by endothelial lipase. *Circulation*, **108**, 2121–2126.

135 Broedl, U. C., Maugeais, C., Marchadier, D., Glick, J. M., Rader, D. J. (2003) Effects of nonlipolytic ligand function of endothelial lipase on high density lipoprotein metabolism in vivo. *J Biol Chem.*, **278**, 40688–40693.

136 Ishida, T., Choi, S. Y., Kundu, R. K., Spin, J., Yamashita, T., Hirata, K., Kojima, Y., Yokoyama, M., Cooper, A. D., Quertermous, T. (2004) Endothelial lipase modulates susceptibility to atherosclerosis in apolipoprotein-E-deficient mice. *J Biol Chem.*, **279**, 45085–45092.

137 Ko, K. W., Paul, A., Ma, K., Li, L., Chan, L. (2005) Endothelial lipase modulates HDL but has no effect on atherosclerosis development in apoE−/− and LDLR−/− mice. *J Lipid Res.*, **46**, 2586–2594.

138 Ahmed, W., Orasanu, G., Nehra, V., Asatryan, L., Rader, D. J., Ziouzenkova, O., Plutzky, J. (2006) High-density lipoprotein hydrolysis by endothelial lipase activates PPARalpha: a candidate mechanism for high-density lipoprotein-mediated repression of leukocyte adhesion. *Circ Res.*, **98**, 490–498.

139 Chawla, A., Lee, C. H., Barak, Y., He, W., Rosenfeld, J., Liao, D., Han, J., Kang, H., Evans, R. M. (2003) PPARdelta is a very low-density lipoprotein sensor in macrophages. *Proc Natl Acad Sci USA*, **100**, 1268–1273.

140 Ziouzenkova, O., Perrey, S., Asatryan, L., Hwang, J., MacNaul, K. L., Moller, D. E., Rader, D. J., Sevanian, A., Zechner, R., Hoefler, G., Plutzky, J. (2003) Lipolysis of triglyceride-rich lipoproteins generates PPAR ligands: evidence for an antiinflammatory role for lipoprotein lipase. *Proc Natl Acad Sci USA*, **100**, 2730–2735.

141 Santamarina-Fojo, S., Gonzalez-Navarro, H., Freeman, L., Wagner, E., Nong, Z. (2004) Hepatic lipase, lipoprotein metabolism, and atherogenesis. *Arterioscler Thromb Vasc Biol.*, **24**, 1750–1754.

142 Zambon, A., Deeb, S. S., Hokanson, J. E., Brown, B. G., Brunzell, J. D. (1998) Common variants in the promoter of the hepatic lipase gene are associated with lower levels of hepatic lipase activity, buoyant LDL, and higher HDL2 cholesterol. *Arterioscler Thromb Vasc Biol.*, **18**, 1723–1729.

143 Dugi, K. A., Amar, M. J., Haudenschild, C. C., Shamburek, R. D., Bensadoun, A., Hoyt, R. F., Jr., Fruchart-Najib, J., Madj, Z., Brewer, H. B. Jr., Santamarina-Fojo, S. (2000) In vivo evidence for both lipolytic and nonlipolytic function of hepatic lipase in the metabolism of HDL. *Arterioscler Thromb Vasc Biol.*, **20**, 793–800.

144 Breckenridge, W. C., Little, J. A., Alaupovic, P., Wang, C. S., Kuksis, A., Kakis, G., Lindgren, F., Gardiner,

G. (1982) Lipoprotein abnormalities associated with a familial deficiency of hepatic lipase. *Atherosclerosis*, **45**, 161–179.

145 Connelly, P. W., Maguire, G. F., Lee, M., Little, J. A. (1990) Plasma lipoproteins in familial hepatic lipase deficiency. *Arteriosclerosis*, **10**, 40–48.

146 Homanics, G. E., deSilva, H. V., Osada, J., Zhang, S. H., Wong, H., Borensztajn, J., Maeda, N. (1995) Mild dyslipidemia in mice following targeted inactivation of the hepatic lipase gene. *J Biol Chem.*, **270**, 2974–2980.

147 Castle, C. K., Kuiper, S. L., Blake, W. L., Paigen, B., Marotti, K. R., Melchior, G. W. (1998) Remodeling of the HDL in NIDDM: a fundamental role for cholesteryl ester transfer protein. *Am J Physiol.*, **274**, E1091–1098.

148 Rashid, S., Trinh, D. K., Uffelman, K. D., Cohn, J. S., Rader, D. J., Lewis, G. F. (2003) Expression of human hepatic lipase in the rabbit model preferentially enhances the clearance of triglyceride-enriched versus native high-density lipoprotein apolipoprotein A-I. *Circulation*, **107**, 3066–3072.

149 Busch, S. J., Barnhart, R. L., Martin, G. A., Fitzgerald, M. C., Yates, M. T., Mao, S. J., Thomas, C. E., Jackson, R. L. (1994) Human hepatic triglyceride lipase expression reduces high density lipoprotein and aortic cholesterol in cholesterol-fed transgenic mice. *J Biol Chem.*, **269**, 16376–16382.

150 Dichek, H. L., Brecht, W., Fan, J., Ji, Z. S., McCormick, S. P., Akeefe, H., Conzo, L., Sanan, D. A., Weisgraber, K. H., Young, S. G., Taylor, J. M., Mahley, R. W. (1998) Overexpression of hepatic lipase in transgenic mice decreases apolipoprotein B-containing and high density lipoproteins. Evidence that hepatic lipase acts as a ligand for lipoprotein uptake. *J Biol Chem.*, **273**, 1896–1903.

151 Dichek, H. L., Johnson, S. M., Akeefe, H., Lo, G. T., Sage, E., Yap, C. E., Mahley, R. W. (2001) Hepatic lipase overexpression lowers remnant and LDL levels by a noncatalytic mechanism in LDL receptor-deficient mice. *J Lipid Res.*, **42**, 201–210.

152 Mezdour, H., Jones, R., Dengremont, C., Castro, G., Maeda, N. (1997) Hepatic lipase deficiency increases plasma cholesterol but reduces susceptibility to atherosclerosis in apolipoprotein E-deficient mice. *J Biol Chem.*, **272**, 13570–13575.

153 Nong, Z., Gonzalez-Navarro, H., Amar, M., Freeman, L., Knapper, C., Neufeld, E. B., Paigen, B. J., Hoyt, R. F., Fruchart-Najib, J., Santamarina-Fojo, S. (2003) Hepatic lipase expression in macrophages contributes to atherosclerosis in apoE-deficient and LCAT-transgenic mice. *J Clin Invest.*, **112**, 367–378.

154 Merkel, M., Eckel, R. H., Goldberg, I. J. (2002) Lipoprotein lipase: genetics, lipid uptake, and regulation. *J Lipid Res.*, **43**, 1997–2006.

155 Stein, Y. and Stein, O. (2003) Lipoprotein lipase and atherosclerosis. *Atherosclerosis*, **170**, 1–9.

156 Eisenberg, S. (1984) High density lipoprotein metabolism. *J Lipid Res.*, **25**, 1017–1058.

157 Evans, V. and Kastelein, J. J. (2002) Lipoprotein lipase deficiency–rare or common?*Cardiovasc Drugs Ther.*, **16**, 283–287.

158 Coleman, T., Seip, R. L., Gimble, J. M., Lee, D., Maeda, N., Semenkovich, C. F. (1995) COOH-terminal disruption of lipoprotein lipase in mice is lethal in homozygotes, but heterozygotes have elevated triglycerides and impaired enzyme activity. *J Biol Chem.*, **270**, 12518–12525.

159 Weinstock, P. H., Bisgaier, C. L., Aalto-Setala, K., Radner, H., Ramakrishnan, R., Levak-Frank, S., Essenburg, A. D., Zechner, R., Breslow, J. L. (1995) Severe hypertriglyceridemia, reduced high

density lipoprotein, and neonatal death in lipoprotein lipase knockout mice. Mild hypertriglyceridemia with impaired very low density lipoprotein clearance in heterozygotes. *J Clin Invest.*, **96**, 2555–2568.

160 Zsigmond, E., Kobayashi, K., Tzung, K. W., Li, L., Fuke, Y., Chan, L. (1997) Adenovirus-mediated gene transfer of human lipoprotein lipase ameliorates the hyperlipidemias associated with apolipoprotein E and LDL receptor deficiencies in mice. *Hum Gene Ther.*, **8**, 1921–1933.

161 Excoffon, K. J., Liu, G., Miao, L., Wilson, J. E., McManus, B. M., Semenkovich, C. F., Coleman, T., Benoit, P., Duverger, N., Branellec, D., Denefle, P., Hayden, M. R., Lewis, M. E. (1997) Correction of hypertriglyceridemia and impaired fat tolerance in lipoprotein lipase-deficient mice by adenovirus-mediated expression of human lipoprotein lipase. *Arterioscler Thromb Vasc Biol.*, **17**, 2532–2539.

162 Liu, G., Ashbourne Excoffon, K. J., Wilson, J. E., McManus, B. M., Rogers, Q. R., Miao, L., Kastelein, J. J., Lewis, M. E., Hayden, M. R. (2000) Phenotypic correction of feline lipoprotein lipase deficiency by adenoviral gene transfer. *Hum Gene Ther.*, **11**, 21–32.

163 Strauss, J. G., Frank, S., Kratky, D., Hammerle, G., Hrzenjak, A., Knipping, G., von Eckardstein, A., Kostner, G. M., Zechner, R. (2001) Adenovirus-mediated rescue of lipoprotein lipase-deficient mice. Lipolysis of triglyceride-rich lipoproteins is essential for high density lipoprotein maturation in mice. *J Biol Chem.*, **276**, 36083–36090.

164 Faustinella, F., Chang, A., Van Biervliet, J. P., Rosseneu, M., Vinaimont, N., Smith, L. C., Chen, S. H., Chan, L. (1991) Catalytic triad residue mutation (Asp156—Gly) causing familial lipoprotein lipase deficiency. Co-inheritance with a nonsense mutation (Ser447—Ter) in a Turkish family. *J Biol Chem.*, **266**, 14418–14424.

165 Groenemeijer, B. E., Hallman, M. D., Reymer, P. W., Gagne, E., Kuivenhoven, J. A., Bruin, T., Jansen, H., Lie, K. I., Bruschke, A. V., Boerwinkle, E., Hayden, M. R., Kastelein, J. J. (1997) Genetic variant showing a positive interaction with beta-blocking agents with a beneficial influence on lipoprotein lipase activity, HDL cholesterol, and triglyceride levels in coronary artery disease patients. The Ser447-stop substitution in the lipoprotein lipase gene. REGRESS Study Group. *Circulation*, **95**, 2628–2635.

166 Gagne, S. E., Larson, M. G., Pimstone, S. N., Schaefer, E. J., Kastelein, J. J., Wilson, P. W., Ordovas, J. M., Hayden, M. R. (1999) A common truncation variant of lipoprotein lipase (Ser447X) confers protection against coronary heart disease: the Framingham Offspring Study. *Clin Genet.*, **55**, 450–454.

167 Wittrup, H. H., Tybjaerg-Hansen, A., Nordestgaard, B. G. (1999) Lipoprotein lipase mutations, plasma lipids and lipoproteins, and risk of ischemic heart disease. A meta-analysis. *Circulation*, **99**, 2901–2907.

168 Ross, C., Meulenberg, J., Twisk, J., JA, K., Liu, G., Miao, F., Moraal, E., Oranje, P., Bakker, A., Hermens, W., Sheenhart-van derMeer, J., Kastelein, J., Hayden, M. (2003) Long-term correction of murine lipoprotein lipase deficiency by single intramuscular administration of AAV1-LPL^{S447X}. *Mol Ther.*, **7**, S22.

169 Rip, J., Nierman, M., Sierts, J., van Raalte, D., Petersen, W., van den Oever, K., Ross, C., Hayden, M. R., Meulenberg, J., Twisk, J., Kastelein, J. J., Kuivenhoven, J. (2003) Working towards clinical application of gene therapy for LPL deficiency. *Circulation*, **108**, IV-133.

170 Ross, C. J., Twisk, J., Bakker, A. C., Miao, F., Verbart, D., Rip, J., Godbey, T., Dijkhuizen, P., Hermens, W. T., Kastelein, J. J., Kuivenhoven, J. A., Meulenberg, J. M., Hayden, M. R. (2006) Correction of feline lipoprotein lipase deficiency with adeno-associated virus serotype 1-mediated gene transfer of the lipoprotein lipase S447X beneficial mutation. *Hum Gene Ther.*, **17**, 487–499.

171 Garner, B., Waldeck, A. R., Witting, P. K., Rye, K. A., Stocker, R. (1998) Oxidation of high density lipoproteins. II. Evidence for direct reduction of lipid hydroperoxides by methionine residues of apolipoproteins AI and AII. *J Biol Chem.*, **273**, 6088–6095.

172 Panzenbock, U., Kritharides, L., Raftery, M., Rye, K. A., Stocker, R. (2000) Oxidation of methionine residues to methionine sulfoxides does not decrease potential antiatherogenic properties of apolipoprotein A-I. *J Biol Chem.*, **275**, 19536–19544.

173 Panzenbock, U. and Stocker, R. (2005) Formation of methionine sulfoxide-containing specific forms of oxidized high-density lipoproteins. *Biochim Biophys Acta.*, **1703**, 171–181.

174 Primo-Parmo, S. L., Sorenson, R. C., Teiber, J., La Du, B. N. (1996) The human serum paraoxonase/arylesterase gene (PON1) is one member of a multigene family. *Genomics.*, **33**, 498–507.

175 Hegele, R. A. (1999) Paraoxonase genes and disease. *Ann Med.*, **31**, 217–224.

176 Draganov, D. I. and La Du, B. N. (2004) Pharmacogenetics of paraoxonases: a brief review. *Naunyn Schmiedebergs Arch Pharmacol.*, **369**, 78–88.

177 Getz, G. S. and Reardon, C. A. (2004) Paraoxonase, a cardioprotective enzyme: continuing issues. *Curr Opin Lipidol.*, **15**, 261–267.

178 Mackness, M. I., Arrol, S., Abbott, C., Durrington, P. N. (1993) Protection of low-density lipoprotein against oxidative modification by high-density lipoprotein associated paraoxonase. *Atherosclerosis.*, **104**, 129–135.

179 Aviram, M., Rosenblat, M., Bisgaier, C. L., Newton, R. S., Primo-Parmo, S. L., La Du, B. N. (1998) Paraoxonase inhibits high-density lipoprotein oxidation and preserves its functions. A possible peroxidative role for paraoxonase. *J Clin Invest.*, **101**, 1581–1590.

180 Jarvik, G. P., Rozek, L. S., Brophy, V. H., Hatsukami, T. S., Richter, R. J., Schellenberg, G. D., Furlong, C. E. (2000) Paraoxonase (PON1) phenotype is a better predictor of vascular disease than is PON1(192) or PON1(55) genotype. *Arterioscler Thromb Vasc Biol.*, **20**, 2441–2447.

181 Jarvik, G. P., Hatsukami, T. S., Carlson, C., Richter, R. J., Jampsa, R., Brophy, V. H., Margolin, S., Rieder, M., Nickerson, D., Schellenberg, G. D., Heagerty, P. J., Furlong, C. E. (2003) Paraoxonase activity, but not haplotype utilizing the linkage disequilibrium structure, predicts vascular disease. *Arterioscler Thromb Vasc Biol.*, **23**, 1465–1471.

182 Costa, L. G., Cole, T. B., Jarvik, G. P., Furlong, C. E. (2003) Functional genomic of the paraoxonase (PON1) polymorphisms: effects on pesticide sensitivity, cardiovascular disease, and drug metabolism. *Annu Rev Med.*, **54**, 371–392.

183 Shih, D. M., Gu, L., Xia, Y. R., Navab, M., Li, W. F., Hama, S., Castellani, L. W., Furlong, C. E., Costa, L. G., Fogelman, A. M., Lusis, A. J. (1998) Mice lacking serum paraoxonase are susceptible to organophosphate toxicity and atherosclerosis. *Nature*, **394**, 284–287.

184 Shih, D. M., Xia, Y. R., Wang, X. P., Miller, E., Castellani, L. W., Subbanagounder, G., Cheroutre, H., Faull, K. F., Berliner, J. A., Witztum, J. L., Lusis, A. J. (2000) Combined serum paraoxonase knockout/apolipoprotein E knockout mice exhibit increased lipoprotein oxidation and atherosclerosis. *J Biol Chem.*, **275**, 17527–17535.

185 Rozenberg, O., Rosenblat, M., Coleman, R., Shih, D. M., Aviram, M. (2003) Paraoxonase (PON1) deficiency is associated with

increased macrophage oxidative stress: studies in PON1-knockout mice. *Free Radic Biol Med.*, **34**, 774–784.

186 Rosenblat, M., Draganov, D., Watson, C. E., Bisgaier, C. L., La Du, B. N., Aviram, M. (2003) Mouse macrophage paraoxonase 2 activity is increased whereas cellular paraoxonase 3 activity is decreased under oxidative stress. *Arterioscler Thromb Vasc Biol.*, **23**, 468–474.

187 Tward, A., Xia, Y. R., Wang, X. P., Shi, Y. S., Park, C., Castellani, L. W., Lusis, A. J., Shih, D. M. (2002) Decreased atherosclerotic lesion formation in human serum paraoxonase transgenic mice. *Circulation*, **106**, 484–490.

188 Bradshaw, G., Gutierrez, A., Miyake, J. H., Davis, K. R., Li, A. C., Glass, C. K., Curtiss, L. K., Davis, R. A. (2005) Facilitated replacement of Kupffer cells expressing a paraoxonase-1 transgene is essential for ameliorating atherosclerosis in mice. *Proc Natl Acad Sci USA*, **102**, 11029–11034.

189 Mackness, B., Quarck, R., Verreth, W., Mackness, M., Holvoet, P. (2006) Human paraoxonase-1 overexpression inhibits atherosclerosis in a mouse model of metabolic syndrome. *Arterioscler Thromb Vasc Biol.*, **26**, 1545–1550.

190 Ng, C. J., Wadleigh, D. J., Gangopadhyay, A., Hama, S., Grijalva, V. R., Navab, M., Fogelman, A. M., Reddy, S. T. (2001) Paraoxonase-2 is a ubiquitously expressed protein with antioxidant properties and is capable of preventing cell-mediated oxidative modification of low density lipoprotein. *J Biol Chem.*, **276**, 44444–44449.

191 Ng, C. J., Bourquard, N., Grijalva, V., Hama, S., Shih, D. M., Navab, M., Fogelman, A. M., Lusis, A. J., Young, S., Reddy, S. T. (2006) Paraoxonase-2 deficiency aggravates atherosclerosis in mice despite lower apolipoprotein-B-containing lipoproteins: anti-atherogenic role for paraoxonase-2. *J Biol Chem.*, **281**, 29491–29500.

192 Ng, C. J., Hama, S. Y., Bourquard, N., Navab, M., Reddy, S. T. (2006) Adenovirus mediated expression of human paraoxonase 2 protects against the development of atherosclerosis in apolipoprotein E-deficient mice. *Mol Genet Metab.*, **89**, 368–373.

193 Reddy, S. T., Wadleigh, D. J., Grijalva, V., Ng, C., Hama, S., Gangopadhyay, A., Shih, D. M., Lusis, A. J., Navab, M., Fogelman, A. M. (2001) Human paraoxonase-3 is an HDL-associated enzyme with biological activity similar to paraoxonase-1 protein but is not regulated by oxidized lipids. *Arterioscler Thromb Vasc Biol.*, **21**, 542–547.

194 Klebanoff, S. J. (2005) Myeloperoxidase: friend and foe. *J Leukoc Biol.*, **77**, 598–625.

195 Niu, X., Zammit, V., Upston, J. M., Dean, R. T., Stocker, R. (1999) Coexistence of oxidized lipids and alpha-tocopherol in all lipoprotein density fractions isolated from advanced human atherosclerotic plaques. *Arterioscler Thromb Vasc Biol.*, **19**, 1708–1718.

196 Hazell, L. J., Arnold, L., Flowers, D., Waeg, G., Malle, E., Stocker, R. (1996) Presence of hypochlorite-modified proteins in human atherosclerotic lesions. *J Clin Invest.*, **97**, 1535–1544.

197 Zheng, L., Nukuna, B., Brennan, M. L., Sun, M., Goormastic, M., Settle, M., Schmitt, D., Fu, X., Thomson, L., Fox, P. L., Ischiropoulos, H., Smith, J. D., Kinter, M., Hazen, S. L. (2004) Apolipoprotein A-I is a selective target for myeloperoxidase-catalyzed oxidation and functional impairment in subjects with cardiovascular disease. *J Clin Invest.*, **114**, 529–541.

198 Shao, B., Oda, M. N., Bergt, C., Fu, X., Green, P. S., Brot, N., Oram, J. F., Heinecke, J. W. (2006) Myeloperoxidase impairs ABCA1-dependent cholesterol efflux through methionine oxidation and site-specific tyrosine chlorination of

apolipoprotein A-I. *J Biol Chem.*, **281**, 9001–9004.

199 Zhang, R., Brennan, M. L., Fu, X., Aviles, R. J., Pearce, G. L., Penn, M. S., Topol, E. J., Sprecher, D. L., Hazen, S. L. (2001) Association between myeloperoxidase levels and risk of coronary artery disease. *Jama*, **286**, 2136–2142.

200 Brennan, M. L., Penn, M. S., Van Lente, F., Nambi, V., Shishehbor, M. H., Aviles, R. J., Goormastic, M., Pepoy, M. L., McErlean, E. S., Topol, E. J., Nissen, S. E., Hazen, S. L. (2003) Prognostic value of myeloperoxidase in patients with chest pain. *N Engl J Med.*, **349**, 1595–1604.

201 McMillen, T. S., Heinecke, J. W., LeBoeuf, R. C. (2005) Expression of human myeloperoxidase by macrophages promotes atherosclerosis in mice. *Circulation*, **111**, 2798–2804.

202 Nikpoor, B., Turecki, G., Fournier, C., Theroux, P., Rouleau, G. A. (2001) A functional myeloperoxidase polymorphic variant is associated with coronary artery disease in French-Canadians. *Am Heart J.*, **142**, 336–339.

203 Asselbergs, F. W., Reynolds, W. F., Cohen-Tervaert, J. W., Jessurun, G. A., Tio, R. A. (2004) Myeloperoxidase polymorphism related to cardiovascular events in coronary artery disease. *Am J Med.*, **116**, 429–430.

204 Castellani, L. W., Chang, J. J., Wang, X., Lusis, A. J., Reynolds, W. F. (2006) Transgenic mice express human MPO -463G/A alleles at atherosclerotic lesions, developing hyperlipidemia and obesity in -463G males. *J Lipid Res.*, **47**, 1366–1377.

205 Brennan, M. L., Anderson, M. M., Shih, D. M., Qu, X. D., Wang, X., Mehta, A. C., Lim, L. L., Shi, W., Hazen, S. L., Jacob, J. S., Crowley, J. R., Heinecke, J. W., Lusis, A. J. (2001) Increased atherosclerosis in myeloperoxidase-deficient mice. *J Clin Invest.*, **107**, 419–430.

III
HDL Formation, Secretion, Removal

High-Density Lipoproteins: From Basic Biology to Clinical Aspects. Edited by Christopher J. Fielding

ISBN: 978-3-527-31717-2

10
Regulation of Genes Involved in the Biogenesis and the Remodeling of HDL*

Dimitris Kardassis, Costas Drosatos, and Vassilis I. Zannis

10.1
Introduction

The initiation of gene transcription is a complex biological process that requires the ordered assembly of multiprotein *transcription initiation complexes* in specific regulatory regions in the promoters of genes [1–4]. All class II genes are transcribed by RNA polymerase II (pol II). This multicomponent enzyme and associated proteins make a complex of more than 70 polypeptides, comparable in size to a ribosome, that is necessary for gene transcription [5]. The factors associated with pol II are general transcription factors, named TFIIA, B, D, E, F, and H, which have diverse functions and participate in the initiation or elongation of transcription [6,7] (Fig. 10.1A). The TFIID factor is a complex of TATA box binding protein (TBP) and at least nine tightly associated polypeptides called TBP-associated Factors (TAFs) [8], which are targets for transcriptional activators. Multiple contacts between different transcription factors and their respective TAFs result in stabilization of the pre-initiation complex and in the synergistic activation of transcription [9,10].

10.2
The Coregulator Complexes

The coregulator complexes are recruited to the promoters through interactions with the transcriptional activation or repression domains of DNA-bound transcription factors. They are also known as coactivators or corepressors, due to their ability to regulate – positively or negatively – the transactivation functions of the transcription factors. Coregulator complexes can modify the histones or the transcription factors either directly or by recruiting other proteins with intrinsic enzymatic activities, including acetylases, methylases, and kinases [1–4]. Histone acetylases enzymatically modify the histones within the regulatory regions of genes and thus facilitate the recruitment of other proteins necessary for gene transcription [11,12]. This process

*Please find a List of Abbreviations at the end of this chapter.

High-Density Lipoproteins. From Basic Biology to Clinical Aspects. Edited by Christopher J. Fielding

ISBN: 978-3-527-31717-2

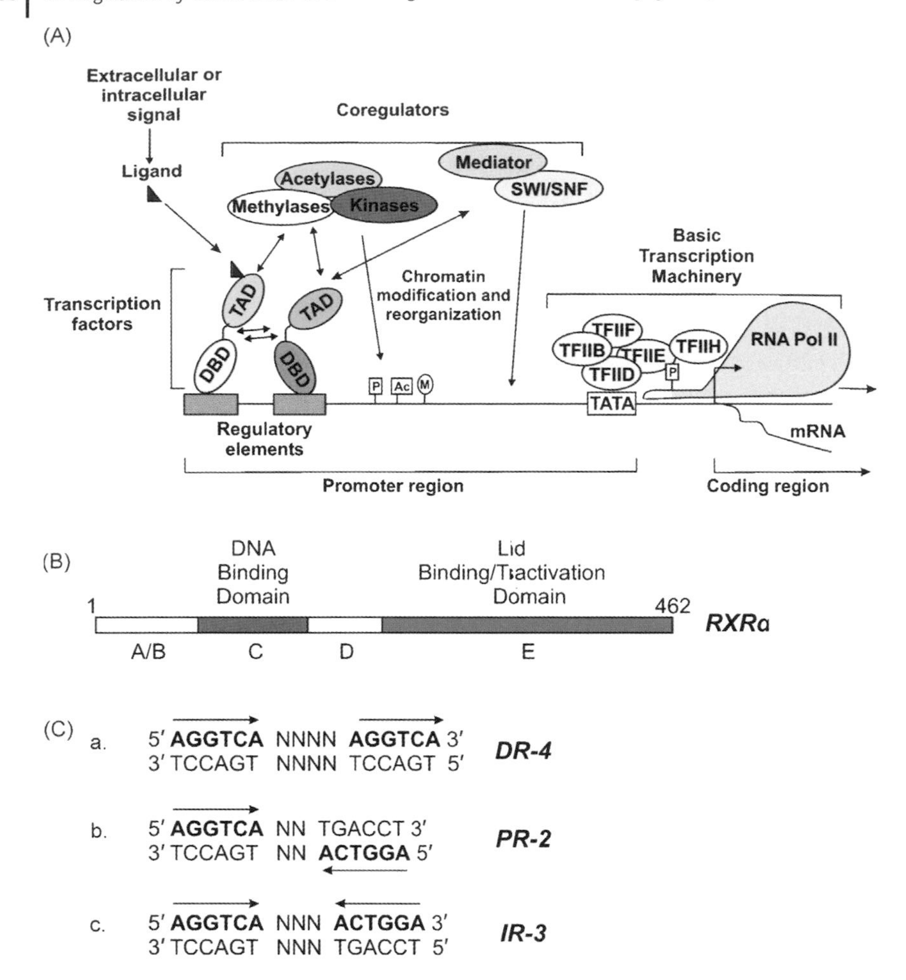

Fig. 10.1 *A*) Schematic representation of the molecular events that take place during transcription initiation in eukaryotic genes. Interactions between the three classes of nuclear proteins that participate in these events – the transcription factors, the coregulators (coactivators, corepressors), and the proteins of the basic transcription apparatus – are indicated by arrows. Transcription factor activation by extracellular signals is also shown. DBD: DNA Binding Domain. TAD: Transactivation Domain. The modifications of the histones elicited by the enzymatic activity of different coregulators are shown with the symbols P (Phosphorylation), M (Methylation), and Ac (Acetylation). *B*) Schematic representation of the functional domains of the nuclear hormone receptor RXRα. *C*) Sequence of a consensus HRE that acts as the DNA binding site of nuclear hormone receptors. a: Direct repeat separated by four nucleotides (DR-4). b: Palindromic repeat separated by two nucleotides (PR-2). c: Inverted repeat separated by three nucleotides (IR-3).

can be reversed by the action of histone deacetylases [11,12]. Coactivators with acetyl transferase activity can also enzymatically modify transcription factors such as the nuclear receptor Hepatocyte Nuclear Factor 4 (HNF-4) and can thus regulate their functions at multiple levels, including their nuclear localization, DNA binding, and interaction with other coactivators [13]. Studies in yeast have identified multicomponent complexes, such as the mediator associated with the carboxy terminal domain of RNA polymerase II [14] and the switch/sucrose nonfermenting complex (SWI/SNF), that remodel the chromatin and make it more accessible to the basal transcription machinery [15,16].

The defining characteristic that discriminates between different promoters in the genome is the unique array of *cis*-regulatory elements that are recognized by sequence-specific DNA-binding transcription factors [1]. Transcription factors may be constitutively active in a cell, or they may be activated by various signal transduction pathways. All known transcription factors are modular in nature, each containing a DNA-binding domain and a transcriptional activation domain. In addition, several factors contain a dimerization domain that permits them to form homodimers and/or heterodimers. A variety of nuclear receptors for steroids, thyroids, retinoids, etc. contain a ligand binding site (Fig. 10.1B) [17,18]. Transcription factors appear to facilitate, through their transcription activation domains, the recruitment of the proteins of the coactivator complex and the basal transcription complex to the transcription initiation site of each gene and thus initiate transcription [19].

The structures and functions of the key transcription factors that regulate the expression of genes implicated in the biogenesis of HDL are summarized below:

10.2.1 Nuclear Hormone Receptors

Nuclear hormone receptors belong to a superfamily of transcription factors that are activated by hormones such as steroids (estrogen, progesterone, glucocorticoids, etc.), retinoids (9-*cis*- and *all-trans*-retinoic acid), thyroid hormone, peroxisome proliferators, and vitamin D and regulate cell differentiation, development, homeostasis, and reproduction [20,21]. The nuclear receptor superfamily also includes numerous orphan receptors [22]. Ligands for some of these orphan receptors have recently been identified; they consist of products of lipid metabolism such as fatty acids, bile acids, prostaglandins, or cholesterol derivatives [22]. The structure of nuclear receptors is highly conserved throughout evolution and consists of an N-terminal transactivation domain called Activation Function 1 (AF-1), a conserved DNA-binding domain that contains two zinc fingers, and a second transactivation domain called AF-2, located close to the ligand-binding domain (LBD) (Fig. 10.1B) [20,21]. Nuclear receptors undergo conformational changes upon ligand binding, and these lead to the recruitment of coactivators [23]. Nuclear receptors may also exert their functions by cross-talk with other signaling pathways or by interaction with other transcription factors [24]. Nuclear receptors that are regulated by steroids form homodimers, whereas nuclear receptors that are regulated by other ligands form heterodimers with the Retinoid X Receptor (RXR) and bind to Hormone Response Elements (HREs) on the promoters of their target genes. The HREs may contain direct repeats (DRs), inverted repeats (IRs), or palindromic repeats (PRs) of the sequence 5′ AG(G/T)TCA 3′. The repeats are separated by one, two, three, four, or five nucleotides and are designated DR1, DR2, etc. (for the direct repeats), IR1, IR2, etc. (for the inverted repeats), and PR1, PR2, etc. (for the palindromic repeats) (Fig. 10.1C).

10.2.2 LXR, FXR, LRH-1, SHP

Liver X receptors LXRα and LXRβ and farnesoid X receptors FXRα and FXRβ are nuclear receptors that activate gene transcription by forming heterodimers with RXR in the presence of their permissive ligands [25,26]. Ligands for LXRα are 22-(R)-hydroxycholesterol and other oxysterols, whereas ligands for FXRα are bile acids such as chenodeoxycholic acid, lithocholic acid, and deoxycholic acid. FXRα and LXRα are expressed in liver, intestine, and other tissues, whereas the β forms are expressed ubiquitously [25,26]. Two other orphan receptors – the Liver Receptor Homolog-1 (LRH-1) [27], which binds as a monomer to DNA, and the Small Heterodimer Partner (SHP) [28], which lacks a DNA binding domain – play an important role in cholesterol homeostasis by regulating the activity of LXRs and FXRs. LXRα knockout mice fed high-cholesterol diets have diminished bile acid production and accumulate cholesterol esters in their livers [29].

10.2.3 PPARα, -β, -γ

Peroxisome proliferator activated receptor α (PPARα) is activated by peroxisome proliferators such as fibrates and various fatty acids and may play a role in the regulation of genes in response to nutrients [30]. It has been shown that 15-deoxy-Δ12,14 prostaglandin J2 is a ligand for PPARγ [31]. Inactivation of PPARα results in viable mice that are refractory to the action of peroxisome proliferators [32]. Thiazolidinediones are PPARγ agonists that are currently used for the treatment of type II diabetes mellitus and other diseases [33].

10.2.4 SF-1

Steroidogenic Factor 1 (SF-1), also called Ad4BP (Adrenal 4 Binding Protein) is a 53 kDa nuclear receptor that is specifically expressed in steroidogenic tissues [34]. SF-1 shows high homology with the drosophila Ftz-F1 transcription factor, which controls fushi tarazu homeotic gene expression [35]. The gene encoding SF-1 generates four different proteins – ELP1, ELP2, ELP3, and SF-1 – by using alternative promoters and splicing [36]. Studies in mice in which the SF-1 gene was inactivated have shown that SF-1 is the key regulator of steroidogenic endocrine development and function [37].

10.2.5 HNF-4

Hepatocyte Nuclear Factor 4 (HNF-4) is a liver-enriched nuclear orphan receptor that regulates the transcription of a large number of liver-specific genes, including the majority of the human apolipoprotein genes [38,39]. It had initially been shown that fatty acyl-CoA thioesters bind to and modulate the transcription activity of HNF-4 [40]. Recent crystallographic studies established that the ligand-binding domain of HNF-4 was occupied by C_{14}–C_{18} fatty acids that lock the receptor in an active conformation [41,42]. Disruption of the HNF-4 gene leads to embryonic lethality [43], and conditional liver-specific inactivation of the HNF-4 gene in mice caused weight loss and lipid abnormalities due to impaired expression of genes involved in lipid and bile acid metabolism and transport [44].

10.2.6
ARP-1, EAR-3

ARP-1 (COUPTFII) and EAR-3 (COUPTFI) are orphan nuclear receptors with sequence homology and common DNA binding specificity with HNF-4 [45]. Both factors have been implicated in embryonic development [46,47] and neuronal cell fate determination [48]. When ARP-1 binds to the same sites as HNF-4 it inhibits transcription [49]; however, in promoters in which HNF-4 binds to unique sites, APR-1 has the capacity to potentiate the activity of HNF-4 through protein–protein interactions [50,51].

10.2.7
SP1 Family of Proteins

SP1 is a ubiquitous transcription factor with a zinc finger-type DNA binding domain that activates several cellular and viral promoters containing the consensus GC-rich binding site 5′ G(T)G(A)GGC(A)GG(T)G(A) 3′ [52,53]. Three other SP1 homologous genes – SP2, SP3, and SP4 – have been identified [52,53]; molecular analysis indicated that these three members may have different functions. Inactivation of SP1 causes embryonic lethality [54].

10.2.8
SREBPs

Sterol Regulatory Element Binding Proteins (SREBPs) are basic-helix-loop-helix-leucine zipper (bHLH-LZ) transcription factors that are synthesized as inactive precursors bound to the endoplasmic reticulum membrane [55]. There are two homologous genes: SREBP-1, which encodes for SREBP-1a and -1c proteins, and SREBP-2, which encodes for the corresponding protein [55,56]. When cholesterol is in excess, membrane protein SREBP cleavage activating protein (SCAP), present in the ER, interacts both with SREBP and with another cholesterol-sensing protein, insig1(2), and all three proteins remain anchored in the ER membrane. Under conditions of cholesterol depletion, insig1(2) dissociates from the complex and allows SREBP and SCAP to move by vesicular transport to the Golgi, where the SREBP proteins are cleaved by two proteases, designated site 1 protease (S1P) and site 2 protease (S2P). S1P cleaves the loop connecting the two transmembrane regions in the Golgi lumen, while S2P cleaves within the first transmembrane domain [57–60]. The cleaved amino-terminal fragment, representing ~480 amino acids of SREBP and containing the bHLHZip motif that includes the DNA binding, and the trans activation domain, translocates to the nucleus and induces transcription of several genes involved in cholesterol or fatty acid biosynthesis and transport [55,56,61].

SREBP-1c is involved in the regulation of the genes of fatty acid synthesis, whereas SREBP-2 in the regulation of the genes of cholesterol metabolism [55].

10.3 Transcriptional Regulation of Genes Involved in HDL Biogenesis and Catabolism

Below we summarize the transcriptional regulatory mechanisms determining the expression of the major genes that contribute to the biogenesis, remodeling, and catabolism of HDL.

10.3.1 Transcriptional Regulation of the Human ApoA-I Gene

Earlier studies established a linkage and a common regulatory mechanism of the apoA-I/apoCIII/apoA-IV gene cluster [62,63]. There are two common features in this cluster. One feature is that the distal regulatory region (−790/−560) of the apoCIII promoter constitutes an enhancer that increases the strength of the neighboring genes *in vitro* [64–67] and *in vivo* [38,68,69]. The other feature is that all the three genes of the cluster have one or more HREs in their proximal promoters [70–72] (Fig. 10.2A).

10.3.1.1 *In vitro* Studies

In humans, the apoA-I gene is expressed abundantly in liver and intestine, and to a lesser extent in other tissues [73]. The proximal human apoA-I promoter contains three regulatory elements, which have been designated D, C, and B (or A, B, and C, respectively, by other investigators [74]) and which are required for transcription in HepG2 cells [71]. Expression in CaCo-2 cells requires a 1.5 Kb 5′ promoter region, the apoCIII enhancer, and the presence of HNF-4 [65]. Elements D and B contain HREs that bind HNF-4 and other members of the nuclear receptor superfamily [71,72]. Mutations in the HREs of elements B or D, which prevent the binding of nuclear receptors to these sites, reduced the promoter strength to approximately 5–7 % of the control [72]. Element C binds CAAT/Enhancer Binding Protein (C/EBP) and Nuclear Factor Y (NFY) [71,75] (Fig. 10.2A).

Although the proximal apoA-I promoter can function independently *in vitro* [71] and *in vivo* [76], it is a very weak promoter. The strength of this promoter increases severalfold, however, when it is linked to the apoCIII enhancer [65]. The enhancer contains two HREs: one is on element G and binds orphan and different combinations of ligand-dependent nuclear receptors, while the other is on element I4 and binds HNF-4, other orphan nuclear receptors, and different combinations of

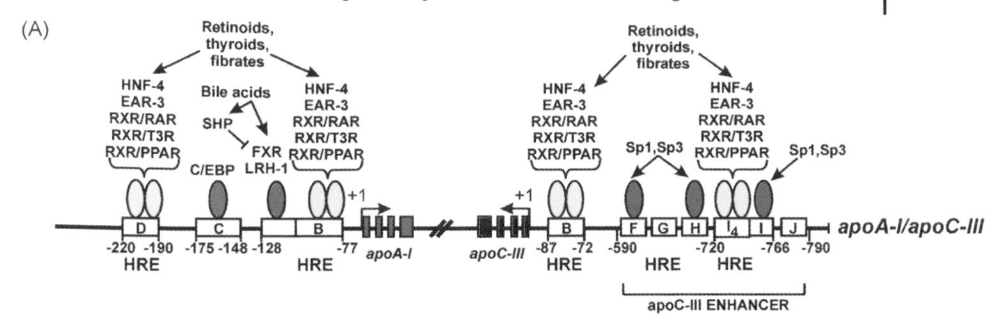

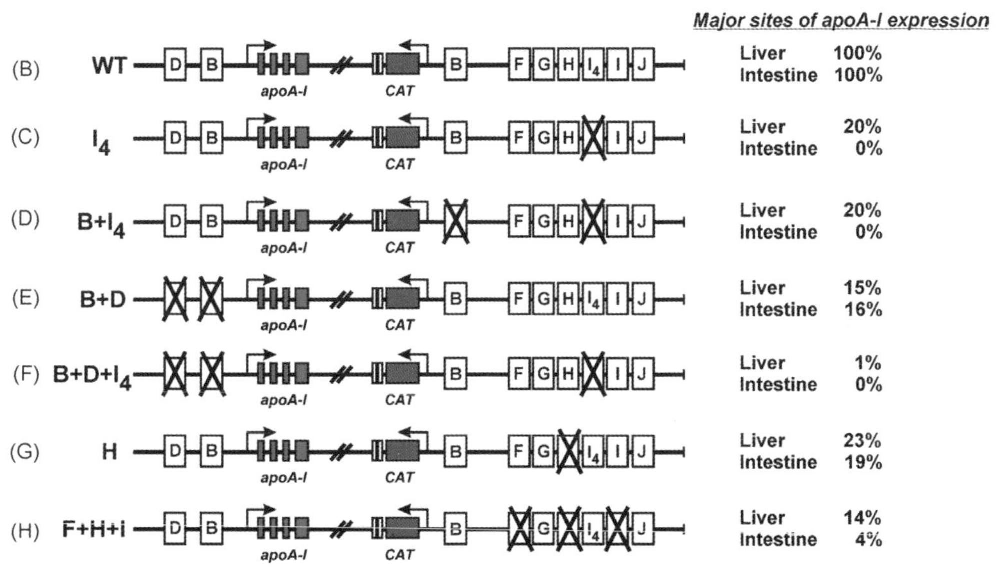

Fig. 10.2 *Panel A:* Regulatory elements and transcription factors governing the expression of the apoA-I/apoCIII/apoA-IV gene cluster. *Panels B–H:* Summary of expression data in transgenic mice carrying the WT apoA-I/apoCIII locus (*Panel B*) or the same locus mutated at the HREs (*Panels C–F*) or the SP1 sites (*Panels G–H*). The mutation sites are indicated by X. The panel illustrates that mutations in the proximal apoA-I promoter or the apoCIII enhancer drastically reduce, but do not eliminate, the hepatic transcription, whereas mutations in the apoCIII enhancer may abolish or diminish the intestinal transcription. Transcription factors are depicted by ovals.

ligand-dependent nuclear receptors. It also contains three Specificity Protein 1 (SP1) binding sites on elements F, H, and I (Fig. 10.2A). *In vitro* mutagenesis of the apoA-I promoter/apoCIII enhancer cluster showed that the enhancer activity was nearly abolished by deletion of the regulatory elements J, I, and H and was reduced to 40–70 % of the control by mutations in the SP1 binding sites [65]. The activity of the apoA-I promoter/apoCIII enhancer was diminished by mutations in the proximal HREs (elements B and D) but was not greatly affected by mutations affecting the binding of C/EBP to regulatory element C. Mutations in various elements of the apoCIII enhancer affected the HNF-4-mediated transactivation of the −1500 apoA-I promoter/apoCIII enhancer cluster in CaCo-2 cells to the same extent as they affected the strength of the −255 to −5 apoA-I promoter/apoCIII enhancer cluster in HepG2 cells [65].

10.3.1.2 *In vivo* Studies

In order for the *in vitro* observations on gene regulation to be valid, the findings also have to apply in an *in vivo* system. For this reason, we generated a variety of transgenic mouse lines expressing the apoA-I/apoCIII cluster. The only change we made in these constructs was that we substituted the apoCIII gene with the Chloramphenicol Acetyl Transferase (CAT) reporter gene (Fig. 10.2B). Three or more mouse lines were studied for each construct; in these mouse lines, individual hormone response elements [68] or individual SP1 binding sites were systematically mutated [69]. The questions addressed with these mouse lines related to the contribution of the hormone response elements and the SP1 sites in the overall expression and tissue-specific expression of the apoA-I gene *in vivo* [38,68,69].

Changes in the Expression of the ApoA-I Gene Resulting from Mutations in the apoA-I Promoter and the apoCIII Enhancer

Analysis of mice expressing the Wild-Type (WT) construct showed that major sites of apoA-I synthesis were the liver and the intestine (Fig. 10.2B). Minor sites of synthesis were the lung, stomach, heart, spleen, and muscle [68].

A mutation in the hormone response element of the apoCIII enhancer abolished the intestinal expression and reduced the hepatic expression to 20 % of the WT control (Fig. 10.2C).

Similar results were obtained by mutations both in the hormone response elements of the apoCIII enhancer and in the proximal apoCIII promoter (Fig. 10.2D) [38,68]. This suggested that the HRE of proximal apoCIII promoter does not affect the transcription of the human apoA-I gene.

Mutations in the hormone response elements of the proximal apoA-I promoter reduced the hepatic and intestinal expression of the apoA-I gene to approximately 15 % of the WT control (Fig. 10.2E). This was unexpected, since earlier cell culture studies had shown that mutations in the proximal hormone response elements inactivate the proximal apoA-I promoter in HepG2 cells [71,72]. The findings suggest that when the proximal promoter is inactivated, other distal regulatory elements localized within the apoCIII enhancer can drive the transcription of the apoA-I gene *in vivo*.

Finally, mutations in all three hormone response elements of the proximal apoA-I promoter and the apoCIII enhancer (the sites that bind HNF-4) abolished the intestinal and hepatic expression of the apoA-I gene [38] (Fig. 10.2F). The findings demonstrate that hormone nuclear receptors are essential for the expression of the apoA-I gene *in vivo*. This conclusion is also supported by other studies, which showed that the expression of the apoA-I/apoCIII/apoA-IV genes was abolished in the fetal livers of mice in which the HNF-4 gene was inactivated by homologous recombination [77].

The mutations in the SP1 sites of the apoCIII enhancer showed that a mutation in one SP1 site on element H of the enhancer reduced the hepatic expression to 23 % of the WT control and intestinal expression to approximately 19 % [38,69] (Fig. 10.2G).

Finally, mutations in all three SP1 sites of the apoCIII enhancer on elements F, H, and I reduced the hepatic expression to 14 % of the WT control and the intestinal expression to 4 % [69] (Fig. 10.2H).

10.3.1.3 Putative Mechanisms of ApoA-I Gene Transcription

Figure 10.3A–D illustrates putative mechanisms of the transcription of the apoA-I gene cluster and of how transcription is affected when different regulatory elements are inactivated. Figure 10.3A shows that a mutation in the hormone nuclear receptor binding site of the enhancer, which binds HNF-4 and other nuclear receptors, abolished the intestinal expression, but that the hepatic expression was maintained at the level of 20 % of the WT control. The data indicate that the apoCIII enhancer is required for the intestinal expression and can enhance the hepatic expression of the apoA-I gene fivefold (Fig. 10.3A). The most interesting observation, however, was that inactivation by mutagenesis of the proximal apoA-I promoter did not eliminate the hepatic and intestinal transcription, but rather reduced it to approximately 15 % relative to the WT control (Fig. 10.3B). This finding indicates that in the absence of the proximal promoter, the apoCIII enhancer alone can independently drive the hepatic and intestinal transcription of the apoA-I gene at lower levels. This information was not borne out by the *in vitro* studies, which showed that mutations in the proximal HRE abolished the apoA-I promoter activity [72].

Finally, when the proximal promoter and the apoCIII enhancer are intact, the proximal promoter alone can contribute 20 % to the transcription, while the enhancer alone can contribute 15 % to the transcription, whereas when both are functional we have 100 % transcription. We observe here that the result is not additive but rather synergistic, so we do not have overall $15 + 20 = 35$ % transcription, but rather 100 % transcription (Fig. 10.3C). These findings established transcriptional synergism *in vivo* for the first time.

Figure 10.3D depicts a putative mechanism that may explain the synergism between the proximal apoA-I promoter and the apoCIII enhancer in the transcription of the genes of apoA-I and the other genes of the cluster (apoCIII, apoA-IV). The transcription factors SP1 and HNF-4 play a major role in the transcriptional activation of the genes of the cluster. When the proximal promoter is inactivated, then factors bound to the enhancer can combine with co-activators and proteins of the basal transcription apparatus, and drive the transcription of the apoA-I at levels of 15 % of the WT control. When the enhancer is inactivated, then the factors bound to the proximal promoter can do the same, and drive transcription at levels of 20 % of the WT control. However, if both the factors that recognize the proximal promoter and the apoCIII enhancer are allowed to bind to their cognate sites, then we suggest, on the basis of the *in vitro* data, that they can cooperate through protein–protein interactions [78]. These interactions may account for transcriptional synergism that increases the transcription at levels of 100 % (Fig. 10.3D).

(A)

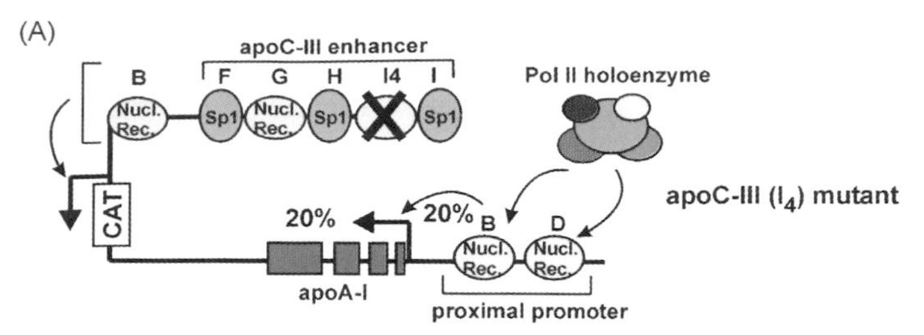

* Reduction in hepatic expression (~20%) of the *apoA-I* gene.
Complete inhibition of the intestinal expression of the *apoA-I* gene.

(B)

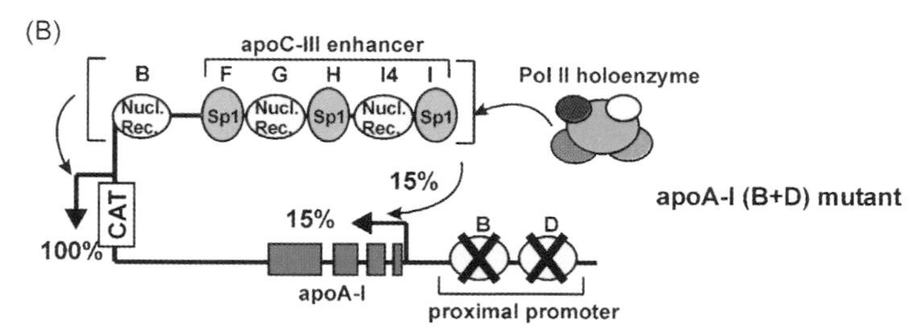

* Reduction in hepatic expression (~15%) and intestinal expression
(16%) of the *apoA-I* gene. No effect on *apoC-III* expression.

(C)

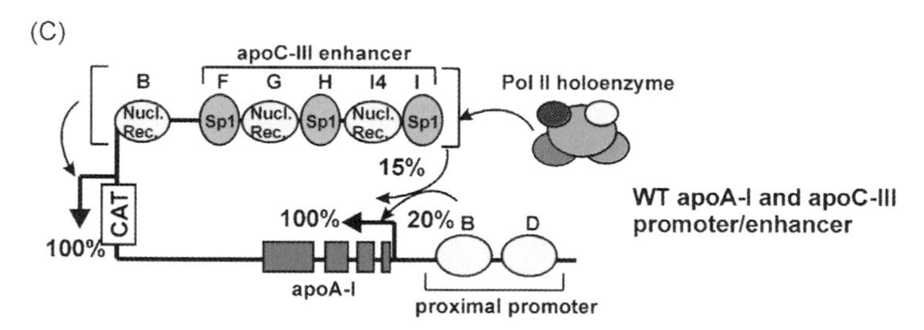

* Normal (100%) hepatic and intestinal expression of the *apoA-I* and *apoC-III* genes.
We observe transcriptional synergism *in vivo*: The contribution of the promoter is 20%
and of the ehnancer is 15%. The result is not additive (35%) but synergistic (100%).

(D)

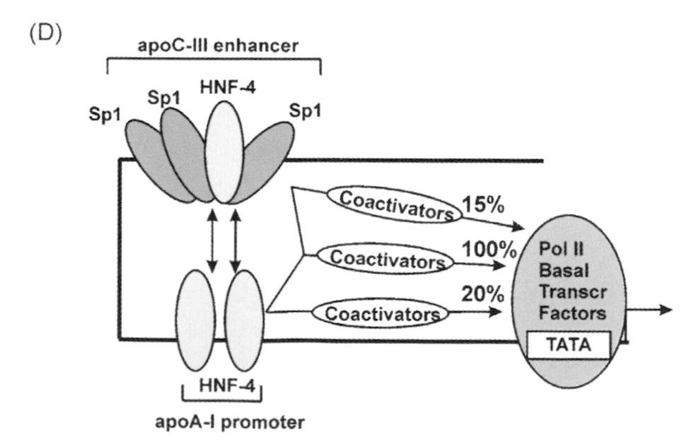

10.3.1.4 *In vivo* Regulation of the ApoA-I Gene in Transgenic and Knockout Animal Models, as well as in Response to Pharmacological and Dietary Treatments

A series of *in vivo* studies have highlighted the importance of the HREs for the transcriptional activity both of apoA-I and of the other genes of the cluster. The expression of the apoA-I/apoCIII/apoA-IV genes is abolished in the fetal livers of mice in which HNF-4 is inactivated by homologous recombination or in hepatic cell cultures infected with recombinant adenoviruses expressing a dominant negative form of HNF-4 [77,79].

Studies in transgenic mice and rabbits showed that fibrates increase human apoA-I gene transcription, and also apoA-I and HDL plasma levels. It was suggested that the increase was mediated by activation of PPARα, which binds to the regulatory element D of apoA-I [74,80] (Fig. 10.2A). The rodent apoA-I gene is repressed by fibrate treatment [74], and it has been proposed that this repression is caused by nucleotide differences in element D of the rodent gene, which prevent binding of PPARα, as well as by the binding of Rev-Erbα adjacent to the TATA box of the rat apoA-I promoter [74]. Inactivation of PPARα in mice is associated with reduced levels of hepatic apoA-I mRNA and reduced plasma apoA-I and HDLc levels [81], whereas the expression of the apoCIII gene is not affected. On the other hand, liver-specific inactivation of the RXRα gene in mice is associated with reduced expression of apoA-I and apoCIII genes [82]. It has been shown that homo- and/or heterodimers of RXRα bind to the HREs present in the proximal promoters of the apoA-I and apoCIII genes as well as to the apoCIII enhancer [70,72] (Fig. 10.2A). Finally, inactivation of the orphan receptor Retinoic acid receptor-related Orphan Receptor α (RORα) in mice decreased intestinal apoA-I mRNA levels [83]. Consistently with this observation, it was found that the orphan receptor RORα binds to the TATA box of the rat and mouse apoA-I gene and increases apoA-I transcription in CaCo-2 cells [83].

Plasma apoA-I and HDLc levels, as well as apoA-I gene transcription, increase in mice with experimental nephrotic syndrome and these changes were associated with a fivefold

Fig. 10.3 Putative transcriptional regulatory mechanisms of the apoA-I gene. *Panel A:* Proposed mechanism of transcription when the enhancer is inactivated as a result of the mutation of the HNF-4 binding site of the enhancer. *Panel B:* Proposed mechanism of transcription when the apoA-I proximal promoter is inactivated as a result of the mutations in HNF-4 binding sites of the proximal promoter. *Panel C:* Proposed mechanism of transcription of the apoA-I gene by the WT promoter and enhancer. *Panel D:* Schematic representation showing both putative independent and the synergistic contributions of protein complexes assembled on the proximal apoA-I promoter and the apoCIII enhancer on the transcription of the apoA-I gene. The diagram is based on the *in vivo* transcription data shown in Panels A–C, as well as on the establishment of physical interactions between HNF-4 and SP1 at the indicated sites of the apoCIII promoter and the apoCIII enhancer [78]. The mechanism involves a simplified version of known protein–protein interactions of the promoter and enhancer complexes and co-activators, with the proteins of the basal transcription complex.

increase in the levels of early growth response factor (EGR-1). In contrast, EGR-1$^{-/-}$ mice had reduced plasma HDLc, apoA-I, and hepatic apoA-I mRNA levels [84]. These findings suggest that EGR-1, which binds to the regulatory element D, contributes both to the basal and the inducible transcription of the human apoA-I gene [84].

Bile acids, which are natural ligands for the nuclear receptor farnesoid X receptor (FXR) inhibit *apoA-I* gene expression both *in vitro* and *in vivo* [85,86]. It was shown that FXR and the monomeric nuclear receptor LRH-1 bind to the apoA-I promoter next to the previously characterized regulatory element B (Fig. 10.2A). FXR inhibits and LRH-1 activates the *apoA-I* promoter. It was proposed that FXR downregulates the *apoA-I* gene transcription both by binding to the apoA-I promoter and by inducing small heterodimer partner (SHP), which in turn represses the activity of LRH-1 [87] (Fig. 10.2A).

10.3.2 Transcriptional Regulation of the ABCA1 Gene

The ABCA1 gene is highly expressed in the liver, testis, small intestine, kidney, adrenal gland, heart, brain, and macrophages and to a lesser extent in other tissues [88–91]. The inducible and tissue-specific expression of the ABCA1 gene is regulated at the level of transcription and is characterized by the utilization of alternative promoters and transcription initiation sites present upstream of the first exon or inside the first intron of the gene (Fig. 10.4A) [92–94]. Based on its function in the efflux of phospholipid and cholesterol, it was anticipated that the expression of the ABCA1 gene is regulated by the levels of intracellular cholesterol. Initial studies had shown that the ABCA1-mediated cholesterol efflux is regulated by oxysterols [95,96]. These oxysterols are natural ligands of Liver X Receptors (LXRs) α and β, a specific subclass of hormone nuclear receptors that are expressed in all tissues. Heterodimers of LXR with RXRα bind to LXR response elements (LXREs) of the DR4 type (direct repeats separated by four nucleotides) [26,97–99]. LXR proteins seem to play an important role in HDL metabolism by coordinately regulating the expression of several genes participating in this process, including ABCA1, with ABCG1, SR-BI, PLTP, and CETP, which are discussed later [26]. The ABCA1 gene contains a thoroughly characterized LXRE at position −65 of the promoter upstream of exon 1, which mediates the activation of ABCA1 gene expression in response to natural or synthetic LXR ligands both *in vitro* and *in vivo* [100]. In addition to serving as an LXRE, the same DR4 element of the upstream ABCA1 promoter has been shown to bind RXR/retinoic acid receptor (RAR) and RXR/thyroid hormone receptor (T3R) heterodimers and to mediate the activation or the repression of ABCA1 transcription by retinoic acid and thyroid hormone, respectively [101,102]. Additional, but not well characterized, LXREs are present at the intron 1 promoter of the ABCA1 gene [94]. Cholesterol depletion leads to activation of the sterol regulatory element binding protein 2 (SREBP-2), which binds to an E-box of the upstream ABCA1 promoter and represses its activity [103]. The same E-box element serves as a binding site for

Upstream Stimulatory Factors (USFs) 1 and 2, which also repress the ABCA1 promoter activity [104,105]. SREBP-1a, a different member of the SREBP family of proteins, binds to several sites present intron 1, although the role of these elements in ABCA1 gene regulation is still unknown (Thymiakou, Zannis, and Kardassis, unpublished observations). The upstream ABCA1 promoter also contains binding sites (GC boxes) for the ubiquitous transcription factor SP1, which is a positive regulator of ABCA1, as well as a binding site for the zinc finger protein ZNF202 (GT box), which acts as a repressor of ABCA1 gene expression [106,107].

Experiments in mice have shown that synthetic LXR ligands upregulate ABCA1 transcription, increase plasma HDL levels, and promote cholesterol efflux from macrophage cultures [95]. This transcriptional activation of the ABCA1 gene depends on the LXRs, as it is lost in macrophages obtained from LXRα,β double-knockout mice [108]. Similarly, LXRs are unable to stimulate cholesterol efflux in fibroblasts from patients with Tangier disease, which lack ABCA1 [109]. However in addition to their important role in cholesterol homeostasis, LXRs are strong inducers of hepatic lipogenesis, due to the transcriptional upregulation of SREBP-1c gene [110,111]. Treatment of mice with synthetic LXR agonists elevates triglyceride levels in the liver and in the plasma [112,113].

In any attempt to increase plasma HDL levels it is important not only to avoid undesirable side effects but also to produce functional HDL particles [114,115]. In this regard, studies of transgenic and knockout mice have established that increased expression of apoA-I and ABCA1 increase HDL levels and protect from atherosclerosis [116–118]. This indicates that the HDL generated by overexpression either of apoA-I or of ABCA1 is functional and has cardioprotective effects. In contrast, inactivation of the SR-BI gene or mutations in the CETP gene increase HDL level but promote atherogenesis, suggesting that the HDL generated under these conditions may not be functional [119,120].

10.3.3 Transcriptional Regulation of the ABCG1 Gene

ABCG1 mRNA is expressed at high or moderate levels in macrophages, spleen, lung, thymus, placenta, brain, and at lower levels in most other tissues, including liver [89,121]. The human ABCG1 gene is expressed from multiple promoters that give rise to alternative transcripts encoding proteins with different amino terminal sequences [122–124]. Like ABCA1, macrophage ABCG1 is regulated by cholesterol loading [96,125]. At least four different LXREs have been identified in the ABCG1 gene (Fig. 10.4B): one LXRE is in the ABCG1 promoter upstream of exon 1 and three are in the region between exons 5 and 8 [124,126]. All these LXREs bind LXR/RXR heterodimers and confer synergistic stimulation of ABCG1 gene transcription by retinoids and oxysterols [124,126]. One of these LXREs (upstream of exon 1) was shown to mediate the inhibition of ABCG1 gene expression by unsaturated fatty acids. It appears that unsaturated fatty acids interfere with the transactivation of the

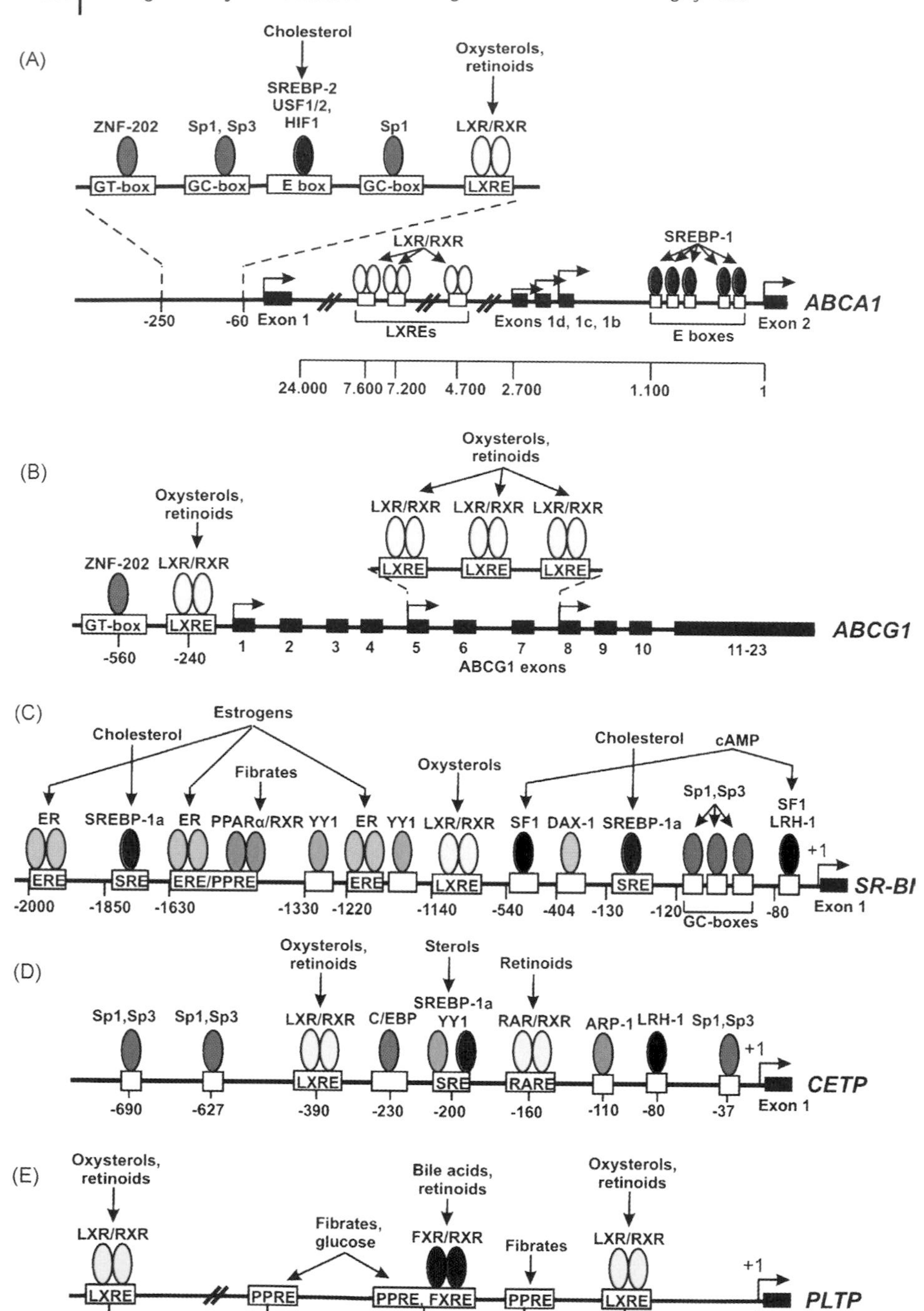

Fig. 10.4 Regulatory elements and factors governing the expression of ABCA1 (*Panel A*), ABCG1 (*Panel B*), SR-BI (*Panel C*), CETP (*Panel D*), and PLTP (*Panel E*). Transcription factors are depicted by ovals.

ABCG1 promoter by LXR/RXR heterodimers in the presence of oxysterols and retinoids [127]. The ABCG1 promoter contains GnT repeats, which serve as binding sites for the zinc-finger factor ZNF202, which is a transcriptional repressor of ABCG1 in HepG2 cells [107].

10.3.4 Transcriptional Regulation of the SR-BI Gene

The human scavenger receptor class B type I (SR-BI) gene is expressed at high levels in the liver and steroidogenic tissues such as adrenal glands and ovaries [128]. The promoters of the human or the rat SR-BI genes contain consensus DNA sequences that bind several positively acting transcription factors (Fig. 10.4C). These factors include:

1. the steroidogenic factor-1 (SF-1), which activates the human and the rat SR-BI promoters and seems to be the key mediator of the cAMP-dependent regulation of the SR-BI gene in response to steroidogenic hormones [129];
2. SREBP-1a, which binds on two sterol responsive elements (SREs) present on the rat SR-BI promoter and regulates SR-BI gene expression in response to intracellular sterol levels [128,130];
3. LXRα, LXRβ, and PPARα, which bind (as heterodimers with the RXRα) to a distal LXRE and a PPARE, respectively, on the human and rat SR-BI promoter as heterodimers and regulate the expression of the human SR-BI gene in response to oxysterols and fibrates [131,132];
4. the LRH-1, a nuclear orphan receptor, which binds to a proximal response element on the human SR-BI promoter (overlapping with the proximal SF-1 binding site) and activates the SR-BI promoter [133];
5. the estrogen receptors α and β (ERα and β), which bind to three different Estrogen Response Elements (EREs) on the rat SR-BI promoter and regulate its activity in response to estrogens in cooperation with SREBP-1a [134];
6. the ubiquitous transcription factors SP1 and SP3, which bind to several GC-rich boxes present on the proximal SR-BI promoter and have been shown to be important for the basal activity and the SREBP-1a-mediated transactivation of the SR-BI promoter [135,136].

In addition to the positively acting transcription factors described above, the SR-BI promoter is also regulated by negatively acting factors. These factors are: *a*) the orphan nuclear receptor dorsal-sensitive sex adrenal hypoplasia congenital critical region on the X-chromosome gene 1 (DAX-1), a protein that plays an important role in adrenal development (DAX-1 represses the rat SR-BI promoter by directly binding to and inhibiting SF-1 and SREBP-1a-mediated transactivation of the SR-BI promoter [137]), and *b*) the Yin Yang-1 (YY-1) transcription factor, which represses the activity of the SR-BI promoter through direct binding to two sites on the SR-BI promoter or by disrupting the SREBP-1a binding to this promoter [135].

Finally, it has been shown that bile acids, which are the natural ligands of the FXR, significantly increase the liver mRNA and protein levels of SR-BI in mice [138]. However, no FXR responsive elements have not yet been identified within the SR-BI promoter regions that have been studied. In another study it was reported that bile acids inhibit SR-BI gene expression in livers of mice and reduce the SR-BI promoter activity [139]. It was proposed that this inhibition was due to the FXR-mediated activation of SHP, which repressed the activity of LRH-1 that binds to the proximal SR-BI promoter [139].

10.3.5 Transcriptional Regulation of the CETP Gene

In humans, cholesterol ester transfer protein (CETP) gene is expressed predominantly in the adipose tissue, liver, and spleen, and at lower levels in small intestine, adrenal glands, kidneys, and heart [140]. CETP mRNA levels are increased in response to high cellular cholesterol content [141–143]. The mechanism of sterol-mediated upregulation of CETP gene expression is complex and is mediated by the interaction of the LXRs and the SREBPs with their cognate regulatory elements on the promoter of the CETP gene (Fig. 10.4D) [144,145]. The human CETP promoter also contains binding sites for several transcription factors that regulate its activity, including the YY1, which binds to the same element as SREBPs and activates the CETP promoter [144], the LRH-1, which potentiates the sterol-mediated induction of the CETP gene by LXRs [146], the orphan nuclear hormone receptor apolipoprotein A-I-regulated protein 1 (ARP-1), which represses CETP promoter activity [147], the RAR, which increases CETP gene expression in response to *all-trans*-retinoic acid [148], and the C/EBP, which is a positive regulator of CETP gene expression [149]. In addition to the above transcription factors, three binding sites for the ubiquitous transcription factors SP1 and SP3 (GC-boxes) have been identified on the CETP gene promoter at positions −690, −629, and −37 and seem to be essential for the basal CETP promoter activity [150].

10.3.6 Transcriptional Regulation of the PLTP Gene

Phospholipid Transfer Protein (PLTP) mRNA is found in numerous tissues, including liver, ovary, thymus, and placenta [151]. The human PLTP promoter contains at least two LXR responsive elements, one in the proximal and one in the distal region, that mediate PLTP gene regulation by oxysterols and retinoids in a synergistic fashion [152,153]. It was shown that LXR ligands increase the steady-state PLTP mRNA levels in various tissues and cell lines, including liver, intestine, macrophages, and adipose tissue [152,153]. PLTP gene expression is also upregulated by bile acids that function as ligands of FXR. Activation is mediated by an FXR responsive element (FXRE) that has been characterized in the proximal PLTP promoter and binds FXR/RXR heterodimers [154]. Fibrates increase PLTP gene expression by activating PPARs, which bind to three PPAR-responsive elements on the PLTP promoter [155,156] (Fig. 10.4E). Two of these PPAR-responsive elements also seem to be responsible for the induction of PLTP expression by high glucose [157].

10.4 Conclusions

The impetus for the study of the transcriptional regulatory mechanisms came from the need to understand what controls the tissue-specific and inducible expression of the genes of an organism. In the case of the genes of the HDL pathway, an additional practical goal was to create new pharmaceuticals to increase plasma HDL levels. Studies of transgenic mice have established that the HNF-4 binding site of the apoCIII enhancer is required for the intestinal expression of apoA-I gene, and synergistically enhances the hepatic expression of the apoA-I gene *in vivo*, acting in concert with other factors that bind to the proximal promoter. The three SP1 sites of the enhancer are also required for intestinal expression and for enhancement of hepatic transcription *in vivo*. Several *in vitro* and a few *in vivo* studies have pointed out the importance of hormone nuclear receptors and specifically the LXRs for the regulation of several of the genes of the HDL pathway. However, attempts to increase HDL levels by specific agonists of LXRs induced lipogenesis due to the transcriptional activation of SREBP-1c. This problem could potentially be solved by the development of more specific ligands that activate genes involved in cholesterol homeostasis and have no effect on genes involved in lipid biosynthesis. Gene-specific ligands have been reported for the ER, which is another member of the hormone nuclear receptor family [158], and recently for LXR [159]. An alternative approach has suggested the development of isoform-specific LXR ligands that selectively activate LXRβ, which does not affect the genes that control lipogenesis [160].

10.5 Acknowledgments

This work was supported by grants from the National Institutes of Health (HL33952 and HL48739) from the General Secretariat for Research and Technology of Greece (programs PENED 2001 and 2003 and the program "RTD collaboration between Greece and USA") and the 6th framework program of the European Union (L$HM–67–2006–037631). We thank Anne Plunkett for preparing the manuscript.

Abbreviations

ARP-1	apolipoprotein A-I-regulatory protein 1
C/EBP	CAAT/enhancer binding protein
CAT	chloramphenicol acetyl transferase
CETP	cholesterol ester transfer protein
DR4 HRE	hormone response element composed of two direct repeats separated by four nucleotides
DAX-1	dorsal-sensitive sex adrenal hypoplasia congenital critical region on the X-chromosome gene 1
EGR-1	early growth response factor
ERα and β	estrogen receptors α and β
ERE	estrogen response elements
FXR	farnesoid X receptor
FXRE	farnesoid X receptor responsive element
HNF-4	hepatocyte nuclear factor-4
HDL	high density lipoprotein
HRE	hormone response element
LRH-1	liver receptor homolog-1
LXRs	liver X receptors
LXRE	LXR response element
NFY	nuclear factor that binds to Y box
PPARα	peroxisome proliferator-activated receptor α
PLTP	phospholipid transfer protein
PPRE	peroxisome proliferator-activated receptor responsive element
RAR	retinoic acid receptor
RORα	retinoic acid receptor-related orphan receptor α
RXRα	retinoid X receptor α
SR-BI	scavenger receptor class B type I
SHP	small heterodimer partner
SP1	specificity protein 1
SF-1	steroidogenic factor-1
SREBP	sterol regulatory element binding protein
SREs	sterol responsive elements
T3R	triiodothyronine (T3) receptor
USFs	upstream stimulatory factors
WT	wild type
YY-1	yin yang-1
ZNF202	zinc finger protein 202

References

1 Tjian, R. and Maniatis, T. (1994) Transcriptional activation: a complex puzzle with few easy pieces. *Cell*, **77**, 5–8.

2 Lemon, B. and Tjian, R. (2000) Orchestrated response: a symphony of transcription factors for gene control. *Genes Dev.*, **14**, 2551–2569.

3 van Holde, K. and Yager, T. (2003) Models for chromatin remodeling: a critical comparison. *Biochem. Cell Biol.*, **81**, 169–172.

4 Torchia, J., Glass, C., Rosenfeld, M. G. (1998) Co-activators and co-repressors in the integration of transcriptional responses. *Curr. Opin. Cell Biol.*, **10**, 373–383.

5 Zawel, L. and Reinberg, D. (1995) Common themes in assembly and function of eukaryotic transcription complexes. *Annu. Rev. Biochem.*, **64**, 533–561.

6 Drapkin, R., Reardon, J. T., Ansari, A., Huang, J. C., Zawel, L., Ahn, K., Sancar, A., Reinberg, D. (1994) Dual role of TFIIH in DNA excision repair and in transcription by RNA polymerase II. *Nature*, **368**, 769–772.

7 Orphanides, G., Lagrange, T., Reinberg, D. (1996) The general transcription factors of RNA polymerase II. *Genes Dev.*, **10**, 2657–2683.

8 Tansey, W. P. and Herr, W. (1997) TAFs: guilt by association?. *Cell*, **88**, 729–732.

9 Sauer, F., Hansen, S. K., Tjian, R. (1995) Multiple TAFIIs directing synergistic activation of transcription. *Science*, **270**, 1783–1788.

10 Chen, J. L., Attardi, L. D., Verrijzer, C. P., Yokomori, K., Tjian, R. (1994) Assembly of recombinant TFIID reveals differential coactivator requirements for distinct transcriptional activators. *Cell*, **79**, 93–105.

11 Struhl, K. (1998) Histone acetylation and transcriptional regulatory mechanisms. *Genes Dev.*, **12**, 599–606.

12 Brown, C. E., Lechner, T., Howe, L., Workman, J. L. (2000) The many HATs of transcription coactivators. *Trends Biochem. Sci.*, **25**, 15–19.

13 Soutoglou, E., Katrakili, N., Talianidis, I. (2000) Acetylation regulates transcription factor activity at multiple levels. *Mol. Cell*, **5**, 745–751.

14 Chadick, J. Z. and Asturias, F. J. (2005) Structure of eukaryotic Mediator complexes. *Trends Biochem. Sci.*, **30**, 264–271.

15 Pazin, M. J. and Kadonaga, J. T. (1997) SWI2/SNF2 and related proteins: ATP-driven motors that disrupt protein-DNA interactions? *Cell*, **88**, 737–740.

16 Sudarsanam, P. and Winston, F. (2000) The Swi/Snf family nucleosome-remodeling complexes and transcriptional control. *Trends Genet.*, **16**, 345–351.

17 Mitchell, P. J. and Tjian, R. (1989) Transcriptional regulation in mammalian cells by sequence-specific DNA binding proteins. *Science*, **245**, 371–378.

18 Collingwood, T. N., Urnov, F. D., Wolffe, A. P. (1999) Nuclear receptors: coactivators, corepressors and chromatin remodeling in the control of transcription. *J. Mol. Endocrinol.*, **23**, 255–275.

19 Farrell, S., Simkovich, N., Wu, Y., Barberis, A., Ptashne, M. (1996) Gene activation by recruitment of the RNA polymerase II holoenzyme. *Genes Dev.*, **10**, 2359–2367.

20 Evans, R. M. (1988) The steroid and thyroid hormone receptor superfamily. *Science*, **240**, 889–895.

21 Mangelsdorf, D. J., Thummel, C., Beato, M., Herrlich, P., Schutz, G., Umesono, K., Blumberg, B., Kastner, P., Mark, M., Chambon, P. (1995) The nuclear receptor superfamily: the second decade. *Cell*, **83**, 835–839.

22 Giguere, V. (1999) Orphan nuclear receptors: from gene to function. *Endocr. Rev.*, **20**, 689–725.

23 Bourguet, W., Germain, P., Gronemeyer, H. (2000) Nuclear

receptor ligand-binding domains: three-dimensional structures, molecular interactions and pharmacological implications. *Trends Pharmacol. Sci.*, **21**, 381–388.

24 Gottlicher, M., Heck, S., Herrlich, P. (1998) Transcriptional cross-talk, the second mode of steroid hormone receptor action. *J. Mol. Med.*, **76**, 480–489.

25 Lu, T. T., Repa, J. J., Mangelsdorf, D. J. (2001) Orphan nuclear receptors as eLiXiRs and FiXeRs of sterol metabolism. *J. Biol. Chem.*, **276**, 37735–37738.

26 Repa, J. J. and Mangelsdorf, D. J. (2002) The liver X receptor gene team: potential new players in atherosclerosis. *Nat. Med.*, **8**, 1243–1248.

27 Fayard, E., Auwerx, J., Schoonjans, K. (2004) LRH-1: an orphan nuclear receptor involved in development, metabolism and steroidogenesis. *Trends Cell Biol.*, **14**, 250–260.

28 Bavner, A., Sanyal, S., Gustafsson, J. A., Treuter, E. (2005) Transcriptional corepression by SHP: molecular mechanisms and physiological consequences. *Trends Endocrinol. Metab*, **16**, 478–488.

29 Peet, D. J., Turley, S. D., Ma, W., Janowski, B. A., Lobaccaro, J. M., Hammer, R. E., Mangelsdorf, D. J. (1998) Cholesterol and bile acid metabolism are impaired in mice lacking the nuclear oxysterol receptor LXR alpha. *Cell*, **93**, 693–704.

30 Lehrke, M. and Lazar, M. A. (2005) The many faces of PPARgamma. *Cell*, **123**, 993–999.

31 Forman, B. M., Tontonoz, P., Chen, J., Brun, R. P., Spiegelman, B. M., Evans, R. M. (1995) 15-Deoxy-delta 12,14-prostaglandin J2 is a ligand for the adipocyte determination factor PPAR gamma. *Cell*, **83**, 803–812.

32 Lee, S. S., Pineau, T., Drago, J., Lee, E. J., Owens, J. W., Kroetz, D. L., Fernandez-Salguero, P. M., Westphal, H., Gonzalez, F. J. (1995) Targeted disruption of the alpha isoform of the peroxisome proliferator-activated receptor gene in mice results in abolishment of the pleiotropic effects of peroxisome proliferators. *Mol. Cell Biol.*, **15**, 3012–3022.

33 Giannini, S., Serio, M., Galli, A. (2004) Pleiotropic effects of thiazolidinediones: taking a look beyond antidiabetic activity. *J. Endocrinol. Invest*, **27**, 982–991.

34 Val, P., Lefrancois-Martinez, A. M., Veyssiere, G., Martinez, A. (2003) SF-1 a key player in the development and differentiation of steroidogenic tissues. *Nucl. Recept.*, **1**, 1–23.

35 Lavorgna, G., Karim, F. D., Thummel, C. S., Wu, C. (1993) Potential role for a FTZ-F1 steroid receptor superfamily member in the control of Drosophila metamorphosis. *Proc. Natl. Acad. Sci. U.S.A*, **90**, 3004–3008.

36 Ninomiya, Y., Okada, M., Kotomura, N., Suzuki, K., Tsukiyama, T., Niwa, O. (1995) Genomic organization and isoforms of the mouse ELP gene. *J. Biochem. (Tokyo)*, **118**, 380–389.

37 Luo, X., Ikeda, Y., Schlosser, D. A., Parker, K. L. (1995) Steroidogenic factor 1 is the essential transcript of the mouse Ftz-F1 gene. *Mol. Endocrinol.*, **9**, 1233–1239.

38 Zannis, V. I., Kan, H. Y., Kritis, A., Zanni, E. E., Kardassis, D. (2001) Transcriptional regulatory mechanisms of the human apolipoprotein genes *in vitro* and *in vivo*. *Curr. Opin. Lipidol.*, **12**, 181–207.

39 Crestani, M., DeFabiani, E., Caruso, D., Mitro, N., Gilardi, F., Vigil Chacon, A. B., Patelli, R., Godio, C., Galli, G. (2004) LXR (liver X receptor) and HNF-4 (hepatocyte nuclear factor-4): key regulators in reverse cholesterol transport. *Biochem. Soc. Trans.*, **32**, 92–96.

40 Hertz, R., Magenheim, J., Berman, I., Bar-Tana, J. (1998) Fatty acyl-CoA thioesters are ligands of hepatic nuclear factor-4alpha. *Nature*, **392**, 512–516.

41 Dhe-Paganon, S., Duda, K., Iwamoto, M., Chi, Y. I., Shoelson, S. E. (2002) Crystal structure of the HNF4 alpha ligand binding domain in complex with endogenous fatty acid ligand. *J. Biol. Chem.*, **277**, 37973–37976.

42 Wisely, G. B., Miller, A. B., Davis, R. G., Thornquest, A. D., Jr., Johnson, R., Spitzer, T., Sefler, A., Shearer, B., Moore, J. T., Miller, A. B., Willson, T. M., Williams, S. P. (2002) Hepatocyte nuclear factor 4 is a transcription factor that constitutively binds fatty acids. *Structure*, **10**, 1225–1234.

43 Chen, W. S., Manova, K., Weinstein, D. C., Duncan, S. A., Plump, A. S., Prezioso, V. R., Bachvarova, R. F., Darnell, J. E., Jr. (1994) Disruption of the HNF-4 gene, expressed in visceral endoderm, leads to cell death in embryonic ectoderm and impaired gastrulation of mouse embryos. *Genes Dev.*, **8**, 2466–2477.

44 Hayhurst, G. P., Lee, Y. H., Lambert, G., Ward, J. M., Gonzalez, F. J. (2001) Hepatocyte nuclear factor 4alpha (nuclear receptor 2A1) is essential for maintenance of hepatic gene expression and lipid homeostasis. *Mol. Cell Biol.*, **21**, 1393–1403.

45 Pereira, F. A., Tsai, M. J., Tsai, S. Y. (2000) COUP-TF orphan nuclear receptors in development and differentiation. *Cell Mol. Life Sci.*, **57**, 1388–1398.

46 Pereira, F. A., Qiu, Y., Tsai, M. J., Tsai, S. Y. (1995) Chicken ovalbumin upstream promoter transcription factor (COUP-TF): expression during mouse embryogenesis. *J. Steroid Biochem. Mol. Biol.*, **53**, 503–508.

47 Qiu, Y., Cooney, A. J., Kuratani, S., DeMayo, F. J., Tsai, S. Y., Tsai, M. J. (1994) Spatiotemporal expression patterns of chicken ovalbumin upstream promoter-transcription factors in the developing mouse central nervous system: evidence for a role in segmental patterning of the diencephalon. *Proc. Natl. Acad. Sci. U.S.A*, **91**, 4451–4455.

48 Qiu, Y., Tsai, S. Y., Tsai, M. J. (1994) COUP-TF: An orphan member of the steroid/thyroid hormone receptor superfamily. *Trends Endocrinol. Metab*, **5**, 234–239.

49 Ladias, J. A., Hadzopoulou-Cladaras, M., Kardassis, D., Cardot, P., Cheng, J., Zannis, V., Cladaras, C. (1992) Transcriptional regulation of human apolipoprotein genes ApoB, ApoCIII, and ApoAII by members of the steroid hormone receptor superfamily HNF-4, ARP-1, EAR-2, and EAR-3. *J. Biol. Chem.*, **267**, 15849–15860.

50 Ktistaki, E. and Talianidis, I. (1997) Chicken ovalbumin upstream promoter transcription factors act as auxiliary cofactors for hepatocyte nuclear factor 4 and enhance hepatic gene expression. *Mol. Cell Biol.*, **17**, 2790–2797.

51 Kardassis, D., Sacharidou, E., Zannis, V. I. (1998) Transactivation of the human apolipoprotein CII promoter by orphan and ligand-dependent nuclear receptors. The regulatory element CIIC is a thyroid hormone response element. *J. Biol. Chem.*, **273**, 17810–17816.

52 Suske, G. (1999) The Sp-family of transcription factors. *Gene*, **238**, 291–300.

53 Philipsen, S. and Suske, G. (1999) A tale of three fingers: the family of mammalian Sp/XKLF transcription factors. *Nucleic Acids Res.*, **27**, 2991–3000.

54 Marin, M., Karis, A., Visser, P., Grosveld, F., Philipsen, S. (1997) Transcription factor Sp1 is essential for early embryonic development but dispensable for cell growth and differentiation. *Cell*, **89**, 619–628.

55 Eberle, D., Hegarty, B., Bossard, P., Ferre, P., Foufelle, F. (2004) SREBP transcription factors: master regulators of lipid homeostasis. *Biochimie*, **86**, 839–848.

56 Rawson, R. B. (2003) The SREBP pathway--insights from Insigs and insects. *Nat. Rev. Mol. Cell Biol.*, **4**, 631–640.

57 Krieger, M. (2003) *Metabolism and movement of lipids*, 5th ed 743–777.

58 Brown, M. S. and Goldstein, J. L. (1997) The SREBP pathway: Regulation of cholesterol metabolism by proteolysis of a membrane-bound transcription factor. *Cell*, **89**, 331–340.

59 Wang, X., Pai, J. T., Wiedenfeld, E. A., Medina, J. C., Slaughter, C. A., Goldstein, J. L., Brown, M. S. (1995) Purification of an interleukin-1 beta

converting enzyme-related cysteine protease that cleaves sterol regulatory element-binding proteins between the leucine zipper and transmembrane domains. *J. Biol. Chem.*, **270**, 18044–18050.

60 Wang, X., Zelenski, N. G., Yang, J., Sakai, J., Brown, M. S., Goldstein, J. L. (1996) Cleavage of sterol regulatory element binding proteins (SREBPs) by CPP32 during apoptosis. *EMBO J.*, **15**, 1012–1020.

61 Goldstein, J. L., Hobbs, H. H., Brown, M. S. (2001) *Familial hypercholesterolemia*, 8th 2863–2913.

62 Kardassis, D., Laccotripe, M., Talianidis, I., Zannis, V. (1996) Transcriptional regulation of the genes involved in lipoprotein transport. The role of proximal promoters and long-range regulatory elements and factors in apolipoprotein gene regulation. *Hypertension*, **27**, 980–1008.

63 Zannis, V. I., Kan, H. Y., Kritis, A., Zanni, E., Kardassis, D. (2001) Transcriptional regulation of the human apolipoprotein genes. *Front Biosci.*, **6**, D456–D504.

64 Talianidis, I., Tambakaki, A., Toursounova, J., Zannis, V. I. (1995) Complex interactions between SP1 bound to multiple distal regulatory sites and HNF-4 bound to the proximal promoter lead to transcriptional activation of liver-specific human APOCIII gene. *Biochemistry*, **34**, 10298–10309.

65 Kardassis, D., Tzameli, I., Hadzopoulou-Cladaras, M., Talianidis, I., Zannis, V. (1997) Distal apolipoprotein C-III regulatory elements F to J act as a general modular enhancer for proximal promoters that contain hormone response elements. Synergism between hepatic nuclear factor-4 molecules bound to the proximal promoter and distal enhancer sites. *Arterioscler. Thromb. Vasc. Biol.*, **17**, 222–232.

66 Ogami, K., Hadzopoulou-Cladaras, M., Cladaras, C., Zannis, V. I. (1990) Promoter elements and factors required for hepatic and intestinal transcription of the human ApoCIII gene. *J. Biol. Chem.*, **265**, 9808–9815.

67 Ktistaki, E., Lacorte, J. M., Katrakili, N., Zannis, V. I., Talianidis, I. (1994) Transcriptional regulation of the apolipoprotein A-IV gene involves synergism between a proximal orphan receptor response element and a distant enhancer located in the upstream promoter region of the apolipoprotein C-III gene. *Nucleic Acids Res.*, **22**, 4689–4696.

68 Kan, H. Y., Georgopoulos, S., Zannis, V. (2000) A hormone response element in the human apolipoprotein CIII (ApoCIII) enhancer is essential for intestinal expression of the ApoA-I and ApoCIII genes and contributes to the hepatic expression of the two linked genes in transgenic mice. *J. Biol. Chem.*, **275**, 30423–30431.

69 Georgopoulos, S., Kan, H. Y., Reardon-Alulis, C., Zannis, V. (2000) The SP1 sites of the human apoCIII enhancer are essential for the expression of the apoCIII gene and contribute to the hepatic and intestinal expression of the apoA-I gene in transgenic mice. *Nucleic Acids Res.*, **28**, 4919–4929.

70 Lavrentiadou, S. N., Hadzopoulou-Cladaras, M., Kardassis, D., Zannis, V. I. (1999) Binding specificity and modulation of the human ApoCIII promoter activity by heterodimers of ligand-dependent nuclear receptors. *Biochemistry*, **38**, 964–975.

71 Papazafiri, P., Ogami, K., Ramji, D. P., Nicosia, A., Monaci, P., Cladaras, C., Zannis, V. I. (1991) Promoter elements and factors involved in hepatic transcription of the human ApoA-I gene positive and negative regulators bind to overlapping sites. *J. Biol. Chem.*, **266**, 5790–5797.

72 Tzameli, I. and Zannis, V. I. (1996) Binding specificity and modulation of the ApoA-I promoter activity by homo- and heterodimers of nuclear receptors. *J. Biol. Chem.*, **271**, 8402–8415.

73 Zannis, V. I., Cole, F. S., Jackson, C. L., Kurnit, D. M., Karathanasis, S. K. (1985) Distribution of apolipoprotein A-I, C-II, C-III, and E mRNA in fetal human tissues. Time-dependent induction of apolipoprotein E mRNA by cultures of human monocyte-macrophages. *Biochemistry*, **24**, 4450–4455.

74 Vu-Dac, N., Chopin-Delannoy, S., Gervois, P., Bonnelye, E., Martin, G., Fruchart, J. C., Laudet, V., Staels, B. (1998) The nuclear receptors peroxisome proliferator-activated receptor alpha and Rev-erbalpha mediate the species-specific regulation of apolipoprotein A-I expression by fibrates. *J. Biol. Chem.*, **273**, 25713–25720.

75 Novak, E. M. and Bydlowski, S. P. (1997) NFY transcription factor binds to regulatory element AIC and transactivates the human apolipoprotein A-I promoter in HEPG2 cells. *Biochem. Biophys. Res. Commun.*, **231**, 140–143.

76 Walsh, A., Ito, Y., Breslow, J. L. (1989) High levels of human apolipoprotein A-I in transgenic mice result in increased plasma levels of small high density lipoprotein (HDL) particles comparable to human HDL3. *J. Biol. Chem.*, **264**, 6488–6494.

77 Li, J., Ning, G., Duncan, S. A. (2000) Mammalian hepatocyte differentiation requires the transcription factor HNF-4alpha. *Genes Dev.*, **14**, 464–474.

78 Kardassis, D., Falvey, E., Tsantili, P., Hadzopoulou-Cladaras, M., Zannis, V. (2002) Direct physical interactions between HNF-4 and Sp1 mediate synergistic transactivation of the apolipoprotein CIII promoter. *Biochemistry*, **41**, 1217–1228.

79 Fraser, J. D., Keller, D., Martinez, V., Santiso-Mere, D., Straney, R., Briggs, M. R. (1997) Utilization of recombinant adenovirus and dominant negative mutants to characterize hepatocyte nuclear factor 4-regulated apolipoprotein AI and CIII expression. *J. Biol. Chem.*, **272**, 13892–13898.

80 Hennuyer, N., Poulain, P., Madsen, L., Berge, R. K., Houdebine, L. M., Branellec, D., Fruchart, J. C., Fievet, C., Duverger, N., Staels, B. (1999) Beneficial effects of fibrates on apolipoprotein A-I metabolism occur independently of any peroxisome proliferative response. *Circulation*, **99**, 2445–2451.

81 Peters, J. M., Hennuyer, N., Staels, B., Fruchart, J. C., Fievet, C., Gonzalez, F. J., Auwerx, J. (1997) Alterations in lipoprotein metabolism in peroxisome proliferator-activated receptor alpha-deficient mice. *J. Biol. Chem.*, **272**, 27307–27312.

82 Wan, Y. J., An, D., Cai, Y., Repa, J. J., Hung-Po, C. T., Flores, M., Postic, C., Magnuson, M. A., Chen, J., Chien, K. R., French, S., Mangelsdorf, D. J., Sucov, H. M. (2000) Hepatocyte-specific mutation establishes retinoid X receptor alpha as a heterodimeric integrator of multiple physiological processes in the liver. *Mol. Cell Biol.*, **20**, 4436–4444.

83 Vu-Dac, N., Gervois, P., Grotzinger, T., DeVos, P., Schoonjans, K., Fruchart, J. C., Auwerx, J., Mariani, J., Tedgui, A., Staels, B. (1997) Transcriptional regulation of apolipoprotein A-I gene expression by the nuclear receptor RORalpha. *J. Biol. Chem.*, **272**, 22401–22404.

84 Zaiou, M., Azrolan, N., Hayek, T., Wang, H., Wu, L., Haghpassand, M., Cizman, B., Madaio, M. P., Milbrandt, J., Marsh, J. B., Breslow, J. L., Fisher, E. A. (1998) The full induction of human apoprotein A-I gene expression by the experimental nephrotic syndrome in transgenic mice depends on cis-acting elements in the proximal 256 base-pair promoter region and the trans-acting factor early growth response factor 1. *J. Clin. Invest*, **101**, 1699–1707.

85 Srivastava, R. A., Srivastava, N., Averna, M. (2000) Dietary cholic acid lowers plasma levels of mouse and human apolipoprotein A-I primarily via a transcriptional mechanism. *Eur. J. Biochem.*, **267**, 4272–4280.

86 Claudel, T., Sturm, E., Duez, H., Torra, I. P., Sirvent, A., Kosykh, V., Fruchart, J. C., Dallongeville, J., Hum, D. W., Kuipers, F., Staels, B. (2002) Bile acid-activated nuclear receptor FXR suppresses apolipoprotein A-I transcription via a negative FXR response element. *J. Clin. Invest*, **109**, 961–971.

87 Delerive, P., Galardi, C. M., Bisi, J. E., Nicodeme, E., Goodwin, B. (2004) Identification of liver receptor homolog-1 as a novel regulator of apolipoprotein AI gene transcription. *Mol. Endocrinol.*, **18**, 2378–2387.

88 Langmann, T., Klucken, J., Reil, M., Liebisch, G., Luciani, M. F., Chimini, G., Kaminski, W. E., Schmitz, G. (1999) Molecular cloning of the human ATP-binding cassette transporter 1 (hABC1): evidence for sterol-dependent regulation in macrophages. *Biochem. Biophys. Res. Commun.*, **257**, 29–33.

89 Langmann, T., Mauerer, R., Zahn, A., Moehle, C., Probst, M., Stremmel, W., Schmitz, G. (2003) Real-time reverse transcription-PCR expression profiling of the complete human ATP-binding cassette transporter superfamily in various tissues. *Clin. Chem.*, **49**, 230–238.

90 Kielar, D., Dietmaier, W., Langmann, T., Aslanidis, C., Probst, M., Naruszewicz, M., Schmitz, G. (2001) Rapid quantification of human ABCA1 mRNA in various cell types and tissues by real-time reverse transcription-PCR. *Clin. Chem.*, **47**, 2089–2097.

91 Wellington, C. L., Walker, E. K., Suarez, A., Kwok, A., Bissada, N., Singaraja, R., Yang, Y. Z., Zhang, L. H., James, E., Wilson, J. E., Francone, O., McManus, B. M., Hayden, M. R. (2002) ABCA1 mRNA and protein distribution patterns predict multiple different roles and levels of regulation. *Lab Invest*, **82**, 273–283.

92 Huuskonen, J., Abedin, M., Vishnu, M., Pullinger, C. R., Baranzini, S. E., Kane, J. P., Fielding, P. E., Fielding, C. J. (2003) Dynamic regulation of alternative ATP-binding cassette transporter A1 transcripts. *Biochem. Biophys. Res. Commun.*, **306**, 463–468.

93 Singaraja, R. R., James, E. R., Crim, J., Visscher, H., Chatterjee, A., Hayden, M. R. (2005) Alternate transcripts expressed in response to diet reflect tissue-specific regulation of ABCA1. *J. Lipid Res.*, **46**, 2061–2071.

94 Singaraja, R. R., Bocher, V., James, E. R., Clee, S. M., Zhang, L. H., Leavitt, B. R., Tan, B., Brooks-Wilson, A., Kwok, A., Bissada, N., Yang, Y. Z., Liu, G., Tafuri, S. R., Fievet, C., Wellington, C. L., Staels, B., Hayden, M. R. (2001) Human ABCA1 BAC transgenic mice show increased high density lipoprotein cholesterol and ApoAI-dependent efflux stimulated by an internal promoter containing liver X receptor response elements in intron 1. *J. Biol. Chem.*, **276**, 33969–33979.

95 Repa, J. J., Turley, S. D., Lobaccaro, J. A., Medina, J., Li, L., Lustig, K., Shan, B., Heyman, R. A., Dietschy, J. M., Mangelsdorf, D. J. (2000) Regulation of absorption and ABC1-mediated efflux of cholesterol by RXR heterodimers. *Science*, **289**, 1524–1529.

96 Venkateswaran, A., Repa, J. J., Lobaccaro, J. M., Bronson, A., Mangelsdorf, D. J., Edwards, P. A. (2000) Human white/murine ABC8 mRNA levels are highly induced in lipid-loaded macrophages. A transcriptional role for specific oxysterols. *J. Biol. Chem.*, **275**, 14700–14707.

97 Fu, X., Menke, J. G., Chen, Y., Zhou, G., Macnaul, K. L., Wright, S. D., Sparrow, C. P., Lund, E. G. (2001) 27-hydroxycholesterol is an endogenous ligand for liver X receptor in cholesterol-loaded cells. *J. Biol. Chem.*, **276**, 38378–38387.

98 Janowski, B. A., Willy, P. J., Devi, T. R., Falck, J. R., Mangelsdorf, D. J. (1996) An oxysterol signalling pathway mediated by the nuclear receptor LXR alpha. *Nature*, **383**, 728–731.

99 Lehmann, J. M., Kliewer, S. A., Moore, L. B., Smith-Oliver, T. A., Oliver, B. B., Su, J. L., Sundseth, S. S., Winegar, D. A., Blanchard, D. E., Spencer, T. A., Willson, T. M. (1997) Activation of the nuclear receptor LXR by oxysterols defines a new hormone response pathway. *J. Biol. Chem.*, **272**, 3137–3140.

100 Costet, P., Luo, Y., Wang, N., Tall, A. R. (2000) Sterol-dependent

transactivation of the ABC1 promoter by the liver X receptor/retinoid X receptor. *J. Biol. Chem.*, **275**, 28240–28245.

101 Costet, P., Lalanne, F., Gerbod-Giannone, M. C., Molina, J. R., Fu, X., Lund, E. G., Gudas, L. J., Tall, A. R. (2003) Retinoic acid receptor-mediated induction of ABCA1 in macrophages. *Mol. Cell Biol.*, **23**, 7756–7766.

102 Huuskonen, J., Vishnu, M., Pullinger, C. R., Fielding, P. E., Fielding, C. J. (2004) Regulation of ATP-binding cassette transporter A1 transcription by thyroid hormone receptor. *Biochemistry*, **43**, 1626–1632.

103 Zeng, L., Liao, H., Liu, Y., Lee, T. S., Zhu, M., Wang, X., Stemerman, M. B., Zhu, Y., Shyy, J. Y. (2004) Sterol-responsive element-binding protein (SREBP) 2 down-regulates ATP-binding cassette transporter A1 in vascular endothelial cells: a novel role of SREBP in regulating cholesterol metabolism. *J. Biol. Chem.*, **279**, 48801–48807.

104 Yang, X. P., Freeman, L. A., Knapper, C. L., Amar, M. J., Remaley, A., Brewer, H. B., Jr., Santamarina-Fojo, S. (2002) The E-box motif in the proximal ABCA1 promoter mediates transcriptional repression of the ABCA1 gene. *J. Lipid Res.*, **43**, 297–306.

105 Langmann, T., Porsch-Ozcurumez, M., Heimerl, S., Probst, M., Moehle, C., Taher, M., Borsukova, H., Kielar, D., Kaminski, W. E., Dittrich-Wengenroth, E., Schmitz, G. (2002) Identification of sterol-independent regulatory elements in the human ATP-binding cassette transporter A1 promoter: role of Sp1/3, E-box binding factors, and an oncostatin M-responsive element. *J. Biol. Chem.*, **277**, 14443–14450.

106 Schmitz, G. and Langmann, T. (2005) Transcriptional regulatory networks in lipid metabolism control ABCA1 expression. *Biochim. Biophys. Acta*, **1735**, 1–19.

107 Porsch-Ozcurumez, M., Langmann, T., Heimerl, S., Borsukova, H., Kaminski, W. E., Drobnik, W., Honer, C., Schumacher, C., Schmitz, G. (2001) The zinc finger protein 202 (ZNF202) is a transcriptional repressor of ATP binding cassette transporter A1 (ABCA1) and ABCG1 gene expression and a modulator of cellular lipid efflux. *J. Biol. Chem.*, **276**, 12427–12433.

108 Tangirala, R. K., Bischoff, E. D., Joseph, S. B., Wagner, B. L., Walczak, R., Laffitte, B. A., Daige, C. L., Thomas, D., Heyman, R. A., Mangelsdorf, D. J., Wang, X., Lusis, A. J., Tontonoz, P., Schulman, I. G. (2002) Identification of macrophage liver X receptors as inhibitors of atherosclerosis. *Proc. Natl. Acad. Sci. U.S.A*, **99**, 11896–11901.

109 Venkateswaran, A., Laffitte, B. A., Joseph, S. B., Mak, P. A., Wilpitz, D. C., Edwards, P. A., Tontonoz, P. (2000) Control of cellular cholesterol efflux by the nuclear oxysterol receptor LXR alpha. *Proc. Natl. Acad. Sci. U.S.A*, **97**, 12097–12102.

110 Horton, J. D., Goldstein, J. L., Brown, M. S. (2002) SREBPs: activators of the complete program of cholesterol and fatty acid synthesis in the liver. *J. Clin. Invest*, **109**, 1125–1131.

111 Repa, J. J., Liang, G., Ou, J., Bashmakov, Y., Lobaccaro, J. M., Shimomura, I., Shan, B., Brown, M. S., Goldstein, J. L., Mangelsdorf, D. J. (2000) Regulation of mouse sterol regulatory element-binding protein-1c gene (SREBP-1c) by oxysterol receptors, LXRalpha and LXRbeta. *Genes Dev.*, **14**, 2819–2830.

112 Schultz, J. R., Tu, H., Luk, A., Repa, J. J., Medina, J. C., Li, L., Schwendner, S., Wang, S., Thoolen, M., Mangelsdorf, D. J., Lustig, K. D., Shan, B. (2000) Role of LXRs in control of lipogenesis. *Genes Dev.*, **14**, 2831–2838.

113 Joseph, S. B., Laffitte, B. A., Patel, P. H., Watson, M. A., Matsukuma, K. E., Walczak, R., Collins, J. L., Osborne, T. F., Tontonoz, P. (2002) Direct and indirect mechanisms for regulation of fatty acid synthase gene expression by liver X receptors. *J. Biol. Chem.*, **277**, 11019–11025.

114 Mineo, C., Yuhanna, I. S., Quon, M. J., Shaul, P. W. (2003) High density lipoprotein-induced endothelial nitric-oxide synthase activation is mediated by Akt and MAP kinases. *J. Biol. Chem.*, **278**, 9142–9149.

115 Rader, D. J. (2002) High-density lipoproteins and atherosclerosis. *Am. J. Cardiol.*, **90**, 62i–70i.

116 Rubin, E. M., Krauss, R. M., Spangler, E. A., Verstuyft, J. G., Clift, S. M. (1991) Inhibition of early atherogenesis in transgenic mice by human apolipoprotein AI. *Nature*, **353**, 265–267.

117 Joyce, C. W., Amar, M. J., Lambert, G., Vaisman, B. L., Paigen, B., Najib-Fruchart, J., Hoyt, R. F., Jr., Neufeld, E. D., Remaley, A. T., Fredrickson, D. S., Brewer, H. B., Jr., Santamarina-Fojo, S. (2002) The ATP binding cassette transporter A1 (ABCA1) modulates the development of aortic atherosclerosis in C57BL/6 and apoE-knockout mice. *Proc. Natl. Acad. Sci. U.S.A*, **99**, 407–412.

118 Zannis, V. I. and Cohen, J. (2000) Old and new players in the lipoprotein system. *Curr. Opin. Lipidol.*, **11**, 101–103.

119 Zhong, S., Sharp, D. S., Grove, J. S., Bruce, C., Yano, K., Curb, J. D., Tall, A. R. (1996) Increased coronary heart disease in Japanese-American men with mutation in the cholesteryl ester transfer protein gene despite increased HDL levels. *J. Clin. Invest*, **97**, 2917–2923.

120 Huszar, D., Varban, M. L., Rinninger, F., Feeley, R., Arai, T., Fairchild-Huntress, V., Donovan, M. J., Tall, A. R. (2000) Increased LDL cholesterol and atherosclerosis in LDL receptor-deficient mice with attenuated expression of scavenger receptorB1. *Arterioscler. Thromb. Vasc. Biol.*, **20**, 1068–1073.

121 Hoekstra, M., Kruijt, J. K., VanEck, M., vanBerkel, T. J. (2003) Specific gene expression of ATP-binding cassette transporters and nuclear hormone receptors in rat liver parenchymal, endothelial, and Kupffer cells. *J. Biol. Chem.*, **278**, 25448–25453.

122 Langmann, T., Porsch-Ozcurumez, M., Unkelbach, U., Klucken, J., Schmitz, G. (2000) Genomic organization and characterization of the promoter of the human ATP-binding cassette transporter-G1 (ABCG1) gene. *Biochim. Biophys. Acta*, **1494**, 175–180.

123 Lorkowski, S., Rust, S., Engel, T., Jung, E., Tegelkamp, K., Galinski, E. A., Assmann, G., Cullen, P. (2001) Genomic sequence and structure of the human ABCG1 (ABC8) gene. *Biochem. Biophys. Res. Commun.*, **280**, 121–131.

124 Kennedy, M. A., Venkateswaran, A., Tarr, P. T., Xenarios, I., Kudoh, J., Shimizu, N., Edwards, P. A. (2001) Characterization of the human ABCG1 gene: liver X receptor activates an internal promoter that produces a novel transcript encoding an alternative form of the protein. *J. Biol. Chem.*, **276**, 39438–39447.

125 Klucken, J., Buchler, C., Orso, E., Kaminski, W. E., Porsch-Ozcurumez, M., Liebisch, G., Kapinsky, M., Diederich, W., Drobnik, W., Dean, M., Allikmets, R., Schmitz, G. (2000) ABCG1 (ABC8), the human homolog of the Drosophila white gene, is a regulator of macrophage cholesterol and phospholipid transport. *Proc. Natl. Acad. Sci. U.S.A*, **97**, 817–822.

126 Sabol, S. L., Brewer, H. B., Jr., Santamarina-Fojo, S. (2005) The human ABCG1 gene: identification of LXR response elements that modulate expression in macrophages and liver. *J. Lipid Res.*, **46**, 2151–2167.

127 Uehara, Y., Miura, S. I., von Eckardstein, A., Abe, S., Fujii, A., Matsuo, Y., Rust, S., Lorkowski, S., Assmann, G., Yamada, T., Saku, K. (2006) Unsaturated fatty acids suppress the expression of the ATP-binding cassette transporter G1 (ABCG1) and ABCA1 genes via an LXR/RXR responsive element. Atherosclerosis.

128 Cao, G., Garcia, C. K., Wyne, K. L., Schultz, R. A., Parker, K. L., Hobbs, H. H. (1997) Structure and localization of the human gene encoding SR-BI/CLA-1. Evidence for transcriptional

control by steroidogenic factor 1. *J. Biol. Chem.*, **272**, 33068–33076.

129 Lopez, D., Sandhoff, T. W., McLean, M. P. (1999) Steroidogenic factor-1 mediates cyclic 3′,5′-adenosine monophosphate regulation of the high density lipoprotein receptor. *Endocrinology*, **140**, 3034–3044.

130 Lopez, D. and McLean, M. P. (1999) Sterol regulatory element-binding protein-1a binds to cis elements in the promoter of the rat high density lipoprotein receptor SR-BI gene. *Endocrinology*, **140**, 5669–5681.

131 Malerod, L., Juvet, L. K., Hanssen-Bauer, A., Eskild, W., Berg, T. (2002) Oxysterol-activated LXRalpha/RXR induces hSR-BI-promoter activity in hepatoma cells and preadipocytes. *Biochem. Biophys. Res. Commun.*, **299**, 916–923.

132 Lopez, D. and McLean, M. P. (2006) Activation of the rat scavenger receptor class B type I gene by PPARalpha. *Mol. Cell Endocrinol.*, **251**, 67–77.

133 Schoonjans, K., Annicotte, J. S., Huby, T., Botrugno, O. A., Fayard, E., Ueda, Y., Chapman, J., Auwerx, J. (2002) Liver receptor homolog 1 controls the expression of the scavenger receptor class B type I. *EMBO Rep.*, **3**, 1181–1187.

134 Lopez, D., Sanchez, M. D., Shea-Eaton, W., McLean, M. P. (2002) Estrogen activates the high-density lipoprotein receptor gene via binding to estrogen response elements and interaction with sterol regulatory element binding protein-1A. *Endocrinology*, **143**, 2155–2168.

135 Shea-Eaton, W., Lopez, D., McLean, M. P. (2001) Yin yang 1 protein negatively regulates high-density lipoprotein receptor gene transcription by disrupting binding of sterol regulatory element binding protein to the sterol regulatory element. *Endocrinology*, **142**, 49–58.

136 Mizutani, T., Yamada, K., Minegishi, T., Miyamoto, K. (2000) Transcrip-tional regulation of rat scavenger receptor class B type I gene. *J. Biol. Chem.*, **275**, 22512–22519.

137 Lopez, D., Shea-Eaton, W., Sanchez, M. D., McLean, M. P. (2001) DAX-1 represses the high-density lipoprotein receptor through interaction with positive regulators sterol regulatory element-binding protein-1a and steroidogenic factor-1. *Endocrinology*, **142**, 5097–5106.

138 Lambert, G., Amar, M. J., Guo, G., Brewer, H. B., Jr., Gonzalez, F. J., Sinal, C. J. (2003) The farnesoid X-receptor is an essential regulator of cholesterol homeostasis. *J. Biol. Chem.*, **278**, 2563–2570.

139 Malerod, L., Sporstol, M., Juvet, L. K., Mousavi, S. A., Gjoen, T., Berg, T., Roos, N., Eskild, W. (2005) Bile acids reduce SR-BI expression in hepatocytes by a pathway involving FXR/RXR, SHP, and LRH-1. *Biochem. Biophys. Res. Commun.*, **336**, 1096–1105.

140 Jiang, X. C., Moulin, P., Quinet, E., Goldberg, I. J., Yacoub, L. K., Agellon, L. B., Compton, D., Schnitzer-Polokoff, R., Tall, A. R. (1991) Mammalian adipose tissue and muscle are major sources of lipid transfer protein mRNA. *J. Biol. Chem.*, **266**, 4631–4639.

141 Quinet, E. M., Agellon, L. B., Kroon, P. A., Marcel, Y. L., Lee, Y. C., Whitlock, M. E., Tall, A. R. (1990) Atherogenic diet increases cholesteryl ester transfer protein messenger RNA levels in rabbit liver. *J. Clin. Invest*, **85**, 357–363.

142 Oliveira, H. C., Chouinard, R. A., Agellon, L. B., Bruce, C., Ma, L., Walsh, A., Breslow, J. L., Tall, A. R. (1996) Human cholesteryl ester transfer protein gene proximal promoter contains dietary cholesterol positive responsive elements and mediates expression in small intestine and periphery while predominant liver and spleen expression is controlled by 5′-distal sequences. Cis-acting sequences mapped in transgenic mice. *J. Biol. Chem.*, **271**, 31831–31838.

143 Jiang, X. C., Agellon, L. B., Walsh, A., Breslow, J. L., Tall, A. (1992) Dietary cholesterol increases transcription of the human cholesteryl ester transfer protein gene in transgenic mice. Dependence on natural flanking sequences. *J. Clin. Invest*, **90**, 1290–1295.

144 Gauthier, B., Robb, M., Gaudet, F., Ginsburg, G. S., McPherson, R. (1999) Characterization of a cholesterol response element (CRE) in the promoter of the cholesteryl ester transfer protein gene: functional role of the transcription factors SREBP-1a, -2, andYY1. *J. Lipid Res.*, **40**, 1284–1293.

145 Luo, Y. and Tall, A. R. (2000) Sterol upregulation of human CETP expression in vitro and in transgenic mice by an LXR element. *J. Clin. Invest*, **105**, 513–520.

146 Luo, Y., Liang, C. P., Tall, A. R. (2001) The orphan nuclear receptor LRH-1 potentiates the sterol-mediated induction of the human CETP gene by liver X receptor. *J. Biol. Chem.*, **276**, 24767–24773.

147 Gaudet, F. and Ginsburg, G. S. (1995) Transcriptional regulation of the cholesteryl ester transfer protein gene by the orphan nuclear hormone receptor apolipoprotein AI regulatory protein-1. *J. Biol. Chem.*, **270**, 29916–29922.

148 Jeoung, N. H., Jang, W. G., Nam, J. I., Pak, Y. K., Park, Y. B. (1999) Identification of retinoic acid receptor element in human cholesteryl ester transfer protein gene. *Biochem. Biophys. Res. Commun.*, **258**, 411–415.

149 Agellon, L. B., Zhang, P., Jiang, X. C., Mendelsohn, L., Tall, A. R. (1992) The CCAAT/enhancer-binding protein trans-activates the human cholesteryl ester transfer protein gene promoter. *J. Biol. Chem.*, **267**, 22336–22339.

150 LeGoff, W., Guerin, M., Petit, L., Chapman, M. J., Thillet, J. (2003) Regulation of human CETP gene expression: role of SP1 and SP3 transcription factors at promoter sites -690, -629, and -37. *J. Lipid Res.*, **44**, 1322–1331.

151 Albers, J. J., Wolfbauer, G., Cheung, M. C., Day, J. R., Ching, A. F., Lok, S., Tu, A. Y. (1995) Functional expression of human and mouse plasma phospholipid transfer protein: effect of recombinant and plasma PLTP on HDL subspecies. *Biochim. Biophys. Acta*, **1258**, 27–34.

152 Laffitte, B. A., Joseph, S. B., Chen, M., Castrillo, A., Repa, J., Wilpitz, D., Mangelsdorf, D., Tontonoz, P. (2003) The phospholipid transfer protein gene is a liver X receptor target expressed by macrophages in atherosclerotic lesions. *Mol. Cell Biol.*, **23**, 2182–2191.

153 Mak, P. A., Kast-Woelbern, H. R., Anisfeld, A. M., Edwards, P. A. (2002) Identification of PLTP as an LXR target gene and apoE as an FXR target gene reveals overlapping targets for the two nuclear receptors. *J. Lipid Res.*, **43**, 2037–2041.

154 Urizar, N. L., Dowhan, D. H., Moore, D. D. (2000) The farnesoid X-activated receptor mediates bile acid activation of phospholipid transfer protein gene expression. *J. Biol. Chem.*, **275**, 39313–39317.

155 Bouly, M., Masson, D., Gross, B., Jiang, X. C., Fievet, C., Castro, G., Tall, A. R., Fruchart, J. C., Staels, B., Lagrost, L., Luc, G. (2001) Induction of the phospholipid transfer protein gene accounts for the high density lipoprotein enlargement in mice treated with fenofibrate. *J. Biol. Chem.*, **276**, 25841–25847.

156 Tu, A. Y. and Albers, J. J. (1999) DNA sequences responsible for reduced promoter activity of human phospholipid transfer protein by fibrate. *Biochem. Biophys. Res. Commun.*, **264**, 802–807.

157 Tu, A. Y. and Albers, J. J. (2001) Glucose regulates the transcription of human genes relevant to HDL metabolism: responsive elements for peroxisome proliferator-activated receptor are involved in the regulation of phospholipid transfer protein. *Diabetes*, **50**, 1851–1856.

158 Gustafsson, J. A. (1998) Therapeutic potential of selective estrogen receptor modulators. *Curr. Opin. Chem. Biol.*, **2**, 508–511.

159 Quinet, E. M., Savio, D. A., Halpern, A. R., Chen, L., Miller, C. P., Nambi, P. (2004) Gene-selective modulation by a synthetic oxysterol ligand of the liver X receptor. *J. Lipid Res.*, **45**, 1929–1942.

160 Alberti, S., Schuster, G., Parini, P., Feltkamp, D., Diczfalusy, U., Rudling, M., Angelin, B., Bjorkhem, I., Pettersson, S., Gustafsson, J. A. (2001) Hepatic cholesterol metabolism and resistance to dietary cholesterol in LXRbeta-deficient mice. *J. Clin. Invest*, **107**, 565–573.

11
ApoA-I Functions and Synthesis of HDL: Insights from Mouse Models of Human HDL Metabolism

Vassilis I. Zannis, Eleni E. Zanni, Angeliki Papapanagiotou, Dimitris Kardassis, Christopher J. Fielding, Angeliki Chroni

11.1
Structures of ApoA-I and High-Density Lipoprotein (HDL)

Human plasma apoA-I is a 243-residue protein that originates from a 249-aa precursor synthesized and secreted primarily by the liver and the intestine [1]. ApoA-I is the major protein component of HDL and is essential for the biogenesis and the functions of HDL. Mice or humans deficient in apoA-I fail to form HDL [2,3]. ApoA-I contains 22- and 11-amino acid repeats [4,5] which, from earlier X-ray crystallography and computer modeling results [5–7], are organized in amphipathic α-helices (Fig. 11.1A).

In the newly derived crystal structure of apoA-I at 2.4 Å resolution, it was shown that under the condition of crystallization the protein consists of a four-helix N-terminal bundle and two C-terminal helices [8] (Fig. 11.1B). From the initial crystal structure of apoA-I[Δ(1–43)], several models to describe the binding of apoA-I to phospholipid bilayer disks have been proposed [9–14]. Such discs can be formed *in vivo* as intermediates in the HDL biogenesis pathway [15–18] and can also be formed *in vitro* as a result of solubilization of multilamellar phospholipid vesicles by apoA-I [19]. In the belt model of discoidal HDL, two apoA-I molecules are wrapped in an antiparallel orientation around the edge of a discoidal lipid bilayer containing 160 lipid molecules. It has been proposed that the orientation of the two apoA-I molecules on the discoidal particle is such that it maximizes intramolecular salt bridge interactions as shown in Fig. 11.1C [9,14].

Molecular dynamics simulation of discoidal recombinant HDL (rHDL) particles containing 160 molecules of phospholipid following gradual removal of the phospholipid leads to intermediates resembling the 95 Å and 78 Å rHDL particles, which have molar phospholipid to apoA-I ratios of 100:2 and 50:2, respectively. As the rHDL is delipidated, the structure of the lipid-bound apoA-I approximates the X-ray crystal structure of lipid-free apoA-I. This model is important for understanding of intermediates in the pathway of biogenesis of HDL [20].

In the ensuing review we describe the various apoA-I mutations and their impact on HDL biogenesis and function with reference to the model given in Fig. 11.1A [5–7]. The domains of apoA-I that affect its various biological functions are shown in color

High-Density Lipoproteins: From Basic Biology to Clinical Aspects. Edited by Christopher J. Fielding

ISBN: 978-3-527-31717-2

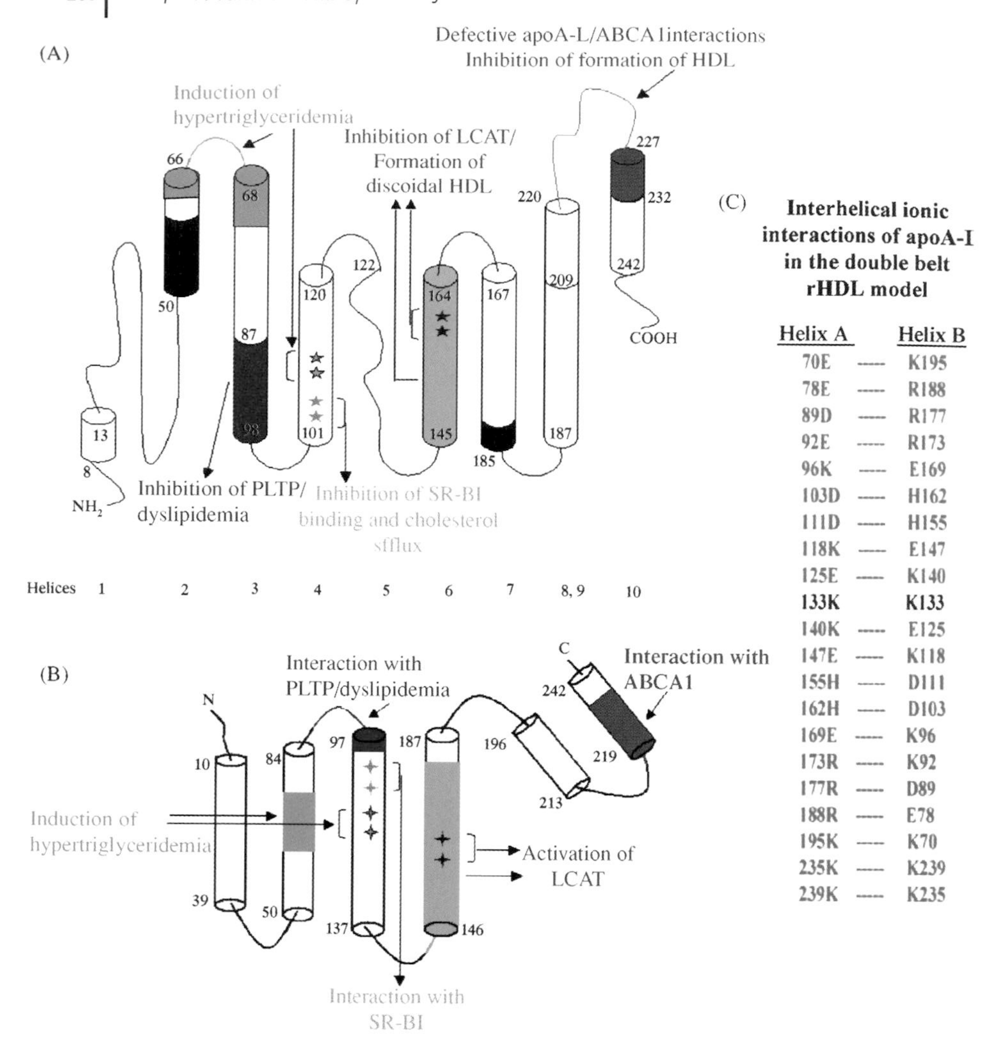

Fig. 11.1 *Summary of structure and functions of apoA-I and HDL. A*) Secondary structure and properties of apoA-I (adapted from [5–7]). Cylinders represent amphipathic α-helices. Predicted amphipathic α-helices are shown in white; additional α-helical regions that have been observed by x-ray crystallography are shown in black. Purple and blue colors indicate regions that affect interactions of apoA-I with ABCA1 and LCAT, respectively. Red indicates region associated with inhibition of PLTP activity and dyslipidemia. Green (and green asterisks) indicate regions and amino acids associated with hypertriglyceridemia. Gray asterisks indicate amino acids that are involved in interactions of apoA-I with SR-BI. *B*) Secondary structure of apoA-I, based on high-resolution x-ray crystallography [8]. The domains of apoA-I that affect its various biological functions are shown in the same colors as described above for Fig. 11.1A. *C*) Proposed interhelical ionic interactions of apoA-I dimers according to the belt model of rHDL [9]. Ionic bonds are shown in green, and repulsion in red. Coordinates for the detailed belt structure can be downloaded from the internet at http://uracil.cmc.uab.edu/Publications.

in Figure 11.1A,B. Details on the structure of apoA-I are provided by Drs. Ryan and Oda (see Chapter 1 review and refs. [14,21]).

11.2
ApoA-I Participates in the Biogenesis and Catabolism of HDL and Interacts with Several Other Proteins of the HDL Pathway

HDL is synthesized through a complex pathway shown in Fig. 11.2 [22], HDL assembly being initiated by an ABCA1-mediated transfer of cellular phospholipids and cholesterol to extracellular lipid-poor apoA-I. The initial lipidation of HDL apoA-I is followed by the remodeling in the plasma compartments of HDL particles through the esterification of cholesterol by the enzyme lecithin:cholesterol acyltransferase (LCAT) and the exchange between HDL and other lipoproteins of apolipoproteins (apoA-I and other less abundant apolipoproteins) and lipids, as well as the putative transfer of additional cellular cholesterol to the growing particles by the scavenger receptor class B, type I (SR-BI) [23] and possibly the cell surface transporter ABCG1 [24]. Finally, hydrolysis of lipids of HDL is mediated by various lipases (lipoprotein lipase, hepatic lipase, endothelial lipase), and exchange of lipids by the phospholipid transfer protein (PLTP) and by the cholesteryl ester transfer protein (CETP). Lipid-free apoA-I secreted by the liver, intestine, and a few other tissues [1] may also interact with ABCA1 present in peripheral cells and tissues and thus contribute to lipid efflux and potentially to HDL biogenesis as described in Fig. 11.2. The regulation of

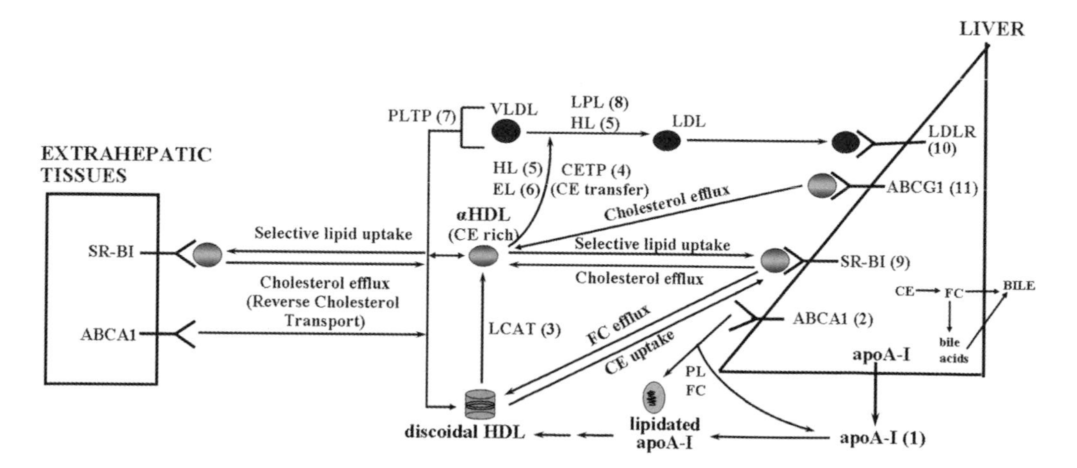

Fig. 11.2 The pathway of biogenesis and catabolism of HDL. The numbers 1–11 indicate key cell membrane or plasma proteins shown to influence HDL levels or composition. They are: 1) apolipoprotein A-I, 2) ATP binding cassette transporter A1, 3) lecithin:cholesterol acyl transferase, 4) cholesteryl ester transfer protein, 5) hepatic lipase, 6) endothelial lipase, 7) phospholipid transfer protein, 8) lipoprotein lipase, 9) scavenger receptor class B type I, 10) LDL receptor, and 11) ABCG1. The figure is modified from Refs. [22,45]. Extensive reviews of several proteins of the HDL pathway are provided in several chapters of this book.

transcription of the genes involved in the biogenesis and remodeling of HDL are reviewed in Chapter 10 by Kardassis and Zannis.

11.3 Functional Interactions between Lipid-Free ApoA-I and the ABCA1 Lipid Transporter Represent the First Step in the Biogenesis of HDL

ABCA1 is a member of the ABC family of transporters. It is a ubiquitous protein of 2261 amino acids expressed abundantly in the liver, macrophages, brain, and various other tissues [25,26] and found on the basolateral surface of the hepatocytes [27]. ABCA1 promotes the efflux of cellular phospholipids and cholesterol to lipid-free apoA-I and other apolipoproteins and amphipathic peptides, but not to spherical HDL particles [28–31]. When cells are depleted of cholesterol, ABCA1 promotes phospholipid efflux to apoA-I, leading to the formation of phospholipid-rich apoA-I particles, which can in turn promote ABCA1-independent cholesterol efflux from cells [32–34]. Inactivating mutations in ABCA1 severely reduced cellular lipid efflux to apoA-I both from fibroblasts of patients with Tangier disease [35–41] and from fibroblasts from $ABCA1^{-/-}$ mice [40]. Tangier disease patients and $ABCA1^{-/-}$ mice fail to form discoidal or spherical HDL particles, but do form preβ1 HDL particles [40–44]. Details on the functions of ABCA1 are given in Ref. [45] and in Chapters 5 by Chimini.

11.3.1 *In vitro* Analysis of the Interactions between ApoA-I and ABCA1 that Lead to Cholesterol and Phospholipid Efflux through the Use of Targeted ApoA-I Mutations

To explore the mechanism of ABCA1-mediated efflux of cellular lipids to apoA-I, we expressed WT apoA-I and numerous variants carrying point mutations or deletions *in vitro*. The apoA-I mutants we generated were then used as lipid acceptors either in cultured cells expressing endogenous WT ABCA1 (J774 macrophages activated by a cAMP analogue) or in cells transfected with a cDNA encoding WT or mutant ABCA1 (HEK293 cells) for *in vitro* ABCA1-mediated lipid efflux and apoA-I/ABCA1 chemical cross-linking experiments. These studies are summarized in Table 11.1A [4,5,15–17,22,46–48], while the alterations in the physicochemical properties of the mutants are shown in Table 11.1B [49,50]. Some of the mutants shown in Tables 11.1A, B were also used in another set of experiments for *in vivo* adenovirus-mediated gene transfer studies in homozygous null apoA-I-deficient mice. The questions addressed were: whether apoA-I interacts directly or indirectly with ABCA1 to promote cholesterol and phospholipid efflux, which domains of apoA-I are required for ABCA1-dependent lipid efflux, and how lipid efflux correlates with apoA-I/ABCA1 binding as assessed by cross-linking.

These studies showed that WT ABCA1-mediated cholesterol and phospholipid efflux was moderately decreased by amino-terminal deletions, was diminished by carboxy-terminal deletions in which residues 220–231 had been removed, and was unexpectedly restored to 80 % of WT control by double deletions of both the amino and the carboxy termini (Tables 11.1A and 11.2) [17,18,22,45]. This finding is

Tab. 11.1A Interactions of apoA-I with ABCA1 (lipid efflux and cross-linking properties), SR-BI, and LCAT in vitro and their ability to form HDL particles in vivo [4,5,15–17,22,46–48].

ApoA-I form	Site of apoA-I mutation (helix)[a]	Relative ABCA1-dependent cholesterol efflux (%)[b]	Relative ABCA1-dependent phospholipid efflux (%)[b]	% cross-linking in the presence of 30-molar excess of competitor[c]	% cross-linking	Activation of LCAT [%]	SR-BI-mediated cholesterol efflux [%]	HDL formation
ApoA-I WT		100	100	13	100	100	100	spherical
ApoA-I[Δ(1–41)]	1 and the region 14–41	86	72	11	120	66.1	–	spherical
ApoA-I[Δ(1–59)]	1, part of 2 and the region 14–50	85	102	6	–[d]	16.7	52	–
ApoA-I[Δ(61–78)]	part of 2 and part of 3	101	108	12	–	47.6	99	spherical
ApoA-I[Δ(89–99)]	lower third part of 3	68	56	9	–	74.5	81	discoidal
ApoA-I (Asp102Ala/Asp103Ala)	4	96	107	6	–	51	21	spherical
ApoA-I (Glu110Ala/Glu111Ala)	4	107	–	–	–	37	53	spherical
ApoA-I (Glu125Lys/Glu128Lys/Lys133Glu/Glu139Lys)	5	109	99	2	–	60.8	114	–
ApoA-I (Leu141Arg)	5	95	104	4	–	0.4	108	few spherical
ApoA-I[Δ(136–143)]	second half of 5	92	93	7	–	30.7	121	–
ApoA-I (Arg160Val/His162Ala)	6	93	91	8	–	0.7	49	discoidal
ApoA-I[Δ(144–165)]	6	71	76	10	–	0.4	116	few spherical
ApoA-I[Δ(165–175)]	first half of 7	79	96	6		62	–	–
ApoA-I[Δ(185–243)]	8, 9, the region 220–227 and 10	20	8	29	67	20.9	59	preβ1 HDL
ApoA-I[Δ(198–243)]	lower two thirds of 8, 9, the region 220–227 and 10	20	–	36	–	38	–	–
ApoA-I[Δ(209–243)]	lower one third of 8,9, the region 220–227 and 10	33	–	32	–	–	–	–

(continued)

Tab. 11.1A *(Continued)*

ApoA-I form	Site of apoA-I mutation (helix)[a]	Relative ABCA1-dependent cholesterol efflux (%)[b]	Relative ABCA1-dependent phospholipid efflux (%)[b]	% cross-linking in the presence of 30-molar excess of competitor[c]	% cross-linking	Activation of LCAT [%]	SR-BI-mediated cholesterol efflux [%]	HDL formation
ApoA-I[Δ(220–243)]	the region 220–227 and 10	9	–	84	46	42	–	preβ1 HDL
ApoA-I[Δ(232–243)]	part of 10	99	–	11	–	14	–	spherical
ApoA-I[Δ(1–41)Δ(185–243)]	1, the region 14–41, 8,9, the region 220–227 and 10	80	77	8	75	14.7	–	discoidal
ApoA-I[Δ(1–59)Δ(185–243)]	1, part of 2, the region 14–50, 8,9, the region 220–227 and 10	78	75	3	–	6.9	45	–
ApoA-I (Asp168Ala/Lys169Ala)	6	–	–	–	–	55.5	87	–
ApoA-I (Lys182Ala/Lys183Ala)	7	–	–	–	–	108.4	125	–
ApoA-I (Arg116Val/Lys118Ala)	4	–	–	–	–	77	98	–
ApoA-I (Glu125Lys/Glu128Lys)	5	–	–	–	–	76.2	–	–
ApoA-I (Glu139Lys)	5	–	–	–	–	116.6	–	–
ApoA-I (Glu191Ala/His193Ala/Lys195Ala)	8	92	–	–	–	107	–	spherical
ApoA-I[Δ(187–197)]	first half of 8	–	–	–	–	43	–	–

[a] Numbers indicate helices as shown in Fig. 11.1A.
[b] Lipid efflux value from J774 macrophages, that have increased expression of ABCA1 after activation with a cAMP analog in the presence of the indicated lipid acceptor relative to the lipid efflux value in the presence of WT apoA-I.
[c] The value that corresponds to the cross-linking of ^{125}I- plasma apoA-I in the presence of 30-fold molar excess of the indicated unlabeled competitor was calculated relative to the value in the absence of competitor, which was set to 100 %.
[d] Not determined.

Tab. 11.1B List of the variant apoA-I forms used in vivo and in vitro studies, expression systems used for their production, their secondary structures and physicochemical properties [17,47,49,50].

		Changes relative to WT apoA-I						
ApoA-I form	Expression system	α-helix [%]	Estimated # of α-helical residue*	Free energy of stabilization ΔG°_D [kcal mol^{-1}]	Effective enthalpy of thermal denaturation ΔH_v [kcal mol^{-1}]	Environment of tryptophan residues	Exposure of hydrophobic surfaces	T ½ [° C]
apoA-I[Δ(1–41)]	baculovirus	~0	−26	−1.1	−12	more polar	increased	
apoA-I[Δ(61–78)]	baculovirus	−8	−29	−1.5	−25	N/A**	greatly increased (↑↑)	
apoA-I[Δ(61–78)] rHDL studies	baculovirus	−8	−31	−1.3				−3
apoA-I[Δ(89–99)]	baculovirus	−6	−21	−0.8	−32	more polar	not studied	
apoA-I[Asp102Ala/Asp103Ala]	baculovirus	−4	−10	−0.4	−13	more polar	increased (↑)	
apoA-I[Asp102Ala/Asp103Ala] rHDL studies	baculovirus	−9	−22	−1.3				−3
apoA-1 [Glu110Ala/Glu111Ala]	adenovirus	−7	−18	−1.0	−18	more polar	greatly increased (↑↑)	
apoA-1 [Glu110Ala/Glu111Ala] rHDL studies	adenovirus	−4	−10	−0.7				–
apoA-I[Arg160Val/His162Ala]	baculovirus		~0	~0	-9	more polar	increased (↑)	
apoA-I[Arg160Val/His162Ala] rHDL studies	baculovirus	~0	~0	~0				~0
apoA-I[Δ(144–165)]	baculovirus	−4	−22	−0.5	−14	more polar	not studied	
apoA-I[Δ(185–243)]	permanent cell lines	~0	−32	~0	~0	no changes	not studied	
apoA-I[Δ(1–41)Δ(185–243)]	baculovirus	−7	−69	−1.1	−11	more polar	increased (↑)	–
apoA-I[Δ(1–41)Δ(185–243)] rHDL studies	baculovirus	+11	−52	−1				

*The number of residues in the helical conformation of WT and each mutant form of apoA-I was estimated by multiplying the total number of residues in the protein by its α-helical content. The changes in the number of residues in the helical conformation of each apoA-1 mutant as compared to that in WT apoA-1 are shown.
**Parameters of Trp intrinsic fluorescence for the apoA-I[Δ(61–78)] were not informative because of the lack of Trp72 in this mutant.

consistent with direct ABCA1/apoA-I interactions involving the central helices of apoA-I [48]. Lipid efflux was either unaffected or moderately reduced (by 10–40 % of WT control) by a variety of point mutations or deletions of internal helices 2–7 as defined in Fig. 11.1A (Tables 11.1A and 11.2) [47,48], suggesting that different combinations of central helices can promote lipid efflux [47,48]. Chemical cross-linking/immunoprecipitation studies showed that the relative abilities of apoA-I mutants to promote ABCA1-mediated lipid efflux correlated with the ability of these mutants to be cross-linked efficiently to ABCA1 [48]. Previous studies also showed that a variety of apolipoproteins and synthetic peptides of D or L configuration can promote cholesterol efflux [17,29,48,51,52], although, as demonstrated by the *in vivo* studies described below, the ability of a peptide or a mutant apoA-I form to cross-link with ABCA1 and promote cholesterol efflux *in vitro* does not necessarily imply that it is capable of promoting synthesis of HDL particles *in vivo*.

Another concept that emerges from our *in vitro* studies and the work of others is that in some instances apoA-I and ABCA1 may bind efficiently, but this interaction may not necessarily lead to the synthesis of HDL. As we discuss later, we call this type of binding "nonproductive" binding. There are two models that illustrate this point. An ABCA1 W590S mutant found in a Tangier patient cross-links more strongly to apoA-I than the WT ABCA1, but fails to promote lipid efflux and to form HDL *in vivo* [53]. When we compared the cross-linking properties of several apoA-I mutants to those of cells expressing either the WT ABCA1 or the W590S mutant, we found significant differences in binding, which raised the possibility that the W590S mutation may have altered the environment in the binding site of ABCA1 in a way that prevented efficient lipid efflux despite the increased binding [48]. Efficient apoA-I/ABCA1 interaction is therefore not in itself sufficient for efficient lipid efflux. A functional relationship between apoA-I/ABCA1 binding and lipid efflux was also seen in the concurrent inhibition both of apoA-I cross-linking to ABCA1 and of lipid efflux by two small-molecule inhibitors: glyburide and BLT4 [1-(2-methoxyphenyl)-3-(naphthalen-2-yl)urea]. These inhibitors increased binding of apoA-I to SR-BI but inhibited efflux [54]. Glyburide is a sulfonyl urea derivative that inhibits the sulfonyl urea receptor-1 (SUR-1) of the KATP channel and also inhibits the ABCA1 mediated lipid efflux. All these *in vitro* data are consistent with a lipid efflux mechanism that involves direct interaction between apoA-I and ABCA1, although other earlier studies had suggested that protein–protein interactions may not be required for lipid efflux [55].

11.3.2
Use of Adenovirus-Mediated Gene Transfer to Study how ApoA-I Mutations Affect the Biogenesis of HDL

These studies initially addressed the question of whether apoA-I mutations that fail to interact properly with ABCA1 would fail to form αHDL particles. In these studies, mice were infected with 1 to 2 × 10^9 pfu of an adenovirus expressing WT apoA-I or different apoA-I mutants that had previously been studied for lipid efflux and ABCA1 cross-linking (Tables 11.1A and 11.2). Four to five days after the gene transfer, plasma lipid analyses, gel filtration chromatography, and two-dimensional gel

Tab. 11.2 In vivo phenotypes of mice expressing apoA-I deletion and point mutants and changes in the in vitro functions of the apoA-I mutants [15–17,47,91–93].

	In vivo phenotypes[a]						Changes in the in vitro functions[b]			
Mutation	Plasma total cholesterol (x-fold)	Plasma triglycerides (x-fold)	VLDL lipolysis	Plasma PLTP activity (%)[d]	Particles in the HDL region	CE/TC of HDL region (%)[e]	Activation of LCAT[f]	ABCA1–mediated cholesterol efflux	SR–BI–mediated cholesterol efflux	HDL subpopulations
ApoA-I[Δ(1–41)][g]	68 % of normal	61 % of normal	–[c]	–	spherical	–	–	86 %	–	–
ApoA-I[Δ(7–43)][h]	82 % of normal	–	–	–	spherical	76	27 %	91 %	–	–
ApoA-I[Δ(7–65)][h]	66 % of normal	–	–	–	discoidal	39	43 %	106 %	–	–
ApoA-I[Δ(1–41)Δ(185–243)][g]	46 % of normal	19 % of normal	–	–	discoidal	–	–	80 %	–	–
ApoA-I [Asp102Ala/Asp103Ala][i]	94 % of normal	62 % of normal	–	–	spherical	103	50 %	96 %	21 %	normal
ApoA-I[Δ(62–78)][j]	2-fold increase	9.5-fold increase	decreased	92	spherical	82	47 %	101 %	96 %	normal
ApoA-I [Glu110Ala/Glu111Ala][k]	1.9-fold increase	6.9-fold increase	decreased	–	spherical	–	37 %	107 %	53 %	normal
ApoA-I[Δ(89–99)][j]	3.2-fold increase	normal	–	32	discoidal	38	77 %	68 %	78 %	increased preβ1/αHDL ratio
ApoA-I [Arg160Val/His162Ala][i]	55 % of normal	normal	–	–	discoidal	44	0.7 %	93 %	49 %	near absence of α1 α2 α3 subpopulation
ApoA-I[Δ(144–165)][i]	25 % of normal	50 % of normal	–	–	spherical	62	0.4 %	71 %	116 %	significant reduction in the α1 α2 α3 subpopulations
ApoA-I[Δ(100–143)][l]	36 % of normal	–	–	–	discoidal, and spherical	61	64 %	–	–	–

(continued)

Tab. 11.2 *(Continued)*

Mutation	In vivo phenotypes[a]						Changes in the in vitro functions[b]			
	Plasma total cholesterol (x-fold)	Plasma triglycerides (x-fold)	VLDL lipolysis	Plasma PLTP activity (%)[d]	Particles in the HDL region	CE/TC of HDL region (%)[e]	Activation of LCAT[f]	ABCA1–mediated cholesterol efflux	SR–BI–mediated cholesterol efflux	HDL subpopulations
ApoA-I[Δ(122–165)][l]	23 % of normal	–	–	–	discoidal	63	13 %	–	–	–
ApoA-I[Δ(144–186)][l]	41 % of normal	–	–	–	discoidal	41	9 %	–	–	–
ApoA-I[Δ(143–164)] Tg mice[m]	29 % of normal	–	–	–	–	59	4 %	–	–	–

[a]Parameters obtained from mice expressing the apoA-I deletion and point mutants in relation to mice expressing the WT apoA-I.
[b]Parameters determined by *in vitro* assays for the apoA-I mutants in relation to the same parameters for the WT apoA-I.
[c]Not determined.
[d]PLTP activity relative to the PLTP activity from plasma of mice expressing WT apoA-I.
[e]CE/TC of particles in the HDL region relative to the CE/TC of the HDL from mice expressing WT apoA-I.
[f]Refers to catalytic efficiency ($Vmax_{app}/Km_{app}$) of LCAT with rHDL particles containing mutant apoA-I forms in relation to rHDL particles containing WT apoA-I.
[g]Ref. [17].
[h]Ref. [92].
[i]Ref. [16].
[j]Ref. [15].
[k]Ref. [47].
[l]Ref. [91].
[m]Ref. [93].

electrophoretic analyses were performed on lipoproteins from the plasma of the apoA-I-deficient mice expressing the WT or mutant apoA-I forms [17,18]. Hepatic apoA-I mRNA levels were also determined, in order to ensure comparable expression of the mutant genes. A set of experiments showed that the WT apoA-I and the amino terminal deletion mutant apoA-I[Δ(1–41)] generated HDL *in vivo*, as determined by FPLC fractionation, consisting of spherical particles as determined by electron microscopy [17] (Fig. 11.3A,B; Table 11.2). Two-dimensional gel electrophoresis showed that the majority of WT apoA-I consisted of particles with α electrophoretic mobility, while a small fraction consisted of particles with β electrophoretic mobility (Fig. 11.3G). Similar analysis showed that the double deletion mutant apoA-I[Δ(1–41)Δ(185–243)] formed discoidal particles (Fig. 11.3C) [17], while the carboxy terminal deletion mutants apoA-I[Δ(185–243)] and apoA-I[Δ(220–243)] formed very little HDL (Fig. 11.3D,E), which consisted exclusively of preβ1-HDL particles (Fig. 11.3H) [17,18,22]. The spherical particles observed in mice expressing the carboxy-terminal deletion mutants apoA-I[Δ(185–243)] and apoA-I[Δ(220–243)] were enriched in apoE (Figs. 10.3F,I). These findings suggested that carboxy-terminal deletions in apoA-I may have prevented the lipidation of apoA-I, thus blocking the first step in the biogenesis of HDL. Preβ-migrating particles similar to those observed in the plasma of apoA-I-deficient mice expressing carboxy-terminal truncated apoA-I forms are also found in the plasma of ABCA1-deficient mice and in the plasma of patients with Tangier disease [41,56,57]. These particles are apparently generated by mechanisms independent of the ABCA1–apoA-I interactions and may be the product of different types of interactions of apoA-I with cells [58,59]. The *in vivo* findings, combined with the relationships of apoA-I mutants to ABCA1 mediated cholestrol efflux and cross linking, determined *in vitro*, permit the following generalizations [17,48]: *1)* ApoA-I mutants that lack the 220–231 domain are defective in the efflux of cellular phospholipids and cholesterol, cross-link poorly to ABCA1, and fail to form discoidal or spherical HDL *in vivo*, but do form preβ HDL particles. *2)* The central region of apoA-I alone, containing helices 2–7 as defined in Fig. 11.1A, has the capacity to promote ABCA1-mediated lipid efflux, to cross-link to ABCA1, and to form discoidal HDL particles *in vivo*. However, the central helices cannot promote lipid efflux and HDL formation if the carboxy-terminal segment of apoA-I is deleted.

Although the kinetics of efflux of cellular cholesterol and phospholipids to apoA-I and other lipid acceptors are indistinguishable [17], it has been proposed on the basis of other evidence that, after ABCA1-catalyzed addition of phospholipid to apoA-I [32–34], the apoA-I/phospholipid complex acquires cholesterol and other lipid components from cholesterol-rich membrane domains such as caveolae and lipid rafts [60]. From the existing data, we suggested a two-step model of lipid efflux that explains Tangier disease and other HDL deficiencies (Fig. 11.3L). The first step is the formation of a tight complex between ABCA1 and its ligand, apoA-I. This step [48,52] appears to be necessary but not sufficient for lipid efflux [52,53]. The second step is the ABCA1-mediated transfer of lipids from the cell to the apoA-I (lipidation). This step appears to require the formation of a productive complex between apoA-I and ABCA1.

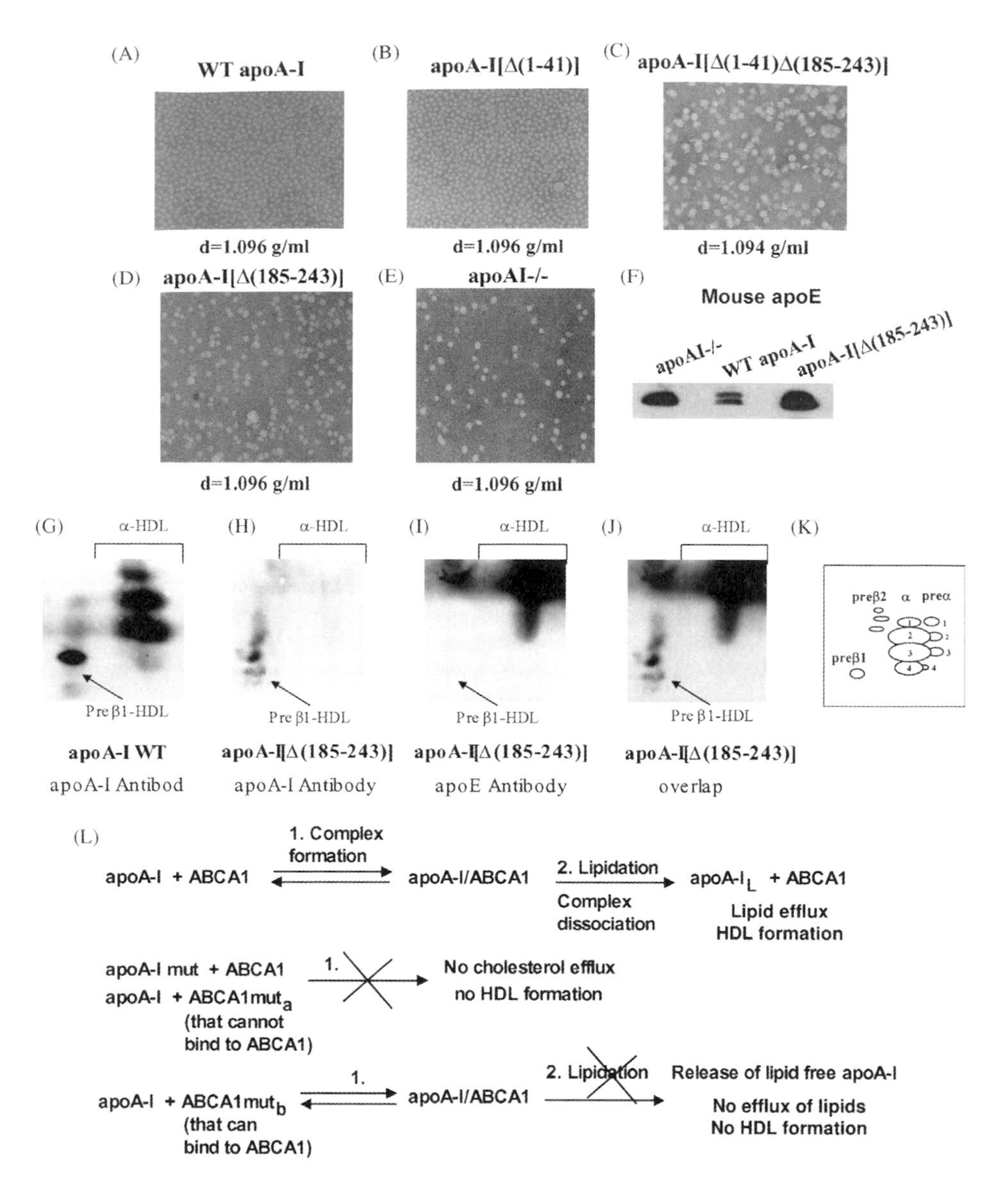

Fig. 11.3 *Analyses of plasma of HDL particles formed in apoA-I$^{-/-}$ mice expressing the WT apoA-I and the apoA-I[Δ(1–41)], apoA-I[Δ(1–41)Δ(185–243)], and apoA-I[Δ(185–243)] deletion mutants.* Schematic representations of subpopulations of HDL and proposed two-step model of lipidation of apoA-I. *A–E*) Electron microscopy of HDL fractions obtained from plasma of apoA-I-deficient mice infected with adenoviruses expressing WT and mutant apoA-I forms as indicated on the top of each panel. *F*) Western blotting of plasma of apoA-I$^{-/-}$ mice that were not infected or apoA-I$^{-/-}$ mice infected with recombinant adenoviruses expressing WT apoA-1 and a carboxy-terminal deletion mutant of apoA-I. Mouse apoE was detected by use of anti-mouse apoE antibodies. *G–J*) Characterization of plasma HDL obtained from mice expressing WT apoA-I or apoA-I[Δ(185–243)] by two-dimensional gel electrophoresis. Detection of particles was achieved with human anti-apoA-I or mouse anti-apoE antibodies. *K*) Schematic representation of the HDL subpopulations obtained by two-dimensional electrophoresis [136]. *L*) Two-step model of ABCA1-mediated cholesterol efflux and apoA-I lipidation.

In the case of the ABCA1[W590S] mutant associated with Tangier disease [53], an apoA-I/ABCA1 complex is formed, but the apoA-I is not lipidated and is subsequently released in a lipid-free form [52], whereas after binding to the WT ABCA1, the apoA-I is released in association with lipids [52]. Lipidation is also not observed when truncated apoA-I forms are used as lipid acceptors (Fig. 11.3H). In principle, it is possible that there may exist currently unidentified point mutations in apoA-I that cause strong binding to ABCA1 but cannot be lipidated because of nonproductive apoA-I/ABCA1 interactions.

11.4
ApoA-I/ABCA1 Interactions: Intracellular Trafficking of the Complex

Intracellular trafficking of apoA-I/ABCA1 complexes has been observed by fluorescence microscopy studies of HeLa cells expressing an ABCA1 green fluorescence fusion protein [27,61]. ABCA1-mediated lipid efflux from macrophages onto apoA-I has been shown to involve binding of apoA-I to ABCA1 in the coated pits, internalization, interaction with intracellular lipid pools, and resecretion of a lipidated particle [62,63]. The recycling and intracellular lipidation of apoA-I is defective in human and murine macrophages that lack ABCA1 [30,31,40]. After a similar pathway, apoA-I is transcytosed through endothelial cells and is secreted from the apical surface in a lipid-bound form [64]. Transcytosis of apoA-I and its release into the subendothelial space in the form of HDL may explain some of the antiatherogenic properties of HDL. Tangier disease fibroblasts that lack ABCA1 have a late endocytic trafficking defect, resulting in the accumulation of cholesterol, sphingomyelin, and Niemann-Pick type C1 protein (NPC1) in late endosomes. This defect of the Tangier fibroblasts could be corrected by ABCA1 gene transfer [65].

Internalization of HDL in hepatocytes is also promoted by interactions of apoA-I with the β chain of the F_1 domain of ATP synthase [66,67]. This topic is reviewed by Collet in Chapter 12.

11.5
Contribution of the HDL Pathway in the Reverse Transport of Cholesterol

The ABCA1-mediated cholesterol efflux to lipid-poor apoA-I represents one of the mechanisms through which peripheral cells may lose their excess cholesterol, which may then be returned to the liver. The return of cholesterol effluxed from peripheral cells via the plasma to the liver for reutilization or excretion in the form of free cholesterol or bile acids has been termed reverse cholesterol transport [58,68,69]. In Tangier patients, who have mutant forms of ABCA1, and in ABCA1$^{-/-}$ mice, cholesterol efflux is inhibited and cholesteryl esters accumulate in macrophages and contribute to atherogenesis [42]. A question that merits consideration is whether the reverse cholesterol transport is a function of the HDL pathway or involves additional processes. Three other mechanisms of

efflux of cholesterol that could carry the excess cholesterol effluxed from the peripheral cells to the liver have been described and involve: *a*) passive diffusion of cholesterol through the aqueous phase between a cholesterol pool at the membrane and albumin, preβ HDL, or other cholesterol acceptors [60,70–72], *b*) protein-facilitated diffusion, which is independent of ATP hydrolysis and may involve surface microdomains with different free cholesterol (FC) and phospholipid (PL) compositions and protein contents [73,74], and *c*) complex mechanisms involving membrane microsolubilization, a process that involves simultaneous removal of cellular cholesterol and phospholipid [75,76].

Most studies have indicated that increases or decreases in HDL due to overexpression or deletion of various genes of the HDL pathway or 7α-hydroxylase, which is the rate-limiting enzyme in bile acid synthesis, did not affect the flux of cholesterol to the liver and/or its excretion into the bile [77–80]. However, simultaneous overexpression of LCAT and SR-BI by adenovirus-mediated gene transfer significantly increased fecal sterol excretion [81]. Through the use of an *in vivo* method that measures the reverse transport of ^{3}H-labeled cholesterol from cholesterol-loaded J774 macrophages via the plasma to the liver and the bile, it has been shown that genes of the HDL pathway may influence this process [82–85]. However, it must be pointed out that this assay measures the reverse transport of cholesterol from the exogenously introduced macrophages and not from the whole peripheral tissues.

Recent data indicate that ABCA1/apoA-I interactions in the liver are essential for the initial lipidation of apoA-I and also determine the subsequent maturation of nascent preβ HDL into spherical αHDL particles [44,86]. When hepatic ABCA1 is inactivated, preβ HDL fails to mature to αHDL and is catabolized rapidly by the kidney [44,86], thus resulting in low HDL levels. It had also been shown previously that cubulin, a 600 kDa membrane protein, binds both apoA and HDL [87] and mediates endocytosis and lysosomal degradation of the HDL particles in the kidney [87,88]. Adenovirus-mediated gene transfer of ABCA1 in total or liver-specific knockout mice for ABCA1 restored the HDL cholesterol levels fully in the liver-specific knockout mice but only partially in the total knockout mice [89]. The apoA-I to phospholipid ratios of the HDL produced by the ABCA1-expressing total knockout and liver-specific knockout mice were similar, but the particles differed in their cholesterol contents [89]. The combined data indicate: *a*) that the liver is the major site for the initial lipidation of apoA-I, which seems to be the rate-limiting step of HDL biogenesis, and that the contribution of the peripheral tissues in this process appears to be small, and *b*) that the ABCA1/apoA-I or ABCA1/preβ1 HDL interactions in the peripheral tissues appear to enrich the initially lipidated particle with cholesterol and to increase its stability. It is possible that when cholesterol levels are low due to defects in apoA-I, ABCA1, or other proteins of the HDL pathway, the various acceptors of cholesterol such as albumin [60] may have other routes independent of the HDL pathway for reverse transport of cholesterol to the liver. In addition, when the HDL pathway is normal, preβ HDL particles generated *de novo* by the liver or preβ1 particles dissociating from αHDL particles in plasma may also contribute to cholesterol efflux from peripheral cells and the reverse transport of cholesterol.

11.6 HDL Biogenesis can be Inhibited by Specific Mutations in ApoA-I that Inhibit Activation of LCAT

ApoA-I that has been lipidated correctly after functional interaction with ABCA1 proceeds through putative intermediate steps to form discoidal particles that are converted into spherical particles by the action of LCAT. Our knowledge of the molecular events that lead to the formation of the discoidal particles and the spherical αHDL subpopulations is at present incomplete. To identify intermediates in later steps of the HDL biogenesis, we performed adenovirus-mediated gene transfer studies of specific apoA-I mutants in apoA-I-deficient mice. This analysis showed that substitution of hydrophobic residues in the 211 to 229 region with either charged or less bulky hydrophobic residues resulted in low levels of HDL and formation of discoidal HDL particles. In contrast, substitution of charged residues 234 to 239 with Ala led to the formation of normal HDL [18], indicating that the hydrophobic residues of the carboxy-terminal region of apoA-I may play an important role in the *in vivo* activation of LCAT. Activation of LCAT was also severely affected by point mutations in helix 6 of apoA-I (apoA-I[Arg-160Val/His162Ala]) as defined in Fig. 11.1A [16]. These point mutations abolished the ability of apoA-I to activate LCAT *in vitro* (Table 11.1A). After adenovirus-mediated gene transfer of the helix 6 point mutant in apoA-I-deficient mice, plasma HDL cholesterol was greatly reduced and HDL cholesterol and apoA-I were distributed in the HDL_3 region (Fig. 11.4A,B). Electron microscopy and two-dimensional gel electrophoresis showed that the helix 6 point mutant predominantly formed high levels of discoidal particles (Fig. 11.4D) and had near absence of α1, α2, and α3 subpopulations (Fig. 11.4H; Table 11.2). The buildup of discoidal particles in the plasma of mice expressing the apoA-I Arg160Val/His162Ala mutant was also associated with a decrease in the CE/TC ratio in HDL sized particles (see Table 11.2) [90]. Coinfection of mice with adenoviruses expressing human LCAT and the helix 6 point mutant dramatically increased plasma HDL and apoA-I levels, converted the discoidal HDL into spherical HDL (Fig. 11.4E), and normalized the ratio of preβ/α-HDL subpopulations (Fig. 11.4I). The findings suggested that the plasma LCAT activity was rate-limiting for esterification of the HDL cholesterol and the conversion of discoidal HDL into spherical HDL. The overall analysis indicated that critical mutations in apoA-I may prevent activation of LCAT *in vivo* and present a phenotype that mimics the phenotype observed in patients with classical LCAT deficiency [90]. Various large deletions in apoA-I had also previously been shown to promote the formation of discoidal HDL particles (Table 11.2) [15,91,92]. An apoA-I deletion mutant apoA-I [Δ(144–165)] designated helix 6Δ, with very low LCAT activation *in vitro*, was also associated with low HDL levels and decreased levels of α1, α2, and α3 subpopulations (Fig. 11.4F,J; Table 11.2). However, the mutant phenotype could not be corrected by coexpression of this mutant and human LCAT, probably because of fast degradation of the product of the initial apoA-I/ABCA1 interactions [93].

11.7
Mutations in ApoA-I may Induce Hypertriglyceridemia and Other Dyslipidemias

By screening the *in vitro* and *in vivo* properties of apoA-I mutants (Tables 11.1A,B and 10.2), we were able to identify three apoA-I mutations that have a profound effect on overall cholesterol and triglyceride homeostasis [15,47]. Two of these – apoA-I [Δ(62–78)] and apoA-I[Glu110Ala/Glu111Ala] (Tables 11.1A,B and 11.2) – caused combined hyperlipidemia, characterized by high plasma cholesterol and severe hypertriglyceridemia (Table 11.2) [15,47]. All the triglycerides and the majority of the excess cholesterol were distributed in apoA-I-enriched VLDL/IDL-sized lipoproteins, suggesting that apoA-I mutants had increased affinity for lower density lipoprotein fractions (Fig. 11.5A,B).

The VLDL/IDL fraction also had decreased levels of apoE and apoCII and increased levels of apoB-48 (Fig. 11.5C) [15,47]. These results raised the possibility that the mutations were associated with inhibition of lipolysis *in vivo*, and this was tested by coinfection of mice with adenoviruses expressing apoA-I[Glu110Ala/Glu111Ala] and

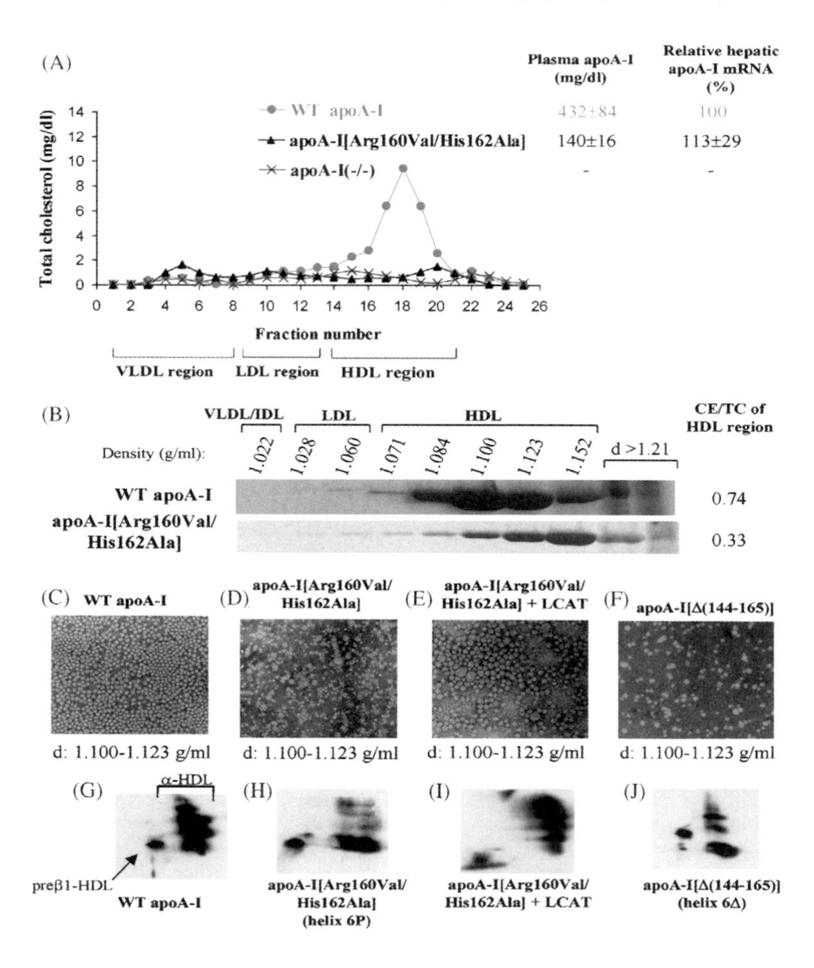

the human lipoprotein lipase. Expression of the lipoprotein lipase reduced plasma and VLDL triglyceride levels but had a smaller effect on plasma cholesterol levels. These observations indicated that some forms of human hypertriglyceridemia or combined dyslipidemia may be due to mutations in apoA-I. The third mutation, apoA-I[Δ(89–99)] [15], caused (relative to WT apoA-I) hypercholesterolemia characterized by increased cholesterol and phospholipids in the VLDL/IDL/LDL size lipoproteins (Fig. 11.5D), normal triglyceride levels, and a substantially decreased CE/TC ratio in HDL and LDL (Fig. 11.5E, Table 11.2). The Δ89–99 deletion also increased apoA-I in the LDL-sized particles (Fig. 11.5E), caused accumulation of discoidal HDL (Fig. 11.5G), and increased the levels of preβ1 relative to the αHDL subpopulation particles (Fig. 5H,I) [15]. Strikingly, mice expressing the apoA-I [Δ(89–99)] mutant exhibited phospholipid transfer protein (PLTP) activity that was only 32 % of that of mice expressing the WT apoA-I [15]. It has been proposed that PLTP functions by linking the donor and acceptor lipoprotein particles, thus facilitating the net transfer of phospholipids to HDL [94]. The potential involvement of the apoA-I[Δ(89–99)] mutant in the inhibition of PLTP activity is also supported by other findings. Previous studies have suggested that apoA-I interacts physically with PLTP [95] and that these interactions may underlie the transfer of the phospholipids from the donor molecule to HDL. PLTP was shown to remodel the HDL and to promote the generation of preβ HDL particles [96–98]. Deficiency of PLTP in mice on a high-fat diet increased the concentration of phospholipids and cholesterol in VLDL and LDL and promoted the formation of discoidal particles [99], so there are remarkable similarities in these phenotypes of mice expressing the apoA-I[Δ(89–99)] mutant and those of PLTP-deficient mice. The dyslipidemic phenotypes of the mice expressing the apoA-I[Δ(89–99)], apoA-I[Δ(62–78)], and apoA-I[Glu110Ala/Glu111Ala] mutants were not clearly correlated with the abilities of these mutants to activate LCAT or to promote ABCA1- and SR-BI-dependent cholesterol efflux (Tables 11.1A and 11.2). As mentioned above, the severe hypertriglyceridemia of the apoA-I[Δ(62–78)] and

Fig. 11.4 *Analyses of plasma apoA-I$^{-/-}$ mice expressing the WT apoA-I and mutant apoA-I forms. A)* FPLC profile of total cholesterol in plasma of apoA-I$^{-/-}$ mice expressing the WT apoA-I or the apoA-I[Arg160Val/His162Ala] mutant or in control apoA-I$^{-/-}$ mice. *B)* SDS-PAGE analysis of density gradient ultracentrifugation fractions of plasma of apoA-I$^{-/-}$ mice expressing the WT apoA-I or the apoA-I[Arg160Val/His162Ala] mutant. The densities of the fractions used are indicated on the top of the figure. ApoA-I was detected by staining with Coomassie Brilliant Blue. The right-hand side of the panel shows the CE/TC ratio from a pool of lipoprotein fractions corresponding to the HDL region. *C,D)* Electron microscopy pictures of HDL fractions obtained from apoA-I$^{-/-}$ mice expressing the WT apoA-I (*C*) or the apoA-I[Arg160Val/His162Ala] mutant (*D*) after density gradient ultracentrifugation of plasma. *E,F)* Similar EM analysis of HDL obtained from mice treated with a mixture of adenovirus expressing the apoA-I [Arg160Val/His162Ala] mutant and human LCAT (*E*) or the apoA-I Δ(144–165) mutant (*F*). The densities of the fractions used are indicated. The photomicrographs were taken at 75 00× magnification and enlarged 3 times. *G,H,I,J)* Analysis of plasma by two-dimensional gel electrophoresis and Western blotting. The plasma was obtained from mice expressing the WT apoA-I (*G*) the apoA-I[Arg160Val/His162Ala] mutant alone (*H*) or combination with human LCAT (*I*), or the apoA-IΔ(144–165)] mutant (*J*).

apoA-I[Glu110Ala/Glu111Ala] mutants was associated with increased presence of apoA-I in the VLDL/IDL fractions and decreased *in vitro* lipolysis of VLDL/IDL (Table 11.2). This observation raised the possibility that the presence of the mutant apoA-I in the VLDL region influenced the activity of lipoprotein lipase *in vivo*. Figure 11.6 shows the normal pathway of biogenesis and catabolism of HDL and various steps where the pathway can be affected and lead to dyslipidemia. These include: *1*) lack of synthesis of HDL due to mutations in ABCA1 or mutations in apoA-I that diminish cholesterol efflux, *2*) induction of hypertriglyceridemia by the apoA-I[Δ(62–78)] and apoA-I[Glu110Ala/Glu111Ala] mutants, *3*) accumulation of discoidal HDL, inhibition of PLTP, and the induction of hypercholesterolemia by the apoA-[Δ(89–99)] mutant, and *4*) the accumulation of discoidal HDL due to the Arg160/Val/His162Ala mutation or other mutations that inhibit LCAT activation. The phenotypes produced by targeted mutagenesis of apoA-I should be valuable for the detection of similar phenotypes in humans and can serve as diagnostic and prognostic markers of dyslipidemia and/or atherosclerosis.

11.8
Remodeling of the HDL Particles after Interactions between HDL-Bound ApoA-I and SR-BI

SR-BI is an 83 kDa membrane glycoprotein containing a large extracellular domain and two transmembrane domains with short cytoplasmic amino- and carboxy-terminal domains [100] that is expressed in the liver, the steroidogenic glands,

Fig. 11.5 *Analyses of plasma of apoA-I$^{-/-}$ mice expressing the WT apoA-I and mutant apoA-I forms.* *A*) FPLC profile of total cholesterol in plasma of apoA-I$^{-/-}$ mice expressing the WT apoA-I or the apoA-I (Glu110Ala/Glu111/Ala) mutant or in control apoA-I$^{-/-}$ mice. *B*) SDS-PAGE analysis of density gradient ultracentrifugation fractions of plasma obtained from apoA-I$^{-/-}$ mice four days post-infection with adenoviruses expressing WT apoA-I or the apoA-I(Glu110Ala/Glu111Ala) mutant. The densities of the fractions are indicated on the top of the figure. ApoA-I was detected by Western blotting with use of a polyclonal antibody to human apoA-I. *C*) Assessment of the apoprotein composition of VLDL/LDL isolated from plasma of mice infected with WT apoA-I and apoA-I(Glu110Alah/Glu111Ala) mutant. The apoproteins were detected by Western blotting with use of polyclonal antibodies to mouse apoB, apoE, and apoCII. *D*) FPLC profile of total plasma cholesterol of apoA-I$^{-/-}$ mice expressing the WT apoA-I or the apoA-I[Δ(89–99)] mutant. *E*) SDS-PAGE analysis of density gradient ultracentrifugation fractions of plasma of apoA-I$^{-/-}$ mice expressing the WT apoA-I or the apoA-I[Δ(89–99)] mutant. The densities of the fractions used are indicated on the top of the figure. ApoA-I was detected by staining with Coomassie Brilliant Blue. The right-hand side of the panel shows the CE/TC ratios from a pool of lipoprotein fractions corresponding to the HDL region. *F,G*) Electron microscopy pictures of HDL fractions obtained from apoA-I$^{-/-}$ mice expressing the WT apoA-I (*F*) or the apoA-I[Δ(89–99)] mutant (*G*) after density gradient ultracentrifugation of plasma. The densities of the fractions used are indicated on the bottom of each picture. The photomicrographs were taken at 75 000× magnification and enlarged 3 times. *H,I*) Analysis of plasma obtained from mice expressing the WT apoA-I (*H*) or the apoA-I[Δ(89–99)] (*I*) mutant after two-dimensional gel electrophoresis and Western blotting.

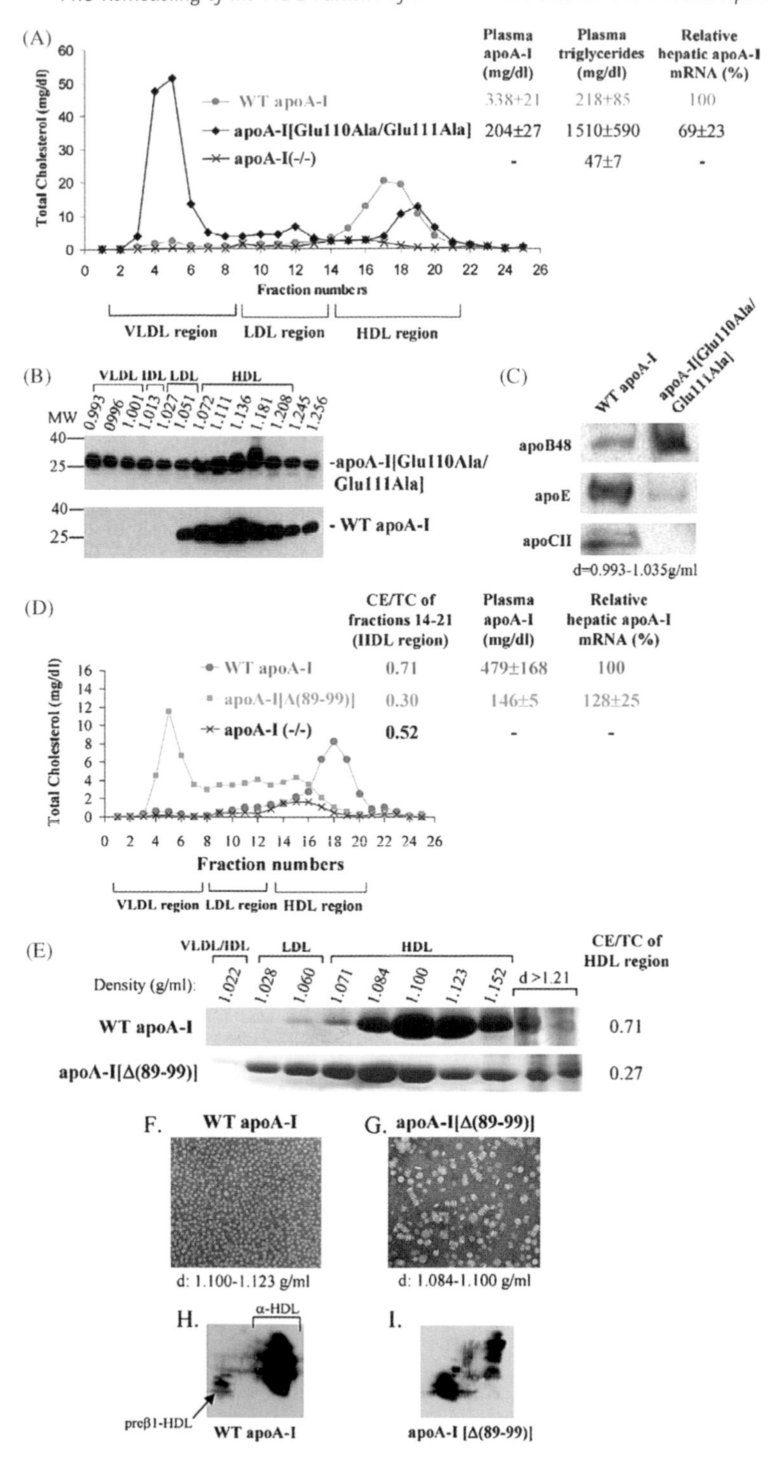

(A)
Total Cholesterol (mg/dl)
Plasma apoA-I (mg/dl)
Plasma triglycerides (mg/dl)
Relative hepatic apoA-I mRNA (%)
WT apoA-I 338±21 218±85 100
apoA-I[Glu110Ala/Glu111Ala] 204±27 1510±590 69±23
apoA-I(-/-) - 47±7 -
Fraction numbers
VLDL region
LDL region
HDL region
(B)
VLDL IDL LDL HDL
MW
0.993 0996 1.001 1.013 1.027 1.051 1.072 1.111 1.136 1.181 1.208 1.245 1.256
-apoA-I[Glu110Ala/Glu111Ala]
- WT apoA-I
(C)
WT apoA-I
apoA-I[Glu110Ala/Glu111Ala]
apoB48
apoE
apoCII
d=0.993-1.035g/ml
(D)
CE/TC of fractions 14-21 (HDL region)
Plasma apoA-I (mg/dl)
Relative hepatic apoA-I mRNA (%)
WT apoA-I 0.71 479±168 100
apoA-I[Δ(89-99)] 0.30 146±5 128±25
apoA-I (-/-) 0.52 - -
Total Cholesterol (mg/dl)
Fraction numbers
VLDL region LDL region HDL region
(E)
VLDL/IDL LDL HDL
Density (g/ml): 1.022 1.028 1.060 1.071 1.084 1.100 1.123 1.152 d >1.21
CE/TC of HDL region
WT apoA-I 0.71
apoA-I[Δ(89-99)] 0.27
F. WT apoA-I
G. apoA-I[Δ(89-99)]
d: 1.100-1.123 g/ml
d: 1.084-1.100 g/ml
H.
α-HDL
I.
preβ1-HDL
WT apoA-I
apoA-I [Δ(89-99)]

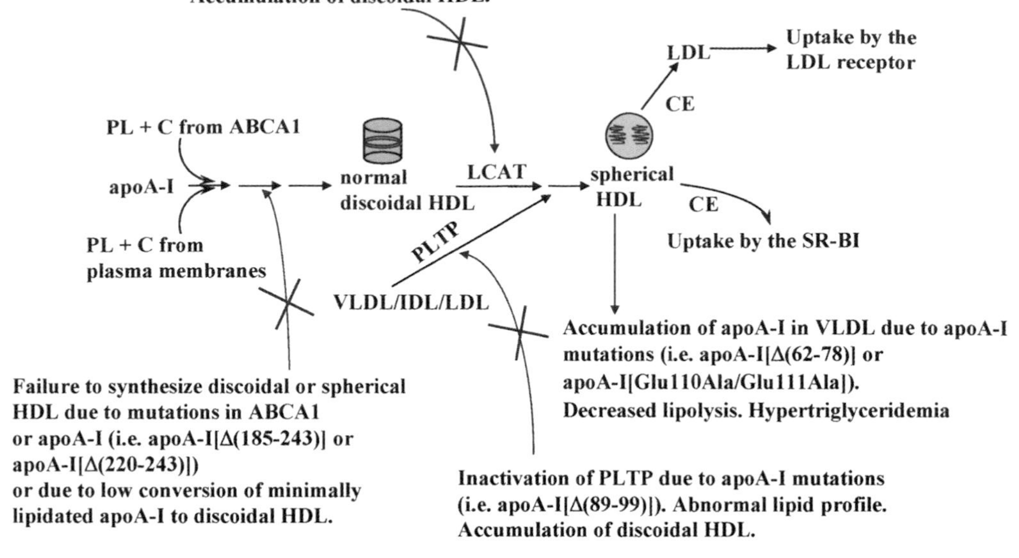

Fig. 11.6 Schematic representation of the defects in the HDL pathway that can account for: *1*) lack of HDL synthesis due to mutation in ABCA1 or C-terminal apoA-I deletions that affect cholesterol efflux, *2*) hypertriglyceridemia induced by the apoA-I[Δ(62–78)] and apoA-IpGlu110Ala/Glu111Ala] mutants, *3*) the high plasma cholesterol levels, inhibition of PLTP activity, and accumulation of discoidal HDL particles induced by the apoA-I[Δ(89–99)] mutant, and *4*) accumulation of discoidal HDL due to LCAT deficiency or the apoA-I[Arg160Val/His162Ala] mutant that inhibits activation of LCAT. PL, phospholipids; C, cholesterol; TC, total cholesterol; CE, cholesteryl esters.

and other tissues [100–102]. SR-BI can bind a variety of ligands [23,100–102], including HDL [101,103], and mediates selective uptake both of cholesteryl ester and of other lipids [101,104,105] from HDL [101,106] and LDL [106,107] to cells as well as bidirectional movement of unesterified cholesterol [108,109]. Numerous *in vivo* studies have provided evidence that SR-BI is a physiologically important HDL receptor. It appears that the interaction of SR-BI with lipid-bound apoA-I, and possibly also apoE, controls the structure, composition, and concentration of plasma HDL [23,110–113], together with the cholesterol content of the adrenal gland, the ovaries, and the bile [112,114,115], and also protects mice from atherosclerosis [116–119]. Details of the *in vitro* and *in vivo* functions of SR-BI are provided in Chapter 7 (McPherson) and in Ref. [45].

In an attempt to understand how the interaction between SR-BI and apoA-I influences SR-BI-mediated lipid transport, we have studied the ability of WT and several mutant forms of apoA-I (incorporated into reconstituted discoidal HDL particles) to bind to SR-BI and to promote cholesterol efflux from LDL receptor-deficient CHO cells (ldlA cells) expressing the WT and two mutant forms (the Met158Arg and the Gln402Arg/Gln418Arg) of SR-BI [120]. The mutations that had the greater cholesterol efflux effects from WT SR-BI were the helix 4 (Asp102Ala/

Asp103/Ala) and the helix 6 (Arg160Val/His162Ala) apoA-I mutants: rHDL containing these mutants had 21 % and 49 % of cholesterol efflux from cells expressing the WT SR-BI [46] (Tables 11.1A,B and 11.2). When the [Met158Arg] and [Gln402Arg/Gln418Arg] mutants of SR-BI were examined, in all but one case the rHDLs bound less tightly and exhibited reduced cholesterol efflux relative to the binding and cholesterol efflux promoted by WT apoA-I rHDL from cells expressing WT SR-BI. Unexpectedly, rHDL containing the helix 6 (Arg160Val/His162Ala) mutant of apoA-I bound almost as tightly to the [Met158Arg] SR-BI mutant as it bound to the cells expressing the WT SR-BI (apparent K_ds of 7 ± 3 and 1.1 ± 0.5 μg protein mL^{-1}, respectively) and the relative cholesterol efflux rates (corrected for binding) were the same [46]. From these and other data, we propose that efficient SR-BI-mediated cholesterol efflux to lipoproteins may require not only direct binding of the lipoprotein to the receptor [109], but also the formation of a "productive complex" [46]. This is also probably the case for selective lipid uptake. A set of small-molecule inhibitors of SR-BI-mediated lipid transport, called BLTs, induced increased HDL binding affinity to SR-BI, but inhibited cholesterol efflux, thus raising the possibility that they may block lipid transport by inducing nonproductive binding [54,121]. In the context of this model, the binding of WT SR-BI to rHDL containing WT apoA-I is high-affinity (K_d 4.4 ± 1.1 μg protein mL^{-1}) and productive. In contrast, the binding of WT SR-BI to rHDL containing mutant apoA-I with the helix 4 mutations is high-affinity K_d = 3.9 ± 0.8 μg protein mL^{-1}), but is not productive. Finally, compensatory mutations in the receptor and ligand can result in high-affinity (K_d 7 ± 3 μg protein mL^{-1}) and productive binding between the helix 6 mutant of apoA-I and the Met158Arg SR-BI mutant. In this case the function of the mutant apoA-I was restored by a corresponding mutation in SR-BI. An alternative model suggested that SR-BI expression alters the cholesterol distribution on the plasma membrane and facilitates its release to the HDL with the need for direct interaction of the HDL ligand with SR-BI *in vivo* [122]. As discussed earlier, the helix 6 Arg160Ala/His162Ala mutation, which inhibits the SR-BI-mediated efflux by 51 % *in vitro* when expressed in apoA-I-deficient mice, inhibited the activation of LCAT and produced a phenotype resembling the phenotype of patients with classical LCAT deficiency (Fig. 11.4H and Table 11.2). Surprisingly, the lipid and lipoprotein phenotypes generated by adenovirus-mediated gene transfer of the helix 4 (Asp102Ala/ Asp103Ala) mutant in apoA-I-deficient mice appeared normal (Table 11.2).

11.9 Remodeling of HDL Particles after Interactions with ABCG1

ABCG1 is a member of the ABC family of half-transporters of approximately 67 kDa molecular mass, expressed in spleen, thymus, lung, and brain [123–125]. ABCG1 is localized in plasma membrane Golgi and recycling endosomes and is induced by LXR agonists or by cholesterol loading in macrophages and in liver [126–128]. Overexpression of ABCG1 in cells promoted cholesterol efflux to HDL but not to lipid-free apoA-I [24,125,128,129], while mutations in the ATP binding motif abolished the ABCG1-dependent cholesterol efflux [129]. Cholesterol efflux did not require direct binding of the acceptor to ABCG1, and it has been suggested that expression of ABCG1 alters the distribution of cholesterol on the cell membranes and allows its removal by HDL [129]. Nascent preβ HDL generated by interactions of apoA-I with ABCA1 promoted ABCG1-mediated cholesterol efflux, thus indicating sequential interaction between ABCA1 and ABCG1 in cholesterol efflux from cells expressing both transporters [130]. ABCG1$^{-/-}$ mice, when placed on high-fat diets, accumulate neutral lipids and phospholipids in hepatocytes and macrophages, whereas lipid accumulation in these cells is prevented in ABCG1 transgenic mice [131]. The combined *in vitro* and *in vivo* data suggest that ABCG1 may play an important role in lipid homeostasis and may contribute to the remodeling of HDL.

11.10 The Downstream Steps of the HDL Pathway

The alterations of HDL by various lipases (hepatic lipase, endothelial lipase) and lipid transfer proteins (CETP, PLTP) are reviewed in Chapter 8 (Ehnholm et al.), Chapter 9 (Chan et al.), Chapter 20 (Barter et al.), and Ref. [45].

11.11 Origins and Functions of HDL Subpopulations

HDL subpopulations have been classified on the bases of different fractionation procedures [132–134]. Two-dimensional electrophoretic separation, which combines agarose gel electrophoresis and nondenaturing polyacrylamide gradient gel electrophoresis [16,57,135,136], separates the HDL into preβ and αHDL subpopulations as shown in Fig. 11.3K. Preβ HDL comprises approximately 5 % of total plasma HDL. It is heterogeneous in size and contains several species of 5–6 nm in diameter [58,137]. The best characterized species are preβ1 and preβ2 [58,138–140]. The concentration of preβ1 HDL is increased in large lymph vessels [141,142] and in aortic intima [143,144] and its cholesterol can be efficiently esterified by LCAT [138].

Important questions that merit additional consideration relate to the origin and the functions of the preβ-HDL subpopulations and their relationship to the αHDL subpopulations. The existing data point to two different routes that lead to the formation of preβ-HDL particles. The first is *de novo* synthesis by the HDL biogenesis pathway (Fig. 11.2), while the second is generation of lipid-poor preβ-HDL particles from αHDL particles.

11.11.1 *De novo* Synthesis of Preβ HDL

Early studies indicated that cellular [^{3}H]cholesterol secreted into a culture medium consisting of plasma was incorporated into preβ-migrating particles, which were later converted into larger preβ particles and eventually into αHDL [145]. Other studies also showed that lipid-free apoA-I added to a culture medium of CHO cells recruits phospholipids and cholesterol, initially to form small 73 Å particles, and subsequently larger apoA-I-containing particles that have a precursor product relationship [146,147]. The LCAT-mediated cholesterol esterification of these particles led to the formation of 8.4 nm diameter particles similar in size to human plasma HDL3a[147].

Cultures of J774 macrophages and human skin fibroblasts produced discoidal particles with α electrophoretic mobility and 9 nm and 12 nm diameters, along with preβ1 particles. The 9 nm and 12 nm αHDL particles contain 2 and 3–4 apoA-I molecules per disc, respectively. These α-migrating particles appear to originate from different cellular domains and there is no precursor product relationship between them [148].

These studies are consistent with sequential lipidation of apoA-I that leads to the formation of preβ, discoidal, and spherical αHDL particles [146–149], but they did not unequivocally clarify the relationship of the different preβ to αHDL subpopulations. Preβ-HDL particles of composition similar to that of the plasma preβ-1 HDL could be also generated by interaction of lipid-free apoA-I with cholesterol-loaded macrophages [150]. As discussed earlier, liver-specific inactivation of ABCA1 indicated that in mice the liver is the major contributor to the initial ABCA1-catalyzed lipidation of apoA-I while the peripheral tissues contribute to maturation of the precursor preβ type of particles into αHDL particles through ABCA1-dependent mechanisms [44,86,89].

A recent study also indicated that a large proportion of apoA-I is secreted from HepG2-, CaCo-2-, and apoA-I-expressing CHO cells as a monomer in lipid-free form, with a Stokes radius of 2.6 nm and preα electrophoretic mobility, that is unable to bind phospholipids or to promote lipid efflux. The 2.6 nm form was converted in an ABCA1-catalyzed reaction into a 3.6 nm monomeric apoA-I form with preβ electrophoretic mobility. This preβ HDL-like particle contained a phospholipids/cholesterol/apoA-I molar ratio of 2:1:1. This particle was able to promote efflux of phospholipids and cholesterol from cells and thus increase its size [149]. Other studies have shown that some types of preβ HDL particles can be formed independently of apoA-I/ABCA1 interactions: the plasma of humans with Tangier disease [41,57] and the plasma of apoA-I-deficient mice infected with adenoviruses expressing truncated apoA-I forms contains preβ HDL but lacks αHDL particles [22] (see Fig. 11.3H). Furthermore, inhibition of ABCA1 in HepG2 cells and macrophage cultures by glyburide inhibited the formation of αHDL particles but did not affect the formation of preβ HDL particles [59]. Although there are no data on the lipid composition of the preβ particles described above, it has been shown that the HDL fraction isolated from the plasma ABCA1$^{-/-}$ mice was highly enriched in phospholipid (10-fold) and sphingomyelin (49-fold), while the plasma PLTP and LCAT activity was approximately 20 % of normal [56].

11.11.2
Generation of Preβ1 HDL from αHDL Particles

Several reactions catalyzed by hepatic lipase [151,152], CETP [153], phospholipid transfer protein [58,98,154,155] and apolipoprotein M [156] generated preβ1 HDL from αHDL species. Details are provided in Chapter 4.

The vascular bed contains increased concentrations of preβ1 HDL, suggesting that these particles are generated locally by gradual lipidation of lipid-poor apoA-I [143,144]. It has been proposed that lipid-free or lipid-poor apoA-I secreted by cells or dissociating from lipoprotein particles might enter the interstitial fluid from the plasma compartment and, through interactions with ABCA1, form preβ1 HDL. Preβ1 HDL particles formed in plasma may also enter the interstitial fluid. These particles may be enlarged further by recruitment of phospholipids and cholesterol from cell membranes and eventually become discoidal HDL particles [58] (Fig. 11.7). Discoidal HDL particles (12–14 nm diameter) containing three to four apoA-I molecules can be generated *in vitro* or in cell cultures [58,157]. When the lymphatic HDL enters the plasma compartment, LCAT activity converts discoidal

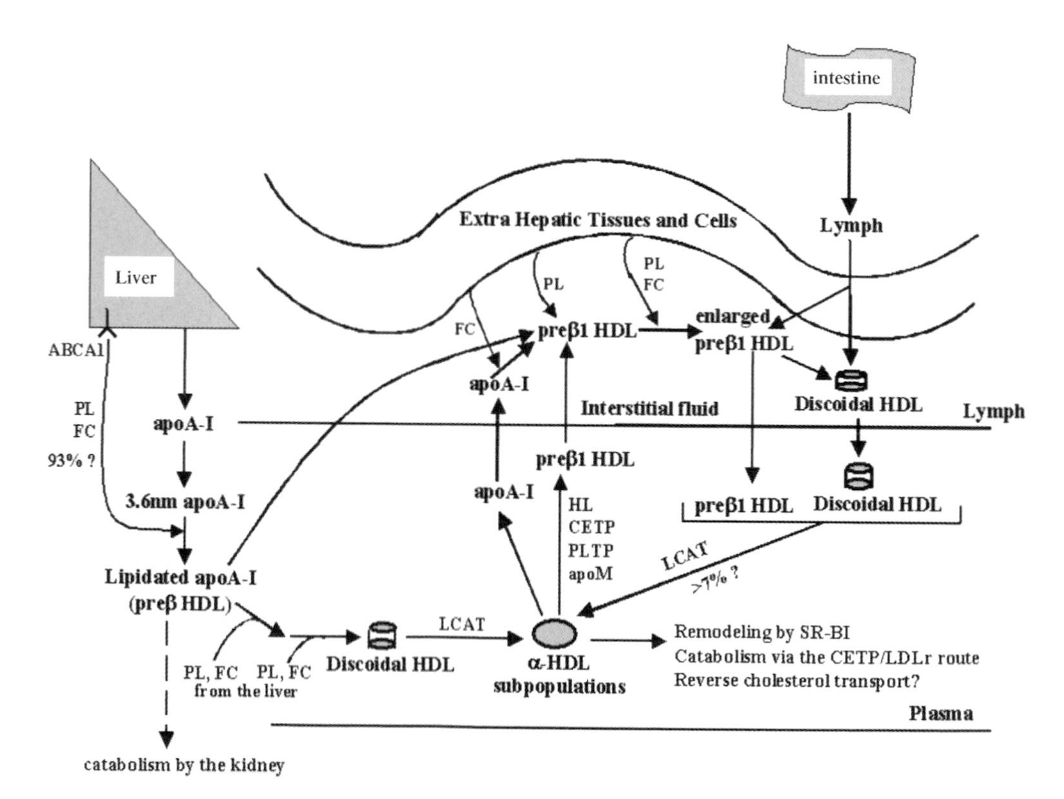

Fig. 11.7 Schematic representation of the contribution of the liver and the peripheral tissues and cells in the biogenesis, maturation, and remodeling of HDL.

HDL into spherical particles. In addition, LCAT esterifies the cholesterol of preβ1 HDL and contributes to its gradual conversion into spherical HDL [138] (Fig. 11.7).

11.12 Importance of HDL Subpopulations

Structural alterations in apoA-I may affect not only its initial interactions with ABCA1, which are essential for the biogenesis of αHDL particles, but also subsequent interactions with other proteins of the HDL pathway that contribute to the maturation and remodeling of HDL (Figures 11.2 and 11.6). Some of these mutations have a profound effect on the concentrations of the preβ and αHDL subpopulations. Accumulation of preβ1 HDL particles may indicate an early blockage of the HDL pathway due to defective apoA-I/ABCA1 interactions (Fig. 11.3H). Alternatively, it may indicate inability of preβ1 particles to be further lipidated and to form discoidal and spherical HDL particles. This pattern was detected in apoA-I$^{-/-}$ mice expressing the helix 3 apoA-I[Δ(89–99)] deletion mutant (Fig. 11.5I). The accumulation of α4 HDL indicates inability of the αHDL particles to interact with LCAT (Fig. 11.4H). Quantitative two-dimensional gel electrophoresis, which can determine the apoA-I contents of the HDL subpopulations, has been used to assess the relative concentrations of the subpopulations of HDL in normal and pathological states, as well as after different pharmacological treatments of human patients [158–163]. Thus, treatment with statins or CETP inhibitors increased the α1 subpopulations [159,161], while preβ1 subpopulations increased and α3 subpopulations decreased in HIV-infected patients after treatment with protease inhibitor-based antiretroviral therapy [163]. In patients with coronary heart disease, the α1 and preα1 and α2 and preα2 levels are decreased and the α3 and preβ1 levels are increased. Decreased α1 and preα1 levels and increased α3 levels were better predictors of new cardiovascular events than HDL cholesterol [162]. Absence of preβ1 HDL occurs in mice with apoM deficiency and this condition predisposes to atherogenesis [156], thus suggesting a beneficial effect of preβ HDL particles.

It has been estimated that preβ1 HDL accounts for 40 % of cholesterol effluxed from cultured cells into the plasma [60] and thus may also contribute to the reverse transport of cholesterol from the periphery to the liver.

11.13 Role of ApoA-I and HDL in Atheroprotection

HDL and apoA-I have been reported to have antioxidant and antiinflammatory properties, can alter prostacyclin levels and platelet function, and may modulate NO release after interaction of HDL with SR-BI [164–170]. All these properties may contribute to the atheroprotective functions of HDL [171], which have been supported by several studies in animal models. Overexpression of apoA-I in different mouse [172,173] or rabbit [174] models protected the animals from atherosclerosis. Unexpectedly, apoA-I$^{-/-}$ mice without other defects do not develop atherosclerosis on normal or atherogenic diets [175]. However, when apoA-I-deficient mice are crossed with apoB transgenic mice they develop more atherosclerosis than the apoB transgenic mice [176]. The atheroprotective functions of apoA-I have also been supported by apoA-I gene transfer studies in apoE$^{-/-}$ or LDLr$^{-/-}$ mice [177–179]. In addition, intravenous administration of proteoliposomes containing apoA-I Milano to apoE$^{-/-}$ mice prevented progression of atherosclerosis [180,181]. Finally, intravenous administration of apoA-I Milano/phospholipid complexes (15 mg kg^{-1}) in five doses at weekly intervals in patients with acute coronary syndrome resulted in significant regression of coronary atherosclerosis, as determined by intravascular ultrasound [182]. Overall, the existing data are consistent with the atheroprotective functions of apoA-I and HDL.

Nevertheless, high levels of HDL cholesterol are not synonymous with atheroprotection; thus human subjects with high HDL levels and documented coronary heart disease attributed to the atherogenic properties of HDL have been identified [183]. Similarly, the increased HDL levels of the SR-BI-deficient mice did not protect, but rather promoted atherosclerosis [116,119], suggesting that the atheroprotective functions of HDL can be influenced by interactions with other proteins of the HDL pathway.

11.14 Conclusions

This review examines the functions of apoA-I in a pathway that leads to the synthesis, remodeling, and catabolism of HDL. Using adenovirus-mediated gene transfer of apoA-I mutants in apoA-I-deficient mice, we were able to identify discrete steps of the pathway where the concentrations of the intermediate accumulate in plasma and formation of mature αHDL is inhibited or impaired. The phenotypes so far identified can be classified into five broad categories:

1. *Deletions in the carboxy-terminal segment of apoA-I that inhibit apoA-I–ABCA1 interactions.* These mutations are characterized by very low HDL levels, formation of preβ HDL, and the complete absence of αHDL.
2. *Mutations that inhibit the activation of LCAT.* These mutations are characterized by low HDL levels, the accumulation of discoidal HDL particles that float in the HDL3 region, and the preponderance of α_4 HDL particles. This phenotype can be corrected *in vivo* by gene transfer of human LCAT.
3. *Mutations that cause hypertriglyceridemia.* These mutations are characterized by the presence of apoA-I in the VLDL/IDL region and can be corrected by gene transfer of human LPL.

4. *Deletion of the Δ(89–99) segment.* This caused dyslipidemia characterized by high cholesterol and normal triglycerides, inhibition of plasma PLTP activity, accumulation of discoidal HDL, and increase in the preβ1/αHDL ratio.
5. *Mutations that cause low HDL levels and increased α4 HDL subpopulations.* These mutations may be the result either of increased intracellular instability of the apoA-I or of rapid catabolism of the secreted preβ particles.

Additional phenotypes that affect the interactions of lipid-bound apoA-I with SR-BI, hepatic lipase, CETP, PLTP, and apoM may emerge in the future. These mutations may affect the structure of αHDL or the formation of preβ HDL particles. New insights into the effects of apoA-I mutations on the biogenesis and maturation of HDL may also emerge in the future, through accurate compositional analysis of individual preβ and αHDL subpopulations of normal and abnormal HDL particles.

Overall, these studies provide new diagnostic parameters for abnormalities in HDL. These parameters include the distribution of apoA-I in HDL3 or in the VLDL/IDL region, the abnormal ratios of preβ/αHDL, and the low cholesteryl ester/total cholesterol ratio of HDL.

The phenotypes generated by targeted mutagenesis of apoA-I should be valuable for the detection of similar phenotypes in humans and should serve as diagnostic and prognostic markers of dyslipidemia and/or atherosclerosis.

11.15 Acknowledgments

This work was supported by grants from the National Institutes of Health (HL48739 and HL67294) and from the General Secretariat for Research and technology of Greece (programs PENED 2003 and the program "RTD Collaboration between Greece and USA") and the 6th framework program of the European Union (LSHM-C1-2006-037631). We thank Anne Plunkett for preparing the manuscript.

References

1 Zannis, V. I., Cole, F. S., Jackson, C. L., Kurnit, D. M., Karathanasis, S. K. (1985) Distribution of apolipoprotein A-I, C-II, C-III, and E mRNA in fetal human tissues. Time-dependent induction of apolipoprotein E mRNA by cultures of human monocyte-macrophages. *Biochemistry*, **24**, 4450–4455.

2 Matsunaga, T., Hiasa, Y., Yanagi, H., Maeda, T., Hattori, N., Yamakawa, K., Yamanouchi, Y., Tanaka, I., Obara, T., Hamaguchi, H. (1991) Apolipoprotein A-I deficiency due to a codon 84 nonsense mutation of the apolipoprotein A-I gene. *Proc. Natl. Acad. Sci. U.S.A*, **88**, 2793–2797.

3 Williamson, R., Lee, D., Hagaman, J., Maeda, N. (1992) Marked reduction of high density lipoprotein cholesterol in mice genetically modified to lack apolipoprotein A-I.

Proc. Natl. Acad. Sci. U.S.A, **89**, 7134–7138.

4 Li, W. H., Tanimura, M., Luo, C. C., Datta, S., Chan, L. (1988) The apolipoprotein multigene family: biosynthesis, structure, structure-function relationships, and evolution. *J. Lipid Res.*, **29**, 245–271.

5 Nolte, R. T. and Atkinson, D. (1992) Conformational analysis of apolipoprotein A-I and E-3 based on primary sequence and circular dichroism. *Biophys. J.*, **63**, 1221–1239.

6 Borhani, D. W., Rogers, D. P., Engler, J. A., Brouillette, C. G. (1997) Crystal structure of truncated human apolipoprotein A-I suggests a lipid-bound conformation. *Proc. Natl. Acad. Sci. U.S.A*, **94**, 12291–12296.

7 Borhani, D. W., Engler, J. A., Brouillette, C. G. (1999) Crystallization of truncated human apolipoprotein A-I in a novel conformation. *Acta Crystallogr. D. Biol. Crystallogr.*, **55** (Pt 9), 1578–1583.

8 Ajees, A. A., Anantharamaiah, G. M., Mishra, V. K., Hussain, M. M., Murthy, H. M. (2006) Crystal structure of human apolipoprotein A-I: insights into its protective effect against cardiovascular diseases. *Proc. Natl. Acad. Sci. U.S.A*, **103**, 2126–2131.

9 Segrest, J. P., Jones, M. K., Klon, A. E., Sheldahl, C. J., Hellinger, M., DeLoof, H., Harvey, S. C. (1999) A detailed molecular belt model for apolipoprotein A-I in discoidal high density lipoprotein. *J. Biol. Chem.*, **274**, 31755–31758.

10 Segrest, J. P. (1977) Amphipathic helixes and plasma lipoproteins: thermodynamic and geometric considerations. *Chem. Phys. Lipids*, **18**, 7–22.

11 Segrest, J. P., Li, L., Anantharamaiah, G. M., Harvey, S. C., Liadaki, K. N., Zannis, V. (2000) Structure and function of apolipoprotein A-I and high-density lipoprotein. *Curr. Opin. Lipidol.*, **11**, 105–115.

12 Panagotopulos, S. E., Horace, E. M., Maiorano, J. N., Davidson, W. S. (2001) Apolipoprotein A-I adopts a belt-like orientation in reconstituted high density lipoproteins. *J. Biol. Chem.*, **276**, 42965–42970.

13 Marcel, Y. L. and Kiss, R. S. (2003) Structure-function relationships of apolipoprotein A-I: a flexible protein with dynamic lipid associations. *Curr. Opin. Lipidol.*, **14**, 151–157.

14 Martin, D. D., Budamagunta, M. S., Ryan, R. O., Voss, J. C., Oda, M. N., (2006.) Apolipoprotein A-I assumes a looped belt conformation on reconstituted high density lipoprotein. *J. Biol. Chem.*,

15 Chroni, A., Kan, H. Y., Shkodrani, A., Liu, T., Zannis, V. I. (2005) Deletions of Helices 2 and 3 of Human ApoA-I Are Associated with Severe Dyslipidemia following Adenovirus-Mediated Gene Transfer in ApoA-I-Deficient Mice. *Biochemistry*, **44**, 4108–4117.

16 Chroni, A., Duka, A., Kan, H. Y., Liu, T., Zannis, V. I. (2005) Point mutations in apolipoprotein A-I mimic the phenotype observed in patients with classical lecithin:cholesterol acyltransferase deficiency. *Biochemistry*, **44**, 14353–14366.

17 Chroni, A., Liu, T., Gorshkova, I., Kan, H. Y., Uehara, Y., vonEckardstein, A., Zannis, V. I. (2003) The central helices of apoA-I can promote ATP-binding cassette transporter A1 (ABCA1)-mediated lipid efflux. Amino acid residues 220-231of the wild-type apoA-I are required for lipid efflux in vitro and high density lipoprotein formation *in vivo*. *J. Biol. Chem.*, **278**, 6719–6730.

18 Reardon, C. A., Kan, H. Y., Cabana, V., Blachowicz, L., Lukens, J. R., Wu, Q., Liadaki, K., Getz, G. S., Zannis, V. I. (2001) In vivo studies of HDL assembly and metabolism using adenovirus-mediated transfer of ApoA-I mutants in ApoA-I-deficient mice. *Biochemistry*, **40**, 13670–13680.

19 Matz, C. E. and Jonas, A. (1982) Micellar complexes of human apolipoprotein A-I with

phosphatidylcholines and cholesterol prepared from cholate-lipid dispersions. *J. Biol. Chem.*, **257**, 4535–4540.

20 Catte, A., Patterson, J. C., Jones, M. K., Jerome, W. G., Bashtovyy, D., Su, Z., Gu, F., Chen, J., Aliste, M. P., Harvey, S. C., Li, L., Weinstein, G., Segrest, J. P. (2006) Novel changes in discoidal high density lipoprotein morphology: a molecular dynamics study. *Biophys. J.*, **90**, 4345–4360.

21 Beckstead, J. A., Block, B. L., Bielicki, J. K., Kay, C. M., Oda, M. N., Ryan, R. O. (2005) Combined N- and C-terminal truncation of human apolipoprotein A-I yields a folded, functional central domain. *Biochemistry*, **44**, 4591–4599.

22 Zannis, V. I., Chroni, A., Kypreos, K. E., Kan, H. Y., Cesar, T. B., Zanni, E. E., Kardassis, D. (2004) Probing the pathways of chylomicron and HDL metabolism using adenovirus-mediated gene transfer. *Curr Opin Lipidol.*, **15**, 151–166.

23 Krieger, M. (2001) Scavenger receptor class B type I is a multiligand HDL receptor that influences diverse physiologic systems. *J. Clin. Invest*, **108**, 793–797.

24 Wang, N., Lan, D., Chen, W., Matsuura, F., Tall, A. R. (2004) ATP-binding cassette transporters G1 and G4 mediate cellular cholesterol efflux to high-density lipoproteins. *Proc. Natl. Acad. Sci. U.S.A*, **101**, 9774–9779.

25 Langmann, T., Klucken, J., Reil, M., Liebisch, G., Luciani, M. F., Chimini, G., Kaminski, W. E., Schmitz, G. (1999) Molecular cloning of the human ATP-binding cassette transporter 1 (hABC1): evidence for sterol-dependent regulation in macrophages. *Biochem. Biophys. Res. Commun.*, **257**, 29–33.

26 Kielar, D., Dietmaier, W., Langmann, T., Aslanidis, C., Probst, M., Naruszewicz, M., Schmitz, G. (2001) Rapid quantification of human ABCA1 mRNA in various cell types and tissues by real-time reverse transcription-PCR. *Clin. Chem.*, **47**, 2089–2097.

27 Neufeld, E. B., Demosky, S. J., Jr., Stonik, J. A., Combs, C., Remaley, A. T., Duverger, N., Santamarina-Fojo, S., Brewer, H. B., Jr. (2002) The ABCA1 transporter functions on the basolateral surface of hepatocytes. *Biochem. Biophys. Res. Commun.*, **297**, 974–979.

28 Wang, N., Silver, D. L., Costet, P., Tall, A. R. (2000) Specific binding of ApoA-I, enhanced cholesterol efflux, and altered plasma membrane morphology in cells expressing ABC1. *J. Biol. Chem.*, **275**, 33053–33058.

29 Remaley, A. T., Stonik, J. A., Demosky, S. J., Neufeld, E. B., Bocharov, A. V., Vishnyakova, T. G., Eggerman, T. L., Patterson, A. P., Duverger, N. J., Santamarina-Fojo, S., Brewer, H. B., Jr. (2001) Apolipoprotein specificity for lipid efflux by the human ABCAI transporter. *Biochem. Biophys. Res. Commun.*, **280**, 818–823.

30 Schmitz, G., Assmann, G., Robenek, H., Brennhausen, B. (1985) Tangier disease: a disorder of intracellular membrane traffic. *Proc. Natl. Acad. Sci. U.S.A*, **82**, 6305–6309.

31 Schmitz, G., Robenek, H., Lohmann, U., Assmann, G. (1985) Interaction of high density lipoproteins with cholesteryl ester-laden macrophages: biochemical and morphological characterization of cell surface receptor binding, endocytosis and resecretion of high density lipoproteins by macrophages. *EMBO J.*, **4**, 613–622.

32 Fielding, P. E., Nagao, K., Hakamata, H., Chimini, G., Fielding, C. J. (2000) A two-step mechanism for free cholesterol and phospholipid efflux from human vascular cells to apolipoprotein A-1. *Biochemistry*, **39**, 14113–14120.

33 Arakawa, R., Abe-Dohmae, S., Asai, M., Ito, J. I., Yokoyama, S. (2000) Involvement of caveolin-1 in cholesterol enrichment of high density lipoprotein during its assembly by apolipoprotein and THP-1 cells. *J. Lipid Res.*, **41**, 1952–1962.

34 Wang, N., Silver, D. L., Thiele, C., Tall, A. R. (2001) ATP-binding cassette transporter A1 (ABCA1) functions as a cholesterol efflux regulatory protein. *J. Biol. Chem.*, **276**, 23742–23747.

35 Francis, G. A., Knopp, R. H., Oram, J. F. (1995) Defective removal of cellular cholesterol and phospholipids by apolipoprotein A-I in Tangier Disease. *J. Clin. Invest*, **96**, 78–87.

36 Bodzioch, M., Orso, E., Klucken, J., Langmann, T., Bottcher, A., Diederich, W., Drobnik, W., Barlage, S., Buchler, C., Porsch-Ozcurumez, M., Kaminski, W. E., Hahmann, H. W., Oette, K., Rothe, G., Aslanidis, C., Lackner, K. J., Schmitz, G. (1999) The gene encoding ATP-binding cassette transporter 1 is mutated in Tangier disease. *Nat. Genet.*, **22**, 347–351.

37 Rust, S., Rosier, M., Funke, H., Real, J., Amoura, Z., Piette, J. C., Deleuze, J. F., Brewer, H. B., Duverger, N., Denefle, P., Assmann, G. (1999) Tangier disease is caused by mutations in the gene encoding ATP-binding cassette transporter 1. *Nat. Genet.*, **22**, 352–355.

38 Brooks-Wilson, A., Marcil, M., Clee, S. M., Zhang, L. H., Roomp, K., vanDam, M., Yu, L., Brewer, C., Collins, J. A., Molhuizen, H. O., Loubser, O., Ouelette, B. F., Fichter, K., Ashbourne-Excoffon, K. J., Sensen, C. W., Scherer, S., Mott, S., Denis, M., Martindale, D., Frohlich, J., Morgan, K., Koop, B., Pimstone, S., Kastelein, J. J., Hayden, M. R. (1999) Mutations in ABC1 in Tangier disease and familial high-density lipoprotein deficiency. *Nat. Genet.*, **22**, 336–345.

39 Lawn, R. M., Wade, D. P., Garvin, M. R., Wang, X., Schwartz, K., Porter, J. G., Seilhamer, J. J., Vaughan, A. M., Oram, J. F. (1999) The Tangier disease gene product ABC1 controls the cellular apolipoprotein-mediated lipid removal pathway. *J. Clin. Invest*, **104**, R25–R31.

40 Orso, E., Broccardo, C., Kaminski, W. E., Bottcher, A., Liebisch, G., Drobnik, W., Gotz, A., Chambenoit, O., Diederich, W., Langmann, T., Spruss, T., Luciani, M. F., Rothe, G., Lackner, K. J., Chimini, G., Schmitz, G. (2000) Transport of lipids from golgi to plasma membrane is defective in tangier disease patients and Abc1-deficient mice. *Nat. Genet.*, **24**, 192–196.

41 Assmann, G., von Eckardstein, A., Brewer, H. B., (2001) Familial analphalipoproteinemia: Tangier disease. In: The Metabolic and Molecular Bases of inherited Disease, Scrives, C.R., Beaudet, A.L., Sly, W.S., Valle, D. (eds). McGraw Hill, New York, 8th edition, 2937–2960.

42 McNeish, J., Aiello, R. J., Guyot, D., Turi, T., Gabel, C., Aldinger, C., Hoppe, K. L., Roach, M. L., Royer, L. J., deWet, J., Broccardo, C., Chimini, G., Francone, O. L. (2000) High density lipoprotein deficiency and foam cell accumulation in mice with targeted disruption of ATP-binding cassette transporter-1. *Proc. Natl. Acad. Sci. U.S.A*, **97**, 4245–4250.

43 Christiansen-Weber, T. A., Voland, J. R., Wu, Y., Ngo, K., Roland, B. L., Nguyen, S., Peterson, P. A., Fung-Leung, W. P. (2000) Functional loss of ABCA1 in mice causes severe placental malformation, aberrant lipid distribution, and kidney glomerulonephritis as well as high-density lipoprotein cholesterol deficiency. *Am. J. Pathol.*, **157**, 1017–1029.

44 Timmins, J. M., Lee, J. Y., Boudyguina, E., Kluckman, K. D., Brunham, L. R., Mulya, A., Gebre, A. K., Coutinho, J. M., Colvin, P. L., Smith, T. L., Hayden, M. R., Maeda, N., Parks, J. S. (2005) Targeted inactivation of hepatic Abca1 causes profound hypoalphalipoproteinemia and kidney hypercatabolism of apoA-I. *J. Clin. Invest*, **115**, 1333–1342.

45 Zannis, V. I., Chroni, A., Krieger, M. (2006) Role of apoA-I, ABCA1, LCAT, and SR-BI in the biogenesis of HDL. *J. Mol. Med.*, **84**, 276–294.

46 Liu, T., Krieger, M., Kan, H. Y., Zannis, V. I. (2002) The effects of

mutations in helices 4 and 6 of apoA-I on scavenger receptor class B type I (SR-BI)-mediated cholesterol efflux suggest that formation of a productive complex between reconstituted high density lipoprotein and SR-BI is required for efficient lipid transport. *J. Biol. Chem.*, **277**, 21576–21584.

47 Chroni, A., Kan, H. Y., Kypreos, K. E., Gorshkova, I. N., Shkodrani, A., Zannis, V. I. (2004) Substitutions of glutamate 110 and 111 in the middle helix 4 of human apolipoprotein A-I (apoA-I) by alanine affect the structure and in vitro functions of apoA-I and induce severe hypertriglyceridemia in apoA-I-deficient mice. *Biochemistry*, **43**, 10442–10457.

48 Chroni, A., Liu, T., Fitzgerald, M. L., Freeman, M. W., Zannis, V. I. (2004) Cross-linking and lipid efflux properties of apoA-I mutants suggest direct association between apoA-I helices and AB CA1. *Biochemistry*, **43**, 2126–2139.

49 Gorshkova, I. N., Liu, T., Zannis, V. I., Atkinson, D. (2002) Lipid-free structure and stability of apolipoprotein A-I: probing the central region by mutation. *Biochemistry*, **41**, 10529–10539.

50 Gorshkova, I. N., Liu, T., Kan, H. Y., Chroni, A., Zannis, V. I., Atkinson, D. (2006) Structure and stability of apolipoprotein a-I in solution and in discoidal high-density lipoprotein probed by double charge ablation and deletion mutation. *Biochemistry*, **45**, 1242–1254.

51 Remaley, A. T., Thomas, F., Stonik, J. A., Demosky, S. J., Bark, S. E., Neufeld, E. B., Bocharov, A. V., Vishnyakova, T. G., Patterson, A. P., Eggerman, T. L., Santamarina-Fojo, S., Brewer, H. B. (2003) Synthetic amphipathic helical peptide mediated efflux of lipid from cells by an ABCA1-dependent and an ABCA1-independent pathway. *J. Lipid Res.*, **44**, 828–836.

52 Fitzgerald, M. L., Morris, A. L., Chroni, A., Mendez, A. J., Zannis, V. I., Freeman, M. W. (2004) ABCA1 and amphipathic apolipoproteins form high-affinity molecular complexes required for cholesterol efflux. *J. Lipid Res.*, **45**, 287–294.

53 Fitzgerald, M. L., Morris, A. L., Rhee, J. S., Andersson, L. P., Mendez, A. J., Freeman, M. W. (2002) Naturally occurring mutations in the largest extracellular loops of ABCA1 can disrupt its direct interaction with apolipoprotein A-I. *J. Biol. Chem.*, **277**, 33178–33187.

54 Nieland, T. J., Chroni, A., Fitzgerald, M. L., Maliga, Z., Zannis, V. I., Kirchhausen, T., Krieger, M. (2004) Cross-inhibition of SR-BI- and ABCA1-mediated cholesterol transport by the small molecules BLT-4 and glyburide. *J. Lipid Res.*, **45**, 1256–1265.

55 Chambenoit, O., Hamon, Y., Marguet, D., Rigneault, H., Rosseneu, M., Chimini, G. (2001) Specific docking of apolipoprotein A-I at the cell surface requires a functional ABCA1 transporter. *J. Biol. Chem.*, **276**, 9955–9960.

56 Francone, O. L., Subbaiah, P. V., vanTol, A., Royer, L., Haghpassand, M. (2003) Abnormal phospholipid composition impairs HDL biogenesis and maturation in mice lacking Abca1. *Biochemistry*, **42**, 8569–8578.

57 Asztalos, B. F., Brousseau, M. E., McNamara, J. R., Horvath, K. V., Roheim, P. S., Schaefer, E. J. (2001) Subpopulations of high density lipoproteins in homozygous and heterozygous Tangier disease. *Atherosclerosis*, **156**, 217–225.

58 Fielding, C. J., Fielding, P. E. (1995) Molecular physiology of reverse cholesterol transport. *J. Lipid Res.*, **36**, 211–228.

59 Krimbou, L., Hajj, H. H., Blain, S., Rashid, S., Denis, M., Marcil, M., Genest, J. (2005) Biogenesis and speciation of nascent apoA-I-containing particles in various cell lines. *J. Lipid Res.*, **46**, 1668–1677.

60 Fielding, C. J. and Fielding, P. E. (2001) Cellular cholesterol efflux.

Biochim. Biophys. Acta, **1533**, 175–189.

61 Neufeld, E. B., Remaley, A. T., Demosky, S. J., Stonik, J. A., Cooney, A. M., Comly, M., Dwyer, N. K., Zhang, M., Blanchette-Mackie, J., Santamarina-Fojo, S., Brewer, H. B., Jr. (2001) Cellular localization and trafficking of the human ABCA1 transporter. *J. Biol. Chem.*, **276**, 27584–27590.

62 Takahashi, Y., Smith, J. D. (1999) Cholesterol efflux to apolipoprotein AI involves endocytosis and resecretion in a calcium-dependent pathway. *Proc. Natl. Acad. Sci. U.S.A*, **96**, 11358–11363.

63 Smith, J. D., Waelde, C., Horwitz, A., Zheng, P. (2002) Evaluation of the role of phosphatidylserine translocase activity in ABCA1-mediated lipid efflux. *J. Biol. Chem.*, **277**, 17797–17803.

64 Rohrer, L., Cavelier, C., Fuchs, S., Schluter, M. A., Volker, W., vonEckardstein, A. (2006) Binding, internalization and transport of apolipoprotein A-I by vascular endothelial cells. *Biochim. Biophys. Acta*, **1761**, 186–194.

65 Neufeld, E. B., Stonik, J. A., Demosky, S. J., Jr., Knapper, C. L., Combs, C. A., Cooney, A., Comly, M., Dwyer, N., Blanchette-Mackie, J., Remaley, A. T., Santamarina-Fojo, S., Brewer, H. B., Jr. (2004) The ABCA1 transporter modulates late endocytic trafficking: insights from the correction of the genetic defect in Tangier disease. *J. Biol. Chem.*, **279**, 15571–15578.

66 Martinez, L. O., Jacquet, S., Esteve, J. P., Rolland, C., Cabezon, E., Champagne, E., Pineau, T., Georgeaud, V., Walker, J. E., Terce, F., Collet, X., Perret, B., Barbaras, R. (2003) Ectopic beta-chain of ATP synthase is an apolipoprotein A-I receptor in hepatic HDL endocytosis. *Nature*, **421**, 75–79.

67 Jacquet, S., Malaval, C., Martinez, L. O., Sak, K., Rolland, C., Perez, C., Nauze, M., Champagne, E., Terce, F., Gachet, C., Perret, B., Collet, X., Boeynaems, J. M., Barbaras, R. (2005) The nucleotide receptor P2Y13 is a key regulator of hepatic high-density lipoprotein (HDL) endocytosis. *Cell Mol. Life Sci.*, **62**, 2508–2515.

68 vonEckardstein, A., Nofer, J. R., Assmann, G. (2001) High density lipoproteins and arteriosclerosis. Role of cholesterol efflux and reverse cholesterol transport. *Arterioscler. Thromb. Vasc. Biol.*, **21**, 13–27.

69 Glomset, J. A. (1968) The plasma lecithins:cholesterol acyltransferase reaction. *J. Lipid Res.*, **9**, 155–167.

70 Bielicki, J. K., Johnson, W. J., Weinberg, R. B., Glick, J. M., Rothblat, G. H. (1992) Efflux of lipid from fibroblasts to apolipoproteins: dependence on elevated levels of cellular unesterified cholesterol. *J. Lipid Res.*, **33**, 1699–1709.

71 Eisenberg, S. (1984) High density lipoprotein metabolism. *J. Lipid Res.*, **25**, 1017–1058.

72 Rothblat, G. H., Mahlberg, F. H., Johnson, W. J., Phillips, M. C. (1992) Apolipoproteins, membrane cholesterol domains, and the regulation of cholesterol efflux. *J. Lipid Res.*, **33**, 1091–1097.

73 Haynes, M. P., Phillips, M. C., Rothblat, G. H. (2000) Efflux of cholesterol from different cellular pools. *Biochemistry*, **39**, 4508–4517.

74 Fielding, C. J., Fielding, P. E. (1981) Evidence for a lipoprotein carrier in human plasma catalyzing sterol efflux from cultured fibroblasts and its relationship to lecithin:cholesterol acyltransferase. *Proc. Natl. Acad. Sci. U.S.A*, **78**, 3911–3914.

75 Gillotte, K. L., Davidson, W. S., Lund-Katz, S., Rothblat, G. H., Phillips, M. C. (1998) Removal of cellular cholesterol by pre-beta-HDL involves plasma membrane microsolubilization. *J. Lipid Res.*, **39**, 1918–1928.

76 Gillotte, K. L., Zaiou, M., Lund-Katz, S., Anantharamaiah, G. M., Holvoet, P., Dhoest, A., Palgunachari, M. N., Segrest, J. P., Weisgraber, K. H., Rothblat, G. H., Phillips, M. C.

(1999) Apolipoprotein-mediated plasma membrane microsolubilization. Role of lipid affinity and membrane penetration in the efflux of cellular cholesterol and phospholipid. *J. Biol. Chem.*, **274**, 2021–2028.

77 Groen, A. K., Bloks, V. W., Bandsma, R. H., Ottenhoff, R., Chimini, G., Kuipers, F. (2001) Hepatobiliary cholesterol transport is not impaired in Abca1-null mice lacking HDL. *J. Clin. Invest*, **108**, 843–850.

78 Alam, K., Meidell, R. S., Spady, D. K. (2001) Effect of up-regulating individual steps in the reverse cholesterol transport pathway on reverse cholesterol transport in normolipidemic mice. *J. Biol. Chem.*, **276**, 15641–15649.

79 Osono, Y., Woollett, L. A., Marotti, K. R., Melchior, G. W., Dietschy, J. M. (1996) Centripetal cholesterol flux from extrahepatic organs to the liver is independent of the concentration of high density lipoprotein-cholesterol in plasma. *Proc. Natl. Acad. Sci. U.S.A*, **93**, 4114–4119.

80 Jolley, C. D., Woollett, L. A., Turley, S. D., Dietschy, J. M. (1998) Centripetal cholesterol flux to the liver is dictated by events in the peripheral organs and not by the plasma high density lipoprotein or apolipoprotein A-I concentration. *J. Lipid Res.*, **39**, 2143–2149.

81 Mucksavage, P., Stoudt, G., Webb, N., Duverger, N., Secreto, A., deBeer, F. C., Glick, J. M., Rothblat, G. H., Rader, D. J. (2001) Co-expression of LCAT and SR-BI markedly increases fecal sterol excretion as a marker of reverse cholesterol transport. *Arterioscler. Thromb. Vasc. Biol.*, **21**, 652.

82 Zhang, Y., Zanotti, I., Reilly, M. P., Glick, J. M., Rothblat, G. H., Rader, D. J. (2003) Overexpression of apolipoprotein A-I promotes reverse transport of cholesterol from macrophages to feces in vivo. *Circulation*, **108**, 661–663.

83 Moore, R. E., Navab, M., Millar, J. S., Zimetti, F., Hama, S., Rothblat, G. H., Rader, D. J. (2005) Increased atherosclerosis in mice lacking apolipoprotein A-I attributable to both impaired reverse cholesterol transport and increased inflammation. *Circ. Res.*, **97**, 763–771.

84 Zhang, Y., Da Silva, J. R., Reilly, M., Billheimer, J. T., Rothblat, G. H., Rader, D. J. (2005) Hepatic expression of scavenger receptor class B type I (SR-BI) is a positive regulator of macrophage reverse cholesterol transport in vivo. *J. Clin. Invest*, **115**, 2870–2874.

85 Naik, S. U., Wang, X., Da Silva, J. S., Jaye, M., Macphee, C. H., Reilly, M. P., Billheimer, J. T., Rothblat, G. H., Rader, D. J. (2006) Pharmacological activation of liver X receptors promotes reverse cholesterol transport in vivo. *Circulation*, **113**, 90–97.

86 Singaraja, R. R., Stahmer, B., Brundert, M., Merkel, M., Heeren, J., Bissada, N., Kang, M., Timmins, J. M., Ramakrishnan, R., Parks, J. S., Hayden, M. R., Rinninger, F. (2006) Hepatic ATP-binding cassette transporter A1 is a key molecule in high-density lipoprotein cholesteryl ester metabolism in mice. *Arterioscler. Thromb. Vasc. Biol.*, **26**, 1821–1827.

87 Kozyraki, R., Fyfe, J., Kristiansen, M., Gerdes, C., Jacobsen, C., Cui, S., Christensen, E. I., Aminoff, M., dela, C. A., Krahe, R., Verroust, P. J., Moestrup, S. K. (1999) The intrinsic factor-vitamin B12 receptor, cubilin, is a high-affinity apolipoprotein A-I receptor facilitating endocytosis of high-density lipoprotein. *Nat. Med.*, **5**, 656–661.

88 Hammad, S. M., Stefansson, S., Twal, W. O., Drake, C. J., Fleming, P., Remaley, A., Brewer, H. B., Jr., Argraves, W. S. (1999) Cubilin, the endocytic receptor for intrinsic factor-vitamin B(12) complex, mediates high-density lipoprotein holoparticle endocytosis. *Proc. Natl. Acad. Sci. U.S.A*, **96**, 10158–10163.

89 Singaraja, R. R., VanEck, M., Bissada, N., Zimetti, F., Collins, H. L.,

Hildebrand, R. B., Hayden, A., Brunham, L. R., Kang, M. H., Fruchart, J. C., vanBerkel, T. J., Parks, J. S., Staels, B., Rothblat, G. H., Fievet, C., Hayden, M. R. (2006) Both Hepatic and Extrahepatic ABCA1 Have Discrete and Essential Functions in the Maintenance of Plasma High-Density Lipoprotein Cholesterol Levels In Vivo. *Circulation*, **114**, 1301–1309.

90 Santamarina-Fojo, S., Hoeg, J. M., Assmann, G., Brewer, H. B., Jr. (2001) Lecithin cholesterol acyltransferase deficiency and fish eye disease. In: The Metabolic and Molecular Bases of inherited Disease, Scrives, C.R., Beaudet, A.L., Sly, W.S., Valle, D. (eds). McGraw Hill, New York, 8th edition, 2817–2834.

91 McManus, D. C., Scott, B. R., Frank, P. G., Franklin, V., Schultz, J. R., Marcel, Y. L. (2000) Distinct central amphipathic alpha-helices in apolipoprotein A-I contribute to the in vivo maturation of high density lipoprotein by either activating lecithin-cholesterol acyltransferase or binding lipids. *J. Biol. Chem.*, **275**, 5043–5051.

92 Scott, B. R., McManus, D. C., Franklin, V., McKenzie, A. G., Neville, T., Sparks, D. L., Marcel, Y. L. (2001) The N-terminal globular domain and the first class A amphipathic helix of apolipoprotein A-I are important for lecithin:cholesterol acyltransferase activation and the maturation of high density lipoprotein in vivo. *J. Biol. Chem.*, **276**, 48716–48724.

93 Sorci-Thomas, M. G., Thomas, M., Curtiss, L., Landrum, M. (2000) Single repeat deletion in ApoA-I blocks cholesterol esterification and results in rapid catabolism of delta6 and wild-type ApoA-I in transgenic mice. *J. Biol. Chem.*, **275**, 12156–12163.

94 Huuskonen, J., Olkkonen, V. M., Jauhiainen, M., Ehnholm, C. (2001) The impact of phospholipid transfer protein (PLTP) on HDL metabolism. *Atherosclerosis*, **155**, 269–281.

95 Pussinen, P. J., Jauhiainen, M., Metso, J., Pyle, L. E., Marcel, Y. L., Fidge, N. H., Ehnholm, C. (1998) Binding of phospholipid transfer protein (PLTP) to apolipoproteins A-I and A-II: location of a PLTP binding domain in the amino terminal region of apoA-I. *J. Lipid Res.*, **39**, 152–161.

96 Jauhiainen, M., Metso, J., Pahlman, R., Blomqvist, S., vanTol, A., Ehnholm, C. (1993) Human plasma phospholipid transfer protein causes high density lipoprotein conversion. *J. Biol. Chem.*, **268**, 4032–4036.

97 Tu, A. Y., Nishida, H. I., Nishida, T. (1993) High density lipoprotein conversion mediated by human plasma phospholipid transfer protein. *J. Biol. Chem.*, **268**, 23098–23105.

98 vonEckardstein, A., Jauhiainen, M., Huang, Y., Metso, J., Langer, C., Pussinen, P., Wu, S., Ehnholm, C., Assmann, G. (1996) Phospholipid transfer protein mediated conversion of high density lipoproteins generates pre beta 1-HDL. *Biochim. Biophys. Acta*, **1301**, 255–262.

99 Jiang, X. C., Bruce, C., Mar, J., Lin, M., Ji, Y., Francone, O. L., Tall, A. R. (1999) Targeted mutation of plasma phospholipid transfer protein gene markedly reduces high-density lipoprotein levels. *J. Clin. Invest*, **103**, 907–914.

100 Krieger, M. (1999) Charting the fate of the "good cholesterol": identification and characterization of the high-density lipoprotein receptor SR-BI. *Annu. Rev. Biochem.*, **68**, 523–558.

101 Acton, S., Rigotti, A., Landschulz, K. T., Xu, S., Hobbs, H. H., Krieger, M. (1996) Identification of scavenger receptor SR-BI as a high density lipoprotein receptor. *Science*, **271**, 518–520.

102 Murao, K., Terpstra, V., Green, S. R., Kondratenko, N., Steinberg, D., Quehenberger, O. (1997) Characterization of CLA-1, a human homologue of rodent scavenger receptor BI, as a receptor for high density lipoprotein and apoptotic

thymocytes. *J. Biol. Chem.*, **272**, 17551–17557.

103 Acton, S. L., Scherer, P. E., Lodish, H. F., Krieger, M. (1994) Expression cloning of SR-BI, a CD36-related class B scavenger receptor. *J. Biol. Chem.*, **269**, 21003–21009.

104 Thuahnai, S. T., Lund-Katz, S., Williams, D. L., Phillips, M. C. (2001) Scavenger receptor class B, type I-mediated uptake of various lipids into cells. Influence of the nature of the donor particle interaction with the receptor. *J. Biol. Chem.*, **276**, 43801–43808.

105 Urban, S., Zieseniss, S., Werder, M., Hauser, H., Budzinski, R., Engelmann, B. (2000) Scavenger receptor BI transfers major lipoprotein-associated phospholipids into the cells. *J. Biol. Chem.*, **275**, 33409–33415.

106 Stangl, H., Hyatt, M., Hobbs, H. H. (1999) Transport of lipids from high and low density lipoproteins via scavenger receptor-BI. *J. Biol. Chem.*, **274**, 32692–32698.

107 Swarnakar, S., Temel, R. E., Connelly, M. A., Azhar, S., Williams, D. L. (1999) Scavenger receptor class B, type I, mediates selective uptake of low density lipoprotein cholesteryl ester. *J. Biol. Chem.*, **274**, 29733–29739.

108 Ji, Y., Jian, B., Wang, N., Sun, Y., Moya, M. L., Phillips, M. C., Rothblat, G. H., Swaney, J. B., Tall, A. R. (1997) Scavenger receptor BI promotes high density lipoprotein-mediated cellular cholesterol efflux. *J. Biol. Chem.*, **272**, 20982–20985.

109 Gu, X., Kozarsky, K., Krieger, M. (2000) Scavenger receptor class B, type I-mediated [3H]cholesterol efflux to high and low density lipoproteins is dependent on lipoprotein binding to the receptor. *J. Biol. Chem.*, **275**, 29993–30001.

110 Wang, N., Arai, T., Ji, Y., Rinninger, F., Tall, A. R. (1998) Liver-specific overexpression of scavenger receptor BI decreases levels of very low density lipoprotein ApoB, low density lipoprotein ApoB, and high density lipoprotein in transgenic mice. *J. Biol. Chem.*, **273**, 32920–32926.

111 Ueda, Y., Royer, L., Gong, E., Zhang, J., Cooper, P. N., Francone, O., Rubin, E. M. (1999) Lower plasma levels and accelerated clearance of high density lipoprotein (HDL) and non-HDL cholesterol in scavenger receptor class B type I transgenic mice. *J. Biol. Chem.*, **274**, 7165–7171.

112 Rigotti, A., Trigatti, B. L., Penman, M., Rayburn, H., Herz, J., Krieger, M. (1997) A targeted mutation in the murine gene encoding the high density lipoprotein (HDL) receptor scavenger receptor class B type I reveals its key role in HDL metabolism. *Proc. Natl. Acad. Sci. U.S.A*, **94**, 12610–12615.

113 Webb, N. R., deBeer, M. C., Yu, J., Kindy, M. S., Daugherty, A., van derWesthuyzen, D. R., deBeer, F. C. (2002) Overexpression of SR-BI by adenoviral vector promotes clearance of apoA-I, but not apoB, in human apoB transgenic mice. *J. Lipid Res.*, **43**, 1421–1428.

114 Mardones, P., Quinones, V., Amigo, L., Moreno, M., Miquel, J. F., Schwarz, M., Miettinen, H. E., Trigatti, B., Krieger, M., VanPatten, S., Cohen, D. E., Rigotti, A. (2001) Hepatic cholesterol and bile acid metabolism and intestinal cholesterol absorption in scavenger receptor class B type I-deficient mice. *J. Lipid Res.*, **42**, 170–180.

115 Ji, Y., Wang, N., Ramakrishnan, R., Sehayek, E., Huszar, D., Breslow, J. L., Tall, A. R. (1999) Hepatic scavenger receptor BI promotes rapid clearance of high density lipoprotein free cholesterol and its transport into bile. *J. Biol. Chem.*, **274**, 33398–33402.

116 Trigatti, B., Rayburn, H., Vinals, M., Braun, A., Miettinen, H., Penman, M., Hertz, M., Schrenzel, M., Amigo, L., Rigotti, A., Krieger, M. (1999) Influence of the high density lipoprotein receptor SR-BI on reproductive and cardiovascular pathophysiology. *Proc. Natl. Acad. Sci. U.S.A*, **96**, 9322–9327.

117 Arai, T., Wang, N., Bezouevski, M., Welch, C., Tall, A. R. (1999) Decreased atherosclerosis in heterozygous low density lipoprotein receptor-deficient mice expressing the scavenger receptor BI transgene. *J. Biol. Chem.*, **274**, 2366–2371.

118 Ueda, Y., Gong, E., Royer, L., Cooper, P. N., Francone, O. L., Rubin, E. M. (2000) Relationship between expression levels and atherogenesis in scavenger receptor class B, type I transgenics. *J. Biol. Chem.*, **275**, 20368–20373.

119 Huszar, D., Varban, M. L., Rinninger, F., Feeley, R., Arai, T., Fairchild-Huntress, V., Donovan, M. J., Tall, A. R. (2000) Increased LDL cholesterol and atherosclerosis in LDL receptor-deficient mice with attenuated expression of scavenger receptorB1. *Arterioscler. Thromb. Vasc. Biol.*, **20**, 1068–1073.

120 Gu, X., Lawrence, R., Krieger, M. (2000) Dissociation of the high density lipoprotein and low density lipoprotein binding activities of murine scavenger receptor class B type I (mSR-BI) using retrovirus library-based activity dissection. *J. Biol. Chem.*, **275**, 9120–9130.

121 Nieland, T. J., Penman, M., Dori, L., Krieger, M., Kirchhausen, T. (2002) Discovery of chemical inhibitors of the selective transfer of lipids mediated by the HDL receptor SR-BI. *Proc. Natl. Acad. Sci. U.S.A*, **99**, 15422–15427.

122 Llera-Moya, M., Rothblat, G. H., Connelly, M. A., Kellner-Weibel, G., Sakr, S. W., Phillips, M. C., Williams, D. L. (1999) Scavenger receptor BI (SR-BI) mediates free cholesterol flux independently of HDL tethering to the cell surface. *J. Lipid Res.*, **40**, 575–580.

123 Savary, S., Denizot, F., Luciani, M., Mattei, M., Chimini, G. (1996) Molecular cloning of a mammalian ABC transporter homologous to Drosophila white gene. *Mamm. Genome*, **7**, 673–676.

124 Croop, J. M., Tiller, G. E., Fletcher, J. A., Lux, M. L., Raab, E., Goldenson, D., Son, D., Arciniegas, S., Wu, R. L. (1997) Isolation and characterization of a mammalian homolog of the Drosophila white gene. *Gene*, **185**, 77–85.

125 Nakamura, K., Kennedy, M. A., Baldan, A., Bojanic, D. D., Lyons, K., Edwards, P. A. (2004) Expression and regulation of multiple murine ATP-binding cassette transporter G1 mRNAs/isoforms that stimulate cellular cholesterol efflux to high density lipoprotein. *J. Biol. Chem.*, **279**, 45980–45989.

126 Klucken, J., Buchler, C., Orso, E., Kaminski, W. E., Porsch-Ozcurumez, M., Liebisch, G., Kapinsky, M., Diederich, W., Drobnik, W., Dean, M., Allikmets, R., Schmitz, G. (2000) ABCG1 (ABC8), the human homolog of the Drosophila white gene, is a regulator of macrophage cholesterol and phospholipid transport. *Proc. Natl. Acad. Sci. U.S.A*, **97**, 817–822.

127 Venkateswaran, A., Repa, J. J., Lobaccaro, J. M., Bronson, A., Mangelsdorf, D. J., Edwards, P. A. (2000) Human white/murine ABC8 mRNA levels are highly induced in lipid-loaded macrophages. A transcriptional role for specific oxysterols. *J. Biol. Chem.*, **275**, 14700–14707.

128 Wang, N., Ranalletta, M., Matsuura, F., Peng, F., Tall, A. R. (2006) LXR-induced redistribution of ABCG1 to plasma membrane in macrophages enhances cholesterol mass efflux to HDL. *Arterioscler. Thromb. Vasc. Biol.*, **26**, 1310–1316.

129 Vaughan, A. M.and Oram, J. F. (2005) ABCG1 redistributes cell cholesterol to domains removable by HDL but not by lipid-depleted apolipoproteins. *J. Biol. Chem.*, **280**, 30150–30157.

130 Gelissen, I. C., Harris, M., Rye, K. A., Quinn, C., Brown, A. J., Kockx, M., Cartland, S., Packianathan, M., Kritharides, L., Jessup, W. (2006) ABCA1 and ABCG1 synergize to mediate cholesterol export to apoA-I.

Arterioscler. Thromb. Vasc. Biol., **26**, 534–540.

131 Kennedy, M. A., Barrera, G. C., Nakamura, K., Baldan, A., Tarr, P., Fishbein, M. C., Frank, J., Francone, O. L., Edwards, P. A. (2005) ABCG1 has a critical role in mediating cholesterol efflux to HDL and preventing cellular lipid accumulation. *Cell Metab*, **1**, 121–131.

132 Chung, B. H., Segrest, J. P., Ray, M. J., Brunzell, J. D., Hokanson, J. E., Krauss, R. M., Beaudrie, K., Cone, J. T. (1986) Single vertical spin density gradient ultracentrifugation. *Methods Enzymol.*, **128**, 181–209.

133 Nichols, A. V., Krauss, R. M., Musliner, T. A. (1986) Nondenaturing polyacrylamide gradient gel electrophoresis. *Methods Enzymol.*, **128**, 417–431.

134 Davidson, W. S., Sparks, D. L., Lund-Katz, S., Phillips, M. C. (1994) The molecular basis for the difference in charge between pre-beta- and alpha-migrating high density lipoproteins. *J. Biol. Chem.*, **269**, 8959–8965.

135 Fielding, C. J., Fielding, P. E. (1996) Two-dimensional nondenaturing electrophoresis of lipoproteins: applications to high-density lipoprotein speciation. *Methods Enzymol.*, **263**, 251–259.

136 Asztalos, B. F., Sloop, C. H., Wong, L., Roheim, P. S. (1993) Two-dimensional electrophoresis of plasma lipoproteins: recognition of new apo A-I-containing subpopulations. *Biochim. Biophys. Acta*, **1169**, 291–300.

137 Nanjee, M. N., Cooke, C. J., Olszewski, W. L., Miller, N. E. (2000) Concentrations of electrophoretic and size subclasses of apolipoprotein A-I-containing particles in human peripheral lymph. *Arterioscler. Thromb. Vasc. Biol.*, **20**, 2148–2155.

138 Nakamura, Y., Kotite, L., Gan, Y., Spencer, T. A., Fielding, C. J., Fielding, P. E. (2004) Molecular mechanism of reverse cholesterol transport: reaction of pre-beta-migrating high-density lipoprotein with plasma lecithin/cholesterol acyltransferase. *Biochemistry*, **43**, 14811–14820.

139 Fielding, P. E., Kawano, M., Catapano, A. L., Zoppo, A., Marcovina, S., Fielding, C. J. (1994) Unique epitope of apolipoprotein A-I expressed in pre-beta-1 high-density lipoprotein and its role in the catalyzed efflux of cellular cholesterol. *Biochemistry*, **33**, 6981–6985.

140 Kunitake, S. T., Chen, G. C., Kung, S. F., Schilling, J. W., Hardman, D. A., Kane, J. P. (1990) Pre-beta high density lipoprotein. Unique disposition of apolipoprotein A-I increases susceptibility to proteolysis. *Arteriosclerosis*, **10**, 25–30.

141 Reichl, D., Hathaway, C. B., Sterchi, J. M., Miller, N. E. (1991) Lipoproteins of human peripheral lymph. Apolipoprotein AI-containing lipoprotein with alpha-2 electrophoretic mobility. *Eur. J. Clin. Invest*, **21**, 638–643.

142 Asztalos, B. F., Sloop, C. H., Wong, L., Roheim, P. S. (1993) Comparison of apo A-I-containing subpopulations of dog plasma and prenodal peripheral lymph: evidence for alteration in subpopulations in the interstitial space. *Biochim. Biophys. Acta*, **1169**, 301–304.

143 Heideman, C. L., Hoff, H. F. (1982) Lipoproteins containing apolipoprotein A-I extracted from human aortas. *Biochim. Biophys. Acta*, **711**, 431–444.

144 Smith, E. B., Ashall, C., Walker, J. E. (1984) High density lipoprotein (HDL) subfractions in interstitial fluid from human aortic intima and atherosclerostic lesions. *Biochem. Soc. Trans.*, **12**, 843–844.

145 Castro, G. R., Fielding, C. J. (1988) Early incorporation of cell-derived cholesterol into pre-beta-migrating high-density lipoprotein. *Biochemistry*, **27**, 25–29.

146 Forte, T. M., Goth-Goldstein, R., Nordhausen, R. W., McCall, M. R. (1993) Apolipoprotein A-I-cell membrane interaction: extracellular assembly of heterogeneous nascent

HDL particles. *J. Lipid Res.*, **34**, 317–324.

147 Forte, T. M., Bielicki, J. K., Goth-Goldstein, R., Selmek, J., McCall, M. R. (1995) Recruitment of cell phospholipids and cholesterol by apolipoproteins A-II and A-I: formation of nascent apolipoprotein-specific HDL that differ in size, phospholipid composition, and reactivity with LCAT. *J. Lipid Res.*, **36**, 148–157.

148 Duong, P. T., Collins, H. L., Nickel, M., Lund-Katz, S., Rothblat, G. H., Phillips, M. C. (2006) Characterization of nascent HDL particles and microparticles formed by ABCA1-mediated efflux of cellular lipids to apoA-I. *J. Lipid Res.*, **47**, 832–843.

149 Chau, P., Nakamura, Y., Fielding, C. J., Fielding, P. E. (2006) Mechanism of prebeta-HDL formation and activation. *Biochemistry*, **45**, 3981–3987.

150 Hara, H. and Yokoyama, S. (1991) Interaction of free apolipoproteins with macrophages. Formation of high density lipoprotein-like lipoproteins and reduction of cellular cholesterol. *J. Biol. Chem.*, **266**, 3080–3086.

151 Clay, M. A., Newnham, H. H., Barter, P. J. (1991) Hepatic lipase promotes a loss of apolipoprotein A-I from triglyceride-enriched human high density lipoproteins during incubation in vitro. *Arterioscler. Thromb.*, **11**, 415–422.

152 Barrans, A., Collet, X., Barbaras, R., Jaspard, B., Manent, J., Vieu, C., Chap, H., Perret, B. (1994) Hepatic lipase induces the formation of pre-beta 1 high density lipoprotein (HDL) from triacylglycerol-rich H DL2.A study comparing liver perfusion to in vitro incubation with lipases. *J. Biol. Chem.*, **269**, 11572–11577.

153 Hennessy, L. K., Kunitake, S. T., Kane, J. P. (1993) Apolipoprotein A-I-containing lipoproteins, with or without apolipoprotein A-II, as progenitors of pre-beta high-density lipoprotein particles. *Biochemistry*, **32**, 5759–5765.

154 Liang, H. Q., Rye, K. A., Barter, P. J. (1994) Dissociation of lipid-free apolipoprotein A-I from high density lipoproteins. *J. Lipid Res.*, **35**, 1187–1199.

155 Settasatian, N., Duong, M., Curtiss, L. K., Ehnholm, C., Jauhiainen, M., Huuskonen, J., Rye, K. A. (2001) The mechanism of the remodeling of high density lipoproteins by phospholipid transfer protein. *J. Biol. Chem.*, **276**, 26898–26905.

156 Wolfrum, C., Poy, M. N., Stoffel, M. (2005) Apolipoprotein M is required for prebeta-HDL formation and cholesterol efflux to HDL and protects against atherosclerosis. *Nat. Med.*, **11**, 418–422.

157 Jonas, A., Steinmetz, A., Churgay, L. (1993) The number of amphipathic alpha-helical segments of apolipoproteins A-I, E, and A-IV determines the size and functional properties of their reconstituted lipoprotein particles. *J. Biol. Chem.*, **268**, 1596–1602.

158 Asztalos, B. F., Roheim, P. S., Milani, R. L., Lefevre, M., McNamara, J. R., Horvath, K. V., Schaefer, E. J. (2000) Distribution of ApoA-I-containing HDL subpopulations in patients with coronary heart disease. *Arterioscler. Thromb. Vasc. Biol.*, **20**, 2670–2676.

159 Asztalos, B. F., Horvath, K. V., McNamara, J. R., Roheim, P. S., Rubinstein, J. J., Schaefer, E. J. (2002) Effects of atorvastatin on the HDL subpopulation profile of coronary heart disease patients. *J. Lipid Res.*, **43**, 1701–1707.

160 Asztalos, B. F., Batista, M., Horvath, K. V., Cox, C. E., Dallal, G. E., Morse, J. S., Brown, G. B., Schaefer, E. J. (2003) Change in alpha1 HDL concentration predicts progression in coronary artery stenosis. *Arterioscler. Thromb. Vasc. Biol.*, **23**, 847–852.

161 Brousseau, M. E., Diffenderfer, M. R., Millar, J. S., Nartsupha, C., Asztalos, B. F., Welty, F. K., Wolfe, M. L., Rudling, M., Bjorkhem, I., Angelin, B., Mancuso, J. P., Digenio,

A. G., Rader, D. J., Schaefer, E. J. (2005) Effects of cholesteryl ester transfer protein inhibition on high-density lipoprotein subspecies, apolipoprotein A-I metabolism, and fecal sterol excretion. *Arterioscler. Thromb. Vasc. Biol.*, **25**, 1057–1064.

162 Asztalos, B. F., Collins, D., Cupples, L. A., Demissie, S., Horvath, K. V., Bloomfield, H. E., Robins, S. J., Schaefer, E. J. (2005) Value of high-density lipoprotein (HDL) subpopulations in predicting recurrent cardiovascular events in the Veterans Affairs HDL Intervention Trial. *Arterioscler. Thromb. Vasc. Biol.*, **25**, 2185–2191.

163 Asztalos, B. F., Schaefer, E. J., Horvath, K. V., Cox, C. E., Skinner, S., Gerrior, J., Gorbach, S. L., Wanke, C. (2006) Protease inhibitor-based HAART, HDL, and CHD-risk in HIV-infected patients. *Atherosclerosis*, **184**, 72–77.

164 Navab, M., Hama, S. Y., Anantharamaiah, G. M., Hassan, K., Hough, G. P., Watson, A. D., Reddy, S. T., Sevanian, A., Fonarow, G. C., Fogelman, A. M. (2000) Normal high density lipoprotein inhibits three steps in the formation of mildly oxidized low density lipoprotein: steps 2 and 3. *J. Lipid Res.*, **41**, 1495–1508.

165 Mineo, C., Yuhanna, I. S., Quon, M. J., Shaul, P. W. (2003) High density lipoprotein-induced endothelial nitric-oxide synthase activation is mediated by Akt and MAP kinases. *J. Biol. Chem.*, **278**, 9142–9149.

166 Yuhanna, I. S., Zhu, Y., Cox, B. E., Hahner, L. D., Osborne-Lawrence, S., Lu, P., Marcel, Y. L., Anderson, R. G., Mendelsohn, M. E., Hobbs, H. H., Shaul, P. W. (2001) High-density lipoprotein binding to scavenger receptor-BI activates endothelial nitric oxide synthase. *Nat. Med.*, **7**, 853–857.

167 Shaul, P. W. (2003) Endothelial nitric oxide synthase, caveolae and the development of atherosclerosis. *J. Physiol*, **547**, 21–33.

168 Navab, M., Hama, S. Y., Cooke, C. J., Anantharamaiah, G. M., Chaddha, M., Jin, L., Subbanagounder, G., Faull, K. F., Reddy, S. T., Miller, N. E., Fogelman, A. M. (2000) Normal high density lipoprotein inhibits three steps in the formation of mildly oxidized low density lipoprotein: step 1. *J. Lipid Res.*, **41**, 1481–1494.

169 Watson, A. D., Navab, M., Hama, S. Y., Sevanian, A., Prescott, S. M., Stafforini, D. M., McIntyre, T. M., Du, B. N., Fogelman, A. M., Berliner, J. A. (1995) Effect of platelet activating factor-acetylhydrolase on the formation and action of minimally oxidized low density lipoprotein. *J. Clin. Invest*, **95**, 774–782.

170 Watson, A. D., Berliner, J. A., Hama, S. Y., LaDu, B. N., Faull, K. F., Fogelman, A. M., Navab, M. (1995) Protective effect of high density lipoprotein associated paraoxonase. Inhibition of the biological activity of minimally oxidized low density lipoprotein. *J. Clin. Invest*, **96**, 2882–2891.

171 Rader, D. J. (2002) High-density lipoproteins and atherosclerosis. *Am. J. Cardiol.*, **90**, 62i–70i.

172 Rubin, E. M., Krauss, R. M., Spangler, E. A., Verstuyft, J. G., Clift, S. M. (1991) Inhibition of early atherogenesis in transgenic mice by human apolipoprotein AI. *Nature*, **353**, 265–267.

173 Paszty, C., Maeda, N., Verstuyft, J., Rubin, E. M. (1994) Apolipoprotein AI transgene corrects apolipoprotein E deficiency-induced atherosclerosis in mice. *J. Clin. Invest*, **94**, 899–903.

174 Emmanuel, F., Caillaud, J. M., Hennuyer, N., Fievet, C., Viry, I., Houdebine, J. C., Fruchart, J. C., Denefle, P., Duverger, N. (1996) Overexpression of human apolipoprotein A-I inhibits atherosclerosis development in Watanabe rabbits. *Circulation*, **94**, I-632.

175 Li, H., Reddick, R. L., Maeda, N. (1993) Lack of apoA-I is not associated with increased susceptibility to atherosclerosis in mice. *Arterioscler. Thromb.*, **13**, 1814–1821.

176 Hughes, S. D., Verstuyft, J., Rubin, E. M. (1997) HDL deficiency in genetically engineered mice requires elevated LDL to accelerate atherogenesis. *Arterioscler. Thromb. Vasc. Biol.*, **17**, 1725–1729.

177 Belalcazar, L. M., Merched, A., Carr, B., Oka, K., Chen, K. H., Pastore, L., Beaudet, A., Chan, L. (2003) Long-term stable expression of human apolipoprotein A-I mediated by helper-dependent adenovirus gene transfer inhibits atherosclerosis progression and remodels atherosclerotic plaques in a mouse model of familial hypercholesterolemia. *Circulation*, **107**, 2726–2732.

178 Benoit, P., Emmanuel, F., Caillaud, J. M., Bassinet, L., Castro, G., Gallix, P., Fruchart, J. C., Branellec, D., Denefle, P., Duverger, N. (1999) Somatic gene transfer of human ApoA-I inhibits atherosclerosis progression in mouse models. *Circulation*, **99**, 105–110.

179 Tangirala, R. K., Tsukamoto, K., Chun, S. H., Usher, D., Pure, E., Rader, D. J. (1999) Regression of atherosclerosis induced by liver-directed gene transfer of apolipoprotein A-I in mice. *Circulation*, **100**, 1816–1822.

180 Shah, P. K., Nilsson, J., Kaul, S., Fishbein, M. C., Ageland, H., Hamsten, A., Johansson, J., Karpe, F., Cercek, B. (1998) Effects of recombinant apolipoprotein A-I(Milano) on aortic atherosclerosis in apolipoprotein E-deficient mice. *Circulation*, **97**, 780–785.

181 Shah, P. K., Yano, J., Reyes, O., Chyu, K. Y., Kaul, S., Bisgaier, C. L., Drake, S., Cercek, B. (2001) High-dose recombinant apolipoprotein A-I(milano) mobilizes tissue cholesterol and rapidly reduces plaque lipid and macrophage content in apolipoprotein e-deficient mice. Potential implications for acute plaque stabilization. *Circulation*, **103**, 3047–3050.

182 Nissen, S. E., Tsunoda, T., Tuzcu, E. M., Schoenhagen, P., Cooper, C. J., Yasin, M., Eaton, G. M., Lauer, M. A., Sheldon, W. S., Grines, C. L., Halpern, S., Crowe, T., Blankenship, J. C., Kerensky, R. (2003) Effect of recombinant ApoA-I Milano on coronary atherosclerosis in patients with acute coronary syndromes: a randomized controlled trial. *JAMA*, **290**, 2292–2300.

183 Ansell, B. J., Navab, M., Hama, S., Kamranpour, N., Fonarow, G., Hough, G., Rahmani, S., Mottahedeh, R., Dave, R., Reddy, S. T., Fogelman, A. M. (2003) Inflammatory/antiinflammatory properties of high-density lipoprotein distinguish patients from control subjects better than high-density lipoprotein cholesterol levels and are favorably affected by simvastatin treatment. *Circulation*, **108**, 2751–2756.

12
Hepatic and Renal HDL Receptors

Laurent O. Martinez, Bertrand Perret, Ronald Barbaras, François Tercé, Xavier Collet

12.1
Historical Background

12.1.1
HDL Receptors: From Myth to Reality

The concept of a lipid transport system between the tissues of the organism through plasma lipoproteins (such as "reverse cholesterol transport") supports the existence of interactions between cells and the lipoprotein particles. Such interactions induce the movement of different metabolites and/or lipoproteins in and out of the cells. The discovery of the LDL receptor in the 1970s and the following major advances in LDL metabolism produced a powerful expectancy of an identical mechanism for HDL metabolism. Unfortunately, though, unlike in the case of LDL, the molecular complexity of HDL particles and poor knowledge of their cellular physiology or related pathology proved to be a significant barrier to the discovery of clear receptors for the HDL, and so the search for clues as to the cell events that regulate HDL metabolism remains an acute issue. Although the discovery in the last ten years of the scavenger receptor class B type I (SR-BI) as an HDL receptor, together with the implication of the ATP binding cassette A1 (ABCA1) in binding to apolipoprotein A-I, mean that the realness of HDL-binding proteins at the cell surface is no longer seriously in question, such an existence had previously been hotly debated. In the 1980s, the major hypothesis for the efflux or uptake of cholesterol by the cells did not require any need for cell surface receptors, and this was supported by different observations: the complexity of the kinetics of cholesterol movement in and out of cells, which involved at least two components [1], and the lack of effect of trypsinisation of HDL on cell interactions [2], for example, were not factors in favor of a receptor. Rather, the cholesterol gradient between plasma membrane and the HDL particles was primarily proposed as the driving force for cholesterol movement. Nowadays, the role for different receptors in the processes of HDL cholesterol exchange with cells has been clearly demonstrated, but the exact mechanisms by which cholesterol moves into or out of cells remains a matter of debate and could still lean on this "gradient" hypothesis.

High-Density Lipoproteins: From Basic Biology to Clinical Aspects. Edited by Christopher J. Fielding

ISBN: 978-3-527-31717-2

HDL binding proteins/receptors have been described in all steps of HDL metabolism: in the synthesis and secretion of small HDL by the liver through ABCA1, at the level of peripheral cells for removal of excess of cellular cholesterol (SR-BI and ABCA1/G1), in the uptake of cholesteryl ester or holo-HDL particles by the liver for cholesterol elimination (SR-BI and ecto-F_1-ATPase), or in the kidney for final catabolism of apolipoprotein A-I (cubilin). The liver thus appears to be a central organ in HDL metabolism, involved in the first steps of synthesis and secretion as well as in the final step of reverse cholesterol transport, leading to bile acid conversion and cholesterol secretion into bile. It is still assumed that this is the only pathway for cholesterol catabolism and elimination from the body, although there are still few data available [3–5]. However, an interesting alternative route of direct elimination of free cholesterol through intestinal secretion from plasma, at least in rodents, has recently been suggested and could represent around 30 % of total cholesterol removal [6].

Actually, three major mechanisms to explain the HDL-dependent cholesterol uptake by the liver have been described:

1. The "selective uptake" of cholesteryl ester by the cells.
   This process, originally observed in the rat by Glass et al. [7], had never been supported by relevant molecular data until the discovery of SR-BI [8]. This pathway is defined as a high-capacity system in which cells internalize more cholesteryl esters than apolipoprotein components of the HDL particle. This is in contrast with the classical LDL (apoB/E) receptor pathway in which the entire particle is internalized through clathrin-coated vesicles for degradation and recycling.
2. The internalization of the holo HDL particle (i.e., protein plus lipid) by the liver through clathrin-coated vesicles [9], which might be the final step of a newly described pathway involving at least a cell surface F_1-ATPase and the $P2Y_{13}$ purinergic receptor [10,11]. These two pathways might today represent the major routes for hepatic cholesterol catabolism.
3. An alternative pathway in which the intracellular uptake of holo-HDL is followed by resecretion of a modified particle depleted in cholesteryl ester, a process called "retro-endocytosis".

12.1.2
To be or not to be an HDL Receptor?

In the 1990s, many attempts were made to identify HDL binding proteins (HBP) at the surfaces of different cell types, including hepatocytes. Most of them were based on the use of ligand blotting with HDL or free apolipoproteins A-I and A-II. Although no clear physiological effect can yet be associated with any of these proteins, they can be differentiated on the basis of their molecular weights, the type of binding site they display (high or low affinity), their differential affinities towards ligands (HDL, apoA-I, apoA-II), their tissue expression, or their regulation by cholesterol. They include two so-called HBP, (HB1-HB2), Gp96/GRP94 or Hsp60. The 110 kDa HBP protein (chicken vigilin) was found in association with the foam cells and smooth muscle cells of atherosclerotic lesions. Also found in endothelial cells, its expression was upregulated by cholesterol loading of macrophage, but the absence of a cell surface signal peptide or transmembrane domain and the weak increase of HDL binding upon overexpression did not favor the hypothesis of a true receptor activity for this protein [12,13]. In contrast, the 65 kDa human HB2 (100 kDa rat HB2), which has 93 % homology with the activated leukocyte-cell adhesion molecule (ALCAM), was downregulated by cholesterol loading in macrophages [14] and clearly increased HDL binding capacity when overexpressed in cells [15]. It has been proposed that interaction of HDL on adhesion molecule HB2 could interfere with the migration of macrophages and thus decrease the development of atherosclerotic lesions. However, overexpression of HB2 was not linked to an increase in cholesteryl ester uptake or cholesterol efflux. These data, together with the lack of any animal models overexpressing or serving as a knockout for HB2, thus do not allow clear identification of this protein as a physiologically relevant HDL receptor.

More recently, a modified ligand blotting method has allowed the identification of a new high-affinity HDL binding protein (95 kDa HBP) specific for HDL, apoA-I, and apoA-II [16]. In contrast with others, this HBP was found not only in fetal and adult hepatocytes, but also in CaCo2 intestinal cells and fibroblasts, and was not affected by deglycosylation of $HepG_2$ membranes. It was found to be different from all other HDL binding proteins including SR-BI or HB2, but no physiological data to support a true role in HDL metabolism are yet available.

Ligand blotting experiments also allowed the subsequent identification of two sets of proteins related to the heat shock protein family as putative HDL binding proteins. The 90 kDa Gp96 (strongly homologous to the stress proteins GRP94 or hsp108) was found in hepatocytes and in endothelial and Kuppfer cells. It was induced by glucose deprivation and its overexpression doubled the binding capacity of HDL. More recently, hsp60 (55 kDa) has been identified as a high-affinity receptor for apoA-II and HDL and to a lesser extent for apoA-I in human hepatocytes, fibroblasts, or macrophages [17]. Interestingly, it is known that human apoA-II overexpression induces HDL conversion into proinflammatory particles, while an anti-hsp60 immunity has been related to the development of atherosclerosis [18]. Hypotheses in which hsp60 might interfere with the dissociation of apolipoproteins from cells and affect their transport in and out of the cells have been proposed.

Beside these more or less characterized HDL receptors, other lipoprotein-binding sites exhibiting broad lipoprotein specificity had been reported earlier in the literature, probably preempting the identification of SR-BI. Bachorik et al. [19] identified a putative receptor in pig hepatocytes – called "lipoprotein binding site" or LBS – that mediated the uptake and degradation of both apoE-free HDL and LDL. They later ascertained that HDL core lipid (mainly cholesteryl ester) was taken up selectively from the holo-particle, a hallmark of SR-BI [20]. Finally, Brisette and Noel also reported an unspecified lipoprotein-binding site in rat liver membranes [21], which suggests that the "LBS" was closely related to SR-BI.

It was in the mid-1990s that Acton et al. used an original approach to detect novel lipoprotein receptors. By screening the binding of LDL to LDL-receptor-deficient CHO cells expressing a cDNA library, they identified the scavenger receptor SR-BI as a protein capable of binding LDL, and which further proved also to bind HDL [8,22]. Generation of SR-BI transgenic mice demonstrated a major role for SR-BI, at least in rodents, in the metabolism of lipoprotein cholesteryl ester. At present, however, the importance of the human homologue of SR-BI (CLA-1 [23]) in HDL cholesteryl ester metabolism in humans, for which the major cholesterol carriers are in the range of the LDL, still needs to be demonstrated. A second protein involved as a high-affinity binding receptor for apoA-I and HDL in hepatocytes was recently found to be the β-chain of human ecto-F_1-ATPase [10]. ApoA-I binding to the ecto-F_1-ATPase leads to cell surface hydrolysis of ATP to ADP, inducing the stimulation of a recently cloned nucleotidic receptor $P2Y_{13}$, followed by holo-HDL particle endocytosis through low-affinity binding sites [11].

In this review we first focus on these two pathways (SR-BI and F_1-ATPase/$P2Y_{13}$), which currently represent the major receptor-dependent routes for hepatic HDL catabolism. The roles of two other well-characterized apoA-I/HDL binding proteins – the ATP binding cassette transporter (ABCA1) and cubilin, involved in hepatic HDL biogenesis and renal apoA-I catabolism, respectively – are also explored. Finally, the relevance of each pathway to human physiopathology is investigated.

12.2
SR-BI-Mediated Selective HDL Cholesterol Uptake

12.2.1
The Selective HDL Cholesteryl Ester Uptake Pathway

As underlined in the introductory section, a process of "selective HDL cholesteryl ester uptake" (referred to below as "selective uptake") was demonstrated in the early 1980s by Glass et al. [7,24]. These pioneering studies revealed that some tissues (mainly hepatic and steroidogenic tissues) internalize more HDL cholesteryl ester (CE) than HDL apolipoprotein. Although the liver contributes to both HDL CE and apolipoprotein uptake, the selective uptake pathway is responsible for the majority of HDL CE uptake by the liver in rodents, which do not express Cholesterol Ester Transfer Protein (CETP). In rabbits, which do possess the CETP-mediated pathway for HDL, CE metabolism (see Chapter 7 for a detailed review about this pathway), 20 % of total HDL CE clearance is still attributed to selective uptake, whereas the holo-HDL particle uptake (cholesterol + apolipoprotein moieties), possibly mediated by the ecto-F_1-ATPase as reported in Section 12.3, could account for 10 % of HDL CE clearance [25]. Even though rabbits have about four times the plasma CETP activity of man, these results suggest that selective uptake may play a substantial role in the clearance of HDL CE in humans. Uptake studies of doubly labeled HDL ([^{3}H]-CE and [^{125}I]-apoA-I) by hepatocytes of different species, however, have reported uptake CE/apoA-I ratios of around 12 in rats [26], 6 in rabbits [27], and between 2 and 4 in humans [26,28], suggesting that major differences in the contributions of selective uptake to the total HDL CE clearance by the liver could exist between species. These studies also revealed that the HDL CE selective uptake process did not involve an endocytosis mechanism [26], neither cholesterol endosomal transport [29] nor retroendocytosis [30], putting into question the requirement for hepatic cell surface receptor(s) involved in HDL CE selective uptake [31].

12.2.2
SR-BI as a Receptor for Selective Lipoprotein Cholesterol Uptake

In the mid-1990s, our understanding of the molecular mechanisms involved in the selective uptake process was dramatically increased by the discovery of rodent scavenger receptor BI (SR-BI) and its human orthologue CLA-1 (CD36 and LIMPII Analogous-1) as lipoprotein receptors. SR-BI, a member of the CD36 family (class B scavenger receptor), is a cell surface membrane glycoprotein made up of 509 amino acids with a molecular mass of 82 kDa. Its predicted secondary structure comprises two transmembranes and two N- and C-terminal cytoplasmic domains, together with a large extracellular loop containing several N-glycosylation sites [32]. SR-BI is mainly expressed in organs that play critical roles in cholesterol metabolism, such as liver, intestine, fat, and steroidogenic tissues (adrenal gland, ovary, testis), but it is also found in various other mammalian tissues and cells, including heart, placenta, kidney, brain, macrophages, smooth muscle cells, and endothelial cells. Although SR-BI/CLA-1 was initially identified as a scavenger receptor for native and modified LDL particles and anionic phospholipids [22,33,34], it was later shown that SR-BI and CLA-1 could also bind HDL and mediate the selective uptake of cholesteryl ester from both LDL and HDL particles [8,35–37], suggesting that SR-BI/CLA-1 may play a role in the metabolism of both HDL and non-HDL particle *in vivo*.

12.2.3
Role of Hepatic SR-BI in Murine Cholesterol Homeostasis and Atherogenesis

Direct evidence of the physiological importance of SR-BI in murine cholesterol metabolism *in vivo* was obtained from studies on genetically engineered mice. Hepatic overexpression of SR-BI is associated with decreased plasma levels of VLDL and of LDL and HDL cholesterol, together with increased biliary cholesterol content, which is correlated with an increase in HDL clearance by the liver [38–41]. Conversely, partial or complete deficiency of SR-BI results in an SR-BI dose-dependent elevation of plasma total and HDL cholesterol with the appearance of large cholesterol-rich HDL particles and a decrease in biliary cholesterol concentration, suggesting that SR-BI deficiency in mice severely disrupts the hepatic selective uptake of HDL cholesterol [42–44]. As a consequence, the expression of SR-BI in mice was expected to be atheroprotective, and this hypothesis was supported by studies of atherosclerosis in several models of SR-BI overexpression or ablation [45].

However, it has been tricky to gain a clear understanding of the liver-specific contribution of SR-BI to protection against atherosclerosis because of the multifunctional properties attributed to this receptor and the broad range of ligands to which SR-BI binds. Indeed, in addition to HDL, LDL, modified LDL, and anionic phospholipids, SR-BI is also reported to bind and/or transport the acute-phase protein serum amyloid A [46,47], vitamin E [48,49], and apoptotic cells, all relevant for atherogenesis. Furthermore, other tissue-specific mechanisms – such as its ability to promote cellular free cholesterol efflux from peripheral tissues [50,51], to activate endothelial nitric oxide synthase (eNOS) [52], and to induce apoptosis in damaged endothelial cells [53] – may also contribute to the antiatherogenic properties of SR-BI. The role of hepatic SR-BI in mice has recently been addressed in a study by Huby et al. performed in liver-specific SR-BI-knockout mice and revealing that these mice develop massive diet-induced atherosclerosis, but not to the same extent as total SR-BI-knockout mice, although very similar plasma lipoprotein cholesterol levels were observed in the two mouse models [54]. These data suggest that peripheral SR-BI expression may protect against atherosclerosis somewhat, through mechanisms that are independent of changes in plasma lipid parameters. Importantly, accumulations of VLDL-sized lipoproteins in high-fat and cholesterol diets were observed in both total and liver-specific SR-BI-knockout mice [54], which might reflect an implication of SR-BI in the uptake of apoB-containing lipoproteins from the circulation, as previously observed in SR-BI transgenic mouse models [39,40].

12.2.4 Regulation of Hepatic SR-BI

As reported above, SR-BI is mostly expressed in liver, adrenal glands, or ovaries for HDL CE uptake, but also in macrophages for free cholesterol efflux or in the intestine for dietary cholesterol uptake. It is thus not surprising that SR-BI expression is highly but differently regulated depending on the cell type. Indeed, SR-BI can be regulated at multiple steps: transcriptional, posttranscriptional, or by the subcellular localization, as reviewed recently ([55] and Chapter 7).

Numerous recently published studies have demonstrated that regulation of SR-BI expression in the liver is at best complex and unclear. Estradiol treatment, for instance, as in ovarian cells, induces regulation at the transcriptional level, decreasing SR-BI expression in rat hepatocyte but increasing it in Kupffer cells [56]. A high-cholesterol diet also decreases SR-BI in rat liver parenchymal cells but induces it in Kupffer cells, confirming the discrepancy between cell lines [56]. In contrast, a cholesterol diet does not change SR-BI level and CE uptake in hamsters [57] or in apoAI-knockout mice (in which HDL level is low), suggesting that plasma HDL level does not control SR-BI hepatic expression [58]. Vitamin E was also found to induce SR-BI expression [59], as were proinflammatory stimuli (lipopolysaccharide, TNF, IL1) [55]. SR-BI is also regulated by transcription factors in the liver, although the data need to be clarified; indeed, natural ligands of PPARs such as fatty acids (FA) display differential effects, with saturated FA decreasing SR-BI [60], but polyunsaturated FA upregulating SR-BI expression and CE uptake in the hamster [61]. Synthetic agonists of PPAR, such as Wy-14643 (PPARα, γ, or δ) or BRL 49653 (PPARγ), also upregulate SR-BI protein levels in rodent hepatocytes or human HepG2 cells [62]. In contrast, fibrate treatment of mice decreases SR-BI protein in a PPARα-dependent way, without affecting mRNA levels, suggesting a posttranscriptional level of regulation through PPAR activation in the mice [63,64]. Interestingly, PPARα/RXR induces SR-BI promoter activation through a putative PPRE motif, further enhanced by fibrate treatment [65], confirming a potential role for PPARs in this regulation. The Liver X receptor (LXR) has also been invoked as a positive regulator in the hepatic expression of SR-BI through its activation by natural oxysterol ligands [66]. In contrast, bile acids, such as chenodeoxycholic acid, activators of Farnesoid X receptor (FXR), reduce SR-BI expression, although no response element for FXR was found in the SR-BI promoter responsive region for the bile acid [67]. This effect could be attributed to a pathway dependent on FXR/RXR, Small Heterodimeric Protein (SHP), and Liver Receptor Homologue-1 (LRH-1) *in vitro* and *in vivo* in rodent or in human [67]. An LRH-1 response element has been found in the SR-BI promoter, while LRH-1 knockout mice display reduced levels of SR-BI [68]. Finally, the effect of FXR was confirmed in FXR-knockout mice in which SR-BI expression is increased [69]. In contradiction with these data, however, a decrease in basal SR-BI expression was found in FXR-knockout mice [70], but could be explained in terms of a dominant positive effect of an LXR/RXR-dependent pathway on the negative FXR effect in a basal regulation status.

Hepatic SR-BI can also be regulated at posttranscriptional level; indeed, the lack of correlation between mRNA and protein levels in hepatocytes [55] and the fact that PPARα agonists reduce SR-BI mouse liver protein but not mRNA [63] suggest such regulation. Interestingly, PPAR activators also induced a decrease in the expression of the SR-BI companion protein PDZK1 [63]. Indeed, PDZK1 is preferentially associated with the baso-lateral membranes in hepatocytes, interacting with the PDZ domain of the SR-BI C-terminus [71]. As a confirmation of this role, PDZ knockout mice do not display cell surface expression of SR-BI, indicating that in the absence of PDZK1, SR-BI is no longer transported to the cell surface and is degraded in the cytosol [72]. Recently, posttranslational regulation of SR-BI through phosphatidylinositol-3-kinase (PI3K) has also been found. Indeed, in fibrate-treated hepatocytes, activation of PI3K increases cell surface expression of SR-BI (and HDL CE uptake) through translocation of SR-BI from intracellular pools to the plasma membrane in a PDZK1-independent way [73].

Finally, chemicals that block lipid transport (BLTs) have been found to inhibit SR-BI activity [74]. These BLTs block SR-BI-mediated cholesterol uptake and efflux, probably by decreasing the dissociation rate of the lipoprotein particle. Interestingly, one of these compounds (BLT-4) has also been found to block ABCA1-mediated cholesterol efflux to lipid-poor apoA-I (see Chapter 5 and Section 12.4.1), suggesting that there may be similarities in the mechanisms of SR-BI- and ABCA1-mediated lipid transport [75].

The regulation of SR-BI expression or activity thus seems to be complex and different pathways might coexist, probably related to tissue origin, cell type or SR-BI function. Anyway, they still need to be clarified to provide access to better knowledge on molecular mechanisms of SR-BI regulation.

12.2.5 Mechanism of SR-BI-Mediated Selective Lipoprotein Cholesterol Uptake

The mechanism of SR-BI-mediated selective lipoprotein cholesterol uptake has mainly been investigated with HDL particles, although, as highlighted in the preceding sections, SR-BI can mediate both HDL and non-HDL cholesterol uptake. The detailed mechanism of SR-BI-mediated selective lipoprotein cholesterol uptake is not yet clear and various theories have been proposed. Firstly, in view of the impressive number of SR-BI heterogeneous ligands (see preceding sections), it is likely that SR-BI interaction with lipoprotein particles does not strictly speaking require a protein–protein interaction. Although an impressive number of independent studies have reported HDL binding to SR-BI, the heterogeneous values obtained in binding isotherm parameters and the lack of correlation of these values with kinetic association/dissociation parameters does not indicate any protein-binding specificity (see [76] for a complete review on this aspect). Consistently with this feature, an elegant study by Williams and colleagues demonstrated an interaction between SR-BI and the class A amphipatic α-helix [77], a common secondary structural motif in the apolipoproteins of HDL. These motifs bind and cross-link to SR-BI only when associated with phosphatidylcholine (DMPC), which suggests that phospholipids play a significant role, probably by maintaining the amphipatic α-helix in a conformation capable of binding SR-BI. This view is consistent with numerous studies reporting that lipid-free apoA-I binds poorly to SR-BI [76] and prompts the conclusion that the lipid content of lipoproteins may be essential for interaction with SR-BI. Cell plasma membrane lipid content in the microenvironment of SR-BI has also been extensively investigated and may regulate SR-BI activity [76]. Indeed, as an example, caveolin-1 is not strictly required for SR-BI activity but, when overexpresed in hepatocytes (naturally poor in caveolin-1), stimulates SR-BI-mediated HDL-CE selective uptake while decreasing LDL-CE selective uptake [78]. Altogether, the data suggest that SR-BI may facilitate spontaneous (non-protein-mediated) bidirectional exchange of lipids by promoting transfer of CE from lipoproteins to the cells [8,79] and an efflux of plasma membrane free cholesterol to lipoproteins and non-lipoprotein acceptors [51]. This process can be called "SR-BI-facilitated diffusion of cholesterol". Furthermore, since CD36 also binds HDL particles without mediating efficient selective CE uptake [80,81], binding does not appear to be a unique requirement for SR-BI activity but, as suggested by Rodrigueza et al., it is more likely that lipoprotein binding to SR-BI allows access of CE molecules to a "channel" formed by the receptor along which CE molecules move down the concentration gradient into the cell plasma membrane [82].

Alternatively, in SR-BI-overexpressing CHO cells it has been reported that SR-BI would promote uptake of HDL-holoparticles, followed by transfer into endosomal compartments and resecretion [83,84]. A similar process of retroendocytosis has been observed in mouse primary hepatocytes and human (HepG$_2$) hepatoma cells [83,85]. However, through the use of different modulators it was recently concluded that the amount of HDL trafficking through the SR-BI-dependent retroendocytic pool was too small to support the SR-BI-mediated CE selective uptake [86] and that hepatic SR-BI-mediated-CE selective uptake was not linked to HDL-holoparticle uptake [87,88]. In our own laboratory, SR-BI would contribute only to a proportion of the low-affinity receptors enabling HDL endocytosis in hepatocytes [89]. The data thus suggest the existence of an SR-BI-independent pathway, involved in uptake and degradation of the holo-HDL particle in hepatocytes [83,88], a process that could be mediated either by SR-BII, an alternative mRNA splicing variant of SR-BI [90], or by the ecto-F$_1$-ATPase pathway [10], as discussed in the next section and illustrated in Fig. 12.1.

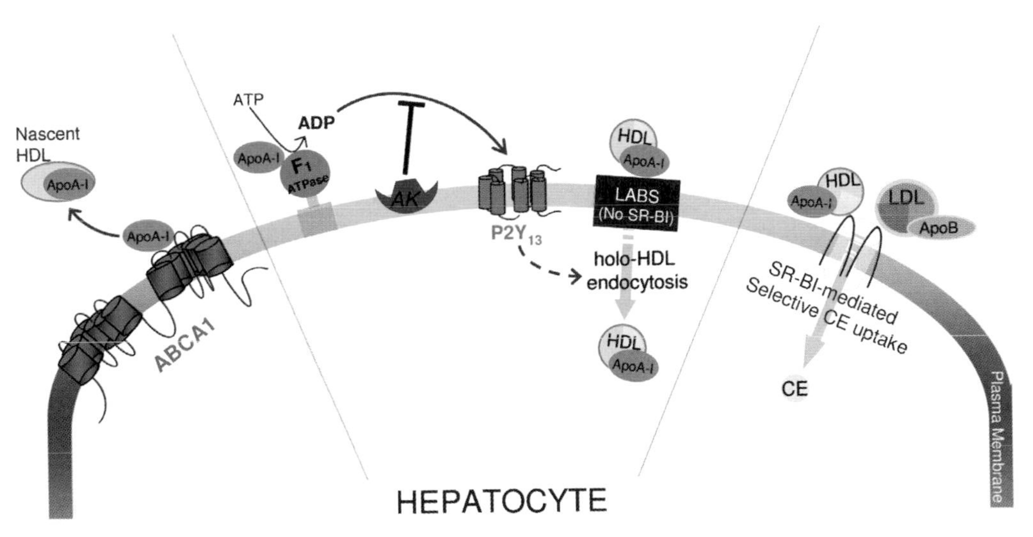

Fig. 12.1 *Pathways of HDL biogenesis and endocytosis by hepatocytes. Left*) Model of HDL biogenesis. Lipid-poor secreted apoA-I can acquire phospholipids and free cholesterol from hepatocytes through an efflux process mediated by ABCA1. *Center*) Model of F$_1$-ATPase/P2Y$_{13}$-mediated HDL endocytosis. Lipid-poor apoA-I binding to the ecto-F$_1$-ATPase stimulates the production of extracellular ADP, which activates the P2Y$_{13}$ nucleotidic receptor and subsequent holo-HDL particle endocytosis through low-affinity binding sites (LABS). Cell surface adenylate kinase (AK) activity downregulates the overall process by consuming the ATP generated by the ecto-F$_1$-ATPase. This holo-HDL endocytosis pathway therefore depends on the balance between the AK activity that removes extracellular ADP and the ecto-F$_1$-ATPase that synthesizes ADP. Activation of ADP synthesis by apoA-I unbalances the pathway and increases HDL endocytosis when required. *Right*) Model of SR-B1-mediated "selective cholesterol uptake". Lipoprotein binding to SR-B1 allows access of cholesteryl ester (CE) molecules to a "channel" formed by the receptor along which CE molecules move down their concentration gradient into the plasma membrane. In this process lipoproteins deliver CE to hepatocytes without the concomitant uptake and degradation of the whole lipoprotein particle.

12.3 Ecto-F_1-ATPase: An Unexpected Role in HDL Endocytosis

12.3.1 Two HDL Binding Sites Allocated to One HDL Endocytosis Pathway

The existence of HDL receptors other than SR-BI on hepatocytes has been suggested by different studies reporting that hepatic HDL uptake in various species occurs both through cholesterol selective uptake mechanisms (a main feature displayed by SR-BI [22]) and through holo-HDL particle endocytosis, a process that is more probably SR-BI-independent [87,88] and may involve the formation of clathrin-coated vesicles [9].

In the last few years our group has made important improvements relating to the identification of receptors and cell surface partners regulating this holo-HDL particle endocytosis pathway. Advances in this field were initiated with the identification and characterization of high- and low-affinity binding sites for HDL (dissociation constants $K_d \approx 10^{-9}$ M and 10^{-7} M, respectively) on rat [91], porcine [89], and human hepatocytes [92]. Interestingly, lipid-free apoA-I (free apoA-I) could only bind to the high-affinity sites, and contribution of SR-BI to these sites has been ruled out mainly because of the poor ability of SR-BI to bind free apoA-I [89,93–95]. In contrast, oxidized LDL (a well known SR-BI ligand [22]) or SR-BI-antibody was able to compete partially with HDL binding to hepatocytes at 4 °C [89], suggesting that the low-affinity binding sites involved at least SR-BI, but also other unknown receptor(s).

The physiological relevance of these high- and low-affinity binding sites and their relative contributions to holo-HDL particle endocytosis by the liver have thus been investigated with regard to metabolic events that affect HDL structure and distribution. Indeed, in the liver, large-size HDL particles, enriched in triglycerides (triglyceride-rich HDL_2 or TG-HDL_2), are preferential substrates for hepatic lipase, acting at the endothelial surfaces of sinusoid capillaries, leading to the formation of a triglyceride and phospholipid-poor "HDL remnants" [96]. TG-HDL_2 displayed only low-affinity binding, whereas the post-lipolysis HDL remnants were able to bind to both low- and high-affinity sites. Moreover, HDL remnants were internalized more rapidly and in greater amounts than their parent TG-HDL_2, suggesting that engagement of high-affinity receptor(s) might stimulate HDL endocytosis occurring through low-affinity binding sites [97]. Through the use of specific ligands for high- and low-binding sites (e.g., free apoA-I and TG-HDL_2 respectively) it was observed that free apoA-I could stimulate TG-HDL_2 endocytosis by hepatocytes by about 30 % after a short co-incubation time of both particles (5 to 10 min), which revealed an important feature of the high-affinity receptor(s): an ability to trigger HDL endocytosis through low-affinity binding sites [10]. This triggering, induced by small quantities of highly specific HDL particles (free apoA-I, pre-β-HDL, or remnant HDL, for instance), may therefore strongly amplify HDL cellular uptake through the low-affinity site. Importantly, *in situ* rat and mouse liver perfusion with lipid-free apoA-I greatly stimulates HDL uptake by the liver [10,11], which may be related to the beneficial effect of recombinant apoA-I/Milano infusion in humans in decreasing coronary artery atheroma [98], and this has given rise to the identification of hepatic high-affinity apoA-I/HDL receptor(s).

12.3.2
Ecto-F_1-ATPase as a High-Affinity ApoA-I Receptor

Two key approaches to fishing out apoA-I binding protein(s) from porcine liver plasma membrane preparations have been developed: apoA-I affinity chromatography purification and Biacore's surface plasmon resonance (SPR) recovery technology. Briefly, SPR technology measures the deviation of the refractive index at the interface between a sensor chip and a flowing solution. This feature, recorded in real time, is converted into a sensogram, which allows calculation of binding parameters. A subsequent process allows specific binding proteins to be eluted and recovered in small volumes for further analysis. Passage of solubilized porcine liver plasma membrane proteins over an immobilized free-apoA-I sensor chip in Biacore's SPR experiments revealed an interaction similar to that observed in human hepatocytes ($K_d \approx 10^{-9}$ M). Subsequently, several rounds of recovery process allowed the purification and identification of a protein with a relative molecular mass of 50 kDa. A larger amount of the 50 kDa apoA-I-binding protein was recovered by apoA-I affinity chromatography and microsequenced by Edman analysis. Unexpectedly, a peptide sequence derived from the 50 kDa apoA-I-binding protein was identical to a segment of the human β-chain of ATP synthase [10].

ATP synthase is classically known as a major protein complex involved in ATP synthesis and localized in the mitochondria. Mitochondrial ATP synthase is a multisubunit complex (~600 kDa) consisting of two major domains: F_1 and F_o [99]. F_1, a globular extra-membrane protein complex composed of five different subunits ($\alpha_3\beta_3\gamma\delta\varepsilon$), contains the catalytic site for ATP synthesis. F_1 is held to the mitochondrial inner membrane through its interaction with F_o, an integral mitochondrial inner membrane complex that contains a transmembrane channel composed of c-subunits (c-ring) through which protons can cross the membrane [100]. The synthesis of ATP requires an electrochemical proton gradient across the c-ring. The collapse or absence (when the F_1 complex is alone, for instance) of the electrochemical proton gradient induces a switch in enzymatic activity from ATP synthesis to ATP hydrolysis. In this case, the F_1 domain (so-called F_1-ATPase) catalyses the hydrolysis of ATP to ADP and phosphate, an activity that is downregulated in mitochondria by the natural inhibitor protein IF_1 [101].

The surprising finding of the presence of the β-chain of F_1-ATPase in hepatocyte plasma membrane preparations led us to demonstrate that it was also present, together with its counterpart α-chain, on the cell surfaces of intact hepatocytes, as previously reported for lymphocytes [102] or human endothelial cells [103]. Furthermore, Beisiegel and Mahley [104,105] had earlier identified the α- and β-chains of the F_1-ATPase as receptors for apolipoprotein E-enriched HDL, but they did not demonstrate the presence of these proteins on the cell surfaces, nor they did propose a role for it. The presence of the α- and β-chains has been further confirmed on the cell surface of $HepG_2$ [10], IHH (immortalized human hepatocytes) [11], and primary human hepatocytes [106], but not on epithelial-like cells such as the CHO cell line [10]. Other studies have supplied further evidence of the α/β complex at the cell surfaces of adipocytes [107], keratinocytes [108], brain cells [109], and a broad range of tumor cells [110]. These observations suggest that the presence of α/β chains at a cell surface is more dependent on the cell type than on the tumorigenic status of the cell, as suggested by Das et al. [102]. In addition, proteomic studies performed on rat liver detected most of the F_1-ATPase subunits in the membrane lipid raft domains [111], strongly suggesting that the entire F_1-ATPase might be present at the cell surface.

Moreover, we identified a cell surface hydrolysis activity of ATP to ADP. Interestingly, the IF_1 protein, which inhibits only the ATP hydrolysis activity of the F_1-ATPase, induced a decrease in the ADP present in the cell medium, demonstrating that the ATP hydrolysis measured is dependent on F_1-ATPase activity. Although numerous ATP and ADP hydrolysis activities at the cell surface have been described, together with those of phosphatases such as members of the ecto-ATPase family [112], the almost complete absence of AMP generated in the time course of our experiments allowed us to exclude implication of these ecto-ATPases. Finally, extracellular ADP generation was strongly stimulated by addition of apoA-I, a process also inhibited by the IF_1 protein. The cell surface ATP hydrolase activity of the F_1-ATPase is therefore downregulated by IF_1 and upregulated by the binding of apoA-I, decreasing and increasing ADP production, respectively.

We further investigated the effect of this ADP production on HDL endocytosis by hepatocytes. It was first found that addition of nanomolar concentrations of ADP to the extracellular medium could stimulate holo-HDL particle endocytosis by hepatocytes to a level comparable to the effect of free apoA-I. This effect of ADP was found to be very specific for HDL endocytosis, as demonstrated by the absence of any effect on LDL endocytosis through its LDL receptor (as a typical cargo receptor), or of EGF receptor endocytosis (as a tyrosine kinase-type receptor). Furthermore, the IF_1 protein, which was shown to inhibit F_1-ATPase-induced ADP production specifically, could dramatically inhibit both basal and apoA-I-induced HDL endocytosis, demonstrating that the F_1-ATPase is constitutively active by providing the ADP required for HDL endocytosis. Interestingly, addition of extracellular ATP failed to stimulate HDL endocytosis, suggesting that the F_1-ATPase substrate (i.e., extracellular ATP) is already present in sufficient amounts at the cell surface and therefore does not appear to be restrictive for F_1-ATPase activity and subsequent HDL endocytosis. This is consistent with numerous studies reporting nanomolar concentrations of ATP in the culture media of different cell type, including hepatocytes [106,113], a feature that results from a dynamic equilibrium between ATP metabolism at the cell surface and constitutive ATP release variously by ATP binding cassette (ABC) transporters [114], cell surface voltage-dependent anion channel VDAC [115], vesicle secretion [116,117], or mitochondrial-associated network [118]. The strict dependence of holo-HDL uptake on extracellular ADP level led us to study events downstream from the production of ADP by the ecto-F_1-ATPase.

12.3.3
The Ecto-F_1-ATPase/$P2Y_{13}$-Mediated HDL Endocytosis Pathway

It is well established that nucleotides act as extracellular signaling molecules that mediate diverse biological effects through cell surface receptors known as nucleotidic receptors [113]. To investigate the molecular mechanisms linking cell surface F_1-ATPase activity and HDL endocytosis, we examined whether nucleotidic receptor could be involved in an extracellular ADP signaling pathway regulating HDL endocytosis, thus investigating the contribution of nucleotide G protein receptors of the P2Y family in HDL endocytosis and in particular focusing our work on the three P2Y receptors preferentially activated by ADP: $P2Y_1$, $P2Y_{12}$, and $P2Y_{13}$ [119]. Whereas the $P2Y_1$ and $P2Y_{12}$ receptors have been implicated in platelet aggregation [120], the physiological function of $P2Y_{13}$ receptor was still unknown [121]. We identified by RT-PCR that $P2Y_1$ and $P2Y_{13}$ – but not $P2Y_{12}$ – messengers were present on human and mouse hepatocytes [11]. We further demonstrated through the use of specific pharmacological inhibitors and siRNA for each P2Y receptor subtype that the effect of ADP on HDL endocytosis by hepatocytes was mediated by the $P2Y_{13}$ receptor and not by $P2Y_1$. Interestingly, we observed that the compound AR-C69931MX, previously designed as a $P2Y_{12}$ inhibitor and currently in clinical development as an antithrombotic agent under the name of *Cangrelor*, was also a partial agonist of $P2Y_{13}$ capable of stimulating HDL endocytosis by $HepG_2$ cells and perfused mouse liver [11]. Therefore, the design of more specific $P2Y_{13}$ agonists or antagonists in the future may give rise to new therapeutic agents for modulation of plasma HDL level in the treatment of atherosclerosis. However, involvement of (an)other unknown P2Y receptor(s) that could work *in vivo* in adjunction to or substitution of $P2Y_{13}$ cannot be excluded. $P2Y_{13}$ knockout mouse models, which should be available shortly, should bring new insight in the physiological importance of $P2Y_{13}$ with regard to the regulation of HDL endocytosis by the liver.

12.3.4
Extracellular ADP Levels as a Crucial Point of Regulation of HDL Endocytosis

As discussed in the previous sections, the limiting factor of this so-called "F_1-ATPase/$P2Y_{13}$-mediated HDL endocytosis pathway" is the availability of extracellular ADP produced by the ecto-F_1-ATPase, so control of extracellular ADP levels may be of great interest in the regulation of this pathway. We therefore investigated the mechanisms controlling extracellular ADP levels in hepatic $HepG_2$ cell lines and primary human hepatocytes in order to determine their impact on HDL endocytosis [106] and identified two nucleotide-converting enzymatic activities, adenylate kinase (AK: 2 ADP $\leftrightarrow$ ATP + AMP) and nucleoside diphosphokinase (NDPK: ADP + NTP $\leftrightarrow$

ATP + NDP), on the cell surfaces of hepatocytes, as already reported for other cell types [108,117,122–125]. Ecto-NDPK activity was able to stimulate HDL endocytosis by hepatocytes only when the enzyme was activated with exogenous substrate (e.g., ATP + GDP) in order to generate extracellular ADP. In contrast, the AK inhibitor diadenosine pentaphosphate (Ap_5A) was able to stimulate HDL endocytosis by human hepatocytes without addition of exogenous nucleotides, and this effect was additive to ADP stimulation. This shows that, unlike NDPK, AK is constitutively active and consumes extracellular ADP (2 ADP $\leftrightarrow$ ATP + AMP), thus inhibiting the ADP-dependent $P2Y_{13}$-mediated HDL endocytosis pathway. Interestingly, the inhibition of AK activity by Ap_5A produces up to ~50 % of the stimulation of HDL endocytosis by apoA-I, which suggests that AK activity counteracts the apoA-I effect by consuming the ADP generated by the interaction of apoA-I on the ecto-F_1-ATPase. Moreover, because Ap_5A did not further stimulate HDL endocytosis when the F_1-ATPase activity was inhibited by IF_1, it appears that most of the ADP constitutively generated by the ecto-F_1-ATPase (ATP $\rightarrow$ ADP + Pi) is consumed by AK activity [106]. Therefore, in the $P2Y_{13}$-mediated pathway of HDL endocytosis, the availability of ADP to stimulate $P2Y_{13}$ and finally HDL uptake is a crucial point of regulation. It may depend on the balance between the constitutive AK activity that removes extracellular ADP and the ecto-F_1-ATPase that generates ADP. Activation of ADP synthesis by apoA-I unbalances the pathway and increases HDL endocytosis when required (Fig. 12.1).

12.3.5
F_1-ATPase as a Moonlighting Protein Complex

The concept of one-gene-one-protein-one-function has become too simple because many proteins are found to fulfill different and apparently unrelated functions. This feature may explain how eukaryotic proteomes, which are limited in size, can achieve organism complexity. The functions of such "moonlighting" or multi-tasking proteins can vary as a consequence of changes in cellular localization or ligand binding, cell type, oligomeric state, or enzymatic activity. From this description, F_1-ATPase may be regarded as a typical illustration of moonlighting protein behavior by using a combination of methods to switch between functions.

Indeed, besides its ability to synthesize ATP when localized in the mitochondrial inner membrane, F_1-ATPase has been identified, depending on the cell type, as a cell-surface receptor for apparently unrelated ligands in the courses of studies carried out on lipid metabolism [10,126], regulation of food intake [127–129], hypertension [130–132], angiogenesis [103,133–135] or innate immunity [102,110].

As described in the above sections describing its ability to bind apoA-I and regulate hepatic HDL uptake, F_1-ATPase may be implicated in lipid metabolism. Another way through which ecto-F_1-ATPase might influence lipid metabolism is through binding of enterostatin, a regulatory pentapeptide acting on satiety with specificity for fat [128,129]. Although the mechanism of action of this peptide is poorly documented, some studies report that enterostatin could act by interacting with the ecto-F_1-ATPase β-chain present in the brain and in the gastrointestinal tract, suggesting both a central and peripheral mechanism of action. It has also been suggested that enterostatin, through its interaction with the ecto-F_1-ATPase, could increase thermogenesis and therefore be linked to the uncoupling activity of the gastrointestinal tract, brown adipose tissue, and insulinoma cells [127].

Other interesting data link the CF_6-subunit of ATP synthase to high blood pressure (HBP), which is often associated with dyslipidemia in the so-called metabolic syndrome [136]. Coupling factor 6 (CF_6 or F_6) is a component of the peripheral stalk of ATP synthase, which acts as a stator by counteracting the tendency of the α/β subunits of F_1 to follow the rotation of the c-subunits of F_o and $\gamma\delta\varepsilon$ complex [137]. CF_6 is found at the cell surfaces of endothelial cells [138]. It has also been demonstrated that CF_6 could be released by into the systemic circulation mechanical forces such as shear stress, which is consistent with physiological and clinical data reporting that high plasma levels of CF_6 are associated with human hypertension [131]. The mechanism of action of CF_6 on vasoconstriction is poorly understood, but data suggest that circulating CF_6 could bind the cell surface F_1-ATPase β-subunit and stimulate its ATP hydrolysis activity. As a hypothesis, this process could lead to vasoconstriction through the following cascade: inhibition of the calcium-dependant phospholipase A_2, inducing a decrease of arachidonate release, and the subsequent inhibition of prostacyclin synthesis [132]. Interestingly, recent data suggest that the gene expression and release of CF_6 are mediated by activation of the NF-κB signaling pathway at the level of I-κB degradation [139], a process that can be induced by TNF-α [140] and blocked by PPAR-γ ligands such as troglitazone and 15-deoxy-$\Delta^{12,14}$-prostaglandin J_2 (15d-PGJ_2), as previously described in other systems [141]. This inhibitory effect of PPAR-γ ligands on circulating CF_6 levels could be associated with the effects on these molecules in lowering blood pressure in diabetic or obese insulin-resistant patients [142,143] and improving the overall lipoprotein profile, thereby reducing the risk of atherosclerosis [144]. Whether circulating CF_6 could act on endothelial cell proliferation or could modulate ecto-F_1-ATPase activity on human hepatocytes and subsequent HDL endocytosis still needs to be investigated.

Expression of the ecto-F_1-ATPase on other cell types may also be associated with a broad range of other metabolic dysfunctions. The binding of angiostatin to the ecto-F_1-ATPase α/β-subunits of endothelial cells (EC), for instance, mediates the antiangiogenic effects of angiostatin (i.e., downregulation of EC proliferation and migration), probably by decreasing extracellular ATP levels (see [135,145] for reviews on this topic). The apoA-I binding sites recently identified on human endothelial cells [146] are possibly carried by the ecto-F_1-ATPase and/or SR-BI [147]. Whatever the receptor involved, this feature could be linked to the stimulatory effect of apoA-I on endothelial cell proliferation and its protective effect on endothelial function, as reported previously [148,149]. The ecto-F_1-ATPase is also expressed on various tumor cell types [102,110] and is recognized in association with apoA-I, by the antigen receptor of circulating cytotoxic lymphocytes of the γδ subset and thus promotes an innate tumor cell recognition and lysis [110]. These recent findings reveal an unanticipated connection between the fields of lipid metabolism and immunology, highlighting the role of apolipoproteins in γδ and NKT cell-mediated innate immunity and antitumor immunosurveillance (see [150] for a detailed review on this topic).

12.3.6
The Checkpoints for Regulation of the F_1-ATPase-Mediated HDL Endocytosis Pathway

From the previous points, it is interesting to highlight new possible targets for the regulation of HDL endocytosis by acting on the already identified steps of this new F_1-ATPase pathway. Indeed, we can access activators of HDL endocytosis at the level of F_1-ATPase by addition of apoA-I or on $P2Y_{13}$ receptor by addition of agonists such as ADP and AR-C69931MX we can also inhibit the pathway through the specific F_1-ATPase inhibitor IF_1 or stimulate the pathway by increasing extracellular ADP levels with aP_5A through the inhibition of the ecto-AK activity. All of these represents potential pharmacological targets.

However, the detailed mechanisms involved in the regulation of the F_1-ATPase pathway have only just begun to be defined. Cholesterol and lipid microenvironments may play important roles since the ecto-F_1-ATPase is localized in the caveolae and/or lipid raft domains of different cell types [107,126,151–153], including hepatocytes [111]. Although no response element for transcription factors and/or nuclear receptors involved in lipid metabolism [144] are currently known in the α- and β-chain gene promoters, it has recently been shown with human endothelial cells that F_1-ATPase cell surface expression is stimulated by cell cholesterol loading [126], which suggests that the enzyme cell surface expression may be linked to cholesterol-sensing processes. On endothelial cells, TNF-α has been reported to modulate the cell surface expression of CF_6, IF_1, and β subunits [140,154], while CF_6 is downregulated by PPAR-γ ligands ([139] and preceding section), but whether TNF-α or PPAR-γ ligands can modulate ecto-F_1-ATPase expression and/or activity on human hepatocytes is currently unknown. Although $P2Y_{13}$ has been shown to be coupled to Gαi on transfected cells [155], the signaling pathway downstream from $P2Y_{13}$ activation, which eventually leads to holo-HDL particle endocytosis through an unknown low-affinity receptor, need to be investigated. Our data suggest that, in human hepatocytes, the low-affinity HDL receptor is not SR-BI/CLA-1 ([10,106] and our unpublished observations) and may involve a clathrin-dependent endocytosis process [9].

Finally, little is known about the targeting of F_1-ATPase to the cell surface. Proteomic studies have revealed that most of the F_1-ATPase subunits are present at the hepatocytes plasma membrane [111], which is consistent with our observations reporting a fully active ecto-F_1-ATPase [10]. Therefore, since the major F_1-ATPase subunits are encoded from the nuclear genome with mitochondrial signal import and are assembled to be fully functional only within the mitochondria, with the requirement for the mitochondrial assembly machinery [156,157], it is likely that the cell surface enzyme is forwarded from the mitochondria towards the plasma membrane. From an impressive recent list of mitochondrial proteins identified at cell surfaces [111,158] and other functions attributed to the ecto-F_1-ATPase on other cell types (see [159] and previous section), the elucidation of the cellular mechanism that takes part in this cell surface targeting of mitochondrial proteins may be relevant to understanding of important physiological processes such as lipid metabolism, angiogenesis, and tumor cell recognition.

12.4
Other Receptors involved in HDL Metabolism

12.4.1
Contribution of ABCA1 to Hepatic ApoA-I Secretion

Tangier disease is a severe HDL deficiency syndrome characterized by a rapid turnover of plasma apolipoprotein A-I, accumulation of sterol in tissue macrophages, and prevalent atherosclerosis. The discovery of the ATP binding cassette transporter ABCA1 as the protein defective in Tangier disease was evidence of a key role for ABCA1 in mediation of cellular phospholipids and cholesterol efflux towards lipid-poor apoA-I [160–163]. Although this feature strongly suggests that ABCA1 could be an excellent apoA-I receptor, the nature of the interaction between ABCA1 and lipid-poor apoA-I is somewhat controversial ([164,165] and [76,166] for extensive reviews on this topic). The binding of apoA-I to fibroblasts from Tangier patients, for instance, has been found to be both normal [167] and disrupted [168–170]. This difference may be related to the various binding phenotypes observed on ABCA1 mutants: the W509S mutant is defective in mediating lipid efflux but still binds to apoA-I, whereas some others mutants fail either to promote lipid efflux or to interact with apoA-I [171]. Finally, the strict energy dependence of ABCA1 ATPase activity on apoA-I binding [165,172], together with the binding affinity variability [76] and the fact that ABCA1 can bind other helical apolipoproteins such as apoA-II, apoA-IV, apoCs, and apoE, and also amphipathic helical motifs [173,174], challenge the classical view of a sequence-based interaction, as it is also observed with SR-BI binding feature (see Section 12.2.5). Whatever the interactions between apoA-I and ABCA1, most investigators agree that the ABCA1-mediated lipid efflux process requires more than a single apoA-I association/dissociation to ABCA1 [165,175].

Although ABCA1 has an important role for mobilizing excess cholesterol from peripheral tissues and macrophages and protecting against atherosclerosis (see Chapter 5), the early observation of its high levels of expression in the liver has led to a search for the role(s) of ABCA1 in this organ. Recent findings in mice with selective inactivation of liver ABCA1 [176] have provided evidence of the role of hepatic ABCA1 as the primary source of plasma HDL-cholesterol, accounting for approximately 80 % of plasma HDL-cholesterol; the remaining fraction of HDL would arise largely from the intestinal pool as found in the intestinal ABCA1 inactivation model [177]. Liver ABCA1 thus seems to be essential in the regulation of the amount of HDL available in the plasma (Fig. 12.1), whereas peripheral cells ABCA1 (in cooperation with ABCG1 [178] and potentially SR-BI) would represent a major actor for cholesterol efflux involved in the maintenance of cellular cholesterol homeostasis. At this point it is important to highlight the fact that, in the process of HDL biogenesis (either by ABCA1 in liver or by ABCA1/ABCG1 in peripheral tissues and macrophages), the part of efflux of cholesterol from macrophages represents only a small fraction of total circulating HDL but that it may be the one responsible for the development of atherosclerosis through a specific "macrophage reverse cholesterol transport" as suggested by Lewis and Rader [179]. This is consistent with data showing that complete ABCA1 deficiency in wild-type, LDLr-knockout, and apoE-knockout mice [180,181] did not alter atherosclerosis, but repopulation of ABCA1-knockout mice with wild-type macrophages [180] or macrophages from ABCA1-overexpressing mice [182] led to significant decreases in atherosclerosis. In contrast, selective inactivation of macrophage ABCA1 in mice increases atherosclerosis [180,183]. Surprisingly, and in contrast with the antiatherogenic role of ABCA1 when expressed in macrophages, ABCA1 overexpression in the livers of LDLr-knockout mice leads to accumulation of proatherogenic lipoproteins and enhanced atherosclerosis [184]. This highlights the fact that tissue-specific regulations of ABCA1 are critical factors that modulate the atheroprotective properties of ABCA1 but also limit the uses of animal models in providing insights in the human lipoprotein and atherosclerosis research fields.

12.4.2
The Role of Cubilin in Renal ApoA-I Catabolism

Cubilin is a 460 kDa glycoprotein, expressed in various polarized absorptive epithelia such as yolk sac, small intestine, and kidney [185,186]. In the kidney cortex, cubulin is present at the apical membranes of epithelial cells in the proximal tubules and at the subcellular level is localized in apical clathrin-coated pits and endosomes [187]. This feature suggests that cubilin should act as a classical receptor involved in renal tubular clearance through endocytosis of protein from the glomerular filtrate, but its structure does not contain any transmembrane domain [188]. Several findings, including some made with the aid of surface plasmon resonance, have demonstrated that cubilin associates with megalin, a 600 kDa multi-ligand receptor belonging to the LDL-receptor family. Colocalization of cubilin and megalin in endosome invaginations argues in favor of a common endocytosis process by both proteins, in which internalization of cubilin would be carried out by megalin [189]. Recently, the 45 kDa transmembrane protein amnionless was identified as a yolk sac and renal protein that interacts with cubilin into a large cubilin/amnionless complex in which cubilin functions as the ligand-binding domain and amnionless functions as an essential entity for trafficking and surface expression of cubilin [190,191].

Among a wide spectrum of ligands [192], cubilin binds apoA-I or HDL [193]. In yolk sac epithelial cells, it has been shown that the megalin and cubilin/amnionless receptor system contributes to uptake and degradation of HDL through holoparticle (apolipoprotein plus lipid) endocytosis in a mechanism similar to that of low density lipoprotein (LDL) and distinct from the SR-BI-mediated selective HDL cholesterol uptake [193,194].

Although the glomerular filtration barrier prevents access of the mature HDL particles to the proximal tubules, the megalin and cubilin/amnionless protein receptor system may be exposed to filtered lipid-free/lipid-poor apoA-I not associated with HDL particles and therefore influence overall lipid metabolism [193,195]. Physiologically, the kidney cortex is a major site of catabolism of apoA-I, either free or lipid-poor. Under normal conditions, this uptake is a result of glomerular filtration, tubular reabsorption, and intracellular degradation of free apoA-I [196–198].

In accordance with a physiological role of cubilin in uptake and degradation process of apoA-I in the proximal tubule, apoA-I, which is normally not detected in urine from healthy subjects, was found in excreted urine from cubilin-deficient patients [193]. Interestingly, this phenotype observed in cubilin-deficient patients was associated with a slight increase in plasma HDL and non-HDL cholesterol concentration and proteinuria [190,193], highlighting the role of cubilin in lipid metabolism and reabsorption of proteins in the proximal tubules.

Whether the megalin and cubilin/amnionless protein receptor system may work in concert with other HDL/apoA-I receptor is not known. It has been suggested that lipid uptake from mature HDL through SR-BI would lead to the formation of lipid-free/lipid-poor apoA-I that would be taken up by cubilin in the kidney [193]. This would explain the observation that transgenic mice overexpressing SR-BI in the liver not only have decreased levels of plasma HDL cholesterol but also show decreased concentrations of apoA-I [38]. Unfortunately, homozygous genetic deletion of both cubilin and amnionless gene in mice are lethal, by causing failure in embryonic development [199,200], and this phenotype therefore does not allow study of the physiological role of cubilin in kidney.

Although the megalin and cubilin/amnionless protein receptor system may have an important role in kidney apoA-I catabolism, its role in hepatic HDL endocytosis has never been demonstrated, probably because of the low level of cubilin expression in liver. Parks and colleagues recently found that hepatic ABCA1, through direct lipidation of hepatic lipid-poor apoA-I, could slow down apoA-I catabolism by the kidney and therefore extend its plasma residence time [176,201], so the phenotype of kidney hypercatabolism of apoA-I observed in liver-specific Abca1-knockout mice may be an alternative model with which to ascertain the physiological role of cubilin in kidney.

12.5 Relevance to Human Physiopathology

Although numerous experiments, both in cell cultures and in genetically modified animals, argue strongly in favor of a major role for SR-BI in the uptake of cholesteryl esters from lipoproteins in rodents (see Section 12.2.3.), evidence of such an implication in humans remains scarce. To gain information on the pathophysiological relevance of SR-BI in humans, the impact of its genetic variations on the lipoprotein phenotype have been investigated. Three polymorphisms with significant frequencies of minor alleles, located in exon 1 (G/A, G2S), exon 8 (C/T, anonymous), and in intron 5, have been studied [202]. In a Caucasian general population, the exon 1 variant (frequency 0.12) was associated with increased HDL-C and lower LDL-C levels in men. On the other hand, the exon 8 rare allele (frequency 0.43) was found associated with lower LDL-C among women only [202]. In the larger Framingham offspring cohort, and following multivariate adjustments, carriers of exon 8 variant, in both genders, displayed higher HDL levels. Moreover, an interaction between exon 1 genotype and type 2 diabetes was described, so that carriage of the minor allele was associated with lower LDL-C only among diabetics [203]. These and other studies would suggest that, in humans, SR-BI may be involved in the metabolism of both LDL and HDL, and that variant alleles might confer a more favorable lipoprotein profile. However, other observations have yielded different results. In a dietary intervention protocol, carriers of the exon 1 minor allele displayed an increased susceptibility of LDL variations in response to saturated fats [204]. On the other hand, in studies on familial hypercholesterolemia, exon 1 and exon 8 variants were found to be associated with significant rises in TG, VLDL-C, and LDL-C [205]. Those different examples suggest that the impact of SR-BI and its variants on VLDL and LDL metabolism is influenced by other genetic or environmental determinants.

Deleterious mutations in the ABCA1 gene have been demonstrated to be the cause of Tangier disease and familial high-density lipoprotein deficiency [162]. These mutations either created a premature stop codon or altered a conserved amino acid residue. Since this discovery, two major questions have been addressed: i) are ABCA1 heterozygous mutations or polymorphisms related to low HDL levels in a general population? ii) what are their impacts on cardiovascular diseases?

Over 50 genetic variations in the ABC-A1 gene have been so far described. In a Danish cohort of >9000 subjects, nine rare mutations (frequency <0.01) were found in about 10 % of subjects displaying HDL-C levels in the lowest percentile of the distribution [206]. In the prospective follow-up of this population, one such mutation, in its heterozygous form, although occurring in only 0.4 % of subjects, conferred a 2.5 times higher risk of coronary artery disease [207]. More common polymorphisms have consistently shown effects on the lipoprotein profile. For instance, the R1587K substitution (a G to A transition, frequency 0.26) resides in a large extracellular loop, important for interactions with apoA-I and for cholesterol efflux. The variant has been repeatedly associated with lower levels of HDL-C and apoA-I [206,208]. Another common variant, R219K (G to A, frequency 0.28), resides in the extracellular loop adjacent to the N terminus. Homozygosity for this variation has been associated with increased HDL-C [209]. This is also the case for two less common variations, V771M and V825I (frequencies around 0.05), affecting the fifth and sixth α-helicoidal transmembrane segments [206]. All those genetic variants would account for modest changes in HDL-C levels, either increases or decreases, ranging between 0.02 and 0.04 g L^{-1}. However, in different cross sectional studies, the R219K and V771M variants were underrepresented in patients presenting coronary artery or cerebrovascular diseases [208,210]. Concordantly, homozygosity for the R219K variation (the AA genotype) was found to be associated with a lower prevalence of coronary calcification in a large study on preclinical atherosclerosis [209]. Altogether, these data support a possible impact of ABCA1 genetic variants on susceptibility to cardiovascular diseases, but a careful examination of published data and the persistence of clinical effects after adjustment on HDL-C suggest that risk predisposition according to ABC-A1 genotype might go beyond effects on HDL-C levels.

Evidence from twin and family studies suggests that genetics might account for about 50 % of HDL-C variability [211]. So far, the available data on candidate HDL receptors indicate only a modest impact of their genetic variants on HDL levels in a general population.

Assessment of the physiological role of the recently discovered membrane F_1-ATPase still requires experimental validation in animal models. Moreover, we have further demonstrated that the $P2Y_{13}$ receptor is the downstream partner of the generated ADP, initiating a signaling cascade that should promote HDL hepatic uptake. More recently, we reported that cell surface adenylate kinase activity plays a key role in controlling the level of extracellular ADP [106]. Elucidation of which step in this new mechanistic sequence plays the major role in regulation of HDL endocytosis should open up new perspectives in the search for genetic variants that might be associated with HDL levels and cardiovascular diseases.

Finally, previous studies from our laboratories and from others had provided evidence that interaction of HDL with hepatocytes is dependent on the prior remodeling of these particles by enzymes and lipid transfer proteins present in the vascular compartment. Of major importance are the coordinated roles of cholesterylester transfer protein (CETP) and of hepatic lipase, in the formation both of a lipid-poor apoA-I and of "HDL remnants", which are avidly taken up by hepatocytes [97]. In several studies, a common variant in the promoter of hepatic lipase, affecting the enzyme synthesis and, *in vivo*, HDL-C levels, has been associated with coronary artery disease [212]. Polymorphisms in the CETP gene have been widely investigated. The common Taq1B polymorphism, affecting CETP mass and activity with reciprocal changes in HDL-C, has consistently been related to the risk of coronary atherosclerosis or to its progression [213].

In summary, multiple candidate genes interact to regulate HDL metabolism and functions. Further clinical studies are needed to evaluate their impacts in vascular pathologies. However, new therapeutic strategies to increase HDL levels or to stimulate their functions are either on the way, such as the CETP inhibitors, or can be foreseen. Increased expression of ABCA1 in macrophages should enhance cholesterol efflux from foam cells and stimulate "macrophage reverse cholesterol transport" [179]. On the other hand, a positive modulation of the hepatic F_1-ATPase/$P2Y_{13}$ pathway should stimulate HDL uptake by liver, one of the last steps of reverse cholesterol transport, and should therefore speed up the processes of HDL turnover and cholesterol removal.

References

1 Rothblat, G. H., Bamberger, M., Phillips, M. C. (1986) Reverse cholesterol transport. *Methods Enzymol.*, **129**, 628–644.

2 Tabas, I., Tall, A. R. (1984) Mechanism of the association of HDL3 with endothelial cells, smooth muscle cells, and fibroblasts. Evidence against the role of specific ligand and receptor proteins. *J Biol Chem.*, **259**, 13897–13905.

3 Chanussot, F. *et al.* (1990) Studies on the origin of biliary phospholipid. Effect of dehydrocholic acid and cholic acid infusions on hepatic and biliary phospholipids. *Biochem J.*, **270**, 691–695.

4 Portal, I. *et al.* (1993) Importance of high-density lipoprotein-phosphatidylcholine in secretion of phospholipid and cholesterol in bile. *Am J Physiol.*, **264**, G1052–1056.

5 Robins, S. J. and Fasulo, J. M. (1997) High density lipoproteins, but not other lipoproteins, provide a vehicle

for sterol transport to bile. *J Clin Invest.*, **99**, 380–384.
6 Kruit, J. K. *et al.* (2005) Increased fecal neutral sterol loss upon liver X receptor activation is independent of biliary sterol secretion in mice. *Gastroenterology*, **128**, 147–156.
7 Glass, C. *et al.* (1983) Dissociation of tissue uptake of cholesterol ester from that of apoprotein A-I of rat plasma high density lipoprotein: selective delivery of cholesterol ester to liver, adrenal, and gonad. *Proc Natl Acad Sci U S A*, **80**, 5435–5439.
8 Acton, S. *et al.* (1996) Identification of scavenger receptor SR-BI as a high density lipoprotein receptor. *Science*, **271**, 518–520.
9 Garcia, A. *et al.* (1996) High density lipoprotein3 (HDL3) receptor-dependent endocytosis pathway in a human hepatoma cell line (HepG2). *Biochemistry*, **35**, 13064–13070.
10 Martinez, L. O. *et al.* (2003) Ectopic beta-chain of ATP synthase is an apolipoprotein A-I receptor in hepatic HDL endocytosis. *Nature*, **421**, 75–79.
11 Jacquet, S., *et al.* (2005) The nucleotide receptor P2Y(13) is a key regulator of hepatic High-Density Lipoprotein (HDL) endocytosis. *Cell Mol Life Sci.*, 62, 2508–2515.
12 McKnight, G. L. *et al.* (1992) Cloning and expression of a cellular high density lipoprotein-binding protein that is up-regulated by cholesterol loading of cells. *J Biol Chem.*, **267**, 12131–12141.
13 Chiu, D. S. *et al.* (1997) High-density lipoprotein-binding protein (HBP)/ vigilin is expressed in human atherosclerotic lesions and colocalizes with apolipoprotein E. *Arterioscler Thromb Vasc Biol.*, **17**, 2350–2358.
14 Matsumoto, A. *et al.* (1997) Cloning and characterization of HB2 a candidate High Density Lipoprotein receptor. *J. Biol. Chem.*, **272**, 16778–16782.
15 Hidaka, H. and Fidge, N. H. (1992) Affinity purification of the hepatic high-density lipoprotein receptor identifies two acidic glycoproteins and enables further characterization of their binding properties. *Biochem J.*, **284** (Pt 1), 161–167.
16 Bocharov, A. V. *et al.* (2001) Characterization of a 95 kDa high affinity human high density lipoprotein-binding protein. *Biochemistry*, **40**, 4407–4416.
17 Bocharov, A. V. *et al.* (2000) Heat shock protein 60 is a high-affinity high-density lipoprotein binding protein. *Biochem Biophys Res Commun.*, **277**, 228–235.
18 Wick, G. (2000) Atherosclerosis–an autoimmune disease due to an immune reaction against heat-shock protein 60. *Herz.*, **25**, 87–90.
19 Bachorik, P. S. *et al.* (1982) High-affinity uptake and degradation of apolipoprotein E free high-density lipoprotein and low-density lipoprotein in cultured porcine hepatocytes. *Biochemistry*, **21**, 5675–5684.
20 Bachorik, P. S., Virgil, D. G., Kwiterovich, P. O., Jr. (1987) Effect of apolipoprotein E-free high density lipoproteins on cholesterol metabolism in cultured pig hepatocytes. *J Biol Chem.*, **262**, 13636–13645.
21 Brissette, L. and Noel, S. P. (1986) The effects of human low and high density lipoproteins on the binding of rat intermediate density lipoproteins to rat liver membranes. *J Biol Chem.*, **261**, 6847–6852.
22 Acton, S. L. *et al.* (1994) Expression cloning of SR-BI, a CD36-related class B scavenger receptor. *J. Biol. Chem.*, **269**, 21003–21009.
23 Calvo, D. *et al.* (1997) CLA-1 an 85-kD plasma membrane glycoprotein that acts as a High-affinity receptor for both native (HDL, LDL, and VLDL) and modified (oxLDL, a,d acLDL) lipoproteins. *Arterioscler. Thromb. Vasc. Biol.*, **17**, 2341–2349.
24 Glass, C. *et al.* (1985) Uptake of high-density lipoprotein-associated apoprotein A-I and cholesterol esters by 16 tissues of the rat *in vivo* and by adrenal cells and hepatocytes *in vitro*. *J Biol Chem.*, **260**, 744–750.
25 Goldberg, D. I., Beltz, W. F., Pittman, R. C. (1991) Evaluation of pathways for

the cellular uptake of high density lipoprotein cholesterol esters in rabbits. *J Clin Invest.*, **87**, 331–346.

26 Pittman, R. C. *et al.* (1987) A nonendocytotic mechanism for the selective uptake of high density lipoprotein-associated cholesterol esters. *J Biol Chem.*, **262**, 2443–2450.

27 Wishart, R. and Mackinnon, M. (1990) Uptake and metabolism of high-density lipoproteins by cultured rabbit hepatocytes. *Biochim Biophys Acta*, **1044**, 375–381.

28 Rinninger, F. and Pittman, R. C. (1988) Regulation of the selective uptake of high density lipoprotein-associated cholesteryl esters by human fibroblasts and Hep G2 hepatoma cells. *J Lipid Res.*, **29**, 1179–1194.

29 Jackle, S. *et al.* (1993) Dissection of compartments in rat hepatocytes involved in the intracellular trafficking of high-density lipoprotein particles or their selectively internalized cholesteryl esters. *Hepatology*, **17**, 455–465.

30 Rinninger, F. *et al.* (1994) Selective uptake of high-density lipoprotein-associated cholesteryl esters by human hepatocytes in primary culture. *Hepatology*, **19**, 1100–1114.

31 Rinninger, F. *et al.* (1993) Selective association of lipoprotein cholesteryl esters with liver plasma membranes. *Biochim Biophys Acta*, **1166**, 284–299.

32 Babitt, J. *et al.* (1997) Murine SR-BI, a high density lipoprotein receptor that mediates selective lipid uptake, is N-glycosylated and fatty acylated and colocalizes with plasma membrane caveolae. *J Biol Chem.*, **272**, 13242–13249.

33 Rigotti, A., Acton, S. L., Krieger, M. (1995) The class B scavenger receptors SR-BI and CD36 are receptors for anionic phospholipids. *J. Biol. Chem.*, **270**, 16221–16224.

34 Calvo, D. *et al.* (1998) Human CD36 is a high affinity receptor for the native lipoproteins HDL, LDL, and VLDL. *J Lipid Res.*, **39**, 777–788.

35 Swarnakar, S. *et al.* (1999) Scavenger receptor class B, type I, mediates selective uptake of low density lipoprotein cholesteryl ester. *J Biol Chem.*, **274**, 29733–29739.

36 Stangl, H., Hyatt, M., Hobbs, H. H. (1999) Transport of lipids from high and low density lipoproteins via scavenger receptor-BI. *J Biol Chem.*, **274**, 32692–32698.

37 Rhainds, D. *et al.* (2003) The role of human and mouse hepatic scavenger receptor class B type I (SR-BI) in the selective uptake of low-density lipoprotein-cholesteryl esters. *Biochemistry*, **42**, 7527–7538.

38 Kozarsky, K. F. *et al.* (1997) Overexpression of the HDL receptor SR-BI alters plasma HDL and bile cholesterol levels. *Nature*, **387**, 414–417.

39 Wang, N. *et al.* (1998) Liver-specific overexpression of scavenger receptor BI decreases levels of very low density lipoprotein ApoB, low density lipoprotein ApoB, and high density lipoprotein in transgenic mice. *J Biol Chem.*, **273**, 32920–32926.

40 Ueda, Y. *et al.* (1999) Lower plasma levels and accelerated clearance of high density lipoprotein (HDL) and non-HDL cholesterol in scavenger receptor class B type I transgenic mice. *J Biol Chem.*, **274**, 7165–7171.

41 Ji, Y. *et al.* (1999) Hepatic scavenger receptor BI promotes rapid clearance of high density lipoprotein free cholesterol and its transport into bile. *J Biol Chem.*, **274**, 33398–33402.

42 Rigotti, A. *et al.* (1997) A targeted mutation in the murine gene encoding the high density lipoprotein (HDL) receptor scavenger receptor class B type I reveals its key role in HDL metabolism. *Proc Natl Acad Sci U S A*, **94**, 12610–12615.

43 Varban, M. L. *et al.* (1998) Targeted mutation reveals a central role for SR-BI in hepatic selective uptake of high density lipoprotein cholesterol. *Proc Natl Acad Sci U S A*, **95**, 4619–4624.

44 Trigatti, B. *et al.* (1999) Influence of the high density lipoprotein receptor SR-BI on reproductive and cardiovascular pathophysiology. *Proc Natl Acad Sci U S A*, **96**, 9322–9327.

45 Krieger, M. and Kozarsky, K. (1999) Influence of the HDL receptor SR-BI on atherosclerosis. *Curr Opin Lipidol.*, **10**, 491–497.

46 Cai, L. *et al.* (2005) Serum amyloid A is a ligand for scavenger receptor class B type I and inhibits high density lipoprotein binding and selective lipid uptake. *J Biol Chem.*, **280**, 2954–2961.

47 Marsche, G., *et al.* (2007) The lipidation status of acute-phase protein serum amyloid A determines cholesterol mobilization via scavenger receptor class B, type I. *Biochem J.*, **402**, 117–124.

48 Mardones, P. *et al.* (2002) Alpha-tocopherol metabolism is abnormal in scavenger receptor class B type I (SR-BI)-deficient mice. *J Nutr.*, **132**, 443–449.

49 Reboul, E. *et al.* (2006) Scavenger receptor class B type I (SR-BI) is involved in vitamin E transport across the enterocyte. *J Biol Chem.*, **281**, 4739–4745.

50 Ji, Y. *et al.* (1997) Scavenger receptor BI promotes high density lipoprotein-mediated cellular cholesterol efflux. *J Biol Chem.*, **272**, 20982–20985.

51 Jian, B. *et al.* (1998) Scavenger receptor class B type I as a mediator of cellular cholesterol efflux to lipoproteins and phospholipid acceptors. *J Biol Chem.*, **273**, 5599–5606.

52 Yuhanna, I. S. *et al.* (2001) High-density lipoprotein binding to scavenger receptor-BI activates endothelial nitric oxide synthase. *Nat Med.*, **7**, 853–857.

53 Li, X. A. *et al.* (2005) A novel ligand-independent apoptotic pathway induced by scavenger receptor class B, type I and suppressed by endothelial nitric-oxide synthase and high density lipoprotein. *J Biol Chem.*, **280**, 19087–19096.

54 Huby, T. *et al.* (2006) Knockdown expression and hepatic deficiency reveal an atheroprotective role for SR-BI in liver and peripheral tissues. *J Clin Invest.*, **116**, 2767–2776.

55 Rigotti, A., Miettinen, H. E., Krieger, M. (2003) The role of the high-density lipoprotein receptor SR-BI in the lipid metabolism of endocrine and other tissues. *Endocr Rev.*, **24**, 357–387.

56 Fluiter, K., vander Westhuijzen, D. R., vanBerkel, T. J. (1998) In vivo regulation of scavenger receptor BI and the selective uptake of high density lipoprotein cholesteryl esters in rat liver parenchymal and Kupffer cells. *J Biol Chem.*, **273**, 8434–8438.

57 Woollett, L. A., Kearney, D. M., Spady, D. K. (1997) Diet modification alters plasma HDL cholesterol concentr-ations but not the transport of HDL cholesteryl esters to the liver in the hamster. *J Lipid Res.*, **38**, 2289–2302.

58 Spady, D. K. *et al.* (1998) Kinetic characteristics and regulation of HDL cholesteryl ester and apolipoprotein transport in the apoA-I-/- mouse. *J Lipid Res.*, **39**, 1483–1492.

59 Witt, W. *et al.* (2000) Regulation by vitamin E of the scavenger receptor BI in rat liver and HepG2 cells. *J Lipid Res.*, **41**, 2009–2016.

60 Loison, C. *et al.* (2002) Dietary myristic acid modifies the HDL-cholesterol concentration and liver scavenger receptor BI expression in the hamster. *Br J Nutr.*, **87**, 199–210.

61 Spady, D. K., Kearney, D. M., Hobbs, H. H. (1999) Polyunsaturated fatty acids up-regulate hepatic scavenger receptor B1 (SR-BI) expression and HDL cholesteryl ester uptake in the hamster. *J Lipid Res.*, **40**, 1384–1394.

62 Malerod, L. *et al.* (2003) Hepatic scavenger receptor class B, type I is stimulated by peroxisome proliferator-activated receptor gamma and hepatocyte nuclear factor 4α. *Biochem Biophys Res Commun.*, **305**, 557–565.

63 Mardones, P. *et al.* (2003) Fibrates down-regulate hepatic scavenger receptor class B type I protein expression in mice. *J Biol Chem.*, **278**, 7884–7890.

64 Fu, T., Kozarsky, K. F., Borensztajn, J. (2003) Overexpression of SR-BI by adenoviral vector reverses the fibrate-induced hypercholesterolemia of apolipoprotein E-deficient mice. *J Biol Chem.*, **278**, 52559–52563.

65 Lopez, D. and McLean, M. P. (2006) Activation of the rat scavenger receptor class B type I gene by PPARα. *Mol Cell Endocrinol.*, **251**, 67–77.

66 Malerod, L. *et al.* (2002) Oxysterol-activated LXRα/RXR induces hSR-BI-promoter activity in hepatoma cells and preadipocytes. *Biochem Biophys Res Commun.*, **299**, 916–923.

67 Malerod, L. *et al.* (2005) Bile acids reduce SR-BI expression in hepatocytes by a pathway involving FXR/RXR, SHP, and LRH-1. *Biochem Biophys Res Commun.*, **336**, 1096–1105.

68 Schoonjans, K. *et al.* (2002) Liver receptor homolog 1 controls the expression of the scavenger receptor class B type I. *EMBO Rep.*, **3**, 1181–1187.

69 Sinal, C. J. *et al.* (2000) Targeted disruption of the nuclear receptor FXR/BAR impairs bile acid and lipid homeostasis. *Cell*, **102**, 731–744.

70 Lambert, G. *et al.* (2003) The farnesoid X-receptor is an essential regulator of cholesterol homeostasis. *J Biol Chem.*, **278**, 2563–2570.

71 Kocher, O. *et al.* (1998) Identification and partial characterization of PDZa novel protein containing PDZ interaction domains. *Lab Invest.*, **78**, 117–125.

72 Silver, D. L. (2002) A carboxyl-terminal PDZ-interacting domain of scavenger receptor B, type I is essential for cell surface expression in liver. *J Biol Chem.*, **277**, 34042–34047.

73 Yesilaltay, A. *et al.* (2006) PDZK1 is required for maintaining hepatic scavenger receptor, class B, type I (SR-BI) steady state levels but not its surface localization or function. *J Biol Chem.*, **281**, 28975–28980.

74 Nieland, T. J. *et al.* (2002) Discovery of chemical inhibitors of the selective transfer of lipids mediated by the HDL receptor SR-BI. *Proc Natl Acad Sci U S A*, **99**, 15422–15427.

75 Nieland, T. J. *et al.* (2004) Cross-inhibition of SR-BI- and ABCA1-mediated cholesterol transport by the small molecules BLT-4 and glyburide. *J Lipid Res.*, **45**, 1256–1265.

76 Martinez, L. O. *et al.* (2004) New insight on the molecular mechanisms of high-density lipoprotein cellular interactions. *CMLS, Cell. Mol. Life Sci.*, **61**, 001–018.

77 Williams, D. L. *et al.* (2000) Binding and cross-linking studies show that scavenger receptor BI interacts with multiple sites in apolipoprotein A-I and identify the class A amphipathic α-helix as a recognition motif. *J Biol Chem.*, **275**, 18897–18904.

78 Truong, T. Q. *et al.* (2006) Opposite effect of caveolin-1 in the metabolism of high-density and low-density lipoproteins. *Biochim Biophys Acta*, **1761**, 24–36.

79 Stangl, H. *et al.* (1998) Scavenger receptor, class B, type I-dependent stimulation of cholesterol esterification by high density lipoproteins, low density lipoproteins, and nonlipoprotein cholesterol. *J Biol Chem.*, **273**, 31002–31008.

80 Gu, X. *et al.* (1998) The efficient cellular uptake of high density lipoprotein lipids via scavenger receptor class B type I requires not only receptor-mediated surface binding but also receptor-specific lipid transfer mediated by its extracellular domain. *J Biol Chem.*, **273**, 26338–26348.

81 Connelly, M. A. *et al.* (1999) Comparison of class B scavenger receptors, CD36 and scavenger receptor BI (SR-BI), shows that both receptors mediate high density lipoprotein-cholesteryl ester selective uptake but SR-BI exhibits a unique enhancement of cholesteryl ester uptake. *J Biol Chem.*, **274**, 41–47.

82 Rodrigueza, W. V. *et al.* (1999) Mechanism of scavenger receptor class B type I-mediated selective uptake of cholesteryl esters from high density lipoprotein to adrenal cells. *J Biol Chem.*, **274**, 20344–20350.

83 Silver, D. L. *et al.* (2001) High density lipoprotein (hdl) particle uptake mediated by scavenger receptor class B type 1 results in selective sorting of hdl cholesterol from protein and polarized cholesterol secretion. *J Biol Chem.*, **276**, 25287–25293.

84 Pagler, T. A. *et al.* (2006) SR-BI-mediated high density lipoprotein (HDL) endocytosis leads to HDL resecretion facilitating cholesterol efflux. *J Biol Chem.*, **281**, 11193–11204.

85 Maxfield, F. R. and Wustner, D. (2002) Intracellular cholesterol transport. *J Clin Invest.*, **110**, 891–898.

86 Sun, B. *et al.* (2006) Quantitative analysis of SR-BI-dependent HDL retroendocytosis in hepatocytes and fibroblasts. *J Lipid Res.*, **47**, 1700–1713.

87 Nieland, T. J. *et al.* (2005) Endocytosis is not required for the selective lipid uptake mediated by murine SR-BI. *Biochim Biophys Acta*, **1734**, 44–51.

88 Harder, C. J. *et al.* (2006) Hepatic SR-BI-mediated cholesteryl ester selective uptake occurs with unaltered efficiency in the absence of cellular energy. *J Lipid Res.*, **47**, 492–503.

89 Martinez, L. O. *et al.* (2000) Characterization of Two High-Density Lipoprotein Binding Sites on Porcine Hepatocyte Plasma Membranes: Contribution of Scavenger Receptor Class B Type I (SR-BI) to the Low-Affinity Component. *Biochemistry*, **39**, 1076–1082.

90 Eckhardt, E. R. *et al.* (2006) High density lipoprotein endocytosis by scavenger receptor SR-BII is clathrin-dependent and requires a carboxyl-terminal dileucine motif. *J Biol Chem.*, **281**, 4348–4353.

91 Morrison, J. R., McPherson, G. A., Fidge, N. H. (1992) Evidence for two sites on rat liver plasma membranes which interact with High Density Lipoprotein3. *J. Biol. Chem.*, **267**, 13205–13209.

92 Barbaras, R. *et al.* (1994) Specific binding of free apolipoprotein A-I to a high-affinity binding site on HepG2 cells: Characterization of two high-density lipoprotein sites. *Biochemistry*, **33**, 2335–2340.

93 Xu, S. *et al.* (1997) Apolipoproteins of HDL can directly mediate binding to the scavenger receptor SR-BI, an HDL receptor that mediates selective lipid uptake. *J Lipid Res.*, **38**, 1289–1298.

94 Liadaki, K. N. *et al.* (2000) Binding of high density lipoprotein (HDL) and discoidal reconstituted HDL to the HDL receptor scavenger receptor class B type I. Effect of lipid association and APOA-I mutations on receptor binding. *J Biol Chem.*, **275**, 21262–21271.

95 Thuahnai, S. T. *et al.* (2003) A quantitative analysis of apolipoprotein binding to SR-BI: multiple binding sites for lipid-free and lipid-associated apolipoproteins. *J Lipid Res.*, **44**, 1132–1142.

96 Guendouzi, K. *et al.* (1999) Biochemical and physical properties of remnant-HDL2 and of pre beta 1- HDL produced by hepatic lipase. *Biochemistry*, **38**, 2762–2768.

97 Guendouzi, K. *et al.* (1998) Remnant high density lipoprotein2 particles produced by hepatic lipase display high-affinity binding and increased endocytosis into a human hepatoma cell line (HEPG2). *Biochemistry*, **37**, 14974–14980.

98 Nissen, S. E. *et al.* (2003) Effect of recombinant ApoA-I Milano on coronary atherosclerosis in patients with acute coronary syndromes: a randomized controlled trial. *Jama*, **290**, 2292–2300.

99 Boyer, P. D. (1997) The ATP synthase – a splendid molecular machine. *Annu Rev Biochem.*, **66**, 717–749.

100 Stock, D., Leslie, A. G., Walker, J. E. (1999) Molecular architecture of the rotary motor in ATP synthase. *Science*, **286**, 1700–1705.

101 Cabezon, E. *et al.* (2003) The structure of bovine F1-ATPase in complex with its regulatory protein. *Nat Struct Biol.*, **10**, 744–750.

102 Das, B. *et al.* (1994) A novel ligand in lymphocyte-mediated cytotoxicity: expression of the beta subunit of H+ transporting ATP synthase on the surface of tumor cell lines. *J Exp Med.*, **180**, 273–281.

103 Moser, T. L. *et al.* (1999) Angiostatin binds ATP synthase on the surface of human endothelial cells. *Proc Natl Acad Sci U S A*, **96**, 2811–2816.

104 Beisiegel, U. *et al.* (1988) Apolipoprotein E-binding proteins isolated from dog and human liver. *Arteriosclerosis*, **8**, 288–297.

105 Mahley, R. W. *et al.* (1989) Chylomicron remnant metabolism. Role of hepatic lipoprotein receptors in mediating uptake. *Arteriosclerosis*, **9**, I14–18.

106 Fabre, A. C. *et al.* (2006) Cell surface adenylate kinase activity regulates the F(1)-ATPase/P2Y (13)-mediated HDL endocytosis pathway on human hepatocytes. *Cell Mol Life Sci.*, **63**, 2829–2837.

107 Kim, B. W. *et al.* (2004) Extracellular ATP is generated by ATP synthase complex in adipocyte lipid rafts. *Exp Mol Med.*, **36**, 476–485.

108 Burrell, H. E. *et al.* (2005) Human keratinocytes release ATP and utilize three mechanisms for nucleotide interconversion at the cell surface. *J Biol Chem.*, **280**, 29667–29676.

109 Berger, K., Winzell, M. S., Erlanson-Albertsson, C. (1998) Binding of enterostatin to the human neuroepithelioma cell line SK-N-MC. *Peptides*, **19**, 1525–1531.

110 Scotet, E. *et al.* (2005) Tumor recognition following Vgamma9Vdelta2 T cell receptor interactions with a surface F1-ATPase-related structure and apolipoprotein A-I. *Immunity*, **22**, 71–80.

111 Bae, T. J. *et al.* (2004) Lipid raft proteome reveals ATP synthase complex in the cell surface. *Proteomics*, **4**, 3536–3548.

112 Plesner, L. (1995) Ecto-ATPases: identities and functions. *Int Rev Cytol.*, **158**, 141–214.

113 Ralevic, V. and Burnstock, G. (1998) Receptors for purines and pyrimidines. *Pharmacol Rev.*, **50**, 413–492.

114 Roman, R. M. *et al.* (1997) Hepatocellular ATP-binding cassette protein expression enhances ATP release and autocrine regulation of cell volume. *J Biol Chem.*, **272**, 21970–21976.

115 Okada, S. F. *et al.* (2004) Voltage-dependent anion channel-1 (VDAC-1) contributes to ATP release and cell volume regulation in murine cells. *J Gen Physiol.*, **124**, 513–526.

116 Lazarowski, E. R., Boucher, R. C., Harden, T. K. (2003) Mechanisms of release of nucleotides and integration of their action as P2X- and P2Y-receptor activating molecules. *Mol Pharmacol.*, **64**, 785–795.

117 Lazarowski, E. R., Boucher, R. C., Harden, T. K. (2000) Constitutive release of ATP and evidence for major contribution of ecto-nucleotide pyrophosphatase and nucleoside diphosphokinase to extracellular nucleotide concentrations. *J Biol Chem.*, **275**, 31061–31068.

118 Beaudoin, A. R., Grondin, G., Gendron, F. P. (1999) Immunolocalization of ATP diphosphohydrolase in pig and mouse brains, and sensory organs of the mouse. *Prog Brain Res.*, **120**, 387–395.

119 Abbracchio, M. P. *et al.* (2003) Characterization of the UDP-glucose receptor (re-named here the P2Y14 receptor) adds diversity to the P2Y receptor family. *Trends Pharmacol Sci.*, **24**, 52–55.

120 Gachet, C. (2001) Identification, characterization, and inhibition of the platelet ADP receptors. *Int J Hematol.*, **74**, 375–381.

121 Communi, D. *et al.* (2001) Identification of a novel human ADP receptor coupled to G(i). *J Biol Chem.*, **276**, 41479–41485.

122 Buckley, K. A. *et al.* (2003) Release and interconversion of P2 receptor agonists by human osteoblast-like cells. *Faseb J.*, **17**, 1401–1410.

123 Donaldson, S. H., Picher, M., Boucher, R. C. (2002) Secreted and cell-associated adenylate kinase and nucleoside diphosphokinase contribute to extracellular nucleotide metabolism on human airway surfaces. *Am J Respir Cell Mol Biol.*, **26**, 209–215.

124 Lazarowski, E. R. *et al.* (1997) Identification of an ecto-nucleoside diphosphokinase and its contribution to interconversion of P2 receptor agonists. *J Biol Chem.*, **272**, 20402–20407.

125 Yegutkin, G. G., Henttinen, T., Jalkanen, S. (2001) Extracellular ATP formation on vascular endothelial cells is mediated by ecto-nucleotide kinase activities via phosphotransfer reactions. *Faseb J.*, **15**, 251–260.

126 Wang, T. *et al.* (2006) Cholesterol loading increases the translocation of ATP synthase beta chain into membrane caveolae in vascular endothelial cells. *Biochim Biophys Acta.*, **1761**, 1182–1190.

127 Berger, K. *et al.* (2002) Mitochondrial ATP synthase-a possible target protein in the regulation of energy metabolism in vitro and in vivo. *Nutr Neurosci.*, **5**, 201–210.

128 Berger, K. *et al.* (2004) Enterostatin and its target mechanisms during regulation of fat intake. *Physiol Behav.*, **83**, 623–630.

129 Park, M. *et al.* (2004) The F1-ATPase beta-subunit is the putative enterostatin receptor. *Peptides*, **25**, 2127–2133.

130 Watts, S. W. (2005) Vasoconstriction caused by the ATP synthase subunit-coupling factor 6: a new function for a historical enzyme. *Hypertension*, **46**, 1100–1102.

131 Osanai, T. *et al.* (2003) Plasma concentration of coupling factor 6 and cardiovascular events in patients with end-stage renal disease. *Kidney Int.*, **64**, 2291–2297.

132 Osanai, T. *et al.* (2005) Intracellular signaling for vasoconstrictor coupling factor 6: novel function of beta-subunit of ATP synthase as receptor. *Hypertension*, **46**, 1140–1146.

133 Moser, T. L. *et al.* (2001) Endothelial cell surface F1-FO ATP synthase is active in ATP synthesis and is inhibited by angiostatin. *Proc Natl Acad Sci U S A*, **98**, 6656–6661.

134 Veitonmaki, N. *et al.* (2004) Endothelial cell surface ATP synthase-triggered caspase-apoptotic pathway is essential for k1-5-induced antiangiogenesis. *Cancer Res.*, **64**, 3679–3686.

135 Wahl, M. L., Moser, T. L., Pizzo, S. V. (2004) Angiostatin and anti-angiogenic therapy in human disease. *Recent Prog Horm Res.*, **59**, 73–104.

136 Executive Summary of The Third Report of The National Cholesterol Education Program (NCEP) Expert Panel on Detection, Evaluation, And Treatment of High Blood Cholesterol In Adults (Adult Treatment Panel III). *Jama* (2001) **285**, 2486–2497.

137 Walker, J. E. and Dickson, V. K. (2006) The peripheral stalk of the mitochondrial ATP synthase. *Biochim Biophys Acta*, **1757**, 286–296.

138 Osanai, T. *et al.* (2001) Mitochondrial coupling factor 6 is present on the surface of human vascular endothelial cells and is released by shear stress. *Circulation*, **104**, 3132–3136.

139 Tomita, H. *et al.* (2005) Troglitazone and 15-deoxy-delta(12,14)-prostaglandin J2 inhibit shear-induced coupling factor 6 release in endothelial cells. *Cardiovasc Res.*, **67**, 134–141.

140 Sasaki, S. *et al.* (2004) Tumor necrosis factor α as an endogenous stimulator for circulating coupling factor 6. *Cardiovasc Res.*, **62**, 578–586.

141 Straus, D. S. *et al.* (2000) 15-deoxy-delta 12,14-prostaglandin J2 inhibits multiple steps in the NF-kappa B signaling pathway. *Proc Natl Acad Sci U S A*, **97**, 4844–4849.

142 Ogihara, T. *et al.* (1995) Enhancement of insulin sensitivity by troglitazone lowers blood pressure in diabetic hypertensives. *Am J Hypertens.*, **8**, 316–320.

143 Nolan, J. J. *et al.* (1994) Improvement in glucose tolerance and insulin resistance in obese subjects treated with troglitazone. *N Engl J Med.*, **331**, 1188–1193.

144 Desvergne, B., Michalik, L., Wahli, W. (2006) Transcriptional regulation of metabolism. *Physiol Rev.*, **86**, 465–514.

145 Wahl, M. L. *et al.* (2005) Angiostatin's molecular mechanism: aspects of specificity and regulation elucidated. *J Cell Biochem.*, **96**, 242–261.

146 Rohrer, L. *et al.* (2006) Binding, internalization and transport of apolipoprotein A-I by vascular endothelial cells. *Biochim Biophys Acta*, **1761**, 186–194.

147 von Eckardstein, A., Hersberger, M., Rohrer, L. (2005) Current understanding of the metabolism and biological actions of HDL. *Curr Opin Clin Nutr Metab Care*, **8**, 147–152.
148 Darbon, J. M. *et al.* (1986) Possible role of protein phosphorylation in the mitogenic effect of high density lipoproteins on cultured vascular endothelial cells. *J Biol Chem.*, **261**, 8002–8008.
149 Meyers, C. D. and Kashyap, M. L. (2004) Pharmacologic elevation of high-density lipoproteins: recent insights on mechanism of action and atherosclerosis protection. *Curr Opin Cardiol.*, **19**, 366–373.
150 Champagne, E. *et al.* (2006) Role of Apolipoproteins in gammadelta and NKT Cell-Mediated Innate Immunity. *Immunol Res.*, **33**, 241–256.
151 Sprenger, R. R. *et al.* (2004) Comparative proteomics of human endothelial cell caveolae and rafts using two-dimensional gel electrophoresis and mass spectrometry. *Electrophoresis*, **25**, 156–172.
152 Bini, L. *et al.* (2003) Extensive temporally regulated reorganization of the lipid raft proteome following T-cell antigen receptor triggering. *Biochem J.*, **369**, 301–309.
153 Li, N. *et al.* (2003) Monocyte lipid rafts contain proteins implicated in vesicular trafficking and phagosome formation. *Proteomics*, **3**, 536–548.
154 Cortes-Hernandez, P. *et al.* (2005) The inhibitor protein of the F1F0-ATP synthase is associated to the external surface of endothelial cells. *Biochem Biophys Res Commun.*, **330**, 844–849.
155 Zhang, F. L. *et al.* (2002) P2Y(13): identification and characterization of a novel Gαi-coupled ADP receptor from human and mouse. *J Pharmacol Exp Ther.*, **301**, 705–713.
156 Wang, Z. G., White, P. S., Ackerman, S. H. (2001) Atp11p and Atp12p are assembly factors for the F(1)-ATPase in human mitochondria. *J Biol Chem.*, **276**, 30773–30778.
157 Wiedemann, N., Frazier, A. E., Pfanner, N. (2004) The protein import machinery of mitochondria. *J Biol Chem.*, **279**, 14473–14476.
158 Soltys, B. J. and Gupta, R. S. (2000) Mitochondrial proteins at unexpected cellular locations: export of proteins from mitochondria from an evolutionary perspective. *Int Rev Cytol.*, **194**, 133–196.
159 Champagne, E. *et al.* (2006) Ecto-F1Fo ATP synthase/F1 ATPase: metabolic and immunological functions. *Curr Opin Lipidol.*, **17**, 279–284.
160 Young, S. G. and Fielding, C. J. (1999) The ABCs of cholesterol efflux. *Nat Genet.* **22**, 316–318.
161 Brooks-Wilson, A. *et al.* (1999) Mutations in ABC1 in Tangier disease and familial high-density lipoprotein deficiency. *Nat Genet.*, **22**, 336–345.
162 Marcil, M. *et al.* (1999) Mutations in the ABC1 gene in familial HDL deficiency with defective cholesterol efflux. *Lancet*, **354**, 1341–1346.
163 Remaley, A. T. *et al.* (1999) Human ATP-binding cassette transporter 1 (ABC1): genomic organization and identification of the genetic defect in the original Tangier disease kindred. *Proc Natl Acad Sci U S A*, **96**, 12685–12690.
164 Wang, N. *et al.* (2000) Specific binding of ApoA-I, enhanced cholesterol efflux, and altered plasma membrane morphology in cells expressing ABC1. *J Biol Chem.*, **275**, 33053–33058.
165 Chambenoit, O. *et al.* (2001) Specific docking of apolipoprotein A-I at the cell surface requires a functional ABCA1 transporter. *J Biol Chem.*, **276**, 9955–9960.
166 Marguet, D. and Chimini, G. (2002) The ABCA1 Transporter and ApoA-I. Obligate or Facultative Partners? *Trends Cardiovasc Med.*, **12**, 294–298.
167 Remaley, A. T. *et al.* (1997) Decreased reverse cholesterol transport from Tangier disease fibroblasts. Acceptor specificity and effect of brefeldin on lipid efflux. *Arterioscler Thromb Vasc Biol.*, **17**, 1813–1821.
168 Francis, G. A., Knopp, R. H., Oram, J. F. (1995) Defective removal of cellular cholesterol and phospholipids

by apolipoprotein A-I in Tangier Disease. *J Clin Invest.*, **96**, 78–87.

169 von Eckardstein, A. *et al.* (1998) Plasma and fibroblasts of Tangier disease patients are disturbed in transferring phospholipids onto apolipoprotein A-I. *J Lipid Res.*, **39**, 987–998.

170 Oram, J. F. *et al.* (1999) Reduction in apolipoprotein-mediated removal of cellular lipids by immortalization of human fibroblasts and its reversion by cAMP: lack of effect with Tangier disease cells. *J Lipid Res.*, **40**, 1769–1781.

171 Fitzgerald, M. L. *et al.* (2002) Naturally occurring mutations in the largest extracellular loops of ABCA1 can disrupt its direct interaction with apolipoprotein A-I. *J Biol Chem.*, **277**, 33178–33187.

172 Wang, N. *et al.* (2001) ATP-binding cassette transporter A1 (ABCA1) functions as a cholesterol efflux regulatory protein. *J Biol Chem.*, **276**, 23742–23747.

173 Remaley, A. T. *et al.* (2001) Apolipoprotein specificity for lipid efflux by the human ABCAI transporter. *Biochem Biophys Res Commun.*, **280**, 818–823.

174 Fitzgerald, M. L. *et al.* (2004) ABCA1 and amphipathic apolipoproteins form high-affinity molecular complexes required for cholesterol efflux. *J Lipid Res.*, **45**, 287–294.

175 Fitzgerald, M. L. *et al.* (2004) ABCA1 and amphipathic apolipoproteins form high affinity molecular complexes required for cholesterol efflux. *J Lipid Res.*, **45**, 287–294.

176 Timmins, J. M. *et al.* (2005) Targeted inactivation of hepatic Abca1 causes profound hypoalphalipoproteinemia and kidney hypercatabolism of apoA-I. *J Clin Invest.*, **115**, 1333–1342.

177 Brunham, L. R. *et al.* (2006) Tissue-specific induction of intestinal ABCA1 expression with a liver X receptor agonist raises plasma HDL cholesterol levels. *Circ Res.*, **99**, 672–674.

178 Curtiss, L. K. (2006) Is two out of three enough for ABCG1? *Arterioscler Thromb Vasc Biol.*, **26**, 2175–2177.

179 Lewis, G. F. and Rader, D. J. (2005) New insights into the regulation of HDL metabolism and reverse cholesterol transport. *Circ Res.*, **96**, 1221–1232.

180 Aiello, R. J. *et al.* (2002) Increased atherosclerosis in hyperlipidemic mice with inactivation of ABCA1 in macrophages. *Arterioscler Thromb Vasc Biol.*, **22**, 630–637.

181 McNeish, J. *et al.* (2000) High density lipoprotein deficiency and foam cell accumulation in mice with targeted disruption of ATP-binding cassette transporter-1. *Proc Natl Acad Sci U S A*, **97**, 4245–4250.

182 Van Eck, M. *et al.* (2006) Macrophage ATP-binding cassette transporter A1 overexpression inhibits atherosclerotic lesion progression in low-density lipoprotein receptor knockout mice. *Arterioscler Thromb Vasc Biol.*, **26**, 929–934.

183 van Eck, M. *et al.* (2002) Leukocyte ABCA1 controls susceptibility to atherosclerosis and macrophage recruitment into tissues. *Proc Natl Acad Sci U S A*, **99**, 6298–6303.

184 Joyce, C. W. *et al.* (2006) ABCA1 overexpression in the liver of LDLr-KO mice leads to accumulation of Pro-atherogenic lipoproteins and enhanced atherosclerosis. *J Biol Chem.*, **281**, 33053–33065.

185 Moestrup, S. K. and Kozyraki, R. (2000) Cubilin, a high-density lipoprotein receptor. *Curr Opin Lipidol.*, **11**, 133–140.

186 Moestrup, S. K. (2006) New insights into carrier binding and epithelial uptake of the erythropoietic nutrients cobalamin and folate. *Curr Opin Hematol.*, **13**, 119–123.

187 Sahali, D. *et al.* (1992) Coexpression in humans by kidney and fetal envelopes of a 280 kDa- coated pit-restricted protein. Similarity with the murine target of teratogenic antibodies. *Am J Pathol.*, **140**, 33–44.

188 Moestrup, S. K. *et al.* (1998) The intrinsic factor-vitamin B12 receptor and target of teratogenic antibodies is a megalin-binding peripheral membrane protein with homology to

developmental proteins. *J Biol Chem.*, **273**, 5235–5242.
189 Kozyraki, R. *et al.* (2001) Megalin-dependent cubilin-mediated endocytosis is a major pathway for the apical uptake of transferrin in polarized epithelia. *Proc Natl Acad Sci U S A*, **98**, 12491–12496.
190 Fyfe, J. C. *et al.* (2004) The functional cobalamin (vitamin B12)-intrinsic factor receptor is a novel complex of cubilin and amnionless. *Blood*, **103**, 1573–1579.
191 Strope, S. *et al.* (2004) Mouse amnionless, which is required for primitive streak assembly, mediates cell-surface localization and endocytic function of cubilin on visceral endoderm and kidney proximal tubules. *Development*, **131**, 4787–4795.
192 Moestrup, S. K. and Verroust, P. J. (2001) Megalin- and cubilin-mediated endocytosis of protein-bound vitamins, lipids, and hormones in polarized epithelia. *Annu Rev Nutr.*, **21**, 407–428.
193 Kozyraki, R. *et al.* (1999) The intrinsic factor-vitamin B12 receptor, cubilin, is a high-affinity apolipoprotein A-I receptor facilitating endocytosis of high-density lipoprotein. *Nat Med.*, **5**, 656–661.
194 Hammad, S. M. *et al.* (1999) Cubilin, the endocytic receptor for intrinsic factor-vitamin B(12) complex, mediates high-density lipoprotein holoparticle endocytosis. *Proc Natl Acad Sci U S A*, **96**, 10158–10163.
195 Moestrup, S. K. and Nielsen, L. B. (2005) The role of the kidney in lipid metabolism. *Curr Opin Lipidol.*, **16**, 301–306.
196 Glass, C. K. *et al.* (1983) Tissue sites of degradation of apoprotein A-I in the rat. *J Biol Chem.*, **258**, 7161–7167.
197 Woollett, L. A. and Spady, D. K. (1997) Kinetic parameters for high density lipoprotein apoprotein AI and cholesteryl ester transport in the hamster. *J Clin Invest.*, **99**, 1704–1713.
198 Braschi, S. *et al.* (2000) Role of the kidney in regulating the metabolism of HDL in rabbits: evidence that iodination alters the catabolism of apolipoprotein A-I by the kidney. *Biochemistry*, **39**, 5441–5449.
199 Smith, B. T. *et al.* (2006) Targeted disruption of cubilin reveals essential developmental roles in the structure and function of endoderm and in somite formation. *BMC Dev Biol.*, **6**, 30.
200 Kalantry, S. *et al.* (2001) The amnionless gene, essential for mouse gastrulation, encodes a visceral-endoderm-specific protein with an extracellular cysteine-rich domain. *Nat Genet.*, **27**, 412–416.
201 Lee, J. Y. *et al.* (2005) HDLs in apoA-I transgenic Abca1 knockout mice are remodeled normally in plasma but are hypercatabolized by the kidney. *J Lipid Res.*, **46**, 2233–2245.
202 Acton, S. *et al.* (1999) Association of polymorphisms at the SR-BI gene locus with plasma lipid levels and body mass index in a white population. *Arterioscler Thromb Vasc Biol.*, **19**, 1734–1743.
203 Osgood, D. *et al.* (2003) Genetic variation at the scavenger receptor class B type I gene locus determines plasma lipoprotein concentrations and particle size and interacts with type 2 diabetes: the framingham study. *J Clin Endocrinol Metab.*, **88**, 2869–2879.
204 Perez-Martinez, P. *et al.* (2003) Polymorphism exon 1 variant at the locus of the scavenger receptor class B type I gene: influence on plasma LDL cholesterol in healthy subjects during the consumption of diets with different fat contents. *Am J Clin Nutr.*, **77**, 809–813.
205 Tai, E. S. *et al.* (2003) Polymorphisms at the SRBI locus are associated with lipoprotein levels in subjects with heterozygous familial hypercholesterolemia. *Clin Genet.*, **63**, 53–58.
206 Frikke-Schmidt, R. *et al.* (2005) Mutation in ABCA1 predicted risk of ischemic heart disease in the Copenhagen City Heart Study Population. *J Am Coll Cardiol.*, **46**, 1516–1520.

207 Frikke-Schmidt, R. *et al.* (2004) Genetic variation in ABC transporter A1 contributes to HDL cholesterol in the general population. *J Clin Invest.*, **114**, 1343–1353.

208 Tregouet, D. A. *et al.* (2004) In-depth haplotype analysis of ABCA1 gene polymorphisms in relation to plasma ApoA1 levels and myocardial infarction. *Arterioscler Thromb Vasc Biol.*, **24**, 775–781.

209 Benton, J. L. *et al.* (2006) Associations between two common polymorphisms in the ABCA1 gene and subclinical atherosclerosis: Multi-Ethnic Study of Atherosclerosis (MESA), in Press. *Atherosclerosis.*

210 Andrikovics, H. *et al.* (2006) Decreased frequencies of ABCA1 polymorphisms R219K and V771M in Hungarian patients with cerebrovascular and cardiovascular diseases. *Cerebrovasc Dis.*, **21**, 254–259.

211 Hunt, S. C. *et al.* (1989) Genetic heritability and common environmental components of resting and stressed blood pressures, lipids, and body mass index in Utah pedigrees and twins. *Am J Epidemiol.*, **129**, 625–638.

212 Baroni, M. G. *et al.* (2003) Genetic study of common variants at the apoE, apoAI, apoCIII, apoB, lipoprotein lipase (LPL) and hepatic lipase (LIPC) genes and coronary artery disease (CAD): variation in LIPC gene associates with clinical outcomes in patients with established CAD. *BMC Med Genet.*, **4**, 8.

213 Boekholdt, S. M. *et al.* (2005) Cholesteryl ester transfer protein TaqIB variant, high-density lipoprotein cholesterol levels, cardiovascular risk, and efficacy of pravastatin treatment: individual patient meta-analysis of 13,677 subjects. *Circulation*, **111**, 278–287.

IV
Biological Activities of HDL

High-Density Lipoproteins: From Basic Biology to Clinical Aspects. Edited by Christopher J. Fielding

ISBN: 978-3-527-31717-2

13
HDL and Inflammation

Mohamad Navab, Srinivasa T. Reddy, Brian J. Van Lenten, Georgette M. Buga, G. M. Anantharamaiah, Alan M. Fogelman

13.1
HDL as Part of the Innate Immune System

Since its proposal by Glomset in 1968, reverse cholesterol transport has been presumed to be the major function of HDL [1]. While reverse cholesterol transport is an important HDL function, however, it is not the only function of HDL. In 1984 Sholomo Eisenberg stated that, "In spite of their large number in plasma, it is difficult to define HDL as a vehicle for lipid transport. Chylomicrons and VLDL are undoubtedly lipoproteins that carry triglycerides from sites of absorption and synthesis to sites of storage and utilization. LDL, a product of VLDL metabolism, is the major cholesterol-carrying lipoprotein in human plasma. HDL cannot be similarly classified and other considerations must apply to this lipoprotein." [2].

HDL appears to have evolved as part of the innate immune system and is a major component of the acute phase response (APR) [3]. The lipids in both LDL and HDL are constantly exchanging, but the main protein in LDL, apoB, is not an exchangeable protein. In contrast, all of the proteins in HDL are exchangeable, so HDL is an ideal platform on which rapidly to assemble and disassemble proteins that participate in the APR. Additionally, HDL is a relatively small protein platform in relation to LDL, allowing HDL readily to enter into extracellular spaces. Unlike LDL, which contains a heparin-binding site in apoB, which results in retention of the particle in proteoglycan-containing extracellular matrices, HDL lacks a heparin-binding site and normally is not retained in extracellular matrices. These characteristics of HDL make it an ideal shuttle platform that can enter and sample tissue spaces and transport lipids and proteins from the tissues to sites at a distance, such as the liver.

A recent proteomic analysis of normal human HDL found 56 HDL-associated proteins [4]. In addition to the exchangeable apolipoproteins and the expected lipid transfer proteins, there were proteins involved in hemostasis and thrombosis, proteins associated with the immune and complement systems, growth factors, receptors, hormone-associated proteins, and many other proteins associated with HDL [4].

High-Density Lipoproteins. From Basic Biology to Clinical Aspects. Edited by Christopher J. Fielding

ISBN: 978-3-527-31717-2

13.2
The Role of Oxidized Phospholipids and HDL-Associated Enzymes

More than a decade ago, Van Lenten et al. [5] reported that HDL from rabbits and humans inhibited the ability of LDL to induce human artery wall cells to produce monocyte chemotactic activity. In contrast, after the induction of APR in both rabbits and humans, HDL enhanced the ability of LDL to induce human artery wall cells to produce monocyte chemotactic activity, which was largely mediated by the production of monocyte chemoattractant protein-1 (MCP-1) [5]. Among the changes in HDL that were associated with the APR were a reduction in the activity of HDL-associated antioxidant enzymes (paraoxonase and platelet activating factor acetylhydrolase) and an increase in HDL-associated serum amyloid A (SAA) and ceruloplasmin [5]. In these studies [5] the APR was induced by croton oil injection in rabbits and by elective surgery in humans.

Evidence that an APR could be induced by hyperlipidemia and produce sustained changes in HDL was subsequently forthcoming in mouse models [6]. On an atherogenic diet, paraoxonase activity (PON) was found to be decreased and the HDL-associated positive acute phase reactant apoJ was found to be increased in atherosclerosis-susceptible C57BL/6J mice, but not in atherosclerosis-resistant C3HeJ mice [6]. ApoE null mice on a chow diet were also found to have a reduction in PON activity and an increase in apoJ. HDL from these apoE null mice stimulated LDL-induced monocyte chemotactic activity in human artery wall cell cocultures, and similar results were found in LDL receptor null mice fed a cholesterol-enriched diet [6]. Rabbits fed an atherogenic diet also demonstrated induction of the positive acute phase reactant apoJ in their livers. Injection of mildly oxidized LDL or injection of oxidized phospholipids (but not the unoxidized phospholipids) into the atherosclerosis-susceptible C57BL/6J mice, but not in the atherosclerosis-resistant C3HeJ mice, induced changes in HDL-associated proteins that were similar to the changes induced by hyperlipidemia [6].

The bioactive molecules in mildly oxidized LDL were found to be oxidized phospholipids containing arachidonic acid in the *sn*-2 position [7–9], formed by the action of reactive oxygen species (ROS) derived in part from products of the lipoxygenase pathway [10]. These ROS could be removed by apoAI [11] from normal HDL, or could be destroyed by the action of enzymes contained in normal HDL [12]. HDL-associated enzymes in normal HDL were also found to be capable of destroying these oxidized phospholipids once they had been formed [12].

The importance of the functional state of HDL-associated enzymes and the lipid hydroperoxide content of HDL, in contradistinction to HDL-cholesterol levels, were established in mice that were transgenic for mouse apoA-II. These mice on a chow diet had elevated HDL-cholesterol levels with increased HDL lipid hydroperoxide content and decreased paraoxonase activity [13]. Those mice observed to have atherosclerosis on a chow diet [14] were found to have proinflammatory HDL, which could be rendered antiinflammatory by *in vitro* supplementation with paraoxonase [13].

Increased levels of the oxidized phospholipids found in mildly oxidized LDL were found in the livers of mice transgenically overexpressing Group II secretory phospholipase A_2, while HDL from these mice was proinflammatory in cultures of human artery wall cells [15].

A direct role for the antioxidant HDL-associated enzyme paraoxonase in determining the inflammatory properties of HDL was established in paraoxonase null mice that were found to have proinflammatory HDL and increased susceptibility to organophosphate toxicity and atherosclerosis [16]. Paraoxonase-3 was found, like paraoxonase-1, to be an HDL-associated enzyme capable of inactivating the oxidized phospholipids found in mildly oxidized LDL, but the expression of paraoxonase-3, unlike that of paraoxonase-1, was not regulated by oxidized lipids [17].

Direct confirmation that HDL inflammatory properties are altered during an infectious process that induces an APR was obtained in a mouse model of viral pneumonia. C57BL/6J mice were infected by intranasal infection with influenza A virus [18]. Despite the fact that no virus was detectable in the plasma in this model, the activities of paraoxonase and platelet activating factor acetylhydrolase in HDL decreased after infection, reaching their lowest levels 7 days after inoculation [18]. While apoA-I levels in HDL did not change, apoJ and ceruloplasmin (two positive acute phase reactants) increased significantly post-infection. As a result of the changes in HDL-associated proteins there was a dramatic shift in HDL from an antiinflammatory state to a proinflammatory state [18].

13.3
A Connection between Reverse Cholesterol Transport and Inflammation

A possible connection between the reverse cholesterol transport pathway and the oxidized phospholipids that are found in mildly oxidized LDL and regulated in part by HDL-associated enzymes was first described by Reddy and colleagues [19]. In these studies it was found that the formation of mildly oxidized LDL by human artery wall cocultures required ATP-binding cassette transporter 1 (ABCA1) [19]. It was also shown that 22-hydroxycholesterol increased ROS within human artery wall cells, contributing to increased formation of mildly oxidized LDL. It was postulated that cholesterol loading of endothelial cells resulted in the formation of oxidized sterols that increased intracellular ROS production and also induced ABCA1 expression, resulting in increased export of the oxidized phospholipids to lipoproteins [19]. Subsequently it was demonstrated that cholesterol loading (as opposed to oxycholesterol loading) was sufficient to increase lipid oxidation within human endothelial cells and to enhance dramatically the formation of mildly oxidized LDL [20].

In cultured macrophages it was demonstrated that activation of Toll-like receptors 3 and 4 by microbial ligands blocks the induction of liver X receptor target genes including ABCA1 and in aortic tissue *in vivo* [21]. As a consequence, ligands for Toll-like receptors 3 and 4 strongly inhibited cholesterol efflux from macrophages [21]. These studies highlighted a common mechanism whereby bacterial and viral pathogens may modulate macrophage cholesterol metabolism [21].

Evidence that the inflammatory properties of HDL are significantly affected by oxidative stress in the tissues was provided by a study of paraoxonase-2. Unlike paraoxonase-1, which is an HDL-associated enzyme, paraoxonase-2 is not found in the plasma but is widely distributed in the tissues [22]. Paraoxonase-2-deficient mice, when challenged with a high-fat, high-cholesterol diet for 15 weeks, had levels of HDL-cholesterol, triglycerides, and glucose that were similar to those in wild-type mice. In contrast, however, serum levels of VLDL/LDL-cholesterol were lower (205.7 ± 46 versus 140.3 ± 21 mg dL^{-1} for wild-type and homozygous deficient paraoxonase-2 mice, respectively; $p < 0.001$) [23]. This was found to be due to lower production of VLDL, which was associated with increased hepatic oxidative stress [23]. Despite the lower circulating levels of apoB, the paraoxonase-2-deficient mice had a 2.7-fold increase in the sizes of their atherosclerotic lesions ($p < 0.05$) and these lesions were particularly rich in macrophages. This was associated with increased intracellular macrophage oxidative stress and enhanced production of inflammatory cytokines in response to bacterial lipopolysaccharide [23]. Interestingly, despite reduced production of VLDL in this animal model of increased oxidative stress, HDL was found to be proinflammatory (i.e., increased LDL-induced monocyte chemotactic activity in cultures of human aortic endothelial cells) [23].

Another example of the potential for the inflammatory properties of HDL to be independent of HDL-cholesterol levels was provided by a study of phospholipid transfer protein null (PLTP null) mice. PLTP mice have decreased atherosclerosis despite decreased HDL-cholesterol levels [24,25]. Interestingly, HDL from PLTP null mice that were also null for the LDL receptor or HDL from PLTP null mice that were also transgenic for the expression of human apoB were antiinflammatory while HDL from control mice without the PLTP deficiency were proinflammatory [26].

A review of a large number of mouse models revealed that, almost without exception, mice that were genetically susceptible to atherosclerosis had proinflammatory HDL while mice that were genetically resistant to atherosclerosis had antiinflammatory HDL [27].

13.4
The Inflammatory Properties of HDL in Humans

To date there have been only small studies of proinflammatory HDL in humans. The first involved 27 subjects with documented coronary heart disease (CHD) and normal total cholesterol, triglycerides, LDL-cholesterol, and HDL-cholesterol. These 27 subjects were not on statins or any hypolipidemic medications. Their HDL was found to be proinflammatory in both a cell-based and a cell-free assay, while HDL from 31 age- and gender-matched controls was found to be antiinflammatory [28]. Interestingly, in these studies it was also established with mice exposed to secondary cigarette smoke that changes in the inflammatory properties of HDL could occur within a matter of hours [28]. However, the 27 patients were non-smokers, were not diabetic, and showed no evidence of an acute illness that would be expected to evoke an APR [28]. The authors [28] concluded that the inflammatory properties of HDL were a form of a "chronic" acute phase response, as had previously been postulated by Gabay and Kushner [29].

Ansell and colleagues studied two groups of patients [30]. The first group included 26 patients with stable CHD or CHD equivalents who had not yet started on a statin that had been recommended by their physicians. The inflammatory properties of HDL were determined before starting and six weeks after treatment with 40 mg daily of simvastatin. Ansell and colleagues constructed an HDL inflammatory index by determining the monocyte chemotactic activity generated by cultured human aortic cells exposed to a standard control LDL in the absence and in the presence of the test HDL. The monocyte chemotactic activity generated in the absence of HDL was normalized to 1.0, so that HDL that promoted an inflammatory response greater than the LDL alone had values greater than 1.0 and were considered proinflammatory. HDL that reduced the inflammatory response to the standard control LDL had values less than 1 and were considered antiinflammatory [30]. Prior to the initiation of statin therapy the patients had an HDL inflammatory index of 1.38 ± 0.91 compared to age- and gender-matched controls with an HDL inflammatory index of 0.38 ± 0.14 ($p = 1.5 \times 10^{-5}$). After simvastain treatment the HDL inflammatory index in the patients had decreased to 1.08 ± 0.71, a significant decrease ($p = 0.002$), but still indicating proinflammatory HDL.

Ansell also studied a second group of 20 patients with documented CHD and high HDL-cholesterol levels. The lowest HDL-cholesterol level in this group of patients was 84 mg dL^{-1} and the highest 145 mg dL^{-1}, with the average value being 95 mg dL^{-1}. The average LDL cholesterol for this group was 108 mg dL^{-1} and the average triglyceride value was 89 mg dL^{-1}. None of these patients were on a statin at the time of the study. The patients had an HDL inflammatory index of 1.28 ± 0.29 by the cell-based assay, in comparison with 0.35 ± 0.11 for the control group, which was age- and gender-matched [30]. Ansell and colleagues obtained similar results with a cell-free assay [30]. These studies demonstrated that the inflammatory properties of HDL could be completely independent of HDL-cholesterol levels.

Women with systemic lupus erythematosus (SLE) have 7–50 times greater risk of CHD. McMahon and colleagues used a cell-based assay to study the HDL from 154 women with systemic lupus erythematosus, 48 women with rheumatoid arthritis (RA), and 72 healthy women [31]. The SLE patients had an HDL inflammatory index of 1.02 ± 0.57, versus 0.68 ± 0.28 in the controls ($p < 0.0001$) and 0.81 ± 0.22 in RA patients ($p = 0.001$ versus SLE patients). A higher proportion of SLE patients had proinflammatory HDL (44.7 % of SLE patients versus 4.1 % of controls), 20.1 % of RA patients had scores >1.0 ($p < 0.006$ between all groups), and SLE patients with CHD had significantly higher scores for their HDL inflammatory index than SLE patients without CHD [31].

13.5 Oxidative Stress and Inflammation

Hedrick and colleagues reported that severe oxidative stress in mice can dramatically alter apoA-I [32]. Two laboratories have demonstrated that human apoA-I can be altered by nitration and chlorination of the amino acids in apoA-I [33–36], the consequences of such alterations having included a loss of its ability to promote ABCA1-mediated cholesterol efflux from macrophages [33–36]. The ability of human HDL to promote ABCA1-mediated cholesterol efflux from human monocyte macrophages was found to be inversely related to the inflammatory property of the HDL (i.e., proinflammatory HDL was relatively poor at promoting such efflux while antiinflammatory HDL was relatively good at promoting such efflux) [27].

Oxidative stress resulting in the formation of oxidized phospholipids can come from a number of pathways [37]. The complexities of the APR with regard to components of HDL (serum amyloid A, apoA-I, and paraoxonase) have recently been elucidated [38,39] and help to explain the changes seen in HDL during an APR. The antiinflammatory properties of HDL are increasingly being recognized as playing an important role in its antiatherogenic potential [40] and the APR appears to be a major negative regulator of these antiinflammatory properties [38,39].

One of the clearest examples of a divergence between HDL-cholesterol levels and the link between reverse cholesterol transport and inflammation was provided by a study of apoA-I null mice that were crossed with LDL receptor null mice and also crossed with mice lacking the editing enzyme for apoB, such that only B-100 particles are formed [41]. HDL-cholesterol levels in these mice were well maintained despite the absence of apoA-I, but the resulting HDL was both proinflammatory and poor in its ability to mediate reverse cholesterol efflux from macrophage *in vivo*, and the mice exhibited increased atherosclerosis [41].

13.6 Apolipoprotein Mimetic Peptides that Modulate HDL Function

The potential to use apoA-I or apoA-I mimetic peptides for the prevention and treatment of atherosclerosis has been a long-term goal of many laboratories [20,42–44].

Datta and colleagues screened a series of class A amphipathic helical peptides on the basis of a number of physical and chemical characteristics and the ability of the peptides to inhibit LDL-induced monocyte chemotactic activity in cultures of human aortic wall cells [45]. Only the latter set of properties proved reliable in predicting which peptides would be effective *in vivo* [46].

The first class A amphipathic helical peptide to be demonstrated to be effective *in vivo* was 5F [47]. This peptide, which contains 18 amino acids, including five phenylalanine residues on the hydrophobic face, significantly reduced diet-induced atherosclerosis in mice after injection without altering plasma lipid profiles [47].

A peptide with four phenylalanine residues on the hydrophobic face (4F) was subsequently synthesized entirely from D-amino acids and was found after oral administration to inhibit atherosclerosis dramatically in apoE null mice on a chow diet and LDL receptor null mice on a Western diet without significantly altering plasma lipid levels [48]. While the peptide did not alter plasma lipid levels, it did significantly improve the antiinflammatory properties of HDL [48]. A major determinant of the antiinflammatory properties of these class A amphipathic helical peptides was found to be related to the positions of aromatic residues on the nonpolar face of the molecule [49].

Administration of 4F to LDL receptor null mice infected with influenza A virus given nasally to cause a viral pneumonia resulted in reduced inflammation associated with the pneumonia, as evidenced by the presence of fewer inflammatory cells and lower levels of IL-6 [50]. The administration of 4F also reduced the viral titers in the lung tissue, prevented HDL from becoming proinflammatory, and dramatically reduced the migration of macrophages into the aortic arch and innominate arteries of the peptide-treated mice [50].

HDL is known to be a contributor to the antiviral activity of human serum. In view of this known function of HDL and the results in the murine viral pneumonia model, Van Lenten and colleagues studied the action of D-4F in human Type II pneumocytes infected with influenza A virus [51]. They found that the cells, upon being infected, responded with a marked secretion of phospholipids into the culture medium. The predominant phospholipids that were secreted were not oxidized, but three oxidized phospholipids that are known to be proinflammatory were also secreted into the medium upon viral infection. The addition of D-4F to the cells did not alter the secretion of the non-oxidized phospholipids, but did significantly reduce the secretion of the oxidized phospholipids. A significant reduction in the production of IL-6 by the cells was concomitant with the reduction in the secretion of the oxidized phospholipids, and there were also significant reductions in caspase activation and in viral titers [51]. The apoA-I mimetic peptide D-4F thus mimicked the action of apoA-I and HDL in this model of human pneumocyte viral infection.

Navab et al. further elucidated the mechanism of action of the peptide D-4F in apoE null mice [52]. They found that, twenty minutes after administration of D-4F to apoE null mice by stomach tube, small cholesterol-containing particles of 7 to 8 nm with pre-β mobility and enriched in apoA-I and paraoxonase activity were found in the plasma. After oral administration of D-4F, HDL became antiinflammatory and demonstrated an increased ability to mediate cholesterol efflux from macrophages *in vitro*. Orally administered D-4F also promoted reverse cholesterol transport from intraperitoneally injected cholesterol-loaded macrophages *in vivo*. Additionally, orally administered D-4F significantly reduced lipoprotein lipid hydroperoxides, except in the case of the pre-β HDL fractions, which actually showed increased lipid hydroperoxide content after D-4F administration. It was concluded that the mechanism of action of orally administered D-4F in apoE null mice involves the formation of pre-β HDL with increased levels of apoA-I and paraoxonase activity. As a result, lipoprotein lipid hydroperoxides are reduced, HDL becomes antiinflammatory, and HDL-mediated cholesterol efflux and reverse cholesterol transport from macrophages are enhanced [52].

As noted above, Ansell and colleagues found that treatment with statins improved the inflammatory properties of HDL, but on average, even after treatment with a statin, the HDL from patients with CHD or CHD equivalents remained proinflammatory [30]. Navab et al. found a remarkable synergy between statins and the apoA-I mimetic peptide D-4F in apoE null mice [53]. They administered oral doses of pravastatin and D-4F, which had been predetermined in pilot studies to be ineffective when given as single agents. The combination of pravastatin and D-4F significantly increased HDL-cholesterol levels, apoA-I levels, and paraoxonase activity, rendered HDL antiinflammatory, prevented lesion formation in young, and caused regression of established lesions in old apoE null mice (i.e., mice receiving the combination for 6 months had lesion areas that were smaller than those before the start of treatment). After six months of combined treatment there was an overall reduction in macrophages of 79 %. Administration of the combination increased intestinal apoA-I synthesis by 60 % ($p = 0.011$). In addition, in monkeys the combination potently rendered HDL antiinflammatory. They concluded that the combination of a statin and an HDL-based therapy may be a particularly potent treatment strategy [53].

Abraham and colleagues [54] reported that administration of D-4F to a rat model of diabetes resulted in a significant preservation of arterial heme oxygenase protein levels and activity and was associated with a significant increase in arterial extracellular superoxide dismutase protein levels. As a result the increase in arterial superoxide anion levels induced by diabetes was significantly ameliorated [54]. Induction of diabetes in this rat model was associated with increased sloughing of endothelial cells, as indicated by the presence of endothelial cells in the circulation and increased circulating fragments of endothelial cells. D-4F treatment significantly reduced the number of circulating endothelial cells and the number of endothelial cell fragments and resulted in a dramatic improvement in vascular reactivity. The authors concluded that an increase in endothelial cell sloughing, superoxide anion, and vasoconstriction in diabetic rats can be prevented by administration of D-4F, which is associated with an increase in heme oxygenase-1 and extracellular superoxide dismutase [54].

Pritchard and colleagues [55] reported that oral D-4F treatment restored endothelium and endothelial nitric oxide synthase-dependent (eNOS-dependent) vasodilation in LDL receptor null mice fed a Western diet in direct relationship to the duration of treatments. They also reported that oral D-4F treatment reduced small artery wall thickness in as little as 2 weeks in vessels with preexisting disease. They found that D-4F had no effect on total or HDL-cholesterol levels in this model but reduced proinflammatory HDL levels. Orally administered D-4F did not alter plasma myeloperoxidase concentrations but did reduce the association of myeloperoxidase with apoA-I and did reduce the content of nitrotyrosine in apoA-I. In LDL receptor null mice lacking apoA-I, D-4F treatment increased endothelium and eNOS-dependent vasodilation without reducing wall thickness. The authors concluded that although D-4F restores vascular endothelial cell and eNOS function to increase vasodilation, HDL containing apoA-I, or at least some critical concentration of apoA-I, is required for D-4F to decrease vessel wall thickness [55].

As noted above, the apoA-I mimetic peptide D-4F was found to be effective in reducing hyperlipidemia-induced inflammation both in large arteries (e.g., aorta) [53,54] and in small arteries (e.g., the facial artery of the mouse) [55]. Buga et al. studied hyperlipidemia-induced inflammation of brain arterioles and found that the macrophages were located on the adventitial side of the vessels rather than between the endothelial and smooth muscle cells as is the case in the larger arteries [56]. However, the production of chemokines by the arterioles was similar to what has been reported for large arteries in diet-induced hyperlipidemia. Treatment of LDL receptor null mice on a Western diet with orally administered D-4F significantly reduced the hyperlipidemia-induced inflammation of these brain arterioles without altering plasma lipids, blood pressure, or arteriole lumen size. The authors found that nonvascular cells in the immediately adjacent areas surrounding brain arterioles were also induced to produce chemokines similar to those in the brain arterioles. However, there was only an increase in macrophage-like cells (microglia) in intimate association with the brain arterioles and not in these immediately adjacent areas. The authors concluded that the chemokines produced by the brain arterioles diffused into the area surrounding the arteriole and stimulated these nonvascular cells. On the Western diet, the mice demonstrated significant deterioration in cognitive function that was significantly improved with orally administered D-4F. The authors concluded that the hyperlipidemia had induced brain arteriole inflammation, resulting in chemokine production by the arterioles, which radiated out from the arterioles to nonvascular brain cells and altered their function, causing the impairment in cognition. The authors suggested that orally administered D-4F prevented the hyperlipidemia-induced deterioration in cognitive function by decreasing chemokine production in the brain arterioles [56].

ApoA-I and D-4F are examples of class A amphipathic peptides. ApoJ is an example of a class G* peptide. Both apoA-I and apoJ are associated with HDL and both have been reported to be antiinflammatory. Navab and colleagues reported on the properties of a peptide synthesized from D-amino acids corresponding to residues 113 to 122 in apoJ [57]. Unlike D-4F, D-[113–122]apoJ showed minimal self-association and helicity in the absence of lipids. While D-4F increased the concentration of apoA-I with pre-β mobility in apoE null mice, D-[113–122]apoJ did not. After an oral dose, D-[113–122]apoJ associated more slowly with lipoproteins and was cleared from plasma much more slowly than D-4F. D-[113–122]apoJ significantly improved the ability of plasma to promote cholesterol efflux and improved the inflammatory properties of HDL for up to 48 hours after a single oral dose in apoE null mice. Oral administration of 125 micrograms of D-[113–122]apoJ per mouse per day reduced atherosclerosis in apoE null mice by approximately 70 %, while in monkeys oral D-[113–122]apoJ rapidly reduced lipoprotein lipid hydroperoxides and improved HDL inflammatory properties. The authors concluded that oral D-[113–122]apoJ significantly improves HDL inflammatory properties in mice and monkeys and inhibits lesion formation in apoE null mice [57].

D-4F and D-[113–122]apoJ are peptides that form amphipathic helixes. Navab and colleagues reported that peptides that are too small to form an amphipathic helix can still demonstrate many of the same properties of the peptides that do [58]. KRES, a peptide that contains only four amino acid residues and is too small to form an amphipathic helix, reduced lipoprotein lipid hydroperoxides, increased paraoxonase activity, increased HDL-cholesterol levels, rendered HDL antiinflammatory, and reduced atherosclerosis in apoE null mice. KRES was orally effective when synthesized either from L- or from D-amino acids, suggesting that peptide–protein interactions were not required. A change in the order of the two central amino acids (from KRES to KERS) resulted in the loss of all biologic activity. Solubility in ethyl acetate and interaction with lipids, as determined by differential scanning calorimetry, indicated significant differences between KRES and KERS, while negative stain electron microscopy revealed that KRES formed organized peptide–lipid structures whereas KERS did not. Another tetrapeptide – FREL – shared many of the physical and chemical properties of KRES and was also biologically active in mice and monkeys when synthesized either from L- or from D-amino acids. After oral administration, KRES and FREL were found associated with HDL, whereas KERS was not. After an oral dose, FREL was associated almost exclusively with HDL as early as 2 hours after the dose and remained associated with HDL for up to 48 hours after a single oral dose. KRES associated with HDL and the nonlipoprotein fractions of plasma after an oral dose and was cleared from HDL by 48 hours after a single oral dose. The authors concluded that the ability of peptides to interact with lipids, to remove lipid hydroperoxides, and to activate antioxidant enzymes associated with HDL determines their antiinflammatory and antiatherogenic properties regardless of their ability to form amphipathic helixes [58].

13.7 Summary

HDL has evolved as part of the innate immune system. In the absence of systemic inflammation it is antiinflammatory, but in the presence of an acute phase response or a "chronic" acute phase response, as is the case in atherosclerosis, HDL becomes proinflammatory. Knowledge of HDL's role in inflammation has led to the development of HDL-associated apolipoprotein mimetic peptides that appear to have potential in the treatment of atherosclerosis and other chronic inflammatory conditions.

13.8 Acknowledgments

This work was supported in part by US Public Health Service grants HL-30568 and HL-34343 and the Laubisch, Castera, and M.K. Grey Funds at UCLA.

References

1 Glomset, J. A. (1968) The plasma lecithin: cholesterol acyltransferase reaction. *Journal of Lipid Research*, **9**, 155–167.

2 Eisenberg, S. (1984) High density lipoprotein metabolism. *Journal of Lipid Research*, **25**, 1017–1058.

3 Navab, M., Berliner, J. A., Subbanagounder, G., Hama, S., Lusis, A. J., Casellani, L. W., Reddy, S., Shih, D., Shi, W., Watson, A. D., Van Lenten, B. J., Vora, D. (2001) Fogelman AM. HDL and the inflammatory response induced by LDL-derived oxidized phospholipids. *Arteriosclerosis Thrombosis, and Vascular Biology*, **21**, 481–488.

4 Rezaee, F., Casetta, B., Levels, J. H. M., Speijer, D., Meijers, J. C. M. (2006) Proteomic analysis of high-density lipoprotein. *Proteomics*, **6**, 721–730.

5 Van Lenten, B. J., Hama, S. Y., deBeer, F. C., Stafforini, D. M., McIntyre, T. M., Prescott, S. M., La Du, B. N., Fogelman, A. M., Navab, M. (1995) Anti-inflammatory HDL becomes pro-inflammatory during the acute phase

response. Loss of protective effect against LDL oxidation in aortic wall cell cocultures. *J Clin Invest*, **96**, 2758–2767.

6 Navab, M., Hama-Levy, S., Van Lenten, B. J., Fonarow, G. C., Cardinez, C. J., Castellani, L. W., Brennan, M.-L., Lusis, A. J., Fogelman, A. M. (1997) Mildly oxidized LDL induces an increased apolipoprotein J/paraoxonase ratio. *J Clin Invest*, **99**, 2005–2019.

7 Watson, A. D., Leitinger, N., Navab, M., Faull, K. F., Horkko, S., Witztum, J. L., Palinski, W., Schwenke, D., Salomon, R. G., Sha, W., Subbanagounder, G., Fogelman, A. M.Berliner, J. A. (1997) Structural identification by mass spectrometry of oxidized phospholipids in minimally oxidized low density lipoprotein that induces monocyte/ endothelial interactions and evidence for their presence in vivo. *J Biol Chem*, **274**, 13597–13607.

8 Watson, A. D., Subbanagounder, G., Welsbie, D. S., Faull, K. F., Navab, M., Jung, M. E., Fogelman, A. M., Berliner, J. A. (1999) *J Biol Chem*, 24787–24798.

9 Subbanagounder, G., Leitinger, N., Schwenke, D. C., Wong, J. W., Lee, H., Rizza, C., Watson, A. D., Faull, K. F., Fogelman, A. M., Berliner, J. A. (2000) Determinants of bioactivity of oxidized phospholipids, specific oxidized fatty acyl groups at the sn-2 position. *Arterioscler Thromb Vasc Biol*, **20**, 2248–2254.

10 Navab, M., Anantharamaiah, G. M., Reddy, S. T., Van Lenten, B. J., Ansell, B. J., Fonarow, G. C., Vahabzadeh, K., Hama, S., Hough, G., Kamranpour, N., Berliner, J. A., Lusis, A. J., Fogelman, A. M. (2004) The oxidation hypothesis of atherogenesis: the role of oxidized phospholipids and HDL. *J Lipid Res*, **45**, 993–1007.

11 Navab, M., Hama, S. Y., Cooke, C. J., Anantharamaiah, G. M., Chaddha, M., Jin, L., Subbanagounder, G., Faull, K. F., Reddy, S. T., Miller, N. E., Fogelman, A. M. (2000) Normal high density lipoprotein inhibits three steps in the formation of mildly oxidized low density lipoprotein: step 1. *J Lipid Res*, **41**, 1481–1494.

12 Navab, M., Hama, S. Y., Anantharamaiah, G. M., Hassan, K. M., Hough, G. P., Watson, A. D., Reddy, S. T., Sevanian, A., Fonarow, G. C., Fogelman, A. M. (2000) Normal high density lipoprotein inhibits three steps in the formation of mildly oxidized low density lipoprotein: steps 2 and 3. *J Lipid Res*, **41**, 1495–1508.

13 Castellani, L. W., Navab, M., Van Lenten, B. J., Hedrick, C. C., Hama, S. Y., Goto, A. M., Fogelman, A. M., Lusis, A. J. (1997) Overexpression of apolipoprotein AII in transgenic mice converts high density lipoproteins to proinflammatory particles. *J Clin Invest*, **100**, 464–474.

14 Warden, C. H., Hedrick, C. C., Qiao, J.-H., Castellani, L. W., Lusis, A. J. (1993) Atherosclerosis in transgenic mice overexpressing apolipoprotein A-II. *Science*, **261**, 469–472.

15 Leitinger, N., Watson, A. D., Hama, S. Y., Ivandic, B., Qiao, J.-H., Huber, J., Faull, K. F., Grass, D. S., Navab, M., Fogelman, A. M., de Beer, F. C., Lusis, A. J., Berliner, J. A. (1999) Role of group II secretory phospholipase A2 in atherosclerosis. 2. Potential involvement of biologically active oxidized phospholipids. *Arterioscler Thromb Vasc Biol*, **19**, 1291–1298.

16 Shih, D. M., Gu, L., Xia, Y.-R., Navab, M., Li, W.-F., Hama, S., Castellani, L. W., Furlong, C. E., Costa, L. G., Fogelman, A. M., Lusis, A. J. (1998) Mice lacking serum paraoxonase are susceptible to organophosphate toxicity and atherosclerosis. *Nature*, **394**, 284–287.

17 Reddy, S. T., Wadleigh, D. J., Grijalva, V., Ng, C., Hama, S., Gangopadhyay, A., Shih, D. M., Lusis, A. J., Navab, M., Fogelman, A. M. (2001) Human paraoxonase-3 is an HDL-associated enzyme with biological activity similar to paraoxonase-1 protein but is not regulated by oxidized lipids. *Arterioscler Thromb Vasc Biol*, **21**, 542–547.

18 Van Lenten, B. J., Wagner, A. C., Nayak, D. P., Hama, S., Navab, M., Fogelman, A. M. (2001) High-density lipoprotein loses its anti-inflammatory properties during acute influenza A infection. *Circulation*, **103**, 2283–2288.

19 Reddy, S. T., Hama, S., Ng, C., Grijalva, V., Navab, M., Fogelman, A. M. (2002) ATP-binding cassette transporter 1 participates in LDL oxidation by artery wall cells. *Arterioscler Thromb Vasc Biol*, **22**, 1877–1883.

20 Reddy, S. T., Anantharamaiah, G. M., Navab, M., Hama, S., Hough, G., Grijalva, V., Garber, D. W., Datta, G., Fogelman, A. M. (2006) Oral amphipathic peptides as therapeutic agents. *Expert Opin Investig Drugs*, **15**, 13–21.

21 Castrillo, A., Joseph, S. B., Vaidya, S. A., Haberland, M., Fogelman, A. M., Cheng, G., Tontonoz, P. (2003) Crosstalk between LXR and Toll-like receptor signaling mediates bacterial and viral antagonism of cholesterol metabolism. *Molecular Cell*, **12**, 805–816.

22 Ng, C. J., Wadleigh, D. J., Gangopadhyay, A., Hama, S., Grijalva, V. R., Navab, M., Fogelman, A. M., Reddy, S. T. (2001) Paraoxonase-2 is a ubiquitously expressed protein with antioxidant properties and is capable of preventing cell-mediated oxidative modification of low density lipoprotein. *J Biol Chem*, **276**, 44444–44449.

23 Ng, C., Bourquard, N., Grijalva, V., Hama, S., Shih, D. M., Navab, M., Fogelman, A. M., Lusis, A. J., Young, S., Reddy, S. T. (2006) Paraoxonase-2 deficiency aggravates atherosclerosis in mice despite lower apolipoprotein-B containing lipoproteins-anti-atherogenic role for paraoxonase-2. *J Biol Chem*, **281**, 29491–29500.

24 Jiang, X. C., Bruce, C., Mar, J., Lin, M., Ji, Y., Francone, O. L., Tall, A. R. (1999) Targeted mutation of plasma phospholipid transfer protein gene markedly reduces high-density lipoprotein levels. *J Clin Invest*, **103**, 907–914.

25 Jian, X. C., Qin, S., Qiao, C., Kawano, K., Lin, M., Skold, A., Xiao, X., Tall, A. R. (2001) Apolipoprotein B secretion and atherosclerosis are decreased in mice with phospholipid-transfer protein deficiency. *Nat Med*, **7**, 847–852.

26 Yan, D., Navab, M., Bruce, C., Fogelman, A. M., Jiang, X.-C. (2004) PLTP deficiency improves the anti-inflammatory properties of HDL and reduces the ability of LDL to induce monocyte chemotactic activity. *J Lipid Res*, **45**, 1852–1858.

27 Navab, M., Anantharamaiah, G. M., Reddy, S. T., Van Lenten, B. J., Ansell, B. J., Hama, S., Hough, G., Bachini, E., Grijalva, V. R., Wagner, A. C., Shaposhnik, Z., Fogelman, A. M. (2005) The double jeopardy of HDL. *Annals of Medicine*, **37**, 173–178.

28 Navab, M., Hama, S. Y., Hough, G. P., Subbanagounder, G., Reddy, S. T., Fogelman, A. M. (2001) A cell-free assay for detecting HDL that is dysfunctional in preventing the formation of or inactivating oxidized phospholipids. *J Lipid Res*, **42**, 1308–1317.

29 Gabay, C. and Kushner, I. (1999) Acute-phase proteins and other systemic responses to inflammation. *N Engl J Med*, **340**, 448–454.

30 Ansell, B. J., Navab, M., Hama, S., Kamranpour, N., Fonarow, G., Hough, G., Rahmani, S., Mottahedeh, R., Dave, R., Reddy, S. T., Fogelman, A. M. (2003) Inflammatory/anti-inflammatory properties of high-density lipoprotein distinguish patients from control subjects better than high-density lipoprotein cholesterol levels and are favorably affected by simvastatin treatment. *Circulation*, **108**, 2751–2756.

31 McMahon, M., Grossman, J., FitzGerald, J., Dahlin-Lee, E., Wallace, D. J., Thong, B. Y., Badsha, H., Kalunian, K., Charles, C., Navab, M., Fogelman, A. M., Hahn, B. H. (2006) Proinflammatory high-density lipoprotein as a biomarker for atherosclerosis in patients with systemic lupus erythematosus and rheumatoid arthritis. *Arthritis & Rheumatism*, **54**, 2541–2549.

32 Hedrick, C. C., Hassan, K., Hough, G. P., Yoo, J.-H., Simzar, S., Quinto, C. R., Kim, S.-M., Dooley, A., Langi, S., Hama, S. Y., Navab, M., Witztum, J. L., Fogelman, A. M. (2000) Short-term feeding of atherogenic diet to mice results in reduction of HDL and paraoxonase that may be mediated by an immune mechanism. *Arterioscler Thromb Vasc Biol*, **20**, 1946–1952.

33 Nicholls, S. J., Zheng, L., Hazen, S. L. (2005) Formation of dysfunctional high-density lipoprotein by myeloperoxidase. *Trends Cardiovasc Med*, **15**, 212–219.

34 Shao, B., Oda, M. N., Oram, J. F., Heinecke, J. W. (2006) Myeloperoxidase: an inflammatory enzyme for generating dysfunctional high density lipoprotein. *Curr Opin Cardiol*, **21**, 322–328.

35 Fogelman, A. M. (2004) When good cholesterol goes bad. *Nat Med*, **10**, 902–903.

36 Navab, M., Anantharamaiah, G. M., Fogelman, A. M. (2005) The role of high-density lipoprotein in inflammation. *Trends Cardiovasc Med*, **15**, 158–161.

37 Fogelman, A. M. (2005) When pouring water on the fire makes it burn brighter. *Cell Metabolism*, **2**, 6–8.

38 Han, C. Y., Chiba, T., Campbell, J. S., Fausto, N., Chaisson, M., Orasanu, G., Pluzky, J., Chait, A. (2006) Reciprocal and coordinate regulation of serum amyloid A versus apolipoprotein A-I and paraoxonase-1 by inflammation in murine hepatocytes. *Arterioscler Thromb Vasc Biol*, **26**, 1806–1813.

39 Van Lenten, B. J., Reddy, S. T., Navab, M., Fogelman, A. M. (2006) Understanding changes in high density lipoproteins during the acute phase response. *Arterioscler Thromb Vasc Biol*, **26**, 1687–1688.

40 Barter, P. J., Nicholls, S., Rye, K.-A., Anantharamaiah, G. M., Navab, M., Fogelman, A. M. (2004) Antiinflammatory properties of HDL. *Circ Res*, **95**, 764–772.

41 Moore, R. E., Navab, M., Millar, J. S., Zimetti, F., Hama, S., Rothblat, G. H., Rader, D. J. (2005) Increased atherosclerosis in mice lacking apolipoprotein A-I attributable to both impaired reverse cholesterol transport and increased inflammation. *Circ Res*, **97**, 763–771.

42 Navab, M., Anantharamaiah, G. M., Reddy, S. T., Van Lenten, B. J., Datta, G., Fogelman, A. M. (2004) Human apolipoprotein A-I and A-I mimetic peptides, potential for atherosclerosis reversal. *Curr Opin Lipidol*, **14**, 645–649.

43 Anantharamaiah, G. M., Navab, M., Reddy, S. T., Garber, D. W., Datta, G., Gupta, H., White, C. R., Handattu, S. P., Palgunachari, M. N., Chaddha, M., Mishra, V. K., Segrest, J. P., Fogelman, A. M. (2006) Synthetic peptides: managing lipid disorders. *Curr Opin Lipidol*, **17**, 233–237.

44 Navab, M., Anantharamaiaha, G. M., Reddy, S. T., Van Lenten, B. J., Datta, G., Garber, D., Fogelman, A. M. (2006) Potential clinical utility of high-density lipoprotein-mimetic peptides. *Curr Opin Lipidol*, **17**, 440–444.

45 Datta, G., Chaddha, M., Hama, S., Navab, M., Fogelman, A. M., Garber, D. W., Mishra, V. K., Epand, R. M., Epand, R. F., Lund-Katz, S., Philips, M. C., Segrest, J. P., Anantharamaiah, G. M. (2001) Effects of increasing hydrophobicity on the physical-chemical and biologic properties of a class A amphipathic helical peptide. *J Lipid Res*, **42**, 1096–1104.

46 Navab, M., Anantharamaiah, G. M., Reddy, S. T., Hama, S., Hough, G., Grijalva, V. R., Yu, N., Ansell, B. J., Datta, G., Garber, D. W., Fogelman, A. M. (2005) Apolipoprotein A-I mimetic peptides. *Arterioscler Thromb Vasc Biol*, **25**, 1325–1331.

47 Garber, D. W., Datta, G., Chaddha, M., Palgunachari, M. N., Hama, S. Y., Navab, M., Fogelman, A. M., Segrest, J. P., Anatharamaiah, G. M. (2001) A new synthetic class A amphipathic peptide analogue protects mice from diet-induced atherosclerosis. *J Lipid Res*, **42**, 545–552.

48 Navab, M., Anantharamaiah, G. M., Hama, S., Garber, D. W., Chaddha, M., Hough, G., Lallone, R., Fogelman,

A. M. (2002) Oral administration of an apoA-I mimetic peptide synthesized from D-amino acids dramatically reduces atherosclerosis in mice independent of plasma cholesterol. *Circulation*, **105**, 290–292.

49 Datta, G., Epand, R. F., Epand, R. M., Chaddha, M., Kirksey, M. A., Garber, D. W., Lund-Katz, S., Phillips, M. C., Hama, S., Navab, M., Fogelman, A. M., Palgunachari, M. N., Segrest, J. P., Anantharamaiah, G. M. (2004) Aromatic residue position on the nonpolar face of class A amphipathic helical peptides determines biological activity. *J Biol Chem*, **279**, 26509–26517.

50 Van Lenten, B. J., Wagner, A. C., Anantharamaiah, G. M., Garber, D. W., Fishbein, M. C., Adhikary, L., Nayak, D. P., Hama, S., Navab, M., Fogelman, A. M. (2002) Influenza infection promotes macrophage traffic into arteries of mice that is prevented by D-4F, an apolipoprotein A-I mimetic peptide. *Circulation*, **106**, 1127–1132.

51 Van Lenten, B. J., Wagner, A. C., Navab, M., Anantharamaiah, G. M., Hui, E.K.-W., Nayak, D. P., Fogelman, A. M. (2004) D-4F, an apolipoprotein A-I mimetic peptide inhibits the inflammatory response induced by influenza A infection of human type II pneumocytes. *Circulation*, **110**, 3252–3258.

52 Navab, M., Anantharamaiah, G. M., Reddy, S. T., Hama, S., Hough, G., Grijalva, V. R., Wagner, A. C., Frank, J. S., Datta, G., Garber, D., Fogelman, A. M. (2004) Oral D-4F causes formation of pre-β high-density lipoprotein and improves high-density lipoprotein-mediated cholesterol efflux and reverse cholesterol transport from macrophages in apolipoprotein E-null mice. *Circulation*, **109**, 3215–3220.

53 Navab, M., Anantharamaiah, G. M., Hama, S., Hough, G., Reddy, S. T., Frank, J. S., Garber, D. W., Handattu, S., Fogelman, A. M. (2005) D-4F and statins synergize to render HDL anti-inflammatory in mice and monkeys and cause lesion regression in old apolipoprotein E-null mice. *Arterioscler Thromb Vasc Biol*, **25**, 1426–1432.

54 Kruger, A. L., Peterson, S., Turkseven, S., Kaminski, P. M., Zhang, F. F., Quan, S., Wolin, M. S., Abraham, N. G. (2005) D-4F induces heme oxygenase-1 and extracellular superoxide dismutase, decreases endothelial cell sloughing, and improves vascular reactivity in rat model of diabetes. *Circulation*, **111**, 3126–3134.

55 Ou, J., Wang, J., Xu, H., Ou, Z., Sorci-Thomas, M. G., Jones, D. W., Signorino, P., Densmore, J. C., Kaul, S., Oldham, K. T., Pritchard, K. A., Jr. (2005) Effects of D-4F on vasodilation and vessel wall thickness in hypercholesterolemic LDL receptor-null and LDL receptor/apolipoprotein A-I double-knockout mice on Western diet. *Circulation Research*, **97**, 1190–1197.

56 Buga, G. M., Frank, J. S., Mottino, G. A., Hendizadeh, M., Hakhamian, A., Tillisch, J. H., Reddy, S. T., Navab, M., Anantharamaiah, G. M., Ignarro, L. J., Fogelman, A. M. (2006) D-4F decreases brain arteriole inflammation and improves cognitive performance in LDL receptor null mice on a Western Diet. *J Lipid Res*, **47**, 2148–2160.

57 Navab, M., Anantharamaiah, G. M., Reddy, S. T., Van Lenten, B. J., Wagner, A. C., Hama, S., Hough, G., Bachini, E., Garber, D. W., Mishra, V. K., Palgunachari, M. N., Fogelman, A. M. (2005) An oral apoJ peptide renders HDL anti-inflammatory in mice and monkeys and dramatically reduces atherosclerosis in apolipoprotein E-null mice. *Arterioscler Thromb Vasc Biol*, **25**, 1932–1937.

58 Navab, M., Anantharamaiah, G. M., Reddy, S. T., Hama, S., Hough, G., Frank, J. S., Grijalva, V. R., Ganesh, V. K., Mishra, V. K., Palgunachari, M. N., Fogelman, A. M. (2005) Oral small peptides render HDL anti-inflammatory in mice and monkeys and reduce atherosclerosis in apoE null mice. *Circulation Research*, **97**, 524–532.

14
HDL in Health and Disease: Effects of Exercise and HIV Infection on HDL Metabolism

Dmitri Sviridov, Michael I. Bukrinsky, Bronwyn A. Kingwell, Jennifer Hoy

14.1
Introduction

HDL is good for you. The negative correlation between HDL-C (HDL cholesterol) and the risk of CAD (coronary artery disease) was first established in the Framingham Heart study [1] and consequently confirmed in a number of cross-sectional and interventional studies [2–7]. An important feature of this relationship is that it is preserved within a wide range of LDL-C (LDL cholesterol) concentrations and in the presence or absence of other cardiovascular risk factors [8], so the atheroprotective effect of HDL is independent of other factors affecting the risk of atherosclerosis. There are two important implications of this conclusion. Firstly, higher HDL levels would probably benefit everybody, including healthy individuals with normal HDL-C levels and without any specific cardiovascular risk factors. Secondly, patients with abnormally low HDL-C levels (hypoalphalipoproteinemia) require active intervention to establish the cause of dyslipidemia, followed by prescription of treatment to increase HDL levels. These considerations establish two strategies for the rapidly emerging "HDL therapy". One is to stimulate the endogenous reverse cholesterol transport pathway, which is already functioning to its natural capacity. Examples of effective interventions are physical exercise [9] and dietary interventions [10]. The second is to diagnose specific defects in the reverse cholesterol transport pathway associated with a given pathological condition and to target this problem. Approaches to reach these two goals may be quite different. An additional issue is that raising HDL-C levels should not be accompanied by adverse changes in the properties of HDL particles that may damage their functionality. Balancing these factors may be difficult and a universal "HDL therapy" may not be possible. Here we examine two conditions that have a profound impact on HDL metabolism: aerobic exercise and HIV infection. These conditions provide a unique insight into reverse cholesterol transport and HDL functionality that may have utility in the development of effective HDL therapies.

High-Density Lipoproteins: From Basic Biology to Clinical Aspects. Edited by Christopher J. Fielding

ISBN: 978-3-527-31717-2

14.2 HDL Metabolism

HDL is a key element of the reverse cholesterol transport (RCT) pathway. Moreover, RCT and HDL metabolism are two pathways with substantial overlap, with only the final step of RCT – the bile acid formation and excretion – being unrelated to HDL. RCT has two physiological components: removal and excretion of excessive cholesterol from the cells and formation of HDL. Which component prevails and has greater impact on protection against atherosclerosis depends on the tissue. Thus, RCT from liver [11] and intestine [12] has a major impact on plasma HDL levels, but probably a lesser effect on tissue cholesterol content. RCT from macrophages, on the other hand, has a profound effect on cell cholesterol content, but little impact on plasma HDL levels [13]. Further, in addition to removing cholesterol from cells by RCT, HDL has a number of other atheroprotective properties that may not be directly related to changes in cellular cholesterol content [4,14]. Changes in RCT may influence not only HDL levels, but also the "quality" or functionality of the HDL particles. Below we briefly describe the reverse cholesterol transport pathway (more detailed information on each step can be found in other chapters of this volume).

14.2.1 HDL Formation

The first step of HDL metabolism is cholesterol efflux, a process underlying the formation of HDL from lipid-free apoA-I and cellular lipids, cholesterol, and/or phospholipids. Cholesterol efflux depends both on the ability of cells to release cholesterol and phospholipids to extracellular acceptors and the ability of plasma to accept these lipids. The initial step of cholesterol efflux occurs at the plasma membranes of the cells as a result of the action of ABCA1 [15]. The products of ABCA1-dependent cholesterol efflux are small, lipid-poor discoid HDL particles, most probably analogous to preβ_1-HDL [16]. These particles are also effective acceptors of cholesterol from cells, although other pathways involving ABCG1 and scavenger receptor B1 (SR-B1) mediate this second step [15]. The products of the second step are large, lipid-rich discoid particles, probably analogous to pre β_2-HDL.

14.2.2 HDL Maturation

Upon reaching the appropriate size and composition, discoid HDL particles become a substrate for lecithin:cholesterol acyltransferase (LCAT). LCAT esterifies free cholesterol and forms the hydrophobic core, transforming HDL from discoid to spherical particles [17]. These particles acquire more cholesterol and apolipoproteins (apoA-I, apoA-II, and apoE) from other lipoproteins in the plasma (such as very low density lipoprotein, VLDL, undergoing lipolysis) and possibly fuse with each other.

The products of this step are large, cholesterol-rich, spherical particles, likely to be analogous to α_1HDL.

14.2.3 HDL Remodeling

Upon reaching the appropriate size and composition, spherical HDL particles become a substrate for a number of enzymes and transfer factors. One set of factors includes SR-B1 and LDL-receptor related protein (LRP) at the surfaces of hepatocytes, which ensure delivery of HDL cholesterol to the liver either by selective cholesteryl ester uptake or through HDL holoparticle uptake, respectively [18–21]. Another important factor is cholesteryl ester transfer protein (CETP), which facilitates exchange of HDL cholesteryl esters for triglycerides in apoB-containing lipoproteins [22,23]. Other factors include phospholipid transfer protein (PLTP) [24] and endothelial [25,26] and hepatic [27] lipases, which transfer and hydrolyze HDL phospholipids and triglycerides. The rate of HDL remodeling and the proportion of cholesterol delivered to the liver versus that delivered to apoB-containing lipoproteins depends on the activities of SR-B1 and LRP on one hand and the levels of TG and activity of CETP on the other. It also depends on HDL composition, as, for example, cholesterol-rich HDL is a better substrate for SR-B1 than TG-rich HDL [28]. The products of this step are small, lipid-poor spherical HDL particles and lipid-free apoA-I released during remodeling. The former join the cycle at the stage of maturation, while the latter can undergo a new cycle of lipidation and HDL formation.

14.2.4 Efficiency of Reverse Cholesterol Transport

The flow of cholesterol through RCT is illustrated in Fig. 14.1A. Identification of a major determinant and/or indicator of the efficiency of this flow is a key issue in the assessment of risk of atherosclerosis and the development of "HDL therapies". Plasma levels of HDL-C were long considered to be such a determinant/indicator, but the concentration of HDL, as well as that of any other intermediate, is not always a true reflection of the flow rate, as this concentration is determined both by the rate of formation and by the rate of remodeling [6]. An example of such a discrepancy is overexpression of one of the HDL receptors, SR-B1, which is associated with low HDL but enhanced RCT [18]. Animal models also support the view that high levels of HDL can be pro- rather than antiatherogenic under certain metabolic conditions [29,30] and there are a number of examples in humans in which high levels of HDL did not provide additional atheroprotection [31]. Conversely, apoA-I_{Milano} mutation leads to lower HDL levels, but enhanced atheroprotection [32]. Finally, HDL has atheroprotective properties not directly related to RCT (see other chapters of this volume) that may contribute significantly to the overall effect.

HDL subfractions, such as preβ_1-HDL or α_1-HDL and α_2-HDL, have been tested as better markers of the efficiency of RCT [33], but are also subject to the same

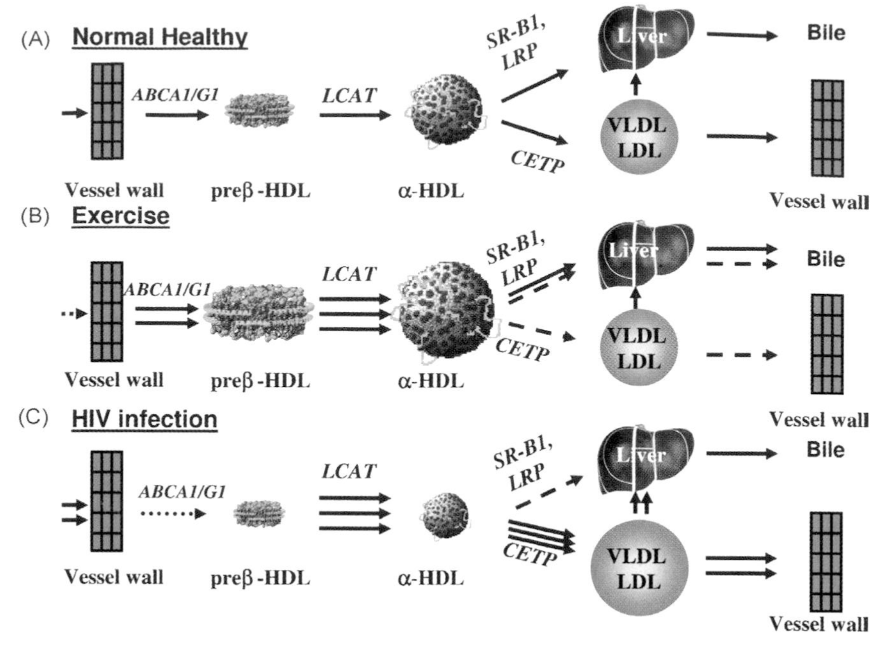

Fig. 14.1 *A) Normal Healthy RCT.* The first step of HDL metabolism is cholesterol efflux: formation of HDL from lipid-free and lipid-poor apoA-I and cellular lipids, cholesterol, and/or phospholipids. Cholesterol efflux occurs at the plasma membrane as a result of action of ABCA1 and ABCG1. The products of ABC-dependent cholesterol efflux are discoid HDL particles, most probably analogous to preβ-HDL. Upon reaching the appropriate size and composition, discoid HDL particles become a substrate for lecithin:cholesterol acyltransferase (LCAT). LCAT esterifies free cholesterol and forms a hydrophobic core, transforming HDL from discoid into spherical particles. These particles acquire more cholesterol and apolipoproteins from other lipoproteins in the plasma, possibly fusing with each other. The products of this step are large, cholesterol-rich, spherical particles, likely to be analogous to α_1HDL. Upon reaching the appropriate size and composition, spherical HDL particles become a substrate for a number of enzymes and transfer factors. One set of factors includes SR-B1 and LDL-receptor related protein (LRP) on the surfaces of hepatocytes, ensuring delivery of HDL cholesterol to the liver either by selective cholesteryl ester uptake or through HDL holoparticle uptake, respectively. Most of this cholesterol will be excreted to the bile either unmodified or after conversion into bile acids. The second factor is cholesteryl ester transfer protein (CETP), facilitating exchange of HDL cholesteryl esters for triglycerides in apoB-containing lipoproteins (VLDL and LDL). Cholesterol transferred to the VLDL and LDL may also be cleared by the liver, but in addition these lipoproteins can interact with LDL receptors (or if modified, with the scavenger receptors) on the cells of vessel wall, delivering cholesterol to these cells and therefore contributing to their cholesterol content. *B) Effects of exercise on RCT.* Formation of HDL is most probably stimulated by exercise, as is HDL maturation, with the overall rate of remodeling not substantially affected. Increased delivery of cholesterol to HDL with an unchanged rate of disposal thus results in elevation of HDL levels. Some evidence indicates that flow of cholesterol to the liver may be partially diverted toward a direct uptake of HDL cholesteryl esters by liver at the expense of the pathway involving CETP and apoB-containing lipoproteins. This could potentially reduce delivery of cholesterol to the cells of the vessel wall. The combination of the increased flow of cholesterol to the reverse cholesterol

limitations. High concentrations of preβ_1-HDL may thus result either from a high level of cholesterol efflux, indicating enhanced efficiency of RCT [34], or from low activity of LCAT [35–37], indicating the opposite. When both cholesterol efflux and LCAT activity are changed, as a result of exercise, for example (see below), it becomes difficult to interpret the quantitative value of preβ_1-HDL. Furthermore, different HDL subfractions may play key roles in different elements of the RCT: small, lipid-poor preβ_1-HDL, for example, is an initial acceptor of cholesterol likely to be active in ABCA1-dependent cholesterol efflux [38,39], while triglyceride-rich HDL is more active in ABCG1-dependent cholesterol efflux [40]. The contributions of each of these pathways to the removal of cholesterol from cells are not clear and may not be the same in all metabolic circumstances, so it is unlikely that the concentration of any specific HDL subfraction could be a reliable indication of "healthy" or "unhealthy" HDL.

The complexity of the reverse cholesterol transport pathway and its dynamic nature argues against a single HDL subfraction, transfer factor, enzyme, or receptor being a universal determinant or reliable indicator of the efficiency of RCT applicable to various metabolic situations. Low HDL levels due to high activity of CETP, for example, were associated with increased risk of CAD in subjects with high levels of triglycerides, but not in subjects with normal and low TGs [41]. The solution to this problem, apart from complex lipoprotein tracer kinetic studies [42], is to consider RCT in its entirety, in an attempt to analyze how changes in activities and concentrations of RCT participants should affect the flow of cholesterol through the pathway in a specific metabolic context. The hypothesis to be tested in such studies is that changes in concentrations of HDL subfractions and/or in activities of enzymes and transfer factors that are consistent with the enhanced flow of cholesterol through RCT would indicate higher efficiency of RCT [6]. Conversely, changes consistent with decreased cholesterol flow would be indicative of the impairment of RCT. Below we present two examples of such analysis of changes in the efficiency of RCT caused by two specific conditions: physical exercise and HIV infection.

Fig. 14.1 (*Continued*)
transport and reduced cholesterol flow back to "forward cholesterol transport" should have antiatherogenic effects. C) *Effects of HIV on RCT.* The initial step of RCT, cholesterol efflux, is severely inhibited by HIV infection. However, as this only affects cells infected with HIV (i.e., macrophages and T-cells), it may have a profound effect on the development of atherosclerosis, but would be unlikely to affect HDL levels in the plasma. Conversion of preβ-HDL into α-HDL and further maturation of HDL is most probably enhanced in HIV-infected patients. However, the combination of hypertriglyceridemia and high activity of CETP diverts the latter steps of RCT from delivery of cholesterol directly to the liver towards delivery of cholesterol to atherogenic VLDL and LDL, so the amount of cholesterol available for modification and uptake by extrahepatic tissues increases, resulting in more cholesterol being delivered to extrahepatic cells in absolute terms. If this situation persists it may contribute significantly to increased risk of atherosclerosis.

14.3 Exercise and HDL Metabolism

The benefits of both acute and chronic exercise for cardiovascular health are well established [43–45]. Not only does exercise reduce the risk of new atherosclerotic plaque development, it also stabilizes existing atherosclerotic plaques in an animal model [46]. Exercise affects multiple aspects of cardiovascular and metabolic physiology related to the development of atherosclerosis, and exercise-related changes in lipoprotein metabolism are likely to make a significant contribution to the atheroprotective effect of physical activity.

The well established effects of exercise on lipoprotein metabolism are lower levels of triglycerides and LDL-C [47] and higher levels of HDL-C [34,48–50]. These changes are antiatherogenic, but the mechanisms linking exercise to the changes in lipoprotein metabolism are not fully established. Below we discuss a potential mechanism explaining the association between exercise and HDL metabolism.

14.3.1 HDL Formation

The effects of acute and chronic exercise on the ability of cells to release their cholesterol have not previously been reported. The only indirect indication that cellular cholesterol efflux may be affected by exercise is preliminary evidence from our laboratory suggesting that leucocytes (but not skeletal muscle cells) from physically fit subjects have elevated expression of ABCA1. Although it is unlikely that expression of ABCA1 in leucocytes is a direct reflection of ABCA1 expression in macrophages in the vessel wall, this finding, if confirmed, would indicate that physical fitness may affect cellular cholesterol metabolism in cells relevant to the development of atherosclerosis. There are also reports that HDL production by the liver is enhanced by exercise [51,52], a finding consistent with exercise increasing cholesterol efflux. Another indirect measure of cholesterol efflux is the plasma concentration of its product: preβ_1-HDL. Physically fit subjects have higher levels of preβ_1-HDL [34,53,54], suggesting that cholesterol efflux is elevated. Furthermore, preβ_1-HDL [53] and mature HDL [55] are released across the leg in humans (arteriovenous flux) suggesting active acute formation of HDL in the muscle (most probably from apoA-I released during lipolysis of VLDL). This process is enhanced by acute exercise [53,55].

The ability of plasma to support cellular cholesterol efflux was increased with greater physical fitness [34,47,48] and even after acute exercise, as long as subjects were physically fit [56]. In most cases this increase was proportional to the levels of apoA-I and/or HDL-C [34,47,48], which indicates that the main cause of increased cholesterol efflux is elevated HDL and that the functionality of HDL is unchanged. The finding that the ability of HDL to prevent LDL oxidation was increased by exercise proportionally to the HDL levels also supports this conclusion [57]. Overall, it can be concluded that formation of HDL is enhanced by exercise training and possibly by acute exercise as well.

14.3.2
HDL Maturation

The key element of HDL maturation is the activity of LCAT. A number of studies have found that LCAT activity in plasma is elevated in physically fit people [34,47,49,58], and this is consistent with the elevation of HDL-C, which mainly reflects the level of cholesteryl ester-rich mature HDL. There are conflicting reports on the effect of acute exercise: it either did not affect LCAT activity and HDL-C levels [59,60] or caused transient elevation of LCAT activity and HDL-C levels in the hours following an exercise session [61–63]. Overall, it is possible to conclude that HDL maturation was enhanced by exercise.

14.3.3
HDL Remodeling

The activity of CETP has been reported variously to be increased [47], to remain unchanged [34,48,50] or to be decreased [64,65] in subjects with higher physical fitness. In association with the lower levels of TG commonly found in physically fit individuals, this would almost certainly lead to lower rates of cholesteryl ester exchanged for TG by the action of CETP. Consequently, less cholesterol would be delivered to apoB-containing lipoproteins and more would be retained in HDL, contributing to the higher HDL-C levels in physically fit subjects. Additional evidence that CETP activity contributes to the hyperalphalipoproteinemia in physically fit subjects is that the effect of exercise on HDL-C levels was dependent on CETP genotype [66]. Furthermore, the effect of exercise on HDL-C was more pronounced in hypertriglyceridemic patients than in patients with normal levels of TG [67], suggesting that the rate of TG exchange plays a major role in determining the effect of exercise on HDL-C levels.

The effect of exercise on the alternate HDL cholesterol disposal mechanism, by direct liver uptake, is unclear as no information is available on the relationship between exercise and expression of SR-B1 or LRP in the liver. Indirect evidence, however, suggests that cholesteryl ester-rich and TG-poor HDL particles are a better substrate for SR-B1 [28], and it would be reasonable to assume that CE will be effectively taken up by the liver and delivered to bile.

There are conflicting reports regarding the effect of exercise on hepatic lipase activity, with some studies reporting elevated levels of HL after physical training [49,68] while others report decreased levels [51,52,69]. A single study reported that the activity of endothelial lipase is reduced by exercise [70]. The effect of exercise on PLTP is not known, and our preliminary studies indicate that it is not affected by exercise.

Overall, in view of the inconsistencies in reports assessing parameters of HDL remodeling, the most likely outcome would be a slight reduction in HDL remodeling or no change at all. This conclusion is consistent with a direct assessment of the fractional catabolic rate of HDL protein, which was reduced by 5 % in fit subjects [52]. Modest reductions in HDL cholesteryl ester metabolism were observed in an animal model [71].

14.3.4
Exercise and the Efficiency of Reverse Cholesterol Transport

The changes in components of HDL metabolism induced by exercise permit prediction of the effect these changes would have on the functionality of RCT, as illustrated in Fig. 14.1A and B. Formation of HDL is, most probably, stimulated by exercise, as is HDL maturation, whereas the overall rate of remodeling is not substantially affected. Increased delivery of cholesterol to HDL with an unchanged rate of disposal would thus result in elevation of HDL levels. Moreover, some evidence points to the possibility that the flow of cholesterol may be partially diverted towards direct uptake of HDL cholesteryl esters by the liver at the expense of the pathway involving CETP and apoB-containing lipoproteins. Although the majority of cholesterol from apoB-containing lipoprotein ends up in the liver, part of it is accessible to extrahepatic cells and such a diversion could potentially reduce the delivery of cholesterol to the extrahepatic cells, including cells of the vessel wall. The combination of the increased flow of cholesterol through reverse cholesterol transport and the reduced cholesterol flow back to "forward cholesterol transport" would have a clear antiatherogenic effect. Importantly, the effect of exercise on HDL metabolism goes well beyond just elevation of HDL-C level; exercise alters the flow of cholesterol through plasma compartments in a way that is clearly beneficial for atheroprotection. Therefore, a similar elevation of HDL-C by other means would only partly emulate the beneficial effects of exercise on lipid metabolism. What remains a completely unexplored area, however, is the biochemical mechanisms connecting exercise and the changes in the parameters of HDL metabolism described above. The prospect that it may be regulated through a simple set of receptors offers the tempting possibility of the existence of a pharmacological agonist that will emulate the benefits of exercise with respect to lipoprotein metabolism. Until the biochemical mechanisms are discovered, however, we will just have to exercise.

14.4
HIV Infection and HDL Metabolism

HIV infection is associated with profound changes in lipid and lipoprotein metabolism [72] including dyslipidemia [73] and lipodystrophy [74]. Dyslipidemia in treatment-naïve HIV-infected patients includes low levels of LDL-C and HDL-C [75,76]. Treatment for HIV infection involves combination therapy with three antiretroviral drugs, also known as Highly Active Antiretroviral Therapy (HAART). HAART has dramatically improved clinical outcomes for HIV-infected individuals, but therapy is commonly associated with elevation of LDL-C levels, while reduced levels of HDL-C persist with most ARV regimens [77,78]. The exception is nevirapine, which consistently caused striking increases in HDL-C (35–49 %) and apoA-I (19 %), as well as HDL particle size (3 %) [79–81]; it is not clear, however, if this effect of nevirapine is related to its antiviral activity. Hypertriglyceridemia is often seen in untreated HIV-infected patients and may be exacerbated with antiretroviral treatment

[82,83]. Consequently, in a majority of cases, persistent hypoalphalipoproteinemia and hypertriglyceridemia associated with elevated levels of total and LDL cholesterol result in a highly atherogenic lipoprotein profile [84], increased risk of development of atherosclerosis [85], at least a threefold increased risk of CAD [86,87], and at least a 26 % increased risk of myocardial infarction [88]. Changes in "forward" cholesterol transport, including elevation of LDL [73] and increased cholesterol uptake through induction of CD36 expression [89], are likely to be caused by HAART [78]. The relative contributions of HAART and the HIV infection itself to HDL metabolism remain to be determined. Preliminary studies from our [90] and other [73,76] laboratories indicate that HIV infection might play a key role in the impairment of HDL metabolism and in increasing the risk of CAD in these individuals.

14.4.1 HDL Formation

Macrophages play a central role in the pathogenesis of atherosclerosis and are also target cells for HIV. Recently, it has been established in our laboratories that HIV infection of macrophages severely impairs ABCA1-dependent cholesterol efflux from these cells [91]. This effect seems to be related to the ability of the viral protein Nef to interact with ABCA1, to reduce the abundance of ABCA1 protein without decreasing ABCA1 expression, and to cause intracellular relocalization of ABCA1 [91]. Tissue-specific knockout of ABCA1 in macrophages causes severe atherosclerosis [13], so it is reasonable to assume that knockout of the ABCA1-dependent pathway by HIV would have the same effect. Indeed, HIV infection and Nef transfection caused accumulation of cholesterol in human and mouse macrophages, transforming them into foam cells, while HIV was also found in foam cells in atherosclerotic plaques of HIV-infected patients [91]. On the other hand, macrophages contribute minimally to the plasma levels of HDL [92], with liver and intestine being the major sources of cholesterol in the HDL fraction [11,12]. Therefore, impairment of ABCA1-dependent cholesterol efflux by HIV would make a significant contribution to the risk of atherosclerosis in HIV-infected patients, but would unlikely to be the cause of HIV-induced hypoalphalipoproteinemia.

When the abilities of plasmas to support cholesterol efflux were evaluated, no difference was found between plasma from HIV-infected patients and plasma from matched HIV-negative subjects. Surprisingly, levels of the likely product of cholesterol efflux – preβ_1-HDL – in HIV-infected subjects were half those found in HIV-negative subjects. The explanation for this paradox appears to be a higher rate of maturation of preβ-HDL (see below), whereas overall changes in HDL formation are unlikely to contribute significantly to the HIV-induced hypoalphalipoproteinemia.

14.4.2 HDL Maturation

The major factor responsible for HDL maturation is LCAT, and we have found that plasma levels of LCAT, as well as its activity, are elevated in HIV-infected patients.

LCAT is also an important determinant of preβ_1-HDL concentration [37], which explains why HIV-infected patients have low levels of preβ_1-HDL despite apparently unchanged rates of HDL formation. Maturation of HDL is thus likely to be increased in HIV-infected patients, which should lead to higher levels of HDL-C in HIV-infected patients. Surprisingly though, the opposite (low HDL-C levels) is commonly observed.

14.4.3 HDL Remodeling

In a cross-sectional study, we found that HIV-infected patients have higher plasma levels and activity of CETP. This finding was confirmed in patients being treated at the time with HAART, patients on an antiretroviral (ARV) treatment interruption for more than 3 months, and in treatment-naïve HIV-infected patients, indicating that the changes in CETP activity are caused by HIV rather then by ARV treatment. It has also been shown that at least some ARV treatment regimens do not affect plasma CETP activity [93]. Higher activity of CETP, especially when associated with hypertriglyceridemia, would increase transfer of cholesteryl esters from HDL to apoB-containing lipoproteins. Consistent with this finding, the ratio of cholesterol to apoA-I in HDL was halved in HIV-infected patients, indicating the presence of cholesterol-poor (and presumably triglyceride-rich) HDL particles, a finding consistent with that of Rimland et al. [94]. In HIV-infected patients, the levels of HDL-C correlated negatively with CETP activity, but not TG levels, while in the HIV-negative group HDL-C negatively correlated with TG levels, but not with CETP activity. One inference of this finding is that increased activity of CETP is an important contributor to the low HDL-C levels in HIV-infected patients. Although the effects of HIV or HAART on the expression and activity of SR-B1 are not known, expression of LRP is inhibited by some ARV regimens [74,95], further inhibiting delivery of cholesterol from HDL to liver. Increased triglyceride content of HDL may also impact on binding and uptake of these particles by liver scavenger receptors, impairing cholesterol delivery through these receptors [96,97] and making HDL particles less protective against atherosclerosis [98].

At least some ARV regimens clearly inhibit lipases [99], including hepatic lipase [100], although endothelial lipase activity in HIV-infected patients has not been investigated. Although it has been hypothesized that changes in lipase activity may be involved in the pathogenesis of lipodystrophy [101] and may be a cause of hypertriglyceriemia, a direct effect of these changes on HDL metabolism is less certain. Indirectly, however, reduced lipolysis of VLDL may reduce the availability of apoA-I to enter the RCT cycle and increase the availability of TG for exchange through CETP (see above); both of these mechanisms may contribute to HIV-induced hypoalphalipoproteinemia.

14.4.4 HIV Infection and the Efficiency of Reverse Cholesterol Transport

The changes in HDL metabolism induced by HIV infection allow one to predict possible effects on the functionality of RCT, as illustrated in Fig. 14.1A and C.

The initial step of RCT, cholesterol efflux, is severely inhibited by HIV infection. However, as this only happens in cells infected with HIV (i.e., macrophages and T-cells), it may have a profound effect on the development of atherosclerosis, but would be unlikely to affect HDL levels in the plasma. Conversion of preβ_1-HDL into α-HDL and further maturation of HDL is most probably enhanced in HIV-infected patients. However, the combination of hypertriglyceridemia and increased activity of CETP diverts the latter steps of RCT from delivery of cholesterol directly to the liver towards delivery of cholesterol to atherogenic VLDL and LDL. Although the liver, through LDL receptors, will still take up the majority of transferred cholesterol, the amount of cholesterol available for modification and uptake by extrahepatic tissues will also increase proportionally, which is likely to result in more cholesterol being delivered to extrahepatic cells. If this situation persists, it may contribute significantly to the increased risk of atherosclerosis in HIV-infected patients.

Changes in HDL remodeling would also lead to the alteration of HDL composition. Although the ability of HDL particles to promote cholesterol efflux is not affected by HIV, the increased triglyceride content of HDL may impact on the uptake of these particles by the liver. Boekholdt et al. [41] and Foger et al. [102] suggested that higher levels of CETP are associated with hypoalphalipoproteinemia and higher risk of CAD, but only in subjects with hypertriglyceridemia. HIV-infected patients may represent such a situation, and the combination of hypertriglyceridemia with significantly elevated CETP activity may contribute considerably to the risk of CAD in these patients. At the same time this model suggests that recently developed CETP inhibitors [103] may be effective in treating dyslipidemia in HIV-infected patients. Although the antiatherogenic potential of CETP inhibitors is yet to be proven in clinical studies, animal models show that inhibition of CETP favorably affects reverse cholesterol transport [104] and is antiatherogenic [105]. This suggestion may have important practical implications, since the efficacy of some statins metabolized by the cytochrome P450 3A4 enzymes is significantly reduced when co-administered with non-nucleoside reverse transcriptase inhibitors, and statin plasma concentrations (and consequently adverse effects) are increased on co-administration with protease inhibitors [106–108].

The observed effects of HIV on plasma HDL metabolism raise an important question regarding how infection-targeting macrophages and T-cells could affect the concentrations and activities of factors that are primarily produced by hepatocytes, which are not infected by HIV. HIV infection is known, however, to have effects on organs not infected by the virus, causing, for example, myocarditis [109] and vasculopathy [110]. One possible explanation is an indirect effect of HIV on hepatocytes through, for example, factors released from HIV-infected hepatic macrophages (Kupfer cells). The answer may also lie in the effects of viral proteins found in the circulation, such as trans-activator protein (Tat) [111]. Recent data also show that HIV depletes CD4 lymphocytes in the gut within 48 hours of acute infection with HIV, which is the cause of long-lasting immune damage [112].

14.5 Fixing HDL Metabolism

In the last 20 years, advances in treatment of dyslipidemia have conclusively established the pivotal role of cholesterol metabolism in the pathogenesis of atherosclerosis and consequently cardiovascular diseases [113]. It has also become clear that a lot can and should be done to reduce the lipid-related risk of CAD further. LDL-related risk is now well controlled, primarily through statin treatment, and this has brought "HDL therapy" to the forefront of research efforts [114–116]. A critical step of such therapy, however, is to develop an understanding of how HDL metabolism is regulated and how it is affected in various metabolic circumstances. Physical exercise and HIV infection both consistently affect HDL metabolism, and detailed analysis of their effects can provide much-needed insight into the regulation of HDL metabolism. Thus, exercise may be a good model of changes in HDL metabolism that could be emulated by HDL-raising therapies. How to extend these benefits further? From the available data it appears that simple elevation of plasma levels of HDL will not fully mimic the effect of exercise on HDL metabolism. Rather, increased HDL production and maturation would more closely emulate the effect of exercise. One can speculate that LXR agonists (if cleared of adverse effects) or similar drugs might be an effective way to enhance formation and maturation of HDL and may more closely simulate the effect, and hopefully the benefits, of exercise with regard to HDL. At the other end of the spectrum is hypoalphalipoproteinemia of metabolic (nonfamilial) origin, such as that associated with HIV infection or type 2 diabetes [117]. It is a clear risk factor for atherosclerosis and requires urgent attention, but not before the specific biochemical mechanism of the particular hypoalphalipoproteinemia is established. In the case of HIV-induced hypoalphalipoproteinemia, CETP inhibitors appear to be the most appropriate type of HDL therapy as they target the exact cause of this condition; other approaches may be required for, say, type 2 diabetes. It seems that systematic analysis of the effects of metabolic conditions and treatments on HDL metabolism and functionality, not just HDL levels, may result in the successful development of "HDL therapy".

References

1 Wilson, P. W., Abbott, R. D., Castelli, W. P. (1988) High density lipoprotein cholesterol and mortality. The Framingham Heart Study. *Arteriosclerosis*, **8**, 737–741.

2 Stein, O. and Stein, Y. (1999) Atheroprotective mechanisms of HDL. *Atherosclerosis*, **144**, 285–301.

3 von Eckardstein, A., Nofer, J. R., Assmann, G. (2001) High density lipoproteins and arteriosclerosis: role of cholesterol efflux and reverse cholesterol transport. *Arterioscler. Thromb. Vasc. Biol.*, **21**, 13–27.

4 Nofer, J. R., Kehrel, B., Fobker, M., Levkau, B., Assmann, G., Eckardstein, A. (2002) HDL and arteriosclerosis: beyond reverse cholesterol transport. *Atherosclerosis*, **161**, 1–16.

5 Young, C. E., Karas, R. H., Kuvin, J. T. (2004) High-density lipoprotein cholesterol and coronary heart disease. *Cardiol Rev*, **12**, 107–119.

6 Sviridov, D. and Nestel, P. (2002) Dynamics of reverse cholesterol transport; protection against atherosclerosis. *Atherosclerosis*, **161**, 245–254.

7 Gordon, D. J., Probstfield, J. L., Garrison, R. J., Neaton, J. D., Castelli, W. P., Knoke, J. D., Jacobs, D. R., Jr., Bangdiwala, S., Tyroler, H. A. (1989) High-density lipoprotein cholesterol and cardiovascular disease. Four prospective American studies. *Circulation*, **79**, 8–15.

8 Boden, W. E. (2000) High-density lipoprotein cholesterol as an independent risk factor in cardiovascular disease: assessing the data from Framingham to the Veterans Affairs High—Density Lipoprotein Intervention Trial. *Am. J. Cardiol.*, **86**, 19L–22L.

9 Leaf, D. A. (2003) The effect of physical exercise on reverse cholesterol transport. *Metabolism*, **52**, 950–957.

10 Nestel, P. (2005) Dietary factors that affect high density lipoprotein. *Curr. Medicinal Chem. – IEMA*, **5**, 361–365.

11 Basso, F., Freeman, L., Knapper, C. L., Remaley, A., Stonik, J., Neufeld, E. B., Tansey, T., Amar, M. J. A., Fruchart-Najib, J., Duverger, N., Santamarina-Fojo, S., Brewer, H. B., Jr. (2003) Role of the hepatic ABCA1 transporter in modulating intrahepatic cholesterol and plasma HDL cholesterol concentrations. *J. Lipid Res.*, **44**, 296–302.

12 Brunham, L. R., Kruit, J. K., Iqbal, J., Fievet, C., Timmins, J. M., Pape, T. D., Coburn, B. A., Bissada, N., Staels, B., Groen, A. K., Hussain, M. M., Parks, J. S., Kuipers, F., Hayden, M. R. (2006) Intestinal ABCA1 directly contributes to HDL biogenesis in vivo. *J. Clin. Invest.*, **116**, 1052–1062.

13 Aiello, R. J., Brees, D., Bourassa, P.-A., Royer, L., Lindsey, S., Coskran, T., Haghpassand, M., Francone, O. L. (2002) Increased Atherosclerosis in Hyperlipidemic Mice With Inactivation of ABCA1 in Macrophages. *Arterioscler. Thromb. Vasc. Biol.*, **22**, 630–637.

14 Barter, P., Kastelein, J., Nunn, A., Hobbs, R. (2003) High density lipoproteins (HDLs) and atherosclerosis; the unanswered questions. *Atherosclerosis*, **168**, 195–211.

15 Cavelier, C., Lorenzi, I., Rohrer, L., von Eckardstein, A. (2006) Lipid efflux by the ATP-binding cassette transporters ABCA1 and ABCG1. *Biochim. Biophys. Acta*, **1761**, 655–666.

16 Sviridov, D., Miyazaki, O., Theodore, K., Hoang, A., Fukamachi, I., Nestel, P. (2002) Delineation of the Role of Pre-{beta}1-HDL in Cholesterol Efflux Using Isolated Pre-{beta}1-HDL. *Arterioscler. Thromb. Vasc. Biol.*, **22**, 1482–1488.

17 Kuivenhoven, J. A., Pritchard, H., Hill, J., Frohlich, J., Assmann, G., Kastelein, J. (1997) The molecular pathology of lecithin:cholesterol acyltransferase (LCAT) deficiency syndromes. *J. Lipid Res.*, **38**, 191–205.

18 Zhang, Y., Da Silva, J. R., Reilly, M., Billheimer, J. T., Rothblat, G. H., Rader, D. J. (2005) Hepatic expression of scavenger receptor class B type I (SR-BI) is a positive regulator of macrophage reverse cholesterol transport in vivo. *J. Clin. Invest.*, **115**, 2870–2874.

19 Trigatti, B. L., Krieger, M., Rigotti, A. (2003) Influence of the HDL Receptor SR-BI on Lipoprotein Metabolism and Atherosclerosis. *Arterioscler. Thromb. Vasc. Biol.*, **23**, 1732–1738.

20 Acton, S., Rigotti, A., Landschulz, K. T., Xu, S., Hobbs, H. H., Krieger, M. (1996) Identification of scavenger receptor SR-BI as a high density lipoprotein receptor. *Science*, **271**, 518–520.

21 Herz, J. and Strickland, D. K. (2001) LRP: a multifunctional scavenger and signaling receptor. *J. Clin. Invest.*, **108**, 779–784.

22 de Grooth, G. J., Klerkx, A. H., Stroes, E. S., Stalenhoef, A. F., Kastelein, J. J., Kuivenhoven, J. A.

(2004) A review of CETP and its relation to atherosclerosis. *J. Lipid Res.*, **45**, 1967–1974.

23 Tall, A. (1995) Plasma lipid transfer proteins. *Annu. Rev. Biochem.*, **64**, 235–257.

24 Huuskonen, J., Olkkonen, V. M., Jauhiainen, M., Ehnholm, C. (2001) The impact of phospholipid transfer protein (PLTP) on HDL metabolism. *Atherosclerosis*, **155**, 269–281.

25 Ishida, T., Choi, S., Kundu, R. K., Hirata, K. -i., Rubin, E. M., Cooper, A. D., Quertermous, T. (2003) Endothelial lipase is a major determinant of HDL level. *J. Clin. Invest.*, **111**, 347–355.

26 Jahangiri, A., Rader, D. J., Marchadier, D., Curtiss, L. K., Bonnet, D. J., Rye, K.-A. (2005) Evidence that endothelial lipase remodels high density lipoproteins without mediating the dissociation of apolipoprotein A-I. *J. Lipid Res.*, **46**, 896–903.

27 Perret, B., Mabile, L., Martinez, L., Terce, F., Barbaras, R., Collet, X. (2002) Hepatic lipase: structure/function relationship, synthesis, and regulation. *J. Lipid Res.*, **43**, 1163–1169.

28 Greene, D. J., Skeggs, J. W., Morton, R. E. (2001) Elevated triglyceride content diminishes the capacity of HDL to deliver cholesteryl esters via the scavenger receptor BI (SR-BI). *J. Biol. Chem.*, **276**, 4804–4811.

29 Berard, A. M., Foger, B., Remaley, A., Shamburek, R., Vaisman, B. L., Talley, G., Paigen, B., Hoyt, R. F., Jr., Marcovina, S., Brewer, H. B., Jr., Santamarina-Fojo, S. (1997) High plasma HDL concentrations associated with enhanced atherosclerosis in transgenic mice overexpressing lecithin-cholesteryl acyltransferase. *Nat Med*, **3**, 744–749.

30 Ansell, B. J., Fonarow, G. C., Fogelman, A. M. (2006) High-density Lipoprotein: Is It Always Atheroprotective? *Curr Atheroscler Rep*, **8**, 405–411.

31 Sviridov, D., Chin-Dusting, J., Nestel, P., Kingwell, B., Hoang, A., Olchawa, B., Starr, J., Dart, A. (2006) Elevated HDL Cholesterol is Functionally Ineffective in Cardiac Transplant Recipients: Evidence for Impaired Reverse Cholesterol Transport. *Transplantation*, **81**, 361–366.

32 Franceschini, G., Calabresi, L., Chiesa, G., Parolini, C., Sirtori, C. R., Canavesi, M., Bernini, F. (1999) Increased cholesterol efflux potential of sera from ApoA-IMilano carriers and transgenic mice. *Arterioscler. Thromb. Vasc. Biol.*, **19**, 1257–1262.

33 Asztalos, B. F., Collins, D., Cupples, L. A., Demissie, S., Horvath, K. V., Bloomfield, H. E., Robins, S. J., Schaefer, E. J. (2005) Value of High-Density Lipoprotein (HDL) Subpopulations in Predicting Recurrent Cardiovascular Events in the Veterans Affairs HDL Intervention Trial. *Arterioscler. Thromb. Vasc. Biol.*, **25**, 2185–2191.

34 Olchawa, B., Kingwell, B. A., Hoang, A., Schneider, L., Miyazaki, O., Nestel, P., Sviridov, D. (2004) Physical Fitness and Reverse Cholesterol Transport. *Arterioscler. Thromb. Vasc. Biol.*, **24**, 1087–1091.

35 Miida, T., Kawano, M., Fielding, C. J., Fielding, P. E. (1992) Regulation of the concentration of pre beta high-density lipoprotein in normal plasma by cell membranes and lecithin-cholesterol acyltransferase activity. *Biochemistry*, **31**, 11112–11117.

36 Miida, T., Miyazaki, O., Hanyu, O., Nakamura, Y., Hirayama, S., Narita, I., Gejyo, F., Ei, I., Tasaki, K., Kohda, Y., Ohta, T., Yata, S., Fukamachi, I., Okada, M. (2003) LCAT-Dependent Conversion of Prebeta1-HDL into alpha-Migrating HDL is Severely Delayed in Hemodialysis Patients. *J Am Soc Nephrol*, **14**, 732–738.

37 Miida, T., Obayashi, K., Seino, U., Zhu, Y., Ito, T., Kosuge, K., Hirayama, S., Hanyu, O., Nakamura, Y., Yamaguchi, T., Tsuda, T., Saito, Y., Miyazaki, O., Okada, M. (2004) LCAT-dependent conversion rate is a determinant of plasma prebeta1-HDL concentration in healthy Japanese. *Clin Chim Acta*, **350**, 107–114.

38 Fielding, C. J. and Fielding, P. E. (2001) Cellular cholesterol efflux. *Biochim. Biophys. Acta*, **1533**, 175–189.

39 Kawano, M., Miida, T., Fielding, C. J., Fielding, P. E. (1993) Quantitation of pre beta-HDL-dependent and nonspecific components of the total efflux of cellular cholesterol and phospholipid. *Biochemistry*, **32**, 5025–5028.

40 Matsuura, F., Wang, N., Chen, W., Jiang, X. -C., Tall, A. R. (2006) HDL from CETP-deficient subjects shows enhanced ability to promote cholesterol efflux from macrophages in an apoE- and ABCG1-dependent pathway. *J. Clin. Invest.*, **116**, 1435–1442.

41 Boekholdt, S. M., Kuivenhoven, J.-A., Wareham, N. J., Peters, R. J. G., Jukema, J. W., Luben, R., Bingham, S. A., Day, N. E., Kastelein, J. J. P., Khaw, K.-T. (2004) Plasma Levels of Cholesteryl Ester Transfer Protein and the Risk of Future Coronary Artery Disease in Apparently Healthy Men and Women: The Prospective EPIC (European Prospective Investigation into Cancer and nutrition)-Norfolk Population Study. *Circulation*, **110**, 1418–1423.

42 Barrett, P. H., Chan, D. C., Watts, G. F. (2006) Thematic review series: patient-oriented research. Design and analysis of lipoprotein tracer kinetics studies in humans. *J. Lipid Res.*, **47**, 1607–1619.

43 Myers, J. (2003) Exercise and Cardiovascular Health. *Circulation*, **107**, 2e–5.

44 Thompson, P. D., Buchner, D., Pina, I. L., Balady, G. J., Williams, M. A., Marcus, B. H., Berra, K., Blair, S. N., Costa, F., Franklin, B., Fletcher, G. F., Gordon, N. F., Pate, R. R., Rodriguez, B. L., Yancey, A. K., Wenger, N. K. (2003) Exercise and Physical Activity in the Prevention and Treatment of Atherosclerotic Cardiovascular Disease: A Statement From the Council on Clinical Cardiology (Subcommittee on Exercise, Rehabilitation, and Prevention) and the Council on Nutrition, Physical Activity, and Metabolism (Subcommittee on Physical Activity). *Circulation*, **107**, 3109–3116.

45 Sandvik, L., Erikssen, J., Thaulow, E., Erikssen, G., Mundal, R., Rodahl, K. (1993) Physical fitness as a predictor of mortality among healthy, middle-aged Norwegian men. *N. Engl. J. Med.*, **328**, 533–537.

46 Napoli, C., Williams-Ignarro, S., de Nigris, F., Lerman, L. O., D'Armiento, F. P., Crimi, E., Byrns, R. E., Casamassimi, A., Lanza, A., Gombos, F., Sica, V., Ignarro, L. J. (2006) Physical training and metabolic supplementation reduce spontaneous atherosclerotic plaque rupture and prolong survival in hypercholesterolemic mice. *PNAS*, **103**, 10479–10484.

47 Gupta, A. K., Ross, E. A., Myers, J. N., Kashyap, M. L. (1993) Increased reverse cholesterol transport in athletes. *Metabolism*, **42**, 684–690.

48 Brites, F., Verona, J., De Geitere, C., Fruchart, J. -C., Castro, G., Wikinski, R. (2004) Enhanced cholesterol efflux promotion in well-trained soccer players. *Metabolism*, **53**, 1262–1267.

49 Lehmann, R., Engler, H., Honegger, R., Riesen, W., Spinas, G. A. (2001) Alterations of lipolytic enzymes and high-density lipoprotein subfractions induced by physical activity in type 2 diabetes mellitus. *Eur J Clin Invest*, **31**, 37–44.

50 Thomas, T. R., Smith, B. K., Donahue, O. M., Altena, T. S., James-Kracke, M., Sun, G. Y. (2004) Effects of omega-3 fatty acid supplementation and exercise on low-density lipoprotein and high-density lipoprotein subfractions. *Metabolism*, **53**, 749–754.

51 Gorski, J., Oscai, L. B., Palmer, W. K. (1990) Hepatic lipid metabolism in exercise and training. *Med. Sci. Sports Exerc.*, **22**, 213–221.

52 Thompson, P. D., Yurgalevitch, S. M., Flynn, M. M., Zmuda, J. M., Spannaus-Martin, D., Saritelli, A., Bausserman, L., Herbert, P. N.

(1997) Effect of prolonged exercise training without weight loss on high-density lipoprotein metabolism in overweight men. *Metabolism*, **46**, 217–223.

53 Sviridov, D., Kingwell, B., Hoang, A., Dart, A., Nestel, P. (2003) Single Session Exercise Stimulates Formation of Prebeta1-HDL in Leg Muscle. *J. Lipid. Res.*, **44**, 522–526.

54 Jafari, M., Leaf, D., MacRae, H., Kasem, J., O'Conner, P., Pullinger, C., Malloy, M., Kane, J. (2003) The effects of physical exercise on plasma prebeta-1 high-density lipoprotein. *Metabolism*, **52**, 437–442.

55 Ruys, T., Sturgess, I., Shaikh, M., Watts, B. G., Nordestgaard, G. F., Lewis, B. (1989) Effects of exercise and fat ingestion on high density lipoprotein production by peripheral tissues. *Lancet*, **2**, 1119–1122.

56 Campaigne, B. N., Fontaine, R. N., Park, M. S., Rymaszewski, Z. J. (1993) Reverse cholesterol transport with acute exercise. *Med. Sci. Sports Exerc.*, **25**, 1346–1351.

57 Brites, F., Zago, V., Verona, J., Muzzio, M. L., Wikinski, R., Schreier, L. (2006) HDL capacity to inhibit LDL oxidation in well-trained triathletes. *Life Sci*, **78**, 3074–3081.

58 Tsopanakis, C., Kotsarellis, D., Tsopanakis, A. (1988) Plasma lecithin: cholesterol acyltransferase activity in elite athletes from selected sports. *Eur J Appl Physiol Occup Physiol*, **58**, 262–265.

59 Thompson, C. E., Thomas, T. R., Araujo, J., Albers, J. J., Decedue, C. J. (1985) Response of HDL cholesterol, apoprotein A-I, and LCAT to exercise withdrawal. *Atherosclerosis*, **54**, 65–73.

60 Thomas, T. R., Adeniran, S. B., Iltis, P. W., Aquiar, C. A., Albers, J. J. (1985) Effects of interval and continuous running on HDL-cholesterol, apoproteins A-1 and B, and LCAT. *Can J Appl Sport Sci*, **10**, 52–59.

61 Dufaux, B., Order, U., Muller, R., Hollmann, W. (1986) Delayed effects of prolonged exercise on serum lipoproteins. *Metabolism*, **35**, 105–109.

62 Frey, I., Baumstark, M. W., Berg, A., Keul, J. (1991) Influence of acute maximal exercise on lecithin:cholesterol acyltransferase activity in healthy adults of differing aerobic performance. *Eur J Appl Physiol Occup Physiol*, **62**, 31–35.

63 Weise, S. D., Grandjean, P. W., Rohack, J. J., Womack, J. W., Crouse, S. F. (2005) Acute changes in blood lipids and enzymes in postmenopausal women after exercise. *J Appl Physiol*, **99**, 609–615.

64 Seip, R. L., Moulin, P., Cocke, T., Tall, A., Kohrt, W. M., Mankowitz, K., Semenkovich, C. F., Ostlund, R., Schonfeld, G. (1993) Exercise training decreases plasma cholesteryl ester transfer protein. *Arterioscler Thromb*, **13**, 1359–1367.

65 Freeman, D. J., Griffin, B. A., Holmes, A. P., Lindsay, G. M., Gaffney, D., Packard, C. J., Shepherd, J. (1994) Regulation of plasma HDL cholesterol and subfraction distribution by genetic and environmental factors. Associations between the TaqI B RFLP in the CETP gene and smoking and obesity. *Arterioscler Thromb*, **14**, 336–344.

66 Wilund, K., Ferrell, R., Phares, D., Goldberg, A., Hagberg, J. (2002) Changes in high-density lipoprotein-cholesterol subfractions with exercise training may be dependent on cholesteryl ester transfer protein (CETP) genotype. *Metabolism*, **51**, 774–778.

67 Couillard, C., Despres, J. -P., Lamarche, B., Bergeron, J., Gagnon, J., Leon, A. S., Rao, D. C., Skinner, J. S., Wilmore, J. H., Bouchard, C. (2001) Effects of Endurance Exercise Training on Plasma HDL Cholesterol Levels Depend on Levels of Triglycerides: Evidence From Men of the Health, Risk Factors, Exercise Training and Genetics (HERITAGE) Family Study. *Arterioscler. Thromb. Vasc. Biol.*, **21**, 1226–1232.

68 Duncan, G. E., Perri, M. G., Theriaque, D. W., Hutson, A. D., Eckel, R. H., Stacpoole, P. W. (2003) Exercise training, without weight

loss, increases insulin sensitivity and postheparin plasma lipase activity in previously sedentary adults. *Diabetes Care*, **26**, 557–562.

69 Thomas, T. R., Horner, K. E., Langdon, M. M., Zhang, J. Q., Krul, E. S., Sun, G. Y., Cox, R. H. (2001) Effect of exercise and medium-chain fatty acids on postprandial lipemia. *J Appl Physiol*, **90**, 1239–1246.

70 Das, U. N. (2005) Long-chain polyunsaturated fatty acids, endothelial lipase and atherosclerosis. *Prostaglandins Leukot Essent Fatty Acids*, **72**, 173–179.

71 von Duvillard, S. P., Foxall, T. L., Davis, W. P., Terpstra, A. H. (2000) Effects of exercise on plasma high-density lipoprotein cholesteryl ester metabolism in male and female miniature swine. *Metabolism*, **49**, 826–832.

72 Dobs, A. and Brown, T. (2002) Metabolic abnormalities in HIV disease and injection drug use. *J Acquir Immune Defic Syndr*, **31** (Suppl 2,), S70–S77.

73 El-Sadr, W., Mullin, C., Carr, A., Gibert, C., Rappoport, C., Visnegarwala, F., Grunfeld, C., Raghavan, S. (2005) Effects of HIV disease on lipid, glucose and insulin levels: results from a large antiretroviral-naive cohort. *HIV Med*, **6**, 114–121.

74 Carr, A., Samaras, K., Chisholm, D. J., Cooper, D. A. (1998) Pathogenesis of HIV-1-protease inhibitor-associated peripheral lipodystrophy, hyperlipidaemia, and insulin resistance. *Lancet*, **351**, 1881–1883.

75 Shor-Posner, G., Basit, A., Lu, Y., Cabrejos, C., Chang, J., Fletcher, M., Mantero-Atienza, E., Baum, M. K. (1993) Hypocholesterolemia is associated with immune dysfunction in early human immunodeficiency virus-1 infection. *Am J Med*, **94**, 515–519.

76 Asztalos, B. F., Schaefer, E. J., Horvath, K. V., Cox, C. E., Skinner, S., Gerrior, J., Gorbach, S. L., Wanke, C. (2006) Protease inhibitor-based HAART, HDL, and CHD-risk in HIV-infected patients. Atherosclerosis, **184**, 72–77.

77 Riddler, S. A., Smit, E., Cole, S. R., Li, R., Chmiel, J. S., Dobs, A., Palella, F., Visscher, B., Evans, R., Kingsley, L. A. (2003) Impact of HIV Infection and HAART on Serum Lipids in Men. *Jama*, **289**, 2978–2982.

78 Carpentier, A., Patterson, B. W., Uffelman, K. D., Salit, I., Lewis, G. F. (2005) Mechanism of highly active anti-retroviral therapy-induced hyperlipidemia in HIV-infected individuals. *Atherosclerosis*, **178**, 165–172.

79 Fisac, C., Virgili, N., Ferrer, E., Barbera, M. J., Fumero, E., Vilarasau, C., Podzamczer, D. (2003) A comparison of the effects of nevirapine and nelfinavir on metabolism and body habitus in antiretroviral-naive human immunodeficiency virus-infected patients: a randomized controlled study. *J Clin Endocrinol Metab*, **88**, 5186–5192.

80 van der Valk, M., Kastelein, J. J., Murphy, R. L., van Leth, F., Katlama, C., Horban, A., Glesby, M., Behrens, G., Clotet, B., Stellato, R. K., Molhuizen, H. O., Reiss, P. (2001) Nevirapine-containing antiretroviral therapy in HIV-1 infected patients results in an anti-atherogenic lipid profile. *Aids*, **15**, 2407–2414.

81 van Leth, F., Phanuphak, P., Stroes, E., Gazzard, B., Cahn, P., Raffi, F., Wood, R., Bloch, M., Katlama, C., Kastelein, J. J., Schechter, M., Murphy, R. L., Horban, A., Hall, D. B., Lange, J. M., Reiss, P. (2004) Nevirapine and Efavirenz Elicit Different Changes in Lipid Profiles in Antiretroviral- Therapy-Naive Patients Infected with HIV-1. *Plos Med*, **1**, e19.

82 Mauss, S., Stechel, J., Willers, R., Schmutz, G., Berger, F., Richter, W. O. (2003) Differentiating hyperlipidaemia associated with antiretroviral therapy. *Aids*, **17**, 189–194.

83 Calza, L., Manfredi, R., Chiodo, F. (2003) Statins and fibrates for the treatment of hyperlipidaemia in HIV-infected patients receiving HAART. *Aids*, **17**, 851–859.

84 Fontas, E., van Leth, F., Sabin, C. A., Friis-Moller, N., Rickenbach, M., d'Arminio Monforte, A., Kirk, O., Dupon, M., Morfeldt, L., Mateu, S., Petoumenos, K., El-Sadr, W., de Wit, S., Lundgren, J. D., Pradier, C., Reiss, P. (2004) Lipid profiles in HIV-infected patients receiving combination antiretroviral therapy: are different antiretroviral drugs associated with different lipid profiles?. *J Infect Dis*, **189**, 1056–1074.

85 Hsue, P. Y., Lo, J. C., Franklin, A., Bolger, A. F., Martin, J. N., Deeks, S. G., Waters, D. D. (2004) Progression of Atherosclerosis as Assessed by Carotid Intima-Media Thickness in Patients With HIV Infection. *Circulation*, **109**, 1603–1608.

86 Hsue, P. Y. and Waters, D. D. (2005) What a Cardiologist Needs to Know About Patients With Human Immunodeficiency Virus Infection. *Circulation*, **112**, 3947–3957.

87 Vittecoq, D., Escaut, L., Chironi, G., Teicher, E., Monsuez, J. J., Andrejak, M., Simon, A. (2003) Coronary heart disease in HIV-infected patients in the highly active antiretroviral treatment era. *Aids*, **17** (Suppl 1,), S70–S76.

88 Friis-Moller, N., Sabin, C. A., Weber, R., d'Arminio Monforte, A., El-Sadr, W. M., Reiss, P., Thiebaut, R., Morfeldt, L., De Wit, S., Pradier, C., Calvo, G., Law, M. G., Kirk, O., Phillips, A. N., Lundgren, J. D. (2003) Combination antiretroviral therapy and the risk of myocardial infarction. *N. Engl. J. Med.*, **349**, 1993–2003.

89 Allred, K. F., Smart, E. J., Wilson, M. E. (2006) Estrogen Receptor-{alpha} Mediates Gender Differences in Atherosclerosis Induced by HIV Protease Inhibitors. *J. Biol. Chem.*, **281**, 1419–1425.

90 Rose, H., Woolley, I., Hoy, J., Dart, A., Bryant, B., Mijch, A., Sviridov, D. (2006) HIV infection and high-density lipoprotein: the effect of the disease vs the effect of treatment. *Metabolism*, **55**, 90–95.

91 Mujawar, Z., Rose, H., Morrow, M. P., Pushkarsky, T., Dubrovsky, L., Mukhamedova, N., Dart, A., Fu, Y., Orenstein, J. M., Bobryshev, Y. V., Bukrinsky, M., Sviridov, D. (2006) Human Immunodeficiency Virus Impairs Reverse Cholesterol Transport from Macrophages. PLoS Biology, **4**, e365.

92 Haghpassand, M., Bourassa, P. A., Francone, O. L., Aiello, R. J. (2001) Monocyte/macrophage expression of ABCA1 has minimal contribution to plasma HDL levels. *J. Clin. Invest.*, **108**, 1315–1320.

93 Petit, J. M., Duong, M., Masson, D., Buisson, M., Duvillard, L., Bour, J. B., Brindisi, M. C., Galland, F., Guiguet, M., Gambert, P., Portier, H., Verges, B. (2004) Serum adiponectin and metabolic parameters in HIV-1-infected patients after substitution of nevirapine for protease inhibitors. *Eur J Clin Invest*, **34**, 569–575.

94 Rimland, D., Guest, J., Hernandez, I., del Rio, C., Le, N., Brown, W. (2005) Antiretroviral therapy in HIV-positive men is associated with increased apolipoprotein CIII in triglyceride-rich lipoproteins. *HIV Med.*, **6**, 326–333.

95 Tran, H., Robinson, S., Mikhailenko, I., Strickland, D. K. (2003) Modulation of the LDL receptor and LRP levels by HIV protease inhibitors. *J Lipid Res*, **44**, 1859–1869.

96 Greene, D. J., Skeggs, J. W., Morton, R. E. (2001) Elevated triglyceride content diminishes the capacity of high density lipoprotein to deliver cholesteryl esters via the scavenger receptor class B type I (SR-BI). *J Biol Chem*, **276**, 4804–4811.

97 Skeggs, J. W., Morton, R. E. (2002) LDL and HDL enriched in triglyceride promote abnormal cholesterol transport 10.1194/

jlr.M100431-JLR200. *J. Lipid Res.*, **43**, 1264–1274.

98 Moore, R. E., M.-a., Kawashiri,Kitajima, K., Secreto, A., Millar, J. S., Pratico, D., Rader, D. J. (2003) Apolipoprotein A-I Deficiency Results in Markedly Increased Atherosclerosis in Mice Lacking the LDL Receptor. *Arterioscler. Thromb. Vasc. Biol.*, **23**, 1914–1920.

99 den Boer, M. A. M., Berbee, J. F. P., Reiss, P., van der Valk, M., Voshol, P. J., Kuipers, F., Havekes, L. M., Rensen, P. C. N., Romijn, J. A. (2006) Ritonavir Impairs Lipoprotein Lipase-Mediated Lipolysis and Decreases Uptake of Fatty Acids in Adipose Tissue 10.1161/ 01.ATV.0000194073.87647.10. *Arterioscler. Thromb. Vasc. Biol.*, **26**, 124–129.

100 Purnell, J. Q., Zambon, A., Knopp, R. H., Pizzuti, D. J., Achari, R., Leonard, J. M., Locke, C., Brunzell, J. D. (2000) Effect of ritonavir on lipids and post-heparin lipase activities in normal subjects. *Aids*, **14**, 51–57.

101 Tershakovec, A. M., Frank, I., Rader, D. (2004) HIV-related lipodystrophy and related factors. *Atherosclerosis*, **174**, 1–10.

102 Foger, B., Ritsch, A., Doblinger, A., Wessels, H., Patsch, J. R. (1996) Relationship of plasma cholesteryl ester transfer protein to HDL cholesterol. Studies in normotriglyceridemia and moderate hypertriglyceridemia. *Arterioscler. Thromb. Vasc. Biol.*, **16**, 1430–1436.

103 Clark, R. W., Sutfin, T. A., Ruggeri, R. B., Willauer, A. T., Sugarman, E. D., Magnus-Aryitey, G., Cosgrove, P. G., Sand, T. M., Wester, R. T., Williams, J. A., Perlman, M. E., Bamberger, M. J. (2004) Raising High-Density Lipoprotein in Humans Through Inhibition of Cholesteryl Ester Transfer Protein: An Initial Multidose Study of Torcetrapib. *Arterioscler. Thromb. Vasc. Biol.*, **24**, 490–497.

104 Zhang, B., Fan, P., Shimoji, E., Xu, H., Takeuchi, K., Bian, C., Saku, K. (2004) Inhibition of Cholesteryl Ester Transfer Protein Activity by JTT-705 Increases Apolipoprotein E-Containing High-Density Lipoprotein and Favorably Affects the Function and Enzyme Composition of High-Density Lipoprotein in Rabbits. *Arterioscler. Thromb. Vasc. Biol.*, **24**, 1910–1915.

105 Okamoto, H., Yonemori, F., Wakitani, K., Minowa, T., Maeda, K., Shinkai, H. (2000) A cholesteryl ester transfer protein inhibitor attenuates atherosclerosis in rabbits. *Nature*, **406**, 203–207.

106 Gerber, J. G., Rosenkranz, S. L., Fichtenbaum, C. J., Vega, J. M., Yang, A., Alston, B. L., Brobst, S. W., Segal, Y., Aberg, J. A. (2005) Effect of Efavirenz on the Pharmacokinetics of Simvastatin, Atorvastatin, and Pravastatin: Results of AIDS Clinical Trials Group 5108 Study. *J Acquir Immune Defic Syndr*, **39**, 307–312.

107 Moro, H., Tsukada, H., Tanuma, A., Shirasaki, A., Iino, N., Nishibori, T., Nishi, S., Gejyo, F. (2004) Rhabdomyolysis after simvastatin therapy in an HIV-infected patient with chronic renal failure. *AIDS Patient Care STDS*, **18**, 687–690.

108 Aberg, J. A., Rosenkranz, S. L., Fichtenbaum, C. J., Alston, B. L., Brobst, S. W., Segal, Y., Gerber, J. G. (2006) Pharmacokinetic interaction between nelfinavir and pravastatin in HIV-seronegative volunteers: ACTG Study A5108. *Aids*, **20**, 725–729.

109 Barbaro, G. (2002) Cardiovascular Manifestations of HIV Infection. *Circulation*, **106**, 1420–1425.

110 Baliga, R. S., Chaves, A. A., Jing, L., Ayers, L. W., Bauer, J. A. (2005) AIDS-related vasculopathy: evidence for oxidative and inflammatory pathways in murine and human AIDS. *Am J Physiol Heart Circ Physiol*, **289**, H1373–H1380.

111 Behnisch, T., Francesconi, W., Sanna, P. P. (2004) HIV secreted protein Tat prevents long-term potentiation in the hippocampal CA1 region. *Brain Res*, **1012**, 187–189.

112 Li, Q., Duan, L., Estes, J. D., Ma, Z. M., Rourke, T., Wang, Y., Reilly, C., Carlis, J., Miller, C. J., Haase, A. T. (2005) Peak SIV replication in resting memory CD4 + T cells depletes gut lamina propria CD4 + T cells. *Nature*, **434**, 1148–1152.

113 Steinberg, D. (2006) Thematic review series: The Pathogenesis of Atherosclerosis. An interpretive history of the cholesterol controversy, part V: The discovery of the statins and the end of the controversy. *J. Lipid Res.*, **47**, 1339–1351.

114 Sviridov, D. (2005) Therapeutic Targeting of HDL and Reverse Cholesterol Transport. *Current Medicinal Chemistry-IEMA*, **5**, 299–309.

115 Linsel-Nitschke, P. and Tall, A. R. (2005) HDL as a target in the treatment of atherosclerotic cardiovascular disease. *Nat Rev Drug Discov*, **4**, 193–205.

116 Duffy, D. and Rader, D. J. (2006) Emerging Therapies Targeting High-Density Lipoprotein Metabolism and Reverse Cholesterol Transport. *Circulation*, **113**, 1140–1150.

117 Berthezene, F. (1996) Non-insulin dependent diabetes and reverse cholesterol transport. *Atherosclerosis*, **124** (Suppl), S39–S42.

15
Endothelial Protection by High-Density Lipoproteins

Monica Gomaraschi, Laura Calabresi, Guido Franceschini

15.1
Endothelial Dysfunction and Cardiovascular Disease

The vascular endothelium is the monolayer of endothelial cells lining the lumen of the vascular beds, and is mechanically and metabolically strategically located, separating the vascular wall from the circulation and the blood components. This monocellular layer is continuously exposed to flowing blood and reacts to even minute changes in local and circulating factors by generating paracrine and autocrine mediators that modulate vascular tone and permeability, inflammatory responses, hemostasis, and the integrity of the endothelium itself. As a major regulator of vascular homeostasis, the endothelium maintains the balance between vasodilatation and vasoconstriction, prevention and stimulation of cell adhesion, and thrombogenesis and fibrinolysis [1]. Damage to the endothelium upsets this tightly regulated balance, leading to endothelial dysfunction, which is pivotal in the development, progression, and rupture of an atherosclerotic lesion and implicated in the clinical manifestations of atherosclerotic cardiovascular disease (Fig. 15.1).

The concept of endothelial dysfunction developed in the mid-1980s, following the discovery that acetylcholine requires the presence of the endothelial cells to relax the underlying vascular smooth muscle [2]. Endothelium-dependent vasodilation was established to occur primarily through the release of endothelium-derived relaxing factor, a humoral mediator that has since been identified as nitric oxide (NO) [3]. Impaired NO-dependent vasodilation soon became the hallmark of endothelial dysfunction. However, during the last two decades it has become evident that endothelial dysfunction extends beyond impaired NO-dependent vasodilation, and reflects a more comprehensive state of endothelial activation. Endothelial cells synthesize and secrete various other vasodilators (e.g., prostacyclin and bradykinin), as well as vasoconstrictors (e.g., endothelin and angiotensin II), and in the presence of endothelial damage, contracting factors predominate. Furthermore, endothelial cells produce a variety of substances that regulate platelet adhesion and aggregation (e.g., NO, prostacyclin, and von Willebrand factor), the clotting cascade (e.g., thrombomodulin and tissue factor), and the fibrinolytic system (e.g., tissue plasminogen activator and plasminogen-activator inhibitor), while damage to the endothelium induces

High-Density Lipoproteins: From Basic Biology to Clinical Aspects. Edited by Christopher J. Fielding

ISBN: 978-3-527-31717-2

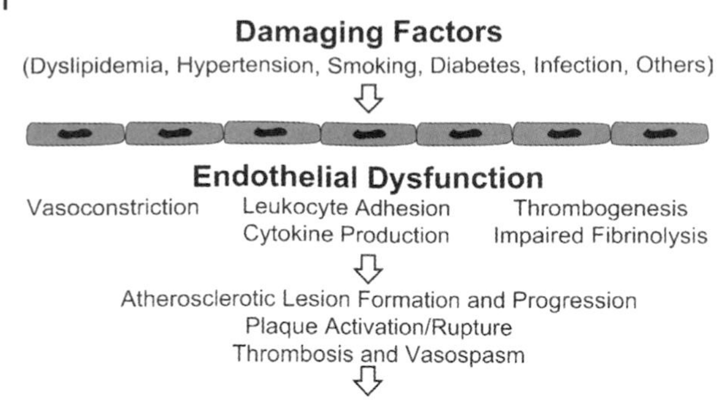

Fig. 15.1 Role of endothelial dysfunction in the pathogenesis of cardiovascular events.

endothelial cells to adopt a prothrombotic phenotype. Finally, when damaged or exposed to certain pathogenetic proinflammatory stimuli, endothelial cells express cellular adhesion molecules (CAMs), leukocyte chemotactic factors (e.g., monocyte chemotactic protein 1), and inflammatory cytokines (e.g., interleukin-1, interleukin-6, and tumor necrosis factor-α) that regulate and direct the inflammatory process [4]. The term "endothelial dysfunction" thus refers to this broad alteration in endothelial phenotype, which develops in response to damaging factors (Fig. 15.1).

Endothelial dysfunction in humans has traditionally been assessed by directly measuring the endothelial cell response to certain stimuli that trigger the release of NO from the vascular endothelium to mediate vasorelaxation. Quantitative angiography has been used to measure changes in coronary diameter in response to the infusion of endothelium-dependent vasodilators, such as acetylcholine, bradykinin, or serotonin [5]. A less invasive method that has gained popularity in recent years uses high-resolution ultrasonography to measure changes in brachial artery diameter in response to reactive hyperemia [6]; the shear stress induced by reactive hyperemia stimulates the release of NO, with subsequent vasodilation. A correlation exists between flow-mediated dilation of the brachial artery and coronary dilation in response to acetylcholine [7], underscoring the systemic nature of endothelial dysfunction and the potential of peripheral vasodilation as a surrogate marker to identify individuals with coronary endothelial dysfunction. An alternative, indirect, and noninvasive approach to the assessment of endothelial dysfunction is the measurement of circulating factors released by damaged and dysfunctional endothelium [8], such as soluble CAMs, proinflammatory cytokines, von Willebrand factor (vWF), tissue factor (TF), and plasminogen activator inhibitor-1 (PAI-1). The major advantage of biomarkers of endothelial function is that their measurement is less expensive than functional tests and provides excellent reproducibility. Elevated plasma levels of such biomarkers, such as intracellular adhesion molecule-1 (ICAM-1), vascular cell adhesion molecule-1 (VCAM-1), vWF, and PAI-1, correlate with impaired NO-dependent vasodilatation [9–12], which validates their use as a clinical tool for the assessment of endothelial dysfunction.

The first evidence of a linkage between endothelial dysfunction and atherosclerosis in humans was provided by assessment of coronary vasodilatory response to acetylcholine. While acetylcholine evoked a NO-mediated vasodilation in healthy subjects, in patients with mild to severe coronary artery disease this effect was blunted, or paradoxical vasoconstriction was observed [5]. Further studies assessing coronary or peripheral NO-mediated vasodilation established that endothelial dysfunction can be detected in patients with coronary risk factors but no angiographic or ultrasound evidence of arterial disease [13,14], indicating that endothelial dysfunction is present in the preclinical stage of atherosclerosis. More convincing evidence for a pathogenetic role of endothelial dysfunction is provided by prospective studies demonstrating that functional or biochemical measures of endothelial dysfunction are not only associated with atherosclerotic vascular disease, but also predict the occurrence of future clinical events, such as cardiac death, myocardial infarction, or ischemic stroke, in patients with or without coronary artery disease [15–19].

Given the relationship between endothelial dysfunction and atherosclerosis, it is not surprising that many of the traditional coronary risk factors, such as hyperlipidemia, hypertension, smoking, diabetes, and family history of premature coronary heart disease are associated with functional [20–24], or biochemical [25–29] measures of endothelial dysfunction. Moreover, the number of risk factors present in a given individual is a strong independent predictor of endothelial dysfunction, as assessed in the coronary or peripheral circulation [1].

Endothelial dysfunction has been reported in patients with primary hypoalphalipoproteinemia [30–32], and a low plasma concentration of high density lipoprotein cholesterol (HDL-C) is an independent predictor of endothelial dysfunction both in healthy individuals and in hyperlipidemic, diabetic, and coronary patients [33–36]. Moreover, elevation of plasma HDL levels by drug treatment, or by infusion of synthetic HDL (sHDL), lead to a significant improvement of endothelial dysfunction [32,37–39]. All together, these findings argue for a protective effect of HDL against the development of endothelial dysfunction, which may well contribute, together with the known HDL function in cell cholesterol removal and transport, to HDL-mediated protection against cardiovascular diseases. In this chapter we report and discuss the results of *in vitro* and *in vivo* investigations that have unequivocally demonstrated that HDL are an autonomous protective factor for vascular endothelium, positively affecting a number of endothelial functions in the regulation of vascular tone, inflammation, and hemostasis, and in the maintenance of endothelium integrity (Fig. 15.2).

15.2 HDL and Regulation of Vascular Tone

15.2.1 Nitric Oxide

NO is a potent vascular relaxing factor that exerts its effect by activating guanylate cyclase, thus increasing cyclic guanosine monophosphate and decreasing cytosolic

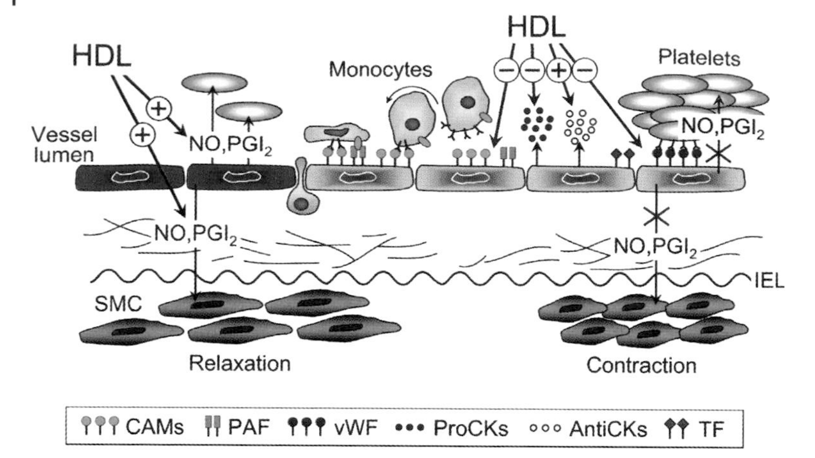

Fig. 15.2 Multiple biological actions of HDL on vascular endothelium. Functional endothelial cells are in dark blue, dysfunctional endothelial cells are in light blue. Abbreviations: antiCKs: antiinflammatory cytokines. CAMs: cell adhesion molecules. NO: nitric oxide. PAF: platelet activating factor. PGI_2: prostacyclin. proCKs: proinflammatory cytokines. TF: tissue factor. vWF: von Willebrand factor.

calcium in vascular smooth muscle cells (SMC). NO is generated in the endothelium by a constitutive endothelial NO synthase enzyme (eNOS). This process uses electrons donated by nicotinamide adenine dinucleotide phosphate to reduce heme iron and to activate molecular oxygen, thereby oxidizing L-arginine to form NO, L-citrulline, and H_2O. Although eNOS is constitutively present in endothelial cells, its expression and activity may be increased through transcriptional or posttranscriptional regulation. Endothelial NOS transcription is increased by physical forces, such as shear stress, cyclic strain, and heat shock, and also by biochemical species such as oxidized low-density lipoproteins (ox-LDL) and their component lysophosphatidylcholine (LPC) [40]. Posttranscriptional palmitoylation of eNOS assists in targeting the enzyme to plasmalemmal caveolae, a prerequisite for its efficient activity [41].

Classified as a calcium/calmodulin-dependent enzyme, eNOS is typically stimulated by its activators through an increase in intracellular calcium. Calcium binding to calmodulin displaces the eNOS inhibitor caveolin-1 from its inhibitory site, thereby increasing eNOS activity [42]. In addition, however, calcium-independent phosphorylation of the serine-1177 residue of eNOS by Akt also enhances enzyme activity [43].

In vitro experiments with cultured endothelial cells have shown HDL to be able to enhance NO production through eNOS stimulation. HDL increase eNOS protein abundance, and offset eNOS downregulation by LDL [37], without affecting mRNA levels or half-life [44]. Instead, HDL prolong the half-life of eNOS protein, in a process that involves extracellular signal-regulated kinases 1 and 2 (ERK1/2), and Akt activation [44]. Lipid-free apolipoprotein A-I (apoA-I) also increases the amount of eNOS, although less effectively than HDL, suggesting that other components of the HDL particle also contribute to this effect [44]. HDL not only increase cellular eNOS content, but also enhance eNOS activity through indirect and direct mechanisms.

HDL prevent the ox-LDL-induced displacement of eNOS from caveolae and its consequent inactivation [45] by preserving the caveolae lipid environment, thanks to the provision of cholesteryl esters mediated by interaction with the scavenger receptor B type I (SR-BI) [46]. In addition, HDL directly activate eNOS by Akt-dependent and -independent mechanisms [47]. The first mechanism involves the binding of apoA-I to SR-BI; however, apoA-I is necessary but not sufficient for eNOS stimulation, since no activation is observed with lipid-free apoA-I [47,48]. The interaction of HDL with SR-BI activates a cascade of kinases in the form of Src family kinase, phosphoinositide 3-kinase (PI3K), and Akt, ultimately leading to eNOS phosphorylation and activation [49]. This process requires cholesterol flux, the C-terminal transmembrane domain of SR-BI that directly binds cholesterol, and the C-terminal PDZ-interacting domain of SR-BI, with SR-BI thus acting as a cholesterol sensor on the plasma membrane [49].

Moreover, HDL cause an increase in the intracellular calcium concentration; the consequent binding of calcium-calmodulin to eNOS positively regulates enzyme activity, independently of Akt-stimulated phosphorylation [47]. HDL lysophospholipids (sphingosine-1-phosphate (S1P), sphingosylphosphorylcholine, and lysosulfatide) are implicated in such calcium-dependent eNOS activation through binding to the S1P receptors [47].

While the interaction of apoA-I and S1P with their receptors may activate distinct signaling pathways, each contributing to eNOS activation, a unifying model for full eNOS activation by plasma HDL has been proposed. In this, SR-BI could provide the necessary spatial proximity for the HDL lysophospholipids to interact effectively with S1P receptors [47].

Ex vivo studies with isolated arteries underscore the complexity of the effects of HDL on NO bioavailability. Early studies with isolated strips of rabbit arteries showed HDL to normalize the impaired NO-dependent vasorelaxation induced by ox-LDL, either by removal of LPC from ox-LDL, thus preventing LPC from acting on the endothelium [50], or by reversal of the suppressive effect of LPC on NO bioavailability [40]. Such effects are mimicked by sHDL, but not by lipid-free apoA-I [51]; notably, sHDL containing the apoA-I_{Milano} (A-I_M) variant are more effective than those containing wild-type apoA-I in restoring endothelial function in aortic strips pretreated with LPC [51], implying a role for the apoA-I structure within HDL in modulation of endothelial NO production. Aortas harvested from apoE-deficient mice chronically treated with A-I_M-containing sHDL display an improved dilator response to acetylcholine in relation to arteries from untreated mice [51]; in this case the NO-dependent vasorelaxation is inversely correlated with the arterial cholesterol content [51]. HDL also cause direct relaxation of phenylephrine-precontracted rings of rat aortas, an effect that is prevented by a selective inhibitor of Akt activation by PI3K [47], is abolished by the eNOS antagonist L-NAME, and does not occur with aortas from eNOS-deficient mice. This effect may again involve the SR-BI receptor, thus providing a link between HDL-mediated cell cholesterol depletion and NO bioavailability.

In vivo studies provide further convincing evidence of the relevant role of HDL in stimulating endothelial NO production, thus enhancing NO-dependent vasorelaxation. Intra-arterial administration of HDL, or of lysophospholipids, lowers arterial blood pressure in rats in which vasoconstriction has been induced by intravenous endothelin infusion [47]. In addition, intravenous administration of HDL enhances myocardial perfusion in mice, an effect evidently mediated by an NO-dependent mechanism, as it is abolished in eNOS-deficient mice [52]. Notably, the administration of S1P to wild-type mice causes an opposite effect, a reduction of myocardial perfusion, which is abolished in mice lacking the SIP_3 receptor [52]. In $S1P_3$-deficient mice, the stimulatory effect of HDL on myocardial perfusion is fully preserved, indicating that HDL increase myocardial perfusion *in vivo* by a NO-dependent and a $S1P_3$-independent mechanism [52].

The importance of the vasodilating properties of HDL in human physiology and pathology was underscored by early investigations showing a positive correlation between plasma HDL-C concentration and NO-dependent coronary vasodilation, as assessed either by quantitative coronary angiography [53] or by intravascular ultrasound [54]. More recently, plasma HDL-C concentration has been found to be an independent predictor of NO-dependent peripheral vasodilation in healthy individuals [55], hyperlipidemic and diabetic patients [34,35], and coronary patients [36]. These findings support the concept that enhancing plasma HDL levels may favorably affect the regulation of the vascular tone. Indeed, short-term treatment with niacin in patients with low HDL causes an elevation of plasma HDL with a parallel increase in NO-mediated vasodilation [37]. Even more strikingly, intravenous infusion of sHDL in subjects with hypercholesterolemia or hypoalphalipoproteinemia rapidly restores the altered endothelium-dependent vasodilation [32,39].

15.2.2
Prostacyclin

Prostacyclin (PGI_2) elicits SMC relaxation by activating a specific cell surface receptor (IP) that is G protein-coupled to adenylyl cyclase and thereby elevates cAMP levels. Endothelial cells synthesize PGI_2 from arachidonate derived either from membrane phospholipids or from exogenous sources (e.g., phospholipids and cholesteryl esters of circulating lipoproteins). Two different cyclooxygenase (Cox) isoforms, a constitutive Cox-1 and an inducible Cox-2, metabolize arachidonate to PGH_2, which is further converted into PGI_2 by prostacyclin synthase. Although both Cox isoforms can mediate PGI_2 production, the synthesis of endothelial PGI_2 seems to be mainly driven by Cox-2 [56]. Endothelial Cox-2 expression is induced by a variety of stimuli, through activation of multiple signaling pathways [57]. Mitogen-activated protein kinase (MAPK), protein kinase C (PKC), and tyrosine kinases have all been reported to be involved in Cox-2 induction in endothelial cells; once activated, these kinases modulate the activity of various transcription factors, such as CRE and NF-κB, involved in Cox-2 expression [58].

Incubation of cultured endothelial cells with HDL causes a dose-dependent increase in PGI_2 release, which is prevented by a non isoform-selective Cox inhibitor, implying an effect on PGI_2 synthesis [59,60]; the two major HDL subfractions, HDL_2 and HDL_3, are equally effective [61]. Delipidated HDL apolipoproteins also enhance PGI_2 production, but to a lower extent than intact HDL [62], suggesting that activation of different mechanisms by HDL lipids and apolipoproteins may ultimately enhance PGI_2 production. Notably, sHDL made with apoA-I and soybean phosphatidylcholine are equally as effective as plasma-derived HDL in enhancing endothelial PGI_2 production, indicating that a lipoprotein particle is needed to achieve full stimulation of Cox-2-mediated PGI_2 synthesis [63]. HDL can stimulate PGI_2 production either by providing endothelial cells with arachidonate for Cox-mediated PGI_2 synthesis [62], or by enhancing Cox-2 expression, or both. As HDL cholesteryl esters are the most efficient donors of arachidonate to endothelial cells [62], it is tempting to speculate that the HDL-induced PGI_2 synthesis may depend on SR-BI-mediated selective uptake and hydrolysis of HDL cholesteryl esters. HDL may also activate a calcium-sensitive, membrane-bound phospholipase [64], thus enhancing the availability of endogenous arachidonate for PGI_2 synthesis [65]. HDL enhance basal and cytokine-induced Cox-2 content of endothelial cells in a time- and dose-dependent manner, through a combination of transcriptional and posttranscriptional mechanisms [63,66]. In unstimulated cells, HDL activate p38 MAPK, one of the major kinases involved in Cox-2 expression; the activation of this signaling pathway promotes Cox-2 mRNA transcription and stabilization, ultimately enhancing Cox-2 protein content [66]. What is less clear is which transcription factor is involved in HDL-induced Cox-2 expression downstream of MAPK. The human Cox-2 promoter contains binding sites for NF-κB and CRE. Early studies in cytokine-activated endothelial cells showed that HDL do not affect NF-κB translocation or transactivation, suggesting that HDL induce Cox-2 expression through an NF-κB-independent pathway [63]. More recent transient transfection experiments in CHO cells show that a mutation in the CRE site completely abolishes the stimulation of Cox-2 promoter by HDL, which is instead reduced by 50 % by a mutation in the NF-κB site [66], indicating that both transcription factors may play a role in Cox-2 induction by HDL.

Ex vivo studies in isolated rabbit and rat heart preparations show that homologous and heterologous HDL increase basal cardiac PGI_2 release [65,67]; moreover, plasma-derived and synthetic HDL further enhance the post-ischemic release of PGI_2 in isolated hearts [67,68], indicating a role for HDL-induced PGI_2 production in strengthening the heart's own defenses against the deleterious effects of ischemia reperfusion.

There is limited information on the effects of HDL on PGI_2 synthesis/release *in vivo* in humans, but the plasma HDL-C level has been found to correlate with the plasma concentration of the stable PGI_2 metabolite 6-keto-$PGF_{1\alpha}$[69,70].

15.3 HDL and Vascular Inflammation

15.3.1 Cellular Adhesion Molecules

CAMs mediate the attachment and transmigration of leukocytes across the endothelial surface in response to pathogenic proinflammatory stimuli, and are hypothesized to play an important role in the initiation of atherosclerosis [71]. Three groups of CAMs have been described: integrins, selectins, and members of the immunoglobulin family. Several selectins – L-selectin, P-selectin, and E-selectin – have been found on the endothelium, the last of these being exclusively expressed on the surfaces of activated endothelial cells. E-selectin binds neutrophils, monocytes, eosinophils, basophils, natural killer cells, and subsets of lymphocytes.

VCAM-1, a member of the immunoglobulin family, is normally expressed at low levels by endothelial cells, smooth muscle cells, tissue macrophages, lymphoid dendritic cells, and renal tubular cells. Several stimuli, including cytokines, modified lipoproteins, advanced glycation end-products, increased blood pressure, and oxidants, can upregulate VCAM-1 expression. ICAM-1, also a member of the immunoglobulin family, is a membrane-bound adhesion molecule receptor for monocytes, lymphocytes, and neutrophils, and also functions as a receptor for soluble fibrinogen. ICAM-1 is expressed by endothelial cells, smooth muscle cells, and monocytes. ICAM-1 levels can be upregulated by inflammatory cytokines, but can also increase after exposure to high glucose levels. Accumulating data imply that selectins mediate initial rolling of leukocytes along the endothelium, and that VCAM-1 and ICAM-1 play important roles in the firm attachment and transendothelial migration of leukocytes.

Soluble forms of E-selectin, VCAM-1, and ICAM-1 are detectable in plasma and have been reported to parallel the expression of the membrane-bound forms on endothelial cells [72].

HDL have been shown to inhibit TNF-α- and IL-1-induced expression of E-selectin, ICAM-1, and VCAM-1 in cultured human umbilical vein endothelial cells [73]. The effect is dose-dependent and is maximal at physiological HDL concentrations. HDL isolated from plasma from different individuals vary in their inhibitory activities, possibly reflecting differences in the concentrations of the major HDL subfractions HDL_2 and HDL_3: the latter is more effective than the former in downregulating CAM expression [74]. sHDL made with apoA-I and phosphatidylcholine (but not phospholipid vesicles, nor the lipid-free apolipoprotein) also prevent TNF-α-induced CAM upregulation [75]. The inhibitory effect of sHDL is unaffected by variations in particle shape or size, or in the apolipoprotein, cholesteryl ester, or triglyceride composition [75,76]. In contrast, the degree of saturation of the fatty acid in the *sn-2* position in the phosphatidylcholine has a marked influence on the ability of sHDL to inhibit VCAM-1 expression by activated endothelial cells [76]. The molecular mechanism/s by which HDL downregulate cytokine-induced CAM expression are not fully understood. The inhibitory effect was initially related to the inhibition of TNF-α-induced activation of sphingosine kinase (SphK) and consequent decreased intracellular S1P accumulation [77]. This prevents ERK activation and the subsequent nuclear translocation and DNA binding of NF-κB [77,78], a prerequisite for TNF-α–induced CAM expression. Notably, SphK activity is not inhibited either by lipid-free apoA-I or by phosphatidylcholine vesicles, while sHDL made with the same components inhibit TNF-α-induced SphK activation to the same degree as plasma-derived HDL [77]. The signaling molecules upstream of SphK have not been investigated. More recently, both SR-BI and the S1P receptors have been found to be involved in attenuation of endothelial NF-κB activation by HDL, with subsequent downregulation of CAM expression [79]. HDL interaction with SR-BI, possibly through apoA-I, and/or with the S1P receptors, through S1P and other lysophospholipids, activates an inhibitory PI3K/NOS pathway, with the consequent reduced translocation of NF-κB into the cell nucleus [79]. This dual-receptor model for downregulation of CAM expression by HDL recalls the unifying model that has been proposed for eNOS activation by HDL [47] (see above); however, the molecular mechanisms linking enhanced NO synthesis and inhibition of CAM expression through the SphK/ERK/NF-κB pathway remain to be established.

While studies are all consistent in showing the ability of HDL to modulate CAM expression in umbilical vein endothelial cells, conflicting results have been obtained with cells that more directly model blood vessels affected by atherosclerosis (i.e., aortic and coronary endothelial cells). HDL and sHDL failed to downregulate TNF-α-induced CAM expression in human aortic and coronary endothelial cells [80], but sHDL inhibited the cytokine-induced expression of E-selectin in porcine aortic endothelial cells [81]. The reasons for these discrepant findings are not known.

Several studies in experimental animals provide evidence for HDL-mediated downregulation of CAM expression *in vivo*. Elevation of plasma HDL concentration by overexpression of apoA-I in apoE-deficient mice significantly inhibits ICAM-1 and VCAM-1 expression on vascular endothelium, and also monocyte recruitment into the arterial wall [82]. Moreover, single or repeated infusions of sHDL reduce endothelial VCAM-1 expression and monocyte recruitment into the arterial walls of apoE-deficient mice [83], inhibit IL-1-induced E-selectin expression in a porcine model of acute inflammation [81], decrease ICAM-1 and P-selectin expression in a rat model of hemorrhagic shock [84], and inhibit the endothelial expression of ICAM-1 and VCAM-1, and the consequent infiltration of neutrophils, induced by a periarterial collar in normocholesterolemic rabbits [85].

In humans, a low plasma HDL concentration is associated with enhanced plasma levels of soluble CAMs [30,86], a surrogate marker of endothelial CAM expression. Moreover, increasing the plasma HDL-C level in low-HDL subjects by fenofibrate treatment results in a significant reduction of soluble CAM concentrations [30]. Finally, the ability of HDL to inhibit CAM expression in cultured endothelial cells is significantly affected by the consumption of single meals with different dietary fat compositions, consistently with the results of experiments with sHDL [76], HDL isolated after a meal enriched in polyunsaturated fat are more effective – and those isolated after a meal enriched in saturated fat less effective – than fasting HDL in downregulating cytokine-induced CAM expression [87].

15.3.2 Cytokines

Endothelial cells, like all the other vascular cells, are able to produce a number of pro- and antiinflammatory cytokines; the proinflammatory repertoire includes IL-1, IL-6, IL-8, and MCP-1, while the antiinflammatory repertoire is represented predominantly by transforming growth factor-β (TGF-β) [88]. Under homeostatic conditions, the endothelium produces little to no proinflammatory factors, which are instead synthesized and released when endothelial cells are exposed to pathogenic proinflammatory stimuli [89].

HDL inhibit the TNF-α-induced production of IL-6 and IL-8 by cultured endothelial cells in a dose-dependent manner [90,91]. Plasma-derived and synthetic HDL are equally effective, while lipid-free apoA-I or phospholipid vesicles are totally ineffective [90]. The inhibitory effect on IL-6 production is mediated through activation of p38 MAPK, the critical kinase in cytokine-induced IL-6 production in endothelial cells [90]. In addition, HDL induce the expression of the antiinflammatory TGF-β2 in unstimulated endothelial cells [92]. In this case, the effect of HDL is mimicked by lysosphingolipids, and is dependent on PI3K/Akt activation; lipid-free apoA-I has only a minimal effect [92].

The ability of HDL to modify endothelial cytokine production has also been demonstrated *in vivo*. TGF-β2 expression is increased in the aortas of transgenic mice with elevated plasma HDL levels due to overexpression of human apoA-I [92]. In humans, a low plasma HDL concentration is associated with increased plasma levels of IL-6 [90].

15.3.3 Platelet-Activating Factor

Platelet-activating factor (PAF) is a highly bioactive phospholipid that exerts a broad range of biological effects, such as stimulation of cell adhesion, vascular permeability, platelet aggregation, and smooth muscle contraction. HDL inhibit, in a dose-dependent manner, the agonist-induced production of PAF by cultured human endothelial cells; both lipid and apolipoprotein components are required for full inhibition [93]. The biological activity of PAF is abolished by hydrolysis of the acetyl residue through the action of intra- and extracellular PAF-acetylhydrolases (PAF-AHs). Human plasma PAF-AH is in part associated with HDL [94], which also bind and transport other PAF-degrading enzymes such as lecithin: cholesterol acyltransferase and paraoxonase [95,96]. Therefore, by limiting PAF production by endothelial cells, and by enhancing its degradation by circulating enzymes, HDL may prevent PAF-induced adhesion of leukocytes to the activated endothelium, which may well contribute to the *in vivo* antiadhesive effects of HDL described above.

15.4 HDL and Hemostasis

15.4.1 Platelet Adhesion and Reactivity

Besides being crucial mediators in the regulation of vascular tone, endothelium-derived NO and PGI_2 have powerful antithrombotic effects, inhibiting platelet aggregation by increasing cGMP and cAMP contents, respectively. PAF instead stimulates platelet aggregation. vWF is another protein expressed by endothelial cells that plays an essential role in platelet adhesion to the dysfunctional endothelium. The circulating vWF levels are inversely correlated with plasma HDL [97], suggesting that HDL may inhibit vWF production. Therefore, by modulating the production/activity of a variety of NO, PGI_2, PAF, and vWF, HDL favorably affect both vascular tone and thrombogenicity. It is noteworthy that high plasma HDL-C levels in humans are associated with reduced *ex vivo* thrombogenic potential [98], and that in an animal model of arterial thrombosis the injection of sHDL significantly reduces thrombus formation on the injured endothelial surface [99].

15.4.2
Coagulation

TF is a membrane-bound protein that initiates the extrinsic coagulation pathway by mediating the activation of factors IX and X by factor VIIa. Healthy endothelial cells express negligible amounts of TF and have low TF activity, which can be upregulated by various stimuli such as thrombin, TNF-α, and IL-1.

Early studies using an isolated *in vitro* reaction assay showed that HDL and the delipidated apolipoprotein fraction could inhibit the activation of factor X by TF and factor VIIa [100]: the active component was later identified as apoA-II, which may act by altering the lipids around TF and reducing its ability to associate functionally with factor VIIa [101]. In addition, HDL and sHDL made with phosphatidylcholine and apoA-I have been shown to inhibit thrombin-induced TF expression in cultured endothelial cells [102]. This effect is mediated by inhibition of RhoA activation and stimulation of PI3K, independently of Akt/NOS activation [102].

Another mechanism by which HDL may inhibit coagulation is through potentiation of endogenous anticoagulant factors, such as activated protein C (APC) and protein S. In an isolated *in vitro* reaction assay, HDL (but not LDL) significantly enhanced inactivation of coagulation factor Va by APC and protein S, and stimulated protein S-dependent proteolytic inactivation of Va by APC, apparently through Va cleavage [103]. When added to plasma, HDL enhance APC/protein S anticoagulant activity in a prothrombin-time clotting assay; both HDL phospholipids and apoA-I are necessary for enhancement of anticoagulant activity. Moreover, in healthy individuals, apoA-I levels correlate with the anticoagulant response to APC/protein S tested by the same clotting assay [103]. The HDL anticoagulant effect is probably mediated through stabilization of the APC/protein S/factor Va complex [103]. In addition to this cell-independent activity, HDL may counteract coagulation by enhancing the endothelial production of endogenous anticoagulants. Repeated infusions of autologous HDL into cholesterol-fed rabbits undergoing balloon denudation of abdominal aorta increased the expression of thrombomodulin, which is essential for protein C activation [104]. This effect is mimicked by sHDL made with phosphatidylcholine and either apoA-I or apoA-II [104]. The observation that the infusion of sHDL into human volunteers limits the procoagulant state associated with endotoxemia [105] further supports the concept that HDL may exert a significant anticoagulant effect *in vivo* in humans.

15.4.3
Fibrinolysis

The endothelium participates in the regulation of fibrinolysis by producing and releasing plasminogen activators, such as tissue plasminogen activator (tPA), and inhibitors, such as PAI-1. An early report indicated that HDL reduce tPA secretion and mRNA levels in unstimulated and stimulated human endothelial cells [106], but this observation has not been confirmed by others [107]. Both studies are consistent in showing no or weak effects of HDL on PAI production. HDL apolipoproteins directly enhanced urokinase-induced plasminogen activation in a purified *in vitro* system, through an unknown mechanism [108]. This activity may lead to the formation of unstable thrombi, with an increased tendency to dissolve spontaneously, as observed in rats injected with sHDL [99].

15.5
HDL and Endothelial Monolayer Integrity

15.5.1
Endothelial Cell Apoptosis

Extended endothelial cell damage through mechanical/biochemical stress can result in endothelial cell apoptosis with loss of the integrity of the endothelial monolayer. Apoptosis is a form of cell death responsible for the selective and controlled elimination of senescent, damaged, infected, or mutated cells, including endothelial cells. Characteristic features of apoptosis include cytoplasm and nucleus shrinkage, plasma membrane blebbing, chromatin condensation, and DNA fragmentation. These events are mediated by caspases, a series of aspartate-specific cysteine proteases that exist as latent zymogens, and, once activated by proteolysis, systematically dismantle the cell by cleaving key cellular proteins [109]. Inflammatory cytokines, ox-LDL, products of lipid peroxidation, and growth factor deprivation can induce excessive and uncontrolled apoptosis of endothelial cells, leading to increased permeability of vascular endothelium, blood cell adhesion to the vessel wall, enhanced thrombogenesis, and SMC proliferation.

Two different apoptotic pathways have been elucidated in endothelial cells. The first is triggered by death receptors, members of the TNF receptor gene superfamily, among which the best characterized are CD95 (also called Fas or Apo1) and TNFR1; the downstream signaling cascade leads to the activation of initiator caspase-8 and effector caspase-3 [109]. Plasma-derived and synthetic HDL have been shown to protect cultured human endothelial cells from TNFα-induced apoptosis [110]. The effect is dose-dependent, it occurs at physiological HDL concentrations, and it is mediated through inhibition of caspase-3 activity; the signaling pathway responsible for this inhibition has not been elucidated [110]. The apolipoprotein composition of sHDL affects antiapoptotic activity, with apoA-I-containing particles being the most effective [110].

The second apoptotic pathway is triggered by growth factor deprivation and is characterized by activation of initiator caspase-9 and consequently of caspase-3; a central role is played by mitochondria, through the release into the cytoplasm of cytochrome c, an essential component of caspase-9 activation machinery [109]. HDL have been shown to suppress growth factor deprivation-induced apoptosis of endothelial cells by inhibiting the dissipation of mitochondrial potential, ROS generation, and cytochrome c release, thus preventing caspase-9 and -3 activation [111]. The effect of HDL is mediated by the activation of the PI3K/Akt pathway, as demonstrated by the use of different PI3K inhibitors and of a dominant negative Akt mutant [111]. HDL apolipoproteins do not inhibit this apoptotic pathway; in contrast, the HDL lipid fraction and, in particular, HDL-carried lysosphingolipids fully mimic the HDL antiapoptotic effect, thus indicating the involvement of an S1P receptor-activated pathway.

Recently, a novel ligand-independent apoptotic pathway has been elucidated in different cell systems, including mouse aortic endothelial cells. This pathway is triggered by SR-BI and involves caspase-8 activation [112]; in this case HDL seem to act as a survival factor that, when present, prevents SR-BI-induced apoptosis [112]. It has been proposed that this effect could be mediated by the preservation of eNOS localization within caveolae [112], but the relevance of this new apoptotic pathway in endothelium homeostasis and the mechanism responsible for the HDL inhibitory effect are still unknown.

15.5.2
Migration and Proliferation of Endothelial Cells

Prompt repair of the damaged endothelium is required to restore vascular wall homeostasis. The restoration of endothelium integrity is achieved through migration of healthy endothelial cells to the site of the lesion and subsequent proliferation. The migration of endothelial cells is mediated by controlled cell adhesion, cytoskeletal reorganization, and localized degradation of the extracellular matrix. Cytoskeletal changes are mainly regulated by members of the Ras superfamily of small GTPases, and among these RhoA appears to be the key element in the induction of endothelial cell migration [113]. The major proliferative stimuli for endothelial cells are vascular endothelial growth factor (VEGF) and fibroblast growth factor (FGF), both acting mainly through the activation of the ERK cascade [113].

Endothelial progenitor cells (EPC) have recently been shown to contribute to endothelium maintenance and repair; these bone marrow-derived cells are recruited to the site of injury by an unknown molecular mechanism and subsequently differentiate into cells that display classical endothelial morphology and characteristics [114]. Incorporation of EPCs has been demonstrated at sites of endothelial injury in various animal models [115–117]. In humans, major cardiovascular risk factors, such as hypercholesterolemia, hypertension, diabetes, and smoking, are associated with reduced EPC number and/or impaired EPC migration [118], and a low level of circulating EPCs predicts the occurrence of cardiovascular events and death from cardiovascular causes in coronary patients [119].

In early studies with cultured bovine aortic endothelial cells (BAECs), HDL were shown to promote endothelial cell migration in a dose-dependent manner; the effect was comparable to that of FGF but occurred through an FGF-independent mechanism [120]. More recently, S1P has been indicated as the active HDL component involved in HDL-mediated migration of human endothelial cells; S1P binds to its receptors $S1P_1$ and $S1P_3$, which in turn activate the PI3K/Akt pathway [121], or the Ras/Raf1/ERK pathway [122], ultimately stimulating cell migration. Nevertheless, BAEC migration is also triggered by the interaction between HDL and SR-BI, with subsequent activation of Src family kinases and PI3K/MAPK [123]. The response to HDL is comparable to that of VEGF, but unlike that of VEGF, does not involve eNOS activation. The effect of plasma-derived HDL can be fully replicated by incubating cells with sHDL made of apoA-I, phosphatidylcholine, and cholesterol, indicating that HDL-stimulated endothelial cell migration, as shown for eNOS activation and inhibition of CAM expression, can occur independently of SP1 and its receptors [123].

HDL also stimulate the proliferation of human [64] and bovine [124,125] endothelial cells. Initially, this HDL-induced proliferation was believed to occur through a PKC-mediated pathway [125]; HDL apolipoproteins were required for this effect [125]. More recent data suggest that the mitogenic effect of HDL is mediated by a rise in intracellular pH and calcium [64], initiated by phospholipase C activation [126]. The lipid fraction of HDL is responsible for the rise in intracellular calcium [126], suggesting that activation of two different signaling pathways by HDL apolipoproteins and lipids may ultimately enhance proliferation of endothelial cells.

HDL have been shown to facilitate the repair of injured endothelium in different animal models. Carotid artery reendothelialization after perivascular electric injury is impaired in apoA-I-deficient mice in relation to wild-type animals, and is restored by reconstitution of apoA-I expression through liver-directed transfer of the human apoA-I gene [123]. Consistently with the *in vitro* findings implying a role for SR-BI in HDL-stimulated endothelial cell migration, carotid reendothelialization after electrical injury is also impaired in SR-BI-deficient mice in relation to wild-type specimens, despite the elevated plasma HDL levels [123]. Direct involvement of HDL in the recruitment of EPC by injured endothelium has been demonstrated in apoE-deficient mice, in which a single infusion of sHDL doubles the number of EPCs incorporated into the endothelial layer without affecting circulating levels of EPC [127]. In humans, the number of circulating EPCs is positively correlated with the plasma HDL-C level [128,129], suggesting that HDL may also improve the survival and the proliferation of EPCs by a still unknown molecular mechanism, which may involve an HDL-mediated increase of eNOS expression in EPCs [129].

15.6
Conclusions and Perspectives

Endothelial dysfunction is a systemic disorder and a critical element in the pathogenesis of atherosclerotic disease and its complications. Growing evidence suggests that the functional and structural integrity of vascular endothelium in a given individual depends on the balance between damaging factors and protective elements. HDL have the capacity to prevent and correct endothelial dysfunction, thus complementing their well known role in cholesterol removal in protecting the arterial wall from the development of atherosclerotic disease. These two mechanisms of vasculoprotection provide an excellent basis for the development of novel therapeutic strategies that favorably influence pathological processes underlying a variety of clinical conditions, such as accelerated atherosclerosis, acute coronary syndromes, and restenosis after coronary angioplasty, through the enhancement of the concentration and/or biological function of HDL.

References

1 Davignon, J. and Ganz, P. (2004) Role of endothelial dysfunction in atherosclerosis. *Circulation*, **109**, III27–III32.

2 Furchgott, R. F. and Zawadzki, J. V. (1980) The obligatory role of endothelial cells in the relaxation of arterial smooth muscle by acetylcholine. *Nature*, **288**, 373–376.

3 Ignarro, L. J., Buga, G. M., Wood, K. S., Byrns, R. E., Chaudhuri, G. (1987) Endothelium-derived relaxing factor produced and released from artery and vein is nitric oxide. *Proc. Natl. Acad. Sci. U.S.A.*, **84**, 9265–9269.

4 Libby, P., Ridker, P. M., Maseri, A. (2002) Inflammation and atherosclerosis. *Circulation*, **105**, 1135–1143.

5 Ludmer, P. L., Selwyn, A. P., Shook, T. L., Wayne, R. R., Mudge, G. H., Alexander, R. W., Ganz, P. (1986) Paradoxical vasoconstriction induced by acetylcholine in atherosclerotic coronary arteries. *N. Engl. J. Med.*, **315**, 1046–1051.

6 Corretti, M. C., Anderson, T. J., Benjamin, E. J., Celermajer, D., Charbonneau, F., Creager, M. A., Deanfield, J., Drexler, H., Gerhard-Herman, M., Herrington, D., Vallance, P., Vita, J., Vogel, R. (2002) Guidelines for the ultrasound assessment of endothelial-dependent flow-mediated vasodilation of the brachial artery: a report of the International Brachial Artery Reactivity Task Force. *J. Am. Coll. Cardiol.*, **39**, 257–265.

7 Anderson, T. J., Uehata, A., Gerhard, M. D., Meredith, I. T., Knab, S., Delagrange, D., Lieberman, E. H., Ganz, P., Creager, M. A., Yeung, A. C. (1995) Close relation of endothelial function in the human coronary and peripheral circulations. *J. Am. Coll. Cardiol.*, **26**, 1235–1241.

8 Verma, S., Buchanan, M. R., Anderson, T. J. (2003) Endothelial function testing as a biomarker of vascular disease. *Circulation*, **108**, 2054–2059.

9 Witte, D. R., Broekmans, W. M., Kardinaal, A. F., Klopping-Ketelaars, I. A., van Poppel, P. G., Bots, M. L., Kluft, C., Princen, J. M. (2003) Soluble intercellular adhesion molecule 1 and flow-mediated dilatation are related to the estimated risk of coronary heart disease independently from each other. *Atherosclerosis*, **170**, 147–153.

10 Lupattelli, G., Lombardini, R., Schillaci, G., Ciuffetti, G., Marchesi, S., Siepi, D., Mannarino, E. (2000) Flow-mediated vasoactivity and circulating adhesion molecules in hypertriglyceridemia: association with small, dense LDL cholesterol particles. *Am. Heart J.*, **140**, 521–526.

11 Felmeden, D. C., Blann, A. D., Spencer, C. G., Beevers, D. G., Lip, G. Y. (2003) A comparison of flow-mediated dilatation and von Willebrand factor as markers of endothelial cell function in health and in hypertension: relationship to cardiovascular risk and effects of

treatment: a substudy of the Anglo-Scandinavian Cardiac Outcomes Trial. *Blood Coagul. Fibrinolysis*, **14**, 425–431.

12 Kathiresan, S., Gona, P., Larson, M. G., Vita, J. A., Mitchell, G. F., Tofler, G. H., Levy, D., Newton-Cheh, C., Wang, T. J., Benjamin, E. J., Vasan, R. S. (2006) Cross-sectional relations of multiple biomarkers from distinct biological pathways to brachial artery endothelial function. *Circulation*, **113**, 938–945.

13 Reddy, K. G., Nair, R. N., Sheehan, H. M., Hodgson, J. M. (1994) Evidence that selective endothelial dysfunction may occur in the absence of angiographic or ultrasound atherosclerosis in patients with risk factors for atherosclerosis. *J. Am. Coll. Cardiol.*, **23**, 833–843.

14 Celermajer, D. S., Sorensen, K. E., Bull, C., Robinson, J., Deanfield, J. E. (1994) Endothelium-dependent dilation in the systemic arteries of asymptomatic subjects relates to coronary risk factors and their interaction. *J. Am. Coll. Cardiol.*, **24**, 1468–1474.

15 Heitzer, T., Schlinzig, T., Krohn, K., Meinertz, T., Munzel, T. (2001) Endothelial dysfunction, oxidative stress, and risk of cardiovascular events in patients with coronary artery disease. *Circulation*, **104**, 2673–2678.

16 von Mering, G. O., Arant, C. B., Wessel, T. R., McGorray, S. P., Bairey Merz, C. N., Sharaf, B. L., Smith, K. M., Olson, M. B., Johnson, B. D., Sopko, G., Handberg, E., Pepine, C. J., Kerensky, R. A. (2004) Abnormal coronary vasomotion as a prognostic indicator of cardiovascular events in women: results from the National Heart, Lung, and Blood Institute-Sponsored Women's Ischemia Syndrome Evaluation (WISE). *Circulation*, **109**, 722–725.

17 Perticone, F., Ceravolo, R., Pujia, A., Ventura, G., Iacopino, S., Scozzafava, A., Ferraro, A., Chello, M., Mastroroberto, P., Verdecchia, P., Schillaci, G. (2001) Prognostic significance of endothelial dysfunction in hypertensive patients. *Circulation*, **104**, 191–196.

18 Ridker, P. M., Hennekens, C. H., Roitman-Johnson, B., Stampfer, M. J., Allen, J. (1998) Plasma concentration of soluble intercellular adhesion molecule 1 and risks of future myocardial infarction in apparently healthy men. *Lancet*, **351**, 88–92.

19 Ridker, P. M., Rifai, N., Stampfer, M. J., Hennekens, C. H. (2000) Plasma concentration of interleukin-6 and the risk of future myocardial infarction among apparently healthy men. *Circulation*, **101**, 1767–1772.

20 Creager, M. A., Cooke, J. P., Mendelsohn, M. E., Gallagher, S. J., Coleman, S. M., Loscalzo, J., Dzau, V. J. (1990) Impaired vasodilation of forearm resistance vessels in hypercholesterolemic humans. *J. Clin. Invest.*, **86**, 228–234.

21 Panza, J. A., Quyyumi, A. A., Brush, J. E.Jr., Epstein, S. E. (1990) Abnormal endothelium-dependent vascular relaxation in patients with essential hypertension. *N. Engl. J. Med.*, **323**, 22–27.

22 Celermajer, D. S., Adams, M. R., Clarkson, P., Robinson, J., McCredie, R., Donald, A., Deanfield, J. E. (1996) Passive smoking and impaired endothelium-dependent arterial dilatation in healthy young adults. *N. Engl. J. Med.*, **334**, 150–154.

23 Schachinger, V., Britten, M. B., Elsner, M., Walter, D. H., Scharrer, I., Zeiher, A. M. (1999) A positive family history of premature coronary artery disease is associated with impaired endothelium-dependent coronary blood flow regulation. *Circulation*, **100**, 1502–1508.

24 Vita, J. A., Treasure, C. B., Nabel, E. G., McLenachan, J. M., Fish, R. D., Yeung, A. C., Vekshtein, V. I., Selwyn, A. P., Ganz, P. (1990) Coronary vasomotor response to acetylcholine relates to risk factors for coronary artery disease. *Circulation*, **81**, 491–497.

25 Lundman, P., Eriksson, M. J., Silveira, A., Hansson, L. O., Pernow,

J., Ericsson, C. G., Hamsten, A., Tornvall, P. (2003) Relation of hypertriglyceridemia to plasma concentrations of biochemical markers of inflammation and endothelial activation (C-reactive protein, interleukin-6, soluble adhesion molecules, von Willebrand factor, and endothelin-1). *Am. J. Cardiol.*, **91**, 1128–1131.

26 Hackman, A., Abe, Y., Insull, W.Jr., Pownall, H., Smith, L., Dunn, K., Gotto, A. M.Jr., Ballantyne, C. M. (1996) Levels of soluble cell adhesion molecules in patients with dyslipidemia. *Circulation*, **93**, 1334–1338.

27 Bermudez, E. A., Rifai, N., Buring, J. E., Manson, J. E., Ridker, P. M. (2002) Relation between markers of systemic vascular inflammation and smoking in women. *Am. J. Cardiol.*, **89**, 1117–1119.

28 Roep, B. O., Heidenthal, E., de Vries, R. R., Kolb, H., Martin, S. (1994) Soluble forms of intercellular adhesion molecule-1 in insulin-dependent diabetes mellitus. *Lancet*, **343**, 1590–1593.

29 Keaney, J. F.Jr., Massaro, J. M., Larson, M. G., Vasan, R. S., Wilson, P. W., Lipinska, I., Corey, D., Sutherland, P., Vita, J. A., Benjamin, E. J. (2004) Heritability and correlates of intercellular adhesion molecule-1 in the Framingham Offspring Study. *J. Am. Coll. Cardiol.*, **44**, 168–173.

30 Calabresi, L., Gomaraschi, M., Villa, B., Omoboni, L., Dmitrieff, C., Franceschini, G. (2002) Elevated soluble cellular adhesion molecules in subjects with low HDL-cholesterol. *Arterioscler. Thromb. Vasc. Biol.*, **22**, 656–661.

31 Vaudo, G., Marchesi, S., Lupattelli, G., Pirro, M., Pasqualini, L., Roscini, A. R., Siepi, D., Schillaci, G., Mannarino, E. (2003) Early vascular damage in primary hypoalphalipoproteinemia. *Metabolism*, **52**, 328–332.

32 Bisoendial, R. J., Hovingh, G. K., Levels, J. H., Lerch, P. G., Andresen, I., Hayden, M. R., Kastelein, J. J., Stroes, E. S. (2003) Restoration of endothelial function by increasing high-density lipoprotein in subjects with isolated low high-density lipoprotein. *Circulation*, **107**, 2944–2948.

33 Chan, N. N., Colhoun, H. M., Vallance, P. (2001) Cardiovascular risk factors as determinants of endothelium-dependent and endothelium-independent vascular reactivity in the general population. *J. Am. Coll. Cardiol.*, **38**, 1814–1820.

34 Lupattelli, G., Marchesi, S., Roscini, A., Siepi, D., Gemelli, F., Pirro, M., Sinzinger, H., Schillaci, G., Mannarino, E. (2002) Direct association between high-density lipoprotein cholesterol and endothelial function in hyperlipemia. *Am. J. Cardiol.*, **90**, 648–650.

35 O'Brien, S. F., Watts, G. F., Playford, D. A., Burke, V., O'Neal, D. N., Best, J. D. (1997) Low-density lipoprotein size, high-density lipoprotein concentration, and endothelial dysfunction in non-insulin-dependent diabetes. *Diabet. Med.*, **14**, 974–978.

36 Zhang, X., Zhao, S. P., Li, X. P., Gao, M., Zhou, Q. C. (2000) Endothelium-dependent and -independent functions are impaired in patients with coronary heart disease. *Atherosclerosis*, **149**, 19–24.

37 Kuvin, J. T., Ramet, M. E., Patel, A. R., Pandian, N. G., Mendelsohn, M. E., Karas, R. H. (2002) A novel mechanism for the beneficial vascular effects of high-density lipoprotein cholesterol: enhanced vasorelaxation and increased endothelial nitric oxide synthase expression. *Am. Heart J.*, **144**, 165–172.

38 Wang, T. D., Chen, W. J., Lin, J. W., Cheng, C. C., Chen, M. F., Lee, Y. T. (2003) Efficacy of fenofibrate and simvastatin on endothelial function and inflammatory markers in patients with combined hyperlipidemia: relations with baseline lipid profiles. *Atherosclerosis*, **170**, 315–323.

39 Spieker, L. E., Sudano, I., Hurlimann, D., Lerch, P. G., Lang, M. G., Binggeli, C., Corti, R., Ruschitzka, F., Luscher, T. F., Noll, G. (2002) High-density lipoprotein restores endothelial function in hypercholesterolemic men. *Circulation*, **105**, 1399–1402.

40 Hirata, K., Miki, N., Kuroda, Y., Sakoda, T., Kawashima, S., Yokoyama, M. (1995) Low concentration of oxidized low-density lipoprotein and lysophosphatidylcholine upregulate constitutive nitric oxide synthase mRNA expression in bovine aortic endothelial cells. *Circ. Res.*, **76**, 958–962.

41 Shaul, P. W. (2002) Regulation of endothelial nitric oxide synthase: location, location, location. *Annu. Rev. Physiol.*, **64**, 749–774.

42 Michel, J. B., Feron, O., Sacks, D., Michel, T. (1997) Reciprocal regulation of endothelial nitric-oxide synthase by Ca2+-calmodulin and caveolin. *J. Biol. Chem.*, **272**, 15583–15586.

43 Dimmeler, S., Fleming, I., Fisslthaler, B., Hermann, C., Busse, R., Zeiher, A. M. (1999) Activation of nitric oxide synthase in endothelial cells by Akt-dependent phosphorylation. *Nature*, **399**, 601–605.

44 Ramet, M. E., Ramet, M., Lu, Q., Nickerson, M., Savolainen, M. J., Malzone, A., Karas, R. H. (2003) High-density lipoprotein increases the abundance of eNOS protein in human vascular endothelial cells by increasing its half-life. *J. Am. Coll. Cardiol.*, **41**, 2288–2297.

45 Blair, A., Shaul, P. W., Yuhanna, I. S., Conrad, P. A., Smart, E. J. (1999) Oxidized low density lipoprotein displaces endothelial nitric-oxide synthase (eNOS) from plasmalemmal caveolae and impairs eNOS activation. *J. Biol. Chem.*, **274**, 32512–32519.

46 Uittenbogaard, A., Shaul, P. W., Yuhanna, I. S., Blair, A., Smart, E. J. (2000) High density lipoprotein prevents oxidized low density lipoprotein-induced inhibition of endothelial nitric-oxide synthase localization and activation in caveolae. *J. Biol. Chem.*, **275**, 11278–11283.

47 Nofer, J. R., van der Giet, G. M., Tolle, M., Wolinska, I., von Wnuck, L. K., Baba, H. A., Tietge, U. J., Godecke, A., Ishii, I., Kleuser, B., Schafers, M., Fobker, M., Zidek, W., Assmann, G., Chun, J., Levkau, B. (2004) HDL induces NO-dependent vasorelaxation via the lysophospholipid receptor S1P3. *J. Clin. Invest.*, **113**, 569–581.

48 Yuhanna, I. S., Zhu, Y., Cox, B. E., Hahner, L. D., Osborne-Lawrence, S., Lu, P., Marcel, Y. L., Anderson, R. G., Mendelsohn, M. E., Hobbs, H. H., Shaul, P. W. (2001) High-density lipoprotein binding to scavenger receptor-BI activates endothelial nitric oxide synthase. *Nat. Med.*, **7**, 853–857.

49 Assanasen, C., Mineo, C., Seetharam, D., Yuhanna, I. S., Marcel, Y. L., Connelly, M. A., Williams, D. L., De, L. L.-M., Shaul, P. W., Silver, D. L. (2005) Cholesterol binding, efflux, and a PDZ-interacting domain of scavenger receptor-BI mediate HDL-initiated signaling. *J. Clin. Invest.*, **115**, 969–977.

50 Matsuda, Y., Hirata, K., Inoue, N., Suematsu, M., Kawashima, S., Akita, H., Yokoyama, M. (1993) High density lipoprotein reverses inhibitory effect of oxidized low density lipoprotein on endothelium-dependent arterial relaxation. *Circ. Res.*, **72**, 1103–1109.

51 Kaul, S., Coin, B., Hedayiti, A., Yano, J., Cercek, B., Chyu, K. Y., Shah, P. K. (2004) Rapid reversal of endothelial dysfunction in hypercholesterolemic apolipoprotein E-null mice by recombinant apolipoprotein A-I(Milano)-phospholipid complex. *J. Am. Coll. Cardiol.*, **44**, 1311–1319.

52 Levkau, B., Hermann, S., Theilmeier, G., van der Giet, G. M., Chun, J., Schober, O., Schafers, M. (2004) High-density lipoprotein stimulates myocardial perfusion in vivo. *Circulation*, **110**, 3355–3359.

53 Kuhn, F. E., Mohler, E. R., Satler, L. F., Reagan, K., Lu, D. Y., Rackley, C. E. (1991) Effects of high-density lipoprotein on acetylcholine-induced coronary vasoreactivity. *Am. J. Cardiol.*, **68**, 1425–1430.

54 Zeiher, A. M., Schachlinger, V., Hohnloser, S. H., Saurbier, B., Just, H. (1994) Coronary atherosclerotic wall thickening and vascular reactivity in humans. Elevated high-density lipoprotein levels ameliorate abnormal vasoconstriction in early atherosclerosis. *Circulation*, **89**, 2525–2532.

55 Kuvin, J. T., Patel, A. R., Sidhu, M., Rand, W. M., Sliney, K. A., Pandian, N. G., Karas, R. H. (2003) Relation between high-density lipoprotein cholesterol and peripheral vasomotor function. *Am. J. Cardiol.*, **92**, 275–279.

56 Caughey, G. E., Cleland, L. G., Penglis, P. S., Gamble, J. R., James, M. J. (2001) Roles of cyclooxygenase (COX)-1 and COX-2 in prostanoid production by human endothelial cells: selective up-regulation of prostacyclin synthesis by COX-2. *J. Immunol.*, **167**, 2831–2838.

57 Eligini, S., Stella, B. S., Cavalca, V., Camera, M., Brambilla, M., De Franceschi, F. M., Tremoli, E., Colli, S. (2005) Diversity and similarity in signaling events leading to rapid Cox-2 induction by tumor necrosis factor-alpha and phorbol ester in human endothelial cells. *Cardiovasc. Res.*, **65**, 683–693.

58 Appleby, S. B., Ristimaki, A., Neilson, K., Narko, K., Hla, T. (1994) Structure of the human cyclo-oxygenase-2 gene. *Biochem. J.*, **302** (Pt 3), 723–727.

59 Fleisher, L. N., Tall, A. R., Witte, L. D., Miller, R. W., Cannon, P. J. (1982) Stimulation of arterial endothelial cell prostacyclin synthesis by high density lipoproteins. *J. Biol. Chem.*, **257**, 6653–6655.

60 Spector, A. A., Scanu, A. M., Kaduce, T. L., Figard, P. H., Fless, G. M., Czervionke, R. L. (1985) Effect of human plasma lipoproteins on prostacyclin production by cultured endothelial cells. *J. Lipid Res.*, **26**, 288–297.

61 Oravec, S., Demuth, K., Myara, I., Hornych, A. (1998) The effect of high density lipoprotein subfractions on endothelial eicosanoid secretion. *Thromb. Res.*, **92**, 65–71.

62 Pomerantz, K. B., Fleisher, L. N., Tall, A. R., Cannon, P. J. (1985) Enrichment of endothelial cell arachidonate by lipid transfer from high density lipoproteins: relationship to prostaglandin I_2 synthesis. *J. Lipid Res.*, **26**, 1269–1276.

63 Cockerill, G. W., Saklatvala, J., Ridley, S. H., Yarwood, H., Miller, N. E., Oral, B., Nithyanathan, S., Taylor, G., Haskard, D. O. (1999) High-density lipoproteins differentially modulate cytokine-induced expression of E-selectin and cyclooxygenase-2. Arterioscler. *Thromb. Vasc. Biol.*, **19**, 910–917.

64 Tamagaki, T., Sawada, S., Imamura, H., Tada, Y., Yamasaki, S., Toratani, A., Sato, T., Komatsu, S., Akamatsu, N., Yamagami, M., Kobayashi, K., Kato, K., Yamamoto, K., Shirai, K., Yamada, K., Higaki, T., Nakagawa, K., Tsuji, H., Nakagawa, M. (1996) Effects of high-density lipoproteins on intracellular pH and proliferation of human vascular endothelial cells. *Atherosclerosis*, **123**, 73–82.

65 Van Sickle, W. A., Wilcox, H. G., Malik, K. U., Nasjletti, A. (1986) High density lipoprotein-induced cardiac prostacyclin synthesis in vitro: relationship to cardiac arachidonate mobilization. *J. Lipid Res.*, **27**, 517–522.

66 Norata, G. D., Callegari, E., Inoue, H., Catapano, A. L. (2004) HDL3 induces cyclooxygenase-2 expression and prostacyclin release in human endothelial cells via a p38 MAPK/CRE-dependent pathway: effects on COX-2/PGI-synthase coupling. *Arterioscler. Thromb. Vasc. Biol.*, **24**, 871–877.

67 Calabresi, L., Rossoni, G., Gomaraschi, M., Sisto, F., Berti, F.,

Franceschini, G. (2003) High-density lipoproteins protect isolated rat hearts from ischemia-reperfusion injury by reducing cardiac tumor necrosis factor-alpha content and enhancing prostaglandin release. *Circ. Res.*, **92**, 330–337.

68 Rossoni, G., Gomaraschi, M., Berti, F., Sirtori, C. R., Franceschini, G., Calabresi, L. (2004) Synthetic high-density lipoproteins exert cardioprotective effects in myocardial ischemia/reperfusion injury. *J. Pharmacol. Exp. Ther.*, **308**, 79–84.

69 Rauramaa, R., Salonen, J. T., Kukkonen-Harjula, K., Seppanen, K., Seppala, E., Vapaatalo, H., Huttunen, J. K. (1984) Effects of mild physical exercise on serum lipoproteins and metabolites of arachidonic acid: a controlled randomised trial in middle aged men. *Br. Med. J.*, **288**, 603–606.

70 Symons, J. D. (1990) Longitudinal and cross-sectional studies of the relationship between 6-keto PGF1 alpha and high density lipoproteins. *Prostag. Leukotr. Ess. Fatty Acids*, **39**, 159–165.

71 Ross, R. (1999) Atherosclerosis – an inflammatory disease. *N. Engl. J. Med.*, **340**, 115–126.

72 Gearing, A. J. and Newman, W. (1993) Circulating adhesion molecules in disease. *Immunol. Today*, **14**, 506–512.

73 Cockerill, G. W., Rye, K. A., Gamble, J. R., Vadas, M. A., Barter, P. J. (1995) High-density lipoproteins inhibit cytokine-induced expression of endothelial cell adhesion molecules. *Arterioscler. Thromb. Vasc. Biol.*, **15**, 1987–1994.

74 Barter, P. J., Baker, P. W., Rye, K. A. (2002) Effect of high-density lipoproteins on the expression of adhesion molecules in endothelial cells. *Curr. Opin. Lipidol.*, **13**, 285–288.

75 Calabresi, L., Franceschini, G., Sirtori, C. R., de Palma, A., Saresella, M., Ferrante, P., Taramelli, D. (1997) Inhibition of VCAM-1 expression in endothelial cells by reconstituted high density lipoproteins. *Biochem. Biophys. Res. Commun.*, **238**, 61–65.

76 Baker, P. W., Rye, K. A., Gamble, J. R., Vadas, M. A., Barter, P. J. (2000) Phospholipid composition of reconstituted high density lipoproteins influences their ability to inhibit endothelial cell adhesion molecule expression. *J. Lipid Res.*, **41**, 1261–1267.

77 Xia, P., Vadas, M. A., Rye, K. A., Barter, P. J., Gamble, J. R. (1999) High density lipoproteins (HDL) interrupt the sphingosine kinase signaling pathway. A possible mechanism for protection against atherosclerosis by HDL. *J. Biol. Chem.*, **274**, 33143–33147.

78 Park, S. H., Park, J. H., Kang, J. S., Kang, Y. H. (2003) Involvement of transcription factors in plasma HDL protection against TNF-alpha-induced vascular cell adhesion molecule-1 expression. *Int. J. Biochem. Cell Biol.*, **35**, 168–182.

79 Kimura, T., Tomura, H., Mogi, C., Kuwabara, A., Damirin, A., Ishizuka, T., Sekiguchi, A., Ishiwara, M., Im, D. S., Sato, K., Murakami, M., Okajima, F. (2006) Role of scavenger receptor class B type I and sphingosine 1-phosphate receptors in high-density lipoprotein-induced inhibition of adhesion molecule expression in endothelial cells. *J. Biol. Chem.*, **281**, 37457–37467.

80 Zhang, W. J., Stocker, R., McCall, M. R., Forte, T. M., Frei, B. (2002) Lack of inhibitory effect of HDL on TNFalpha-induced adhesion molecule expression in human aortic endothelial cells. *Atherosclerosis*, **165**, 241–249.

81 Cockerill, G. W., Huehns, T. Y., Weerasinghe, A., Stocker, C., Lerch, P. G., Miller, N. E., Haskard, D. O. (2001) Elevation of plasma high-density lipoprotein concentration reduces interleukin-1-induced expression of E-selectin in an in vivo model of acute inflammation. *Circulation*, **103**, 108–112.

82 Theilmeier, G., De Geest, B., Van Veldhoven, P. P., Stengel, D., Michiels, C., Lox, M., Landeloos, M., Chapman, M. J., Ninio, E., Collen,

D., Himpens, B., Holvoet, P. (2000) HDL-associated PAF-AH reduces endothelial adhesiveness in apoE–/– mice. *Faseb J.*, **14**, 2032–2039.

83 Shah, P. K., Yano, J., Reyes, O., Chyu, K. Y., Kaul, S., Bisgaier, C. L., Drake, S., Cercek, B. (2001) High-dose recombinant apolipoprotein A-I_{Milano} mobilizes tissue cholesterol and rapidly reduces plaque lipid and macrophage content in apolipoprotein E-deficient mice: potential implications for acute plaque stabilization. *Circulation*, **103**, 3047–3050.

84 Cockerill, G. W., McDonald, M. C., Mota-Filipe, H., Cuzzocrea, S., Miller, N. E., Thiemermann, C. (2001) High density lipoproteins reduce organ injury and organ dysfunction in a rat model of hemorrhagic shock. *Faseb J.*, **15**, 1941–1952.

85 Nicholls, S. J., Dusting, G. J., Cutri, B., Bao, S., Drummond, G. R., Rye, K. A., Barter, P. J. (2005) Reconstituted high-density lipoproteins inhibit the acute pro-oxidant and proinflammatory vascular changes induced by a periarterial collar in normocholesterolemic rabbits. *Circulation*, **111**, 1543–1550.

86 Soro-Paavonen, A., Westerbacka, J., Ehnholm, C., Taskinen, M. R. (2006) Metabolic syndrome aggravates the increased endothelial activation and low-grade inflammation in subjects with familial low HDL. *Ann. Med.*, **38**, 229–238.

87 Nicholls, S. J., Lundman, P., Harmer, J. A., Cutri, B., Griffiths, K. A., Rye, K. A., Barter, P. J., Celermajer, D. S. (2006) Consumption of saturated fat impairs the anti-inflammatory properties of high-density lipoproteins and endothelial function. *J. Am. Coll. Cardiol.*, **48**, 715–720.

88 Tedgui, A. and Mallat, Z. (2006) Cytokines in atherosclerosis: pathogenic and regulatory pathways. *Physiol Rev.*, **86**, 515–581.

89 Widlansky, M. E., Gokce, N., Keaney, J. F.Jr., Vita, J. A. (2003) The clinical implications of endothelial dysfunction. *J. Am. Coll. Cardiol.*, **42**, 1149–1160.

90 Gomaraschi, M., Calabresi, L., Franceschini, G. (2006) High-density lipoproteins: a therapeutic target for atherosclerotic cardiovascular disease. *Expert. Opin. Ther. Targets*, **10**, 561–572.

91 Cockerill, G. W. and Reed, S. (1999) High-density lipoprotein: multipotent effects on cells of the vasculature. *Int. Rev. Cytol.*, **188**, 257–297.

92 Norata, G. D., Callegari, E., Marchesi, M., Chiesa, G., Eriksson, P., Catapano, A. L. (2005) High-density lipoproteins induce transforming growth factor-beta2 expression in endothelial cells. *Circulation*, **111**, 2805–2811.

93 Sugatani, J., Miwa, M., Komiyama, Y., Ito, S. (1996) High-density lipoprotein inhibits the synthesis of platelet-activating factor in human vascular endothelial cells. *J. Lipid Mediat. Cell Signal.*, **13**, 73–88.

94 Tselepis, A. D. and Chapman, M. J. (2002) Inflammation, bioactive lipids and atherosclerosis: potential roles of a lipoprotein-associated phospholipase A2, platelet activating factor-acetylhydrolase. *Atheroscler. Suppl.*, **3**, 57–68.

95 Liu, M. and Subbaiah, P. V. (1994) Hydrolysis and transesterification of platelet-activating factor by lecithin-cholesterol acyltransferase. *Proc. Natl. Acad. Sci. U.S.A.*, **91**, 6035–6039.

96 Rodrigo, L., Mackness, B., Durrington, P. N., Hernandez, A., Mackness, M. I. (2001) Hydrolysis of platelet-activating factor by human serum paraoxonase. *Biochem. J.*, **354**, 1–7.

97 Conlan, M. G., Folsom, A. R., Finch, A., Davis, C. E., Sorlie, P., Marcucci, G., Wu, K. K. (1993) Associations of factor VIII and von Willebrand factor with age, race, sex, and risk factors for atherosclerosis. The Atherosclerosis Risk in Communities (ARIC) Study. *Thromb. Haemost.*, **70**, 380–385.

98 Naqvi, T. Z., Shah, P. K., Ivey, P. A., Molloy, M. D., Thomas, A. M.,

Panicker, S., Ahmed, A., Cercek, B., Kaul, S. (1999) Evidence that high-density lipoprotein cholesterol is an independent predictor of acute platelet-dependent thrombus formation. *Am. J. Cardiol.*, **84**, 1011–1017.

99 Li, D., Weng, S., Yang, B., Zander, D. S., Saldeen, T., Nichols, W. W., Khan, S., Mehta, J. L. (1999) Inhibition of arterial thrombus formation by Apo A1 Milano. *Arterioscler. Thromb. Vasc. Biol.*, **19**, 378–383.

100 Carson, S. D. (1981) Plasma high density lipoproteins inhibit the activation of coagulation factor X by factor VIIa and tissue factor. *FEBS Lett.*, **132**, 37–40.

101 Carson, S. D. (1987) Tissue factor (coagulation factor III) inhibition by apolipoprotein A-II. *J. Biol. Chem.*, **262**, 718–721.

102 Viswambharan, H., Ming, X. F., Zhu, S., Hubsch, A., Lerch, P., Vergeres, G., Rusconi, S., Yang, Z. (2004) Reconstituted high-density lipoprotein inhibits thrombin-induced endothelial tissue factor expression through inhibition of RhoA and stimulation of phosphatidylinositol 3-kinase but not Akt/endothelial nitric oxide synthase. *Circ. Res.*, **94**, 918–925.

103 Griffin, J. H., Kojima, K., Banka, C. L., Curtiss, L. K., Fernandez, J. A. (1999) High-density lipoprotein enhancement of anticoagulant activities of plasma protein S and activated protein C. *J. Clin. Invest.*, **103**, 219–227.

104 Nicholls, S. J., Cutri, B., Worthley, S. G., Kee, P., Rye, K. A., Bao, S., Barter, P. J. (2005) Impact of short-term administration of high-density lipoproteins and atorvastatin on atherosclerosis in rabbits. *Arterioscler. Thromb. Vasc. Biol.*, **25**, 2416–2421.

105 Pajkrt, D., Lerch, P. G., van der Poll, T., Levi, M., Illi, M., Doran, J. E., Arnet, B., Van den Ende, A., ten Cate, J. W., van Deventer, S. J. (1997) Differential effects of reconstituted high-density lipoprotein on coagulation, fibrinolysis and platelet activation during human endotoxemia. *Thromb. Haemost.*, **77**, 303–307.

106 Levin, E. G., Miles, L. A., Fless, G. M., Scanu, A. M., Baynham, P., Curtiss, L. K., Plow, E. F. (1994) Lipoproteins inhibit the secretion of tissue plasminogen activator from human endothelial cells. *Arterioscler. Thromb.*, **14**, 438–442.

107 Kaneko, T., Wada, H., Wakita, Y., Minamikawa, K., Nakase, T., Mori, Y., Deguchi, K., Shirakawa, S. (1994) Enhanced tissue factor activity and plasminogen activator inhibitor-1 antigen in human umbilical vein endothelial cells incubated with lipoproteins. *Blood Coagul. Fibrinolysis*, **5**, 385–392.

108 Saku, K., Ahmad, M., Glas-Greenwalt, P., Kashyap, M. L. (1985) Activation of fibrinolysis by apolipoproteins of high density lipoproteins in man. *Thromb. Res.*, **39**, 1–8.

109 Grutter, M. G. (2000) Caspases: key players in programmed cell death. *Curr. Opin. Struct. Biol.*, **10**, 649–655.

110 Sugano, M., Tsuchida, K., Makino, N. (2000) High-density lipoproteins protect endothelial cells from tumor necrosis factor-alpha-induced apoptosis. *Biochem. Biophys. Res. Commun.*, **272**, 872–876.

111 Nofer, J. R., Levkau, B., Wolinska, I., Junker, R., Fobker, M., von Eckardstein, A., Seedorf, U., Assmann, G. (2001) Suppression of endothelial cell apoptosis by high density lipoproteins (HDL) and HDL-associated lysosphingolipids. *J. Biol. Chem.*, **276**, 34480–34485.

112 Li, X. A., Guo, L., Dressman, J. L., Asmis, R., Smart, E. J. (2005) A Novel Ligand-independent Apoptotic Pathway Induced by Scavenger Receptor Class B, Type I and Suppressed by Endothelial Nitric-oxide Synthase and High Density Lipoprotein. *J. Biol. Chem.*, **280**, 19087–19096.

113 Munoz-Chapuli, R., Quesada, A. R., Angel, M. M. (2004) Angiogenesis and signal transduction in

endothelial cells. *Cell Mol. Life Sci.*, **61**, 2224–2243.

114 Urbich, C. and Dimmeler, S. (2004) Endothelial progenitor cells functional characterization. *Trends Cardiovasc. Med.*, **14**, 318–322.

115 Walter, D. H., Rittig, K., Bahlmann, F. H., Kirchmair, R., Silver, M., Murayama, T., Nishimura, H., Losordo, D. W., Asahara, T., Isner, J. M. (2002) Statin therapy accelerates reendothelialization: a novel effect involving mobilization and incorporation of bone marrow-derived endothelial progenitor cells. *Circulation*, **105**, 3017–3024.

116 Hess, D. C., Hill, W. D., Martin-Studdard, A., Carroll, J., Brailer, J., Carothers, J. (2002) Bone marrow as a source of endothelial cells and NeuN-expressing cells after stroke. *Stroke*, **33**, 1362–1368.

117 Kalka, C., Masuda, H., Takahashi, T., Kalka-Moll, W. M., Silver, M., Kearney, M., Li, T., Isner, J. M., Asahara, T. (2000) Transplantation of ex vivo expanded endothelial progenitor cells for therapeutic neovascularization. *Proc. Natl. Acad. Sci. U.S.A.*, **97**, 3422–3427.

118 Werner, N. and Nickenig, G. (2006) Influence of cardiovascular risk factors on endothelial progenitor cells: limitations for therapy?*Arterioscler. Thromb. Vasc. Biol.*, **26**, 257–266.

119 Werner, N., Kosiol, S., Schiegl, T., Ahlers, P., Walenta, K., Link, A., Bohm, M., Nickenig, G. (2005) Circulating endothelial progenitor cells and cardiovascular outcomes. *N. Engl. J. Med.*, **353**, 999–1007.

120 Murugesan, G., Sa, G., Fox, P. L. (1994) High-density lipoprotein stimulates endothelial cell movement by a mechanism distinct from basic fibroblast growth factor. *Circ. Res.*, **74**, 1149–1156.

121 Kimura, T., Sato, K., Malchinkhuu, E., Tomura, H., Tamama, K., Kuwabara, A., Murakami, M., Okajima, F. (2003) High-density lipoprotein stimulates endothelial cell migration and survival through sphingosine 1-phosphate and its receptors. *Arterioscler. Thromb. Vasc. Biol.*, **23**, 1283–1288.

122 Miura, S., Tanigawa, H., Matsuo, Y., Fujino, M., Kawamura, A., Saku, K. (2003) Ras/Raf1-dependent signal in sphingosine-1-phosphate-induced tube formation in human coronary artery endothelial cells. *Biochem. Biophys. Res. Commun.*, **306**, 924–929.

123 Seetharam, D., Mineo, C., Gormley, A. K., Gibson, L. L., Vongpatanasin, W., Chambliss, K. L., Hahner, L. D., Cummings, M. L., Kitchens, R. L., Marcel, Y. L., Rader, D. J., Shaul, P. W. (2006) High-density lipoprotein promotes endothelial cell migration and reendothelialization via scavenger receptor-B type I. *Circ. Res.*, **98**, 63–72.

124 Cohen, D. C., Massoglia, S. L., Gospodarowicz, D. (1982) Correlation between two effects of high density lipoproteins on vascular endothelial cells. The induction of 3-hydroxy-3-methylglutaryl coenzyme A reductase activity and the support of cellular proliferation. *J. Biol. Chem.*, **257**, 9429–9437.

125 Darbon, J. M., Tournier, J. F., Tauber, J. P., Bayard, F. (1986) Possible role of protein phosphorylation in the mitogenic effect of high density lipoproteins on cultured vascular endothelial cells. *J. Biol. Chem.*, **261**, 8002–8008.

126 Honda, H. M., Wakamatsu, B. K., Goldhaber, J. I., Berliner, J. A., Navab, M., Weiss, J. N. (1999) High-density lipoprotein increases intracellular calcium levels by releasing calcium from internal stores in human endothelial cells. *Atherosclerosis*, **143**, 299–306.

127 Tso, C., Martinic, G., Fan, W. H., Rogers, C., Rye, K. A., Barter, P. J. (2006) High-density lipoproteins enhance progenitor-mediated endothelium repair in mice. *Arterioscler. Thromb. Vasc. Biol.*, **26**, 1144–1149.

128 Pellegatta, F., Bragheri, M., Grigore, L., Raselli, S., Maggi, F. M.,

Brambilla, C., Reduzzi, A., Pirillo, A., Norata, G. D., Catapano, A. L. (2006) In vitro isolation of circulating endothelial progenitor cells is related to the high density lipoprotein plasma levels. *Int. J. Mol. Med.*, **17**, 203–208.

129 Noor, R., Shuaib, U., Wang, C. X., Todd, K., Ghani, U., Schwindt, B., Shuaib, A. (2007) High-density lipoprotein cholesterol regulates endothelial progenitor cells by increasing eNOS and preventing apoptosis. *Atherosclerosis*, **192**, 92–99.

16
The Effect of Nutrients on Apolipoprotein A-I Gene Expression

Arshag D. Mooradian

16.1
Introduction

It is generally accepted that one of the major promoters of cardiovascular disease (CVD) is the consumption of an atherogenic diet [1,2]. The biochemical pathways by which dietary components contribute to atherosclerosis are multiple and often interlinked, and include changes in plasma lipid profile, postprandial excursion of blood glucose, and deficiency or excess of a specific nutrient that may variously alter oxidant and antioxidant balance, increase atherogenic lipids, or downregulate cardioprotective lipid and lipoprotein fractions. Although excess caloric content of the diet eventually leads to obesity along with its attendant increased risk of CVD, the precise components of the diet that are atherogenic remain controversial [3,4]. Most current dietary recommendations are based on population studies, metabolic ward experiments in limited numbers of subjects, and often on expert opinion. Population studies are limited by interindividual variability of the plasma lipid changes in response to dietary changes. In addition, since a change in one component of the diet is bound to be associated with a change in another component, it is often difficult to attribute changes in plasma lipids to a specific nutrient. Species-specificity of response to dietary manipulations and peculiarities of cell culture experiments limit the clinical applicability of the findings of these studies. Nevertheless, these studies may provide valuable mechanistic information that can be utilized in designing and interpreting studies in human subjects, so it is necessary that the totality of evidence from various types of studies be reviewed before any conclusions can be drawn as to the effect of a particular diet on lipid metabolism.

In this communication the effects of various dietary components on high-density lipoprotein (HDL), its major apolipoproteins – notably apolipoprotein A-I (apoA-I) – and HDL cholesterol (HDLc) are discussed.

High-Density Lipoproteins: From Basic Biology to Clinical Aspects. Edited by Christopher J. Fielding

ISBN: 978-3-527-31717-2

16.2
Effect of Caloric Intake

One of the hallmarks of insulin resistance associated with obesity is reduced plasma concentration of HDLc and low plasma levels of apoA-I [5–7]. The reduced serum HDLc is mostly attributed to increased fractional clearance of HDL as a result of depletion of HDL cholesterol following its transfer to triglyceride-rich lipoproteins [6] . However, subjects with low HDLc without hypertriglyceridemia had significantly reduced HDLc production rates without any change in fractional clearance rates [8]. Approximately 50 % of obese normotriglyceridemic individuals have low HDLc levels [9] and the degree of adiposity and insulin resistance continues to be a major predictor of low HDLc in this subgroup. Obese individuals with low HDLc levels were more likely than those with normal HDLc levels to have elevated serum TNFα and IL-6 levels [9]. TNFα is implicated in obesity-related insulin resistance [10] and is found to downregulate the apoA-I gene expression and to reduce serum HDLc levels [11,12]. Reduced caloric intake during a weight reducing program is associated with a temporary decline in HDLc, though once a stable weight has been achieved, HDLc is increased by 0.35 $mg\,dL^{-1}$ per kilogram of weight reduced [13]. In addition to increased HDLc levels, lipoprotein lipase levels and lecithin:cholesterol acyltransferase activity are also increased and may contribute to the increased cholesterol esterification and reverse cholesterol transport [14].

16.3
Effects of Various Nutrients on HDLc and ApoA-I

The effects of select nutrients on HDLc or apoA-I levels are summarized in Table 16.1. There are multiple mechanisms by which various nutrients alter HDLc or apoA-I levels [15], but the two general determinants of HDLc levels are the rate of transfer of cholesterol to and from the HDL and the production and degradation of the major apoprotein constituents, notably of apoA-I. Several enzymes and transporters are involved in the regulation of HDLc turnover and are discussed in the other chapters of the book.

Expression of the apoA-I gene is regulated at both the transcriptional and the posttranscriptional level [15]. The apoA-I gene promoter contains several *cis* elements that regulate expression of the gene in either a positive or a negative manner in response to various hormonal or metabolic signals (Figure 16.1) [15,16].

The myriad of changes that may occur in apoA-I gene expression related to various nutrients is summarized in Table 16.2.

16.3.1
Effects of Lipids

The relationship of dietary fat to increased prevalence of cardiovascular disease has been studied extensively. The current consensus is that limiting dietary saturated fat

Tab. 16.1 The effects of select nutrients on HDLc or apoA-I levels measured in various experimental models. NC: no change. Inc.: increase. Dec.: decrease.?: data not available. PUFA: polyunsaturated fatty acids; MUFA; monounsaturated fatty acids; Trans–FA: trans fatty acids; MCFA: medium chain fatty acids; w–3–FA: omega 3 fatty acids. CHO: carbohydrates. ARP-1: apoA-I regulatory protein -1 (Adapted from Mooradian AD, Haas MJ, Wong NCW. Endocr Rev. 2006, 27, 2–16 with permission of the Endocrine Society).

Nutrient	Cell cultures	Animal studies	Human studies	Predominant mechanism
I. Dietary fat				
Saturated fat	NC	Inc.	Inc.	Decreased clearance; Gender specific
PUFA	NC or Dec.	Dec.	Dec.	Increased clearance; Dec. apoA-I mRNA; Dec. apoA-I secretion
MUFA	NC	Inc.	Dec. HDLc NC in apoA-I	Increased apo A-I mRNA; Species Specific
Trans-FA	NC	Dec.	Dec. or NC	Dec. apoA-I promoter
MCFA	?	?	Dec.	Modest effects
ω-3-FA	Dec.	Dec.	Dec. or NC	Increased clearance; Dec. apo A-I mRNA
Cholesterol	NC	NC	NC	
II. CHO or CHO metabolites				
Glucose	Dec.	Dec.	Dec.	Dec. gene expression
Fructose	?	Inc.	Dec.	Insulin resistance
Glucosamine	Inc.	?	NC	Stabilize apoA-I mRNA
III. protein	?	?	NC or Dec.	Phytoestrogens in soy proteins increase apo A-I and HDLc (see below).
IV. Micronutrients				
1) Vitamins				
A, C, E, D	Dec.	Dec.	NC or Dec. Inc. in diabetics	High doses or deficiency dec. gene expression
2) Minerals				
Zn deficiency	Dec.	Dec.	NC or Dec.	High doses or deficiency dec. gene expression
Cu deficiency	Inc.	Inc.	Inc.	Inc. apo A I synthesis; apo A-I mRNA inc. or NC.
Cr supplement	Dec.	Inc.	NC or Inc.	Insulinomimetic; High doses dec. gene expression
V. Others				
Soy proteins	Inc.	Inc.	Inc.	Inc. gene expression
Alcohol	Inc.	Inc.	Inc.	Inc. gene expression

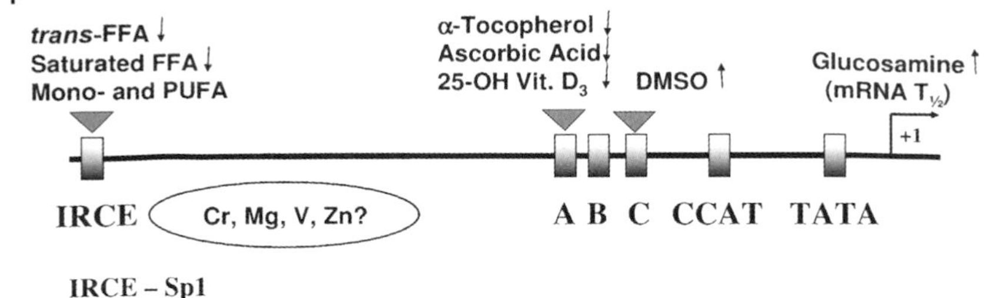

IRCE – Sp1

Site A – TR, HNF-4, RARα,β, RXRα, C/EBP, ARP-1 PPARα, Rev-erbα,

Site B - HNF-3, glucocorticoid receptor, estrogen receptor

Site C - HNF-4, ARP-1

TATA – Basal transcriptional apparatus, nTRE, pH-RE

Fig. 16.1 Organization of regulatory elements within the apoA-I gene promoter. IRCE: insulin response core element that binds specific protein 1 (Sp1) and is also a carbohydrate response element. This is the region where various fatty acids and, possibly, high concentrations of minerals such as chromium (Cr), magnesium (Mg), vanadate (V), and zinc (Zn) alter promoter activity. Site A contains several response elements and is also the site where α tocopherol, ascorbic acid, and vitamin D modulate the promoter activity. TR: thyroid hormone receptor. nTRE: negative thyroid hormone response element. pH-RE: pH response element. A nTRE and a pH-RE are located in the vicinity at the TATA box. ↑: increases promoter activity. ↓: decreases promoter activity. (Reproduced from Mooradian AD, Haas MJ, Wong NCW. Endocr Rev. 2006;27: 2–16 with permission of the Endocrine Society).

intake will reduce the risk of atherosclerosis [17,18]. A large study in women over the age of 50 years found that reducing total fat in the diet does not reduce the risk of cardiovascular events over 8 years of observation [4]. Limiting the consumption of fat often results in decreasing HDLc levels [19]. In a study of sedentary men and women, consumption of a low-fat diet (19 % of total calories from fat) was associated with lower HDLc (54 vs. 63 mg dL^{-1}) and apoA-I levels (118 vs. 127 mg dL^{-1}) than were observed in a high-fat diet (50 % of calories from fat) [20]. This effect is gender-specific [21,22]. HDLc and apoA-I levels decreased more in hypercholesterolemic women than in men ingesting a National Cholesterol Education Program (NCEP) Step II diet [21]. In another study, the low-fat diet resulted in greater reductions in total cholesterol and plasma apoB concentrations in men than in women. Post-prandial triacylglycerol and LpAI:AII concentrations were reduced in men, but not in women, while similar decreases in LpAI concentrations and LDL and HDL particle size were observed in men and women [22]. These data are consistent with the concept that men may have a more favorable lipoprotein response to a low-fat, low-cholesterol diet than postmenopausal women.

Both the amount of fat and qualitative changes in fat composition, such as altered proportions of saturated in relation to unsaturated fatty acid content, are important determinants of the changes in HDLc and apoA-I levels [23].

Tab. 16.2 Effects of various nutrients on apoA- I gene expression (adapted from Mooradian AD, Haas MJ, Wong NCW. Endocr. Rev. 2006, 27, 2–16 with permission of the Endocrine Society).

Nutrient	Effect on ApoA-I gene expression	References
free fatty acids	1) most of the effects of saturated fatty acids are posttranscriptional.	[36]
	2) suppression of insulin-mediated induction of apoA-I gene expression *in vitro* by saturated fatty acids and *trans*-fatty acids; *cis*-unsaturated fatty acids had no effect	[44]
dextrose	suppression of apoA-I gene expression in cell cultures and in vivo through inhibition of transcription	[83]
fructose feeding	1) may decrease HDL cholesterol in hyperinsulinemic men	[88]
	2) induction of apoA-I gene expression in vivo	[87]
glucosamine	increase in apoA-I in cell cultures through stabilization of apoA-I mRNA; no change in humans	[96,97]
ketones	1) ketosis is associated with suppression in apoA-I expression	[68]
	2) ketones do not have a direct effect	[69]
	3) high-fat ketogenic diet increases apoA-I mRNA	[69,70]
vitamins	α-tocopherol, ascorbate and vitamins D and A suppress apoA-I gene expression at the transcriptional level;	[109,117]
	vitamin A does not alter apoA-I protein level.	[120]
minerals	zinc, chromium, magnesium, and vanadate in high concentrations suppress apoA-I gene transcription; deficiency of zinc is also associated with reduced apoA-I expression	[124–127]
soy proteins	induction of apoA-I gene transcription	[103–105]
alcohol	induction of apoA-I gene transcription	[132]

16.3.1.1 Effect of Fatty Acid Saturation

Dietary intake of saturated fatty acids increases both plasma HDLc level and the apoA-I level [19]. The decreased plasma HDLc levels associated with the recommended heart-healthy diet may reduce the overall cardioprotective efficacy of these diets. However, it is likely that the HDLc capacity to reduce the risk of atherosclerosis depends on the lipid and apoprotein composition, which is in turn determined by the changes in lipid transfer and the expression and turnover of the apoproteins. Thus, the modestly decreased HDLc level associated with low dietary fat consumption may have more cardioprotective effects than the modestly increased HDLc that occurs following consumption of a high-fat diet.

It is noteworthy that diets enriched in saturated fatty acids, unlike diets enriched in unsaturated fatty acids, are associated with increased HDLc and in apoA-II levels but not in apoA-I levels [24]. This observation suggests that saturated fatty acids may have a suppressive effect on apoA-I expression, since the overall turnover of HDL molecules is reduced with consumption of saturated fatty acids.

Relative to saturated fatty acids, dietary polyunsaturated fatty acids upregulate hepatic SR-B1 expression and increase HDL cholesterol ester transport to the liver, as a consequence of which plasma HDLc level is reduced [25].

In golden Syrian hamsters, feeding diets enriched in canola and soybean oils significantly lowered serum HDLc concentrations relative to butter (rich in saturated fat) [26]. Canola oil produced higher exogenous LCAT activity than butter did, while both soybean and canola oils significantly increased hepatic apoA-I and SR-B1 mRNA abundance. Butter, relative to margarine, coconut, and soybean oils, significantly increased serum non-HDL cholesterol concentrations. Endogenous and exogenous LCAT, CETP, and PLTP activities did not differ in hamsters fed margarine or saturated fat diets, despite lower hepatic LCAT, apoA-I, and SR-B1 mRNA abundance, suggesting that changes in available substrate and/or modification to the LCAT protein may have been involved in lipoprotein changes. The lower HDLc concentrations resulting from canola and soybean oil feeding may not be detrimental, because of increases in components involved in the reverse cholesterol transport in these hamsters, and may retard the progression of atherosclerosis [26]. The physiologic roles of LCAT, CETP, PLTP, and SR-BI are discussed in Chapters 6–8.

Multiple mechanisms account for the changes in plasma HDLc in response to alterations in dietary fat. Reducing dietary intake of saturated fat and cholesterol is associated with reduced plasma cholesterol content of various lipoproteins including low density lipoprotein (LDL), very low density lipoprotein (VLDL), and HDL [19,24,25,27]. The reduced cholesterol content of HDL accelerates its clearance. Dietary fat may also alter the expression of apoproteins. Several studies have documented major *in vivo* and *in vitro* interactions between various nutrients and the expression of lipoprotein genes [28]. It is noteworthy that genetic variations influence the response of plasma lipids to dietary intervention [29–31].

In male rats, a corn oil diet (enriched with polyunsaturated fatty acids) lowered both HDLc and LDL-cholesterol (LDLc). In comparison with control or high-cholesterol diets, plasma apoA-I concentrations were significantly increased in coconut and 40 % olive oil diets. Hepatic apoA-I mRNA concentrations were not altered after the consumption of coconut oil or corn oil diets. However, 40 % and 10 % olive oil diets induced a significant increase in the expression of this mRNA in relation to the cholesterol diet. These observations suggest that diets containing different fatty acids regulate apoA-I through different mechanisms [32].

In rabbits, coconut oil feeding for 4 weeks caused a 220 % increase in the hepatic apoA-I transcription rate in relation to controls [33]. However, treatment of rabbit liver cells in culture with various saturated fatty acids and sera from chow-fed and coconut oil-fed rabbits did not alter apoA-I mRNA levels as observed *in vivo*[33], so caution should be exercised on extrapolation of the results of cell culture studies to the intact organism. In addition, the changes reported are often species-specific. Unlike the observations made in rabbits, increased apoA-I production by increased dietary fat in mice was not associated with any increase in hepatic or intestinal apoA-I mRNA, suggesting that the mechanism of the dietary fat effect was posttranscriptional, involving either increased translatability of the apoA-I mRNA or decreased intracellular apoA-I degradation [34]. Another potential mechanism may be through increasing the fraction of apoA-I mRNA in the translating pool [35]; this was supported by additional studies by Srivastava [36], who concluded that transcriptional regulation of the apoA-I gene was not responsible for increased plasma apoA-I levels in mice and rats on high-fat and fat/cholesterol combination diets [36]. In his studies, the rat hepatic apoA-I mRNA was decreased in high-fat and fat/cholesterol combination diets but the translational efficiency of apoA-I on isolated polysomes was significantly increased [36]. The high-fat diet increased HDLc levels by 20 % in rats and 55 % in mice and although the high-cholesterol diet did not alter plasma HDLc levels, a combination of high saturated fat and high-cholesterol diet raised plasma HDLc levels by 36 and 67 % in rats and mice, respectively. Plasma apoA-I levels increased in parallel with HDLc concentrations. These results suggest that additional regulation through impaired catabolism of HDL particles by high-fat diet feeding may be another pathway by which HDL levels are increased. Unlike apoA-I mRNA, the mRNA of the other HDL apoproteins apoA-II and apoA-IV were increased by high-fat and combination diet feeding. These results suggest that saturated fatty acids regulate plasma HDL levels by translational and posttranslational mechanisms [36]. Posttranslational mechanisms contributing to the changes in HDLc level were also demonstrated in another study of rats in which triolein, but not tripalmitin or menhaden oil, increased the secretion of HDL and its components (apoA-I and lecithin:cholesterol acetyl transferase) and stimulated the production of hepatic SR-B1 receptor protein [37]. These changes associated with triolein feeding may promote reverse cholesterol transport [37].

The metabolic effects of polyunsaturated fatty acids (PUFA) on HDLc and apoA-I are similar, but not identical, to the effects of monounsaturated fatty acids (MUFA). In a crossover study of 38 healthy young adults, the effects of two fat-reduced diets, one rich in MUFAs, the other rich in PUFAs, were compared [38]. Both test diets led to significant reductions in serum cholesterol, LDLc, and HDLc levels. However, apoA-I was significantly higher on the MUFA-rich diet than on the PUFA-rich diet. [38]. In a study of the effects of canola, corn, and olive oils on postprandial plasma lipoproteins in people on NCEP step 2 diet, Lichtenstein et al found that plasma HDLc concentrations declined after consumption of the canola (7 %) and corn (9 %) oil enriched diets (high in polyunsaturated fatty acids) [39]. The olive oil enriched diet (high in monounsaturated fatty acids) did not alter HDLc significantly. Plasma cholesterol concentrations declined after consumption of diets enriched in the test oils. Changes in apoB concentrations paralleled those of LDLc. Switching from the baseline to the vegetable oil enriched diets had no significant effect on plasma triglyceride, apoA-I, or lipoprotein (a) concentrations or the total cholesterol to HDLc ratio. ApoB to apoA-I ratios were significantly reduced when the subjects consumed the vegetable oil enriched diets [39]. In general, a diet enriched in MUFA does not cause a significant change in HDLc [40], yet it may limit the carbohydrate loads that in certain populations may adversely affect the serum lipid profile [41].

Differences in the effects of MUFA- and PUFA-enriched diets are also species-specific. In a study of cynomolgus monkeys, dietary MUFAs were comparable to PUFAs in their effects on hepatic lipid and apoA-I mRNA levels [42]. In African green monkeys, isocaloric substitution of polyunsaturated fat (PUFA) for saturated fat reduced the concentrations of total plasma cholesterol and HDLc, while plasma apoA-I concentration was reduced by 16% [43]. A PUFA diet reduced the hepatic apoA-I mRNA concentrations and the rate of hepatic apoA-I secretion, while intestinal apoA-I mRNA concentrations were not altered. Plasma apoA-II and hepatic apoA-II mRNA concentrations were also not altered by the type of dietary fat. Dietary polyunsaturated fat can therefore selectively alter the expression of the apoA-I gene in a tissue-specific manner [43].

The effect of the degree of fat saturation on apoA-I transcription has not been well studied. In Hep G2 cells, various concentrations of FFA do not alter basal apoA-I mRNA levels or its protein secretion in culture media. However, treatment of cells with saturated FFA such as stearic acid (C_{18}), myristic acid (C_{14}), and palmitic acid (C_{16}) abolishes both insulin and exogenously delivered Sp-1-stimulated activation of the apoA-I promoter [44]. Unlike saturated FFA, three unsaturated fatty acids – oleic ($C_{18:1}$), linoleic ($C_{18:2}$), or linolenic acid ($C_{18:3}$) – had no effect on Sp1-mediated induction of the apoA-I promoter. These results suggest that the ability of FFA to repress Sp1-mediated induction of apoA-I promoter activity is dependent on the degree of saturation but not necessarily on chain length. Additional experiments involving Sp1–DNA binding have indicated that stearic acid must affect Sp1 function through a mechanism unrelated to regulation of Sp1–DNA binding, most probably by altering posttranslational modification [44].

Monounsaturated FFA cannot prevent the downregulation of apoA-I transcription by saturated FFA. It is noteworthy that the effect of mixing monounsaturated or polyunsaturated FFA with saturated FFA may vary depending on the precise combinations and proportions. The differential effects of saturated and unsaturated fatty acids on the transcriptional process have been recognized previously [45].

Overall, it appears that the main effect of fatty acids on apoA-I synthesis occurs only when Sp1 is activated, while the effects are modest at best during basal conditions when Sp1 is not overexpressed. Unlike saturated fatty acids, the unsaturated fatty acids do not appear to have this potential deleterious effect on apoA-I gene expression.

16.3.1.2 **Effects of Oxidized Fatty Acids**

Significant amounts of oxidized lipids are found in processed foods, while the oxidation state of the FFA may also affect their ability to modulate gene expression. It appears that oxidized fatty acids, unlike non-oxidized fatty acids, may enhance intestinal apoA-I gene expression through peroxisomal proliferation activating receptor gamma (PPAR-γ) [46]. In preliminary studies, oxidized lipids in the diet were associated with elevated levels of plasma apoA-I in humans. In Caco-2 cells, oxidized linoleic acid [ox-linoleic acid, 13-hydroperoxylinoleic acid (13-HPODE)], caused a dose-dependent increase in the levels of apoA-I protein and mRNA. ApoB production was not increased. PPAR-γ receptors are expressed in these cells and could increase apoA-I secretion [46].

16.3.1.3 Effect of *trans*-Fatty Acids

Potential adverse effects of *trans* fat on plasma lipid profiles are well recognized. Relative to saturated fatty acids, *trans*-fatty acids/hydrogenated fat-enriched diets increase LDLc levels and either decrease or have no effect on HDLc levels [47]. Although overall HDLc level may not change, *trans* fat consumption was associated with a relative decrease in HDL particles containing apoA-I only. This adverse effect could not be attributed to differences in CETP, phospholipid transfer protein activity, or the fractional esterification rate of cholesterol in HDL [47]. The observed differences in HDL could have been related to the saturated fatty acid component of the diet, rather than the *trans*- or polyunsaturated fatty acid component [47]. In another study of 80 healthy subjects, it was found that, in relation to a dairy fat diet, stearic acid and *trans*-fatty acids decreased serum total cholesterol concentrations similarly (by 13 % and 12 %, respectively) but the *trans*-fatty acid diet decreased HDLc (17 %) and apoA-I (15 %) significantly more than the stearic acid diet did (11 % and 12 %, respectively) [48]. In comparison with a *cis*-monounsaturated fatty acid (namely oleic acid), *trans*-monounsaturated fatty acids (TFA) increase plasma cholesterol level to a degree equivalent to that seen with saturated fatty acids (FA) [49]. However, it is unknown to what extent these effects may be due to cholesterol lowering by oleic acid rather than elevation by saturated or *trans*-fatty acid.

It appears that *trans*-fatty acids may accelerate apoA-I catabolism and may also decrease LDL apoB-100 catabolism [50]. In a study of eight postmenopausal hypercholesterolemic women, the subjects were provided in random order with three diets for 5 week periods. Two-thirds of the fat was soybean oil (unsaturated fat), stick margarine (hydrogenated fat), or butter (saturated fat). Total and LDL-C levels were highest after the saturated diet, whereas HDL-C levels were lowest after the hydrogenated diet. Plasma apoA-I levels and pool size (PS) were lower, whereas apoA-I fractional catabolic rate (FCR) was higher after the hydrogenated diet than after the saturated variant. LDL apoB-100 levels and PS were significantly higher, whereas LDL apoB-100 FCR was lower, with the saturated and hydrogenated fat diets relative to the unsaturated fat diet [50]. There was no significant difference among diets in apoA-I or B-100 production rates or apoB-48 kinetic parameters.

In another study of 26 men and 36 healthy normolipidemic women, the effects of linoleic acid (*cis, cis*-C18:2(*n*-6)) and its hydrogenation products elaidic (*trans*-C18:1(*n*-9)) and stearic acid (C18:0) on serum lipoprotein levels were compared [51]. The HDLc was decreased by 2 mg dL^{-1} with the stearate diet and by 4 mg dL^{-1} with the *trans* diet. Hydrogenation of linoleic acid to either stearic or *trans*-fatty acids produces fatty acids that may increase LDLc and decrease HDLc relative to linoleic acid itself [52].

According to a report by the US FDA, replacement of *trans* fat with PUFA (1 % of energy) would reduce CV risk by 2.96 % when taking into account changes in both LDL and HDL cholesterol [53].

The effect of *trans*-fatty acids on apoA-I expression has also been evaluated in Hep G2 cells. Whereas apoB accumulation in the medium increases with oleic (*cis* 18:1) and elaidic acids (*trans* 18:1), apoA-I accumulation does not change [54]. A subsequent study confirmed the lack of effect of *trans*-fatty acids on apoA-I secretion in Hep G2 cells [44]. However, *trans*-fatty acids, like saturated fatty acids, blunt Sp1 upregulation of apoA-I promoter activity. The reduction in apoA-I promoter activity in the presence of *trans*-fatty acids is consistent with the literature on the effect of dietary *trans*-fatty acids on lowering HDLc levels [55,56].

16.3.1.4 **Effects of Medium-Chain Fatty Acids**

There are limited studies on the effects of medium-chain fatty acids (MCFA) on apoA-I. Clinical trials have suggested that MCFA may raise LDLc concentrations slightly and affect the apoA-I to apoB ratio unfavorably in relation to oleic acid [57]. In a study of thirty healthy adults (16 men and 14 postmenopausal women) with LDLc concentrations over 3.37 mM (130 mg dL^{-1}) subjected to 5 weeks of experimental diets, it was found that an interesterified mixture of butter, medium-chain triacylglycerol, and safflower oils had no appreciable effect on plasma cholesterol concentrations in relation to native butter, but was associated with a modest rise in plasma triacylglycerol [58].

Overall, the effects of MCFA on plasma lipids and lipoproteins were modest.

16.3.1.5 **Effects of ω-3-Fatty Acids**

The triglyceride-lowering effect of *n*–3 polyunsaturated (poly) fat consumption is well established [59,60]. In a systematic review of randomized controlled trials that evaluated the effect of consumption of fish oil and plant-source oil ω-3 fatty acids, it was found that there is a dose-dependent beneficial effect of fish oil on serum triglycerides, particularly among people with more elevated levels [61]. Fish oil consumption also modestly improves HDLc and increases LDLc levels, but does not appear to affect glucose homeostasis adversely. The evidence regarding plant-source oil ω-3 fatty acids is inconclusive [61]. Fish consumption shifts lipoprotein subfractions to a less atherogenic pattern. In a study by Li *et al.*, consumption of the high-fish NCEP Step 2 diet was associated with a significant reduction in medium and small VLDL in relation to the average American diet, whereas the low-fish diet did not affect VLDL subfractions [62]. Both high-fish and low-fish NCEP 2 diets significantly reduced LDLc and lowered HDLc concentrations, but the high-fish diet significantly lowered only the HDL fraction containing both apolipoprotein (apo) A-I and A-II (LpAI:AII) whereas the low-fish diet significantly lowered the HDL fraction containing only apoA-I (LpAI) [62]. Thus within the context of a diet restricted in fat and cholesterol, a higher fish content favorably affects VLDL and HDL subspecies [62].

In several studies the lowering of plasma triglyceride by *n*–3 poly fat is often associated with a reduction in HDLc. Since ordinarily triglyceride lowering is typically associated with an increase in HDLc [63], it is likely that *n*–3 fatty acids directly downregulate HDLc levels. In African green monkeys this change has been attributed to increased fractional catabolic rate of medium-sized HDL particles [64].

In rats, a significant decrease in liver apoA-I and apoA-II mRNA levels was observed after fish oil (*n*-3 poly) feeding, in relation to rats fed isocaloric diets enriched in saturated fat (hydrogenated coconut oil) or *n*–6 polys (safflower oil). In primary hepatocytes, the addition of the *n*–3 polys, docosanohexaenoic acid and eicosanopentaenoic acid or the fatty acid derivative α-bromopalmitate, decreased apoA-I mRNA in a dose- and time-dependent manner while apoA-II mRNA did not change significantly [65]. These observations suggest that fish oil feeding reduces rat liver apoA-I gene expression, through direct transcriptional action [65].

Native Alaskans who consume a diet rich in *n*–3 polyunsaturated fatty acids have high HDLc [66]. However the elevated HDLc in this population may be independent of their dietary habits. In people with type 2 diabetes, treatment with dietary fish oil was associated with a decrease in plasma triglyceride levels, while the total cholesterol, HDLc, and plasma apoA-I levels remained unchanged [67]. Interestingly, although the apoA-I fractional catabolic rate was reduced after fish oil feeding, a concomitant decrease in the apoA-I absolute production rate kept the plasma apoA-I levels unchanged. These observations raise the possibility that *n*–3 poly-rich fish oil may have reduced apoA-I synthesizing machinery. The effect of *n*–3 fatty acids on transcriptional regulation of the apoA-I gene has not been well studied.

16.3.1.6 **Effect of Fat Metabolites: Ketones and Prostanoids**

Diabetic ketoacidosis is often associated with reductions in plasma HDLc levels and there are potentially multiple mechanisms underlying this change. In experimental models, ketoacidosis profoundly suppresses apoA-I gene expression. It appears that acidosis is a required element of apoA-I gene suppression by ketoacidosis and that ketones have a minor effect, if any, on apoA-I levels [68–70].

The effect of acidosis on apoA-I promoter activity occurs through a pH-responsive element (pH-RE) located within the promoter [70]. Acidosis increases the specific DNA-binding activity of a putative repressor protein, and the response element is either close to or overlaps a negative thyroid hormone response element (nTRE) that is located 3′ and overlaps the apoA-I TATA element [69,70]. Repression of apoA-I promoter action by acidosis does not require *de novo* protein or mRNA synthesis [70].

Prostanoids are metabolites of arachidonic acid and are known to have important roles in regulating the expression of various genes. Prostanoids may have a role in modulating apoA-I expression through their interference with insulin action [71,72] or through interaction with PPAR receptors [73–76], although treatment of cells with arachidonic acid or a select group of its prostanoid metabolites including prostaglandin I_2, thromboxane B_2, (±)-5-HETE or (±)-12-HETE, and prostaglandins E_1 and E_2 did not alter apoA-I expression [77]. However, inhibition of cyclooxygenase (COX) activity with indomethacin or acetylsalicylic acid downregulates apoA-I protein and mRNA expression at the transcriptional level [77], which raises the possibilities either that a prostanoid that was not evaluated is involved in apoA-I expression or that the effect of indomethacin on apoA-I gene expression is independent of COX activity. It is noteworthy that aspirin, but not indomethacin, has been shown to decrease the expression of apolipoprotein (a) (apoa) in human hepatocytes at the transcriptional level [78]. This effect was also independent of COX inhibition. Apo(a) is a protein of unknown function with a high sequence homology with plasminogen. Lipoprotein (a) [Lp(a)] consists of large LDL particles in which the apoB protein is covalently bonded to apo(a). Elevated Lp(a) ($>30\,\mathrm{mg\,dL^{-1}}$) is usually observed in patients with premature CVD, and some ethnic or racial groups (e.g., African American population) appear to have higher levels of Lp(a) [79].

16.4 Effects of Carbohydrates or their Metabolites

Low-fat diets are often enriched with carbohydrates, and these may also have independent effects on HDLc and apoA-I production.

16.4.1 Effect of Glucose

Glucose is the most abundantly absorbed carbohydrate after the digestion of starch or table sugar (sucrose). In addition to being an important metabolic substrate, glucose is also an important regulator of gene expression [80,81]. A sucrose-rich diet stimulates hepatic lipogenesis and the abundance of cellular apoA-IV mRNA without altering apoA-I and apoC-III mRNA levels [82]. If hyperglycemia occurs, however, such as in diabetes, hepatic expression of apoA-I protein and its messenger RNA (mRNA) are reduced [83]. In isolated rat hepatocytes or in human hepatoma cell line HepG2, dextrose in the culture media is associated with concentration-dependent downregulation of apoA-I secretion [83,84]. Treatment of cells with $100\,\mathrm{uU\,mL^{-1}}$ of insulin causes a twofold increase in apoA-I mRNA levels [83]. The transcriptional activity of apoA-I promoter is also suppressed by dextrose and stimulated by insulin in a dose-dependent fashion [83]. Insulinomimetics, such as bisperoxo (1,10-phenathroline) oxovanadate (bpv) and the protein kinase C activator phorbol-12,13-dibutyrate (PDBu), upregulate apoA-I promoter [85,86]. Insulin modulates apoA-I promoter activity through at least two signaling pathways. One is through Ras-raf activation of MAPK and the other is PI3-K-dependent activation of PKC kinases [85].

A single *cis*-acting element within the promoter located between nucleotides −425 to −376 regulates the effects of both insulin and glucose on transcriptional activity of the apoA-I gene. However, only insulin – not dextrose – alters insulin response core element (IRCE) binding to transcriptional factor Sp1; this can be modified by the degree of phosphorylation [85]. There appears to be additional carbohydrate-responsive elements remaining to be characterized, and the IRCE may have only a permissive role for the inhibitory effects of glucose [83]. It is also possible that a unique carbohydrate-responsive element overlaps the IRCE and interferes with its ability to respond to insulin.

16.4.2
Effects of Fructose

Feeding of experimental rats with large quantities of fructose is associated with insulin resistance, hyperlipidemia, and secondarily high HDLc [87], while feeding of rats with a sucrose-rich_diet did not affect the abundance of cellular and nuclear apoA-I and apoC-III mRNA or the transcriptional activity of these genes in the liver, although apoA-IV mRNA increased significantly [82]. The effect of a modest increase in fructose content of the diet in humans is less dramatic. Reiser *et al.* found that a diet enriched with fructose increases the levels of risk factors associated with heart disease, especially in hyperinsulinemic men [88]. However, in a study of 13 poorly controlled type 2 diabetic patients, 60 gm fructose a day did not significantly alter the fasting serum lipids, lipoproteins, and apolipoproteins A-1 and B-100 levels [89].

In HepG2 cells treated with fructose, no significant changes in apoA-I expression could be demonstrated (unpublished data), so the *in vivo* effect of a high-fructose diet is likely to be secondary to hyperinsulinemia or to increased intracellular glucosamine levels (see below). Alternatively, increased apoA-I in fructose-fed rats may have been secondary to increased hepatic triglyceride content. Indeed, it has been shown in studies of cebus monkeys that hepatic apoA-I mRNA concentrations increase as liver lipid content increases [90].

16.4.3
Effect of Glucosamine

Glucosamine is not a major nutrient and the amount in commonly consumed diets is small enough not to cause significant changes in HDLc or apoA-I levels. The intracellular glucosamine level increases in periods of hyperglycemia when glutamine:fructose-6-phosphate amidotransferase (GFAT) is increased and causes insulin resistance [91,92]. Conversion of glucosamine-6-phosphate into UDP-N-acetylglucosamine provides a substrate necessary for glycosylation of proteins, including those that regulate transcriptional activity [93,94]. One such transcription factor implicated in apoA-I gene transcription is Sp1. It has recently been demonstrated that Sp1 transcriptional activity and half-life are regulated by glycosylation on serine and threonine residues with N-acetylglucosamine, the primary end-product of elevated glucosamine levels [93,95]. However, treatment of HepG2 cells with varying concentrations of glucosamine had no effect on apoA-I gene transcription rate, although the amount of apoA-I protein and apoA-I mRNA levels were increased [96]. ApoA-I mRNA turnover studies have shown that treatment of cells with glucosamine is associated with increased apoA-I mRNA half-life, from 7.6 to 16.6 h, so increased apoA-I gene expression by glucosamine occurs primarily through apoA-I mRNA stabilization, while the changes in Sp1 are not significant enough to alter apoA-I gene transcription [96]. The clinical implications of these observations are not clear. In a recent study of diet supplementation with 1.5 gm of glucosamine daily for 6 weeks, there was no significant change in insulin sensitivity or any change in serum apoA-I levels in lean or obese individuals [97].

16.5
Effects of Protein

There is some discrepancy between observational and interventional studies as to the effect of proteins on cardiovascular disease. Whereas observational studies suggested a negative association between vegetable protein and cardiovascular disease risk, interventional studies have been less consistent.

In general, when body weight is stable, the type of dietary protein or individual amino acids have little effect on lipoprotein patterns [98,99]. In the OmniHeart trial, a diet rich in carbohydrates, a diet rich in protein, about half from plant sources, and a diet rich in unsaturated fat, predominantly monounsaturated fat, were compared in 164 adults with pre-hypertension or stage 1 hypertension [100]. Blood pressure, LDLc, and estimated coronary heart disease risk were lower in relation to baseline on each diet. In comparison with the carbohydrate-rich diet, the protein-rich diet further decreased mean systolic blood pressure by 1.4 mm Hg among those with hypertension and decreased LDLc by 3.3 mg dL^{-1}, HDLc by 1.3 mg dL^{-1}, and triglycerides by 15.7 mg dL^{-1}. Whereas the carbohydrate-rich diet was associated with its well established HDLc-lowering effects in relation to the high-fat diet, the small reduction in HDLc of those on the protein-rich diet (2.6 % vs. carbohydrate; 4.7 % vs. unsaturated fat) was unexpected [100].

The effect of amount or source of proteins on HDLc levels is not well studied, though certain proteins such as whey protein has been shown to enhance pancreatic insulin secretion through potentiation of incretins [101]. This process is expected to have anabolic effects on apoA-I and on HDLc levels. The effect of whey protein on incretin secretion is believed to be a direct function of the peptide fragments generated through digestion. In contrast, the effect of soy proteins appears to be mostly related to the other components of soy, notably isoflavones, which contribute to the hypocholesterolemic effect of soy [102–104]. A large intake of soy protein ($\geq$25 g per day) is required to have a small (5 %) hypocholesterolemic effect. However, consumption of isoflavone in a study of gerbils did not contribute to the hypocholesterolemic effect of soy [105]. These effects may be species-specific and gender-specific.

In addition, because of structural similarities between isoflavones and estrogen, soy phytoestrogens have been shown to increase plasma levels of HDLc and apoA-I in animal and in some human studies. In Hep G2 cells, both genistein and daidzein increase apoA-I secretion in a dose-dependent fashion [103]. The effect of genistein on apoA-I secretion is similar to that observed with 17-β-estradiol (E2) and it increases transcriptional activity of the apoA-I gene through the mitogen activated protein (MAP) kinase pathway [104].

16.6
Effects of Micronutrients

Although micronutrients are essential ingredients of healthy diets, supplementation with pharmacologic doses of vitamins without a clear clinical indication is fraught with therapeutic uncertainties. Supplementation with antioxidants has been suggested as a potential intervention to reduce cardiovascular disease, and yet interventional trials have failed to show clear benefit [106]. The effect of antioxidants on HDLc levels depends on the population of patients studied and the quantities of vitamins used. In a randomized trial of 69 type 2 diabetic subjects, supplementation with magnesium, zinc, and vitamins C and E significantly increased HDLc and apoA-I levels [107]. However, another small interventional trial found that antioxidants may partially blunt HDLc induction by simvastatin and niacin combination therapy [108]. In cell culture studies, vitamins with antioxidative potential downregulate apoA-I gene expression. [109]. However, a high dose (1 gm per day) vitamin C supplementation *in vivo* does not have consistent effects on plasma lipids or lipoprotein levels [110–113]. It therefore appears that optimal concentrations of these vitamins are needed for apoA-I gene expression.

As in the cases of vitamins C and E, optimal vitamin D levels are necessary to support adequate apoA-I levels. In a Belgian study, those with high vitamin D levels also had the highest plasma apoA-I levels [114]. However, vitamin D supplementation in postmenopausal women has recently been shown to be associated with adverse changes in serum lipoproteins [115,116]. In Hep G2 cells, treatment with 1,25-$(OH)_2$ D_3 suppresses apoA-I secretion and mRNA levels, along with a decrease in apoA-I promoter activity [117].

Vitamin A appears to have suppressive effects on hepatic but not intestinal apoA-I gene transcription, although apoA-I protein content does not change, suggesting additional regulation of posttranscriptional events [118–120]. The abundance of hepatic apoA-I mRNA of vitamin A-deficient rats was 2.2 to 6 times that of sufficient rats [118]. Treatment of co-cultured rat hepatocytes with retinoic acid resulted in a specific decrease in apoA-I mRNA levels, whereas no marked difference in apoA-II mRNA levels was observed [119]. Recently, retinol binding protein 4 (RBP4), a specific carrier protein for retinol, has been implicated in insulin resistance [121]. Serum levels of RBP4, a protein secreted by adipocytes, are increased in insulin-resistant states and these levels correlate with the magnitude of insulin resistance in subjects with obesity, impaired glucose tolerance, or type 2 diabetes, and also in non-obese, nondiabetic subjects with a strong family history of type 2 diabetes [121]. At the present time it is not known whether retinol can alter the biology of RBP4 and whether some of the effects of vitamin A on HDLc are mediated through its effect on RBP4 and insulin sensitivity.

Mineral supplementation of diets can also affect HDLc and apoA-I levels. Excessive zinc supplementation has been associated with copper deficiency, leading to hypertriglyceridemia and low HDLc. In cell cultures, however, low cellular copper status can enhance apoA-I mRNA production and increase apoA-I synthesis [122,123]. This increase was attributed to an elevated level of the regulatory factor HNF-4 [122]. In rats, copper deficiency increases hepatic apoA-I synthesis and secretion, but does not alter hepatic total cellular apoA-I mRNA abundance [123].

Some minerals such as chromium, vanadium, magnesium, and zinc either have insulinomimetic effects or have permissive effects on insulin action [124–127]. In addition, these minerals have direct effects on apoA-I expression. In particular, zinc deficiency causes downregulation of apoA-I expression [126], while supraphysiologic concentrations of zinc, as well as chromium or vanadium, downregulate apoA-I promoter activity [127].

Future clinical trials need to address the safety and efficacy of pharmacologic doses of micronutrient supplementation. Until then, consumption of micronutrients, with a few possible exceptions, should be limited to the Recommended Daily Allowances.

16.7 Effects of Alcohol

It is generally accepted that a moderate amount of alcohol (ethanol) consumption (1–2 drinks daily) has favorable effects on cardiovascular health. The cardioprotective effect of alcohol is partly secondary to improved plasma lipid profile, notably increased plasma HDLc levels. In general, alcohol consumption is associated with increases in serum levels of both LpA-I and LpA-I:A-II particles [128]. In a study of postmenopausal women, plasma HDLc and apoA-I levels increased significantly after daily consumption of 30 g alcohol while plasma apoB decreased [129]. It is estimated that consumption of 1 fluid oz (30 g) of alcohol per day increases HDLc by approximately 4 $mg\,dL^{-1}$ [130]. It is noteworthy that this favorable effect may be overshadowed by alcohol-induced hypertriglyceridemia in individuals who have metabolic syndrome. The increasing plasma triglyceride levels will dampen or prevent the rise in HDLc.

Alcohol raises HDLc levels through multiple mechanisms: it increases cellular cholesterol efflux and cholesterol esterification [131] and also has direct effects on apoA-I expression in HepG2 and Hep3B cells [132]. Treatment of cells with ethanol increases apoA-I and apoB levels without significant changes in apoC-II, apoC-III, and apoE. In these experiments, 10 mM ethanol increased HDLc by 40–70 %, while at 25 mM or greater concentrations, apoB was increased and VLDL and LDL triglyceride were increased by 20 % to 25 % over the control level. Thus, in cell cultures the effects of ethanol on apoA-I and HDLc levels were more pronounced than its effects on apoB-containing lipoproteins [132].

16.8 Conclusions

Dietary factors are important determinants of recent changes in the prevalence of cardiovascular disease. There are multiple ways in which diet can contribute to the emergence of coronary heart disease, one potential component of this process being the effect of various nutrients on plasma lipid profiles.

The currently recommended low saturated fat diet reduces atherogenic LDLc, but it can also reduce cardioprotective HDLc. However, the overall clinical effect of such diets remains positive and the benefits of limiting saturated fat in the diet exceed the expectations based on modest changes in lipid profile. The reduced plasma apoA-I levels related to consumption of a low-fat diet are mostly attributable to increased fractional clearance of HDLc [19]. However, it is possible that the mechanism underlying the changes in HDLc is an important determinant of its biologic functions. Reduced serum apoA-I levels secondary to changes in apoA-I gene expression induced by various nutrients may have more detrimental effects than reduced HDLc levels secondary to increased turnover. Understanding the effects of nutrients on apoA-I gene expression should help in the rational design of dietary recommendations.

References

1 Tavazzi, L. (1999) Clinical epidemiology of acute myocardial infarction. *Am. Heart J.*, **138**, S48–S54.

2 Keys, A. (1997) Coronary heart disease in seven countries. *Nutrition*, **13**, 250–252.

3 Hooper, L., Summerbell, C. D., Higgins, J. P., Thompson, R. L., Clements, G., Capps, N., Davey, S., Riemersma, R. A., Ebrahim, S. (2001) Reduced or modified dietary fat for preventing cardiovascular disease. *Cochrane Database Syst Rev.*, (3), CD002137

4 Howard, B. V., Van Horn, L., Hsia, J., Manson, J. E., *et al.* (2006) Low-fat dietary pattern and risk of cardiovascular disease: the Women's Health Initiative Randomized Controlled Dietary Modification Trial. *JAMA*, **295**, 655–66.

5 Lopez-Candales, A. (2001) Metabolic syndrome X: a comprehensive review of the pathophysiology and recommended therapy. *J Med*, **32**, 283–300.

6 Vajo, Z., Terry, J. G., Brinton, E. A. (2002) Increased intra-abdominal fat may lower HDL levels by increasing the fractional catabolic rate of Lp A-I in postmenopausal women. *Atherosclerosis*, **160**, 495–501.

7 Duvillard, L., Pont, F., Florentin, E., Gambert, P., Verges, B. (2000) Inefficiency of insulin therapy to correct apolipoprotein A-I metabolic abnormalities in non-insulin-dependent diabetes mellitus. *Atherosclerosis*, **152**, 229–37.

8 Le, N. A. and Ginsburg, H. N. (1988) Heterogeneity of apolipoprotein A-I turnover in subjects with reduced concentrations of plasma high density lipoprotein cholesterol. *Metabolism*, **37**, 614–7.

9 Mooradian, A. D., Hass, M. J., Albert, S. G. (2007) Low serum HDL cholesterol in obese subjects with normal serum triglycerides: The role of insulin resistance and inflammatory cytokines. *Diabetes, Obesity and Metabolism* 9: 441–443.

10 Mooradian, A. D. (2001) Obesity: A rational target for managing diabetes mellitus. *Growth Hormone & IGF research (suppl. A)*, S1–S5.

11 Haas, M. J., Horani, M., Mreyoud, A., Plummer, B., Wong, N.C.W., Mooradian, A. D. (2003) Suppression of apolipoprotein AI gene expression in Hep G2 cells by TNF α and IL-1 β. *Biochem. Biophys. Acta*, **1623**, 120–128.

12 Beers, A., Haas, M. J., Wong, N.C.W., Mooradian, A. D. (2006) Inhibition of apolipoprotein AI gene expression by tumor necrosis factor α: Roles for MEK/ERK and JNK Signaling. *Biochemistry*, **45**, 2408–2413.

13 Dattilo, A. M., Kris-Etherton, P. M. (1992) Effects of weight reduction on blood lipids and lipoproteins: a meta-analysis. *Am J Clin Nutr*, **56**, 320–328.

14 Weisweiler, P. (1987) Plasma lipoproteins and lipase and lecithin:cholesterol acyltransferase activities in obese subjects before and after weight reduction. *J Clin Endocrinol Metab*, **65**, 969–973.

15 Mooradian, A. D., Haas, M. J., Wong, N. C. W. (2006) The effect of select nutrients on serum high density lipoprotein cholesterol and apolipoprotein A-I levels. *Endocrine Reviews*, **27**, 2–16.

16 Mooradian, A. D., Haas, M. J., Wong, N.C.W. (2004) Transcriptional control of apolipoprotein A-I gene expression in diabetes mellitus. *Diabetes*, **53**, 513–520.

17 Lichtenstein, A. H., Appel, L. J., Brands, M., Carnethon, M., Daniels, S., Franch, H. A., Franklin, B., Kris-Etherton, P., Harris, W. S., Howard, B., Karanja, N., Lefevre, M., Rudel, L., Sacks, F., Van Horn, L., Winston, M., Wylie-Rosett, J., Diet and lifestyle recommendations revision 2006: a scientific statement from the American Heart Association Nutrition Committee. *Circulation* **114**, 82–96.

18 Bantle, J. P., Wylie-Rosett, J., Albright, A. L., Apovian, C. M., Clark, N. G., Franz, M. J., Hoogwerf, B. J., Lichtenstein, A. H., Mayer-Davis, E., Mooradian, A. D., Wheeler, M. L. (2006) Nutrition recommendations and interventions for diabetes–2006: a position statement of the American Diabetes Association. *Diabetes Care*, **29**, 2140–2157.

19 Mensink, R. P., Zock, P. L., Kester, A. D., Katan, M. B. (2003) Effects of dietary fatty acids and carbohydrates on the ratio of serum total to HDL cholesterol and on serum lipids and apolipoproteins: a meta-analysis of 60 controlled trials. *Am J Clin Nutr.*, **77**, 1146–1155.

20 Meksawan, K., Pendergast, D. R., Leddy, J. J., Mason, M., Horvath, P. J., Awad, A. B. (2004) Effect of low and high fat diets on nutrient intakes and selected cardiovascular risk factors in sedentary men and women. *J Am Coll Nutr*, **23**, 131–140.

21 Walden, C. E., Retzlaff, B. M., Buck, B. L., Wallick, S., McCann, B. S., Knopp, R. H. (2000) Differential effect of National Cholesterol Education Program (NCEP) Step II diet on HDL cholesterol, its subfractions, and apoprotein A-I levels in hypercholesterolemic women and men after 1 year: the beFIT Study. *Arterioscler Thromb Vasc Biol*, **20**, 1580–1587.

22 Li, Z., Otvos, J. D., Lamon-Fava, S., Carrasco, W. V., Lichtenstein, A. H., McNamara, J. R., Ordovas, J. M., Schaefer, E. J. (2003) Men and women differ in lipoprotein response to dietary saturated fat and cholesterol restriction. *J Nutr*, **133**, 3428–3433.

23 Vessby, B., Boberg, J., Gustafsson, I. B., Karlstrom, B., Lithell, H., Ostlund-Linqvist, A. M. (1980) Reduction of high density lipoprotein cholesterol and apoliproprotein A-I concentrations by a lipid-lowering diet. *Atherosclerosis*, **35**, 21–27.

24 Srivastava, R. A., Tang, J., Krul, E. S., Pfleger, B., Kitchens, R. T., Schonfeld, G. (1992) Dietary fatty acids and dietary cholesterol differ in their effect on the in vivo regulation of apolipoprotein A-I and A-II gene expression in inbred strains of mice. *Biochim Biophys Acta.*, **1125**, 251–261.

25 Spady, D. K., Kearney, D. M., Hobbs, H. H. (1999) Polyunsaturated fatty acids up-regulate hepatic scavenger receptor B1 (SR-BI) expression and HDL cholesteryl ester uptake in the hamster. *J Lipid Res.*, **40**, 1384–1394.

26 Dorfman, S. E., Wang, S., Vega-Lopez, S., Jauhiainen, M., Lichtenstein, A. H. (2005) Dietary fatty acids and cholesterol differentially modulate HDL cholesterol metabolism in Golden-Syrian hamsters. *J Nutr.*, **135**, 492–8.

27 Sanchez-Muniz, F. J., Merinero, M. C., Rodriguez-Gil, S., Ordovas, J. M., Rodenas, S., Cuesta, C. (2002) Dietary fat saturation affects apolipoprotein AII levels and HDL composition in postmenopausal women. *J Nutr.*, **132**, 50–54.

28 Talmud, P. J., Waterworth, D. M. (2000) In-vivo and in-vitro nutrient-gene interactions. *Curr Opin Lipidol.*, **11**, 31–36.

29 Talmud, P. J., Hawe, E., Robertson, K., Miller, G. J., Miller, N. E., Humphries, S. E. (2002) Genetic and environmental determinants of plasma high density lipoprotein cholesterol and apolipoprotein AI concentrations in healthy middle-aged men. *Ann Hum Genet*, **66** (Pt 2), 111–124.

30 Blangero, J., MacCluer, J. W., Kammerer, C. M., Mott, G. E., Dyer, T. D., McGill, H. C., Jr. (1990) Genetic analysis of apolipoprotein A-I in two dietary environments. *Am J Hum Genet*, **47**, 414–428.

31 Ordovas, J. M., Corella, D., Cupples, L. A., Demissie, S., Kelleher, A., Coltell, O., Wilson, P. W., Schaefer, E. J., Tucker, K. (2002) Polyunsaturated fatty acids modulate the effects of the APOA1 G-A polymorphism on HDL-cholesterol concentrations in a sex-specific manner: the Framingham Study. *Am J Clin Nutr*, **75**, 38–46.

32 Calleja, L., Trallero, M. C., Carrizosa, C., Mendez, M. T., Palacios-Alaiz, E., Osada, J. (2000) Effects of dietary fat amount and saturation on the regulation of hepatic mRNA and plasma apolipoprotein A-I in rats. *Atherosclerosis*, **152**, 69–78.

33 Schwab, D. A., Rea, T. J., Hanselman, J. C., Bisgaier, C. L., Krause, B. R., Pape, M. E. (2000) Elevated hepatic apolipoprotein A-I transcription is associated with diet-induced hyperalphalipoproteinemia in rabbits. *Life Sci*, **66**, 1683–1694.

34 Hayek, T., Ito, Y., Azrolan, N., Verdery, R. B., Aalto-Setala, K., Walsh, A., Breslow, J. L. (1993) Dietary fat increases high density lipoprotein (HDL) levels both by increasing the transport rates and decreasing the fractional catabolic rates of HDL cholesterol ester and apolipoprotein (Apo) A-I. Presentation of a new animal model and mechanistic studies in human Apo A-I transgenic and control mice. *J Clin Invest*, **91**, 1665–1671.

35 Azrolan, N., Odaka, H., Breslow, J. L., Fisher, E. A. (1995) Dietary fat elevates hepatic apo A-I production by increasing the fraction of apolipoprotein A-I mRNA in the translating pool. *J Biol Chem*, **270**, 19833–19838.

36 Srivastava, R. A. (1994) Saturated fatty acid, but not cholesterol, regulates apolipoprotein AI gene expression by posttranscriptional mechanism. *Biochem Mol Biol Int*, **34**, 393–402.

37 Hatahet, W., Cole, L., Kudchodkar, B. J., Fungwe, T. V. (2003) Dietary fats differentially modulate the expression of lecithin:cholesterol acyltransferase, apoprotein-A1 and scavenger receptor b1 in rats. *J Nutr.*, **133**, 689–694.

38 Wahrburg, U., Martin, H., Sandkamp, M., Schulte, H., Assmann, G. (1992) Comparative effects of a recommended lipid-lowering diet vs a diet rich in monounsaturated fatty acids on serum lipid profiles in healthy young adults. *Am J Clin Nutr*, **56**, 678–683.

39 Lichtenstein, A. H., Busman, L. M., Carrasco, W., Jenner, J. L., Gualtieri, L. J., Goldin, B. R., Ordovas, J. M., Schaefer, E. J. (1993) Effects of canola, corn, and olive oils on fasting and postprandial plasma lipoproteins in humans as part of a National Cholesterol Education Program Step 2 diet. *Arterioscler Thromb.*, **13**, 1533–42.

40 Pelkman, C. L., Fishell, V. K., Maddox, D. H., Pearson, T. A., Mauger, D. T., Kris-Etherton, P. M. (2004) Effects of moderate-fat (from monounsaturated fat) and low-fat weight loss diets on the serum lipid profile in overweight and obese men and women. *Am J Clin Nutr*, **79**, 204–212.

41 Syper, A., Jurva, J., Pleuss, J., Hoffman, R., Gutterman, D. (2005) Influence of glycemic load on HDL cholesterol in youth. *Am J Clin Nutr*, **81**, 376–379.

42 Brousseau, M. E., Ordovas, J. M., Osada, J., Fasulo, J., Robins, S. J., Nicolosi, R. J., Schaefer, E. J. (1995) Dietary monounsaturated and polyunsaturated fatty acids are

comparable in their effects on hepatic apolipoprotein mRNA abundance and liver lipid concentrations when substituted for saturated fatty acids in cynomolgus monkeys. *J Nutr,* **125**, 425–436.

43 Sorci-Thomas, M., Prack, M. M., Dashti, N., Johnson, F., Rudel, L. L., Williams, D. L. (1989) Differential effects of dietary fat on the tissue-specific expression of the apolipoprotein A-I gene: relationship to plasma concentration of high density lipoproteins. *J Lipid Res,* **30**, 1397–1403.

44 Haas, M. J., Horani, M. H., Wong, N. C., Mooradian, A. D. (2004) Induction of the apolipoprotein AI promoter by Sp1 is repressed by saturated fatty acids. *Metabolism,* **53**, 1342–1348.

45 Lee, J. Y., Sohn, K. H., Rhee, S. H., Hwang, D. (2001) Saturated fatty acids, but not unsaturated fatty acids, induce the expression of cyclooxygenase-2 mediated through Toll-like receptor 4. *J Biol Chem,* **276**, 16683–166899.

46 Rong, R., Ramachandran, S., Penumetcha, M., Khan, N., Parthasarathy, S. (2002) Dietary oxidized fatty acids may enhance intestinal apolipoprotein A-I production. *J Lipid Res,* **43**, 557–564.

47 Lichtenstein, A. H., Jauhiainen, M., McGladdery, S., Ausman, L. M., Jalbert, S. M., Vilella-Bach, M., Ehnholm, C., Frohlich, J., Schaefer, E. J. (2001) Impact of hydrogenated fat on high density lipoprotein subfractions and metabolism. *J Lipid Res,* **42**, 597–604.

48 Aro, A., Jauhiainen, M., Partanen, R., Salminen, I., Mutanen, M. (1997) Stearic acid, trans fatty acids, and dairy fat: effects on serum and lipoprotein lipids, apolipoproteins, lipoprotein A, and lipid transfer proteins in healthy subjects. *Am J Clin Nutr,* **65**, 1419–1426.

49 Judd, J. T., Baer, D. J., Clevidence, B. A., Kris-Etherton, P., Muesing, R. A., Iwane, M. (2002) Dietary *cis* and *trans* monounsaturated and saturated FA and plasma lipids and lipoproteins in men. *Lipids,* **37**, 123–131.

50 Methane, N. R., Welty, F. K., Barrett, P. H., Harausz, C., Dolnikowski, G. G., Parks, J. S., Eckel, R. H., Schaefer, E. J., Lichtenstein, A. H. (2004) Dietary hydrogenated fat increases high-density lipoprotein apoA-I catabolism and decreases low-density lipoprotein apoB-100 catabolism in hypercholesterolemic women. *Arterioscler Thromb Vasc Biol.,* **24**, 1092–7.

51 Zock, P. L., Katan, M. B. (1992) Hydrogenation alternatives: effects of *trans* fatty acids and stearic acid versus linoleic acid on serum lipids and lipoproteins in humans. *J Lipid Res.,* **33**, 399–410.

52 Mensink, R. P., Katan, M. B. (1990) Effect of dietary *trans* fatty acids on high-density and low-density lipoprotein cholesterol levels in healthy subjects. *N. Engl. J. Med.,* **323**, 439–445.

53 Food and Drug Administration 2006. 21 CFR Part 101. Food Labeling; *Trans* Fatty Acids in Nutrition Labeling; [Docket No. 94P-0036]. RIN 0910-AB66. Food Labeling: *Trans* Fatty Acids in Nutrition Labeling, Nutrient Content Claims, and Health Claims. Federal Register: July 11, 2003 (Volume 68, Number 133)].http://vm.cfsan.fda.govw~lrd/fr03711a.html accessed July 7, 2006.

54 Dashti, N., Feng, Q., Franklin, F. A. (2000) Long-term effects of cis and *trans* monounsaturated (18:1) and saturated (16:0) fatty acids on the synthesis and secretion of apolipoprotein A-I- and apolipoprotein B-containing lipoproteins in HepG2 cells. *J Lipid Res,* **41**, 1980–1990.

55 deRoos, N. M., Bots, M. L., Katan, M. B. (2001) Replacement of dietary saturated fatty acids by *trans* fatty acids lowers serum HDL cholesterol and impairs endothelial function in healthy men and women. *Arterioscler Thromb Vasc Biol.,* **21**, 1233–1237.

56 Nicolosi, R. J., Wilson, T. A., Rogers, E. J., Kritchevsky, D. (1998) Effects of specific fatty acids (8:0, 14:0, *cis*-18:1, *trans*-18:1) on plasma lipoproteins, early atherogenic potential, and LDL oxidative properties in the hamster. *J Lipid Res.*, **39**, 1972–1980.

57 Temme, E. H., Mensink, R. P., Hornstra, G. (1997) Effects of medium chain fatty acids (MCFA), myristic acid, and oleic acid on serum lipoproteins in healthy subjects. *J Lipid Res*, **38**, 1746–1754.

58 Mascioli, E. A., McLennan, C. E., Schaefer, E. J., Lichtenstein, A. H., Hoy, C. E., Christensen, M. S., Bistrian, B. R. (1999) Lipidemic effects of an interesterified mixture of butter, medium-chain triacylglycerol and safflower oils. *Lipids*, **34**, 889–94.

59 Harris, W. S. (1996) *n*-3 Fatty acids and lipoproteins: comparison of results from human and animal studies. *Lipids*, **31**, 243–252.

60 Sanders, T. A., Sullivan, D. R., Reeve, J., Thompson, G. R. (1985) Triglyceride-lowering effect of marine polyunsaturates in patients with hypertriglyceridemia. *Arteriosclerosis*, **5**, 459–465.

61 Balk, E. M., Lichtenstein, A. H., Chung, M., Kupelnick, B., Chew, P., Lau, J. (2006) Effects of omega-3 fatty acids on serum markers of cardiovascular disease risk: A systematic review. *Atherosclerosis.*, **189**, 19–30.

62 Li, Z., Lamon-Fava, S., Otvos, J., Lichtenstein, A. H., Velez-Carrasco, W., McNamara, J. R., Ordovas, J. M., Schaefer, E. J. (2004) Fish consumption shifts lipoprotein subfractions to a less atherogenic pattern in humans. *J Nutr.*, **134**, 1724–8.

63 Ginsberg, H. N. (1996) Diabetic dyslipidemia: basic mechanisms underlying the common hypertriglyceridemia and low HDL cholesterol levels. *Diabetes*, **45** (Suppl 3), S27–S30.

64 Huggins, K. W., Colvin, P. L., Burleson, E. R., Kelley, K., Sawyer, J. K., Barrett, P. H., Rudel, L. L., Parks, J. S. (2001) Dietary n-3 polyunsaturated fat increases the fractional catabolic rate of medium-sized HDL particles in African green monkeys. *J Lipid Res*, **42**, 1457–1466.

65 Berthou, L., Saladin, R., Yaqoob, P., Branellec, D., Calder, P., Fruchart, J. C., Denefle, P., Auwerx, J., Staels, B. (1995) Regulation of rat liver apolipoprotein A-I, apolipoprotein A-II and acyl-coenzyme A oxidase gene expression by fibrates and dietary fatty acids. *Eur J Biochem*, **232**, 179–187.

66 Gerasimova, E., Perova, N., Ozerova, I. *et al.* (1991) The effect of dietary n-3 polyunsaturated fatty acids on HDL-cholesterol in Chukot residents vs. Muscovites. *Lipids*, **26**, 261–265.

67 Frenais, R., Ouguerram, K., Maugeais, C., Mahot, P., Charbonnel, B., Magot, T., Krempf, M. (2001) Effect of dietary omega-3 fatty acids on high-density lipoprotein apolipoprotein AI kinetics in type II diabetes mellitus. *Atherosclerosis*, **157**, 131–135.

68 Haas, M. J., Pun, K., Reinacher, D., Wong, N.C.W., Mooradian, A. D. (2000) Effects of ketoacidosis on rat apolipoprotein A1 gene expression: A link with acidosis but not with ketones. *J. Mol. Endocrinol*, **25**, 129–139.

69 Haas, M. J., Reinacher, D., Pun, K., Wong, N.C.W., Mooradian, A. D. (2000) Induction of the apolipoprotein A1 gene by fasting: A relationship with ketosis but not with ketone bodies. *Metabolism*, **49**, 1572–1578.

70 Haas, M. J., Reinacher, D., Li, J. P., Wong, N.C.W., Mooradian, A. D. (2001) Regulation of ApoA1 gene expression with acidosis: Requirement for a transcriptional repressor. *J. Mol. Endocrinol*, **27**, 43–57.

71 Wasner, H. K., Weber, S., Partke, H. J., Amini-Hadi-Kiashar, H. (1994) Indomethacin treatment causes loss of insulin action in rats: involvement of prostaglandins in the mechanism

of insulin action. *Acta Diabetologica*, **31**, 175–182.

72 Christensen, J. R., Hammond, B. J., Smith, G. D. (1990) Indomethacin inhibits endocytosis and degradation of insulin. *Biochem Biophys Res. Commun.*, **173**, 127–133.

73 Willson, T. M., Brown, P. J., Sternbach, D. D., Heake, B. R. (2000) The PPARS: From orphan receptors to drug discovery. *J Med Chem*, **43**, 527–550.

74 Dussault, I., Forman, B. M. (2000) Prostaglandins and fatty acids regulate transcriptional signaling via the peroxisome proliferator activated receptor nuclear receptors. *Prostaglandins & other Lipid Mediators*, **62**, 1–13.

75 Lefebre, A. M., Peinado-0nsurke, J., Leitersdorf, I., Briggs, M. R., Paterniti, J. R., Fruchart, J. C., Fievet, C., Auwerx, J., Staels, B. (1997) Regulation of lipoprotein metabolism by thiazolidinediones occurs through a distinct but complementary mechanism relative to fibrates. *Arterio Thromb Vasc Biol*, **17**, 1756–1764.

76 Vu-Dac, N., Chopin-Delannoy, S., Gervois, P., Bonnelye, E., Martin, G., Fruchart, J-C., Laudet, V., Staels, B. (1998) The nuclear receptors peroxisome proliferator-activated receptor α and Rev-erb α mediate the species-specific regulation of apolipoprotein A1 expression by fibrates. *J Biol Chem*, **273**, 25713–25720.

77 Horani, M., Gobal, F., Haas, M. J., Wong, N.C.W., Mooradian, A. D. (2004) Cyclooxygenase (COX) inhibition is associated with downregulation of apolipoprotein AI (Apo-AI) promoter activity in cultured hepatoma cell line-Hep G2. *Metabolism*, **53**, 174–181.

78 Kagawa, A., Azuma, H., Akaike, M., Kanagawa, Y., Matsumoto, T. (1999) Aspirin reduces apolipoprotein (a) (apo(a) production in human hepatocytes by suppression of apo (a) gene transcription. *J. Biol. Chem.*, **274**, 34111–34115.

79 Hachem, S., Mooradian, A. D. (2006) Familial dyslipidemias: An overview of pathophysiology and management. *Drugs* 66: 1949–1969.

80 Mooradian, A. D., Albert, S. G. (1999) The age-related changes in lipogenic enzymes: the role of dietary factors and thyroid hormone responsiveness. *Mech Ageing Dev.*, **108**, 139–149.

81 Mooradian, A. D., Mariash, C. N. (1987) Effects of insulin and glucose on cultured rat hepatocyte gene expression. *Diabetes*, **36**, 938–943.

82 Radosavljevic, M., Lin-Lee, Y. C., Soyal, S. M., Strobl, W., Seelos, C., Gotto, A. M., Jr., Patsch, W. (1992) Effect of sucrose diet on expression of apolipoprotein genes A-I, C-III and A-IV in rat liver. *Atherosclerosis*, **95**, 147–156.

83 Murao, K., Wada, Y., Nakamura, T., Taylor, A. H., Mooradian, A. D., Wong, N.C.W. (1998) Effects of glucose and insulin on rat apolipoprotein A-I gene expression. *J Biol Chem*, **273**, 18959–18965.

84 Zheng, X. L., Matsubara, S., Diao, C., Hollenberg, M. D., Wong, N. C. (2000) Activation of apolipoprotein AI gene expression by protein kinase A and kinase C through transcription factor, Sp1. *J Biol Chem*, **275**, 31747–31754.

85 Samson, S. L. and Wong, N. C. (2002) Role of Sp1 in insulin regulation of gene expression. *J Mol Endocrinol.*, **29**, 265–279.

86 Lam, J. K., Matsubara, S., Mihara, K., Zheng, X. L., Mooradian, A. D., Wong, N. C. (2003) Insulin induction of apolipoprotein AI, role of Sp1. **42**, 2680–90.

87 Mooradian, A. D., Wong, N. C. W., Shah, G. N. (1997) Apolipoprotein A1 expression in young and aged rats is modulated by dietary carbohydrates. *Metabolism*, **46**, 1132–1136.

88 Reiser, S., Powell, A. S., Scholfield, D. J., Panda, P., Ellwood, K. C., Canary, J. J. (1989) Blood lipids, lipoproteins, apoproteins, and uric acid in men fed diets containing

fructose or high-amylose cornstarch. *Am J Clin Nutr*, **49**, 832–839.

89 Osei, K., Bossetti, B. (1989) Dietary fructose as a natural sweetener in poorly controlled type 2 diabetes: a 12-month crossover study of effects on glucose, lipoprotein and apolipoprotein metabolism. *Diabet Med.*, **6**, 506–511.

90 Hennessy, L. K., Osada, J., Ordovas, J. M., Nicolosi, R. J., Stucchi, A. F., Brousseau, M. E., Schaefer, E. J. (1992) Effects of dietary fats and cholesterol on liver lipid content and hepatic apolipoprotein A-I, B, and E and LDL receptor mRNA levels in cebus monkeys. *J Lipid Res*, **33**, 351–360.

91 McClain, D. A. (2001) Hexosamines as mediators of nutrient sensing: relevance to obesity, insulin resistance, and diabetes. *Curr Opin Endocrinol Diabetes*, **8**, 186–191.

92 Sakai, K. and Clemmons, D. R. (2003) Glucosamine induces resistance to insulin-like growth factor I (IGF-I) and insulin in Hep G2 cell cultures: Biological significance of IGF-I/insulin hybrid receptors. *Endocrinology*, **144**, 2388–2395.

93 Han, I. and Kudlow, J. E. (1997) Reduced O-glycosylation of Sp1 is associated with increased proteasome susceptibility. *Mol Cell Biol*, **17**, 2550–2558.

94 Comer, F. I. and Hart, G. W. (1999) O-GlcNAc and the control of gene expression. *Biochim Biophys Acta*, **1473**, 161–171.

95 Roos, M. D., Su, K., Baker, J. R., Kudlow, J. E. (1997) O glycosylation of an Sp1-derived peptide blocks known Sp1 protein interactions. *Mol Cell Biol*, **17**:6472–6480.

96 Haas, M. J., Wong, N. C., Mooradian, A. D. (2004) Effect of glucosamine on apolipoprotein AI mRNA stabilization and expression in HepG2 cells. *Metabolism*, **53**, 766–771.

97 Muniyappa, R., Karne, R. J., Hall, G., Crandon, S. K., Bronstein, J. A., Ver, M. R., Hortin, G. L., Quon, M. J. (2006) Oral glucosamine for 6 weeks at standard doses does not cause or worsen insulin resistance or endothelial dysfunction in lean or obese subjects. *Diabetes 55*, 3142–3150.

98 Lichtenstein, A. H. (2006) Thematic review series: patient-oriented research. Dietary fat, carbohydrate, and protein: effects on plasma lipoprotein patterns. *J Lipid Res.*, **47**, 1661–7.

99 Vega-Lopez, S. and Lichtenstein, A. H. (2005) Dietary protein type and cardiovascular disease risk factors. *Prev Cardiol.*, **8**, 31–40.

100 Appel, L. J., Sacks, F. M., Carey, V. J., Obarzanek, E., Swain, J. F., Miller, E. R. 3rd, Conlin, P. R., Erlinger, T. P., Rosner, B. A., Laranjo, N. M., Charleston, J., McCarron, P., Bishop L.M. OmniHeart Collaborative Research Group, (2005) Effects of protein, monounsaturated fat, and carbohydrate intake on blood pressure and serum lipids: results of the OmniHeart randomized trial. *JAMA.*, **294**, 2455–64.

101 Gunnarson, P. T., Winzell, M. S., Deacon, C. F., Larsen, M. O., Jelic, K., Carr, R. D., Ahren, B. (2006) Glucose-induced incretin hormone release and inactivation are differently modulated by oral fat and protein in mice. *Endocrinology*, **147**, 3173–3180.

102 Potter, S. M. (1995) Overview of proposed mechanism for the hypocholesterolemic effect of soy. *J. Nutr*, **125**, 606S–611S.

103 Lamon-Fava, S. (2000) Genistein activates apolipoprotein A-I gene expression in the human hepatoma cell line Hep G2. *J Nutr*, **130**, 2489–2492.

104 Lamon-Fava, S. and Micherone, D. (2004) Regulation of apoA-I gene expression: mechanism of action of estrogen and genistein. *J Lipid Res*, **45**, 106–112.

105 Tovar-Palacio, C., Potter, S. M., Hafermann, J. C., Shay, N. F. (1998) Intake of soy protein and soy protein extracts influences lipid metabolism

and hepatic gene expression in Gerbils. *J Nutr*, **128**, 839–842.

106 Mooradian, A. D. (2006) Antioxidants and diabetes. *Nestle Nutr Workshop Ser Clin Perform Programme.*, **11**, 107–22.

107 Farvid, M. S. Siassi, F. Jalali, M. Hosseini, M. Saadat, N. (2004) The impact of vitamin and/or mineral supplementation on lipid profiles in type 2 diabetes. *Diabetes Research & Clinical Practice.*, **65**, 21–28.

108 Cheung, M. C., Zhao, X-Q., Chait, A., Albers, J. J., Brown, G. (2001) Antioxidant supplements block the response of HDL to simvastatin-niacin therapy in patients with coronary artery disease and low HDL. *Arterioscler Thromb Vasc Biol*, **21**, 1320–1326.

109 Mooradian, A. D., Haas, M. J., Wadud, K. (2006) Ascorbic acid and α-tocopherol down regulate apolipoprotein AI gene expression in HepG2 and Caco-2 cell lines. *Metabolism: Clinical and Experimental*, **55**, 159–167.

110 Wahlberg, G. and Walldius, G. (1982) Lack of effect of ascorbic acid on serum lipoprotein concentrations in patients with hypertriglyceridaemia. *Atherosclerosis.*, **43**, 283–288.

111 Buzzard, I. M. McRoberts, M. R. Driscoll, D. L. Bowering, J. (1982) Effect of dietary eggs and ascorbic acid on plasma lipid and lipoprotein cholesterol levels in healthy young men. *Am J Clinic Nutr*, **36**, 94–105.

112 Johnson, G. E. and Obenshain, S. S. (1981) Nonresponsiveness of serum high-density lipoprotein-cholesterol to high dose ascorbic acid administration in normal men. *Am J Clinic Nutr*, **34**, 2088–2091.

113 Shidfar, F. Keshavarz, A. Jallali, M. Miri, R. Eshraghian, M. (2003) Comparison of the effects of simultaneous administration of vitamin C and omega-3 fatty acids on lipoproteins, apo A-I, apo B, and malondialdehyde in hyperlipidemic patients. *Int J Vit & Nutr Res*, **73**, 163–170.

114 Auwerx, J., Bouillon, R., Kesteloot, H. (1992) Relation between 25-hydroxyvitamin D3, apoprotein A-I, and high-density lipoprotein cholesterol. *Arterioscler Throm*, **12**, 671–674.

115 Heikkinen, A-M., Tuppurainen, M. T., Niskanen, L., Komulainen, M., Penttila, I., Saarikoski, S. (1997) Long-term vitamin D3 supplementation may have adverse effects on serum lipids during postmenopausal hormone replacement therapy. *Eur J Endocrinol*, **137**, 495–502.

116 Takeo, S., Anan, M., Fujioka, K., Kajihara, T., Hiraga, S., Miyake, K., Tanonaka, K., Minematsu, R., Mori, H., Taniguchi, Y. (1989) Functional changes of aorta with massive accumulation of calcium. *Atheroscler*, **77**, 175–181.

117 Wehmeier, K., Beers, A., Haas, M. J., Wong, N. C. W., Steinmeyer, A., Zugel, U., Mooradian, A. D. (2005) Inhibition of apolipoprotein AI gene expression by 1, 25-dihydroxyvitamin D3. *Biochim. Biophys. Acta - Molecular and Cell Biology of Lipids.*, **1737**, 16–26.

118 Zolfaghari, R. and Ross, A. C. (1994) Effect of vitamin A deficiency and retinoic acid repletion on intestinal and hepatic apolipoprotein A-I mRNA levels of adult rats. *J Lipid Res*, **35**, 1985–1992.

119 Nagasaki, A., Kikuchi, T., Kurata, K., Masushige, S., Hasegawa, T., Kato, S. (1994) Vitamin A regulates the expression of apolipoprotein AI and CIII genes in the rat. *Biochem Biophys Res Commun*, **205**, 1510–1517.

120 Berthou, L., Langouet, S., Grude, P., Denefle, P., Branellec, D., Guillouzo, A. (1998) Negative regulation of Apo A-I gene expression by retinoic acid in rat hepatocytes maintained in a coculture system. *Biochim Biophys Acta*, **1391**, 329–336.

121 Graham, T. E., Yang, Q., Bluher, M., Hammarstedt, A., Ciaraldi, T. P., Henry, R. R., Wason, C. J., Oberbach, A., Jansson, P. A., Smith, U., Kahn,

B. B. (2006) Retinol-binding protein 4 and insulin resistance in lean, obese, and diabetic subjects. *N Engl J Med.*, **354**, 2552–63.

122 Zhang, J. J., Wang, Y., Lei, K. Y. (1995) Apolipoprotein A-I synthesis and secretion are increased in Hep G2 cells depleted of copper by cupruretic tetramine. *J Nutr.*, **125**, 172–182.

123 Hoogeveen, R. C., Reaves, S. K., Lei, K. Y. (1995) Copper deficiency increases hepatic apolipoprotein A-I synthesis and secretion but does not alter hepatic total cellular apolipoprotein A-I mRNA abundance in rats. *J Nutr,* **125**, 2935–2944.

124 Wu, J. Y., Reaves, S. K., Wang, Y. R., Wu, Y., Lei, P. P., Lei, K. Y. (1998) Zinc deficiency decreases plasma level and hepatic mRNA abundance of apolipoprotein A-I in rats and hamsters. *Am J Physiol,* **275** (6 Pt 1), C1516–C1525.

125 Wu, J. Y., Zhang, J. J., Wang, Y., Reaves, S. K., Wang, Y. R., Lei, P. P., Lei, K. Y. (1997) Regulation of apolipoprotein A-I gene expression in Hep G2 cells depleted of Cu by cupruretic tetramine. *Am J Physiol,* **273** (4 Pt 1), C1362–C1370.

126 Wu, J. Y., Wu, Y., Reaves, S. K., Wang, Y. R., Lei, P. P., Lei, K. Y. (1999) Apolipoprotein A-I gene expression is regulated by cellular zinc status in hep G2 cells. *Am. J. Physiol.*, **277**, C537–C544.

127 Haas, M. J., Sawaf, R., Horani, M. H., Gobal, F., Wong, N.C.W., Mooradian, A. D. (2003) Effect of chromium on apolipoprotein A-I expression in HepG2 cells. *Nutrition*, **19**, 353–357.

128 Branchi, A., Rovellini, A., Tomella, C., Sciariada, L., Torri, A., Molgora, M., Sommariva, D. (1997) Association of alcohol consumption with HDL subpopulations defined by apolipoprotein A-I and apolipoprotein A-II content. *Eur J Clin Nutr,* **51**, 362–365.

129 Baer, D. J., Judd, J. T., Clevidence, B. A., Muesing, R. A., Campbell, W. S., Brown, E. D., Taylor, P. R. (2002) Moderate alcohol consumption lowers risk factors for cardiovascular disease in postmenopausal women fed a controlled diet. *Am J Clin Nutr,* **75**, 593–599.

130 Rimm, E. B.. Williams, P., Fosher, K., Criqui, M., Stampfer, M. J. (1999) Moderate alcohol intake and lower risk of coronary heart disease: meta-analysis of effects on lipids and haemostatic factors. *BMJ,* **319**, 1523–1528.

131 van der Gaag, M. S., van Tol, A., Vermunt, S. H., Scheek, L. M., Schaafsma, G., Hendriks, H.F.J. (2001) Alcohol consumption stimulates early steps in reverse cholesterol transport. *J Lipid Res,* **42**, 2077–83.

132 Tam, S. P. (1992) Effect of ethanol on lipoprotein secretion in two human hepatoma cell lines, HepG2 and Hep3B. *Alcohol Clin Exp Res,* **16**, 1021–1028.

17
Nutritional Factors and High-Density Lipoprotein Metabolism

Ernst J. Schaefer, Stefania Lamon-Fava, Bela F. Asztalos

17.1
Introduction

Low levels of plasma high-density lipoprotein (HDL) cholesterol ($<40\,mg\,dL^{-1}$ in men and $<50\,mg\,dL^{-1}$ in women) have been shown to be an important independent risk factor for coronary heart disease (CHD) [1]. Raising HDL cholesterol with resins, fibrates, statins, and niacin has been associated with decreased CHD risk and reduced CHD progression in randomized trials [2–6]. HDL are a polydisperse collection of lipoproteins in the plasma density range of $1.063–1.21\,g\,mL^{-1}$. By weight, HDL contains approximately 50 % protein, 25 % phospholipids, 20 % cholesterol (mainly esterified), and 5 % triglyceride [7]. The major proteins of HDL are apolipoproteins (apo) A-I and A-II, present in an approximately 3:1 concentration ratio, while minor protein constituents include apoA-IV, apoA-V, apoC-I, apoC-II, apoC-III, apoD, apoE, apoJ, and amyloid A protein. Levels of HDL apoA-I are regulated by alterations in both production and clearance, with enhanced fractional clearance being the most common cause of low HDL cholesterol and apoA-I, usually together with increased levels of triglyceride-rich lipoproteins [7–10].

An overview of plasma lipoprotein metabolism, based on human stable isotope methodology using endogenous labeling of apolipoproteins with deuterated leucine in the fed state carried out in our laboratory [11], is shown in Fig. 17.1. Intestinal chylomicron apoB-48 enters plasma with a secretion rate of $1–2\,mg\,kg^{-1}\,day^{-1}$, and after lipolysis due to lipoprotein lipase (LPL) and hepatic lipase (HL), chylomicrons pick up apoE and cholesteryl ester in exchange for triglyceride through cholesteryl ester transfer protein (CETP) from HDL. Chylomicron apoB-48 has a plasma residence time of 4–5 hours, and is cleared by apoE-mediated liver receptor uptake. HDL-derived apoA-I, apoA-IV, C apolipoproteins, and apoE can recycle between TRL and HDL multiple times [11–16]. In the case of apoA-I there are on average 20 recyclings between TRL apoB-48 particles and HDL, while for apoA-IV there are on average about 8 recyclings [15,16]. The transfer of apoA-I and apoA-IV from TRL apoB-48 particles to HDL occurs rapidly in plasma as TRL apoB-48 particles undergo lipolysis [12,13]. In our view the transfer of apoA-I and apoA-IV from HDL back to

High-Density Lipoproteins: From Basic Biology to Clinical Aspects. Edited by Christopher J. Fielding

ISBN: 978-3-527-31717-2

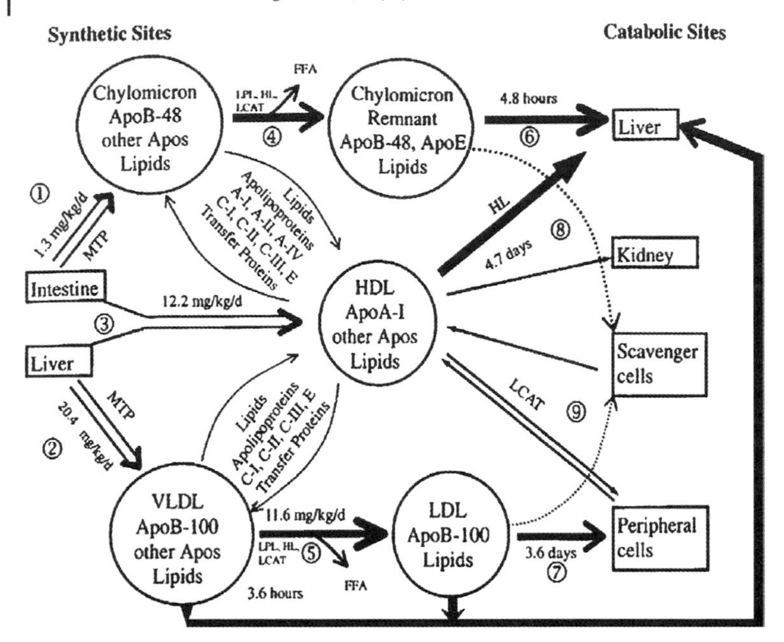

Fig. 17.1 An overview of lipoprotein metabolism based on stable isotope studies in humans in the fed state [11]. Abbreviations: Apolipoproteins: apo. CETP: cholesteryl ester transfer protein. FFA: free fatty acids. HL: hepatic lipase. LCAT: lecithin:cholesterol acyltransferase. LPL: lipoprotein lipase. MTP: microsomal transfer protein. Open arrows indicate synthesis pathways; closed arrows indicate catabolic pathways.

newly synthesized TRL apoB-48 particles occurs in the extra-plasma space, mainly the lymph [11–16].

Hepatic VLDL apoB-100 has a secretion rate of about 20 mg kg^{-1} day^{-1} in the fed state, and as VLDL undergoes lipolysis, it loses much of its triglyceride, and picks up cholesteryl ester and apolipoproteins from HDL in exchange for triglyceride [17]. ApoE and the C apolipoproteins can recycle between newly formed TRL and HDL particles multiple times. In the fed state, about 50 % of VLDL apoB-100 is converted into LDL, which takes about 3–4 hours, while the remainder is directly catabolized. VLDL apoB-100 synthesis is increased by age, obesity, insulin resistance, and diabetes. LDL apoB-100 is cleared from the plasma with a residence time of about 3 days [11]. LDL apoB-100 fractional clearance is decreased by aging, dietary cholesterol, and dietary saturated fat [18–20]. Conversely it is increased by restriction of saturated fat and cholesterol, in exchange for polyunsaturated fat [11,20].

Apo(a) made in the liver binds to newly formed VLDL apoB-100 and can either become lipoprotein (a) in the LDL–HDL density region (1.019–1.125 g mL^{-1}) or be cleaved off from VLDL that is directly catabolized and can reassociate with newly formed VLDL apoB-100. Its synthesis is about 0.5 mg kg^{-1} day^{-1}, and its residence is about 5 days [21]. On average apo(a) recycles twice on an apoB-100 TRL particle, and in plasma it exists mainly as Lp(a) in the LDL density region. While Lp(a) and apo(a) levels are elevated due to increased secretion, patients with increased LDL levels due to delayed clearance of LDL also have Lp(a) elevations, due to delayed apo(a) fractional catabolism [21–23].

HDL apoA-I is synthesized in both the liver and intestine and has a secretion rate of approximately 12 mg kg^{-1} day^{-1} and a plasma residence time of 4–5 days. In our view apoA-I enters plasma after being secreted in the liver and intestine mainly as small, hardly lipidated preβ-1 HDL [11,24–26]. Its metabolism is more fully discussed below.

Common genetic lipoprotein disorders associated with premature coronary heart disease and low HDL cholesterol (<40 mg dL^{-1}) include familial combined hyperlipidemia (elevated low density lipoproteins and TRL) and familial dyslipidemia (elevated TRL), each observed in about 15 % of families, while isolated low HDL (familial hypoalphalipoproteinemia) is less commonly observed (about 5 %) [27]. Subjects selected for CHD and low HDL are often overweight or obese and usually have elevated insulin levels [28].

HDL is generally assessed in plasma or serum by measuring HDL cholesterol by standardized automated enzymatic methods after removal of other lipoprotein particles. An HDL cholesterol level of less than 40 mg dL^{-1} in men and less than 50 mg dL^{-1} in women has been defined as decreased by the third Adult Treatment Panel of the National Cholesterol Education Program, and serves as one of the five criteria for the metabolic syndrome [1]. HDL can also be assessed plasma or serum apoA-I by immunoassay. After puberty, women have higher concentrations of plasma apoA-I than men, due to increased estrogen-mediated production [7,29,30]. At the time of menopause, HDL apoA-I remains unchanged in women because of loss of both estrogen and progestin [31]. Hormonal replacement with estrogen increases HDL apoA-I production, while progestin treatment has an opposing effect by decreasing production [30]. Anabolic steroid treatment in men also lowers HDL by decreasing HDL apoA-I production, as well as enhancing clearance [32]. Subjects with hypertriglyceridemia have decreased levels of HDL, which is triglyceride-rich and cholesteryl ester poor [7–10]. In this setting there is enhanced transfer of cholesteryl ester to triglyceride-rich lipoproteins (TRL) in exchange for triglyceride, and enhanced HDL apoA-I fractional catabolism [7–10]. In contrast, patients with isolated low HDL tend to have decreased apoA-I production [10].

17.2
High Density Lipoprotein Particles and their Metabolism

Methodology for examining HDL particles includes measurement of HDL_3 cholesterol after precipitation of apoB-containing lipoproteins and HDL_2, gradient gel electrophoresis, nuclear magnetic resonance, and two-dimensional gel electrophoresis. This last technology was originally developed by Fielding and colleagues [25,26], and the method was further refined by Asztalos and Roheim and has been applied to samples from populations and a variety of dyslipidemic states by Asztalos and Schaefer (see Fig. 17.2) [33–36]. HDL in whole plasma or

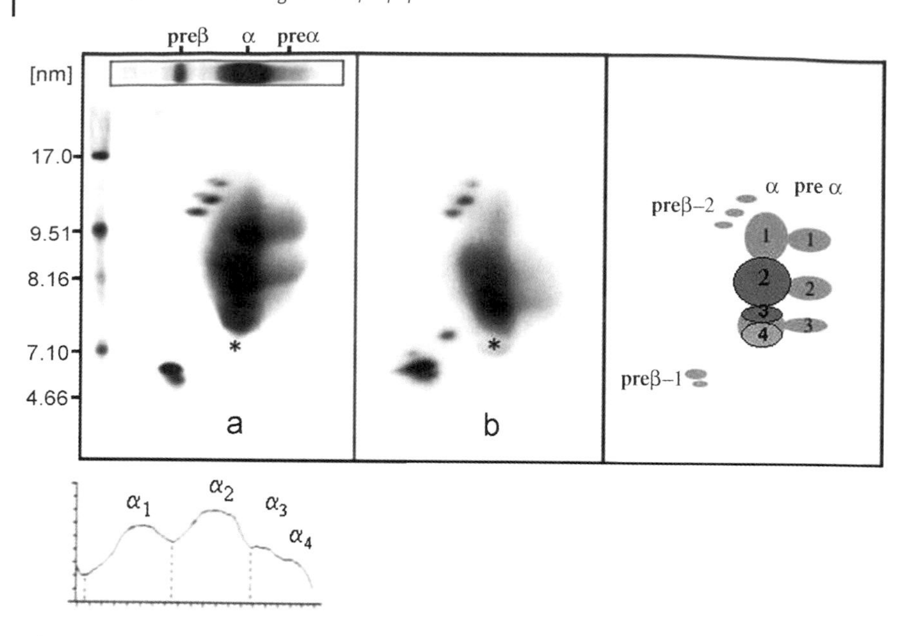

Fig. 17.2 Two-dimensional gel patterns of apoA-I-containing HDL particles as detected in whole plasma as shown in a) a normal subject, and b) a patient with premature coronary artery disease with modest increases in preβ-1 HDL and decreases in α-1 and α-2 HDL and preα-1 and preα-2 HDL particles. Locations of apoA-I containing HDL particles are shown (c). Only α-2 and α-3 HDL particles contain both apoA-I and apoA-II.

serum can be examined by non-denaturating two-dimensional gel electrophoresis followed by immunoblotting with apoA-I antibody (see Fig. 17.2). This technique separates HDL by size in the vertical dimension and by charge in the horizontal dimension. The major apoA-I-containing HDL subspecies have α mobility on electrophoresis, ranging from large spherical (or α-1) to intermediate α-2 and α-3 particles also containing apoA-II, to small, discoidal α-4 HDL particles (containing apoA-I without apoA-II). Adjacent to these α particles are corresponding preα HDL particles, containing apoA-I, but no apoA-II. In addition, there are small, discoidal preβ-1 and large preβ-2 migrating apoA-I-containing HDL (see Fig. 17.2). In collaboration with Dr. George Rothblat's group, we have documented that both preβ-1 and α-2 HDL particles serve as acceptors of free cholesterol by the ATP binding cassette protein A1 (ABCA-I) pathway, while the large α-1 and α-2 particles interact with SRB1 to promote cholesterol efflux from liver cells [37]. Since this is a bidirectional process we hypothesize that these particles can also serve as very efficient cholesteryl ester donors.

Research in patients with rare inborn errors of HDL metabolism has given us a framework for HDL particle metabolism (see Fig. 17.3) [24]. The first step in HDL metabolism is apoA-I secretion by the liver and intestine, and its binding to phospholipids to form small discoidal preβ-1 HDL. When no apoA-I is secreted as a result of rare mutations affecting the apoA-I gene, no apoA-I-containing HDL is

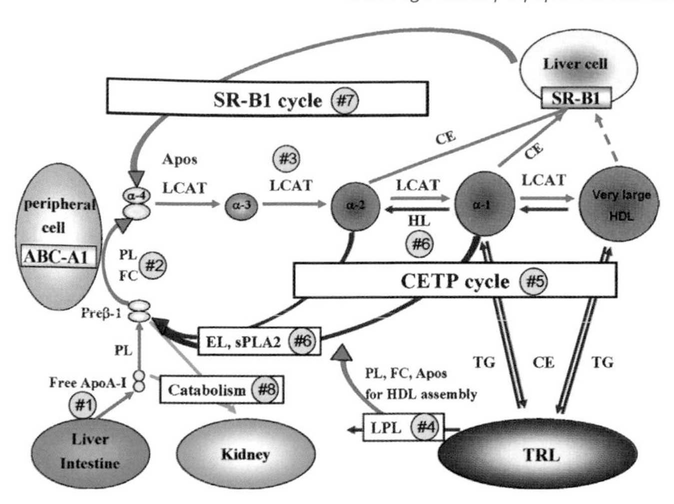

Fig. 17.3 An overview of HDL particle metabolism based on studies using the two-dimensional gel technique shown in Figure 17.2 in normal subjects, in patients with premature CHD, and in patients with rare inborn errors of metabolism including apoA-I deficiency, Tangier disease, LCAT deficiency, LPL deficiency, HL deficiency, and CETP deficiency [24,34–55].

present in plasma, HDL cholesterol is less than 5 mg dL^{-1}, triglyceride and LDL cholesterol levels are normal, and severe premature coronary artery disease occurs [38–43]. The second step in HDL metabolism is the efflux of cellular free cholesterol and phospholipids via ABCA1 to convert pre-β1 HDL into small discoidal α-4 HDL. In Tangier patients with defective ABCAI-mediated cellular cholesterol efflux, only preβ-1 HDL is present in plasma, and this is accompanied by mild hypertriglyceridemia and low LDL cholesterol, cholesterol ester deposition in macrophages in the liver, spleen, bone marrow, and other tissues, and significant premature CAD [44–53]. While intestinal cholesterol absorption is normal in these patients, their LDL is about 50 % of normal, because of lack of cholesterol transfer from HDL to apoB-containing lipoproteins, and the fact that the resultant small, cholesterol-poor, and triglyceride-rich LDL is hypercatabolized [44–53].

The third step in HDL metabolism is the esterification of free cholesterol (an HDL surface component) with the transfer of a fatty acid from the phospholipid lecithin or phosphatidylcholine to free cholesterol through the action of lecithin:cholesterol acyltransferase (LCAT). This process results in the movement of cholesteryl ester to the HDL core, with the formation of intermediate, large, and very large spherical αHDL particles of α-3, α-2, and α-1 type and even larger mobility. Adjacent to each αHDL particle is a preα HDL particle that in similar to its corresponding α particle, except that it lacks apoA-II, due to apoA-II being removed, and combining with newly formed α-3 and α-2 HDL particles. In the absence of LCAT only preβ-1 and discoidal α-4 HDL particles are present in plasma, and apoA-I and apoA-II are not observed on the same HDL particle [54].

The fourth step in HDL metabolism is the lipolysis of triglyceride on all lipoprotein particles, especially TRL, by lipoprotein lipase. This process results in the transfer of surface components of TRL to HDL, namely the C apolipoproteins and apoE, as well as phospholipids and free cholesterol. In the absence of lipoprotein lipase, severe hypertriglyceridemia is observed, only preβ-1 and α-4 HDL particles are present, and apoA-I and apoA-II are not seen on the same HDL particles.

The fifth step in HDL metabolism is the exchange of core cholesteryl ester in the core of larger αHDL particles (α-2, α-1, and very large HDL) for triglyceride from TRL through the action of cholesteryl ester transfer protein (CETP), which is essential for the formation of α-1 HDL and preα-1 HDL particles containing apoA-I without apoA-II. During this process some apoA-I is shed from HDL and recycles back to preβ-HDL, while apoA-II is shed from HDL and recycles to combine with newly formed α-3 and α-2 HDL particles, while apoE is shed to form its own large preβ-HDL particles, which do not contain apoA-I or apoA-II. In the absence of CETP, HDL cholesterol is markedly elevated, and mainly very large spherical αHDL are present in plasma. These particles contain apoA-I, apoA-II, and apoE, and are not seen in normal plasma [55].

The sixth step in HDL metabolism represents another option for cholesteryl ester on large HDL particles of α-1 and α-2 mobility, and is a critical step in reverse cholesterol transport. This is the selective bidirectional transport of cholesteryl ester between HDL particles and the liver through the action of scavenger receptor B1 (SR-B1) for ultimate excretion of cholesterol into the bile. During this process apoA-I is shed from HDL to form small discoidal α-4 HDL particles. Humans with SR-B1 deficiency have not been described, but we hypothesize that such patients exist, have premature CHD, markedly elevated HDL cholesterol, and an excess of α-1 and α-2 HDL particles.

The seventh step in HDL metabolism involves the lipolysis of phospholipids on HDL mediated by hepatic lipase, endothelial lipase, and secretory phospholipase with the conversion of α-1 to α-2 HDL particles. In the absence of hepatic lipase there is a lack of α-2 HDL particle formation.

The eighth step in HDL metabolism is the rapid clearance of preβ-1 HDL through the kidney, instead of its picking up free cholesterol from cells to become α-4 HDL. This process is markedly accentuated in Tangier disease (ABCA1 deficiency) [44,46]. HDL apoA-I and apoA-II may also be directly catabolized by the liver [24].

In both the Framingham Offspring Study and the Veterans Affairs High Density Lipoprotein Intervention Trial (VA-HIT) we have clearly documented that low levels of α-1 and α-2 HDL apoA-I levels and elevated levels of pre-β-1 HDL are much better predictors than HDL cholesterol of the presence of CHD [34,35]. We have also collaborated with Dr. Greg Brown to examine samples from the HDL Atherosclerosis Treatment Study (HATS), and have documented that the substantial increases (mean 115 %) in the large α-1 particles with the niacin/simvastatin combination were correlated with less progression or more regression of coronary atherosclerosis in the study [36]. Niacin is currently the most effective agent for raising HDL cholesterol by 20–30 %, by decreasing HDL apoA-I fractional clearance. Use of niacin has been associated with CHD risk reduction, and a large-scale clinical trial is currently underway in CHD patients with low HDL in order to determine whether niacin treatment adds benefit above and beyond that seen with statin alone. The use of niacin has been limited by its side effect profile, especially flushing.

Inhibition of cholesterol ester transfer activity with available inhibitors (JTT 705 and torcetrapib) can raise HDL cholesterol by 50 % or more, and is well tolerated [24,56–58]. Torcetrapib raises HDL cholesterol by preventing transfer of its cholesteryl ester to TRL, resulting in delayed apoA-I fraction catabolism, as well as enhanced fractional clearance of apoB-100 within VLDL, IDL, and LDL [57,58]. Torcetrapib can normalize HDL particles in patients with HDL deficiency and raise α-1 HDL by 150 % [24,57]. Clinical outcome studies to define the potential clinical benefit are underway. These studies were terminated in early December of 2006 of excess mortality and congestive heart failure. This drug was found to form a "non productive complex" with HDL and CETP and to raise systolic blood pressure, in some cases, significantly. Other CETP inhibitors do not have this effect, and are still being tested.

17.3 Effects of Dietary Cholesterol

Stucchi et al. and Hennessy et al. have examined the effects of isolated increased dietary cholesterol in cebus monkeys [19,59]. They documented that high cholesterol intake was associated with significant increases, relative to low-cholesterol diets, in: 1) hepatic cholesterol content, 2) hepatic apoA-I mRNA levels, 3) apoA-I production as assessed by tracer studies, and 4) HDL cholesterol and apoA-I levels in plasma [18,59]. In human studies, Lichtenstein et al. have documented that increasing dietary cholesterol from less than 200 mg day^{-1} to over 400 mg day^{-1} results in modest, but significant increases in both LDL cholesterol and HDL cholesterol, by about 10 % in each case [60]. Studies by Velez-Carrasco have shown that diets restricted in cholesterol and saturated fat in humans are associated with significant reductions in HDL cholesterol and apoA-I levels in plasma, due to decreased apoA-I production rates [61]. The overall data are consistent with the concept that dietary cholesterol increases apoA-I secretion by directly stimulating increases in hepatic apoA-I mRNA levels, which in turn were linked to increased hepatic cellular cholesteryl ester content [18,59–61]. These data suggest that there is upregulation of hepatic apoA-I mRNA when there is an increased need for cellular cholesterol efflux.

17.4 Effects of Dietary Fatty Acids

Schaefer et al. reported marked reductions in both LDL cholesterol and HDL cholesterol in humans when saturated fat was replaced by polyunsaturated fat [62]. Lichtenstein et al. reported similar results when saturated fat-rich oil was replaced by polyunsaturated fat-rich oil in humans [63]. Brousseau et al. examined the effects of diets high in saturated fat, monounsaturated fat, and polyunsaturated fat (constant dietary cholesterol content) on plasma HDL cholesterol and apoA-I levels, hepatic apoA-I mRNA levels, and plasma apoA-I kinetics in cynomolgus monkeys [64,65]. HDL cholesterol and apoA-I levels were lowest on the high polyunsaturated fat diet and highest on the saturated fat diet, with intermediate levels on the high monounsaturated fat diet. There were no differences in hepatic cholesterol or apoA-I mRNA levels, and kinetic studies revealed significant differences in plasma apoA-I fractional catabolic rates, accounting for these differences [64,65]. The data discussed pertain mainly to the *n*6 fatty acids, particularly linoleic acid (18:2*n*6) and arachidonic acid (20:4*n*6)

Spady et al. have examined the effects of a control chow diet, a high saturated fat (coconut) diet, a high polyunsaturated fat (safflower oil) diet, and then the same diets, but also high in cholesterol (0.08 %), in hamsters [66]. Animals on the high polyunsaturated fat diet had significantly lower HDL cholesterol levels and about 50 % higher liver SR-BI mRNA and protein levels, higher delivery of HDL cholesteryl esters to the liver, and higher liver cholesterol content than animals on the high saturated fat diets. There were no significant differences in extrahepatic cholesterol content [66]. Their major conclusion was that substitution of polyunsaturated fatty acids for saturated fat lowered HDL cholesterol, and increased HDL cholesterol delivery to the liver and liver cholesterol content, but had no effect on overall reverse cholesterol transport in this animal model under the conditions tested [66]. The overall data are consistent with the concept that replacing saturated fatty acids with ω-6 polyunsaturated fatty acids in the diet lowers HDL levels, by upregulating HDL apoA-I fractional catabolic rate and SR-B1 gene expression, which may, along with LDL reduction, account for the beneficial effects of polyunsaturated fats in CHD risk reduction [67–69].

The effects of ω-3 fatty acids on HDL metabolism appear to be somewhat different from those of *n*6 fatty acids. Frenais et al. reported in six diabetic subjects that six one gram ω-3 fatty acid capsules (MaxEpa) per day significantly lowered plasma triglyceride levels, with no effect on apoA-I or apoA-II plasma levels, but with significant reductions both in the production of HDL apoA-I and apoA-II, as well as in their fractional catabolism [70]. Chan et al. reported similar findings for fish oil in men with increased waist circumference studied at baseline and then on placebo ($n = 12$), fish oil (Omacor 4 capsules day^{-1}) ($n = 12$), atorvastatin (40 mg day^{-1}) ($n = 13$), and on the combination ($n = 11$) [71]. Atorvastatin treatment had no effect on the pool size, production, or fractional clearance of HDL apoA-I and apoA-II. These data indicate that, in contrast to *n*6 fatty acids, which enhance HDL apoA-I fractional clearance, *n*3 fatty acids have the opposite effect of delaying fractional clearance, but also lowering apoA-I production.

17.5 Effects of Exchange of Dietary Fat and Carbohydrate

Schaefer et al. examined the effects of a virtually fat-free, high-carbohydrate diet in humans relative to an average American diet in humans, and reported that the low-fat, high-carbohydrate diet lowered both LDL cholesterol and HDL cholesterol, each by approximately 30 % [62]. Schaefer et al. subsequently reported similar finding with diets restricted in total fat, saturated fat, and cholesterol, even along with weight loss [72–75]. Dansinger et al. have reported that calory-restricted diets for promotion of weight loss will raise HDL cholesterol provided that there is carbohydrate restriction [76].

Blum et al. carried out kinetic studies in humans and reported that high-carbohydrate, low-fat diets reduced HDL cholesterol and apoA-I levels significantly, in association with significant increases in the HDL apoA-I fractional catabolic rate, with no effect on HDL apoA-I production [77]. Brinton et al., in contrast, reported that low-fat, high-carbohydrate diets lowered HDL apoA-I by decreasing production [78], though that diet was also low in cholesterol [80]. Velez-Carrasco et al. reported that a National Cholesterol Education Program step 2 diet containing less than 200 mg day^{-1} of cholesterol, less than 7 % of calories as saturated fat, and less than 30 % of calories as total fat, significantly lowered plasma apoA-I levels relative to an average American diet (higher in cholesterol, saturated fat, and total fat), due to decreased production [61].

Woolett et al. examined the effects of control chow diets, diets high in carbohydrate and low in cholesterol and saturated fat, the same diets supplemented with fiber, as well as a diet high in saturated fat and cholesterol (Western diet) in hamsters [79,80]. In these studies the Western diet was associated with 46 % higher plasma HDL cholesteryl esters and 25 % higher apoA-I levels than the control diet, and 86 % and 45 % higher values than the high-fiber diet. These differences were all due to differences in HDL cholesteryl ester and apoA-I entry into plasma versus differences in plasma clearance rates [80]. Woolett et al. concluded that the dietary modifications studied altered plasma HDL cholesterol levels through changes in production, and not through alterations in the transport of HDL cholesteryl esters to the liver [80]. The combined data suggest that when there is no change in dietary cholesterol or fatty acid content in the diet, and when a substantial amount of fat in the diet is replaced isocalorically by carbohydrate, there are usually elevations in TRL, while HDL apoA-I decreases due to enhanced fractional catabolism, associated with increased HDL cholesteryl ester transfer to TRL. These data are consistent with observations that HDL apoA-I fractional catabolic rates are increased in patients with hypertriglyceridemia [7–9]. In contrast, when there is only modest replacement of fat with carbohydrate, and significant restriction of dietary cholesterol and saturated fat, the primary mechanism by which plasma apoA-I is reduced is by decreased apoA-I production, with little or no change in TRL concentration. This effect is probably mediated by less need for reverse cellular cholesterol efflux in the context of dietary cholesterol restriction.

17.6 Effects of Diets Restricted in Cholesterol and Saturated Fat

Schaefer et al. has reported that diets restricted in cholesterol and saturated fat in humans reduce HDL cholesterol almost as much as LDL cholesterol [72–74]. Velez Carrasco et al. related such changes to decreased HDL apoA-I production, as did Brinton et al. [61,78], while Asztalos et al. reported that such diets were associated with reductions in α-1 and α-2 HDL apoA-I concentrations only in those with normal HDL cholesterol concentrations, and not those with decreased HDL cholesterol values [81]. As previously mentioned, the plasma apoA-I levels in this context are reduced because of decreased apoA-I production, and it may be that subjects with normal levels of HDL on an atherogenic diet have been able to upregulate apoA-I production, while those with low HDL are unable to do so [81].

17.7 Effects of Alcohol Intake

It is well known that alcohol intake increases HDL cholesterol and apoA-I levels [82,83]. Clevidence has reported increases in both HDL2 and HDL3 cholesterol, as well as increases in HDL particles containing both apoA-I and apoA-II, and also in particles containing apoA-I without apoA-II [82]. De Oliveira e Silva et al. have reported that alcohol (vodka, mean 0.45 $ml\,kg^{-1}\,day^{-1}$) increased HDL cholesterol, apoA-I, and apoA-II plasma mass by 18 %, 10 %, and 17 %, respectively, in association with significant increases in apoA-I and apoA-II production rates of 21 % and 17 %, respectively [83]. These effects may account for the cardioprotective effect of alcohol intake when used in moderation [84–86].

17.8 Effects of Weight Loss

Weight loss can have variable effects on HDL cholesterol and apoA-I levels. If there is no change in dietary composition, and only restriction in calories, HDL cholesterol and apoA-I levels decrease during the hypocaloric phase, and then increase significantly above the baseline value when the subjects are in a new steady state with substantially lower body weight [75,76]. These effects are generally associated with lower plasma TRL, less HDL cholesteryl ester transfer to TRL, and delayed HDL apoA-I fractional clearance [7,57]. Dattilo et al. have developed precise equations for predicting the rise in HDL cholesterol based on weight reduction, once a new lower stable body weight has been achieved [87].

17.9 Effects of Exercise

Exercise has variable effects on HDL cholesterol and apoA-I levels. If there is no significant change in body composition or weight, increased exercise has little effect on HDL levels [88–91]. However, if the exercise intervention is substantial and prolonged, and if there is associated weight loss, with increased muscle mass and decreased body fat, HDL cholesterol levels can increase quite significantly, along with lowering of TRL, reduction in HDL cholesteryl ester transfer to TRL, and a decrease in HDL apoA-I fractional catabolic rate [90]. Intensive aerobic exercise over seven years of follow-up has been associated with significant benefit in terms of heart disease morbidity and mortality in patients sustaining a myocardial infarction [91].

17.10 Conclusions

Analysis of HDL subspecies by non-denaturing two-dimensional gel electrophoresis followed by immunoblotting for apoA-I and quantitation can provide substantial information about CHD risk, as well as important clues as to metabolic defects in various rare familial HDL deficiency or HDL excess states [24]. Increasing efforts are now being directed at the development of better strategies to raise HDL and to reduce CHD risk, including better niacin preparations and inhibitors of cholesteryl ester transfer protein [6,24,36,56]. Dietary restriction of saturated fat and its replacement with polyunsaturated fat rich in polyunsaturated fatty acids remains the cornerstone of CHD risk reduction and disease prevention, and such interventions have been shown to reduce CHD morbidity and mortality, despite HDL cholesterol lowering [67–69]. Lowering dietary cholesterol is also important in this regard, even though this intervention reduces apoA-I production [59,61]. In contrast, exchanging saturated fat with polyunsaturated fat lowers HDL cholesterol by enhancing HDL cholesteryl ester delivery to the liver by upregulation of hepatic SR-BI mRNA and protein [64,66]. Replacement of total fat with carbohydrate also enhances apo-A-I fractional catabolism [77]. Weight loss with calorie restriction and exercise will raise HDL cholesterol and apoA-I significantly, because of lowering of TRL, reduction of transfer of HDL cholesteryl ester to TRL, and delayed HDL apoA-I fractional catabolism. Alcohol intake raises HDL cholesterol and apoA-I by increasing apoA-I production [83], with moderate alcohol intake being associated with decreased CHD risk [84–86]. The most effective strategies for raising HDL cholesterol and apoA-I are caloric restriction and weight loss if indicated, restriction of sugars and refined high glycemic index carbohydrates, and increased physical activity, as well as moderate alcohol intake. While restriction of dietary saturated fat and cholesterol, as well as increased intake of essential fatty acids, will lower HDL cholesterol, such interventions have been shown to reduce CHD morbidity and mortality substantially. Equations to quantitate dietary effects on HDL cholesterol have been developed, but there is a wide variability in response [92,93]. Moreover, those individuals who cannot increase their HDL levels on an atherogenic diet may be at the greatest risk [81].

References

1 Expert Panel. (2001) Executive summary of the third report of the National Cholesterol Education Program (NCEP) Expert Panel on Detection, Evaluation, and Treatment of High Blood Cholesterol in Adults (Adult Treatment Panel III). *J. Am. Med. Assoc.*, **285**, 2486–2497.

2 The Lipid Clinics Coronary Primary Prevention Trial results: II. (1984) The relationship of reduction in incidence of coronary heart disease to cholesterol

lowering. *J. Am. Med. Assoc.*, **251**, 365–374.

3 Manninen, V., Elo, O., Frick, H. H. *et al.* (1988) Lipid alterations and decline in the incidence of coronary heart disease in the Helsinki Heart Study. *J. Am. Med. Assoc.*, **260**, 641–651.

4 Robins, S. J., Collins, D., Wittes, J. T., Papademetriou, V., Deedwania, P. C., Schaefer, E. J., McNamara, J. R., Kashyap, M. L., Hershman, J. M., Wexler, L. F., Rubins, H. B. (2001) for the VA-HIT Study Group.Relation of gemfibrozil treatment and lipid levels with major coronary events. VA-HIT: A randomized controlled trial. *J. Am. Med. Assoc.*, **285**, 1585–91.

5 Pedersen, T. R., Olsson, A. G., Faergeman, O. *et al.* (1998) Lipoprotein changes and reduction: the incidence of major coronary heart disease events in the Scandinavian Simvastatin Survival Study (4S). *Circulation*, **97**, 1453–1460.

6 Brown, G. B., Zhao, X. Q., Chait, A., Fisher, L. D., Cheung, M., Morse, J. S., Dowdy, A. A., Marino, E. K., Bolson, E. L., Alaupovic, P., Frohlich, J., Albers, J. J. (2001) Simvastatin and niacin, antioxidant vitamins, or the combination for the prevention of coronary artery disease. *N. Engl. J. Med.*, **345**, 1583–1592.

7 Schaefer, E. J., Zech, L. A., Jenkins, L. L., Aamodt, R. A., Bronzert, T. J., Rubalcaba, E. A., Lindgren, F. T., Brewer, H. B.Jr: (1982) Human apolipoprotein A-I and A-II metabolism. *J. Lipid Res.*, **23**, 850–862.

8 Schaefer, E. J., Ordovas, J. M., (1986) Metabolism of the apolipoproteins A-I, A-II, and A-IV. In *Methods in Enzymology, Plasma Lipoproteins, Part B: characterization, Cell Biology and Metabolism* (J. Segrest, J. Albers, eds.). Academic Press. **129**, 420–442.

9 Brinton, E. A., Eisenberg, S., Breslow, J. L. (1994) Human HDL cholesterol levels are determined by apoA-I fractional catabolic rate, which correlates inversely with estimates of HDL particle size: effects of gender, hepatic and lipoprotein lipases, triglyceride and insulin levels, and body fat distribution. *Arterioscler. Thromb.*, **14**, 707–720.

10 Velez-Carrasco, W., Lichtenstein, A. H., Welty, F. K., Li, Z., Lamon-Fava, S., Dolnikowski, G. G., Schaefer, E. J. (1999) Dietary restriction of saturated fat and cholesterol decreases HDL apoA-I secretion. *Arterioscler. Thromb. Vasc. Biol.*, **19**, 918–924.

11 Schaefer, E. J., McCollum, E. V. (2002) Award Lecture: Lipoproteins, nutrition, and heart disease. *Am. J. Clin. Nutr.* **75**, 191–212.

12 Schaefer, E. J., Jenkins, L. L., Brewer, H. B.Jr. (1978) Human chylomicron apolipoprotein metabolism. *Biochem. Biophys. Res. Commun.*, **80**, 405–412.

13 Schaefer, E. J., Wetzel, M. G., Bengtsson, G., Scow, R. O., Brewer, H. B.Jr, Olivercrona, T. (1982) Transfer of human lymph chylomicron constituents to other lipoprotein density fractions during in vitro lipolysis. *J. Lipid Res.*, **3**, 1259–1273.

14 Anderson, D. W., Schaefer, E. J., Bronzert, T. J., Lindgren, F. T., Forte, T., Starzl, T. B., Niblack, G. D., Zech, L. A., Brewer, H. B.Jr. (1981) Transport of apolipoprotein A-I and A-II by human thoracic duct lymph. *J. Clin. Invest.*, **67**, 857–866.

15 Velez-Carrasco, W., Lichtenstein, A. H., Barrett, P. H. R., Sun, Z., Dolnikowski, G. G., Welty, F. K., Schaefer, E. J. (1999) Human apolipoprotein A-I kinetics within triglyceride-rich lipoproteins and high density lipoproteins. *J. Lipid Res.*, **40**, 1695–1700.

16 Sun, Z., Lichtenstein, A. H., Dolnikowski, G. G., Welty, F. K., Schaefer, E. J. (2001) Human apolipoprotein A-IV metabolism within triglyceride-rich lipoproteins and plasma. *Atherosclerosis*, **156**, 363–372.

17 Welty, F. K., Lichtenstein, A. H., Barrett, P. H. R., Dolnikowski, G. G., Schaefer, E. J. (1999) Human apolipoprotein (Apo) B-48 and Apo B-100 kinetics with stable isotopes. *Arterioscler. Thromb. Vasc. Biol.*, **19**, 2966, 2974.

18 Nicolosi, R. J., Stucchi, A. F., Kowala, M. C., Hennessy, L. K., Hegsted, D. M., Schaefer, E. J. (1990) Effect of dietary fat saturation and cholesterol on low density lipoprotein composition and metabolism. I. In vivo studies of receptor and non-receptor mediated catabolism of LDL in Cebus monkeys. *Arteriosclerosis*, **10**, 119–128.

19 Hennessy, L. K., Osada, J., Ordovas, J. M., Nicolosi, R. J., Brousseau, M. E., Schaefer, E. J. (1992) Effects of dietary fatty acids and cholesterol on liver lipid content and hepatic apolipoprotein A-I, B and E and LDL receptor mRNA levels in Cebus monkeys. *J. Lipid Res.*, **33**, 351–360.

20 Millar, J. S., Lichtenstein, A. H., Cuchel, M., Dolnikowski, G. G., Hachey, D. L., Cohn, J. S., Schaefer, E. J. (1995) Impact of age on the metabolism of VLDL, IDL, and LDL apolipoprotein B-100. *J. Lipid Res.*, **36**, 1155–1167.

21 Jenner, J. L., Ordovas, J. M., Lamon-Fava, S., Schaefer, M. M., Wilson, P. W. F., Castelli, W. P., Schaefer, E. J. (1993) Effects of age, gender, and menopausal status on plasma lipoprotein (a) levels: The Framingham Offspring Study. *Circulation*, **87**, 135–1141.

22 Schaefer, E. J., Lamon-Fava, S., Jenner, J. L., Ordovas, J. M., Davis, C. E., Lippel, K., Levy, R. I. (1994) Lipoprotein(a) levels predict coronary heart disease in the lipid research clinics coronary prevention trial. *JAMA*, **271**, 999–1003.

23 Jenner, J. L., Seman, L. J., Millar, J. S., Lamon-Fava, S., Welty, F. K., Dolnikowski, G. G., Marcovina, S. M., Lichtenstein, A. H., Barrett, P. H. R., deLuca, C., Schaefer, E. J. (2005) The metabolism of apolipoproteins (a) and B-100 within plasma lipoprotein(a) in human beings. *Metabolism*, **54**, 361–369.

24 Schaefer, E. J. and Asztalos, B. F. (2006) Cholesteryl ester transfer protein inhibition, high-density lipoprotein metabolism, and heart disease risk reduction. *Current Opinion in Lipidology.*, **17**, 394–398.

25 Ishida, B. Y., Frohlich, J., Fielding, C. J. (1987) Pre-beta migrating HDL: quantitation in normal and hyperlipidemic plasma by solid phase radioimmunoassay following electrophoretic transfer. *J. Lipid Res.*, **28**, 778–786.

26 Chau, P., Nakamura, Y., Fielding, C. J., Fielding, P. E. (2006) Mechanisms of pre-beta HDL formation and activation. *Biochemistry*, **45**, 3981–3987.

27 Genest, J. J., Martin-Munley, S., McNamara, J. R., Ordovas, J. M., Jenner, J., Meyers, R., Wilson, P. W. F., Schaefer, E. J. (1992) Prevalence of familial lipoprotein disorders in patients with premature coronary artery disease. *Circulation*, **85**, 2025–2033.

28 Rubins, H. B., Robins, S. J., Collins, D., Nelson, D. B., Elam, M. B., Schaefer, E. J., Faas, F. H., Anderson, J. W. (2002) For the VA-HIT Study Group. Diabetes, plasma insulin, and cardiovascular disease. Subgroup Analysis From the Department of Veterans Affairs High-Density Lipoprotein Intervention Trial (VA-HIT). *Arch. Intern. Med.*, **162**, 2597–2604.

29 Schaefer, E. J., Foster, D. A., Zech, L. A., Brewer, H. B., Jr, Levy, R. I. (1983) The effect of estrogen administration on plasma lipoprotein metabolism in premenopausal females. *J. Clin. Endocr. Metab.*, **57**, 262–270.

30 Lamon-Fava, S., Postfai, B., Diffenderfer, M., DeLuca, C., O'Connor, J., Welty, F. K., Dolnikowski, G. G., Barrett, P. H. R., Schaefer, E. J. (2006) Role of estrogen and progestin in hormonal replacement therapy on apolipoprotein A-I kinetics in post-menopausal women. *Arterioscler. Thromb. Vasc. Biol.*, **26**, 385–391.

31 Schaefer, E. J., Lamon-Fava, S., Ordovas, J. M., Cohn, S. D., Schaefer, M. M., Castelli, W. P., Wilson, P. W. F. (1994) Factors associated with low and elevated plasma high density lipoprotein cholesterol and apolipoprotein A-I levels in the Framingham Offspring Study. *J. Lipid Res.*, **35**, 871–882.

32 Haffner, S. M., Kushwaha, R. S., Foster, D. M., Applebaum-Bowden, D., Hazzard, W. R. (1983) Studies on the metabolic mechanism of reduced HDL during anabolic steroid therapy. *Metabolism*, **32**, 413–420.

33 Asztalos, B. F., Sloop, C. H., Wong, L., Roheim, P. S. (1993) Two dimensional electrophoresis of plasma lipoproteins: recognition of new apoA-I containing subpopulations. *Biochim. Biophys. Acta*, **1169**, 291–300.

34 Asztalos, B. F., Cupples, L. A., Demissie, S., Horvath, K. V., Cox, C. E., Batista, M. C., Schaefer, E. J. (2004) High-density lipoprotein subpopulation profile and coronary heart disease prevalence in male participants in the Framingham Offspring Study. *Arterioscler. Thromb. Vasc. Biol.*, **24**, 2181–2187.

35 Asztalos, B. F., Collins, D., Cupples, L. A., Demissie, S., Horvath, K. V., Bloomfield, H. E., Robins, S. J., Schaefer, E. J. (2005) Value of high density lipoprotein (HDL) subpopulations in predicting recurrent cardiovascular events in the Veterans Affairs HDL Intervention Trial. *Arterioscler. Thromb. Vasc. Biol.*, **25**, 2185–2191.

36 Asztalos, B. F., Batista, M., Horvath, K. V., Cox, C. E., Dallal, G. E., Morse, J. S., Brown, G. B., Schaefer, E. J. (2003) Change in alpha 1 HDL concentration predicts progression in coronary artery stenosis. *Arterioscler. Thromb. Vasc. Biol.*, **23**, 847–852.

37 Asztalos, B. F., De la Llera-Moya, M., Dallal, G. E., Horvath, K. V., Schaefer, E. J., Rothblat, G. H. (2005) Differential effects of HDL subpopulations on cellular ABCA1 and SRB1-mediated cholesterol efflux. *J. Lipid Res.*, **46**, 2246–2253.

38 Schaefer, E. J., Heaton, W. H., Wetzel, M. G., Brewer, H. B., Jr. (1982) Plasma apolipoprotein A-I absence associated with marked reduction of high density lipoproteins and premature coronary artery disease. *Arteriosclerosis*, **2**, 16–26.

39 Norum, R. A., Lakier, J. B., Goldstein, S., Angel, A., Goldberg, R. B., Block, W. D., Noffze, D. K., Dolphin, P. J., Edelglass, J., Bogorad, D. D., Alaupovic, P. (1982) Familial deficiency of apolipoproteins A-I and C-III and precocious coronary artery disease. *N. Engl. J. Med.*, **306**, 1513–1519.

40 Schaefer, E. J., Ordovas, J. M., Law, S., Ghiselli, G., Kashyap, M. L., Srivastava, L. S., Heaton, W. H., Albers, J. J., Connor, W. E., Lemeshev, Y., Segrest, J., Brewer, H. B., Jr. (1985) Familial apolipoprotein A-I and C-III deficiency, variant II. *J. Lipid Res.*, **26**, 1089–1101.

41 Lamon-Fava, S., Ordovas, J. M., Mandel, G., Forte, T. M., Goodman, R. H., Schaefer, E. J. (1987) Secretion of apolipoprotein A-I in lipoprotein particles following transfection of the human apolipoprotein A-I gene into 3T3 cells. *J. Biol. Chem.*, **262**, 8944–8947.

42 Ordovas, J. M., Cassidy, D. K., Civeira, F., Bisgaier, C. L., Schaefer, E. J. (1989) Familial apolipoprotein A-I, C-III, and A-IV deficiency with marked high density lipoprotein deficiency and premature atherosclerosis due to a deletion of the apolipoprotein A-I, C-III, and A-IV gene complex. *J. Biol. Chem.*, **264**, 16339–16342.

43 Ng, D. S., Leiter, L. A., Vezina, C., Connelly, P. W., Hegele, R. A. (1994) Apolipoprotein A-I Q[-2]X causing isolated apolipoprotein A-I deficiency in a family with analphalipoproteinemia. *J. Clin. Invest.*, **93**, 223–229.

44 Schaefer, E. J., Blum, C. B., Levy, R. I., Jenkins, L. L., Alaupovic, P., Foster, D. M., Brewer, H. B., Jr. (1978) Metabolism of high density lipoprotein apolipoproteins in Tangier disease. *N. Eng. J. Med.*, **299**, 905–910.

45 Schaefer, E. J., Zech, L. A., Schwartz, D. S., Brewer, H. B., Jr. (1980) Coronary heart disease prevalence and other clinical features in familial high density lipoprotein deficiency (Tangier disease). *Ann. Int. Med.*, **93**, 261–266.

46 Schaefer, E. J., Anderson, D. W., Zech, L. A., Lindgren, F. T., Bronzert, T. J., Rubalcaba, E. A., Brewer, H. B., Jr.

(1981) Metabolism of high density lipoprotein subfractions and constituents in Tangier disease following the infusion of high density lipoproteins. *J. Lipid Res.*, **22**, 217–226.

47 Alaupovic, P., Schaefer, E. J., McConathy, W. J., Fesmire, J. D., Brewer, H. B., Jr. (1981) Plasma apolipoprotein concentrations in familial apolipoprotein A-I and A-II deficiency (Tangier disease). *Metabolism*, **30**, 805–809.

48 Serfaty-Lacrosniere, C., Lanzberg, A., Civeira, F., Isaia, P., Berg, J., Janus, E. D., Smith, M. P., Pritchard, P. H., Frohlich, J., Lees, R. S., Ordovas, J. M., Schaefer, E. J. (1994) Homozygous Tangier disease and cardiovascular disease. *Atherosclerosis*, **107**, 85–98.

49 Brousseau, M. E., Bodzioch, M., Schaefer, E. J., Goldkamp, A. L., Kielar, D., Probst, M., Ordovas, J. M., Aslanidas, C., Lackner, K., Rubins, H. B., Collins, D., Robins, S. J., Wilson, P. W. F., Schmitz, G. (2001) Common variants in the gene encoding ATP-binding cassette transporter 1 in men with low HDL cholesterol levels and coronary heart disease. *Atherosclerosis*, **154**, 607–611.

50 Brousseau, M. E., Eberhart, G. P., Dupuis, J., Asztalos, B. F., Goldkamp, A. L., Schaefer, E. J., Freeman, M. W. (2000) Cellular cholesterol efflux in heterozygotes for Tangier disease is markedly reduced and correlates with high density lipoprotein cholesterol concentration and particle size. *J. Lipid Res.*, **41**, 1125–1135.

51 Brousseau, M. E., Schaefer, E. J., Dupuis, J., Eustace, B., Van Eerdewegh, P., Goldkamp, A. L., Thurston, L. M., FitzGerald, M. G., Yasek-McKenna, D., O'Neill, G., Eberhart, G. P., Weiffenbach, B., Ordovas, J. M., Freeman, M. W., Brown, R. H., Jr., Gu, J. Z. (2000) Novel mutations in the gene encoding ATP-binding cassette 1 in four Tangier disease kindreds. *J. Lipid Res.*, **41**, 433–441.

52 Schaefer, E. J., Brousseau, M. E., Diffenderfer, M. R., Cohn, J. S., Welty, F. K., O'Connor, J., Dolnikowski, G. G., Wang, J., Hegele, R. A., Jones, P. J. (2001) Cholesterol and apolipoprotein B metabolism in Tangier disease. *Atherosclerosis*, **159**, 231–236.

53 Asztalos, B. F., Brousseau, M. E., McNamara, J. R., Horvath, K. V., Roheim, P. S., Schaefer, E. J. (2001) Subpopulations of high-density lipoproteins in homozygous and heterozygous Tangier disease. *Atherosclerosis*, **156**, 217–225.

54 Asztalos, B. F., Schaefer, E. J., Horvath, K. V., Yameshita, S., Miller, M., Franceschini, G., Calabresi, L. (2001) Role of LCAT in HDL Remodeling: Investigation of HDL Deficiency States. *J. Lipid Res.* **48**, 597–599.

55 Asztalos, B. F., Horvath, K. V., Kajinami, K., Nartsupha, C., Cox, C. E., Batista, M., Schaefer, E. J., Inazu, A., Mabuchi, H. (2004) Apolipoprotein composition of HDL in cholesteryl ester transfer protein deficiency. *J. Lipid Res.*, **45**, 448–455.

56 Brousseau, M. E., Schaefer, E. J., Wolfe, M. L., Bloedan, L. T., Digenio, A. G., Clark, R. W., Mancuso, J. P., Rader, D. J. (2004) Effects of an inhibitor of cholesteryl ester transfer protein on HDL cholesterol. *N. Engl. J. Med.*, **350**, 1505–1515.

57 Brousseau, M. E., Diffenderfer, M. R., Millar, J. S., Nartsupha, C., Asztalos, B. F., Welty, F. K., Wolfe, M. L., Rudling, M., Bjorkhem, I., Angelin, B., Mancuso, J. P., DiGenio, A., Rader, D. J., Schaefer, E. J. (2005) Effects of cholesteryl ester transfer protein inhibition on high-density lipoprotein subspecies, apolipoprotein A-I metabolism, and fecal sterol excretion. *Arterioscler. Thromb. Vasc. Biol.*, **25**, 1057–1064.

58 Millar, J. S., Brousseau, M. E., Diffenderfer, M. R., Barrett, P. H., Welty, F. K., Faruqi, A., Wolfe, M. L., Nartsupha, C., Digenio, A. G., Mancuso, J. P., Dolnikowski, G. G., Schaefer, E. J., Rader, D. J. (2006) Effects of the cholesteryl ester transfer protein inhibitor torcetrapib on

apolipoprotein B100 metabolism. *Arterioscler Thromb Vasc. Biol.*, **26**, 1350–1356.

59 Stucchi, A. F., Hennessy, L. K., Vespa, D. B., Weiner, E. J., Osada, J., Ordovas, J. M., Schaefer, E. J., Nicolosi, R. J. (1991) Effect of corn and coconut oil-containing diets with and without cholesterol on high density lipoprotein apoprotein A-I metabolism and hepatic apoprotein A-I mRNA levels in Cebus monkeys. *Arterioscler. Thromb.*, **11**, 1719–1729.

60 Lichtenstein, A. H., Ausman, L. M., Carrasco, W., Jenner, J. L., Ordovas, J. M., Schaefer, E. J. (1994) Hypercholesterolemic effect of dietary cholesterol in diets enriched in polyunsaturated and saturated fat. *Arterioscler. Thromb.*, **14**, 168–175.

61 Velez-Carrasco, W., Lichtenstein, A. H., Welty, F. K., Li, Z., Lamon-Fava, S., Dolnikowski, G. G., Schaefer, E. J. (1999) Dietary restriction of saturated fat and cholesterol decreases HDL apoA-I secretion. *Arterioscler. Thromb. Vasc. Biol.*, **19**, 918–924.

62 Schaefer, E. J., Levy, R. I., Ernst, N. D., Van Sant, F. D., Brewer, H. B., Jr (1981) The effect of low cholesterol, high polyunsaturated fat, and low fat diets on plasma lipid and lipoprotein cholesterol levels in normal and hypercholesterolemic subjects. *Am. J. Clin. Nutr.*, **34**, 1758–1763.

63 Lichtenstein, A. H., Carrasco, W., Jenner, J. L., Ausman, L., Ordovas, J. M., Schaefer, E. J. (1993) Effects of canola, corn, olive, and rice bran oil on fasting and post-prandial lipoproteins in humans as part of a National Cholesterol Education Program Step 2 diet. *Arteriosclerosis and Thrombosis*, **13**, 1533–1542.

64 Brousseau, M. E., Schaefer, E. J., Stucchi, A. F., Osada, J., Vespa, D. B., Ordovas, J. M., Nicolosi, R. J. (1995) Diets enriched in unsaturated fatty acids enhance apolipoprotein A-I catabolism but do not affect either its production or hepatic mRNA abundance in cynomolgus monkeys. *Atherosclerosis*, **15**, 107–119.

65 Brousseau, M. E., Ordovas, J., Osada, J., Fasulo, J., Robins, S., Nicolosi, R., Schaefer, E. J. (1995) Dietary monounsaturated and polyunsaturated fatty acids are comparable in their effects on hepatic apolipoprotein mRNA abundance and liver lipid concentrations when substituted for saturated fatty acids in cynomolgus monkeys. *J. Nutrition*, **125**, 425–436.

66 Spady, D. K., Kearney, D. M., Hobbs, H. H. (1999) Polyunsaturated fatty acids upregulate hepatic scavenger receptor B1 (SR-B1) expression and HDL cholesteryl ester uptake in the hamster. *J. Lipid Res.*, **40**, 1384–1394.

67 Leren, P. (1966) The effect of plasma cholesterol lowering diet in male survivors of myocardial infarction. *Acta Med. Scand.*, **466**, 1–92.

68 Dayton, S., Pearce, M. L., Goldman, H. *et al.* (1968) Controlled trial of a diet high in unsaturated fat for prevention of atherosclerotic complications. *Lancet*, **2**, 1060–1062.

69 Turpeinen, O. (1979) Effect of a cholesterol lowering diet on mortality from coronary heart diseases and other causes. *Circulation*, **59**, 1–7.

70 Frenais, R., Ouguerram, K., Maugeais, C., Mahot, P., Charbonnel, B., Magot, T., Krempf, M. (2001) Effect of dietary omega-3 fatty acids on high density lipoprotein apolipoprotein AI kinetics in type II diabetes mellitus. *Atherosclerosis*, **157**, 131–135.

71 Chan, D. C., Watts, G. F., Barrett, P. H. R. (2006) Factorial study of the effect of n-3 fatty acid supplementation and atorvastatin on the kinetics of HDL apolipoproteins A-I and A-II in men with abdominal obesity. *Am. J. Clin. Nutr.*, **84**, 37–43.

72 Schaefer, E. J., Lichtenstein, A. H., Lamon-Fava, S., Contois, J. H., Li, Z., Rasmussen, H., McNamara, J. R., Ordovas, J. M. (1995) Efficacy of a National Cholesterol Education Program Step 2 Diet in normolipidemic and hyperlipidemic middle aged and elderly men and women. *Arterioscler. Thromb. & Vasc. Biol.*, **15**, 1079–1085.

73 Schaefer, E. J., Lichtenstein, A. H., Lamon-Fava, S., Contois, J. H., Li, Z., Goldin, B. R., Rasmussen, H., McNamara, J. R., Ordovas, J. M. (1996) Effects of National Cholesterol Education Program Step 2 diets relatively high or relatively low in fish-derived fatty acids on plasma lipoproteins in middle-aged and elderly subjects. *Am. J. Clin. Nutr.*, **63**, 234–241.

74 Schaefer, E. J., Lamon-Fava, S., Ausman, L. M., Ordovas, J. M., Clevidence, B. A., Judd, J. T., Goldin, B. R., Woods, M., Gorbach, S., Lichtenstein, A. H. (1997) Individual variability in lipoprotein cholesterol response to National Cholesterol Education Program Step 2 diets. *Am. J. Clin. Nutr.*, **65**, 823–830.

75 Schaefer, E. J., Lichtenstein, A. H., Lamon-Fava, S., McNamara, J. R., Schaefer, M. M., Rasmussen, H., Ordovas, J. M. (1995) Body weight and low-density lipoprotein cholesterol changes after consumption of a low fat ad libitum diet. *JAMA*, **274**, 1450–1455.

76 Dansinger, M. L., Gleason, J. A., Griffith, J. L., Selker, H. P., Schaefer, E. J. (2005) Comparison of the Atkins, Ornish, Weight Watchers, and Zone diets for weight loss and heart disease risk reduction. *JAMA*, **293**, 43–53.

77 Blum, C. B., Levy, R. I., Eisenberg, S., Hall, M, 3rd, Goebel, R. H., Berman, M. (1977) High density lipoprotein metabolism in man. *J. Clin. Invest.*, **60**, 795–807.

78 Brinton, E. A., Eisenberg, S., Breslow, J. L. (1990) A low-fat diet decreases high density lipoprotein (HDL) cholesterol by decreasing HDL apolipoprotein transport rates. *J. Clin. Invest.*, **85**, 144–151.

79 Woollett, L. A. and Spady, D. K. (1997) Kinetic parameters for high density lipoprotein apoprotein AI and cholesteryl ester transport in the hamster. *J. Clin. Invest.*, **99**, 1704–1713.

80 Woollett, L. A., Kearney, D. M., Spady, D. K. (1997) Diet modification alters HDL cholesterol, but not the transport of HDL cholesteryl ester to the liver in the hamster. *J. Lipid Res.*, **38**, 2289–2302.

81 Asztalos, B. F., Lefevre, M., Wong, L., Foster, T. A., Tully, R., Windhauser, M., Zhang, W., Roheim, P. S. (2000) Differential response to low fat diet between low and normal HDL cholesterol subjects. *J. Lipid Res.*, **41**, 321–328.

82 Clevidence, B. A., Reichman, M. E., Judd, J. T., Muesing, R. A., Schatzkin, A., Schaefer, E. J., Li, Z., Jenner, J., Brown, C. C., Sunkin, M., Campbell, W. S., Taylor, P. R. (1995) Effects of alcohol consumption on lipoproteins of premenopausal women. *Arterioscler. Thromb. Vasc. Biol.*, **15**, 179–184.

83 De Oliveira e Silva, E. R., Foster, D., Harper, M. M., Seidman, C. E., Smith, J. D., Breslow, J. L., Brinton, E. A. (2000) Alcohol consumption raises HDL cholesterol by increasing the transport rate of apolipoproteins A-I and A-II. *Circulation*, **102**, 2347–2352.

84 Stampfer, M. J., Colditz, G. A., Willett, W. C. *et al.* (1988) A prospective study of moderate alcohol intake and risk of coronary disease and stroke in women. *N. Engl. J. Med.*, **319**, 267–273.

85 Gaziano, J. M., Buring, J. E., Breslow, J. L. *et al.* (1993) Moderate alcohol intake, increased levels of high density lipoprotein and its subfractions, and decreased risk of myocardial infarction. *N. Engl. J. Med.*, **329**, 1829–1834.

86 Hegsted, M. and Ausman, L. M. (1987) Diet, alcohol, and coronary heart disease in men. *J. Nutr.*, **46**, 941–948.

87 Dattilo, A. M. and Kris-Etherton, P. M. (1992) Effects of weight reduction on blood lipids and lipoproteins: a meta analysis. *Am. J. Clin. Nutr.*, **56**, 320–328.

88 Lipson, L. C., Bonow, R. O., Schaefer, E. J., Brewer, H. B., Jr, Lindgren, F. T. (1980) Effect of exercise conditioning on plasma high density lipoproteins (HDL) and other lipoproteins. *Atherosclerosis*, **37**, 529–538.

89 Lamon-Fava, S., McNamara, J. R., Farber, H. W., Hill, N. S., Schaefer,

E. J. (1989) Acute changes in lipid, lipoprotein, apolipoprotein and low density lipoprotein particle size after an endurance triathlon. *Metabolism*, **38**, 921–925.

90 Thompson, P. D., Jurqalevitch, S. M., Flynn, M. M., Zmuda, J. M., Spannauser-Martin, D., Saritelli, A., Bausserman, L., Herbert, P. N. (1997) Effect of prolonged exercise training without weight loss on HDL metabolism in overweight men. *Metabolism*, **46**, 217–223.

91 Steffen-Battey, L., Nichaman, M. Z., Goff, D. C., Jr, Frankowski, R. F., Hanis, C. L., Ramsey, D. J., Labarthe, D. (2000) Change in physical activity and risk of all cause mortality: the Corpus Christi Heart Project. *Circulation*, **102**, 2204–2209.

92 Mensink, R. P. and Katan, M. B. (1992) Effect of dietary fatty acids on serum lipids and lipoproteins: a meta analysis of 27 trials. *Arterioscler. Thromb.*, **12**, 911–919.

93 Kris-Etherton, P. M. and Yu, S. (1995) Individual fatty acid effects on plasma lipids and lipoproteins: human studies. *Am. J. Clin. Nutr.*, **65** (Suppl), 1628S–1644S.

V
Clinical Aspects of HDL

High-Density Lipoproteins: From Basic Biology to Clinical Aspects. Edited by Christopher J. Fielding

ISBN: 978-3-527-31717-2

18
HDL and Metabolic Syndrome

Clara Cavelier, Arnold von Eckardstein

18.1
Introduction

The concept of metabolic syndrome originates from the observation that certain metabolic abnormalities associated with an increased risk of diabetes and cardiovascular disease tend to cluster within individuals. In 1988, Reaven described syndrome X as the tendency of hyperinsulinaemia and glucose intolerance, hypertension, low HDL-C, and raised triglycerides (TG) to occur within the same individual [1]. To date, three definitions of the metabolic syndrome have found the broadest acceptance and application: namely those of the World Health Organization (WHO) [2], the National Cholesterol Education Program, Adult Treatment Panel III (NCEP: ATP III) [3], and the International Diabetes Federation (IDF) [4] (Table 18.1). All definitions include glucose intolerance, obesity, hypertension, and dyslipidaemia as major components of the metabolic syndrome, which by any definition is very common in adults worldwide, typically being found in 10–30 % of the adult population [5]. The risk of cardiovascular disease in subjects with the metabolic syndrome is two to four times higher than in subjects without it [6], though it is controversial whether the metabolic syndrome represents an etiological entity [7]. Hence, the ultimate importance of defining the metabolic syndrome is that it helps to identify individuals at high risk of type 2 diabetes and cardiovascular diseases.

18.2
Altered Lipoprotein Metabolism in the Metabolic Syndrome

Dyslipidaemia is considered to be the key mediator of the atherogenicity in the metabolic syndrome and is characterized by increased plasma concentrations of triglycerides (TG), decreased plasma levels of HDL cholesterol (HDL-C), and by the occurrence of small, dense LDL particles. In addition, the HDL pool is constituted of small, cholesterol-poor, TG-rich particles, and furthermore, the relative decrease in plasma levels of HDL subclass LpA-I/A-II is more pronounced than that of HDL subclass LpA-I. These changes in HDL-C plasma levels and HDL subclasses

High-Density Lipoproteins. From Basic Biology to Clinical Aspects. Edited by Christopher J. Fielding

ISBN: 978-3-527-31717-2

Tab. 18.1 Definitions of metabolic syndrome.

WHO 1999	NCEP:ATP III 2001	IDF 2005
glucose intolerance and/or insulin resistance and two or more of the following:	three or more of the following:	obesity: waist circumference ethnic specificity* and two or more of the following:
1. obesity: WHR>0.9 (men) WHR>0.85 (women) or BMI>30 kg m^{-2}	1. obesity: waist circumference > 102 cm (men) waist circumference > 88 cm (women)	
2. dyslipidaemia: triglycerides ≥ 1.7 mmol L^{-1} HDL-C < 0.9 mmol L^{-1} (men) HDL-C < 1 mmol L^{-1} (women)	2. hypertriglyceridaemia: triglycerides ≥ 1.7 mmol L^{-1}	1. hypertriglyceridaemia: triglycerides ≥ 1.7 mmol L^{-1}
	3. low HDL-C: HDL-C < 1.03 mmol L^{-1} (men) HDL-C < 1.29 mmol L^{-1} (women)	2. low HDL-C: HDL-C < 1.03 mmol L^{-1} (men) HDL-C < 1.29 mmol L^{-1} (women)
3. hypertension: blood pressure ≥ 140/90 mmHg	4. hypertension: blood pressure ≥ 130/85 mmHg	3. hypertension: blood pressure ≥ 130/85 mmHg
4. microalbuminuria: albumin excretion 20 ug min^{-1} or albumin:creatine ratio ≥ 30 mg g^{-1}	5. fasting plasma glucose: ≥6.1 mmol L^{-1}	4. fasting plasma glucose: ≥5.6 mmol L^{-1} or diagnosed type 2 diabetes

*Europids: men ≥ 94 cm, women ≥ 80 cm; South Asian and Chinese: men ≥ 90 cm, women ≥ 80 cm; Japanese: men ≥ 85 cm, women ≥ 90 cm; Sub Saharan Africans, eastern Mediterranean and middle east orign populations: Europid values until specific date are availlable; Ethnic central and south Americans: south Asian values until specific data are available.

distribution are mostly due to enhanced HDL and apolipoprotein A-I (apoA-I) catabolism. Indeed, in patients with type 2 diabetes mellitus [8,9] and metabolic syndrome [10–13], low plasma levels of HDL and apoA-I coincide with enhanced HDL and apoA-I catabolism. No *in vivo* lipoprotein turnover study in individuals with the metabolic syndrome has revealed a reduced apoA-I production; actually, a recent turnover study even demonstrated enhanced LpA-I production but normal LpA-I/A-II production. In conjunction with the also observed enhanced catabolism of both LpA-I and LpA-I/A-II, this finding is in good agreement with the more pronounced decrease in LpA-I/A-II and apoA-II levels in relation to LpA-I and apoA-I levels [11].

Insulin resistance is considered a key etiological factor of low HDL cholesterol in patients with the metabolic syndrome, either directly or indirectly through increasing of the plasma concentrations of free fatty acids (FFA) and TG, which in turn have a strong impact on HDL metabolism. Quantitatively the indirect effects of insulin resistance appear to be more important than its direct effects, since apoA-I expression has previously been reported to be only slightly stimulated by insulin [14,15]. In contrast, FFA were found to regulate several important genes and proteins of HDL metabolism at the transcriptional and posttranslational levels [16]. As well as this, triglycerides, notably in VLDL, have a strong impact on HDL-C levels, since they stimulate cholesteryl ester transfer protein (CETP) activity (see below). Insulin is an important regulator of FFA production in the adipose tissue, since it suppresses lipolysis by inhibiting both acyltrigylceride lipase (ATGL) and hormone-sensitive lipase. Insulin and FFA are also important regulators of TG plasma levels, insulin inhibiting hepatic VLDL production whereas FFA stimulate it. Insulin also regulates the activity of lipoprotein lipase (LPL), which degrades TG in VLDL and chylomicrons. In insulin resistance, hepatic TG and VLDL production are hence increased whereas the LPL-mediated lipolysis of VLDL (and chylomicrons) is decreased [17].

Finally, since most patients with metabolic syndrome are overweight or obese, it is important that the adipose tissue produces proteins that are involved in lipoprotein metabolism as well as hormones (i.e., adipokines) that may exert direct effects on HDL metabolism. For example, lipoprotein lipase (LPL), phospholipid transfer protein (PLTP), and CETP are expressed by the adipose tissue [18]. Any change in adipocyte mass and distribution will hence influence the plasma concentration and activity of these enzymes and lipid transfer proteins. Moreover, plasma concentrations of adiponectin are more closely related to HDL-C levels than any other anthropometric or biochemical measure of excess weight, such as body mass index, waist circumference, or leptin levels. In fact, in a multivariate analysis of data from a turnover study, adiponectin plasma levels were correlated with the fractional catabolic rate of apoA-I independently of any other measure including insulin resistance and explained more than 40 % of the variance in apoA-I catabolism [19]. However, no target gene of adiponectin in HDL metabolism has so far been identified [20].

18.3 Dysregulated Genes and Proteins of HDL Metabolism

Impaired insulin action, increased concentrations of FFA, and hypertriglyceridemia are typically observed in the metabolic syndrome and affect the production and maturation of HDL by altering the expression of apoA-I, ABCA1, and ABCG1 (Fig. 18.1), as well as the remodeling of HDL particles, which involves LPL, CETP, PLTP, and HL (Figure 18.2).

Apolipoprotein A-I ApoA-I is the major protein constitusent in HDL; its absence prevents the formation of typical HDL particles [21]. Insulin had previously been shown to stimulate apoA-I gene expression in rat liver [14,15] so that insulin

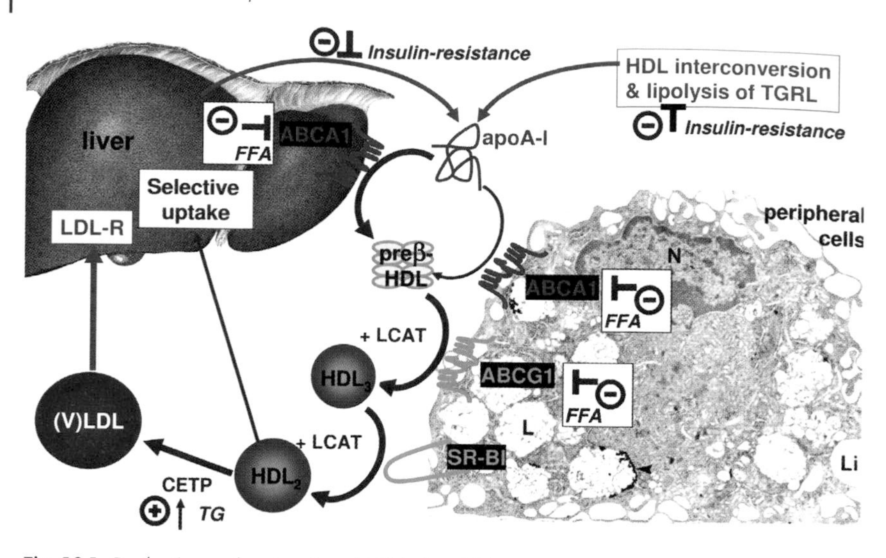

Fig. 18.1 *Production and maturation of HDL.* ApoA-I is synthesized by the liver. ABCA1, ABCG1, and SR-BI play a critical role in HDL production and maturation by loading apoA-I with lipids. In the condition of metabolic syndrome both HDL production and maturation are modulated by FFA and insulin resistance.

resistance would decrease apoA-I production. However, turnover studies in men rather found increased apoA-I and LpA-I production, so the relevance of the *in vitro* findings is questionable [11].

ATP Binding Cassette Transporter A1 (ABCA1) ABCA1 mediates the efflux of phospholipids and cholesterol from many cells onto lipid-free apoA-I and thereby helps to generate discoidal HDL precursors, so-called nascent HDL [22] (see also Chapter 5 by Chimini and Chapter 10 by Zannis). The importance of ABCA1 for regulating HDL cholesterol levels and cholesterol homeostasis in macrophages is highlighted by the virtual absence of HDL from plasma and the accumulation of cholesterol in macrophages of men and mice lacking ABCA1 [21]. ABCA1 is expressed in many organs but shows the highest activity in the liver and macrophages. In a mouse model of diabetes mellitus, ABCA1 gene expression was reduced in both the liver and peritoneal macrophages [23]. *In vitro*, unesterified long chain and unsaturated fatty acids were found to decrease the expression of ABCA1 in macrophages and hepatocytes [23], this effect being most pronounced if the cells were stimulated with oxysterols. Therefore, and also because mutagenesis of the DR4 domain of the ABCA1 promoter, which binds the oxysterol activated transcription factor LXR, abrogates this effect [24], free fatty acids are considered antagonists of oxysterols. In addition to the transcriptional inhibition of ABCA1, FFA limit the availability of ABCA1 at the posttranslational level. Notably, after stimulation with cyclic AMP, FFA inhibit the translocation of ABCA1 from intracellular

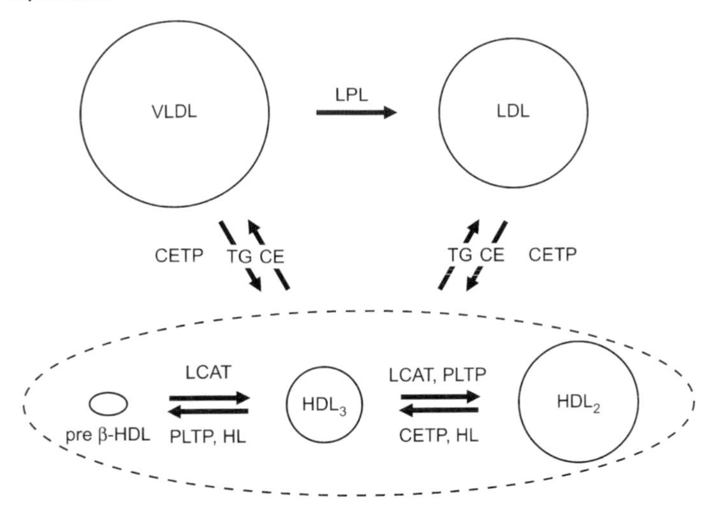

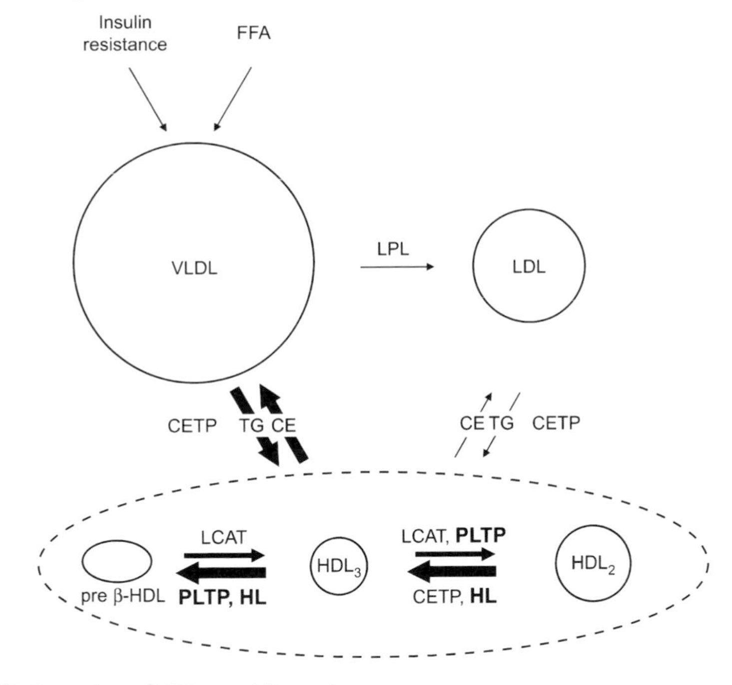

Fig. 18.2 *Comparison of HDL remodeling under normolipidemia and metabolic syndrome.* HDL remodeling, which involves LPL, LCAT, HL, PLTP and CETP, is impaired in the condition of metabolic syndrome.

compartments to the plasma membrane and stimulate proteasomal degradation of ABCA1 [16]. Both the transcriptional and posttranslational regulation of ABCA1 by FFA limit the abundance of ABCA1 and thereby the maturation of HDL.

Cholesteryl Ester Transfer Protein (CETP) CETP mediates the transfer of cholesteryl ester (CE) and TG between TG-rich lipoproteins and HDL (see Chapter 6 by McPherson). CETP-deficient patients have very high plasma levels of HDL [25]. Expression of CETP in mice, which are naturally CETP-deficient, results in markedly reduced HDL-C levels [26], so CETP activity has the net effect of reducing HDL-C. In insulin resistance, CETP activity was found to be unchanged [27]. However, the mass transfer of TG and CE catalyzed by CETP is highly dependent on the pool size of the TG-rich lipoproteins in plasma [28], so hypertrigleridaemia as observed in metabolic syndrome enhances the exchange of neutral lipids between TG-rich particles and HDL, resulting in the intermediary production of TG-rich HDL, which have been shown to be thermodynamically unstable. Indeed, enrichment of HDL with TG enhances apoA-I clearance *in vivo* [29]. Mechanistically, it has been proposed that the decrease in the CE/TG ratio decreases the thermodynamic stability and structural integrity of the particles [30], while furthermore, it has been observed that *in vitro* incubation of HDL with CETP and VLDL (to form TG-enriched HDL) produces a time-dependent dissociation of apoA-I from HDL [31]. This, however, may not be the only mechanism leading to the low HDL-C concentrations observed in metabolic syndrome; indeed, some investigators have proposed that the lipolysis of TG-rich HDL by HL further contributes to low HDL-C concentrations [32].

In addition to TG, FFA also regulate CETP activity. On the one hand, unsaturated long chain fatty acids were found to inhibit CETP gene expression, which is induced by sterols [33,34]. Since CETP, like ABCA1 is upregulated by LXR, this inhibitory effect probably occurs through competition with oxysterols for the activation of this nuclear receptor. On the other hand, *in vitro* studies showed that FFA stimulate the transfer of CE from HDL to VLDL and LDL [35]. These opposite effects on CETP expression and CETP activity may explain why FFA concentrations showed little correlation with CETP activity in patients with metabolic syndrome or type 2 diabetes.

Phospholipid Transfer Protein (PLTP) PLTP facilitates the transfer of phospholipids from TG-rich lipoproteins to HDL during lipolysis and accounts for most phospholipid transfer in plasma (see Chapter 8 by Ehnholm et al.). Both knockout and overexpression of PLTP in mice lead to reduced HDL-C levels [36,37]; however, PLTP transgenic mice – but not PLTP knockout mice – have increased plasma levels of lipid-poor preβ-HDL [37].

Controversial findings have been reported on PLTP activity in patients with diabetes mellitus or metabolic syndrome: PLTP activity was found to be elevated in diabetes mellitus (both type 1 and type 2), obesity, and insulin resistance [38,39]. In particular, PLTP-mediated conversion of HDL into larger and smaller particles is enhanced in HDL that are enriched with TG [40]. In another case-control study, PLTP activity was much lower in overweight patients with metabolic syndrome relative to controls with normal weight and lipoprotein profile [11]. The controversial results may reflect different effects of various components of the metabolic syndrome on PLTP. On the one hand, PLTP is an LXR target gene, so that FFA may suppress PLTP gene expression [41], while on the other hand, since both type 1 and type 2 diabetic subjects show elevated PLTP activity and since insulin treatment lowers PLTP activity, insulin may also exert direct regulatory effects on PLTP [38]. Whether PLTP activity is high or low may hence depend on the relative activity of FFA, insulin, and insulin sensitivity.

Hepatic Lipase Hepatic lipase (HL) promotes the catabolism of TG-rich HDL, which are formed as intermediate products in metabolic syndrome (see Chapter 7 by Chan). It has been demonstrated that *in vitro* incubation of TG-rich HDL with HL leads to a reduction in HDL size and the loss of apoA-I from HDL [42]. Several studies have shown that insulin resistance is associated with enhanced post-heparin plasma HL activity [43–45]. Actually, it seems that only chronic – not acute – insulin administration induces HL activity [46]. Two processes may play a role in the lowering of HDL-C and apoA-I plasma levels in metabolic syndrome: firstly, apoA-I shedding and formation of preβ-HDL may enhance apoA-I clearance from the circulation, and secondly, the HDL particles produced by the lipolytic processes described above may be more rapidly taken up by the liver because of enhanced receptor-mediated uptake [47,48].

Lipoprotein Lipase (LPL) LPL is predominantly expressed in the adipose tissue, the skeletal muscle, and the heart, and hydrolyses TG contained in TG-rich lipoproteins. During this process, redundant surface lipids (CE and phospholipids) and apolipoproteins are transferred from TG-rich particles to HDL [49]. HDL-C concentrations correlate positively with LPL plasma activity; indeed, LPL deficiency in humans is associated with reduced plasma HDL concentration [49]. HDL levels in LPL knockout mice are also low [50], while antibody-mediated LPL inhibition in monkeys results in reduced HDL levels associated with enhanced HDL and apoA-I clearance [51]. Interestingly, post-heparin plasma LPL activity is decreased in insulin-resistant subjects without diabetes [52], reflecting the stimulatory effect of insulin on LPL gene transcription and the inhibitory effect of insulin on the transcription of angiopoietin-like proteins 3 and 4, (i.e., endogenous inhibitors of lipoprotein lipase) [53,54]. The shedding of lipids and apolipoproteins from the surfaces of TG-rich lipoproteins during LPL-mediated lipolysis is thus diminished, contributing to lower plasma HDL-C and apoA-I concentration. Reduction in LPL activity also aggravates the enrichment in TG-rich lipoproteins [55], which, as described above, provokes a compositional change of the HDL pool involving CETP, PLTP, and HL, ultimately enhancing HDL-clearance from the circulation. Relative LPL deficiency in metabolic syndrome thus contributes both directly and indirectly to low HDL plasma levels.

HDL metabolism is therefore strongly disturbed in metabolic syndrome, which results in a reduced HDL-pool consisting of small, TG-rich, and CE-poor particles. This modified HDL composition enhances the risk of cardiovascular disease. However, it has also been suggested that HDL atheroprotective properties are altered in metabolic syndrome, contributing further to the enhanced risk of cardiovascular disease.

18.4 Biological Properties of Impaired HDL

HDL and apoA-I exert diverse potentially atheroprotective functions (see Chapters 13, 15 and 20 by Fogelman, Franceschini, and Barter, respectively): for example, they reduce oxidative damage, correct endothelial dysfunction, inhibit inflammation, and mediate reverse cholesterol transport, which might be distorted in the metabolic syndrome.

Impaired Reverse Cholesterol Transport Reverse cholesterol transport involves the removal of excess cholesterol from peripheral tissues, including foam cells of the arterial wall, and its delivery to the liver for biliary excretion [56]. This process is affected, at least in theory, both by altered HDL composition and by dysregulated cellular lipid transporters. Brites et al. demonstrated that serum samples from hypertriglyceridemic patients have a lower capacity to promote cholesterol efflux from hepatocytes and macrophages, this impaired cholesterol efflux being attributable not only to low HDL levels but also to changes in HDL particle composition [57]. It has also been suggested that glycation of HDL apolipoproteins can interfere with cholesterol efflux in diabetes type 2 [58], while cholesterol efflux to apoA-I is also inhibited in the presence of TG-rich lipoproteins [59]. However, other investigators have reported normal phospholipid and cholesterol efflux to HDL of hypertriglyceridemic subjects from hepatocytes, smooth muscle cells, and fibroblasts [60–62]. In addition, the metabolic syndrome may interfere with the activity of cellular factors that regulate cholesterol efflux; as discussed above, for example, ABCA1 expression is reduced in macrophages and hepatocytes treated with unsaturated FFA [23]. ABCG1, which mediates cholesterol efflux to lipidated HDL particles, is also suppressed by FFA in an LXR-dependent manner [24], and unsaturated FFA were also found to induce the phosphorylation, and hence the destabilization, of ABCA1 in a phospholipase D2-dependent manner [16]. Both the transcriptional and the posttranslational suppression of ABCA1 by FFA reduce apoA-I-mediated cholesterol efflux. In addition, advanced glycation end products that are formed in diabetic patients impair ABCA1- and SR-BI-mediated cholesterol removal from cells [63,64], while the changes in the remodeling of HDL described above may also interfere with reverse cholesterol transport. Furthermore, it has also been suggested that the uptake of HDL-derived cholesterol by the liver is impaired in metabolic syndrome [65], and finally, the excretion of cholesterol into the bile by ABC transporters G5 and G8 is also regulated by FFA and may be suppressed in metabolic syndrome [66,67]. Taken together, these results suggest that reverse cholesterol transport is compromised in patients with metabolic syndrome. However, as the whole question of the physiological importance of reverse cholesterol transport is still a matter of debate, this hypothesis awaits validation by *in vivo* studies [56,68,69].

Impaired Antioxidant Actions The metabolic syndrome is characterized by elevated systemic oxidative stress and, in particular, the ability of HDL to protect LDL from oxidation was found to be diminished in these patients [70]. It has also been established that the antioxidative activity of small, dense HDL particles is impaired in subjects with metabolic syndrome [71], with this impairment of HDL antioxidative activity in the metabolic syndrome being partly related to the enrichment of small HDL in TG and their depletion in CE, to the replacement of apoA-I by serum amyloid A, and to glycation and oxidation of apoA-I [72]. In addition, HDL from type 2 diabetic subjects without coronary heart disease were found to have a decreased ability to metabolize oxidized phospholipids, which could lead to increased susceptibility to development of cardiovascular disease [73]. Dysfunctional HDL in the metabolic syndrome may thus be responsible for elevated oxidative stress, which is a key component of endothelial dysfunction and thereby of increased cardiovascular disease risk.

HDL and β-cell Homeostasis The identification of low HDL-C and hypertriglyceridemia as risk factors of diabetes is traditionally regarded as an early symptom of insulin resistance that later progresses to diabetes mellitus. However, initial lines of evidence indicate that low HDL-C and hypertriglyceridemia are also markers of a β-cell-toxic metabolic situation and may even contribute causally to β-cell failure and hence to the manifestation of diabetes mellitus. Insulin resistance turns into manifest type 2 diabetes mellitus when β-cells do not produce enough insulin to overcome insulin resistance. Among others, glucose, fatty acids, and various inflammatory cytokines derived either from the adipose tissue (e.g., leptin and TNF-α) or the innate immune system (e.g., IL-1β) have previously been found to impair β-cell function and survival [74]. Since HDL cholesterol is frequently low in patients with hyperglycemia and hyperacylemia or hyperleptinemia, low HDL cholesterol can be considered an indirect marker of a metabolic situation that favors β-cell death. Interestingly, Roehrich and colleagues provided support for a direct pathogenetic contribution of lipoproteins to β-cell death and survival [75]. The islet cell line βTC3, as well as primary murine islet cells, were found to express the lipoprotein receptors SR-BI, LDLR, VLDLR, and apoER2 and to bind and internalize VLDL, LDL, and HDL. In parallel, LDL and VLDL dose-dependently inhibited the expression of insulin mRNA and β-cell proliferation and induced apoptosis. The β-cells of LDL receptor-deficient mice were found to be partially resistant to the pro-apoptotic effect of LDL, while the pro-apoptotic effect of VLDL involved caspase-3 cleavage and reduction of cellular levels of the c-Jun N-terminal kinase-interacting protein-1 and was dose-dependently antagonized by HDL. The anti-apoptotic effect of HDL was mediated in part by inhibition of caspase-3 cleavage and activation of protein kinase B (Akt) [75]. Our laboratory has also obtained evidence from experiments in β-cells from human islets that HDL inhibits apoptosis induced either by high glucose or by IL-1β and stimulates proliferation (*Rütti, Rohrer, von Eckardstein, Donath*; unpublished data). Overall, HDL is decreased in the prediabetic state, hampering its antiapoptotic and hence β-cytoprotective signals, which again aggravates the pro-apoptotic and β-cytotoxic signals of insulin resistance (VLDL, FFA, proinflammatory cytokines). This puts HDL into the center of a vicious cycle in which insulin resistance decreases HDL and thereby exacerbates a factor that might protect from β-cell loss and from the manifestation of type 2 diabetes mellitus (Fig. 18.3).

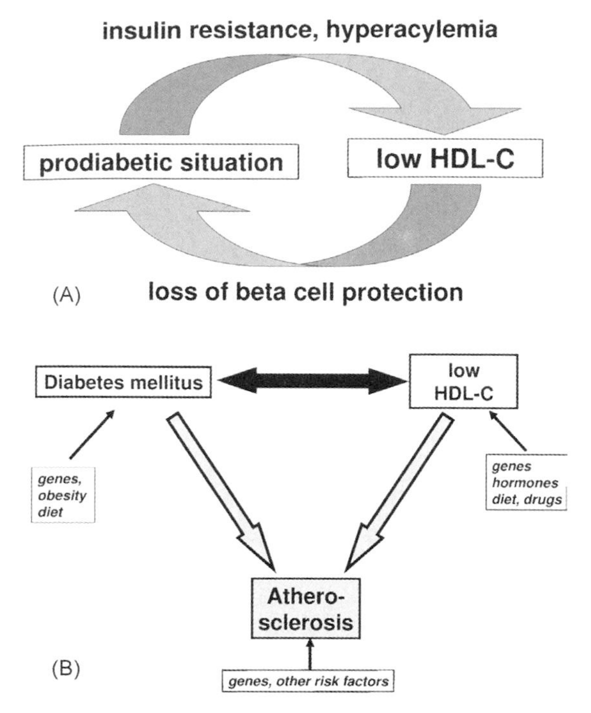

Fig. 18.3 *Possible interplay of HDL, diabetes, and atherosclerosis. A)* Insulin resistance and free fatty acids contribute to low HDL-C in the prediabetic situation (e.g., metabolic syndrome). Low or dysfunctional HDL favor the loss of beta cells and thereby the progression to manifest diabetes mellitus type 2. *B)* Both diabetes and HDL-C are proatherogenic. Because of the interplay shown in A, (pre)diabetes mellitus type 2 and low HDL-C mutually promote their pathogenesis.

18.5
Treatment of Low HDL-C in Metabolic Syndrome

Therapeutic Lifestyle Changes The initial intervention recommended in order to increase HDL-C levels is therapeutic lifestyle changes (see also Chapter 14 by Sviridov). These include smoking cessation, weight loss, and exercise. All three interventions have been shown to improve the cardiovascular risk profile, including an increase in HDL-C, and to reduce cardiovascular morbidity and mortality [76,77]. Smoking cessation was found to increase HDL-C by a mean of 4 mg dL^{-1} (0.10 mmol L^{-1}), more so in women than in men and in persons with elevated baseline HDL-C levels (>47 mg dL^{-1} [1.21 mmol L^{-1}]) [78]. In a meta-analysis, one kilogram of weight reduction was associated with an increase in HDL-C of 0.35 mg dL^{-1} (0.009 mmol L^{-1}) in subjects who achieved a stabilized, reduced weight ($P \leq 0.01$) [79]. Regular aerobic exercise was also found to increase HDL-C level by about 5 percent in healthy, sedentary persons [80].

Endocannabinoid Receptor 1 Antagonists Rimonabant, a selective endocannabinoid type 1 receptor blocker, was primarily developed to reduce appetite and thereby help overweight and obese patients to reduce bodyweight (see also Chapter 21 by Rader). In addition, it was found to improve cardiovascular and metabolic risk factors in nondiabetic and diabetic patients with excess weight or obesity. In both nondiabetic and diabetic overweight individuals, rimonabant was found to increase HDL-C dose-dependently by up to 20 % [81,82]. This increase was more than would be expected merely from the reduction in body weight, probably because endocannabinoid type 1 receptors are also expressed outside of the central nervous system – namely in liver, pancreas and adipose tissue – and mediate direct metabolic regulatory effects [83].

Niacin Niacin (nicotinic acid) is the most effective agent currently available for increasing HDL-C (15–35 %) [3] (see chapter 21 by Rader). It reduces apoA-I catabolism by interfering with hepatocyte uptake of HDL [84] and by stimulating ABCA1 [85]. Niacin also decreases TG level, slightly reduces LDL level, and increases LDL particle size [86]. Trials have shown that treatment with nicotinic acid reduces progression of atherosclerosis and of clinical events and mortality from coronary heart disease [87].

Peroxisome Proliferator-Activated Receptor (PPAR) Agonists The PPAR family is composed of three members: PPARα, PPARγ, and PPARδ, which are activated by several ligands including eicosanoids, fibrates, glitazones, and FFA (see Chapter 21 by Rader). PPARα activation results in upregulation of genes involved in the uptake of FFA and their oxidation, as a consequence of which TG biosynthesis is reduced, while activation of PPARs also triggers the upregulation of apoA-I and apoA-II expression [88]. ABCA1-mediated cholesterol efflux from macrophages is also promoted upon PPARα activation [89]. Both effects result in an increase in HDL-C levels. Fibrates, the classical PPARα agonists, can raise HDL by 10–20 % [3] and are even more potent in lowering triglycerides. In large controlled trials, the effect of fibrates on cardiovascular outcomes was controversial: in VA-HIT, gemfibrozil was found to reduce cardiovascular event rates, whereas in BIP and FIELD bezafibrate and fenofibrate, respectively, did not lower cardiovascular event rates [90–92]. In post-hoc analyses of two of these trials, patients with metabolic syndrome and diabetes mellitus type 2 were found to benefit from treatment with gemfibrozil and bezafibrate [93,94]. Furthermore, glitazones, which were primarily developed as insulin sensitizers, were found to suppress FFA through the inhibition of lipolysis in the adipose tissue and showed antiinflammatory, antioxidant, and antiprocoagulant properties [95]. In patients with type 2 diabetes mellitus or metabolic syndrome, the PPAR-γ agonist pioglitazone increased HDL-C by about 15 % [96], while in the PROactive study pioglitazone treatment also increased HDL-C by about 10 % and reduced combined endpoint of all-cause mortality and myocardial infarction by 15 % [97]. Surprisingly, another PPAR-γ agonist, rosiglitazone, only increased HDL-C in patients with diabetes mellitus, but not in patients with metabolic syndrome without diabetes [98]. Since patients with metabolic syndrome present both impaired glucose and lipid metabolism, combined PPAR-α and -γ agonists – so called dual agonists – have been developed. Dual agonists, as well as PPAR-δ agonists and pan-agonists, are currently being evaluated in clinical trials and appear to be more efficient in raising HDL-C than the first generation of selective PPAR-α or -γ agonists [99,100].

CETP Inhibitors CETP-mediated exchange of CE and TG between HDL and TG-rich lipoproteins is enhanced in patients with metabolic syndrome. As high plasma CETP correlates with low HDL, a new strategy to increase HDL-C levels may be to inhibit CETP [101]. JTT-705 and torcetrapib have been shown to raise HDL-C levels in patients with low HDL-C [102,103] (see also Chapter 21 by Rader). It seems that the increase in plasma HDL and apoA-I levels is associated with a reduction in the catabolism of HDL and apoA-I [104]. In theory, CETP inhibitors could provide a powerful treatment for people with metabolic syndrome, although the consequences for reverse cholesterol transport remain unknown. The development of torcetrapib was previously stopped, because it had no effect on atherosclerosis as assessed by vascular ultrasound examination and caused an excess of cardiovascular events (105,106). Further analyses will have to show whether this adverse effect is related to the compound and its side effects or to the mode of CETP inhibition in general [105,106].

18.6 Conclusions

In the metabolic syndrome, HDL-C is decreased, HDL composition changed, and HDL metabolism altered. In particular, HDL production by ABCA1 and HDL maturation involving LPL, CETP, PLTP, and HL are impaired. HDL possess various antiatherogenic properties, which are or may be impaired in the metabolic syndrome. Altered HDL metabolism and function may thus contribute causally to the high cardiovascular risk of patients with metabolic syndrome. To raise HDL-C concentrations to reduce the atherogenic risk in patients with metabolic syndrome may be a particularly successful measure, but drugs targeting HDL effectively still have to be developed and evaluated in controlled clinical trials.

References

1 Reaven, G. M. (1988) Banting lecture 1988. Role of insulin resistance in human disease. *Diabetes*, **37** (12), 1595–1607.
2 Definition, diagnosis, and classification of diabetes mellitus and its complications. Report of a WHO consultation. Part 1: Diagnosis and classification of diabetes mellitus (1999) Geneva: World Health Organisation.
3 Executive Summary of The Third Report of The National Cholesterol Education Program (NCEP) Expert Panel on Detection, Evaluation, And Treatment of High Blood Cholesterol In Adults (Adult Treatment Panel III). *Jama*, (2001) **285** (19), 2486–2497.
4 Alberti, K. G., Zimmet, P., Shaw, J. (2005) The metabolic syndrome – a new worldwide definition. *Lancet*, **366** (9491), 1059–1062.
5 Unwin, N. (2006) The metabolic syndrome. *J. R. Soc. Med.*, **99** (9), 457–462.

6 Bonora, E. (2006) The metabolic syndrome and cardiovascular disease. *Ann. Med.*, **38** (1), 64–80.
7 Kahn R., Buse J., Ferrannini E., Stern M.; American Diabetes Association; European Association for the Study of Diabetes. (2005) The metabolic syndrome: time for a critical appraisal: joint statement from the American Diabetes Association and the European Association for the Study of Diabetes. *Diabetes Care*, **28** (9), 2289–2304.
8 Duvillard, A., Pont F., Florentin E., Galland-Jos C, Gambert P., Vergès B. (2000) Inefficiency of insulin therapy to correct apolipoprotein A-I metabolic abnormalities in non-insulin-dependent diabetes mellitus. *Atherosclerosis*, **152** (1), 229–237.
9 Frénais R., Ouguerram K., Maugeais C, Mahot P., Maugère P., Krempf M., Magot T. (1997) High density lipoprotein apolipoprotein AI kinetics in NIDDM: a stable isotope study. *Diabetologia*, **40** (5), 578–583.
10 Pietzsch J., Julius U., Nitzsche S., Hanefeld M. (1998) In vivo evidence for increased apolipoprotein A-I catabolism in subjects with impaired glucose tolerance. *Diabetes*, **47** (12), 1928–1934.
11 Ji J., Watts G. F., Johnson A. G., Chan D. C, Ooi E. M., Rye K. A., Serone A. P., Barrett P. H. (2006) High-density lipoprotein (HDL) transport in the metabolic syndrome: application of a new model for HDL particle kinetics. *J. Clin. Endocrinol. Metab.*, **91** (3), 973–979.
12 Pont F., Duvillard L., Florentin E., Gambert P., Verges B. (2002) High-density lipoprotein apolipoprotein A-I kinetics in obese insulin resistant patients. An in vivo stable isotope study. *Int. J. Obes. Relat. Metab. Disord.*, **26** (9), 1151–1158.
13 Watts G. F., Barrett P. H., Ji J., Serone A. P., Chan D. C, Croft K. D., Loehrer F., Johnson A. G. (2003) Differential regulation of lipoprotein kinetics by atorvastatin and fenofibrate in subjects with the metabolic syndrome. *Diabetes*, **52** (3), 803–811.
14 Murao K., Wada Y., Nakamura T., Taylor A. H., Mooradian A. D., Wong N. C. (1998) Effects of glucose and insulin on rat apolipoprotein A-I gene expression. *J. Biol. Chem.*, **273** (30), 18959–18965.
15 Zheng X. L., Matsubara S., Diao C, Hollenberg M. D., Wong N. C. (2001) Epidermal growth factor induction of apolipoprotein A-I is mediated by the Ras-MAP kinase cascade and Sp1. *J. Biol. Chem.*, **276** (17), 13822–13829.
16 Wang, Y. and Oram, J. F. (2005) Unsaturated fatty acids phosphorylate and destabilize ABCA1 through a phospholipase D2 pathway. *J. Biol. Chem.*, **280** (43), 35896–35903.
17 Adiels M., Olofsson S. O, Taskinen M. R, Boren J. (2006) Diabetic dyslipidaemia. *Curr. Opin. Lipidol.*, **17** (3), 238–246.
18 Dusserre, E., Moulin, P., Vidal, H. (2000) Differences in mRNA expression of the proteins secreted by the adipocytes in human subcutaneous and visceral adipose tissues. *Biochim. Biophys. Acta*, **1500** (1), 88–96.
19 Verges B., Petit J. M., Duvillard L., Dautin G., Florentin E., Galland F., Gambert P. (2006) Adiponectin is an important determinant of apoA-I catabolism. *Arterioscler. Thromb. Vasc. Biol.*, **26** (6), 1364–1369.
20 Neumeier M., Sigruener A., Eggenhofer E., Weigert J., Weiss T. S., Schaeffler A., Schlitt H. J., Aslanidis C, Piso P., Langmann T., Schmitz G., Scholmerich J., Buechler C. (2007) High molecular weight adiponectin reduces apolipoprotein B and E release in human hepatocytes. *Biochem. Biophys. Res. Commun.*, **352** (2), 543–548.
21 von Eckardstein, A. (2006) Differential diagnosis of familial high density lipoprotein deficiency syndromes. *Atherosclerosis*, **186** (2), 231–239.
22 Cavelier C, Lorenzi I., Rohrer L., von Eckardstein A. (2006) Lipid efflux by

the ATP-binding cassette transporters ABCA1 and ABCG1. *Biochim. Biophys. Acta*, **1761** (7), 655–666.

23 Uehara Y., Engel T., Li Z., Goepfert C, Rust S., Zhou X., Langer C, Schachtrup C, Wiekowski J., Lorkowski S., Assmann G., von Eckardstein A. (2002) Polyunsaturated fatty acids and acetoacetate downregulate the expression of the ATP-binding cassette transporter A1. *Diabetes*, **51** (10), 2922–2928.

24 Uehara Y, Miura S, von Eckardstein A, Abe S, Fujii A, Matsuo Y, Rust S, Lorkowski S, Assmann G, Yamada T, Saku K. (2007) Unsaturated fatty acids suppress the expression of the ATP-binding cassette transporter G1 (ABCG1) and ABCA1 genes via an LXR/RXR responsive element. *Atherosclerosis*, **191** (1), 11-21.

25 Inazu A., Brown M. L., Hesler C. B., Agellon L. B., Koizumi J., Takata K., Maruhama Y., Mabuchi H., Tall A. R. (1990) Increased high-density lipoprotein levels caused by a common cholesteryl-ester transfer protein gene mutation. *N. Engl. J. Med.*, **323** (18), 1234–1238.

26 Agellon L. B., Walsh A., Hayek T., Moulin P., Jiang X. C, Shelanski S. A., Breslow J. L., Tall A. R. (1991) Reduced high density lipoprotein cholesterol in human cholesteryl ester transfer protein transgenic mice. *J. Biol. Chem.*, **266** (17), 10796–10801.

27 Riemens S. C, Van Tol A., Stulp B. K., Dullaart R. P. (1999) Influence of insulin sensitivity and the TaqIB cholesteryl ester transfer protein gene polymorphism on plasma lecithin:cholesterol acyltransferase and lipid transfer protein activities and their response to hyperinsulinemia in non-diabetic men. *J. Lipid Res.*, **40** (8), 1467–1474.

28 Bagdade J. D., Lane J. T., Subbaiah P. V., Otto M. E., Ritter M. C. (1993) Accelerated cholesteryl ester transfer in noninsulin-dependent diabetes mellitus. *Atherosclerosis*, **104** (1–2), 69–77.

29 Lamarche B., Uffelman K. D., Carpentier A., Cohn J. S., Steiner G., Barrett P. H., Lewis G. F. (1999) Triglyceride enrichment of HDL enhances in vivo metabolic clearance of HDL apo A-I in healthy men. *J. Clin. Invest.*, **103** (8), 1191–1199.

30 Sparks D. L., Davidson W. S., Lund-Katz S., Phillips M. C. (1995) Effects of the neutral lipid content of high density lipoprotein on apolipoprotein A-I structure and particle stability. *J. Biol. Chem.*, **270** (45), 26910–26917.

31 Liang, H. Q., Rye, K. A., Barter, P. J. (1994) Dissociation of lipid-free apolipoprotein A-I from high density lipoproteins. *J. Lipid Res.*, **35** (7), 1187–1199.

32 Rashid S., Uffelman K. D., Barrett P. H., Vicini P., Adeli K., Lewis G. F. (2001) Triglyceride enrichment of HDL does not alter HDL-selective cholesteryl ester clearance in rabbits. *J. Lipid Res.*, **42** (2), 265–271.

33 Hirano R., Igarashi O., Kondo K., Itakura H., Matsumoto A. (2001) Regulation by long-chain fatty acids of the expression of cholesteryl ester transfer protein in HepG2 cells. *Lipids*, **36** (4), 401–406.

34 Cheema S. K., Agarwal-Mawal A., Murray C. M., Tucker S. (2005) Lack of stimulation of cholesteryl ester transfer protein by cholesterol in the presence of a high-fat diet. *J. Lipid Res.*, **46** (11), 2356–2366.

35 Lagrost L., Florentin E., Guyard-Dangremont V., Athias A., Gandjini H., Lallemant C, Gambert P. (1995) Evidence for nonesterified fatty acids as modulators of neutral lipid transfers in normolipidemic human plasma. *Arterioscler. Thromb. Vasc. Biol.*, **15** (9), 1388–1396.

36 Jiang X. C., Bruce C., Mar J., Lin M., Ji Y., Francone O. L., Tall A. R. (1999) Targeted mutation of plasma phospholipid transfer protein gene markedly reduces high-density lipoprotein levels. *J. Clin. Invest.*, **103** (6), 907–914.

37 Jiang X., Francone O. L., Bruce C., Milne R., Mar J., Walsh A., Breslow J. L., Tall A. R. (1996) Increased

prebeta-high density lipoprotein, apolipoprotein AI, and phospholipid in mice expressing the human phospholipid transfer protein and human apolipoprotein AI transgenes. *J. Clin. Invest.*, **98** (10), 2373–2380.

38 Riemens S. C., van Tol A., Sluiter W. J., Dullaart R. P. (1998) Plasma phospholipid transfer protein activity is related to insulin resistance: impaired acute lowering by insulin in obese Type II diabetic patients. *Diabetologia*, **41** (8), 929–934.

39 van Tol, A. (2002) Phospholipid transfer protein. *Curr. Opin. Lipidol.*, **13** (2), 135–139.

40 Rye K. A., Jauhiainen M., Barter P. J., Ehnholm C. (1998) Triglyceride-enrichment of high density lipoproteins enhances their remodelling by phospholipid transfer protein. *J. Lipid Res.*, **39** (3), 613–622.

41 Cao G., Beyer T. P., Yang X. P., Schmidt R. J., Zhang Y., Bensch W. R., Kauffman R. F., Gao H., Ryan T. P., Liang Y., Eacho P. I., Jiang X. C. (2002) Phospholipid transfer protein is regulated by liver X receptors in vivo. *J. Biol. Chem.*, **277** (42), 39561–39565.

42 Clay, M. A., Newnham, H. H., Barter, P. J. (1991) Hepatic lipase promotes a loss of apolipoprotein A-I from triglyceride-enriched human high density lipoproteins during incubation in vitro. *Arterioscler. Thromb.*, **11** (2), 415–422.

43 Despres J. P., Ferland M., Moorjani S., Nadeau A., Tremblay A., Lupien P. J., Theriault G., Bouchard C. (1989) Role of hepatic-triglyceride lipase activity in the association between intra-abdominal fat and plasma HDL cholesterol in obese women. *Arteriosclerosis*, **9** (4), 485–492.

44 Laakso M., Sarlund H., Ehnholm C., Voutilainen E., Aro A., Pyorala K. (1987) Relationship between postheparin plasma lipases and high-density lipoprotein cholesterol in different types of diabetes. *Diabetologia*, **30** (9), 703–706.

45 Lewis G. F., Murdoch S., Uffelman K., Naples M., Szeto L., Albers A,. Adeli K., Brunzell J.D. (2004) Hepatic lipase mRNA, protein, and plasma enzyme activity is increased in the insulin-resistant, fructose-fed Syrian golden hamster and is partially normalized by the insulin sensitizer rosiglitazone. *Diabetes*, **53** (11), 2893–2900.

46 Knauer T. E., Woods J. A., Lamb R. G., Fallon H. J. (1982) Hepatic triacylglycerol lipase activities after induction of diabetes and administration of insulin or glucagon. *J. Lipid Res.*, **23** (4), 631–637.

47 Barrans A., Collet X., Barbaras R., Jaspard B., Manent J., Vieu C., Chap H., Perret B. (1994) Hepatic lipase induces the formation of pre-beta 1 high density lipoprotein (HDL) from triacylglycerol-rich H DL2. A study comparing liver perfusion to in vitro incubation with lipases. *J. Biol. Chem.*, **269** (15), 11572–11577.

48 Guendouzi K., Jaspard B., Barbaras R., Motta C., Vieu C., Marcel Y., Chap H., Perret B., Collet X. (1999) Biochemical and physical properties of remnant-HDL2 and of pre beta 1-HDL produced by hepatic lipase. *Biochemistry*, **38** (9), 2762–2768.

49 Merkel, M., Eckel, R. H., Goldberg, I. J. (2002) Lipoprotein lipase: genetics, lipid uptake, and regulation. *J. Lipid Res.*, **43** (12), 1997–2006.

50 Weinstock P. H., Bisgaier C. L., Aalto-Setala K., Radner H., Ramakrishnan R., Levak-Frank S., Essenburg A. D., Zechner R., Breslow J. L. (1995) Severe hypertriglyceridemia, reduced high density lipoprotein, and neonatal death in lipoprotein lipase knockout mice. Mild hypertriglyceridemia with impaired very low density lipoprotein clearance in heterozygotes. *J. Clin. Invest.*, **96** (6), 2555–2568.

51 Goldberg I. J., Blaner W. S., Vanni T. M., Moukides M., Ramakrishnan R. (1990) Role of lipoprotein lipase in the regulation of high density lipoprotein apolipoprotein metabolism. Studies in normal and

lipoprotein lipase-inhibited monkeys. *J. Clin. Invest.*, **86** (2), 463–473.
52 Knudsen P., Eriksson J., Lahdenpera S., Kahri J., Groop L., Taskinen M. R. (1995) Changes of lipolytic enzymes cluster with insulin resistance syndrome. Botnia Study Group. *Diabetologia*, **38** (3), 344–350.
53 Shimamura M., Matsuda M., Ando Y., Koishi R., Yasumo H., Furukawa H., Shimomura I. (2004) Leptin and insulin down-regulate angiopoietin-like protein 3, a plasma triglyceride-increasing factor. *Biochem. Biophys. Res. Commun.*, **322** (3), 1080–1085.
54 Yamada T., Ozaki N., Kato Y., Miura Y., Oiso Y. (2006) Insulin downregulates angiopoietin-like protein 4 mRNA in 3T3-L1 adipocytes. *Biochem. Biophys. Res. Commun.*, **347** (4), 1138–1144.
55 Rinninger F., Kaiser T., Mann W. A., Meyer N., Greten H., Beisiegel U. (1998) Lipoprotein lipase mediates an increase in the selective uptake of high density lipoprotein-associated cholesteryl esters by hepatic cells in culture. *J. Lipid Res.*, **39** (7), 1335–1348.
56 Cuchel, M. and Rader, D. J. (2006) Macrophage reverse cholesterol transport: key to the regression of atherosclerosis?*Circulation*, **113** (21), 2548–2555.
57 Brites F. D., Bonavita C. D., De Geitere C, Cloes M., Delfly B., Yael M. J., Fruchart J., Wikinski R. W., Castro G. R. (2000) Alterations in the main steps of reverse cholesterol transport in male patients with primary hypertriglyceridemia and low HDL-cholesterol levels. *Atherosclerosis*, **152** (1), 181–192.
58 Duell, P. B., Oram, J. F., Bierman, E. L. (1991) Nonenzymatic glycosylation of HDL and impaired HDL-receptor-mediated cholesterol efflux. *Diabetes*, **40** (3), 377–384.
59 Palmer, A. M., Murphy, N., Graham, A. (2004) Triglyceride-rich lipoproteins inhibit cholesterol efflux to apolipoprotein (apo) A1 from human macrophage foam cells. *Atherosclerosis*, **173** (1), 27–38.
60 Alenezi M. Y., Marcil M., Blank D., Sherman M., Genest J. (2004) Is the decreased high-density lipoprotein cholesterol in the metabolic syndrome due to cellular lipid efflux defect? *J. Clin. Endocrinol. Metab.*, **89** (2), 761–764.
61 Foumier N., Atger V., Cogny A., Vedie B., Giral P., Simon A., Moatti N., Paul J. L. (2001) Analysis of the relationship between triglyceridemia and HDL-phospholipid concentrations: consequences on the efflux capacity of serum in the Fu5AH system. *Atherosclerosis*, **157** (2), 315–323.
62 Uint L., Sposito A., Brandizzi L. I., Yoshida V. M., Maranhao R. C., Luz P. L. (2003) Cellular cholesterol efflux mediated by HDL isolated from subjects with low HDL levels and coronary artery disease. *Arq. Bras. Cardiol.*, p. 39–41, 35–38.
63 Passarelli M., Tang C., McDonald T. O., O'Brien K. D., Gerrity R. G., Heinecke J. W., Oram J. F. (2005) Advanced glycation end product precursors impair ABCA1-dependent cholesterol removal from cells. *Diabetes*, **54** (7), 2198–2205.
64 Ohgami N., Nagai R., Miyazaki A., Ikemoto M., Arai H., Horiuchi S., Nakayama H. (2001) Scavenger receptor class B type I-mediated reverse cholesterol transport is inhibited by advanced glycation end products. *J. Biol. Chem.*, **276** (16), 13348–13355.
65 Borggreve, S. E., DeVries, R., Dullaart, R. P. (2003) Alterations in high-density lipoprotein metabolism and reverse cholesterol transport in insulin resistance and type 2 diabetes mellitus: role of lipolytic enzymes, lecithin:cholesterol acyltransferase and lipid transfer proteins. *Eur. J. Clin. Invest.*, **33** (12), 1051–1069.
66 Lally S., Tan C. Y., Owens D., Tomkin G. H. (2006) Messenger RNA levels of genes involved in dysregulation of postprandial lipoproteins in type 2 diabetes: the role of Niemann-Pick C1-like 1, ATP-binding cassette, transporters G5 and

G8, and of microsomal triglyceride transfer protein. *Diabetologia*, **49** (5), 1008–1016.

67 Bloks V. W., Bakker-Van Waarde W. M., Verkade H. J., Kema I. P., Wolters H., Vink E., Groen A. K., Kuipers F. (2004) Down-regulation of hepatic and intestinal Abcg5 and Abcg8 expression associated with altered sterol fluxes in rats with streptozotocin-induced diabetes. *Diabetologia*, **47** (1), 104–112.

68 Dietschy, J. M. and Turley, S. D. (2002) Control of cholesterol turnover in the mouse. *J. Biol. Chem.*, **277** (6), 3801–3804.

69 Groen A. K., Oude Elferink R. P., Verkade H. J., Kuipers F. (2004) The ins and outs of reverse cholesterol transport. *Ann. Med.*, **36** (2), 135–145.

70 Hansel B., Giral P., Nobecourt E., Chantepie S., Bruckert E., Chapman M. J., Kontush A. (2004) Metabolic syndrome is associated with elevated oxidative stress and dysfunctional dense high-density lipoprotein particles displaying impaired antioxidative activity. *J. Clin. Endocrinol. Metab.*, **89** (10), 4963–4971.

71 Nobécourt E., Jacqueminet S., Hansel B., Chantepie S., Grimaldi A., Chapman M. J., Kontush A. (2005) Defective antioxidative activity of small dense HDL3 particles in type 2 diabetes: relationship to elevated oxidative stress and hyperglycaemia. *Diabetologia*, **48** (3), 529–538.

72 Hansel B., Kontush A., Bonnefont-Rousselot D., Bruckert E., Chapman M. J. (2006) Alterations in lipoprotein defense against oxidative stress in metabolic syndrome. *Curr. Atheroscler. Rep.*, **8** (6), 501–509.

73 Mastorikou, M., Mackness, M., Mackness, B. (2006) Defective metabolism of oxidized phospholipid by HDL from people with type 2 diabetes. *Diabetes*, **55** (11), 3099–3103.

74 Donath MY, Storling J., Maedler K., Mandrup-Poulsen T. (2003) Inflammatory mediators and islet beta-cell failure: a link between type 1 and type 2 diabetes. *J. Mol. Med.*, **81** (8), 455–470.

75 Roehrich M. E., Mooser V., Lenain V., Herz J., Nimpf J., Azhar S., Bideau M., Capponi A., Nicod P., Haefliger J. A., Waeber G. (2003) Insulin-secreting beta-cell dysfunction induced by human lipoproteins. *J. Biol. Chem.*, **278** (20), 18368–18375.

76 Reaven, G. M. (2005) The insulin resistance syndrome: definition and dietary approaches to treatment. *Annu. Rev. Nutr.*, **25**: 391–406.

77 Lakka, T. A. and Bouchard, C. (2005) Physical activity, obesity and cardiovascular diseases. *Handb. Exp. Pharmacol.*, (170), 137–163.

78 Maeda, K., Noguchi, Y., Fukui, T. (2003) The effects of cessation from cigarette smoking on the lipid and lipoprotein profiles: a meta-analysis. *Prev. Med.*, **37** (4), 283–290.

79 Dattilo, A. M. and Kris-Etherton, P. M. (1992) Effects of weight reduction on blood lipids and lipoproteins: a meta-analysis. *Am. J. Clin. Nutr.*, **56** (2), 320–328.

80 King A. C., Haskell W. L., Young D. R., Oka R. K., Stefanick M. L. (1995) Long-term effects of varying intensities and formats of physical activity on participation rates, fitness, and lipoproteins in men and women aged 50 to 65 years. *Circulation*, **91** (10), 2596–2604.

81 Scheen A. J., Finer N., Hollander P., Jensen M. D., Van Gaal L. F.; RIO-Diabetes Study Group. (2006) Efficacy and tolerability of rimonabant in overweight or obese patients with type 2 diabetes: a randomised controlled study. Lancet, 368 (9548), 1660–1672.

82 Despres, J. P., Golay, A., Sjostrom, L. (2005) Effects of rimonabant on metabolic risk factors in overweight patients with dyslipidemia. *N. Engl. J. Med.*, **353** (20), 2121–2134.

83 Matias, I.and Di Marzo, V. (2007) Endocannabinoids and the control of energy balance. *Trends Endocrinol. Metab.*, **18** (1), 27–37.

84 Jin, F. Y., Kamanna, V. S., Kashyap, M. L. (1997) Niacin decreases removal of high-density lipoprotein apolipoprotein A-I but not cholesterol ester by Hep G2 cells. Implication for reverse cholesterol transport. *Arterioscler. Thromb. Vasc. Biol.*, **17** (10), 2020–2028.

85 Rubic, T., Trottmann, M., Lorenz, R. L. (2004) Stimulation of CD36 and the key effector of reverse cholesterol transport ATP-binding cassette A1 in monocytoid cells by niacin. *Biochem. Pharmacol.*, **67** (3), 411–419.

86 Pan J., Lin M., Kesala R. L., Van J., Charles M. A. (2002) Niacin treatment of the atherogenic lipid profile and Lp (a) in diabetes. *Diabetes Obes. Metab.*, **4** (4), 255–261.

87 Carlson, L. A. (2005) Nicotinic acid: the broad-spectrum lipid drug. A 50th anniversary review. *J. Intern. Med.*, **258** (2), 94–114.

88 Staels B., Dallongeville J., Auwerx J., Schoonjans K., Leitersdorf E., Fruchart J. C. (1998) Mechanism of action of fibrates on lipid and lipoprotein metabolism. *Circulation*, **98** (19), 2088–2093.

89 Chinetti G., Lestavel S., Bocher V., Remaley A. T., Neve B., Torra I. P., Teissier E., Minnich A., Jaye M., Duverger N., Brewer H. B., Fruchart J. C., Clavey V., Staels B. (2001) PPAR-alpha and PPAR-gamma activators induce cholesterol removal from human macrophage foam cells through stimulation of the ABCA1 pathway. *Nat. Med.*, **7** (1), 53–58.

90 Keech A., Simes R. J., Barter P., Best J., Scott R., Taskinen M. R., Forder P., Pillai A., Davis T., Glasziou P., Drury P., Kesaniemi Y. A., Sullivan D., Hunt D., Colman P., d'Emden M., Whiting M., Ehnholm C., Laakso M.; FIELD study investigators. (2005) Effects of long-term fenofibrate therapy on cardiovascular events in 9795 people with type 2 diabetes mellitus (the FIELD study): randomised controlled trial. *Lancet*, **366** (9500), 1849–1861.

91 Tanne D., Koren-Morag N., Graff E., Goldbourt U. (2001) Blood lipids and first-ever ischemic stroke/transient ischemic attack in the Bezafibrate Infarction Prevention (BIP) Registry: high triglycerides constitute an independent risk factor. *Circulation*, **104** (24), 2892–2897.

92 Robins S. J., Collins D., Wittes J. T., Papademetriou V., Deedwania P. C., Schaefer E. J., McNamara J. R., Kashyap M. L., Hershman J. M., Wexler L. F., Rubins H. B.; VA-HIT Study Group. Veterans Affairs High-Density Lipoprotein Intervention Trial. (2001) Relation of gemfibrozil treatment and lipid levels with major coronary events: VA-HIT: a randomized controlled trial. *Jama*, **285** (12), 1585–1591.

93 Tenenbaum A., Motro M., Fisman E. Z., Tanne D., Boyko V., Behar S. (2005) Bezafibrate for the secondary prevention of myocardial infarction in patients with metabolic syndrome. *Arch. Intern. Med.*, **165** (10), 1154–1160.

94 Rubins H. B., Robins S. J., Collins D., Nelson D. B., Elam M. B., Schaefer E. J., Faas F. H., Anderson J. W. (2002) Diabetes, plasma insulin, and cardiovascular disease: subgroup analysis from the Department of Veterans Affairs high-density lipoprotein intervention trial (VA-HIT). *Arch. Intern. Med.*, **162** (22), 2597–2604.

95 Staels, B. and Fruchart, J. C. (2005) Therapeutic roles of peroxisome proliferator-activated receptor agonists. *Diabetes*, **54** (8), 2460–2470.

96 Szapary P. O., Bloedon L. T., Samaha F. F., Duffy D., Wolfe M. L., Soffer D., Reilly M. P., Chittams J., Rader D. J. (2006) Effects of pioglitazone on lipoproteins, inflammatory markers, and adipokines in nondiabetic patients with metabolic syndrome. *Arterioscler. Thromb. Vasc. Biol.*, **26** (1), 182–188.

97 Dormandy J. A., Charbonnel B., Eckland D. J., Erdmann E., Massi-Benedetti M., Moules I. K., Skene A. M., Tan M. H., Lefebvre P. J., Murray G. D., Standl E., Wilcox R. G., Wilhelmsen L., Betteridge J., Birkeland K., Golay A., Heine R. J., Koranyi L., Laakso M., Mokan M.,

Norkus A., Pirags V., Podar T., Scheen A., Scherbaum W., Schernthaner G., Schmitz O., Skrha J., Smith U., Taton J.; PROactive investigators. (2005) Secondary prevention of macrovascular events in patients with type 2 diabetes in the PROactive Study (PROspective pioglitAzone Clinical Trial In macroVascular Events): a randomised controlled trial. *Lancet*, **366** (9493), 1279–1289.

98 Samaha F. F., Szapary P. O., Iqbal N., Williams M. M., Bloedon L. T., Kochar A., Wolfe M.L., Rader D.J. (2006) Effects of rosiglitazone on lipids, adipokines, and inflammatory markers in nondiabetic patients with low high-density lipoprotein cholesterol and metabolic syndrome. *Arterioscler. Thromb. Vasc. Biol.*, **26** (3), 624–630.

99 Kendall D. M., Rubin C. J., Mohideen P., Ledeine J. M., Belder R., Gross J., Norwood P., O'Mahony M., Sail K., Sloan G., Roberts A., Fiedorek F. T., DeFronzo R. A. (2006) Improvement of glycemic control, triglycerides, and HDL cholesterol levels with muraglitazar, a dual (alpha/gamma) peroxisome proliferator-activated receptor activator, in patients with type 2 diabetes inadequately controlled with metformin monotherapy: A double-blind, randomized, pioglitazone-comparative study. *Diabetes Care*, **29** (5), 1016–1023.

100 Furnsinn, C., Willson, T. M., Brunmair, B. (2007) Peroxisome proliferator-activated receptor-delta, a regulator of oxidative capacity, fuel switching and cholesterol transport. *Diabetologia*, **50** (1), 8–17.

101 van der Steeg W. A., Kuivenhoven J. A., Klerkx A. H., Boekholdt S. M., Hovingh G. K., Kastelein J. J. (2004) Role of CETP inhibitors in the treatment of dyslipidemia. *Curr. Opin. Lipidol.*, **15** (6), 631–636.

102 Kuivenhoven J. A., de Grooth G. J., Kawamura H., Klerkx A. H., Wilhelm F., Trip M. D., Kastelein J. J. (2005) Effectiveness of inhibition of cholesteryl ester transfer protein by JTT-705 in combination with pravastatin in type II dyslipidemia. *Am. J. Cardiol.*, **95** (9), 1085–1088.

103 Brousseau M. E., Schaefer E. J., Wolfe M. L., Bloedon L. T., Digenio A. G., Clark R. W., Mancuso J. P., Rader D. J. (2004) Effects of an inhibitor of cholesteryl ester transfer protein on HDL cholesterol. *N. Engl. J. Med.*, **350** (15), 1505–1515.

104 Brousseau M. E., Diffenderfer M. R., Millar J. S., Nartsupha C., Asztalos B. F., Welty F. K., Wolfe M. L., Rudling M., Bjorkhem I., Angelin B., Mancuso J. P., Digenio A. G., Rader D. J., Schaefer E. J. (2005) Effects of cholesteryl ester transfer protein inhibition on high-density lipoprotein subspecies, apolipoprotein A-I metabolism, and fecal sterol excretion. *Arterioscler. Thromb. Vasc. Biol.*, **25** (5), 1057–1064.

105 Nissen S. E., Tardif J. C., Nicholls S. J., Revkin J. H., Shear C. L., Duggan W. T., Ruzyllo W., Bachinsky W. B., Lasala G. P., Tuzcu E. M.; ILLUSTRATE Investigators. (2007) Effect of torcetrapib on the progression of coronary atherosclerosis. N Engl J Med. **356** (13): 1304–1316.

106 Kastelein J. J., van Leuven S. I., Burgess L., Evans G. W., Kuivenhoven J. A., Barter P. J., Revkin J. H., Grobbee D. E., Riley W. A., Shear C. L., Duggan W. T., Bots M. L.; RADIANCE 1 Investigators. (2007) Effect of torcetrapib on the carotid atherosclerosis in familial hypercholesterolemia. N Engl J Med. **356** (16): 1620–1630

19
Genetics of High-Density Lipoproteins

Jacques Genest, Zari Dastani, James C. Engert, Michel Marcil

19.1
Introduction

Genetic lipoprotein disorders are frequently seen in patients with premature coronary artery disease (CAD), one of the most common lipoprotein disorders being low levels of high-density lipoprotein cholesterol (HDL-C) [1]. Isolated familial HDL deficiency is seen in approximately 4 % of patients with premature CAD [2,3]. In most cases low HDL-C is a consequence, rather than a cause, of multiple metabolic disorders, such as abdominal obesity, elevated blood pressure, hypertriglyceridemia, and insulin resistance, commonly seen as part of the metabolic syndrome [4]. The estimation of genetic risk for CAD has been examined in the prospective Framingham Heart Study, where the relative risk of CAD in a patient with an affected first degree relative is approximately double (depending on the presence of premature CAD in one or both parents and correction for conventional cardiovascular risk factors) [5]. While this estimate represents the average for an entire cohort, the genetic contribution for an individual's HDL-C subjects may vary considerably. Examples of strong genetic predisposition are familial hypercholesterolemia, where a single gene defect (in the low-density lipoprotein receptor) contributes to most of the familial expression of CAD and Tangier disease and familial HDL deficiency, in which defects at the ABCA1 gene locus confer a dominant (and probably dominant negative) form of HDL deficiency [6–8]. Conversely, lifestyle, diabetes, dyslipidemia, cigarette smoking, and hypertension contribute to most of the population attributable risk in the INTERHEART study [9]. Remarkably, the apolipoproteins B to AI ratio (apoB/AI) contributes 49 % of the population attributable risk, underlying the relative and absolute importance of lipoproteins in the pathogenesis of CAD. The cholesterol/HDL-C ratio remains the best lipoprotein predictor of CVD [9]. In the majority of cases of low HDL-C encountered clinically, the decrease in HDL-C is the result of altered metabolism, as outlined in the next chapter.

The identification of single gene disorders may lead the way to better understanding of complex metabolic pathways. Because of their importance in lipoprotein metabolism, vascular endothelial function [10,11], and prevention of atherosclerosis

High-Density Lipoproteins: From Basic Biology to Clinical Aspects. Edited by Christopher J. Fielding

ISBN: 978-3-527-31717-2

[12] (see also Chapter 21), understanding of the genes that regulate HDL metabolism may pave the way for novel therapeutic approaches. This approach has proven remarkably successful in the case of the statin class of medications, where specific inhibitors of cellular sterol synthesis were targeted for drug discovery. New approaches in the treatment of HDL have been emphasized by two novel therapies in man for the treatment of CAD: the infusion of apoA-I-containing proteoliposomes, through the use of a mutant form of apoA-I, apoA-I_{Milano}, in patients with acute coronary syndromes [13], and the long-term treatment of subjects with low HDL-C with the cholesteryl ester transfer protein (CETP) inhibitor Torcetrapib© [14] (see Chapter 21 by D. Rader). A novel therapeutic approach using agonists of the LxR/RxR pathway is based on the molecular cellular physiology of cellular cholesterol transport and the role of the ATP binding cassette A1 (ABCA1) transporter in the genesis of HDL particles [15].

This chapter examines the genetic etiology of HDL-C. We first look at monogenic disorders associated with HDL, including diseases in humans in which low HDL-C is a secondary manifestation of a rare disorder. We will then examine genome-wide studies in which a level of HDL-C is treated as a dichotomous trait and those in which HDL-C is treated as a quantitative trait in the search for quantitative trail loci (QTL). We then review genetic polymorphism studies of candidate genes associated with HDL-C.

19.2 Background

19.2.1 Epidemiology of HDL and Risk of Coronary Artery Disease CAD (See also Chapter 21)

Disorders of lipoprotein metabolism are associated with atherosclerosis, leading to ischemic manifestations in all vascular beds, including the coronary arteries (CAD) (leading to acute coronary syndromes and myocardial infarction), the cerebral circulation (causing strokes), and the peripheral circulation (causing intermittent claudication and gangrene). Ischemic vascular diseases, especially myocardial infarctions, represent the largest cause of death worldwide [16]. Plasma (or serum) level of HDL-C is a continuous and graded negative cardiovascular risk factor [17–19]. Canadian and American CAD prevention guidelines define HDL-C as a categorical risk factor and the absolute value of HDL-C is used in a multivariate model to predict cardiovascular risk and to determine the need for and intensity of preventive therapies [4,20]. It is estimated that approximately 40 % of patients with premature CAD have low HDL-C [2,3], and this represents the most common lipoprotein disorder in patients with CAD. We have shown, as have others, that most patients with low HDL-C have multiple cardiovascular risk factors and features of the metabolic syndrome, with obesity (predominantly abdominal), elevated plasma triglyceride levels, high blood pressure and hyperglycemia, insulin resistance, or diabetes [4]. Despite the strong association between metabolic disorders and HDL-C, plasma levels of HDL-C are in part genetically determined.

19.2.2
HDL Metabolism [10,21] (See also Chapter 21)

The purpose of lipoproteins is to shuttle hydrophobic molecules (cholesteryl esters and triglycerides) from one organ to another in the aqueous environment of the plasma. Lipoproteins are spherical macromolecules consisting of an envelope of phospholipids and free (unesterified) cholesterol and a core of triglycerides and cholesteryl esters. Several proteins – the apolipoproteins (apo) – are arranged circumferentially and provide structural integrity to the lipoprotein, are involved in its assembly and secretion, serve as ligands for specific receptors, and have enzymatic co-activator functions (Chapters 1 and 2). Lipoproteins are separated by hydrated density and by apolipoprotein content. Triglyceride-rich lipoproteins (chylomicrons and very low density lipoproteins – VLDL), low-density lipoproteins (LDL), and HDL are the major lipoprotein classes. Reviews of lipoprotein metabolism and HDL functions have recently been published. [10,21] (Chapters 11 and 12).

The liver and intestine secrete apoA-I as a lipid-free or lipid-poor protein. ApoA-I acquires phospholipids and cholesterol from triglyceride-rich lipoprotein hydrolysis and from cell-mediated phospholipid and cholesterol efflux. The ABCA1 transporter mediates the transfer of phospholipids and cholesterol onto apoA-I to form a nascent HDL particle in hepatic and peripheral cells, including macrophages [22]. The hepatic expression of ABCA1 is necessary and essential for the formation of nascent HDL and the formation of mature forms of HDL [23]. Once in circulation, HDL particles continue to remove cellular cholesterol. The free cholesterol is acylated to form cholesteryl esters, through the action of lecithin:cholesterol acyltransferase (LCAT). Once formed, these cholesteryl esters move to more thermodynamically stable positions in the core of the HDL particle, which assumes a spherical configuration. The formation of cholesteryl esters allows the maintenance of a gradient of cholesterol from the plasma membrane to HDL particles, to ensure a net movement of cholesterol from the cell to HDL. The growing, cholesterol-rich HDL_2 particle is further modified in plasma by the combined actions of CETP and phospholipid transfer protein (PLTP), which mediate the exchange of HDL cholesteryl esters for triglycerides in VLDL. This step is part of the reverse cholesterol transport pathway, a mechanism by which some cholesterol is returned to the liver for elimination into bile. The triglyceride-enriched HDL_{2b} is hydrolyzed by endothelial lipase (EL) and hepatic lipase (HL), while phospholipids are altered by the secretion of phospholipase A2 (sPLA2) and sphingomyelinase (Smase). HDL particles can bind to at least two receptors: the scavenger receptor B1 (SR-B1) [24] and the ABCA1 transporter [22]. All the genes coding for the proteins mentioned above have been implicated in disorders of HDL in humans or in animal models.

High-density lipoproteins have many roles (Chapter 21). The best studied, and therefore considered the most important, is reverse cholesterol transport, though there is a remarkable paucity of evidence that reverse cholesterol transport constitutes a major role of HDL. Indeed, recent data from genetically altered mice indicate that the major determinant of plasma levels of HDL-C is hepatic ABCA1-mediated production of HDL particles [23], with a relatively minor (20 %) contribution from the intestine [25]. Conditional inactivation of ABCA1 in macrophages does not lead to significant changes in plasma HDL-C levels [26].

19.3 Genetic Approaches

In the past two and a half decades, techniques for the identification of genes associated with biological traits evolved dramatically as the complexity of the genome became more obvious and as the tools for genomic research increased the analytical throughput by several orders of magnitude. There are approximately 20–25 000 genes in the human genome, probably coding for over 1 000 000 proteins. After the publication of the apolipoprotein AI (apoA-I) gene in 1983 [27], the candidate gene approach dominated the field of genetics until techniques for performing genome-wide scans for study of Mendelian traits such as Tangier disease (a severe deficiency of HDL-C) became more readily available [28]. The tandem evolution of the fields of computational genetics and high-throughput genomics allowed the identification of many chromosomal regions that may harbor genes related to HDL. Innovative techniques based on sequence homology between species and high-throughput sequencing of large chromosomal regions allowed the identification of further novel genes related to lipoprotein metabolism in general, influencing HDL-C levels [29], as was the case for the identification of apoA5, on chromosome 11q23, part of the apoA-I/CIII/AIV gene complex [30,31]

19.3.1 Monogenic Disorders, Candidate Genes

Family studies of extreme disorders of HDL (severe deficiency or HDL-C excess) have allowed the discovery of monogenic defects in several proteins that modulate HDL-C levels (Figure 19.1). The roles of these genes and their products have been examined and confirmed in genetically modified animal models. These disorders, however, probably account for only a small portion of all individuals displaying low HDL-C.

Genes encoding proteins associated with HDL particles or involved in its metabolism have been studied in great detail. HDL metabolism, however, is very complex and not fully understood. While the major apolipoprotein (apoA-I), LCAT, and ABCA1 account for the majority of known monogenic disorders of low HDL-C and CETP for cases of hyperalphalipoproteinemia, in the majority of cases of extremes of HDL-C the molecular basis remains unknown. These have been

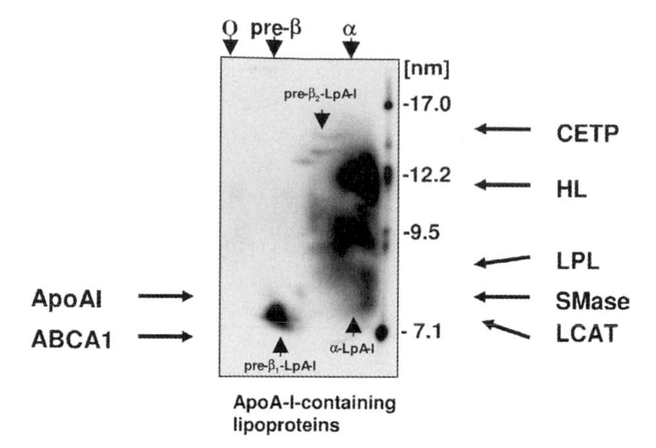

Fig. 19.1 Monogenic causes of low HDL-C in humans (Table 1A, 1B, 1D, and 1E).

characterized over the past four decades firstly on the basis of clinical manifestations, biochemical characterization, and metabolic disorders and finally by identification of the gene coding for the defective protein.

Over 100 proteins have been identified as associated with, or involved in, HDL metabolism (Table 19.1A to G). From proteinomic study of HDL proteins, at least 56 different proteins have been identified, by a combination of one- and two-dimension electrophoresis, MALDI-TOF, and isotope-coded affinity tag (ICAT) [32]. All of the apolipoproteins (with the exception of apoB) are found in HDL, as are several enzymes involved in lipoprotein processing, inflammation and free-radical metabolism as well as many other protein whose function is yet to be characterized [30]. Several other proteins, including apoAV, apoF, and α-2 macroglobulin, have been found associated with HDL particles and should be added to this list, and to these must also be added a nearly equal number of proteins, enzymes, receptors, and transcription factors known to be involved in HDL metabolism [29]. Wang and Paigent reported 54 genes involved in HDL metabolism and function; of these, apoA-I, CETP, LCAT, LIPC, LIPG, LPL, ABCA1, and ABCG5/8 have been shown to regulate HDL in humans [29].

More recently, Vaisar et al. [33] have identified several protein associated with the inflammatory response and acute phase reactants. Importantly, they have identified several protease inhibitors, suggesting a potential role for HDL to protect against proteolytic degradation of the atherosclerotic fibrous plaque.

For the present review, genes coding for proteins associated with HDL (Table 19.1A), HDL related enzymes and transfer proteins (Table 19.1B), HDL-associated inflammatory and acute phase proteins (Table 19.1C) plasma and cellular enzymes involved in HDL metabolism and remodeling (Table 19.1D), cellular receptors, transporters and adaptor proteins (Table 19.1E), transcription factors (Table 19.1F) and other proteins (Table 19.1G) are indicated.

Tab. 19.1 A–G Candidate genes for HDL metabolism and functions (modified from Wang [29]. Rezaee F. et al. [32] and Vaisar T et al. J. Clin Invest. 2007;117:746.)* indicates proteins isolated in the HDL fraction of human serum or plasma. [a] OMIM database. √ Hum mut: mutation identified in humans, causing disorders of HDL-C.

*A) HDL-associated apolipoprotein**			
Gene symbol[a]	***Gene name***	***hChr***	***Hum mut***
*apolAI**	apolipoprotein A-I	11q23.3	√
*apolAII**	apolipoprotein A-II	1q23.3	
*apolAIV**	apolipoprotein A-IV	11q23.3	
*apolAV**	apolipoprotein A-V	11q23.3	
*apolCII**	apolipoprotein C-II	19q13.32	√
*apolCIII**	apolipoprotein C-III	11q23.3	
*apolD**	apolipoprotein D	3q29	
*apolE**	apolipoprotein E	19q13.32	√
*apolF**	apolipoprotein F	12q13.3	
*apolJ**	apolipoprotein J (clusterin)	8p21	
*apolL1**	apolipoprotein L1	—	
*apolM**	apolipoprotein M	621.3	

B) HDL-associated enzymes and transfer proteins			
Gene symbol[a]	***Gene name***	***hChr***	***Mut***
*Cetp**	cholesteryl ester transfer protein	16q13	√
*CRP**	C-reactive protein	1q21	
*Lcat**	lecithin:cholesterol acyltransferase	16q22.1	√
*Pafah1b1**	platelet-activating factor (PAF) acetylhydrolase, isoform 1b, beta1 subunit	17p13.3	
*Pafah1b2**	PAF acetylhydrolase, isoform 1b, α2 subunit	11q23.3	
*Pafah1b3**	PAF acetylhydrolase, isoform 1b, α1 subunit	19q13.2	
*Pltp**	phospholipid transfer protein	20q13.12	
*Pon1**	paraoxonase 1	7q21.3	
*Pon2**	paraoxonase 2	7q21.3	
*Pon3**	paraoxonase 3	7q21.3	

C) HDL-associated inflammatory proteins			
Gene Symbol[a]	***Protein Name***	***HChr***	***Mut***
C3	Complement component C3	19p13.3-p13.2	
ORM2	α-1-acid glycoprotein 2	9q34.1-q34.3	
AHSG	α-2-HS glycoprotein	3q27	
PI	α-1 antitrypsin	14q32.1	
A1BG	α-1B glycoprotein	19cen-q13.2	
FGA	Fibrinogen (α chain)	4q28	
TF	Serotransferrin	3q21	
HPR	Haptoglobin-related protein	16q22.1	
SAA1	Serum amyloid A1	11p15.1	
SAA2	Serum amyloid A2	11p15.1	
SAA3P	Serum amyloid A3	11p15.1	

Tab. 19.1 A–G *(Continued)*

C) HDL-associated inflammatory proteins

Gene Symbol[a]	*Protein Name*	*HChr*	*Mut*
TTR	Transthyretin	18q11.2-q12.1	
GC	Vitamin D binding protein	4q12–q13	
ALB	Serum albumin	4q11-q13	
C4B	Complement component C4B	6p21.3	
C4A	Complement component C4A	6p21.3	
C9	Complement component C9	5p13	
VTN	Vitronectin	17q11	
AAP	α-2 antiplasmin	17pter-p12	
AMBP	α-1 microglobulin/bikunin	9q32-q33	
ITIH4	Inter-α-trypsin inhibitor heavy chain H4	3p21.2-p14.1	
AGT	Angiotensinogen	1q42.2	
SERPINF1	Serpin peptidase inhibitor (Clade F, member 1)	17p13.3	
KNG1	Kininogen 1	3q27.3	
RBP4	Plasma-retinol binding protein 4	10q23.33	
PCYOX1	Prenylcysteine oxidase 1	2p14	
HPX	Hemopexin	11p15.5-p15.4	

D) Plasma- and cell-associated enzymes

Gene symbol[a]	*Gene name*	*hChr*	*Mut*
Dgat1	diacylglycerol *O*-acyltransferase 1	8q24.3	
HL	hepatic lipase	15q21.3	√
HSL	hormone-sensitive lipase	19q13.2	
EL	endothelial lipase	18q21.1	
LPL	lipoprotein lipase	8p21.3	√
Pcyt1a	phosphate cytidylyltransferase 1, choline, α isoform	3q29	
Pemt	phosphatidylethanolamine *N*-methyltransferase	17p11.2	
Soat1	sterol *O*-acyltransferase 1	1q25.2	
Soat2	sterol *O*-acyltransferase 2	12q13.13	
SMPD-1	sphingomyelinase phosphodiasterase 1	11p15.4	√

E) Cellular receptors, their adaptors, and transporters

Gene symbol[a]	*Gene name*	*hChr*	*Mut*
Abca1	ATP-binding cassette, subfamily A, member 1	9q31.1	√
Abca7	ATP-binding cassette, subfamily A, member 7	19p13.3	
Abcb4	ATP-binding cassette, subfamily B, member 4	7q21.12	
Abcb11	ATP-binding cassette, subfamily B, member 11	2q24.3	
Abcc6	ATP-binding cassette, subfamily C, member 6	16p13.11	
Abcg1	ATP-binding cassette, subfamily G, member 1	21q22.3	
Abcg4	ATP-binding cassette, subfamily G, member 4	11q23.3	
Abcg5/8	ATP-binding cassette, subfamily G, member 5/8	2p21	√
Cav	caveolin	7q31.2	
Cav2	caveolin 2	7q31.2	

(continued)

Tab. 19.1 A–G *(Continued)*

E) Cellular receptors, their adaptors, and transporters

Gene symbol[a]	*Gene name*	*hChr*	*Mut*
Cd36	CD36 antigen	7q21.11	
Cubulin	cubilin (intrinsic factor-cobalamin receptor)	10p13	
Lrp2	low-density lipoprotein receptor-related protein 2 (megalin)	2q31.1	
Map17	membrane-associated protein 17	1p33	
NPC1	Niemann–Pick disease type C protein	18q11	√
Pdzk1	PDZ domain containing 1	1q21.1	
Scarb1	scavenger receptor class B type 1 (SR-BI)	12q24.31	

F) Transcription factors

Gene symbol[a]	*Gene name*	*hChr*	*Mut*
Hnf4	hepatic nuclear factor 4	20q13.12	
N-fkb1	nuclear factor of κlight chain gene enhancer in B-cells 1	4q24	
Nr1h2	nuclear receptor subfamily 1, group H, member 2 (LXRβ)	19q13.33	
Nr1h3	nuclear receptor subfamily 1, group H, member 3 (LXRα)	11p11.2	
Nr1h4	nuclear receptor subfamily 1, group H, member 4 (FXR)	12q23.1	
Ppara	peroxisome proliferator–activated receptor alpha	22q13.31	
Ppard	peroxisome proliferator–activated receptor delta	6p21.31	
Pparg	peroxisome proliferator–activated receptor gamma	3p25.2	
Rxra	retinoid X receptor α (RXRα)	9q34.2	
Rxrb	retinoid X receptor β (RXRβ)	6p21.32	
Tcf1	transcription factor 1	12q24.31	
Znf202	zinc finger 202		

G) Other Proteins

Gene symbol[a]	*Gene name*	*hChr*	*Mut*
	ceruloplasmin		
	haptoglobin	16q22.1	
	serum amyloid A1	11p15.1	
	prosaponin	10q22.1	
	transthyretin	18q11.2	
	transferrin	3q21	
	complement component inhibitor 1		
	macrophage stimulator factor 1		
	lymphocyte antigen		
	HAL-A protein		
	meningioma expressed antigen 5		
	NOTCH1		
	sialic acid binding Ig-like lectin 5		

Tab. 19.1 A–G *(Continued)*

G) Other Proteins			
Gene symbol[a]	*Gene name*	*hChr*	*Mut*
	C-type lectin superfamily member 1		
	H factor 1		
	complement component 1		
	Insulinoma-associated protein 1A-6		
	latent TGF-β		
	growth arrest-specific gene-6		
	LTBP-2		
	ryanodine receptor-1		
	POU 5 domain protein		
	plasma kallikrein B1		
	TFPI		
	CIP interacting ZFP		
	TAT interactive protein 72 kDa		
	protein tyrosine phosphatase		
	desmocollin		
	coagulation factor VIII associated protein		
	IgG		
	HAS		
	α-1β glycoprotein		
	a-2 macroglobulin		
	11 other unnamed proteins [32]		

Some of these candidate genes leading to genetic HDL deficiency or excess are discussed below.

- *ApoA-I*: ApoA-I constitutes the major apolipoprotein in HDL, accounting for approximately 70 % of the protein mass within HDL. The role of apoA-I is structural; the recent characterization of the apoA-I crystal structure [34] and molecular modeling of the lipidated apoA-I peptide [35] have shed considerable light on the structure/function of apoA-I domains critical for nascent HDL particle assembly. ApoA-I also activates LCAT (see Chapter 6) and binds ABCA1 [22]. Mutations within the apoA-I gene can lead to altered apoA-I and HDL-C levels, but not all apoA-I mutations lead to decreased apoA-I. Still other mutations lead to amyloidosis [35]. At least 47 mutations within the apoA-I gene have been identified [36,37], and most of the 18 reported mutations that affect apoA-I and HDL-C levels appear to cluster between residues ~100 and 200 (i.e., within amphipathic helices 5, 6 and 7, critical for HDL assembly). Two of these mutations, apoA-I_{Milano} (apoA-I_{R173C}) and apoA-I_{Paris} (apoA-I_{R151C}), are associated with low HDL-C, but, paradoxically, with no increase in the incidence of heart disease [38]. In a proof-of-concept clinical study, A-I_{Milano} proteoliposomes were intravenously injected weekly in 47 patients with acute coronary syndromes for a period of 5 weeks. At the end of the study, a small, but significant regression of atheroma volume (−4.2 %) was documented by intravascular ultrasound [13]. Mutations within the apoA-I gene clustering in the amino terminus are associated with amyloidosis, a disease that has multi-systemic effects but is not associated with atherosclerosis. Other mutations, scattered throughout the apoA-I gene, are not associated with low HDL-C or amyloidosis; their physiological significance is unknown. Most cases of apoA-I mutations have been described in single families and account for a minority of HDL deficiency cases seen in humans.

- *ABCA1*: The ABCA1 gene codes for a phospholipid (and possibly cholesterol) cellular transporter. ApoA-I – and other apolipoproteins that contain amphipathic α-helical motifs – bind to ABCA1 and mediate the transport of cellular lipids onto acceptor proteins through signaling molecules (protein kinase A) [39]. A defect in ABCA1 results in Tangier disease [6–8] and familial HDL deficiency in the homozygous and heterozygous forms, respectively [40], and multiple mutations of ABCA1 have been identified in Tangier disease and in severe HDL deficiency. Data on the tertiary structure of ABCA1 suggest that the functional unit of ABCA1 is oligomeric in nature and that a tetrameric structure is the minimal functional unit of the ABCA1 efflux complex [41]. This would explain the clinical observation that some heterozygous mutations may behave in a dominant negative fashion. At least three studies have examined the prevalence of ABCA1 mutations in patients with low HDL-C. Cohen et al. [42] examined patients from the US and Canada and identified an ABCA1 mutation or gene variant in 10–16 % of patients with low HDL-C, and a similar finding was reported in Germany, with approximately 10 % of low HDL-C being associated with ABCA1 mutations [43]. Alrassadi et al. reported a 20 % prevalence of ABCA1 mutation in subjects of French-Canadian descent [44], possibly reflecting a Founder effect. In a study of patients with heterozygous ABCA1 mutations, the risk of coronary artery disease was increased 3.5-fold [45,46]. The risk of vascular disease in Tangier disease remains a matter of debate. Despite very low plasma HDL-C levels, not all patients with Tangier disease develop severe premature atherosclerosis; this may be due, at lease in part, to the very low plasma levels of LDL-C seen in Tangier disease. Some patients develop a severe demyelinating neuropathy that contributes to increased morbidity and mortality in Tangier disease.

- *LCAT*: Mutations in this gene cause Norum disease and Fish-eye disease, so called because of the pathogneumonic corneal infiltrations seen with LCAT deficiency. The pathogenesis of LCAT deficiency stems from an inability to form cholesteryl esters in plasma and thus an inability to form mature HDL particles. Clinically, patients with LCAT deficiency may present with a normochromic anemia, proteinuria, corneal opacifications, and increased foam cell formation in bone marrow and the kidneys [48–50]. While patients with LCAT deficiency have a marked decrease in HDL-C, there appears to be no increase in premature coronary artery disease.
- *CETP*: Mutations in the CETP gene are associated with increases in HDL-C [51]. CETP mediates the equimolar transfer of cholesteryl esters from HDL particles to triglyceride-rich (VLDL and IDL) particles in exchange for triglycerides [52–54]. A deficiency of CETP increases the cholesteryl ester content of HDL and the formation of large, buoyant HDL_2 particles. It has recently been found that these particles maintain their functional abilities in terms of promoting cellular cholesterol efflux [55].
- *HL*: In reports relating to subjects with hepatic lipase deficiency, the biochemical phenotype is characterized by triglyceride enrichment of LDL and HDL. The LDL-C is normal or mildly elevated and HDL-C is elevated. Heterozygote carriers do not display significant lipoprotein abnormalities; postprandial clearance of triglycerides is delayed. Premature atherosclerosis has been reported in affected individuals [56–58].
- *LPL and ApoCII*: Deficiencies in LPL and its activator ApoCII cause severe hypertriglyceridemia and hyperchylomicronemia with a secondary HDL deficiency. The inability to hydrolyze triglycerides within chylomicrons and VLDL leads to accumulation of these particles in plasma. Heterozygous mutations in the LPL gene are associated with elevations in plasma triglycerides, a reduction in HDL-C, and an increased risk of CAD [59–61].
- *SMPD1*: Sphingomyelinase deficiency results from mutations in the gene coding for sphingomyelin phosphodiasterase-1 causes Neimann–Pick disease type 1 (subtypes A and B – NPD A and B) [62]. While the clinical manifestations of the Niemann–Pick group of diseases encompass pulmonary infiltrates, hepato-splenomegaly, the formation of lipid-laden histiocytes in the bone marrow, and mental retardation, a low HDL-C is a frequent biochemical feature [63]. In NPD A and B, the

deficiency in SMPD1 impairs the formation of nascent HDL particles and LCAT-mediated cholesterol esterification [64,65].
- *NPC-1*: Niemann–Pick disease type C is caused by mutations at the Niemann–Pick type C gene (*NPC-1*). In NPC disease, impaired cholesterol transport from the late endoplasmic pathway impairs the regulation of ABCA1 and cellular cholesterol efflux [66].
- *ApoE*: Patients homozygous for the apo$E_{2/2}$ genotype are at risk of developing type III hyperlipidemia, also known as dysbetalipoproteinemia or "broad-beta" dyslipidemia, so called because of the accumulation of remnant lipoproteins on ultracentrifugation of lipoprotein electrophoresis. The hypertriglyceridemia identified in affected individuals is associated with a low HDL-C [67].

19.3.2
Genome-Wide Studies

While much progress has been made in the elucidation of the genes and molecules that play a role in HDL-C, there are surely novel contributors left to be discovered. In order to identify these novel loci, whole genome linkage scans have been performed on many populations, with use of a variety of methodologies and phenotype definitions.

Genome-wide linkage analyses of families for HDL-C as a continuous quantitative trait have been performed by several research groups (reviewed in [29]). Some studies used probands that were selected on the basis of specific lipoprotein disorders. The most famous example of this are the well defined families that were used to identify the ABCA1 gene locus for Tangier disease [28]. Other studies have used unselected family studies in which multiple phenotypes are examined in the general population. These are summarized in Table 19.2.

Many investigators have used threshold values applied to HDL-C for linkage analysis. Use of an arbitrary cut-off point for HDL-C (0.9 mmol L^{-1} or 35 mg dL^{-1}, for instance) to define hypoalphalipoproteinemia does not take account of gender or age or the presence of concomitant metabolic disorders. This approach may reduce the power to detect genes related to HDL-C. To treat HDL-C as a discrete trait we favor the selection of patients with an HDL-C $< 5^{th}$ percentile of subjects age and gender-matched by use of the Lipid Research Clinical database reference manual [68] and the study of single families or groups of families with careful phenotypic assessment for the identification of genes that modulate HDL-C levels. The genetic heterogeneity of disorders of HDL-C makes analysis of multiple families problematic. We have performed a genome-wide scan on 12 families with severe HDL-C deficiency and reported evidence of linkage to chromosomes 4 and 16 [69]. Candidate gene analysis in the same cohort allowed the identification of a mutation within the apoA-I gene (apoA-I_{E136X}) [37] that was not identified in the aggregate genome scan data; hence the imperative necessity of providing precise clinical, biochemical, and cellular phenotyping. In addition, a complete map of the human

Tab. 19.2 Genome-wide studies of the genetic determinants of HDL-C.

Author/year [ref]	*Population*	*n families*	*Subjects*	*Selection*	*Results*
Elbein SC. 2002 [102]		42	576	type II DM	Chr1q21–q23,
Almasy L 1999 [82]	San Antonio	10	477	none†	Chr 8 (HDL cholesterol), 15
Kort EN 2000 [83]	Utah	105	3116	CHD, HDL	Chr 11q23; ZNF 202
Imperatore G 2000 [98]	Puma Indians	188	547	none	Chr 3
Peacock JM 2001 [80]	FHS, ARIC, Utah	101	1027	none	Chr 5, 13
Arya R 2002 [95]	Mexican Am	27	415	diabetes	Chr 9p
Klos KL 2001 [99]	Rochester Mn	232	1484	none	Chr 12, Chr 17
Coon H 2001 [97]	HyperGEN		1271†	hypertension	Chr 1
Soro A 2002 [79]	Finland	25	176	HDL-C <10%	Chr 8q23; 16q24.1; 20q13.11
Mahaney MC 2003 [100]	San Antonio	10	472	none	Chr 16q
Pajukanta P 2004 [78]	Finland + Dutch	53	341	FCH	Chr 16q
Dastani Z 2006 [69]	Canada	13	362	HDL-C <5%	Chr 4, 12, 14, 15, 16
Lin JP 2003 [101]	FHS	333			Chr 1, 6
Bosse Y 2004 [96]	French Canada	292	930		Chr 12
Harrap SB 2006 [103]	Australia	233	932	CV risk factors	Chr 4
Shoulders C 2004 [93]	review	–	–	FCH–HDL-C	Chr 16q22–24

†Same cohort of subjects.

genome is essential; the publication of the mouse genome and single-nucleotide polymorphisms and the results of the HapMap project in man should make the laborious work of gene identification easier [70]. The apparent lack of consistency observed in the studies that follow can, in part, be explained by the oligogenetic and multivariate nature of HDL regulation, pleiotropy, and selection criteria. Population-based studies that treat HDL-C as a parametric variable and use family studies to identify a quantitative trait locus (QTL) rely on unbiased assumptions and objective analysis. Unfortunately, such analyses are often not based on sound physiological principles and potential genetic effects are diluted by the complexity of HDL metabolism and by the complex gene–environment interactions. Such interactions are characteristic of the metabolic syndrome, where low HDL-C is a diagnostic criterion. Unexpectedly, genes not necessarily associated with HDL metabolism may lead to novel discoveries (Table 19.1F and G). The recent identification of ZNF202 (Zinc finger 202) as a modulator of lipoprotein metabolism highlights the importance of transcriptional factors in lipid metabolism [71].

19.3.3 Segregation of HDL-C

Segregation analysis is a general method for evaluating the transmission of a trait within pedigrees. It is used to assess the mode of inheritance and can also provide evidence for a major gene effect in a population. Different modes of inheritance have been observed in different segregation studies, probably as a result of the complexity of HDL-C metabolism and population specific differences. Friedlander et al. [72] performed a segregation analysis, under the mixed model, in a sample of nuclear families from Israel and showed evidence of a recessive major gene for low HDL-C. A genetic study of hypoalphalipoproteinemia by Byard et al. [73] also identified a recessive mode of inheritance for low HDL-C. Segregation of the age- and sex-adjusted HDL-C values in the Bogalusa Heart Study, however, provided evidence of the influence of major locus with neither dominant nor recessive expression, but this study was conducted on only one family of 200 individuals [74]. A similar study by Feitosa et al. [75] showed evidence for a major gene for HDL-C, but could not confirm a mode of inheritance for hypoalphalipoproteinemia. In addition, genetic analysis of the Stanford LRC family to assess the mode of inheritance of HDL-C as a continuously distributed trait is more consistent with some degree of multifactorial transmission or major gene model with additive allelic effects and similar allele frequencies [76]. Although segregation analysis of HDL-C did not indicate an allele for a major gene locus associated with low HDL-C in pedigrees recruited for the NHLBI Family Heart Study and Utah for early CHD, the authors did find some evidence for a recessive major gene that associates with high HDL-C [77]. These inconsistent results are probably due to the etiological heterogeneity of HDL-C, but in addition differences in the specific protocols for ascertaining probands, and differences in the methodology could also lead to discrepant results.

Several linkage studies of HDL-C levels, using HDL-C levels as a discrete trait, have been performed on the basis of this information. Some investigators dichotomize continuous variables, such as HDL-C, prior to linkage analysis. Parametric linkage studies resulted in the identification of several proposed chromosomal regions that might harbor genes that influence HDL-C levels. Combined analysis of genome scans of Dutch and Finnish families, ascertained through a preband with Amilial combined Hyperlipidemia (FCHL), revealed a susceptibility locus for HDL-C on chromosome 16q24.1 [77]. Parametric two-point linkage analyses under the assumption of a recessive mode of inheritance on the 13 multigenerational French Canadian families, whose probands were ascertained by HDL-C level < the 5th age/sex, identified a LOD score of 4.6 on chromosome 4.31–21 [69]. In both studies, subjects were coded as affected or unaffected on the basis of 10th age/sex specific percentiles for HDL-C. A similar study by Soro et al. revealed a locus on chromosomes 8q23 [79]. The inconsistency of separate linkage analysis could be explained by different underlying segregating variants conferring susceptibility to low levels of plasma HDL-C between the families and indeed between populations.

19.3.4
Heritability of HDL-C

In examinations of quantitative genetic contribution to the determination of HDL-C levels, there have been at least nine published twin studies and 14 family studies [reviewed in reference [75]. Estimates of the heritability of plasma HDL-C levels vary widely from 0.24 to 0.83, but are most often quoted in the 40 % to 60 % range [81,82]. In our study of families with HDL deficiency we determined that the overall heritability of the low HDL-C trait is 0.58 ($n = 699$ subjects), with ascertainment correction, with age and sex being the major co-variates. In subjects with known ABCA1 mutations, the heritability index increases to 0.68 ($n = 91$ subjects).

Some recent attempts to map genes for HDL-C have used the HDL-C measurement as a quantitative trait instead of to define affection status. Quantitative Trait Loci (QTL) analysis is an efficient approach used to uncover the genetic determinants of a trait such as HDL-C that can underlie complex human diseases. Generally, quantitative trait analysis should have more power to identify loci controlling these traits than use of dichotomized traits. However, ascertainment criteria for families may lead to restricted variances and thus reduce the power of QTL analyses if not accounted for.

Roughly speaking, there are three possible ascertainment strategies for analysis of quantitative traits. Firstly, under the affected ascertainment, a proband based on a threshold value for a measured quantitative trait is chosen. The second strategy is extreme the selection of sib pairs exhibiting discordance (families in which one sib is in the lower tail of the distribution of the quantitative trait and another sib is in the upper end of the distribution of the same trait are chosen). The last scheme is a random selection of families from a population regardless of the affection status.

In humans, at least 21 studies, identifying over 45 QTLs for HDL-C, have been published. As some of the QTLs map to similar chromosomal localizations, it is possible that they represent the same genes. In their review, Wang and Paigen (2005) estimated that 30 QTLs with LOD scores of more than 1.4 for HDL-C have been identified in humans. Most of these QTLs (>90 %) are concordant with mice QTLs in syntenic regions. The high degree of concordance between mouse and human QTLs suggests a strong likelihood that the underlying genes will be the same [29]; these authors therefore argue that an examination of HDL-C QTLs in mice will be a more efficient way to identify potential chromosomal regions of interest in the human genome.

The only reproducible QTL reported in different independent studies is 16q22–24. The apparent lack of consistency observed in these studies can be explained, in part, by the polygenetic and multivariate nature of HDL regulation, pleiotropy, and selection criteria. In addition, potential genetic effects can be diluted by the complexity of HDL metabolism and by complex gene–environment interactions [83]; such interactions are characteristic of the metabolic syndrome where low HDL cholesterol is a diagnostic criterion. Many independent studies, sometimes with different statistical methods, are therefore unable to replicate observed QTLs.

The initial stage of a whole genome scan typically identifies a large (10–20 cM) genomic region containing hundreds of genes. The major obstacle in identifying the causal gene is to narrow the linked region; this step has proven to be more arduous than originally anticipated. The publication of the mouse genome and mouse single-nucleotide polymorphisms [29] and the results of the HapMap project [70] in humans should make the laborious work of gene identification easier by enabling the analysis of a more efficient set of SNPs and haplotypes in humans and allowing for a finer-scale comparison of mouse and human homologous QTLs.

Linkage studies have good power for discovering less frequent alleles, but association studies should have more power to discover common alleles. It has recently been proposed that a significant portion of the risk for common diseases (such as low HDL-C) are the result of common genetic variants (the common-disease common-variant (CDCV) hypothesis). While there is currently some controversy in the field [84,85], this theoretical work has been matched by recent technological advances in high-throughput genotyping, which has led to the possibility of whole genome association (WGA) studies. Until recently, most association studies searching for common variants related to common diseases utilized a candidate gene approach and genome-wide association studies were limited by the cost and availability of the technology. Utilization of information relating to the nonrandom association of alleles at nearby loci (known as linkage disequilibrium or LD) generated from the International HAPMAP project allows the use of the highly informative tagged SNPs (tSNPs), and reduces the number of markers required to scan the entire genome [86]. Genome-wide panels of 300–500 000 markers are now available at reasonable cost, making such genome-wide association studies a relatively cost-efficient method to test the common-disease common-variant hypothesis. Furthermore, with a genome-wide approach no assumptions are made regarding the candidate genes and variants, obviating the need to "guess" at potential causal genes. Results of this methodology for a variety of diseases including low HDL-C should be forthcoming in the near future.

19.3.5
Genetic Association Studies

In the 25 years since the apoA-I gene was sequenced, genetic association studies between polymorphisms at candidate genes and a specific trait (HDL-C or CAD, for the most part) have been performed worldwide. Many such studies have been limited to specific ethnic or population subgroups and do not have sufficient statistical power to allow significant conclusions or to lead to clinically useful diagnostic tests. One exception is the apoE genotype or phenotype, where the effect of the apoE allele (especially the E4 allele) has been consistently been associated with elevated triglycerides and the apoE2/2 homozygocity, with the occurrence of type III dysbetalipoproteinemia [67].

Interestingly, recent data have highlighted new paradigms in the genetic basis of specific traits and cardiovascular risk. These are outlined below.

- *ApoA-I*: The apoA-I gene is part of a gene complex on chromosome 11 comprising the apoA-I/CIII/AIV gene cluster and the apoA5 gene. Genetic polymorphisms within this region have been examined in various populations for association with plasma lipid and lipoprotein lipids as well as CAD. In a recent examination of the Physician's Health study, two apoA-I polymorphisms (Xmi1 and Sst1 restriction fragment-length polymorphisms) were not associated with risk of myocardial infarction, despite a weak association with triglyceride levels [87].
- *ABCA1*: DNA sequencing of the ABCA1 gene in populations from the US, Canada, and Germany has shown that ABCA1 gene variability accounts for 10–20 % of cases of HDL-C deficiency, making ABCA1 the major cause of genetic HDL deficiency [38–40]. Subjects with heterozygous ABCA1 mutations have a 3.5 times greater risk of CAD than control subjects [Wellington]. Polymorphisms of ABCA1 have also been found to be associated with CAD [47,49,88–90].
- *LPL*: A meta-analysis of LPL gene polymorphisms has identified several mutations within the LPL gene as being associated with elevated triglycerides and low HDL-C and increased risk of CAD [60]. These findings have since been confirmed in separate populations [48,59]. Although the LPL mutations are associated with alterations in lipid and lipoprotein lipid levels, the authors conclude that the association with CAD may in part be independent of the changes in lipid parameters.
- *CETP*: While mutations at the CETP gene locus are associated with elevated HDL-C levels, the potential cardioprotective effect of this effect remains to be determined. In a large population study performed in the Netherlands, the presence of the –629A promoter SNP was associated with increased HDL-C (by 0.08 mmol L^{-1}), but with a 26 % increased risk of

CAD ($p = 0.1$ after adjustment for HDL-C), in relation to the −629CC genotype [91].

- *Paraoxonase*: Paraoxonase-1 (PON1 – Table 19.0B) and HDL are tightly associated in plasma, and some studies have found a close association of PON1 levels and HDL-C. The C-107T SNP was found to increase PON1 expression and correlates with HDL-C levels. In patients with CAD this relationship was altered, suggesting that differential expression of PON1 in CAD cases is not associated with a concomitant change in HDL-C [92].

19.4 Conclusions

Low HDL-C is a risk factor for cardiovascular disease the number one cause of death in North America. The polygenic nature and environmental interaction result in the complexity of HDL metabolism, with many independent studies with different statistical methods having therefore been unable to replicate the observed QTLs and yield to identify different susceptibility regions on different chromosomes. The only QTL to have been reported in different independent studies is 16q22–24, which is more likely to harbor causative genetic variants [89] (see Table 19.2).

In spite of the importance role of QTLs in this kind of continuous trait, the major obstacle to identification of the causal genes seems to be narrowing down the QTL to a few candidate genes, which is time-consuming and requires the sequencing of many genes.

The publication of the mouse genome and single-nucleotide polymorphisms, together with the results of the HapMap project [70] in man, should make the laborious work of gene identification easier. Use of the HapMap data to choose more efficient set of SNPs should help researchers in designing and analysis of genetics association to look for more associated SNPs in the case cohort versus control cohort and subsequently to help them to identify causal genes underlying QTL. Such an approach has been used, in a proof-of-concept study, to identify SNPs that are associated with HDL-C levels. Using a method to derive haplotypes information, Hinds et al. [94] identified 7283 SNPs in 71 candidate gene regions associated with low HDL-C levels. The CETP gene was identified as a potential important modulator of HDL-C levels in this population.

19.5 Future Considerations

The identification of single-gene disorders causing low HDL-C is likely to be coming to an end; few clinical syndromes representing extremes of clinical or biochemical phenotypes in the lipoprotein field remain to be elucidated in molecular terms, although the possibility that novel, rare disorders of HDL deficiency or excess might be encountered in the future cannot be excluded. The next step is the search for genes that modulate HDL. The identification of novel genes involved in HDL metabolism may help in the identification of novel metabolic pathways and possibly provide novel therapeutic targets for modulation of HDL-C in humans.

References

1 Genest, J. Jr., Martin-Munley, S. S., McNamara, J. R., Ordovas, J. M., Jenner, J. L., Meyers, R. H., Silberman, S. R., Wilson, P.W.F., Salem, D. N., Schaefer, E. J. (1992) Familial Lipoprotein Disorders in Patients With Premature Coronary Artery Disease. *Circulation*, **85**, 2025–2033.

2 Despres, J. P., Lemieux, I., Dagenais, G. R., Cantin, B., Lamarche, B. (2000 Dec) HDL-cholesterol as a marker of coronary heart disease risk: the Quebec cardiovascular study. *Atherosclerosis*, **153** (2), 263–272.

3 Genest, J., Jr.,McNamara, J. R., Salem, D. N., Ordovas, J. M., Jenner, J. L., Millar, J. S., Silberman, S. R., Wilson, P. F. W., Schaefer, E. J. (1992) Lipoprotein cholesterol, apolipoproteins A-I and B, and Lipoprotein(a) in men with premature coronary artery disease. *J. Am. Coll. Cardiol.*, **19**, 782–802.

4 Grundy, S. M., Cleeman, J. I., Merz, C. N., Brewer, H. B., Jr.,Clark, L. T., Hunninghake, D. B., Pasternak, R. C., Smith, S. C., Jr.,Stone, N. J.National Heart, Lung, and Blood Institute; American College of Cardiology Foundation; American Heart Association. (2004 Jul 13) Implications of recent clinical trials for the National Cholesterol Education Program Adult Treatment Panel III guidelines. *Circulation*, **110** (2), 227–239.

5 Lloyd-Jones, D. M., Nam, B. H., D'Agostino, R. B., Sr.,Levy, D., Murabito, J. M., Wang, T. J., Wilson, P. W., O'Donnell, C. J. (2004 May 12) Parental cardiovascular disease as a risk factor for cardiovascular disease in middle-aged adults: a prospective study of parents and offspring. *JAMA*, **291** (18), 2204–2211.

6 Brooks-Wilson, A., Marcil, M., Clee, S., Zhang, L. H., Romp, C., van Dam, M., Yu, L., Brewer, C., Collins, J., Molhuizen, H. O. F., Loubser, O., Francis-Ouelette, B. F., Fichter, K., Ashbourne-Excoffon, K., Sensen, C., Scherer, S., Mott, S., Martindale, D., Frohlich, J., Morgan, K., Koop,Pimstone, S., Genest, J., Jr.,Kastelein, J., Hayden, M. (1999) Mutations in the ATP binding cassette (ABC1) transporter gene in Tangier disease and Familial HDL deficiency. *Nature Genet.*, **22**, 336–345.

7 Bodzioch, M., Orso, E., Klucken, J., Langmann, T., Bottcher, A., Diederich, W., Drobnik, W., Barlage, S., Buchler, C., Porsch-Ozcurumez, M., Kaminski, W. E., Hahmann, H. W., Oette, K., Rothe, G., Aslanidis, C., Lackner, K. J., Schmitz, G. (1999) The gene encoding ATP-binding cassette transporter 1 is mutated in Tangier disease. *Nat. Genet.*, **22**, 347–351.

8 Rust, S., Rosier, M., Funke, H., Real, J., Amoura, Z., Piette, J.-C., Deleuze, J.-F., Brewer, H. B., Duverger, N., Denefle, P., Assman, G. (1999) Tangier disease is caused by mutations in the gene encoding ATP-binding cassette transporter 1. *Nature Genet.*, **22**, 352–355.

9 Yusuf, S., Hawken, S., Ounpuu, S., Dans, T., Avezum, A., Lanas, F., McQueen, M., Budaj, A., Pais, P., Varigos, J., Lisheng, L.INTERHEART Study Investigators. (2004 Sep 11) Effect of potentially modifiable risk

factors associated with myocardial infarction in 52 countries (the INTERHEART study): case-control study. *Lancet*, **364** (9438), 937–952.

10 Assmann, G. and Nofer, J. R. (2003) Atheroprotective effects of high-density lipoproteins. *Annu. Rev. Med.*, **54**, 321–341.

11 O'Connell, B. and Genest, J.Jr. (2001) High Density Lipoproteins and endothelial function. *Circulation*, **104**, 1978–1983.

12 Barter, P. J. (2005 Jul) Cardioprotective effects of high-density lipoproteins: the evidence strengthens. *Arterioscler. Thromb. Vasc. Biol.*, **25** (7), 1305–1306.

13 Nissen, S. E., Tsunoda, T., Tuzcu, E. M., Schoenhagen, P., Cooper, C. J., Yasin, M., Eaton, G. M., Lauer, M. A., Sheldon, W. S., Grines, C. L., Halpern, S., Crowe, T., Blankenship, J. C., Kerensky, R. (2003) Nov 5 Effect of recombinant ApoA-I Milano on coronary atherosclerosis in patients with acute coronary syndromes: a randomized controlled trial. *JAMA*, **290** (17), 2292–2300.

14 Brousseau, M. E., Schaefer, E. J., Wolfe, M. L., Bloedon, L. T., Digenio, A. G., Clark, R. W., Mancuso, J. P., Rader, D. J. (2004) Apr 8 Effects of an inhibitor of cholesteryl ester transfer protein on HDL cholesterol. *N. Engl. J. Med.*, **350** (15), 1505–1515.

15 Miao, B., Zondlo, S., Gibbs, S., Cromley, D., Hosagrahara, V. P., Kirchgessner, T. G., Billheimer, J., Mukherjee, R. (2004 Aug) Raising HDL cholesterol without inducing hepatic steatosis and hypertriglyceridemia by a selective LXR modulator. *J. Lipid Res.*, **45** (8), 1410–1417.

16 Yusuf, S., Reddy, S., Ounpuu, O., Anand, S. (2001) Global burden of cardiovascular diseases. *Circulation*, **104**, 2746–2753, *Circulation*, (2001) 2855–64.

17 Assmann, G., Schulte, H., von Eckardstein, A., Huang, Y. (1996) High-density lipoprotein cholesterol as a predictor of coronary heart disease risk. The PROCAM experience and pathophysiological implications for reverse cholesterol transport. *Atherosclerosis*, **124**, S11–S20.

18 Gordon, T., Castelli, W. P., Hjortland, M. C., Kannel, W. B., Dawber, T. R. (1977) High density lipoprotein as a protective factor against coronary heart disease. The Framingham Study. *Am. J. Med.*, **62**, 707–714.

19 Wilson, P. W. (1990 Sep 4) High-density lipoprotein, low-density lipoprotein and coronary artery disease. *Am. J. Cardiol*, **66** (6), 7A–10A. Review.

20 McPherson, R., Frohlich, J., Fodor, G., Genest, J. (2006) Canadian Cardiovascular Society position statement: recommendations for the diagnosis and treatment of dyslipidemias and prevention of cardiovascular disease. *Can. J. Cardiol.*, **22**, 913–927.

21 Lewis, G. F. and Rader, D. J. (2005) New insights into the regulation of HDL metabolism and reverse cholesterol transport. *Circulation Research*, **96**, 1221.

22 Denis, M., Haidar, B., Marcil, M., Krimbou, L., Genest, J. (2004) Molecular and cellular physiology of apolipoprotein A-I lipidation by the ATP biding cassette A1 (ABCA1). *J. Biol. Chem.*, **279**, 7384–7394.

23 Timmins, J. M., Lee, J.-Y., Boudyguina, E. *et al.* Apr(2005). Targeted inactivation of hepatic ABCA1 causes profound hypoalphalipoproteineinemia and kidney hypercatabolism of apoA-I. *J. Clin. Invest.*, online pub.

24 Wang, N., Weng, W., Breslow, J. L., Tall, A. R. (1996) Scavenger receptor B1 (SR-B1) is up-regulated in adrenal glands an apolipoprotein AI and hepatic lipase knock-out mice as a response to depletion of cholesterol stores. In vivo evidence that SB-B1 is a functional high density lipoprotein receptor under feedback control. *J. Biol. Chem.*, **271**, 21001–21004.

25 Brunham, L. R., Kruit, J. K., Pape, T. D., Parks, J. S., Kuipers, F., Hayden, M. R. (2006 Sep 29) Tissue-specific

induction of intestinal ABCA1 expression with a liver X receptor agonist raises plasma HDL cholesterol levels. *Circ. Res.*, **99** (7), 672–674.

26 Joyce, C. W., Amar, M. J., Lambert, G., Vaisman, B. L., Paigen, B., Najib-Fruchart, J., Hoyt, R. F., Jr.,Neufeld, E. D., Remaley, A. T., Fredrickson, D. S., Brewer, H. B., Jr.,Santamarina-Fojo, S. (2002)Jan 8 The ATP binding cassette transporter A1 (ABCA1) modulates the development of aortic atherosclerosis in C57BL/6 and apoE-knockout mice. *Proc. Natl. Acad. Sci. U. S. A.*, **99** (1), 407–412.

27 Karathanasis, S. K., Zannis, V. I., Breslow, J. L. (1983)Oct Isolation and characterization of the human apolipoprotein A-I gene. *Proc. Natl. Acad. Sci. U. S. A.*, **80** (20), 6147–6151.

28 Rust, S., Walter, M., Funke, H., vonEckardstein, A., Cullen, P., Kroes, H. Y., Hordijk, R., Geisel, J., Kastelein, J., Molhuizen, H. O., Schreiner, M., Mischke, A., Hahmann, H. W., Assmann, G. (1998)Sep Assignment of Tangier disease to chromosome 9q31 by a graphical linkage exclusion strategy. *Nat. Genet.*, **20** (1), 96–98.

29 Wang, X. and Paigen, B. (2005) Genetics variations in HDL cholesterol in humans and mice. *Circ. Res.*, **96**, 27–42.

30 Pennacchio, L. A. and Rubin, E. M. (2003)Apr 1 Apolipoprotein A5, a newly identified gene that affects plasma triglyceride levels in humans and mice. *Arterioscler. Thromb. Vasc. Biol.*, **23** (4), 529–534.

31 O'Brien, P. J., Alborn, W. E., Sloan, J. H., Ulmer, M., Boodhoo, A., Knierman, M. D., Schultze, A. E., Konrad, R. J. (2005)Feb The novel apolipoprotein A5 is present in human serum, is associated with VLDL, HDL, and chylomicrons, and circulates at very low concentrations compared with other apolipoproteins. *Clin. Chem.*, **51** (2), 351–359.

32 Rezaee, F., Cassetta, B., Levels, J.H.M., Speijer, D., Meijers, J.C.M. (2006) Proteionomics analysis of high-density lipoproteins. *Proteinomics*, **6**, 721–730.

33 Vaisar, T., Pennathur, S., Green, P. S., Gharib, S. A., Hoofnagle, A. N., Cheung, M. C., Byun, J., Vuletic, S., Kassim, S., Singh, P., Chea, H., Knopp, R. H., Brunzell, J., Geary, R., Chait, A., Zhao, X. Q., Elkon, K., Marcovina, S., Ridker, P., Oram, J. F., Heinecke, J. W. (2007) Mar 1 Shotgun proteomics implicates protease inhibition and complement activation in the antiinflammatory properties of HDL. *J Clin Invest.*, **117** (3), 746–756.

34 Ajees, A. A., Anantharamaiah, G. M., Mishra, V. K., Hussain, M. M., Murthy, H. M. (2006) Crystal structure of human apolipoprotein A-I: insights into its protective effect against cardiovascular diseases. *Proc. Natl. Acad. Sci. U. S. A.*, **103** (7), 2126–2131.

35 Catte, A., Patterson, J. C., Jones, M. K., Jerome, W. G., Bashtovyy, D., Su, Z., Gu, F., Chen, J., Aliste, M. P., Harvey, S. C., Li, L., Weinstein, G., Segrest, J. P. (2006) Novel changes in discoidal high density lipoprotein morphology: a molecular dynamics study. *Biophys. J.*, **90** (12), 4345–4360.

36 Sorci-Thomas, M. G. and Thomas, M. J. (2002 Apr) The effects of altered apolipoprotein A-I structure on plasma HDL concentration. *Trends Cardiovasc. Med.*, **12** (3), 121–128.

37 Dastani, Z., Dangoisse, C., Boucher, B., DesBiens, K., Dufour, R., Hegele, R., Paivi Pajukanta,Jamie Engert,Krimbou, L., Genest, J., Marcil, M. (2006) A Novel Mutation of Apolipoprotein AI (ApoA-IE136X) Causes a low HDL-C in French Canadians. *Atherosclerosis*, **185** (1), 127–136.

38 Sirtori, C. R., Calabresi, L., Franceschini, G., Baldassarre, D., Amato, M., Johansson, J., Salvetti, M., Monteduro, C., Zulli, R., Muiesan, M. L., Agabiti-Rosei, E. (2001) Apr 17 Cardiovascular status of carriers of the apolipoprotein A-I(Milano) mutant: the Limone sul

Garda study. *Circulation*, **103** (15), 1949–1954.

39 Haidar, B., Denis, M., Marcil, M., Krimbou, L., Genest, J. (2004) Apolipoprotein A-I activates cellular camp signalling through the ABCA1 receptor. Evidence for molecular interactions between ABCA1 receptor and G protein. *J. Biol. Chem.*, **279**, 9963–9969.

40 Marcil, M., Brooks-Wilson, A., Clee, S. M., Roomp, K., Zhang, L. H., Yu, L., Collins, J. A., van Dam, M., Molhuizen, H. O., Loubster, O., Ouellette, B. F., Sensen, C. W., Fichter, K., Mott, S., Denis, M., Boucher, B., Pimstone, S., Genest, J., Jr.,Kastelein, J. J., Hayden, M. R. (1999) Mutations in the ABC1 gene in familial HDL deficiency with defective cholesterol efflux. *Lancet*, **354**, 1341–1346.

41 Denis, M., Haidar, B., Marcil, M., Bouvier, M., Krimbou, L., Genest, J. (2004) Characterization of oligomeric human ATP binding cassette transporter A1.Potential implication for determining the structure of nascent HDL particles. [CIHR MOP 15042][*]. *J. Biol. Chem.*, **279**, 41529–41536.

42 Cohen, J. C., Kiss, R. S., Pertsemlidis, A., Marcel, Y. L., McPherson, R., Hobbs, H. H. (2004) Multiple rare alleles contribute to low plasma levels of HDL cholesterol. *Science*, **305**, 869–872., 1198–1205.

43 Frikke-Schmidt, R. and Brge, G. (2004) Nordestgaard, Gorm B. Jensen and Anne Tybjærg-Hansen, Genetic variation in ABC transporter A1 contributes to HDL cholesterol in the general population. *J. Clin. Invest.*, **114**, 1343–1353.

44 Alrasadi, K., Ruel, I., Marcil, M., Genest, J. (2006) Functional mutations in the ABCA1 gene in subjects of French Canadian descent with HDL deficiency [CIHR MOP 62834] [*]. *Atherosclerosis*, **188**, 281–291.

45 Clee, S. M., Kastelein, J. J., van Dam, M., Marcil, M., Roomp, K., Zwarts, K. Y., Collins, J. A., Roelants, R., Tamasawa, N., Stulc, T., Suda, T., Ceska, R., Boucher, B., Rondeau, C., DeSouich, C., Brooks-Wilson, A., Molhuizen, H. O., Frohlich, J., Genest, J., Jr.,Hayden, M. R. (2000) Age and residual cholesterol efflux affect HDL cholesterol levels and coronary artery disease in ABCA1 heterozygotes. *J. Clin. Invest.*, **106**, 1263–1270.

46 Brousseau, M. E., Bodzioch, M., Schaefer, E. J., Goldkamp, A. L., Kielar, D., Probst, M., Ordovas, J. M., Aslanidis, C., Lackner, K. J., Bloomfield, R. H., Collins, D., Robins, S. J., Wilson, P. W., Schmitz, G. (2001) Common variants in the gene encoding ATP-binding cassette transporter 1 in men with low HDL cholesterol levels and coronary heart disease. *Atherosclerosis.*, **154**, 607–611.

47 Clee, S. M., Zwinderman, A. H., Engert, J. C., Zwarts, K. Y., Molhuizen, H. O., Roomp, K., Jukema, J. W., van Wijland, M., van Dam, M., Hudson, T. J., Brooks-Wilson, A., Genest, J., Jr.,Kastelein, J. J., Hayden, M. R. (2001) Common genetic variation in ABCA1 is associated with altered lipoprotein levels and a modified risk for coronary artery disease. *Circulation*, 103.

48 Brousseau, M. E. (2004) Common Variation in Genes Involved in HDL Metabolism Influences Coronary Heart Disease Risk at the Population Level. *Rev Endocr Metab Disord.*, **5**, 343–349.

49 Evans, D. and Beil, F. U. (2003) The association of the R219K polymorphism in the ATP-binding cassette transporter 1 (ABCA1) gene with coronary heart disease and hyperlipidaemia. *J Mol Med.*, **81**, 264–270.

50 Kuivenhoven, J. A., Pritchard, H., Hill, J., Frohlich, J., Assmann, G., Kastelein, J. (1997 Feb) The molecular pathology of lecithin:cholesterol acyltransferase (LCAT) deficiency syndromes. *J. Lipid Res.*, **38** (2), 191–205.

51 Kuivenhoven, J. A., de Knijff, P., Boer, J. M., Smalheer, H. A., Botma,

G. J., Seidell, J. C., Kastelein, J. J., Pritchard, P. H. (1997 Mar) Heterogeneity at the CETP gene locus. Influence on plasma CETP concentrations and HDL cholesterol levels. *Arterioscler. Thromb. Vasc. Biol.*, **17** (3), 560–568.

52 Thompson, J. F., Durham, L. K., Lira, M. E., Shear, C., Milos, P. M. (2005) CETP polymorphisms associated with HDL cholesterol may differ from those associated with cardiovascular disease. *Atherosclerosis*, **181**, 45–53.

53 Kondo, I., Berg, K., Drayna, D., Lawn, R. (1989 Jan) DNA polymorphism at the locus for human cholesteryl ester transfer protein (CETP) is associated with high density lipoprotein cholesterol and apolipoprotein levels. *Clin. Genet.*, **35** (1), 49–56.

54 Inazu, A., Jiang, X. C., Haraki, T., Yagi, K., Kamon, N., Koizumi, J., Mabuchi, H., Takeda, R., Takata, K., Moriyama, Y. *et al.* (1994 Nov) Genetic cholesteryl ester transfer protein deficiency caused by two prevalent mutations as a major determinant of increased levels of high density lipoprotein cholesterol. *J. Clin. Invest.*, **94** (5), 1872–1882.

55 Matsuura, F., Wang, N., Chen, W., Jiang, X. C., Tall, A. R. (2006) HDL from CETP-deficient subjects shows enhanced ability to promote cholesterol efflux from macrophages in an apoE- and ABCG1-dependent pathway. *J. Clin. Invest.*, **116** (5), 1435–1442.

56 Hegele, R. A., Little, J. A., Vezina, C., Maguire, G. F., Tu, L., Wolever, T. S., Jenkins, D. J., Connelly, P. W. (1993 May) Hepatic lipase deficiency. Clinical, biochemical, and molecular genetic characteristics. *Arterioscler Thromb.*, **13** (5), 720–728.

57 Connelly, P. M. and Hegele, R. A. (1998) Hepatic lipase deficiency. *Crit. Rev. Clin. Lab. Sci.*, **35**, 547–572.

58 Guerra, R., Wang, J., Grundy, S. M., Cohen, J. C. (1997)Apr 29 A hepatic lipase (LIPC) allele associated with high plasma concentrations of high density lipoprotein cholesterol. *Proc. Natl. Acad. Sci. U.S.A.*, **94** (9), 4532–4537.

59 Wittrup, H. H., Andersen, R. V., Tybjaerg-Hansen, A., Jensen, G. B., Nordestgaard, B. G. (2006 Apr) Combined analysis of six lipoprotein lipase genetic variants on triglycerides, high-density lipoprotein, and ischemic heart disease: cross-sectional, prospective, and case-control studies from the Copenhagen City Heart Study. *J. Clin. Endocrinol. Metab.*, **91** (4), 1438–1445.

60 Wittrup, H. H., Tybjaerg-Hansen, A., Nordestgaard, B. G. (1999) Lipoprotein lipase mutations, plasma lipids and lipoproteins, and risk of ischemic heart disease. *Circulation*, **99**, 2901–2907.

61 Brousseau, M. E., Goldkamp, A. L., Collins, D., Demissie, S., Connelle, A. C., Cupples, L. A., Ordovas, J. M., Bloomfield, H. E., Robbins, S. J., Schaefer, E. J. (2004) Polymorphisms in the gene encoding lipoprotein lipase in men with low HDL-C and coronary artery disease: the Veterans Affairs HDL Intervention Trail. *J. Lipid Res.*, **45**, 1885–1891.

62 Schuchman, E. H. and Miranda, S. R. (1997) Niemann–Pick disease: mutation update, genotype/phenotype correlations, and prospects for testing. *Genet. Test.*, **113**, 329–334.

63 Viana, M. B., Giugliani, R., Leite, V.H.R., Barth, M. L., Lekhwani, C., Slade, C. M., Fensom, A. (1990) Very low levels of high density lipoprotein cholesterol in four sibs of a family with non-neuropathic Niemann–Pick disease and sea-blue histiocytosis. *J. Med. Genet.*, **27**, 499–504.

64 Lee, C. Y., Krimbou, L., Boucher, B., Bernard, C., Laramée, P., Genest, J., Marcil, M. (2003) Compound Heterozygocity at the Sphingomyelin Phosphodiesterase-1 (SMPD-1) gene is Associated with a low HDL cholesterol. *Hum. Genet.*,

65 Lee, C.-Y., Vincent, J., Lesimple, A., Larsen, Å., Laliberté, J., Denis, M., Mamer, O., Krimbou, L.Genest, J.,

Marcil, M. (2006) Increased sphingomyelin content of HDL impairs HDL biogenesis and maturation in human Niemann-Pick Disease Type B. *J. Lipid Res.*, **47**, 322–332.

66 Choi, H. Y., Karten, B., Chan, T., Vance, J. E., Greer, W. L., Heidenreich, R. A., Garver, W. S., Francis, G. A. (2003 Aug 29) Impaired ABCA1-dependent lipid efflux and hypoalphalipoproteinemia in human Niemann-Pick type C disease. *J. Biol. Chem.*, **278** (35), 32569–32577.

67 Walden, C. C. and Hegele, R. A. (1994 Jun 15) Apolipoprotein E in hyperlipidemia. *Ann. Intern. Med.*, **120** (12), 1026–1036.

68 N.I.H., Washington D.C. Lipid Research Clinics population studies databook. [Book, volume, 1], 28–81, 1980. Department of health and human services, Public Health Service, Washington DC. NIH publication 80–1527.

69 Dastani, Z., Quiogue, L., Engert, J. C., Plaisier, C., Marcil, M., Genest, J., Pajukanta, P. (2006) Evidence of a gene influencing HDL-C on chromosome 4q31.21. *Atheroscler. Thromb. Vasc. Biol.*, **26** (2), 392–397.

70 Altshuler, D., Brooks, L. D., Chakravarti, A., Collins, F. S., Daly, M. J., Donnelly, P. (2005 Oct 27) International HapMap Consortium. A haplotype map of the human genome. *Nature*, **437** (7063), 1299–1320.

71 Wagner, S., Hess, M. A., Ormonde-Hanson, P. *et al.* (2000) A broad role for the zinc finger protein ZNF 202 in human lipid metabolism. *J. Biol. Chem.*, **275**, 15685–15690.

72 Friedlander, Y. and Kark, J. D. (1987) Complex segregation analysis of plasma lipid and lipoprotein variables in a Jerusalem sample of nuclear families. *Hum. Hered.*, **37** (1), 7–19.

73 Byard, P. J., Borecki, I. B., Glueck, C. J., Laskarzewski, P. M., Third, J. L., Rao, D. C. (1984) A genetic study of hypoalphalipoproteinemia. *Genet. Epidemiol.*, **1** (1), 43–51.

74 Amos, C. I., Wilson, A. F., Rosenbaum, P. A., Srinivasan, S. R., Webber, L. S., Elston, R. C., Berenson, G. S. (1986) An approach to the multivariate analysis of high-density-lipoprotein cholesterol in a large kindred: the Bogalusa Heart Study. *Genet. Epidemiol.*, **3** (4), 255–267.

75 Feitosa, M. F., Province, M. A., Heiss, G., Arnett, D. K., Myers, R. H., Pankow, J. S., Hopkins, P. N., Borecki, I. B. (2006 Mar 7) Evidence of QTL on 15q21 for high-density lipoprotein cholesterol: The National Heart, Lung, and Blood Institute Family Heart Study (NHLBI FHS). *Atherosclerosis*,

76 Karlin, S., Williams, P. T., Haskell, W. L., Wood, P. D. (1981 Mar) Genetic analysis of the Stanford LRC family study data. II. Structured exploratory data analysis of lipids and lipoproteins. *Am. J. Epidemiol.*, **113** (3), 325–337.

77 Kronenberg, F., Coon, H., Ellison, R. C., Borecki, I., Arnett, D. K., Province, M. A., Eckfeldt, J. H., Hopkins, P. N., Hunt, S. C. (2002) Jun Segregation analysis of HDL cholesterol in the NHLBI Family Heart Study and in Utah pedigrees. *Eur. J. Hum. Genet.*, **10** (6), 367–374.

78 Pajukanta, P., Allayee, H., Krass, K. L., Kuraishy, A., Soro, A., Lilja, H. E., Mar, R., Taskinen, M. R., Nuotio, I., Laakso, M., Rotter, J. I., de Bruin, T. W., Cantor, R. M., Lusis, A. J., Peltonen, L. (2003 Apr) Combined analysis of genome scans of Dutch and Finnish families reveals a susceptibility locus for high-density lipoprotein cholesterol on chromosome16q. *Am. J. Hum. Genet.*, **72** (4), 903–917.

79 Soro, A., Pajukanta, P., Lilja, H. E., Ylitalo, K., Hiekkalinna, T., Perola, M., Cantor, R. M., Viikari, J. S., Taskinen, M. R., Peltonen, L. (2002 May) Genome scans provide evidence for low-HDL-C loci on chromosomes 8q23, 16q24. 1-24.2, and 20q13.11 in Finnish families. *Am. J. Hum. Genet.*, **70** (5), 1333–1340.

80 Peacock, J. M., Arnett, D. K., Atwood, L. D., Myers, R. H., Coon, H., Rich, S. S., Province, M. A., Heiss, G. (2001 Nov), Genome scan for quantitative trait loci linked to high-density lipoprotein cholesterol: The NHLBI Family Heart Study. *Arterioscler. Thromb. Vasc. Biol.*, **21** (11), 1823–1828.

81 Breslow, J. L. (2001) Familial disorders of high density lipoprotein metabolism, Scriver, C. R.Beaudet, A. L.Sly, W. S.ValleD.*The Molecular and Metabolic Basis of Inherited Diseases*, McGraw-Hill, New York.

82 Almasy, L., Hixson, J. E., Rainwater, D. L., Cole, S., Williams, J. T., Mahaney, M. C., VandeBerg, J. L., Stern, M. P., MacCluer, J. W., Blangero, J. (1999 Jun) Human pedigree-based quantitative-trait-locus mapping: localization of two genes influencing HDL-cholesterol metabolism. *Am. J. Hum. Genet.*, **64** (6), 1686–1693.

83 Kort, E. N., Ballinger, D. G., Ding, W. *et al.* (2000) Evidence of linkage of familial hypoalphalipoproteinemia to a novel locus on chromosome 11q23. *Am. J. Hum. Genet.*, **66**, 1845–1856.

84 Reich, D. E. and Lander, E. S. (2001) Sep On the allelic spectrum of human disease. *Trends Genet.*, **17** (9), 502–510.

85 Pritchard, J. K. (2001 Jul) Are rare variants responsible for susceptibility to complex diseases?. *Am. J. Hum. Genet.*, **69** (1), 124–137.

86 Stram, D. O. (2004 Dec) Tag SNP selection for association studies. *Genet. Epidemiol.*, **27** (4), 365–374.

87 Liu, S., Song, Y., Hu, F. B., Niu, T., Ma, J., Gaziano, M., Stampfer, M. J. (2004) A prospective study of the apoA-I XmnI and apoC3 SstI polymorphisms in the apoA-I/C3/A4 gene cluster and risk of incident myocardial infarction in men. *Atherosclerosis*, **177**, 119–126.

88 Srinivasan, S. R., Li, S., Chen, W., Boerwinkle, E., Berenson, G. S. (2003) R219K polymorphism of the ABCA1 gene and its modulation of the variations in serum high-density lipoprotein cholesterol and triglycerides related to age and adiposity in white versus black young adults. *The Bogalusa heart study. Metabolism.*, **52**, 930–934.

89 Tregouet, D. A., Ricard, S., Nicaud, V., Arnould, I., Soubigou, S., Rosier, M., Duverger, N., Poirier, O., Mace, S., Kee, F., Morrison, C., Denefle, P., Tiret, L., Evans, A., Deleuze, J. F., Cambien, F. (2004) In-depth haplotype analysis of ABCA1 gene polymorphisms in relation to plasma ApoA1 levels and myocardial infarction. *Arterioscler. Thromb. Vasc. Biol.*, **24**, 775–781.

90 Wang, J., Burnett, J. R., Near, S., Young, K., Zinman, B., Hanley, A. J., Connelly, P. W., Harris, S. B., Hegele, R. A. (2000) Common and rare ABCA1 variants affecting plasma HDL cholesterol. *Arterioscler. Thromb. Vasc. Biol.*, **20**, 1983–1989.

91 Borggreve, S. E., Hillege, H. L., Wolfenbuttel, B. H., de Jong, P. E., Zuurman, M. W., van der Stege, G., van Tol, A., Dullart, R. p. (2006) PREVEND Study Group. An increase coronary risk is paradoxically associated with common cholesteryl ester transfer protein gene variation that relate to higher high-density lipoprotein cholesterol: a population-based study. *Clin. Endocrinol. Metab.*, **91**, 3382–3388.

92 Blatter, G., Morne, X., James, R. W. (2006) Paraoxonase-1 and serum concentrations of HDL-cholesterol and apoA-I. *J. Lipid Res.*, **47**, 515–520.

93 Shoulders, C. C., Jones, E. L., Naoumova, R. P. (2004) Genetics of familial combined hyperlipidemia and risk of CAD. *Hum. Mol. Genetics*, **13**, R149–R160.

94 Hinds, D. A., Seymour, A. B., Durham, L. K., Banarjee, P., Ballinger, D. G., Milos, P. M., Cox, D. R., Thompson, J. F., Frazer, K. A. (2004) Application of pooled genotyping to scan candidate regions for association with HDL cholesterol levels. *Hum. Genomics*, **1**, 421–434.

95 Arya, R., Duggirala, R., Almasy, L., Rainwater, D. L., Mahaney, M. C.,

Cole, S., Dyer, T. D., Williams, K., Leach, R. J., Hixson, J. E., MacCluer, J. W., O'Connell, P., Stern, M. P., Blangero, J. (2002) Linkage of high-density lipoprotein-cholesterol concentrations to a locus on chromosome 9p in Mexican Americans. *Nat. Genet.*, **30** (1), 102–105.

96 Bosse, Y., Chagnon, Y. C., Despres, J. P., Rice, T., Rao, D. C., Bouchard, C., Perusse, L., Vohl, M. C. (2004) Genome-wide linkage scan reveals multiple susceptibility loci influencing lipid and lipoprotein levels in the Quebec Family Study. *J. Lipid Res.*, **45**, 419–426.

97 Coon, H., Leppert, M. F., Eckfeldt, J. H., Oberman, A., Myers, R. H., Peacock, J. M., Province, M. A., Hopkins, P. N., Heiss, G. (2001 Dec) genome-wide linkage analysis of lipids in the Hypertension Genetic Epidemiology Network (HyperGEN) Blood Pressure Study. *Arterioscler. Thromb. Vasc. Biol.*, **21** (12), 1969–1976.

98 Imperatore, G., Knowler, W. C., Pettitt, D. J., Kobes, S., Fuller, J. H., Bennett, P. H., Hanson, R. L. (2000) A locus influencing total serum cholesterol on chromosome 10p: results from an autosomal scan of serum lipid concentrations in Pima Indians. *Arterioscler. Thromb. Vasc. Biol.*, **20**, 2651–2656.

99 Klos, K. L., Kardia, S. L., Ferrell, R. E., Turner, S. T., Boerwinkle, E., Sing, C. F. (2001 Jun) Genome-wide linkage analysis reveals evidence of multiple regions that influence variation in plasma lipid and apolipoprotein levels associated with risk of coronary heart disease. *Arterioscler. Thromb. Vasc. Biol.*, **21** (6), 971–978.

100 Mahaney, M. C., Almasy, L., Rainwater, D. L., Vandeberg, J. L., Cole, S. A., Hixon, J. E., Blangero, J., MacCluer, J. W. (2003) A quantitative trait locus on chromosome 16q influences variations in plasma HDL-C in Mexican Americans. *Atheroscler. Thromb. Vasc. Biol.*, **23**, 339–345.

101 Lin, J. P. (2003 Dec 31) genome-wide scan on plasma triglyceride and high density lipoprotein cholesterol levels, accounting for the effects of correlated quantitative phenotypes. *BMC Genet.*, **4** Suppl(1), S47.

102 Elbein, S. C. and Hasstedt, S. J. (2002) Feb Quantitative trait linkage analysis of lipid-related traits in familial type 2 diabetes: evidence for linkage of triglyceride levels to chromosome 19q. *Diabetes*, **51** (2), 528–535.

103 Harrap, S. B., Wong, Z.Y.H., Scurrah, K., Lamantia, A. (2006) genome-wide linkage analysis of population variation in high-density lipoprotein cholesterol. *Hum. Genet.*, **119**, 541–546.

20
HDL and Atherosclerosis

Philip J. Barter, Kerry-Anne Rye

20.1
Introduction

An inverse relationship between the concentration of cholesterol in high-density lipoproteins (HDL) and the development of premature coronary heart disease (CHD) has been observed in many large-scale prospective studies [1–5], in several of which the level of HDL cholesterol has been the single most powerful lipid predictor of future CHD events. On the basis of the population studies it has been concluded that for every 1 mg dL^{-1} (0.025 mmol L^{-1}) increase in HDL cholesterol, the CHD risk is reduced by 2–3 % [6]. In one large study, the concentration of HDL cholesterol was also an inverse predictor of total mortality [7]. It is noteworthy that a low level of HDL cholesterol remains predictive of future CV risk even when the concentration of cholesterol in low density lipoproteins (LDL) has been taken to low levels by treatment with statins [8–10].

In support of these human population studies, there are numerous intervention studies in animals showing that an increase in the concentration of HDL inhibits the development of atherosclerosis [11–14]. In humans, however, the evidence for a protective effect of HDL is still circumstantial and, although strengthening, awaits direct confirmation. This chapter discusses potential mechanisms that underlie a protective role of HDL and then examines in detail the evidence that interventions that raise HDL levels are indeed antiatherogenic.

20.2
HDL and Atherosclerosis: Cause and Effect or an Epiphenomenon?

It has been argued that HDL may not be directly involved as a protector against atherosclerosis but rather that a low level of HDL cholesterol is the reflection of some other factor or factors that causes the disease. Low levels of HDL cholesterol are common in people with hypertriglyceridemia, for example [15]. Since some populations of triglyceride-rich lipoproteins are proatherogenic [15], it is possible that the observation of an increase in CHD in people with low levels of HDL cholesterol may

High-Density Lipoproteins: From Basic Biology to Clinical Aspects. Edited by Christopher J. Fielding

ISBN: 978-3-527-31717-2

reflect no more than an accompanying higher level of triglyceride-rich lipoproteins. Furthermore, patients with low levels of HDL cholesterol are often obese and insulin-resistant [16,17], conditions known in their own right to be associated with an increased risk of CHD.

However, there is robust evidence that a low concentration of HDL cholesterol predicts CHD events, independently of the levels of LDL cholesterol, plasma triglyceride, body weight, and the presence of diabetes [18]. As described in greater detail below, there is also growing understanding of potential mechanisms by which HDL protect, with evidence mounting that HDL-raising interventions will indeed translate into a reduced risk of future cardiovascular events.

20.3 Potential Antiatherogenic Properties of HDL

There are several properties of HDL that have the potential to protect against the development of atherosclerosis. Most of these have been described in detail in earlier chapters of this book and will only be summarized here. The best documented is the ability of HDL to promote the efflux of cholesterol from cells in the artery wall [19]. However, HDL have a number of additional potentially antiatherogenic properties that may be unrelated to their role in plasma lipid transport. For example, HDL bind lipopolysaccharide [20] and promote endothelial repair by enhancing the migration of cells from neighboring undamaged tissue [21] and by recruiting progenitor endothelial cells from plasma into damaged endothelium [22]. HDL also inhibit the synthesis of platelet-activating factor by endothelial cells [23] and are antithrombotic [24], while they also modulate endothelial function, probably by stimulating endothelial NO production [25], and also possess antioxidant and antiinflammatory activities [26]. The degree to which any or all of these nonlipid transport functions of HDL contribute to protection against atherosclerosis is still uncertain, although evidence is mounting that at least some of them may be of substantial importance.

Potential antiatherogenic properties of HDL are:

- promotion of cholesterol efflux from macrophages
- antioxidant properties
- antiinflammatory properties
- antithrombotic properties
- enhancement of endothelial repair
- enhancement of endothelial function

20.3.1 HDL and Cholesterol Efflux

The best known of the potentially protective functions of HDL is their role in promoting the efflux of cholesterol from macrophage foam cells in the artery wall [27], thus inhibiting the progression (and even promoting the regression) of atherosclerosis. This efflux is the first step in the pathway known as reverse cholesterol transport, in which HDL transport cholesterol from extrahepatic tissues to the liver for recycling or for excretion from the body through the bile.

There are at least four documented processes that promote the efflux of cholesterol from cell membranes to acceptors in the extracellular space. One involves ABCA1, another ABCG1, a third involves SR-B1, and a fourth involves passive diffusion. While all of these processes have been well documented in studies conducted with cells growing in tissue culture, their relative contributions to cholesterol efflux *in vivo* remain to be determined.

The ABCA1 transporter translocates phospholipids and cholesterol from the inner leaflets of cell membranes to the outer leaflets, where they are picked up by apolipoprotein A-I (apoA-I) (or other HDL apolipoproteins) in the extracellular space [28]. This interaction is limited to apolipoproteins that contain no or very little lipid and results in the formation of discoidal HDL complexes containing apolipoproteins, phospholipids, and unesterified cholesterol. The ABCG1 transporter promotes the unidirectional transfer of cholesterol from cells, including macrophages, to preformed HDL in the extracellular space [29,30]. Another mechanism involves binding of pre-formed HDL to the scavenger receptor-B1 (SR-BI) on the surfaces of cells [27] in a process that promotes bidirectional transfers of unesterified cholesterol between cells and HDL. This process promotes a net efflux from the cell only if there is a concentration gradient of unesterified cholesterol from the donor cell to the acceptor HDL. In contrast with ABCA1-mediated efflux, which favors lipid-poor apolipoproteins, both ABCG1 and SR-B1 promote an efflux of cholesterol to larger, spherical, highly lipidated HDL. Both discoidal and spherical HDL in the extracellular space are also capable of accepting unesterified cholesterol from cells in a process of passive aqueous diffusion that is apparently independent of the processes driven by ABCA1, ABCG1, or SR-B1 [27]. In this diffusion process, unesterified cholesterol in cell membranes is spontaneously released into the extracellular fluid, where it collides with and incorporates into any preformed HDL particles that are present. This is a bidirectional process in which unesterified cholesterol exchanges between HDL and cell membranes. However, a net transfer of cholesterol into HDL may result when LCAT-mediated esterification of cholesterol on the HDL surface generates a concentration gradient between the cell surface and the HDL particle, down which cholesterol flows into the HDL. The relative contributions of these different processes to the efflux of cholesterol from macrophages in the artery wall are not known.

Regardless of the mechanism, once cellular cholesterol has been transferred to HDL in the extracellular space, it may be transported to the liver for elimination from the body by several pathways.

20.3.2
Antioxidant Properties of HDL

HDL have been shown by many investigators to possess antioxidant properties [31]. Given that oxidative modification of LDL is proatherogenic, the capacity of HDL to inhibit the process is potentially antiatherogenic. In studies conducted *in vitro*, HDL inhibit the transmigration of monocytes induced by oxidized LDL [32], the cytotoxicity induced by oxidized LDL [33], and the oxidized LDL-induced adhesion of monocytes to endothelial cells [34]. The exact mechanism of the inhibition is uncertain. One possibility is an involvement of the paraoxonase that is transported in plasma by HDL [35], although it should be noted that some of the major HDL apolipoproteins have antioxidant properties independently of paraoxonase [36]. The relevance of the antioxidant properties of HDL to their antiatherogenic function *in vivo* remains to be determined.

20.3.3
Antiinflammatory Properties of HDL

Inflammation has been implicated in the genesis, progression, and instability of atherosclerotic plaques, and there is growing circumstantial evidence that HDL have the capacity to inhibit this inflammation; the cytokine-induced expression of VCAM-1, ICAM-1, and E-selectin in endothelial cells growing in tissue culture, for example, has been shown to be inhibited by HDL in a concentration-dependent manner [37] in a process involving an HDL-mediated inhibition of endothelial cell sphingosine kinase [38]. The inhibition of VCAM-I and E-selectin protein expression by HDL is paralleled by significant reductions in the steady-state mRNA levels of these adhesion molecules [37], suggesting that transcription may be suppressed by the lipoprotein. These *in vitro* antiinflammatory properties of HDL are dependent on the presence of phospholipids in the particles [39] and are also influenced by the species of phospholipid that predominates in the particle [40]. HDL also inhibit binding of monocytes [41] and neutrophils [42] to endothelial cells growing in culture.

There is evidence that antiinflammatory properties of HDL also operate *in vivo*, although in most cases this has been demonstrated in a setting of hypercholesterolemia and atherosclerosis. For example, intravenous infusion of reconstituted HDL (rHDL) reduces the *in vivo* expression of endothelial adhesion molecules induced by insertion of carotid periarterial cuffs in cholesterol-fed, apoE knockout mice [43]. In another study of apoE knockout mice, the increase in HDL concentration accompanying an overexpression of the human apoA-I gene reduced macrophage accumulation in the aortic root by more than threefold [44], this being associated with reduced *in vivo* oxidation of β-VLDL, lower ICAM-1 and VCAM-1 expression, and diminished *ex vivo* leukocyte adhesion. In another study, conducted in rabbits in which aortic atherosclerosis was induced by a balloon injury followed by 17 weeks of a high-cholesterol diet, as little as two intravenous injections of relatively small amounts of HDL given during the last week of the study markedly inhibited the extent of inflammation in the aortic wall [45].

However, there are also examples of *in vivo* antiinflammatory effects of HDL in the absence of hypercholesterolemia and atherosclerosis. In a porcine model, for example, injection of rHDL inhibits the development of a local inflammatory infiltrate after the subcutaneous administration of interleukin-1 [46], while pretreatment with rHDL in studies of experimental stroke in rats significantly and substantially reduces the brain necrotic area in a process possibly related to an rHDL-induced reduction in levels of reactive oxygen species [47]. Furthermore, in a study of hemorrhagic shock in rats, the resulting multiple organ dysfunction syndrome was largely abolished by a single injection of human HDL given 90 minutes after the hemorrhage and one min before resuscitation. In that model, injection of HDL prevented both the severe disruption of tissue architecture and the extensive cellular infiltration into the affected tissues [48]. In other *in vivo* studies it was found that infusion of rHDL containing 25 mg apoA-I into normo-cholesterolemic rabbits on three consecutive days markedly inhibited the infiltration of neutrophils into the carotid arterial wall in response to application of a silastic peri-arterial collar. This was associated by inhibition of collar-induced increases in the generation of reactive oxygen species by the vascular wall, as well as by expression of adhesion molecules and chemokines on the endothelial surface [49].

There is also circumstantial evidence that HDL have acute antiinflammatory effects *in vivo* in humans. In one study, single intravenous infusions of rHDL into hypercholesterolemic humans normalized endothelium-dependent vasodilation, possibly by increasing NO bioavailability [50]. In a second human study a single injection of rHDL corrected the endothelial dysfunction associated with low levels of HDL in ABCA1 heterozygotes [51]. Whether these antiinflammatory properties of HDL contribute to their ability to protect against atherosclerosis remains to be determined.

20.3.4 Antithrombotic Properties of HDL

Antithrombotic effects of HDL are secondary to an attenuation of the expression of tissue factor and selectins, to downregulation of thrombin generation by the protein C pathway, and to direct and indirect blunting of platelet activation [24]. There is also evidence that HDL protect against both arterial and venous thrombosis in processes that may involve the activation of prostacyclin synthesis [24].

20.3.5 HDL and Endothelial Repair

The disruption of endothelial monolayer integrity is an important contributing factor in vascular disorders and its repair plays a fundamental role in the ultimate outcome. HDL have the capacity to enhance endothelial repair by at least two distinct mechanisms. Firstly, in studies conducted *in vitro* in a model of endothelial injury, HDL have been shown to stimulate endothelial cell migration in a nitric oxide-independent manner through SR-BI-mediated activation of Rac GTPase [21]. This process is dependent on the activation of Src kinases, phosphatidylinositol 3-kinase, and p44/42 mitogen-activated protein kinases. Paralleling the *in vitro* findings, re-endothelialization of carotid arteries after perivascular electric injury has been shown to be blunted in apoA-I knockout mice; reconstitution of apoA-I expression promotes restoration of the endothelium but only in mice expressing SR-B1 [21]. In other studies conducted in mice, it has been shown that endothelial progenitor cell engraftment into the endothelial layer also promotes endothelial repair in a process that is enhanced by HDL [22].

20.4 HDL Subpopulations and Atherosclerosis

The HDL fraction in human plasma is heterogeneous in terms of shape, size, density, composition, and surface charge [52]. When isolated on the basis of density by ultracentrifugation, human HDL separate into two major subfractions: HDL_2 and HDL_3. Nondenaturing polyacrylamide gradient gel electrophoresis separates HDL by particle size into at least five distinct subpopulations.

HDL can also be divided into two main subpopulations by their apolipoprotein composition. One subpopulation comprises HDL that contain apoA-I but no apoA-II (A-I HDL), while another comprises particles that contain both apoA-I and apoA-II (A-I/A-II HDL. ApoA-I is divided approximately equally between A-I HDL and A-I/A-II HDL in most subjects, while almost all of the apoA-II is in A-I/A-II HDL. When subjected to agarose gel electrophoresis, HDL have either α, preα, preβ, or γ migration [52]. The α-migrating particles are spherical lipoproteins and account for the major proportion of HDL in plasma. They include the HDL_2 and HDL_3 subfractions, as well as A-I HDL and A-I/A-II HDL subpopulations. Preβ HDL are either lipid-poor apoA-I or discoidal particles consisting two or three molecules of apoA-I complexed with phospholipids and possibly a small amount of unesterified cholesterol. γ-HDL contain apoE and no apoA-I.

Results of some human population studies and some transgenic animal studies have raised the possibility that A-I HDL may be superior to A-I/A-II HDL in their ability to protect against atherosclerosis [53,54], although other studies have suggested that the protection conferred by A-I HDL and by A-I/A-II HDL is comparable [55]. It has been reported that populations of larger HDL are more protective than those of smaller HDL [56], while it has been suggested by others that minor subpopulations of discoidal, preβ-migrating HDL are superior to spherical, α-migrating HDL in their ability to inhibit atherosclerosis because such particles are the preferred acceptors of cholesterol released from cells by the ABCA1 transporter. However, the discovery that another transporter, ABCG1, promotes cholesterol efflux from cells to large HDL particles supports the epidemiological finding that larger HDL particles are also protective.

Overall, it may be concluded that the evidence linking protection against CHD to specific HDL subpopulations in humans is conflicting and confusing, and so it remains unknown whether the cardioprotective effects of HDL are influenced by their apolipoprotein composition, their size, density, or their electrophoretic mobility.

20.5 Evidence that Raising the Concentration of HDL Protects against Atherosclerosis

20.5.1 Intervention Studies in Animals

There have been many studies investigating the effects of increasing HDL on atherosclerosis animal models. It should be emphasized, however, that all of these models have their limitations and that none is a true model for human disease. With this reservation, however, the animal studies have provided powerful evidence that increasing the concentration of HDL does protect against atherosclerosis.

HDL raising interventions in animals that protect against atherosclerosis are:

- increased synthesis of apoA-I
- intravenous infusion of native or reconstituted HDL
- overexpression of LCAT (in rabbits)
- inhibition of CETP (in rabbits)

Badimon et al. were the first to report direct antiatherogenic effects of HDL [11]. Using a model of experimental atherosclerosis in rabbits, they showed that weekly infusions of HDL significantly reduced the development of aortic fatty streaks. A similar beneficial effect has been observed in rabbits infused with rHDL containing either apoA-I_{Milano} [57] or native apoA-I [45]. The most compelling evidence, however, has come from studies in a range of genetically modified animals.

Overexpression of the human apoA-I gene in transgenic rabbits [58] and mice [13,14] results in an increased concentration of HDL cholesterol, which is accompanied by protection against atherosclerosis. Overexpression of lecithin:cholesterol acyltransferase (LCAT) increases the concentrations of HDL cholesterol in both rabbits [59] and mice [60]. This is associated with an inhibition of atherosclerosis in rabbits [59] but an increase in atherosclerosis in mice [60], possibly a reflection of the fact that mice, unlike rabbits, are deficient in CETP and may thus be unable to dispose of the cholesteryl esters that accumulate in HDL after overexpression of LCAT. A high level of activity of CETP in rabbits may be one reason for the susceptibility of this species to development of atherosclerosis. As described below, this suggestion is supported by the observation that inhibiting CETP in rabbits inhibits atherosclerosis.

In cholesterol-fed rabbits, the inhibition of CETP by injection of antisense oligodeoxynucleotides (ODNs) against CETP resulted in a reduction in CETP mRNA and mass in the liver, a reduction in plasma total and LDL cholesterol, and an increased concentration of HDL cholesterol [61]. These changes were accompanied by a marked reduction in aortic cholesterol content as a marker of the extent of atherosclerosis. A vaccine approach has been used to generate auto-antibodies against CETP *in vivo* in rabbits. Cholesterol-fed rabbits, animals immunized against CETP, had a reduced plasma activity of CETP, an increase in the concentration of HDL cholesterol, a modest decrease in LDL cholesterol concentration, and a significant reduction in aortic atherosclerotic lesions [61].

A chemical inhibitor of CETP has been also been used in cholesterol-fed rabbits [62]. The inhibitor reduced CETP activity in rabbits by more than 90 %, almost doubled the level of HDL cholesterol, and decreased the non-HDL cholesterol by about 50 %. There was an accompanying 70 % reduction in atherosclerotic lesions in the aortas of these animals.

20.5.2
Intervention Studies in Humans

There have been few reports of clinical trials designed specifically to determine whether raising the level of the total HDL fraction, or of specific HDL subpopulations, results in a reduced incidence of CHD. There are, however, several human intervention studies in which drug-induced elevations of HDL cholesterol are associated with a reduction in atherosclerosis.

In the Lipid Research Clinics Primary Prevention Trial, cholestyramine was used as the active agent. A reduction in CHD events correlated positively with changes in LDL cholesterol levels and negatively with changes in HDL cholesterol. For every 1 % increase in the concentration of HDL cholesterol there was a 0.6 % reduction in CHD events that was independent of the changes in LDL cholesterol levels [63].

The relationship between changes in HDL cholesterol and CHD events in the statin trials is unclear, possibly because it is obscured by the major reduction in LDL cholesterol. The HDL cholesterol increase with simvastatin in 4S was a significant (although weak) predictor of benefit [8], while in the AFCAPS/TexCAPS study with lovastatin, the level of apoA-I at one year was predictive of benefit [64]. However, the increase in HDL cholesterol induced by pravastatin in WOSCOPS, CARE, and LIPID did not correlate significantly with the reduction in CHD events [8,9,65].

The results of fibrate trials have been mixed. In the Helsinki Heart Study, major CHD events were significantly reduced from 4.1 % in the placebo group to 2.7 % in the gemfibrozil group. In this study, a 1 % increase in HDL cholesterol was associated with a 2–3 % decrease in CHD events, which was independent of changes in levels of LDL cholesterol [66]. However, most of this benefit was found in subjects with elevated levels of plasma triglyceride in combination with low levels of HDL cholesterol. This was especially apparent in those who were also overweight [67].

In the VA-HIT study the primary endpoint (non-fatal MI or coronary death) was significantly reduced from 21.7 % in the placebo group to 17.3 % in the gemfibrozil group [68]. The on-treatment HDL cholesterol level was predictive of CHD events in both the active and placebo groups. Multivariate regression analysis showed that, of all of the variables measured, the increase in HDL cholesterol was the only one that predicted benefit [69], but the HDL cholesterol increase accounted for only about one quarter of the benefit. The obvious question that arises is what explains the other three quarters of the benefit, suggesting beneficial effects of gemfibrozil beyond their ability to increase the HDL cholesterol concentration.

The BIP study used bezafibrate as the active agent. In this study there was no significant effect of bezafibrate on the primary outcome (the combined incidence of non-fatal MI or death from CHD) despite an 18 % increase in concentration of HDL cholesterol in the treated group [70]. Post-hoc analysis, however, indicated a significant benefit in the subset of patients who entered the trial with elevated levels of plasma triglyceride [70]. There was also a benefit in those with other features of the metabolic syndrome [71]. Thus, while this study added support to a proposition that fibrates are effective in reducing cardiovascular risk in people with features of the metabolic syndrome, it was not possible to conclude that the benefit was secondary to the increase in concentration of HDL.

The more recently reported FIELD Study, using fenofibrate in people with type 2 diabetes, added little to the argument, since fenofibrate treatment in this study resulted in an HDL cholesterol increase of less than 2 % and a nonsignificant reduction in the primary endpoint of the study [72].

Studies with niacin lend strong circumstantial support to the view that raising the level of HDL cholesterol is protective. Niacin has long been used as a lipid-modifying agent: it lowers plasma triglyceride by 40–50 %, lowers LDL cholesterol by up to 10–15 %, and increases HDL cholesterol by up to 30 % [73]. When coadministered with statins, niacin promotes significant angiographic regression of atheromatous plaque and a reduction in clinical cardiovascular events [74,75].

The most compelling evidence of a direct benefit of HDL raising in humans has been provided by a small study in which humans with documented coronary atherosclerosis received infusions of rHDL containing apoA-I Milano complexed with phospholipid [76]. Subjects received intravenous injections of the rHDL preparation at weekly intervals for just five weeks, which resulted in a statistically significant reduction in the atheroma burden in the coronary arteries as assessed by intravascular ultrasound. While the study included only a small number of subjects, the result was consistent with a profound protective action of HDL and has provided a powerful incentive to conduct further research.

20.6 Conclusions

The epidemiological evidence demonstrating an inverse relationship between the risk of having a cardiovascular event and the plasma concentration of HDL cholesterol is overwhelming. The likelihood that the relationship is causal is supported by the observation that HDL possess a number of properties with antiatherogenic potential. Furthermore, raising the level of HDL reduces or even reverses atherosclerosis in virtually all animal models that have been studied. The challenge ahead is to demonstrate conclusively that raising the plasma concentration of HDL also inhibits the development of atherosclerosis (or even reverses the process) in humans. Several studies addressing this issue are currently underway and their results are awaited with great interest.

References

1 Gordon, D. J., Knoke, J., Probstfield, J. L., Superko, R., Tyroler, H. A. (1986) High-density lipoprotein cholesterol and coronary heart disease in hypercholesterolemic men: the Lipid Research Clinics Coronary Primary Prevention Trial. *Circulation*, **74**, 1217–1225.

2 Enger, S. C., Hjermann, I., Foss, O. P., Helgeland, A., Holme, I., Leren, P., Norum, K. R. (1979) High density lipoprotein cholesterol and myocardial infarction or sudden coronary death: a prospective case-control study in middle-aged men of the Oslo study. *Artery.*, **5**, 170–181.

3 Miller, N. E., Thelle, D. S., Forde, O. H., Mjos, O. D. (1977) The Tromso heart-study. High-density lipoprotein and coronary heart-disease: a prospective case-control study. *Lancet.*, **1**, 965–968.

4 Miller, M., Seidler, A., Kwiterovich, P. O., Pearson, T. A. (1992) Long-term predictors of subsequent cardiovascular events with coronary artery disease and 'desirable' levels of plasma total cholesterol. *Circulation*, **86**, 1165–1170.

5 Pekkanen, J., Linn, S., Heiss, G., Suchindran, C. M., Leon, A., Rifkind, B. M., Tyroler, H. A. (1990) Ten-year mortality from cardiovascular disease in relation to cholesterol level among men with and without preexisting cardiovascular disease. *N Engl J. Med.*, **322**, 1700–1707.

6 Gordon, D. J., Probstfield, J. L., Garrison, R. J., Neaton, J. D., Castelli, W. P., Knoke, J. D., Jacobs, D. R., Jr., Bangdiwala, S., Tyroler, H. A. (1989) High-density lipoprotein cholesterol and cardiovascular disease. Four prospective American studies. *Circulation*, **79**, 8–15.

7 Jacobs, D. R., Jr., Mebane, I. L., Bangdiwala, S. I., Criqui, M. H., Tyroler, H. A. (1990) High density lipoprotein cholesterol as a predictor of cardiovascular disease mortality in men and women: the follow-up study of the Lipid Research Clinics Prevalence Study. *Am J. Epidemiol.*, **131**, 32–47.

8 Pedersen, T. R., Olsson, A. G., Faergeman, O., Kjekshus, J., Wedel, H., Berg, K., Wilhelmsen, L., Haghfelt, T., Thorgeirsson, G., Pyorala, K., Miettinen, T., Christophersen, B., Tobert, J. A., Musliner, T. A., Cook, T. J. (1998) Lipoprotein changes and reduction in the incidence of major coronary heart disease events in the Scandinavian Simvastatin Survival Study (4S). *Circulation*, **97**, 1453–1460.

9 Sacks, F. M., Tonkin, A. M., Shepherd, J., Braunwald, E., Cobbe, S., Hawkins, C. M., Keech, A., Packard, C., Simes, J., Byington, R., Furberg, C. D. (2000) Effect of pravastatin on coronary disease events in subgroups defined by coronary risk factors: the Prospective Pravastatin Pooling Project. *Circulation*, **102**, 1893–1900.

10 Simes, R. J., Marschner, I. C., Hunt, D., Colquhoun, D., Sullivan, D., Stewart, R. A., Hague, W., Keech, A., Thompson, P., White, H., Shaw, J., Tonkin, A. (2002) Relationship between lipid levels and clinical outcomes in the Long-term Intervention with Pravastatin in Ischemic Disease (LIPID) Trial: to what extent is the reduction in coronary events with pravastatin explained by on-study lipid levels?*Circulation*, **105**, 1162–1169.

11 Badimon, J. J., Badimon, L., Fuster, V. (1990) Regression of atherosclerotic lesions by high density lipoprotein plasma fraction in the cholesterol-fed rabbit. *J. Clin. Invest.*, **85**, 1234–1241.

12 Paszty, C., Maeda, N., Verstuyft, J., Rubin, E. M. (1994) Apolipoprotein AI transgene corrects apolipoprotein E deficiency-induced atherosclerosis

in mice. *J. Clin. Invest.*, **94**, 899–903.

13 Rubin, E. M., Krauss, R. M., Spangler, E. A., Verstuyft, J. G., Clift, S. M. (1991) Inhibition of early atherogenesis in transgenic mice by human apolipoprotein AI. *Nature*, **353**, 265–267.

14 Plump, A. S., Scott, C. J., Breslow, J. L. (1994) Human apolipoprotein A-I gene expression increases high density lipoprotein and suppresses atherosclerosis in the apolipoprotein E-deficient mouse. *Proc. Natl. Acad. Sci. U.S.A.*, **91**, 9607–9611.

15 Austin, M. A. (1991) Plasma triglyceride and coronary heart disease. *Arterioscler. Thromb.*, **11**, 2–14.

16 Despres, J. P., Moorjani, S., Ferland, M., Tremblay, A., Lupien, P. J., Nadeau, A., Pinault, S., Theriault, G., Bouchard, C. (1989) Adipose tissue distribution and plasma lipoprotein levels in obese women. Importance of intra-abdominal fat. *Arteriosclerosis*, **9**, 203–210.

17 Haffner, S. M., Valdez, R. A., Hazuda, H. P., Mitchell, B. D., Morales, P. A., Stern, M. P. (1992) Prospective analysis of the insulin-resistance syndrome (syndrome X). *Diabetes*, **41**, 715–722.

18 Gordon, T., Castelli, W. P., Hjortland, M. C., Kannel, W. B., Dawber, T. R. (1977) High density lipoprotein as a protective factor against coronary heart disease. The Framingham Study. *Am. J. Med.*, **62**, 707–714.

19 Duffy, D. and Rader, D. J. (2006) Emerging therapies targeting high-density lipoprotein metabolism and reverse cholesterol transport. *Circulation*, **113**, 1140–1150.

20 Levine, D. M., Parker, T. S., Donnelly, T. M., Walsh, A., Rubin, A. L. (1993) In vivo protection against endotoxin by plasma high density lipoprotein. *Proc. Natl. Acad. Sci. U.S.A.*, **90**, 12040–12044.

21 Seetharam, D., Mineo, C., Gormley, A. K., Gibson, L. L., Vongpatanasin, W., Chambliss, K. L., Hahner, L. D., Cummings, M. L., Kitchens, R. L., Marcel, Y. L., Rader, D. J., Shaul, P. W. (2006) High-density lipoprotein promotes endothelial cell migration and reendothelialization via scavenger receptor-B type I. *Circ. Res.*, **98**, 63–72.

22 Tso, C., Martinic, G., Fan, W. H., Rogers, C., Rye, K. A., Barter, P. J. (2006) High-density lipoproteins enhance progenitor-mediated endothelium repair in mice. *Arterioscler. Thromb. Vasc. Biol.*, **26**, 1144–1149.

23 Sugatani, J., Miwa, M., Komiyama, Y., Ito, S. (1996) High-density lipoprotein inhibits the synthesis of platelet-activating factor in human vascular endothelial cells. *J. Lipid Mediat. Cell Signal.*, **13**, 73–88.

24 Mineo, C., Deguchi, H., Griffin, J. H., Shaul, P. W. (2006) Endothelial and antithrombotic actions of HDL. *Circ. Res.*, **98**, 1352–1364.

25 Bisoendial, R. J., Hovingh, G. K., Levels, J. H., Lerch, P. G., Andresen, I., Hayden, M. R., Kastelein, J. J., Stroes, E. S. (2003) Restoration of endothelial function by increasing high-density lipoprotein in subjects with isolated low high-density lipoprotein. *Circulation*, **107**, 2944–2948.

26 Barter, P. J., Nicholls, S., Rye, K. A., Anantharamaiah, G. M., Navab, M., Fogelman, A. M. (2004) Antiinflammatory properties of HDL. *Circ. Res.*, **95**, 764–772.

27 Yancey, P. G., Bortnick, A. E., Kellner-Weibel, G., de la Llera-Moya, M., Phillips, M. C., Rothblat, G. H. (2003) Importance of different pathways of cellular cholesterol efflux. *Arterioscler. Thromb. Vasc. Biol.*, **23**, 712–719.

28 Yokoyama, S. (2006) ABCA1 and biogenesis of HDL. *J. Atheroscler. Thromb.*, **13**, 1–15.

29 Wang, N., Lan, D., Chen, W., Matsuura, F., Tall, A. R. (2004) ATP-binding cassette transporters G1 and G4 mediate cellular cholesterol efflux to high-density lipoproteins. *Proc.*

Natl. Acad. Sci. U.S.A., **101**, 9774–9779.

30 Nakamura, K., Kennedy, M. A., Baldan, A., Bojanic, D. D., Lyons, K., Edwards, P. A. (2004) Expression and regulation of multiple murine ATP-binding cassette transporter G1 mRNAs/isoforms that stimulate cellular cholesterol efflux to high density lipoprotein. *J. Biol. Chem.*, **279**, 45980–45989.

31 Mackness, M. I., Durrington, P. N., Mackness, B. (2004) The role of paraoxonase 1 activity in cardiovascular disease: potential for therapeutic intervention. *Am. J. Cardiovasc. Drugs*, **4**, 211–217.

32 Navab, M., Imes, S. S., Hama, S. Y., Hough, G. P., Ross, L. A., Bork, R. W., Valente, A. J., Berliner, J. A., Drinkwater, D. C., Laks, H. et al. (1991) Monocyte transmigration induced by modification of low density lipoprotein in cocultures of human aortic wall cells is due to induction of monocyte chemotactic protein 1 synthesis and is abolished by high density lipoprotein. *J. Clin. Invest.*, **88**, 2039–2046.

33 Suc, I., Escargueil-Blanc, I., Troly, M., Salvayre, R., Negre-Salvayre, A. (1997) HDL and ApoA prevent cell death of endothelial cells induced by oxidized LDL. *Arterioscler. Thromb. Vasc. Biol.*, **17**, 2158–2166.

34 Maier, J. A., Barenghi, L., Pagani, F., Bradamante, S., Comi, P., Ragnotti, G. (1994) The protective role of high-density lipoprotein on oxidized-low-density-lipoprotein-induced U937/endothelial cell interactions. *Eur. J. Biochem.*, **221**, 35–41.

35 Mackness, M. I. and Durrington, P. N. (1995) HDL, its enzymes and its potential to influence lipid peroxidation. *Atherosclerosis*, **115**, 243–253.

36 Garner, B., Witting, P. K., Waldeck, A. R., Christison, J. K., Raftery, M., Stocker, R. (1998) Oxidation of high density lipoproteins. I. Formation of methionine sulfoxide in apolipoproteins AI and AII is an early event that accompanies lipid peroxidation and can be enhanced by alpha-tocopherol. *J. Biol. Chem.*, **273**, 6080–6087.

37 Cockerill, G. W., Rye, K. A., Gamble, J. R., Vadas, M. A., Barter, P. J. (1995) High-density lipoproteins inhibit cytokine-induced expression of endothelial cell adhesion molecules. *Arterioscler. Thromb. Vasc. Biol.*, **15**, 1987–1994.

38 Xia, P., Vadas, M. A., Rye, K. A., Barter, P. J., Gamble, J. R. (1999) High density lipoproteins (HDL) interrupt the sphingosine kinase signaling pathway. A possible mechanism for protection against atherosclerosis by HDL. *J. Biol. Chem.*, **274**, 33143–33147.

39 Baker, P. W., Rye, K. A., Gamble, J. R., Vadas, M. A., Barter, P. J. (1999) Ability of reconstituted high density lipoproteins to inhibit cytokine-induced expression of vascular cell adhesion molecule-1 in human umbilical vein endothelial cells. *J. Lipid Res.*, **40**, 345–353.

40 Baker, P. W., Rye, K. A., Gamble, J. R., Vadas, M. A., Barter, P. J. (2000) Phospholipid composition of reconstituted high density lipoproteins influences their ability to inhibit endothelial cell adhesion molecule expression. *J. Lipid Res.*, **41**, 1261–1267.

41 Navab, M., Imes, S. S., Hama, S. Y., Hough, G. P., Ross, L. A., Bork, R. W., Valente, A. J., Berliner, J. A., Drinkwater, D. C., Laks, H., Fogelman, A. M. (1991) Monocyte transmigration induced by modification of low density lipoprotein in cocultures of human aortic wall cells is due to induction of monocyte chemotactic protein 1 synthesis and is abolished by high density lipoprotein. *J. Clin. Invest.*, **88**, 2039–2046.

42 Moudry, R., Spycher, M. O., Doran, J. E. (1997) Reconstituted high density lipoprotein modulates adherence of polymorphonuclear leukocytes to human endothelial cells. *Shock.*, **7**, 175–181.

43 Dimayuga, P., Zhu, J., Oguchi, S., Chyu, K-Y., Xu, X-O.H., Yano, J., Shah, P. K., Nilsson, J., Cercek, B. (1999) Reconstituted HDL containing human apolipoprotein A-I reduces VCAM-1 expression and neointima formation following periadventitial cuff-induced carotid injury in apoE null mice. *Biochem. Biophys. Res. Comm.*, **264**, 465–468.

44 Rong, J. X., Li, J., Reis, E. D., Choudhury, R. P., Dansky, H. M., Elmalem, V. I., Fallon, J. T., Breslow, J. L., Fisher, E. A. (2001) Elevating high-density lipoprotein cholesterol in apolipoprotein E-deficient mice remodels advanced atherosclerotic lesions by decreasing macrophage and increasing smooth muscle cell content. *Circulation*, **104**, 2447–2452.

45 Nicholls, S. J., Cutri, B., Worthley, S. G., Kee, P., Rye, K. A., Bao, S., Barter, P. J. (2005) Impact of short-term administration of high-density lipoproteins and atorvastatin on atherosclerosis in rabbits. *Arterioscler. Thromb. Vasc. Biol.*, **25**, 2416–2421.

46 Cockerill, G. W., Huehns, T. Y., Weerasinghe, A., Stocker, C., Lerch, P. G., Miller, N. E., Haskard, D. O. (2001) Elevation of plasma high-density lipoprotein concentration reduces interleukin-1-induced expression of E-selectin in an *in vivo* model of acute inflammation. *Circulation*, **103**, 108–112.

47 Paterno, R., Ruocco, A., Postiglione, A., Hubsch, A., Anderson, I., Lang, M. G. (2004) Reconstituted high-density lipoprotein exhibits neuroprotection in two rat models of stroke. *Cerebrovasc. Dis.*, **17**, 204–211.

48 Cockerill, G. W., McDonald, M. C., Mota-Filipe, H., Cuzzocrea, S., Miller, N. E., Thiemermann, C. (2001) High density lipoproteins reduce organ injury and organ dysfunction in a rat model of hemorrhagic shock. *Faseb J.*, **15**, 1941–1952.

49 Nicholls, S. J., Dusting, G. J., Cutri, B., Bao, S., Drummond, G. R., Rye, K. A., Barter, P. J. (2005) Reconstituted high-density lipoproteins inhibit the acute pro-oxidant and proinflammatory vascular changes induced by a periarterial collar in normocholesterolemic rabbits. *Circulation*, **111**, 1543–1550.

50 Spieker, L. E., Sudano, I., Hurlimann, D., Lerch, P. G., Lang, M. G., Binggeli, C., Corti, R., Ruschitzka, F., Luscher, T. F., Noll, G. (2002) High-density lipoprotein restores endothelial function in hypercholesterolemic men. *Circulation*, **105**, 1399–1402.

51 Bisoendial, R. J., Hovingh, G. K., Levels, J. H. M., Lerch, P. G., Andresen, I., Hayden, M. R., Kastelein, J. J. P., Stroes, E. S. G. (2003) Restoration of Endothelial Function by Increasing High-Density Lipoprotein in Subjects With Isolated Low High-Density Lipoprotein. *Circulation*, **107**, 2944–2948.

52 Rye, K. A., Clay, M. A., Barter, P. J. (1999) Remodelling of high density lipoproteins by plasma factors. *Atherosclerosis*, **145**, 227–238.

53 Warden, C. H., Hedrick, C. C., Qiao, J. H., Castellani, L. W., Lusis, A. J. (1993) Atherosclerosis in transgenic mice overexpressing apolipoprotein A-II. *Science*, **261**, 469–472.

54 Amouyel, P., Isorez, D., Bard, J. M., Goldman, M., Lebel, P., Zylberberg, G., Fruchart, J. C. (1993) Parental history of early myocardial infarction is associated with decreased levels of lipoparticle AI in adolescents. *Arterioscler. Thromb.*, **13**, 1640–1644.

55 Tailleux, A., Bouly, M., Luc, G., Castro, G., Caillaud, J. M., Hennuyer, N., Poulain, P., Fruchart, J. C., Duverger, N., Fievet, C. (2000) Decreased susceptibility to diet-induced atherosclerosis in human apolipoprotein A-II transgenic mice. *Arterioscler. Thromb. Vas.c Biol.*, **20**, 2453–2458.

56 Miller, N. E. (1987) Associations of high-density lipoprotein subclasses and apolipoproteins with ischemic heart disease and coronary atherosclerosis. *Am. Heart J.*, **113**, 589–597.

57 Chiesa, G., Monteggia, E., Marchesi, M., Lorenzon, P., Laucello, M., Lorusso, V., Di Mario, C., Karvouni, E., Newton, R. S., Bisgaier, C. L., Franceschini, G., Sirtori, C. R. (2002) Recombinant apolipoprotein A-I (Milano) infusion into rabbit carotid artery rapidly removes lipid from fatty streaks. *Circ Res.*, **90**, 974–980.

58 Duverger, N., Kruth, H., Emmanuel, F., Caillaud, J. M., Viglietta, C., Castro, G., Tailleux, A., Fievet, C., Fruchart, J. C., Houdebine, L. M., Denefle, P. (1996) Inhibition of atherosclerosis development in cholesterol-fed human apolipoprotein A-I-transgenic rabbits. *Circulation*, **94**, 713–717.

59 Hoeg, J. M., Santamarina-Fojo, S., Berard, A. M., Cornhill, J. F., Herderick, E. E., Feldman, S. H., Haudenschild, C. C., Vaisman, B. L., Hoyt, R. F., Jr., Demosky, S. J., Jr., Kauffman, R. D., Hazel, C. M., Marcovina, S. M., Brewer, H. B., Jr. (1996) Overexpression of lecithin:cholesterol acyltransferase in transgenic rabbits prevents diet-induced atherosclerosis. *Proc. Natl. Acad. Sci. U.S.A.*, **93**, 11448–11453.

60 Berard, A. M., Foger, B., Remaley, A., Shamburek, R., Vaisman, B. L., Talley, G., Paigen, B., Hoyt, R. F., Jr., Marcovina, S., Brewer, H. B., Jr., Santamarina-Fojo, S. (1997) High plasma HDL concentrations associated with enhanced atherosclerosis in transgenic mice overexpressing lecithin-cholesteryl acyltransferase. *Nat Med.*, **3**, 744–749.

61 Sugano, M., Makino, N., Sawada, S., Otsuka, S., Watanabe, M., Okamoto, H., Kamada, M., Mizushima, A. (1998) Effect of antisense oligonucleotides against cholesteryl ester transfer protein on the development of atherosclerosis in cholesterol-fed rabbits. *J. Biol. Chem.*, **273**, 5033–5036.

62 Okamoto, H., Yonemori, F., Wakitani, K., Minowa, T., Maeda, K., Shinkai, H. (2000) A cholesteryl ester transfer protein inhibitor attenuates atherosclerosis in rabbits. *Nature*, **406**, 203–207.

63 The Lipid Research Clinics Coronary Primary Prevention Trial results, (1984) I. Reduction in incidence of coronary heart disease. *Jama.*, **251**, 351–364.

64 Gotto, A. M., Jr., Whitney, E., Stein, E. A., Shapiro, D. R., Clearfield, M., Weis, S., Jou, J. Y., Langendorfer, A., Beere, P. A., Watson, D. J., Downs, J. R., de Cani, J. S. (2000) Relation between baseline and on-treatment lipid parameters and first acute major coronary events in the Air Force/Texas Coronary Atherosclerosis Prevention Study (AFCAPS/TexCAPS). *Circulation*, **101**, 477–484.

65 (1998) Influence of pravastatin and plasma lipids on clinical events in the West of Scotland Coronary Prevention Study (WOSCOPS). *Circulation*, **97**, 1440–1445.

66 Manninen, V., Tenkanen, L., Koskinen, P., Huttunen, J. K., Manttari, M., Heinonen, O. P., Frick, M. H. (1992) Joint effects of serum triglyceride and LDL cholesterol and HDL cholesterol concentrations on coronary heart disease risk in the Helsinki Heart Study. Implications for treatment. *Circulation*, **85**, 37–45.

67 Tenkanen, L., Manttari, M., Manninen, V. (1995) Some coronary risk factors related to the insulin resistance syndrome and treatment with gemfibrozil. Experience from the Helsinki Heart Study. *Circulation*, **92**, 1779–1785.

68 Rubins, H. B., Robins, S. J., Collins, D., Fye, C. L., Anderson, J. W., Elam, M. B., Faas, F. H., Linares, E., Schaefer, E. J., Schectman, G., Wilt, T. J., Wittes, J. (1999) Gemfibrozil for the secondary prevention of coronary heart disease in men with low levels of high-density lipoprotein cholesterol. Veterans Affairs High-Density Lipoprotein Cholesterol Intervention Trial Study Group. *N. Engl. J. Med.*, **341**, 410–418.

69 Robins, S. J., Collins, D., Wittes, J. T., Papademetriou, V., Deedwania, P. C., Schaefer, E. J., McNamara, J. R.,

Kashyap, M. L., Hershman, J. M., Wexler, L. F., Rubins, H. B. (2001) Relation of gemfibrozil treatment and lipid levels with major coronary events: VA-HIT: a randomized controlled trial. *Jama.*, **285**, 1585–1591.

70 (2000) Secondary prevention by raising HDL cholesterol and reducing triglycerides in patients with coronary artery disease: the Bezafibrate Infarction Prevention (BIP) study. *Circulation*, **102**, 21–27.

71 Tenenbaum, A., Motro, M., Fisman, E. Z., Tanne, D., Boyko, V., Behar, S. (2005) Bezafibrate for the secondary prevention of myocardial infarction in patients with metabolic syndrome. *Arch Intern Med.*, **165**, 1154–1160.

72 Keech, A., Simes, R. J., Barter, P., Best, J., Scott, R., Taskinen, M. R., Forder, P., Pillai, A., Davis, T., Glasziou, P., Drury, P., Kesaniemi, Y. A., Sullivan, D., Hunt, D., Colman, P., d'Emden, M., Whiting, M., Ehnholm, C., Laakso, M. (2005) Effects of long-term fenofibrate therapy on cardiovascular events in 9795 people with type 2 diabetes mellitus (the FIELD study): randomised controlled trial. *Lancet*, **366**, 1849–1861.

73 Carlson, L. A. (2004) Niaspan, the prolonged release preparation of nicotinic acid (niacin), the broad-spectrum lipid drug. *Int. J. Clin. Pract.*, **58**, 706–713.

74 Brown, G., Albers, J. J., Fisher, L. D., Schaefer, S. M., Lin, J. T., Kaplan, C., Zhao, X. Q., Bisson, B. D., Fitzpatrick, V. F., Dodge, H. T. (1990) Regression of coronary artery disease as a result of intensive lipid-lowering therapy in men with high levels of apolipoprotein B. *N. Engl. J. Med.*, **323**, 1289–1298.

75 Brown, B. G., Zhao, X. Q., Chait, A., Fisher, L. D., Cheung, M. C., Morse, J. S., Dowdy, A. A., Marino, E. K., Bolson, E. L., Alaupovic, P., Frohlich, J., Albers, J. J. (2001) Simvastatin and niacin, antioxidant vitamins, or the combination for the prevention of coronary disease. *N. Engl. J. Med.*, **345**, 1583–1592.

76 Nissen, S. E., Tsunoda, T., Tuzcu, E. M., Schoenhagen, P., Cooper, C. J., Yasin, M., Eaton, G. M., Lauer, M. A., Sheldon, W. S., Grines, C. L., Halpern, S., Crowe, T., Blankenship, J. C., Kerensky, R. (2003) Effect of recombinant ApoA-I Milano on coronary atherosclerosis in patients with acute coronary syndromes: a randomized controlled trial. *Jama.*, **290**, 2292–2300.

21
Therapeutic Targeting of High-Density Lipoproteins

Daniel J. Rader

21.1
Introduction

Low plasma concentrations of high-density lipoprotein (HDL) cholesterol, as well as of its major protein apolipoprotein A-I (apoA-I), have been demonstrated in many observational studies to be independent risk factors for atherosclerotic cardiovascular events [1,2]. Many controlled clinical trials have also demonstrated that treating patients who have low HDL-C with various lipid modifying therapies, including statins, fibrates, niacin, or combination therapy, can reduce major coronary events. The National Cholesterol Education Program (NCEP) Adult Treatment Panel (ATP) III guidelines recognized low HDL-C (<40 mg dL^{-1}) as an independent major risk factor for CHD, as a component of the metabolic syndrome, and as a potential target for therapeutic intervention [3]. The primary target of the NCEP ATP III guidelines is to lower levels of low-density cholesterol (LDL-C), while an update to the ATP III guidelines recommends even more aggressive treatment of LDL-C in the light of new clinical trial data [4]. However, even in high-risk patients treated to aggressive LDL-C goals, coronary events still occur at a high rate [5,6] and low HDL-C is a major risk factor in patients treated aggressively for LDL-C. Therefore, a natural next step in the search for therapies to reduce cardiovascular morbidity and mortality further involves raising HDL-C levels and/or improving HDL function.

21.2
Current Clinical Approaches to Patients with Low HDL-C

Therapeutic lifestyle changes are first-line therapy in patients with low HDL-C levels. Smoking cessation is associated with modest increases in HDL-C [7,8], while aerobic exercise can raise HDL-C levels modestly or, if accompanied by weight loss, more substantially [9,10]. The impact of dietary composition on HDL metabolism is complex. Diets high in saturated fat raise HDL-C levels, diets high in polyunsaturated fat reduce HDL-C levels, and monounsaturated fats tend to be neutral with regard to HDL-C [11]. Alcohol use has substantial HDL-C raising effects that are

High-Density Lipoproteins: From Basic Biology to Clinical Aspects. Edited by Christopher J. Fielding

ISBN: 978-3-527-31717-2

dose-dependent [8] but is not generally considered to be a clinical HDL-raising approach.

The HDL-C raising effects of currently available drugs translate only into modest effects on HDL-C levels. Statins raise HDL-C by only 5–10 % [3] and the Cholesterol Treatment Trialists' Collaborators meta-analysis of 14 statin trials showed no difference in benefit from statin therapy by HDL-C subgroup [12]. However, a meta-analysis of statins studies using coronary intravascular ultrasound to measure coronary atheroma suggested that statin therapy was associated with regression of coronary atherosclerosis when LDL-C was substantially reduced and HDL-C was increased by more than 7.5 % [13]. Fibrates, agonists of PPARα, raise HDL-C by 5–20 % [3] depending on triglyceride levels. While some clinical trials with fibrates have shown significant reductions in major CHD events [14,15], others have been disappointing [108] Keech, 2005 #4558}. Whether the HDL-raising effect of fibrates contributes to their cardiovascular benefits is uncertain. Thiazolidinediones, agonists of PPARγ used for the treatment of type 2 diabetes, also have a modest effect in increasing HDL-C by 5–15 % [16,17]. Nicotinic acid (niacin) is the most effective HDL-raising drug currently available, with increases of up to 35 % [3]. The only fully powered outcome trial is the Coronary Drug Project, which showed that niacin reduced coronary events in men with CHD [18]. Thus, while some therapies to raise HDL-C are available their effects are modest and new therapies targeted to HDL are needed.

21.3 Approaches to Raising Levels of HDL-C and/or ApoA-I

21.3.1 Increasing HDL/ApoA-I Production and Maturation

The biosynthesis of HDL involves multiple components and is an interesting target for the development of novel therapies. The major HDL apolipoprotein apoA-I constitutes approximately 70 % of HDL protein and is present in virtually all HDL particles [19]. ApoA-I is synthesized in both the intestine and the liver, and after secretion it quickly acquires phospholipids and cholesterol through the ATP binding cassette A1 (ABCA1) transporter [20,21]. This nascent HDL acquires lipids from other tissues, and free cholesterol is esterified to cholesteryl ester by the enzyme lecithin:cholesterol acyltransferase (LCAT), forming the mature HDL particle [19].

Hepatic overexpression of apoA-I inhibits progression [22,23] and induces regression [24] of atherosclerosis in mice, and transcriptional upregulation of apoA-I expression would be a potentially effective approach to therapeutic targeting of HDL. There has been great interest in understanding the transcriptional regulation of apoA-I [25] with the goal of developing small molecules that promote the process. PPARα agonists, such as fibrates, upregulate transcription of apoA-I in vitro and in mice [26] but have not been conclusively demonstrated to increase apoA-I production in humans. The orphan nuclear receptor liver receptor homolog-1 (LRH-1) was shown to upregulate apoA-I transcription through direct binding to the apoA-I promoter [27], suggesting another potential therapeutic strategy. Agonism of the farnesoid X receptor (FXR) inhibits apoA-I transcription [28], suggesting that FXR antagonism could be yet another approach to apoA-I transcriptional upregulation. Because increasing apoA-I production is perhaps the most validated HDL-targeted approach in preclinical models, this area is of substantial interest, and entry of new molecules into clinical development would be welcomed with anticipation.

Hepatic overexpression of LCAT substantially increases HDL-C levels in mice [29] and rabbits [30], and reduces atherosclerosis in rabbits [31]. Upregulation of LCAT activity through the use of a small molecule might thus be expected to increase HDL-C levels and could be considered another target, although small-molecule LCAT activators may be difficult to develop (though an apoA-I mimetic peptide that activates LCAT is in clinical trials; see below). Furthermore, the impact of increased LCAT activity on reverse cholesterol transport and atherosclerosis remains uncertain.

21.3.2 Inhibiting ApoA-I Catabolism

The pathways of HDL catabolism are even more complex than the biosynthetic process. Like their biosyntheses, catabolism of HDL apolipoproteins and of HDL cholesterol are, at least in part, distinct processes, and the catabolism of apoA-I is an important determinant of plasma apoA-I concentrations [19]. Studies in animals using trapped ligands [32] established that approximately one-third of apoA-I is catabolized by the kidneys, and the rest by the liver. Lipid-poor apoA-I is glomerularly filtered and then catabolized by proximal renal tubular epithelial cells, where it is internalized and degraded by cubilin [33,34]. However, functional cubilin deficiency in animals or humans does not cause elevated apoA-I levels [35,36], probably because the rate-limiting step in renal catabolism is glomerular filtration, so inhibition of cubilin is not considered a major target of therapy.

Because only lipid-poor apoA-I can be glomerularly filtered, there is interest in the factors that regulate the generation of lipid-poor apoA-I. One of the key processes is the remodeling of HDL, resulting in the generation of lipid-poor apoA-I. The hydrolysis of HDL triglycerides and phospholipids by lipolytic enzymes is an important HDL remodeling process and an important determinant of renal apoA-I catabolism. Hepatic lipase (HL) has the ability to hydrolyze both HDL triglycerides and phospholipids but is probably most active on triglyceride-enriched HDL [19]. HDL becomes enriched with triglycerides after exchange of its cholesteryl ester for triglyceride in a process mediated by the cholesteryl ester transfer protein (CETP; see below). HL is then especially effective in hydrolyzing HDL triglycerides, resulting in the shedding of lipid-poor apoA-I from HDL and increased apoA-I filtration and renal catabolism. Indeed, apoA-I associated with TG-enriched HDL is catabolized significantly more rapidly than non-TG-enriched HDL [37]. Pharmacologic inhibition of HL would therefore be expected to slow apoA-I catabolism and to increase apoA-I levels, but HL also plays a role in the lipolysis of apoB-containing remnant particles, so there is concern that inhibition of HL could increase plasma concentrations of atherogenic lipoproteins. As a result, HL has not been a major focus of drug development.

Endothelial lipase (EL) is a close relative of HL [38,39], having relatively more phospholipase activity than HL and being even more active in hydrolyzing HDL phospholipids (109). Overexpression of EL in mice results in reduction in HDL-C and apoA-I levels [38,40], due to increased catabolism, primarily through the kidneys [41]. Conversely, antibody inhibition [42] or gene deletion [40,43] of EL results in increased HDL-C and apoA-I levels. Subjects with high HDL-C levels have potentially functional mutations in the EL gene [44], while plasma EL levels in humans are significantly inversely associated with HDL-C levels and other components of the metabolic syndrome [45]. Atherosclerosis studies in EL knockout mice have been conflicting [46,47], but plasma EL levels in humans are significantly associated with coronary atherosclerosis [45]. Pharmacologic inhibition of EL would therefore be expected to slow apoA-I catabolism and to increase apoA-I and HDL-C levels, and EL is indeed considered a *bona fide* target for the development of small-molecule inhibitors for raising HDL-C.

A substantial amount of apoA-I is catabolized by the liver [32], but the mechanisms are poorly understood. ApoE-rich HDL is catabolized by the LDL receptor and probably accounts for some HDL apoA-I catabolism [48]. Other pathways of HDL apoA-I uptake and degradation by the liver probably exist, and could be of potential therapeutic interest; niacin has in fact been shown to reduce uptake of HDL apoA-I in hepatocytes *in vitro* [49] and to reduce the catabolic rate of apoA-I in humans *in vivo* [50]. The niacin receptor GPR109A (aka HM74A) is a G protein-coupled receptor primarily expressed on adipocytes, where it mediates suppression of triglyceride lipolysis and release of free fatty acids [51]. Whether this receptor mediates the effects of niacin on the liver uptake of HDL remains to be determined; indeed, the mechanisms by which niacin raises HDL-C levels are still poorly understood. In any case, GPR109A is considered a major target for new drug development, though whether synthetic agonists of GPR109A will raise HDL-C levels remains to be determined.

21.3.3 Inhibiting Cholesteryl Ester Transfer

The cholesteryl ester transfer protein (CETP) transfers cholesteryl esters (CE) from HDL to apoB-containing lipoproteins in exchange for triglycerides, thus resulting in CE depletion of HDL and CE enrichment of apoB-containing lipoproteins. Humans genetically deficient in CETP have markedly increased HDL-C levels [52,53], confirming the critical role CETP plays in HDL metabolism in humans. CETP-deficient persons have high HDL levels and slow turnover of apoA-I [54], confirming that lipid transfer by CETP influences the catabolism of the associated apoA-I. Whether CETP deficiency causes reduced, or increased, risk of CHD is not definitively established. Individuals heterozygous for CETP mutations display only modest increases in HDL-C levels and have no major difference in cardiovascular risk [55], while additional genetic studies of CETP polymorphisms in other populations [56] and measurement of CETP in large prospective observational studies [57] are conflicting.

The fact that CETP-deficient persons have very high HDL-C levels has raised interest in CETP inhibition as a novel approach to raise plasma HDL-C levels. The CETP inhibitor JTT-705 was shown to raise HDL-C levels both as monotherapy [58] and in combination with pravastatin [59]. Another CETP inhibitor, torcetrapib, increased HDL-C levels in healthy volunteers [60] and in subjects with low baseline HDL-C levels [61]. Dose-dependent increases in HDL-C levels and modest reductions in LDL-C levels were seen with torcetrapib monotherapy as well as in combination with atorvastatin [62]. Mice lack CETP, but inhibition of CETP in rabbits reduces atherosclerosis [63]. On the basis of promising preclinical atherosclerosis data and HDL-C raising in humans, torcetrapib was advanced into phase III trials involving both atherosclerosis imaging (intravascular ultrasound of the coronaries and intima-media thickness of the carotids), as well as a large cardiovascular outcome trial [64], but in December 2006 the outcome trial was halted because of excess mortality in the torcetrapib group [65]. At this time it is unclear as to whether this result is due to inhibition of CETP or to molecule-specific effects (such as elevation in blood pressure) [65].

The effect of CETP inhibition on reverse cholesterol transport is of interest. Injection of HDL labeled with a CE tracer in humans demonstrated that the labeled cholesterol that was excreted into bile was mostly firstly transferred to apoB-containing lipoproteins [66], suggesting a role for CETP in reverse cholesterol transport. Administration of torceptrapib resulted in significantly slower catabolism of apoA-I and no significant effect on fecal neutral sterol excretion [67]. While an early report suggested that HDL from CETP-deficient subjects was "dysfunctional" in its ability to promote macrophage cholesterol efflux, a recent study indicated that CETP-deficient HDL is more effective than control HDL in promoting macrophage cholesterol efflux by the ABCG1 pathway [68]. More research is needed to determine whether CETP inhibition will be antiatherogenic in humans, but this question has become considerably more complicated to answer in view of the clinical outcome results with torcetrapib.

21.4 Approaches to Promoting Reverse Cholesterol Transport

Reverse cholesterol transport (RCT) is the process by which excess cholesterol is effluxed from peripheral tissues to HDL and returned to the liver for excretion in the bile and ultimately the feces [69]. The concept that novel therapies that promote this process might be developed is appealing. In particular, there is major interest in promoting the first step of this process: namely upregulation of cholesterol efflux from macrophages. The best understood pathway for macrophage cholesterol efflux is the ABCA1 transporter, which promotes cholesterol efflux to lipid-poor apoA-I. Overexpression of ABCA1 in macrophages increases cholesterol efflux to lipid-poor apoA-I and reduces atherosclerosis in mice [70]; conversely, macrophages from ABCA1-knockout mice display substantially reduced cholesterol efflux to lipid-poor apoA-I, and transplantation of bone marrow from ABCA1-knockout mice into atherosclerosis-prone mice results in significantly increased atherosclerosis [71]. The major regulators of ABCA1 gene expression are the nuclear receptor liver X receptors α and β (LXRα and LXRβ), which act as heterodimers with their partner the retinoid X receptor (RXR) [72]. LXR is activated by cholesterol loading and upregulates ABCA1 in that context. Synthetic LXR agonists upregulate ABCA1 transcription in macrophages and result in increased cholesterol efflux to lipid-poor apoA-I.

Mature HDL is also capable of promoting cholesterol efflux from macrophages through the transporter ABCG1 [73,74]. ABCG1-deficient macrophages have reduced cholesterol efflux to HDL, while ABCG1-knockout mice have accumulation of lipid in macrophages, particularly in the lung [74]. Paradoxically, macrophage deficiency of ABCG1 was shown to result in reduced atherosclerosis [75,76], possibly due to compensatory upregulation of macrophage ABCA1 and apoE [75] or increased susceptibility of ABCG1-deficient macrophages to apoptosis induced by oxidized LDL [76]. Like ABCA1, ABCG1 is upregulated by cholesterol loading and synthetic LXR agonists, which also result in translocation from intracellular pools to the plasma membrane [77].

LXR agonists are therefore of obvious interest as an approach to promoting macrophage cholesterol efflux and RCT. Indeed, a synthetic LXR agonist promoted macrophage cholesterol efflux and RCT *in vivo* despite having little effect on plasma HDL-C levels [78]. Synthetic LXR agonists have been shown to inhibit atherosclerosis progression [79,80] and even to promote atherosclerosis regression [81] in mice, despite having little effect on plasma HDL-C levels. However, some LXR agonists have been found to cause hepatic steatosis and hypertriglyceridemia in animals, which is believed to be due to induction of hepatic expression of SREBP1c, which in turn induces expression of fatty acid synthesis genes [82]. Furthermore, some LXR agonists increased LDL-C levels in animals that express CETP [83]. It would be desirable to develop LXR modulators that were relatively selective for specific tissues or for specific genes (i.e., ABCA1/G1 over SREBP1c). One LXR agonist has been reported to be partially selective and to induce less hepatic steatosis [84]. Alternatively, selective modulation of LXRβ but not LXRα (which is more abundant in liver) might be a viable approach. There remains substantial interest in LXR agonism as a therapeutic approach, and the hope remains that more selective LXR modulators may have fewer hepatic adverse effects. It will be important to investigate systematically the effects of LXR agonists in human on HDL and LDL metabolism, RCT, and hepatic lipid.

Macrophages express PPARβ/δ, and a synthetic PPARβ/δ agonist was shown to raise HDL-C levels in a rhesus monkey model of the metabolic syndrome and to promote macrophage cholesterol efflux *ex vivo*[85]. Whether PPARβ/δ agonists promote macrophage cholesterol efflux remains controversial, however. PPARβ/δ agonists are in clinical development, and early human studies suggest relatively modest effects on HDL-C levels [86]. Whether PPARβ/δ agonists promote RCT or have any role in modulation of plasma lipids or prevention of atherosclerotic disease remains to be determined.

In rodents, there is substantial "selective uptake" of HDL cholesterol by the liver, mediated by the scavenger receptor class BI (SR-BI). Hepatic overexpression of SR-BI in mice markedly reduces plasma HDL-C levels [87] but also reduces atherosclerosis [88]; conversely, SR-BI knockout mice have elevated HDL-C levels [89,90] but markedly increased atherosclerosis [91–93]. This apparent paradox was resolved by the demonstration that hepatic SR-BI overexpression increased RCT whereas SR-BI deficiency reduced RCT [94], which demonstrates the concept that "flux" of HDL cholesterol is more important than steady-state concentrations with regard to effects on atherosclerosis. Pharmacologic upregulation of hepatic SR-BI might very well be an effective antiatherogenic approach, but would be complicated by reduced plasma HDL-C levels. Furthermore, there remain questions regarding the physiologic importance of the hepatic SR-BI pathway for uptake of HDL cholesterol in humans, as studies in humans suggest that relatively little HDL-cholesteryl ester (HDL-CE) is directly taken up by the liver and targeted to bile [66]. Thus, while theoretically interesting, SR-BI upregulation is unlikely to be developed as a therapeutic strategy.

21.5 Full-Length ApoA-I and ApoA-I Mimetic Peptides

Parenteral infusion of full-length apoA-I might be expected to have beneficial effects that, with repeated infusions, could favorably impact on atherosclerosis. Acute bolus intravenous infusion of apoA-I in humans demonstrates relatively little increase in HDL-C levels, but measurable increases in apoA-I and phospholipids [95]. Four subjects with heterozygous familial hypercholesterolemia infused with a single bolus of proapoA-I had a 39 % increase in fecal sterol excretion, suggesting promotion of reverse cholesterol transport [96]. ApoA-I Milano is one of several rare, naturally occurring point mutations in apoA-I that cause low HDL-C levels but do not necessarily increase cardiovascular risk [97]. Studies in animals have shown that apoA-I Milano expression is associated with reduced atherosclerosis [97]. In a small clinical trial, five weekly infusions of recombinant apoA-I Milano complexed with phospholipids resulted in no change in HDL-C or apoA-I levels but a reduction from baseline in coronary atheroma volume as measured by intravascular ultrasound [98]. While this was a small study with some caveats [99], it provided some degree of proof-of-concept that parenteral infusion of apoA-I can impact atherosclerosis and in a relatively short time. Selective delipidation of HDL *ex vivo*, which generates lipid-poor apoA-I, is under development as an alternative strategy; in this model the delipidated HDL is reinfused in an autologous fashion with the hope of influencing atherosclerosis [100].

ApoA-I contains ten amphipathic helices that are adapted to bind lipids on one face and to interact with the aqueous environment on the other. Small amphipathic peptides of 18–22 amino acids based loosely on the apoA-I sequence have similar properties to apoA-I, including the abilities to promote cellular cholesterol efflux and to activate LCAT [101]. One major advantage of these apoA-I mimetic peptides over full-length apoA-I is that they are much smaller and therefore much cheaper and easier to make as a therapeutic molecule [102], but because they are targets of peptidases in the gut they must be parenterally administered. Repeated injection of the prototypical peptide L-5F reduced the progression of atherosclerosis in mice [103], while the apoA-I mimetic peptide ETC-642 (also known as RLT peptide) promotes LCAT activation and is also in clinical development. Several other apoA-I mimetic peptides have been developed and tested in cell and animal models [102], and parenterally administered apoA-I mimetic peptides may ultimately prove to be an effective method of treated acute coronary syndromes in humans.

One oral apoA-I mimetic peptide in clinical development is an oral apoA-I mimetic peptide called D-4F. D-4F is an 18 amino acid peptide almost identical in sequence to the L-5F mentioned above, but differs in that it is has one phenylalanine residue fewer on the hydrophobic face and is composed of all D-amino acids. As a result, it is not recognized by gut peptidases in the same way as naturally occurring proteins containing L-amino acids are, and is therefore not degraded in the gut and is orally bioavailable. Oral administration of D-4F has been shown to reduce atherosclerosis in mice without raising levels of HDL-C [104]; although its mechanism remains uncertain, some data suggest that its major effect may be to enhance the antiinflammatory function of HDL [105]. D-4F has also been shown to promote macrophage RCT *in vivo* in mice [105]. D-4F may provide a test of the hypothesis that enhancing HDL function without increasing plasma HDL-C levels can reduce atherosclerosis or cardiovascular risk.

21.6 Improving HDL Function

Several potentially antiatherogenic properties of HDL independent of cholesterol efflux have been described. These include inhibition of LDL oxidation, inhibition of endothelial inflammation, promotion of endothelial nitric oxide production, promotion of prostacylin bioavailability, and inhibition of platelet aggregation and coagulation [106,107]. Their molecular bases are poorly understood and the importance of these mechanisms to the atheroprotective effects of HDL remains uncertain, but they could be therapeutically important. It may prove possible to develop therapies that enhance these functions of HDL or directly mimic these effects on the vessel wall.

21.7 Conclusions

Low levels of HDL-C remain a major unsolved medical problem in cardiovascular medicine, but raising HDL-C levels alone may not provide protection from atherosclerosis, as plasma steady-state HDL-C levels are not necessarily reflective of the rate of reverse cholesterol transport or HDL function. Indeed, RCT and other aspects of HDL function are probably more important than HDL-C concentration. There is a great need for the development of novel assays of RCT and HDL function in humans.

There are relatively few validated targets for targeting HDL. CETP inhibition has advanced the furthest in clinical development, but is now in question as a result of the increased mortality with the CETP inhibitor torcetrapib, so an urgent issue in this field is to determine whether this outcome was related to inhibition of CETP or to molecule-specific effects of torcetrapib. Because torcetrapib is known to raise blood pressure, an effect that is not mechanism-based, it is plausible that molecule-specific effects were the major factor in the increased mortality. Some of the additional current targets for small molecule development, such as LXR agonists, niacin receptor agonists, and EL inhibitors, are promising, but data in humans are needed to assess their potential. Better understanding of the molecular regulation of HDL metabolism is required, and efforts to identify additional genes contributing to variation in HDL-C levels in humans are needed. Previously unsuspected genes involved in HDL metabolism should be identified by genome-wide studies, and some of these will probably become therapeutic targets.

In addition to small molecules, biological approaches may also become an important area for HDL therapeutics. Infusion of wild-type or modified full-length apoA-I, or more probably of smaller apoA-I mimetic peptides, could become a form of "induction therapy" for acute coronary disease, bringing with it major paradigm changes in the management of acute coronary syndromes. Furthermore, the concept of oral apoA-I mimetic peptides that are absorbed and influence HDL function is exciting and has the potential to be a chronic biologic-based therapy.

In summary, the coming phase is likely to be a time when the HDL hypothesis undergoes definitive interrogation with new therapeutic approaches, and if confirmed, will see an exciting expansion of the therapeutic approaches for raising HDL-C levels or improving its function in order to reduce the risk of atherosclerotic cardiovascular disease.

References

1 Gordon, D. J. and Rifkind, B. M. (1989) High-density lipoproteins – the clinical implications of recent studies. *N. Engl. J. Med.*, **321**, 1311–1316.

2 Boden, W. E. (2000) High-density lipoprotein cholesterol as an independent risk factor in cardiovascular disease: assessing the data from Framingham to the Veterans Affairs High-Density Lipoprotein Intervention Trial. *Am. J. Cardiol.*, **86**, 19L–22L.

3 2001. Executive Summary of The Third Report of The National Cholesterol Education Program (NCEP) Expert Panel on Detection, Evaluation, And Treatment of High Blood Cholesterol In Adults (Adult Treatment Panel III). JAMA 285, 2486–2497.

4 Grundy, S. M., Cleeman, J. I., Merz, C. N., Brewer, H. B., Jr., Clark, L. T., Hunninghake, D. B., Pasternak, R. C., Smith, S. C., Jr., Stone, N. J. (2004) Implications of recent clinical trials for the National Cholesterol Education Program Adult Treatment Panel III guidelines. *Circulation*, **110**, 227–239.

5 Cannon, C. P., Braunwald, E., McCabe, C. H., Rader, D. J., Rouleau, J. L., Belder, R., Joyal, S. V., Hill, K. A., Pfeffer, M. A., Skene, A. M. (2004) Intensive versus moderate lipid lowering with statins after acute coronary syndromes. *N. Engl. J. Med.*, **350**, 1495–1504.

6 LaRosa, J. C., Grundy, S. M., Waters, D. D., Shear, C., Barter, P., Fruchart, J. C., Gotto, A. M., Greten, H., Kastelein, J. J., Shepherd, J. *et al.* (2005) Intensive lipid lowering with atorvastatin in patients with stable coronary disease. *N. Engl. J. Med.*, **352**, 1425–1435.

7 Garrison, R. J., Kannel, W. B., Feinleib, M., Castelli, W. P., McNamara, P. M., Padgett, S. J. (1978) Cigarette smoking and HDL cholesterol: the Framingham offspring study. *Atherosclerosis*, **30**, 17–25.

8 Ellison, R. C., Zhang, Y., Qureshi, M. M., Knox, S., Arnett, D. K., Province, M. A. (2004) Lifestyle determinants of high-density lipoprotein cholesterol: the National Heart, Lung, and Blood Institute Family Heart Study. *Am. Heart J.*, **147**, 529–535.

9 Thompson, P. D. and Rader, D. J. (2001) Does exercise increase HDL cholesterol in those who need it the most? *Arterioscler. Thromb. Vasc. Biol.*, **21**, 1097–1098.

10 Dattilo, A. M. and Kris-Etherton, P. M. (1992) Effects of weight reduction on blood lipids and lipoproteins: a meta-analysis. *Am. J. Clin. Nutr.*, **56**, 320–328.

11 Lichtenstein, A. H. (2003) Dietary fat and cardiovascular disease risk: quantity or quality? *J Womens Health (Larchmt)*, **12**, 109–114.

12 Baigent, C., Keech, A., Kearney, P. M., Blackwell, L., Buck, G., Pollicino, C., Kirby, A., Sourjina, T., Peto, R., Collins, R. *et al.* (2005) Efficacy and safety of cholesterol-lowering treatment: prospective meta-analysis of data from 90,056 participants in 14 randomised trials of statins. *Lancet*, **366**, 1267–1278.

13 Nicholls, S. J., Tuzcu, E. M., Sipahi, I., Grasso, A. W., Schoenhagen, P., Hu, T., Wolski, K., Crowe, T., Desai, M. Y., Hazen, S. L. *et al.* (2007) Statins, high-density lipoprotein cholesterol, and regression of coronary atherosclerosis. *Jama*, **297**, 499–508.

14 Manninen, V., Elo, M. O., Frick, M. F., Haapa, K., Heinonen, O. P., Heinsalmi, P., Helo, P., Huttunen, J. K., Kaitaniemi, P., Koskinen, P. *et al.* (1988) Lipid alterations and decline in the incidence of coronary heart disease in the lelsinki heart study. *JAMA*, **260**, 641–651.

15 Rubins, H. B., Robins, S. J., Collins, D., Fye, C. L., Anderson, J. W., Elam, M. B., Faas, F. H., Linares, E., Schaefer, E. J., Schectman, G. *et al.* (1999) Gemfibrozil for the secondary prevention of coronary heart disease in men with low levels of high-density lipoprotein cholesterol. *N. Engl. J. Med.*, **341**, 410–418.

16 Yki-Jarvinen, H. (2004) Thiazolidinediones. *N. Engl. J. Med.*, **351**, 1106–1118.

17 Szapary, P. O., Bloedon, L. T., Samaha, F. F., Duffy, D., Wolfe, M. L., Soffer, D., Reilly, M. P., Chittams, J., Rader, D. J. (2006) Effects of pioglitazone on lipoproteins, inflammatory markers, and adipokines in nondiabetic patients with metabolic syndrome. *Arterioscler. Thromb. Vasc. Biol.*, **26**, 182–188.

18 Canner, P. L., Berge, K. G., Wenger, N. K., Stamler, J., Friedman, L., Prineas, R. J., Friedewald, W. (1986) Fifteen year mortality in Coronary Drug Project patients: long-term benefit with niacin. *Journal of the American College of Cardiology*, **8**, 1245–1255.

19 Lewis, G. F. and Rader, D. J. (2005) New insights into the regulation of HDL metabolism and reverse cholesterol transport. *Circ Res*, **96**, 1221–1232.

20 Timmins, J. M., Lee, J. Y., Boudyguina, E., Kluckman, K. D., Brunham, L. R., Mulya, A., Gebre, A. K., Coutinho, J. M., Colvin, P. L., Smith, T. L. *et al.* (2005) Targeted inactivation of hepatic Abca1 causes profound hypoalphalipoproteinemia and kidney hypercatabolism of apoA-I. *J. Clin. Invest.*, **115**, 1333–1342.

21 Brunham, L., Kruit, J. K., Pape, T. D., Havinga, R., Fievet, C., Timmins, J. M., Bissada, N., Staels, B., Groen, A. K., Park, J. S. *et al.* (2005) Intestinal ABCA1 is a Significant Contributor to Plasma HDL-C and ApoB Levels in Vivo. *Circulation*, **112** (Suppl), II–170.

22 Rubin, E., Krauss, R., Spangler, E., Verstuyft, J., Clift, S. (1991) Inhibition of early atherogenesis in transgenic mice by human apolipoprotein AI. *Nature*, **353**, 265–267.

23 Plump, A., Scott, C., Breslow, J. (1994) Human apolipoprotein A-I

gene expression increases high density lipoprotein and suppresses atherosclerosis in the apolipoprotein E-deficient mouse. *Proc. Natl. Acad. Sci. USA*, **91**, 9607–9611.

24 Tangirala, R. K., Tsukamoto, K., Chun, S. H., Usher, D., Pure, E., Rader, D. J. (1999) Regression of atherosclerosis induced by liver-directed gene transfer of apolipoprotein A-I in mice [see comments]. *Circulation*, **100**, 1816–1822.

25 Mooradian, A. D., Haas, M. J., Wong, N. C. (2004) Transcriptional control of apolipoprotein A-I gene expression in diabetes. *Diabetes*, **53**, 513–520.

26 Berthou, L., Duverger, N., Emmanuel, F., Langouet, S., Auwerx, J., Guillouzo, A., Fruchart, J. C., Rubin, E., Denefle, P., Staels, B. *et al.* (1996) Opposite regulation of human versus mouse apolipoprotein A-I by fibrates in human apolipoprotein A-I transgenic mice. *J. Clin. Invest*, **97**, 2408–2416.

27 Delerive, P., Galardi, C. M., Bisi, J. E., Nicodeme, E., Goodwin, B. (2004) Identification of liver receptor homolog-1 as a novel regulator of apolipoprotein AI gene transcription. *Mol. Endocrinol.*, **18**, 2378–2387.

28 Claudel, T., Staels, B., Kuipers, F. (2005) The Farnesoid X Receptor. A Molecular Link Between Bile Acid and Lipid and Glucose Metabolism. *Arterioscler. Thromb. Vasc. Biol.* **25**, 2020–2030.

29 Francone, O. L., Gong, E. L., Ng, D. S., Fielding, C. J., Rubin, E. M. (1995) Expression of human lecithin-cholesterol acyltransferase in transgenic mice. Effect of human apolipoprotein AI and human apolipoprotein all on plasma lipoprotein cholesterol metabolism. *J. Clin. Invest.*, **96**, 1440–1448.

30 Hoeg, J. M., Vaisman, B. L., Demosky, S. J., Jr., Meyn, S. M., Talley, G. D., Hoyt, R. F., Jr., Feldman, S., Berard, A. M., Sakai, N., Wood, D. *et al.* (1996) Lecithin: cholesterol acyltransferase overexpression generates hyperalpha-lipoproteinemia and a nonatherogenic lipoprotein pattern in transgenic rabbits. *J. Biol. Chem.*, **271**, 4396–4402.

31 Hoeg, J. M., Santamarina-Fojo, S., Berard, A. M., Cornhill, J. F., Herderick, E. E., Feldman, S. H., Haudenschild, C. C., Vaisman, B. L., Hoyt, R. F., Jr., Demosky, S. J., Jr. *et al.* (1996) Overexpression of lecithin: cholesterol acyltransferase in transgenic rabbits prevents diet-induced atherosclerosis. *Proc. Natl. Acad. Sci. USA*, **93**, 11448–11453.

32 Glass, C., Pittman, R. C., Weinstein, D. B., Steinberg, D. (1983) Dissociation of tissue uptake of cholesterol ester from that of apoprotein A-I of rat plamsa high density lipoprotein: selective delivery of cholesterol ester to liver, adrenal, and gonad. *Proc. Natl. Acad. Sci. USA*, **80**, 5435–5439.

33 Hammad, S. M., Stefansson, S., Twal, W. O., Drake, C. J., Fleming, P., Remaley, A., Brewer, H. B., Jr., Argraves, W. S. (1999) Cubilin, the endocytic receptor for intrinsic factor-vitamin B12 complex, mediates high-density lipoprotein holoparticle endocytosis. *Proc. Natl. Acad. Sci. USA*, **96**, 10158–10163.

34 Barth, J. L. and Argraves, W. S. (2001) Cubilin and megalin: partners in lipoprotein and vitamin metabolism. *Trends Cardiovasc Med*, **11**, 26–31.

35 Christensen, E. I. and Gburek, J. (2004) Protein reabsorption in renal proximal tubule-function and dysfunction in kidney pathophysiology. *Pediatr Nephrol*, **19**, 714–721.

36 He, Q., Madsen, M., Kilkenney, A., Gregory, B., Christensen, E. I., Vorum, H., Hojrup, P., Schaffer, A. A., Kirkness, E. F., Tanner, S. M. *et al.* (2005) Amnionless function is required for cubilin brush-border expression and intrinsic factor-cobalamin (vitamin B12) absorption in vivo. *Blood*, **106**, 1447–1453.

37 Lamarche, B., Uffelman, K. D., Carpentier, A., Cohn, J. S., Steiner, G., Barrett, P. H., Lewis, G. F. (1999)

Triglyceride enrichment of HDL enhances in vivo metabolic clearance of HDL apo A-I in healthy men. *J. Clin. Invest.*, **103**, 1191–1199.

38 Jaye, M., Lynch, K. J., Krawiec, J., Marchadier, D., Maugeais, C., Doan, K., South, V., Amin, D., Perrone, M., Rader, D. J. (1999) A novel endothelial-derived lipase that modulates HDL metabolism. *Nature Genetics*, **21**, 424–428.

39 Hirata, K., Diechek, H. L., Cioffi, J. A., Choi, S. Y., Leeper, N. J., Quintana, L., Kronmal, G. S., Cooper, A. D., Quertermous, T. (1999) Cloning of a unique lipase from endothelial cells extends the lipase gene family. *J. Biol. Chem.*, **274**, 14170–14175.

40 Ishida, T., Choi, S., Kundu, R. K., Hirata, K., Rubin, E. M., Cooper, A. D., Quertermous, T. (2003) Endothelial lipase is a major determinant of HDL level. *J. Clin. Invest.*, **111**, 347–355.

41 Maugeais, C., Tietge, U. J., Broedl, U. C., Marchadier, D., Cain, W., McCoy, M. G., Lund-Katz, S., Glick, J. M., Rader, D. J. (2003) Dose-dependent acceleration of high-density lipoprotein catabolism by endothelial lipase. *Circulation*, **108**, 2121–2126.

42 Jin, W., Millar, J. S., Broedl, U., Glick, J. M., Rader, D. J. (2003) Inhibition of endothelial lipase causes increased HDL cholesterol levels in vivo. *J. Clin. Invest.*, **111**, 357–362.

43 Ma, K., Cilingiroglu, M., Otvos, J. D., Ballantyne, C. M., Marian, A. J., Chan, L. (2003) Endothelial lipase is a major genetic determinant for high-density lipoprotein concentration, structure, and metabolism. *Proc. Natl. Acad. Sci. USA*, **100**, 2748–2753.

44 deLemos, A. S., Wolfe, M. L., Long, C. J., Sivapackianathan, R., Rader, D. J. (2002) Identification of genetic variants in endothelial lipase in persons with elevated high-density lipoprotein cholesterol. *Circulation*, **106**, 1321–1326.

45 Badellino, K. O., Wolfe, M. L., Reilly, M. P., Rader, D. J. (2006) Endothelial lipase concentrations are increased in metabolic syndrome and associated with coronary atherosclerosis. *PLoS Med.*, **3**, e22.

46 Ishida, T., Choi, S. Y., Kundu, R. K., Spin, J., Yamashita, T., Hirata, K., Kojima, Y., Yokoyama, M., Cooper, A. D., Quertermous, T. (2004) Endothelial lipase modulates susceptibility to atherosclerosis in apolipoprotein-E-deficient mice. *J. Biol. Chem.*, **279**, 45085–45092.

47 Ko, K. W., Paul, A., Ma, K., Li, L., Chan, L. (2005) Endothelial lipase modulates HDL but has no effect on atherosclerosis development in apoE-/- and LDLR-/- mice. *J. Lipid Res.*, **46**, 2586–2594.

48 Mahley, R. W., Huang, Y., Weisgraber, K. H. (2006) Putting cholesterol in its place: apoE and reverse cholesterol transport. *J. Clin. Invest.*, **116**, 1226–1229.

49 Meyers, C. D., Kamanna, V. S., Kashyap, M. L. (2004) Niacin therapy in atherosclerosis. *Curr Opin Lipidol*, **15**, 659–665.

50 Carlson, L. A. (2005) Nicotinic acid: the broad-spectrum lipid drug. A 50th anniversary review. *J Intern Med*, **258**, 94–114.

51 Tunaru, S., Kero, J., Schaub, A., Wufka, C., Blaukat, A., Pfeffer, K., Offermanns, S. (2003) PUMA-G and HM74 are receptors for nicotinic acid and mediate its anti-lipolytic effect. *Nat. Med.*, **9**, 352–355.

52 Brown, M. L., Inazu, A., Hesler, C. B., Agellon, L. B., Mann, C., Whitlock, M. E., Marcel, Y. L., Milne, R. W., Koizumi, J., Mabuchi, H. *et al.* (1989) Molecular basis of lipid transfer protein deficiency in a family with increased high-density lipoproteins. *Nature*, **342**, 448–451.

53 Inazu, A., Brown, M. L., Hesler, C. B., Agellon, L. B., Koizumi, J., Takata, K., Maruhama, Y., Mabuchi, H., Tall, A. R. (1990) Increased high-density lipoprotein levels caused by a common cholesteryl-ester transfer

protein gene mutation. *N. Engl. J. Med.*, **323**, 1234–1238.

54 Ikewaki, K., Rader, D. J., Sakamoto, T., Nishiwaki, M., Wakimoto, N., Schaefer, J. R., Ishikawa, T., Fairwell, T., Zech, L. A., Nakamura, H. (1993) Delayed catabolism of high density lipoprotein apolipoproteins A-I and A-II in human cholesteryl ester transfer protein deficiency. *J. Clin. Invest.*, **92**, 1650–1658.

55 Curb, J. D., Abbott, R. D., Rodriguez, B. L., Masaki, K., Chen, R., Sharp, D. S., Tall, A. R. (2004) A prospective study of HDL-C and cholesteryl ester transfer protein gene mutations and the risk of coronary heart disease in the elderly. *J. Lipid Res.*, **45**, 948–953.

56 Boekholdt, S. M., and Thompson, J. F. (2003) Natural genetic variation as a tool in understanding the role of CETP in lipid levels and disease. *J. Lipid Res.*, **44**, 1080–1093.

57 Boekholdt, S. M., Kuivenhoven, J. A., Wareham, N. J., Peters, R. J., Jukema, J. W., Luben, R., Bingham, S. A., Day, N. E., Kastelein, J. J., Khaw, K. T. (2004) Plasma levels of cholesteryl ester transfer protein and the risk of future coronary artery disease in apparently healthy men and women: the prospective EPIC (European Prospective Investigation into Cancer and nutrition)-Norfolk population study. *Circulation*, **110**, 1418–1423.

58 de Grooth, G. J., Kuivenhoven, J. A., Stalenhoef, A. F., de Graaf, J., Zwinderman, A. H., Posma, J. L., van Tol, A., Kastelein, J. J. (2002) Efficacy and safety of a novel cholesteryl ester transfer protein inhibitor, JTT-705, in humans: a randomized phase II dose-response study. *Circulation*, **105**, 2159–2165.

59 Kuivenhoven, J. A., deGrooth, G. J., Kawamura, H., Klerkx, A. H., Wilhelm, F., Trip, M. D., Kastelein, J. J. (2005) Effectiveness of inhibition of cholesteryl ester transfer protein by JTT-705 in combination with pravastatin in type II dyslipidemia. *Am. J. Cardiol.*, **95**, 1085–1088.

60 Clark, R. W., Sutfin, T. A., Ruggeri, R. B., Willauer, A. T., Sugarman, E. D., Magnus-Aryitey, G., Cosgrove, P. G., Sand, T. M., Wester, R. T., Williams, J. A. *et al.* (2004) Raising high-density lipoprotein in humans through inhibition of cholesteryl ester transfer protein: an initial multidose study of torcetrapib. *Arterioscler. Thromb. Vasc. Biol.*, **24**, 490–497.

61 Brousseau, M. E., Schaefer, E. J., Wolfe, M. L., Bloedon, L. T., Digenio, A. G., Clark, R. W., Mancuso, J. P., Rader, D. J. (2004) Effects of an inhibitor of cholesteryl ester transfer protein on HDL cholesterol. *N. Engl. J. Med.*, **350**, 1505–1515.

62 Davidson, M. H., McKenney, J. M., Shear, C. L., Revkin, J. H. (2006) Efficacy and safety of torcetrapib, a novel cholesteryl ester transfer protein inhibitor, in individuals with below-average high-density lipoprotein cholesterol levels. *J. Am. Coll. Cardiol.*, **48**, 1774–1781.

63 Rader, D. J. (2004) Inhibition of Cholesteryl Ester Transfer Protein Activity: A New Therapeutic Approach to Raising High-density Lipoprotein. *Curr Atheroscler Rep*, **6**, 398–405.

64 (2005) http://www.clinicaltrials.gov/ct/gui/show/NCT00134264.

65 Tall, A. R., Yvan-Charvet, L., Wang, N. (2007) The failure of torcetrapib: was it the molecule or the mechanism? *Arterioscler. Thromb. Vasc. Biol.*, **27**, 257–260.

66 Schwartz, C. C., VandenBroek, J. M., Cooper, P. S. (2004) Lipoprotein cholesteryl ester production, transfer, and output in vivo in humans. *J. Lipid Res.*, **45**, 1594–1607.

67 Brousseau, M. E., Diffenderfer, M. R., Millar, J. S., Nartsupha, C., Asztalos, B. F., Welty, F. K., Wolfe, M. L., Rudling, M., Bjorkhem, I., Angelin, B. *et al.* (2005) Effects of cholesteryl ester transfer protein inhibition on high-density lipoprotein subspecies, apolipoprotein A-I

metabolism, and fecal sterol excretion. *Arterioscler. Thromb. Vasc. Biol.*, **25**, 1057–1064.

68 Matsuura, F., Wang, N., Chen, W., Jiang, X. C., Tall, A. R. (2006) HDL from CETP-deficient subjects shows enhanced ability to promote cholesterol efflux from macrophages in an apoE- and ABCG1-dependent pathway. *J. Clin. Invest.*, **116**, 1435–1442.

69 Cuchel, M., Rader, D. J. (2006) Macrophage reverse cholesterol transport: key to the regression of atherosclerosis? *Circulation*, **113**, 2548–2555.

70 Joyce, C. W., Amar, M. J., Lambert, G., Vaisman, B. L., Paigen, B., Najib-Fruchart, J., Hoyt, R. F., Jr., Neufeld, E. D., Remaley, A. T., Fredrickson, D. S. *et al.* (2002) The ATP binding cassette transporter A1 (ABCA1) modulates the development of aortic atherosclerosis in C57BL/6 and apoE-knockout mice. *Proc. Natl. Acad. Sci. USA*, **99**, 407–412.

71 Haghpassand, M., Bourassa, P. A., Francone, O. L., Aiello, R. J. (2001) Monocyte/macrophage expression of ABCA1 has minimal contribution to plasma HDL levels. *J. Clin. Invest.*, **108**, 1315–1320.

72 Repa, J. J., Turley, S. D., Lobaccaro, J. A., Medina, J., Li, L., Lustig, K., Shan, B., Heyman, R. A., Dietschy, J. M., Mangelsdorf, D. J. (2000) Regulation of absorption and ABC1-mediated efflux of cholesterol by RXR heterodimers. *Science*, **289**, 1524–1529.

73 Wang, N., Lan, D., Chen, W., Matsuura, F., Tall, A. R. (2004) ATP-binding cassette transporters G1 and G4 mediate cellular cholesterol efflux to high-density lipoproteins. *Proc. Natl. Acad. Sci. USA*, **101**, 9774–9779.

74 Kennedy, M. A., Barrera, G. C., Nakamura, K., Baldan, A., Tarr, P., Fishbein, M. C., Frank, J., Francone, O. L., Edwards, P. A. (2005) ABCG1 has a critical role in mediating cholesterol efflux to HDL and preventing cellular lipid accumulation. *Cell Metab.*, **1**, 121–131.

75 Ranalletta, M., Wang, N., Han, S., Yvan-Charvet, L., Welch, C., Tall, A. R. (2006) Decreased atherosclerosis in low-density lipoprotein receptor knockout mice transplanted with Abcg1-/- bone marrow. *Arterioscler. Thromb. Vasc. Biol.*, **26**, 2308–2315.

76 Baldan, A., Pei, L., Lee, R., Tarr, P., Tangirala, R. K., Weinstein, M. M., Frank, J., Li, A. C., Tontonoz, P., Edwards, P. A. (2006) Impaired development of atherosclerosis in hyperlipidemic Ldlr-/- and ApoE-/- mice transplanted with Abcg1-/- bone marrow. *Arterioscler. Thromb. Vasc. Biol.*, **26**, 2301–2307.

77 Wang, N., Ranalletta, M., Matsuura, F., Peng, F., Tall, A. R. (2006) LXR-induced redistribution of ABCG1 to plasma membrane in macrophages enhances cholesterol mass efflux to HDL. *Arterioscler. Thromb. Vasc. Biol.*, **26**, 1310–1316.

78 Naik, S. U., Wang, X., Da Silva, J. S., Jaye, M., Macphee, C. H., Reilly, M. P., Billheimer, J. T., Rothblat, G. H., Rader, D. J. (2006) Pharmacological activation of liver X receptors promotes reverse cholesterol transport in vivo. *Circulation*, **113**, 90–97.

79 Terasaka, N., Hiroshima, A., Koieyama, T., Ubukata, N., Morikawa, Y., Nakai, D., Inaba, T. (2003) T-0901317, a synthetic liver X receptor ligand, inhibits development of atherosclerosis in LDL receptor-deficient mice. *FEBS Lett*, **536**, 6–11.

80 Joseph, S. B., McKilligin, E., Pei, L., Watson, M. A., Collins, A. R., Laffitte, B. A., Chen, M., Noh, G., Goodman, J., Hagger, G. N. *et al.* (2002) Synthetic LXR ligand inhibits the development of atherosclerosis in mice. *Proc. Natl. Acad. Sci. USA*, **99**, 7604–7609.

81 Levin, N., Bischoff, E. D., Daige, C. L., Thomas, D., Vu, C. T., Heyman, R. A., Tangirala, R. K., Schulman, I. G. (2005) Macrophage liver X receptor is required for

antiatherogenic activity of LXR agonists. *Arterioscler. Thromb. Vasc. Biol.*, **25**, 135–142.

82 Li, A. C. and Glass, C. K. (2004) PPAR- and LXR-dependent pathways controlling lipid metabolism and the development of atherosclerosis. *J. Lipid Res.*, **45**, 2161–2173.

83 Groot, P. H., Pearce, N. J., Yates, J. W., Stocker, C., Sauermelch, C., Doe, C. P., Willette, R. N., Olzinski, A., Peters, T., d'Epagnier, D. *et al.* (2005) Synthetic LXR agonists increase LDL in CETP species. *J. Lipid Res.*, **46**, 2182–2191.

84 Miao, B., Zondlo, S., Gibbs, S., Cromley, D., Hosagrahara, V. P., Kirchgessner, T. G., Billheimer, J., Mukherjee, R. (2004) Raising HDL cholesterol without inducing hepatic steatosis and hypertriglyceridemia by a selective LXR modulator. *J. Lipid Res.*, **45**, 1410–1417.

85 Oliver, W. R., Jr., Shenk, J. L., Snaith, M. R., Russell, C. S., Plunket, K. D., Bodkin, N. L., Lewis, M. C., Winegar, D. A., Sznaidman, M. L., Lambert, M. H. *et al.* (2001) A selective peroxisome proliferator-activated receptor delta agonist promotes reverse cholesterol transport. *Proc. Natl. Acad. Sci. USA*, **98**, 5306–5311.

86 Sprecher, D.L., Massien, C., Pearce, G., Billin, A.N., Perlstein, I., Willson, T.M., Hassall, D.G., Ancellin, N., Patterson, S.D., Lobe, D.C., et al. 2007. Triglyceride:high-density lipoprotein cholesterol effects in healthy subjects administered a peroxisome proliferator activated receptor delta agonist. *Arterioscler Thromb Vasc Biol*, **27**, 359–365.

87 Kozarsky, K. F., Donahee, M. H., Rigotti, A., Iqbal, S., Edelman, E. R., Krieger, M. (1997) Overexpression of the HDL receptor SR-B1 alters plasma HDL and bile cholesterol levels. *Nature*, **387**, 414–417.

88 Kozarsky, K. F., Donahee, M. H., Glick, J. M., Krieger, M., Rader, D. J. (2000) Gene transfer and hepatic overexpression of the HDL receptor SR-BI reduces atherosclerosis in the cholesterol-fed LDL receptor-deficient mouse. *Arterioscler. Thromb. Vasc. Biol.*, **20**, 721–727.

89 Rigotti, A., Trigatti, B. L., Penman, M., Rayburn, H., Herz, J., Krieger, M. (1997) A targeted mutation in the murine gene encoding the high density lipoprotein (HDL) receptor scavenger receptor class B type I reveals its key role in HDL metabolism. *Proc. Natl. Acad. Sci. USA*, **94**, 12610–12615.

90 Brundert, M., Ewert, A., Heeren, J., Behrendt, B., Ramakrishnan, R., Greten, H., Merkel, M., Rinninger, F. (2005) Scavenger receptor class B type I mediates the selective uptake of high-density lipoprotein-associated cholesteryl ester by the liver in mice. *Arterioscler. Thromb. Vasc. Biol.*, **25**, 143–148.

91 Trigatti, B., Rayburn, H., Vinals, M., Braun, A., Miettinen, H., Penman, M., Hertz, M., Schrenzel, M., Amigo, L., Rigotti, A. *et al.* (1999) Influence of the high density lipoprotein receptor SR-BI on reproductive and cardiovascular pathophysiology. *Proc. Natl. Acad. Sci. USA*, **96**, 9322–9327.

92 Braun, A., Trigatti, B. L., Post, M. J., Sato, K., Simons, M., Edelberg, J. M., Rosenberg, R. D., Schrenzel, M., Krieger, M. (2002) Loss of SR-BI expression leads to the early onset of occlusive atherosclerotic coronary artery disease, spontaneous myocardial infarctions, severe cardiac dysfunction, and premature death in apolipoprotein E-deficient mice. *Circ. Res*, **90**, 270–276.

93 Covey, S. D., Krieger, M., Wang, W., Penman, M., Trigatti, B. L. (2003) Scavenger receptor class B type I-mediated protection against atherosclerosis in LDL receptor-negative mice involves its expression in bone marrow-derived cells. *Arterioscler. Thromb. Vasc. Biol.*, **23**, 1589–1594.

94 Zhang, Y., Da Silva, J. R., Reilly, M., Billheimer, J. T., Rothblat, G. H., Rader, D. J. (2005) Hepatic expression of scavenger receptor class B type I (SR-BI) is a positive regulator of macrophage reverse

cholesterol transport in vivo. *J. Clin. Invest.*, **115**, 2870–2874.

95 Nanje, M. N., Crouse, J. R., King, J. M., Hovorka, R., Rees, S. E., Carson, E. R., Morgenthaler, J. J., Lerch, P., Miller, N. E. (1996) Effects of intravenous infusion of lipid-free apo A-I in humans. *Arterioscler. Thromb. Vasc. Biol.*, **16**, 1203–1214.

96 Eriksson, M., Carlson, L. A., Miettinen, T. A., Angelin, B. (1999) Stimulation of fecal steroid excretion after infusion of recombinant proapolipoprotein A-I: Potential reverse cholesterol transport in humans. *Circulation*, **100**, 594–598.

97 Chiesa, G. and Sirtori, C. R. (2003) Apolipoprotein A-IMilano: current perspectives. *Curr. Opin. Lipidol.*, **14**, 159–163.

98 Nissen, S. E., Tsunoda, T., Tuzcu, E. M., Schoenhagen, P., Cooper, C. J., Yasin, M., Eaton, G. M., Lauer, M. A., Sheldon, W. S., Grines, C. L. *et al.* (2003) Effect of recombinant ApoA-I Milano on coronary atherosclerosis in patients with acute coronary syndromes: a randomized controlled trial. *Jama*, **290**, 2292–2300.

99 Rader, D. J. (2003) High-density lipoproteins as an emerging therapeutic target for atherosclerosis. *Jama*, **290**, 2322–2324.

100 Shab, P.K. 2007. Emerging HDL-based Therapies for Atherothrombotic Vascular Disease. *Curr Treat Options Cardiovasc Med*, **9**, 60–70.

101 Navab, M., Anantharamaiah, G. M., Reddy, S. T., Hama, S., Hough, G., Grijalva, V. R., Yu, N., Ansell, B. J., Datta, G., Garber, D. W. *et al.* (2005) Apolipoprotein A-I mimetic peptides. *Arterioscler. Thromb. Vasc. Biol.*, **25**, 1325–1331.

102 Navab, M., Anantharamaiah, G. M., Reddy, S. T., Fogelman, A. M. (2006) Apolipoprotein A-I mimetic peptides and their role in atherosclerosis prevention. *Nat. Clin. Pract. Cardiovasc. Med.*, **3**, 540–547.

103 Garber, D. W., Datta, G., Chaddha, M., Palgunachari, M. N., Hama, S. Y., Navab, M., Fogelman, A. M., Segrest, J. P., Anantharamaiah, G. M. (2001) A new synthetic class A amphipathic peptide analogue protects mice from diet-induced atherosclerosis. *J. Lipid Res.*, **42**, 545–552.

104 Navab, M., Anantharamaiah, G. M., Hama, S., Garber, D. W., Chaddha, M., Hough, G., Lallone, R., Fogelman, A. M. (2002) Oral administration of an Apo A-I mimetic peptide synthesized from D-amino acids dramatically reduces atherosclerosis in mice independent of plasma cholesterol. *Circulation*, **105**, 290–292.

105 Navab, M., Anantharamaiah, G. M., Reddy, S. T., Hama, S., Hough, G., Grijalva, V. R., Wagner, A. C., Frank, J. S., Datta, G., Garber, D. *et al.* (2004) Oral D-4F causes formation of pre-beta high-density lipoprotein and improves high-density lipoprotein-mediated cholesterol efflux and reverse cholesterol transport from macrophages in apolipoprotein E-null mice. *Circulation*, **109**, 3215–3220.

106 Barter, P. J., Nicholls, S., Rye, K. A., Anantharamaiah, G. M., Navab, M., Fogelman, A. M. (2004) Antiinflammatory properties of HDL. *Circ. Res.*, **95**, 764–772.

107 Mineo, C., Deguchi, H., Griffin, J. H., Shaul, P. W. (2006) Endothelial and antithrombotic actions of HDL. *Circ. Res.*, **98**, 1352–1364.

108 Keech, A., Simes, R.J., Barter, P., Best, J., Scott, R., Taskinen, M.R., Forder, P., Pillai, A., Davis, T., Glasziou, P., et al. 2005. Effects of long-term fenofibrate therapy on cardiovascular events in 9795 people with type 2 diabetes mellitus (the FIELD study): randomised controlled trial. *Lancet*, **366**, 1849–1861.

109 McCoy, M.G., Sun, G.S., Marchadier, D., Maugeais, C., Glick, J.M., and Rader, D.J. 2002. Characterization of the lipolytic activity of endothelial lipase. *J Lipid Res*, **43**, 921–929.

Index

a
ABC, *see* ATP-binding cassette
ACAT 62
activated leukocyte-cell adhesion molecule (ALCAM) 309
activated protein C (APC) 385
activation function 1 (AF-1) 239
Ad4BP (adrenal 4 binding protein) 240
adeno-associated virus (AAV) 210ff.
–rep protein-mediated integration 209
adenoviral vector 212
–helper-dependent (HDAd) 212
adenovirus (Ad) 212
ADP 319
adrenoleukodystrophy 118f.
AIIABI 25
Alzheimer's disease (AD) 105, 191
Akt/PKB 453
alcohol 414
–intake 434
alternative splicing factor/splicing factor 2 (ASF/SF2) 27
amphipathic peptide 349
–class A 349
amyloid
–plaque 5
–serum amyloid A (SAA) 154, 342, 453
antiatherogenic property 490
antiinflammatory property 492
antioxidant 452, 491
antithrombotic property 493
antitumor immunosurveillance 321
apoI 386
apoA-I 3ff., 143ff., 189, 413, 432f., 466ff., 491ff., 506
–ABCA1 279, 323
–antibody 8
–catabolism 507
–function 4ff., 267ff.
–gene transcription 242ff., 399ff.
–HDL-bound 284
–hepatic ApoA-I secretion 323
–hypertriglyceridemia 283
–lipid-bound 11ff.
–lipid-free 3ff., 270, 378
–mimetic peptide 348, 511
–molecular dynamic simulation 15
–mutation 270ff., 474
–nutrient 399ff.
–receptor 317
–recycling 154ff.
–renal ApoA-I catabolism 324
–secretion 429
–structure 3ff.
–X-ray structure 9ff.
apoA-II 25ff.
–level 33
–metabolism 32
–mutation 27
–pathophysiology 33
–plasma level regulation 32
–polymorphism 27
–secretion 32
–SNP haplotype 39
–structure 25ff.
–synthesis 32
–transcriptional regulation 25
apoC 426
apoCII 473
apoCIII enhancer 244
apoE 55ff., 164, 426, 473
–HDL 67
–isoform 56
–recycled 72

High-Density Lipoproteins: From Basic Biology to Clinical Aspects. Edited by Christopher J. Fielding

ISBN: 978-3-527-31717-2

apoE recycling 66ff.
–regulation 70
apoE2/2 homozygocity 479
apoM 89ff., 154
–amino acid sequence 91
–antiatherogenic property 89ff.
–cellular expression 95
–characterization 90
–gene 91
–hormonal regulation 102f.
–protein structure 93f.
–regulation 99
apolipoprotein 3ff., 310ff., 356
–apoa 410, 426
–apoA-I, *see* apoA-I
–apoA-II, *see* apoA-II
–apoE, *see* apoE
–apoM, *see* apoM
–mimetic peptide 346ff.
–regulatory element 1 (ARP-1) 26, 241
ARV (antiretrovirus) 362ff.
atherogenesis 41, 311
atherosclerosis 73, 98, 171, 193f., 216ff., 346, 489ff., 512
–HDL 163, 489ff.
ATP-binding cassette (ABC)
–A family 120f.
–A1, *see* ATP-binding cassette A1
–A7 126f.
–class 118f.
–D1 118
–G family 128ff.
–G1, *see* ATP-binding cassette G1
–G4 163
–G5 167
–G8 167
–signature 116
–subfamily 119
–transporter 113ff., 318
ATP-binding cassette G1 (ABCG1) 68ff., 119, 129, 163, 452
–HDL 287
–transcriptional regulation 249
ATP-binding cassette A1 (ABCA1) 60, 72, 121ff., 145ff., 248, 269ff., 343, 448ff., 464ff., 493, 506
–ApoA-I interaction 279
–HDL-binding 310
–hepatic ApoA-I secretion 323
–lipid transporter 270
–nascent HDL 448
–pathway 70
–transcriptional regulation 248
ATPase 315ff.

b

bacterial attachment site (*attB*) 209
basic-helix-loop-helix-leucine zipper (bhLH-LZ) transcription factor 241
belt-like confirmation 12
bisperoxo (1,10-phenanthroline) oxovanadate (bpv) 410
block lipid transport (BLT) 313
BPI 184

c

c-jun N-terminal kinase-interacting protein-1 453
CAD, *see* coronary artery disease *or* coxackievirus and adenovirus receptor
caloric intake 400
canola oil 404
carbohydrate 432
carcinoma 101
cardiovascular disease (CVD) 183ff., 375ff., 399ff.
caspase 453
CCAAT-enhancer-binding protein (C/EBP) 25, 186
CD95 386
cell
–membrane 146
cellular adhesion molecule-1 (CAM) 376ff.
ceruloplasmin 342
chenodeoxycholic acid (CDCA) 186
cholesterol 279
–dietary 431
–efflux 62ff., 119ff., 190, 347, 490
–free (FC) 143ff., 429
–HDL, *see* HDL-C
–HL deficiency 216
–recognition amino acid consensus (CRAC) motif 150
–secretion 169
–selective HDL cholesterol uptake 310
–selective lipoprotein cholesterol uptake 311ff.
–transport 279
–VLDL/LDL 219
cholesterol transport 35, 68
–reverse, *see* reverse cholesterol transport
–macrophage reverse cholesterol transport 36, 323
cholesteryl ester (CE) 36, 62, 149ff., 308ff., 450f., 508
–selective uptake 308ff.
–transfer (CET) 508

cholesteryl ester transfer protein (CETP) 34, 63, 152, 163ff., 184, 269, 327, 364f., 404, 447ff., 465ff., 495, 508
- –gene 252
- –HDL remodeling 163ff., 357ff.
- –inhibitor 456, 464, 509

chylomicron 57, 217, 465
CLA, *see* linoleic acid
CLA-1 310f.
coagulation 385
coconut oil 432
common disease-common variant (CDCV) hypothesis 478
coregulator complex 237
coronary artery disease (CAD) 32, 163ff., 218, 363ff., 463ff.
coronary heart disease (CHD) 425ff., 508
coupling factor 6 (CF6) 321f.
COUPTF 241
coxsackievirus and adenovirus receptor (CAD) 212
CRAC, *see* cholesterol
CRE 170
CTFR, *see* cystic fibrosis transmembrane regulator gene
cubilin 310, 324
cyclooxygenase (Cox) 380, 409
cystic fibrosis transmembrane regulator gene (CTFR) 27
cytokine 383

d

D-[113-122]apoJ 349
D-4F 347ff., 512
Dax-1 252
diabetes 38, 104
direct repeat (DR) 239
DNA
- –delivery 208
- –nonviral 208

DR4 (direct repeats separated by 4 nucleotide) 248, 448
dysbetalipoproteinemia type III 479
dyslipidemia 38, 283

e

E-selectin 382, 492
EAR-3 241
ecto-F1-ATPase 315ff.
endocannabinoid receptor 1 antagonist 455
endocytosis 315ff.
- –HDL 315ff.

endosomal recycling compartment (ERC) 72
endothelial cell 387
endothelial cell apoptosis 386
endothelial dysfunction 375ff.
endothelial lipase (EL) 35, 214, 508
endothelial nitric oxide synthase (eNOS) 312, 378ff.
endothelial progenitor cell (EPC) 387
endothelial protection 375ff.
endothelial repair 493
ERK 382
ETC-642 511
exonic splicing enhancer (ESE) 26

f

F1-ATPase 317ff.
factor Va 385
familial combined hyperlipidemia (FCHL) 40, 476
farnesoid X-activated receptor (FXR) 186, 240ff., 313
- –responsive element (FXRE) 253

Fas 386
fatty acid (FA) 33, 407
- –ω-3 408
- –dietary 431
- –free (FFA) 37, 406, 447f.
- –high-monounsaturated (MUFA) 33, 405f.
- –medium-chain (MCFA) 408
- –metabolism 37
- –oxidized 406
- –polyunsaturated (PUFA) 405f.
- –saturation 403
- –*trans* 407
- – *trans*-monounsaturated (TFA) 407

FC, *see* cholesterol
fibrinolysis 386
fibroblast growth factor (FGF) 387
first-generation Ad vector (FGAd) 212
fish-eye disease (FED) 150
foamy viral vector (FV) 214
forward cholesterol transport 362
fractional catabolic rate (FCR) 33
free fatty acid (FFA) 37, 406
fructose 411
FXR, *see* farnesoid X-activated receptor
FXRE, *see* farnesoid X-activated receptor

g

gene regulation 237ff.
gene transfer 208ff.

- –adenovirus-mediated 274
- –HDL remodelling 208ff.

genetic association study 478
glucosamine 411
glucose 410
- –homeostasis 38
- –metabolism 37

Gp96/GRP94 309
GPR109A 508
growth factor 100

h

HA-PLTP, *see* phospholipid transfer protein
hairpin model 13
HB1-HB2 309
HBP 309
HBV, *see* hepatitis B virus
HCC, *see* hepatocellular carcinoma
HDL, *see* high-density lipoprotein
HDL particle
- –containing apoA-I (LpA-I) 32
- –containing apoA-II (LpA-II) 32

HDL-C (HDL cholesterol) 61f., 360ff., 377, 400ff., 431f., 446ff., 476, 496, 505ff.
- –heritability 477
- –low 505
- –selective uptake 510

hemostasis 384
heparin sulfate proteoglycan (HSPG) matrix 172
hepatic cholesterol catabolism 308
hepatic HDL receptor 307ff.
hepatic lipase (HL) 35, 164, 214ff., 451, 473, 507
hepatic lipid metabolism 97
hepatitis 101
hepatitis B virus (HBV) 101
hepatocellular carcinoma (HCC) 101
hepatocyte 315
hepatocyte nuclear factor
- –1α (HNF-1α) 99ff.
- –4 (HNF-4) 238ff.

high-density lipoprotein (HDL) 6ff., 33ff., 57ff., 98, 214ff., 267ff., 375ff.
- –2 434
- –3 434
- –α-HDL 12, 151, 280ff., 365, 430
- –α1 HDL 428
- –α2 HDL 428
- –α4 HDL 292, 428f.
- –A-I HDL 494
- –A-I/A-II HDL 494
- –adaptor protein 467ff.
- –antiatherogenic property 490
- –antiinflammatory property 492
- –antioxidant property 491
- –antithrombotic property 493
- –apoA-I 284, 427f.
- –apoA-I maturation 506
- –apoE-enriched 62
- –apolipoprotein mimetic peptide 346
- –associated enzyme 214
- –associated protein 467f.
- –atheroprotection 163, 291
- –atherosclerosis 489ff.
- –biogenesis 124, 237ff., 270ff.
- –catabolism 216, 242
- –cellular receptor 467ff.
- –cholesterol, *see* HDL-C
- –cholesteryl ester 433
- –deficiency 277
- –discoidal 11f., 151, 267, 291, 428
- –disease 355ff.
- –endocytosis 315ff.
- –enzyme 467ff.
- –epidemiology 464
- –formation 60, 356ff.
- –function 346, 512
- –gene transfer 208ff.
- –genetics 463ff.
- –health 355ff.
- –impaired 452
- –inflammation 341ff., 381ff.
- –inflammatory protein 467f.
- –lysophospholipid 379
- –maturation 60, 356ff.
- –metabolic syndrome 445ff.
- –metabolism 33, 60, 269ff., 323, 355ff., 425ff., 447, 465
- –nascent 12, 448
- –nutrient 400ff.
- –particle 427
- –pathway 269ff.
- –pre-β-HDL 12, 143ff., 280ff., 450
- –pre-β1-HDL 289, 360ff., 427ff.
- –pre-β2-HDL 427
- –receptor 307ff.
- –recombinant (rHDL) 267
- –reconstituted (rHDL) 492
- –remodeling 33, 163ff., 189, 208ff., 237ff., 284ff., 357ff.
- –selective HDL cholesteryl ester uptake 308ff.
- –spherical 16, 291, 428

- –structure 267
- –subclass 11, 445
- –subfraction 357
- –subpopulation 290, 494
- –synthesis 269ff.
- –synthetic (sHDL) 377ff.
- –therapy 355ff.
- –transcription factor 467ff.
- –transfer protein 467f.
- –transporter 467ff.
- –triglyceride 507
- –UC-enriched 65f.
- –vascular inflammation 381ff.

highly active antiretroviral therapy (HAART) 362f.
hinge domain 14
hinge segment hypothesis 15
histone deacetylase 237
HIV infection 362ff.
HL, *see* hepatic lipase
HNF, *see* hepatocyte nuclear factor
homeostasis 453
hormone response element (HRE) 239
Hsp60 (heat shock protein) 309
HSPG, *see* heparin sulfate proteoglycan matrix
13-hydroperoxylinoleic acid (13-HPODE) 406
hyperchylomicronemia 217
hyperlipoproteinemia
- –type I 217
- –type III 216
- –type V 37

hypertriglyceridemia 283ff.
hypoalphalipoproteinemia 474

i

ICAM-1, *see* intracellular cell adhesion molecule-1
IDL 190, 283
IL1 312, 382ff., 453
immune response 91ff.
immune system 341
- –innate 341

immunity
- –innate 320f.

immunosurveillance
- –antitumor 321

induction therapy 513
inflammation 101, 341ff.
insulin 103
- –response core element (IRCE) 410

internalization 308
intervention study
- –animal 495
- –human 496

intracellular cell adhesion molecule-1 (ICAM-1) 376ff., 492
intron splicing enhancer (ISE) 27
intronic splicing silencer (ISS) 27
inverted repeat (IR) 239
inverted terminal repeat (ITR) 210

j

JTT 705 431, 456, 509
c-jun N-terminal kinase-interacting protein-1 453

k

KATP channel 274
KERS 349
6-keto-PGF$_{1\alpha}$ 381
ketone 409
kidney
- –function 98
- –hypercatabolism 325

KRES 349

l

L-5F 512
L-NAME 379
LA-PLTP, *see* phospholipid transfer protein
LBP 184
LCAT, *see* lecithin:cholesterol acyltransferase
LD (linkage disequilibrium) 478
LDLR, *see* low density lipoprotein
lecithin:cholesterol acyltransferase (LCAT) 6f., 34, 61, 143ff., 164, 356ff., 429, 465ff., 506ff.
- –ApoA-I 281
- –HDL remodeling 269
- –mechanism 149f.
- –pre-β1-HDL complex 152ff.

lecithin:cholesterol acyltransferase-dependent LpA-I/LpA-II particle fusion process 32
lentiviral vector 213
leptin 102
ligand-binding domain (LBD) 239
linoleic acid (CLA) 33
lipase 214ff.
LIPC 164
lipid 400
- –efflux 277
- –intracellular trafficking 55
- –transport 97, 118
- –transport protein (LTP) 184

lipidation 277
lipocalin 93
lipopolysaccharide (LPS) 184ff., 312, 344
lipoprotein 169, 356
– assembly 56
– catabolism 58
– metabolism 98
– plasma 94
– processing 58
– secretion 56
– triglyceride-rich (TRL) 427ff.
lipoprotein lipase (LPL) 37, 183, 214ff., 447ff., 473ff.
lipoprotein lipid hydroperoxide 347
lipoprotein metabolism 56, 445
– triglyceride-rich 56
liver receptor homolog-1 (LRH-1) 240, 313, 506
liver X-activated receptor (LXR) 70, 100, 122, 240ff., 510
– response element (LXRE) 187
low density lipoprotein (LDL) 33, 152, 404, 426, 465
– cholesterol (LDLc) 404
– oxidized (ox-LDL) 378
– receptor (LDLR) 55ff.
– receptor related protein 1 (LRP1) 59ff., 164
LpA-I 32ff., 414, 445
LpA-II 32ff., 414, 445
γ-LpE 60ff.
LPL, *see* lipoprotein lipase
LRH, *see* liver receptor homolog
LRP1, *see* low density lipoprotein
LXR, *see* liver X-activated receptor
LXRE, *see* liver X-activated receptor

m

macrophage 363
macrophage reverse cholesterol transport 36, 323
major histocompatibility complex (MHC) 89ff.
major urinary protein (MUP) 93
maturity onset diabetes in the young (MODY3) 99ff.
MCFA, *see* fatty acid
membrane
– cell 146
metabolic syndrome 445ff.
methyl β-cyclodextrin (MβCD) 169
microdomain 170
micronutrient 412
microsolubilization 147
migration 387
mitogen-activated protein kinase (MAPK) 380f., 493
MODY3, *see* maturity onset diabetes in the young
monocyte 492
monocyte chemoattractant protein-1 (MCP-1) 342
monogenic disorder 466
MUFA, *see* fatty acid
multidrug transporter 118
myeloperoxidase (MPO) 219
myristic acid 406

n

nephrotic syndrome 248
neutrophil 492
NF-κB (nuclear factor) 381ff.
niacin 455
– receptor 508
Niemann-Pick type C1 protein (NPC1) 279, 473
nitric oxide (NO) 375ff., 493
nitric oxide synthase
– endothelial (eNOS) 312
nuclear hormone receptor 239
nuclear orphan factor 122
nuclear transcription factor 99
nucleotide/ATP-binding domain (NBD) 115ff.
nutrient 399ff.
nutritional factor 425ff.

o

obesity 104
olive oil 405
oncoretroviral vector 213
orphan receptor 248, 506
ox-linoleic acid 406
oxidative stress 345

p

$P2Y_{13}$ 319
p44/42 mitogen-activated protein kinase 493
p5 integration efficiency element (p5IEE) 210
P-glycoprotein 118
PAF, *see* platelet-activating factor
PAF-AH, *see* platelet-activating factor acetylhydrolase
palindromic repeat (PR) 239
palmitic acid 406

paraoxonase (PON) 36, 61, 218f., 479
peroxisome proliferator 239
peroxisome proliferator-activated receptor (PPAR) 99
- –α 26, 186, 240, 313
- –β 240
- –γ 129, 240, 313ff., 406
- –δ 313
- –agonist 455, 506

phage
- –ΦC31 209
- –attachment site (*attP*) 209
- –integrase 209

phorbol-12,13-dibutyrate (PDBu) 410
phosphatidylcholine (PC) 145ff.
phosphatidylinositol 3-kinase (Pi3K) 379, 493
phosphatidylserine 125
phosphoinositide 3-kinase (Pi3K) 379, 493
phospholipase A2 (sPLA2) 465
phospholipid 63, 124f., 451, 507
- –efflux 270
- –transfer activity 187

phospholipid transfer protein (PLTP) 34, 73, 151, 184ff., 269, 357, 447ff., 465
- –atherosclerosis 193, 344
- –function 187
- –gene expression 186
- –high-activity (HA) 192
- –low-activity (LA) 192
- –plasma 147, 183ff., 404
- –tissue 190
- –transcriptional regulation 252

physiopathology 325
picket fence 12
PKB, *see* protein kinase B
plasma lipoprotein 94
plasma phospholipid transfer protein 147, 183ff., 404
plasma recycling 55
plasminogen activator inhibitor-1 (PAI-1) 376
platelet-activating factor (PAF) 384
platelet-activating factor acetylhydrolase (PAF-AH) 36f., 384
platelet adhesion 384
PLTP, *see* phospholipid transfer protein
pneumonia 347
PON, *see* paraoxonase
PPAR, *see* peroxisome proliferator-activated receptor
PR, *see* palindromic repeat
proliferation 387
prostacyclin (PGI_2) 380
prostanoid 409
protein 412
protein kinase B (PKB/Akt) 453
protein kinase C (PKC) 380
protein S 385
PUFA, *see* fatty acid

q

quantitative trait locus approach (QLT) 466ff.

r

RBP, *see* retinol binding protein
RCT, *see* reverse cholesterol transport
reactive oxygen species (ROS) 342
receptor associated protein (RAP) 172
renal function 97
renal HDL receptor 307ff.
rep protein 210
retinoic acid receptor (RAR) 249
retinoic acid receptor-related orphan receptor (RORα) 248
retinoid 239
retinoid X receptor (RXR) 186, 239ff.
- –RXRα 26, 100

retinol binding protein (RBP) 93
retro-endocytosis 308
reverse cholesterol transport (RCT) 35, 55ff., 143ff., 343, 357ff., 509ff.
- –impaired 452
- –macrophage-specific 36, 323

rHDL, *see* HDL
RhoA 387
RLT peptide 510
RNA viral vector 213
ROR, *see* retinoic acid receptor-related orphan receptor
ROS, *see* reactive oxygen species
RXR, *see* retinoid X receptor

s

$S1P_1$ 388
$S1P_3$ 388
SAA, *see* amyloid
safflower oil 432
scavenger receptor class BI (SR-BI) 65ff., 163ff., 251, 287, 310ff., 379, 452, 465
- –gene 251
- –HDL remodeling 284
- –hepatic 311ff.
- –transport 430

selective HDL cholesteryl ester uptake 308ff.
selective lipoprotein cholesterol uptake 311ff.
serine-arginine-rich (SR) protein 27
SF-1, *see* steroidogenic factor 1
site 1 protease (S1P) 241
small heterodimer partner (SHP) 240, 313
SMPD 1 473
specificity protein 1 (SP1) 241
sphingomyelin 279
sphingomyelinase (Smase) 465
sphingosine kinase (SphK) 382
sphingosine-1-phosphate (S1P) 379
  – receptor 379
sphingosylphosphorylcholine 379
splicing component 35 (SC35) 27
Src kinase 493
stearic acid 406
steroid 239
steroidogenic factor 1 (SF-1) 240
sterol regulatory element-binding protein (SREBP) 241ff.
  – SREBP1 26, 170, 186
sterol response element 170
sterol trafficking 128
sulfonyl urea receptor-1 (SUR-1) 274
switch/sucrose nonfermenting complex (SWI/SNF) 239

t
Tangier disease 37, 124, 277, 429f.
TFA, *see* fatty acid
TG, *see* triglyceride
tissue factor (TF) 376
thyroid hormone 239
  – receptor (T3R) 249
TNF, *see* tumor necrosis factor
α-tocopherol 188
torcetrapib® 431, 464
transfer
  – two-step 147
transforming growth factor β (TGFβ) 122, 383f.
transmembrane domain (TMD) 115ff.
triglyceride (TG) 37, 217, 316, 445ff.
TRL, *see* lipoprotein
tumor necrosis factor (TNF) 122, 312ff., 382ff., 453

u
unesterified cholesterol (UC) 60ff.
  – HDL 65f.
upstream stimulatory factor (USF) 25, 249

v
vascular cell adhesion molecule-1 (VCAM-1) 376ff., 492
vascular endothelial growth factor (VEGF) 387
vascular tone 377
vasodilation 376, 493
  – NO-dependent 376
VDAC, *see* voltage-dependent anion channel
vector
  – nonviral 209
  – viral 210
very low density lipoprotein (VLDL) 37, 56ff., 152, 217ff., 283, 356, 404ff., 426, 447ff., 465
  – hepatic 426
vesicle secretion 318
virus
  – adeno-associated (AAV) 210ff.
  – adenovirus (Ad) 212
  – antiretrovirus (ARV) 362ff.
  – DNA viral vector 210
  – hepatitis B (HBV) 101
  – HIV infection 362ff.
vitamin
  – A 413
  – C 413
  – D 239, 413
  – E 312, 413
voltage-dependent anion channel (VDAC) 318
von Willebrand factor (vWF) 376

w
Walker motif 116
weight loss 434

y
Yin Yang-1 (YY-1) transcription factor 252

z
zinc finger protein ZNF202 249, 476
Zellweger syndrome 119

Related Titles

Kaput, J., Rodriguez, R. L. (eds.)

Nutritional Genomics

Discovering the Path to Personalized Nutrition

469 pages
2005
Hardcover
ISBN: 978-0-471-68319-3

Tamm, L. K. (ed.)

Protein-Lipid Interactions

From Membrane Domains to Cellular Networks

470 pages with 82 figures and 3 tables
2005
Hardcover
ISBN: 978-3-527-31151-4

Storey, K. B.

Functional Metabolism

Regulation and Adaptation

approx. 624 pages
2004
Hardcover
ISBN: 978-0-471-41090-4

Östman, J., Britton, M., Jonsson, E. (eds.)

Treating and Preventing Obesity

An Evidence Based Review

372 pages with 12 figures and 42 tables
2004
Hardcover
ISBN: 978-3-527-30818-7

Krauss, G.

Biochemistry of Signal Transduction and Regulation

558 pages with 270 figures and 15 tables
2003
Softcover
ISBN: 978-3-527-30591-9